Eco-friendly and Smart Polymer Systems

Hamid Mirzadeh · Ali A Katbab
Editors

Eco-friendly and Smart Polymer Systems

Springer

Editors
Hamid Mirzadeh
Department of Polymer Engineering
Amirkabir University of Technology
Teheran, Iran

Ali A Katbab
Department of Polymer Engineering
Amirkabir University of Technology
Teheran, Iran

ISBN 978-3-030-45084-7 ISBN 978-3-030-45085-4 (eBook)
https://doi.org/10.1007/978-3-030-45085-4

This Springer imprint is published by the registered company Springer Nature Switzerland AG
The registered company address is: Gewerbestrasse 11, 6330 Cham, Switzerland

Preface

The 13th International Seminar on Polymer Science and Technology (ISPST 2018) was well organized and took place in Amirkabir University of Technology (Tehran Polytechnic), Department of Polymer Engineering, Tehran, during November 10–22, 2018. This forum was the culmination of more than 3 decades of academic and industrial activities of Iranian scholars and professionals and participation of many notable international scientists in covering various important subjects of concern in Iran and the world on polymers including polymer syntheses, processing, wide range of properties and notably issues concerning polymer degradation, stability and environmental aspects.

About ISPST

The 13th International Seminar on Polymer Science and Technology (ISPST 2018) is the continuation of the ISPST past successful series, which is aimed to provide a forum for presentation, discussion and dissemination of current and future trends in polymeric materials and to facilitate information exchange among researchers with diverse interests, in various polymeric fields. The previous 12 seminars were as follows:

- ISPST 2016, Islamic Azad University, Tehran, Iran
- ISPST 2014, Iran Polymer and Petrochemical Institute, Tehran, Iran
- ISPST 2012, Amirkabir University of Technology, Tehran, Iran
- ISPST 2009, Iran Polymer and Petrochemical Institute, Tehran, Iran
- ISPST 2007, Sharif University of Technology, Tehran, Iran
- ISPST 2005, Amirkabir University of Technology, Tehran, Iran
- ISPST 2003, Iran Polymer and Petrochemical Institute, Tehran, Iran
- ISPST 2000, Amirkabir University of Technology, Tehran, Iran
- ISPST 1997, Iran Polymer and Petrochemical Institute, Tehran, Iran
- ISPST 1994, Shiraz University, Shiraz, Iran
- ISPST 1990, Amirkabir University of Technology, Tehran, Iran
- ISPST 1986, Amirkabir University of Technology, Tehran, Iran

ISPST 2018 was held in Amirkabir University of Technology, during 19–22 November, 2018. The main topics of the seminar with respect to the environmental aspects were as follows:

- Polymers for smart and sustainable future
- Polymer recycling to protect environment
- Biocompatible and biodegradable polymers
- Processing and reactive processing of environmental polymers
- Nanoassemblies for sustainable energy and environmental applications
- Green coatings and engineering in polymer science
- Natural bio-based polymers and composites
- Pathways for degradation of plastic polymers floating in the marine environment
- Other novel aspects of polymer science and technology

Four plenary lecturers and 11 international distinguished keynote speakers deliver their latest research works in the seminar fields. About 500 papers were accepted after peer-reviewing, in which 110 papers were presented orally and the rest in the form of poster. Receiving this number of the articles and attracting more than 600 participants show the status and importance of this scientific event.

Forewords

It is for nearly half a century that human health, progress for better living condition and, in the last few decades specifically, environmental pollution have shaped and forced the scientific activities toward creation of smart materials, compatible with ecosystems and human body and urged scientists and technologists to make sound efforts to furnish research using natural sustainable sources.

Indeed, large sections of Iranian society and academic community have taken important steps with heavy responsibility to confront problems made by and potentially solved by polymers and plastics.

For this, we organizers wish that we will have the positive response made by the scientific and industrial people by working together to solve many crucial problems.

And last but not least, it is the main responsibility of polymer scientists to enhance and apply their knowledge and experience to improve the position of polymers in the life, and at the same time, they should pave the way for a solution against pollution of the ecosystems. We should accelerate innovation trends in polymer for solving societal challenges. This is our main message at ISPST 2018.

Hamid Mirzadeh
ISPST 2018 Chairman

Ali A Katbab
Scientific Committee Head

Mehdi Nekoomanesh
IPS & IPPI President

Organization

Seminar Chairman

Hamid Mirzadeh — Department of Polymer Engineering, Amirkabir University of Technology (Tehran Polytechnic), Teheran, Iran
mirzadeh@aut.ac.ir

Scientific Committee Head

Ali A Katbab — Department of Polymer Engineering, Amirkabir University of Technology (Tehran Polytechnic), Teheran, Iran
katbab@aut.ac.ir

Plenary Lecturers

Hamid Mirzadeh	Amirkabir University of Technology, Iran
Ali A Katbab	Amirkabir University of Technology, Iran
Mehdi Nekoomanesh	Iran Polymer and Petrochemical Institute, Iran
Naser Mohammadi	Amirkabir University of Technology, Iran

Keynote Speakers

Volker Altstädt	Werkstoffe Universität, Germany
Tayebeh Ameri	Montan universitaet, Austria
Charles Dubois École	Polytechnique de Montréal, Canada
Fatemeh Goharpey	Amirkabir University of Technology, Iran
Vahid Haddadi-Asl	Amirkabir University of Technology, Iran
Clemens Holzer	University of Munich, Germany

Organizing Committee

Scientific Committee

S. Bonakdar	Pasteur Institute of Iran, Iran
H. Borsi	Leibniz Universität Hannover, Germany
C. Dubois	École Polytechnique de Montréal, Canada
M. Ebrahimi	AUT, Iran
M. Ehsani	IPPI, Iran
M. H. N. Famili	Tarbiat Modares University, Iran
S. Farhangzadeh	IPPI, Iran
M. Frounchi	Sharif University of Technology, Iran
M. Ghaemy	University of Mazandaran, Iran
S. R. Ghaffarian	AUT, Iran
F. Goharpey	AUT, Iran
I. Ghasemi	IPPI, Iran
V. Haddadi-Asl	AUT, Iran
C. Holzer	Montan Universitaet, Austria
E. Jabbari	University of South Carolina, USA
S. H. Jafari	University of Tehran, Iran
M. Jamshidi	Iran University of Science and Technology, Iran
A. Jalali Arani	AUT, Iran
A. Javadi	AUT, Iran
B. Kaffashi	University of Tehran, Iran
H. A. Khonakdar	IPPI, Iran
M. Kokabi	Tarbiat Modares University, Iran
M. M. Lakouraj	University of Mazandaran, Iran
A. R. Mahdavian	IPPI, Iran
S. Mallakpour	Sharif University of Technology, Iran
M. Mehranpour	Islamic Azad University, Iran
M. Mirabedini	IPPI, Iran
G. Mirmohammad Sadeghi	AUT, Iran
A. Mirmohseni	University of Tabriz, Iran
M. R. Moghbeli	Iran University of Science and Technology, Iran
N. Mohammadi	AUT, Iran
M. R. Mohammadzade-Attar	AUT, Iran
S. M. Mohseni Bozorgi	AUT, Iran
M. Morshed	Isfahan University of Technology, Iran
R. Naderi Mahmoudi	University of Tehran, Iran
H. Namazi	University of Tabriz, Iran
A. H. Navarchian	University of Isfahan, Iran
H. Nazokdast	AUT, Iran
B. Nyström	University of Oslo, Norway
A. Olad	University of Tabriz, Iran
B. Pourabas	Sahand University of Technology, Iran
A. Pourjavadi	Sharif University of Technology, Iran
S. Pourmahdian	AUT, Iran
A. Ramazani Saadat-Abadi	Sharif University of Technology, Iran
Z. Ranjbar	Institute for Color Science and Technology, Iran

M. K. Razavi Aghjeh	Sahand University of Technology, Iran
M. Razzaghi	Tarbiat Modares University, Iran
A. R. Shakeri	University of Tehran, Iran
F. Sharif	AUT, Iran
A. Shojaei Segherloo	Sharif University of Technology, Iran
A. Solouk	AUT, Iran
F. Tammadon	Lund University, Sweden
E. Vasheghani Farahani	Tarbiat Modares University, Iran
A. A. Yousefi	IPPI, Iran
M. Zandi	IPPI, Iran
G. H. Zohuri	Ferdowsi University of Mashhad, Iran

Executive Committee

Hamid Mirzadeh	AUT, Iran
Salwa Farhangzadeh	IPPI, Iran
Shadab Bagheri	AUT, Iran
Morteza Ehsani	IPPI, Iran
Mitra Hashemi Tabatabaei	IPPI, Iran
Azizeh Javadi	AUT, Iran
Arezoo Mirghasemi	IPS, Iran
Nafiseh Olov	AUT, Iran
Hadi Mirhoseini Moghadam	IPPI, Iran
Atefeh Solouk	AUT, Iran
Hakimeh Mohammadzad	IPS, Iran
Ashraf Tatlari	IPPI, Iran
Tahereh Khaleghjoo	IPPI, Iran
Sarvenaz Alikhani	IAU, Iran
Talieh Sakhamanesh	IAU, Iran
Ahmad Ahmadi	AUT, Iran
Maryam Amjadi	AUT, Iran
Alireza Kouhkhil	AUT, Iran

Seminar Reviewers

Abbasi, Farhang
Abdouss, Majid
Ahmadi, Mostafa
Ahmadi, Shervin
Ahmadi, Zahed
Ahmadjo, Saeid
Akbari, Somaye
Fahimeh Askari
Mohammad Atai
Ehsan Avazverdi
Reza Bagheri
Shadab Bagheri-Khoulenjani
Mehdi Barikani
Jalal Barzin
Mohamad Hosain Beheshty
Soheil Dariushi

Organizers and Sponsors

Organizers

Iran Polymer Science and Technology Society (IPSTS)
Amirkabir University of Technology (AUT)
Iran Polymer and Petrochemical Institute (IPPI)

Sponsors

Ecole Polytechnic de Montreal
Barez Industrial Group
Jam Petrochemical Company
Iranian Association of Polymer and Chemical Engineers
Avrand Plastic Company
Iran Rheology Society
Baspar Magazin

Acknowledgement

We would like to acknowledge and give our special thanks to Salwa Farhangzadeh, Mitra Hashemi Tabatabaei and Mehdi Vafayan all from Iran Polymer and Petrochemical Institute. Those who have had a vital role as like as co-editors to generate this book and without their significant contribution this work would not be completed.

Contents

Green Coatings and Engineering in Polymer Science

Other Aspects of Polymer Science and Technology

Processing and Reactive Processing of Environmental Polymers

Appendix I. Plenary Lecture

Appendix II. Keynote Lectures

About the Editors

Professor Dr. Hamid Mirzadeh (BSc, MSc, Ph.D) received his Ph.D degree in Polymer Engineering from University of New South Wales, Sydney Australia in 1994. Since 1994, he is working as a full professor at Amirkabir University of Technology (AUT-Tehran Polytechnic), Department of Polymer Science and Engineering with a second affiliation to the Faculty of Medical Engineering at AUT and adjunct professor at Tehran University of Medical Sciences (TUMS). He is the Coordinator of Tissue Engineering and Biomaterials, Iranian Bunch (TEBIB). Professor Hamid Mirzadeh has extensive experience as a published researcher, lecturer, and editor in the field of biomaterials, tissue engineering and regenerative medicine, drug delivery systems, polymer science and technology. He is Editor-in-Chief of Progress in Biomaterials (Springer/Nature). Based on Scopus and Google Scholar, his h-index is 38 and 44, respectively. He was President of seven International Seminar on Polymer Science and Technology (ISPST) out of 13 ISPST. He also serves as a reviewer of 20 international journals and editorial member of six international journals. Additionally, he has published 199 peer-reviewed papers and published/edited 11 books and book chapters. So far, 107 MSc and Ph.D students have been graduated under his supervision and 14 postgraduate students are currently working at Biopolymer Laboratory under his supervision as well. His ORCID ID is: https://orcid.org/0000-0002-9821-1759.

Professor Dr. Ali Asghar Katbab received MSc and Ph.D degrees in polymer science and engineering from Aston University in Birmingham, England, in 1977 and 1980, respectively. He started an academic activity in 1983 at chemical engineering and then at polymer engineering departments of Amirkabir University in Tehran, Iran. He spent his first sabbatical leave in NSW University Australia and second in McMaster University, Canada. He is now a full-time professor at Polymer Engineering Department of Amirkabir University. His research activities have been focused on rubber engineering and processing, nanocomposite polymers, bio-nanocomposite polymers, and since years ago, he has been involved in smart and functional polymer systems such as shape memory polymers, piezoresistive, piezoelectric polymer nanocomposites, for both industrial and tissue engineering applications. He has graduated many Ph.D students in the field of EMI shielding polymer materials and anti-acoustic nanostructured foamed polymer systems.

Biocompatible and Biodegradable Polymers

Relationship Between Curing Characteristics and Mechanical Properties of HTV Silicone Rubber

Hamidreza Heidari and Azam Jalali-Arani(✉)

Department of Polymer Engineering and Color Technology, Amirkabir University of Technology, Tehran, Iran
ajalali@aut.ac.ir

Abstract. Curing characteristics of rubbers through their effects on crosslinking density affect the final properties of vulcanized rubbers. In this study, compounds of high temperature vulcanized silicone rubber (HTV-SR) containing different amounts of dicumyl peroxide (DCP) as curing agent were prepared and the effect of DCP concentration on the curing characteristics and mechanical properties of the vulcanized SR were investigated. The obtained results showed that by increasing the DCP amount the difference between maximum and minimum torque of rheometry was increased, however the tensile strength was increased for the DCP content up to 1 phr and by further increasing DCP and increasing the crosslink density it was unexpectedly decreased gradually. Increasing the DCP amount increased the hardness and decreased the elongation at break of the compounds.

Keywords: Silicone rubber · Curing characteristics · Mechanical properties

1 Introduction

Vulcanization is a process involving the conversion of linear rubber molecules into a network by the formation of chemical crosslinks [1]. The properties of cured rubbers are influenced by the type and amount of the used curing agent as well as the processing parameters [2, 3]. Among the rubbers, silicone rubber (SR) due to its excellent properties including high heat resistance, radiation resistance, excellent electrical insulation, thermal oxidative stability, corrosion resistance, superior fatigue resistance under extreme temperatures, fracture toughness, flexibility, biocompatibility, nontoxicity, and low cost has been widely used in various industrials fields, especially in bio applications. Depending on its structure, silicone rubber can be cured at room (RTV grade) or hot temperatures (HTV grade) [4, 5]. In this regard, dicumyl peroxide (DCP) is commonly used as the curing agent for the HTV grade SR.

In this work, different amounts of DCP were used as the curing agent in silicone rubber matrix, then the relation between curing characteristics and mechanical properties of the samples was studied.

H. Mirzadeh and A. A. Katbab (Eds.): ISPST 2018, *Eco-friendly and Smart Polymer Systems*, pp. 3–6, 2020.
https://doi.org/10.1007/978-3-030-45085-4_1

2 Experimental

2.1 Materials

Silicone rubber (HTV-30, consist of 24 wt% reinforcing silica filler) was supplied by Wacker Company (Germany) and dicumyl peroxide (DCP) (99%, AR) as the curing agent was purchased from a local company.

2.1.1 Preparation of Samples

The compounds which contained various contents, 0.5, 1, 1.5, 2, and 3 phr of DCP, were mixed on a two roll mill at temperature of 50 °C. Then the samples were cured on a hot press (Schwabenthan 400 hot-press (Germany)) at temperature of 160 °C. The cured samples were shelf aged for 24 h at room temperature before testing.

2.2 Characterization

Curing characteristics of the samples were evaluated at 160 °C using an oscillating disk rheometer (Monsanto 100S, England). Tensile tests were performed at room temperature on dumbbell-shaped specimens by using a tensile testing machine (HIWA 200, Iran) according to ASTM D412 test method.

The hardness of the samples was also determined with a Frank durometer (Germany).

3 Results and Discussion

The curing characteristics of the samples are illustrated in Table 1. This table shows that the scorch time of the samples was decreased by increasing the amount of DCP; however the cure time is slightly increased. It can be explained that the presence of a greater amount of DCP allows that the free radicals to be created faster and reduces the scorch time. This table shows that the value of M_H-M_L is increased by increasing the DCP amount. This can be attributed to the more crosslinks formed between the rubber chains in the compounds with higher amount of DCP.

Table 1. Curing characteristics of the samples with different contents of DCP (phr)

DCP phr	0.5	1	1.5	2	3
M_H-M_L (dN.m)	19.79	22.86	23.37	25.07	27.50
Scorch time (min)	2:04	1:35	1:12	1:03	0:56
Curing time (min)	6:50	7:24	7:40	7:55	11

The stress-strain curves of the prepared samples are shown in Fig. 1. As shown in this figure, by increasing the DCP amount up to 1 phr, the values of tensile strength and modulus are increased, while the elongation at break is decreased. As seen in Table 1, the increase in DCP content leads to an increase in crosslink density, by which the mobility of the rubber chains is decreased and concludes to decrement of elongation at break [1].

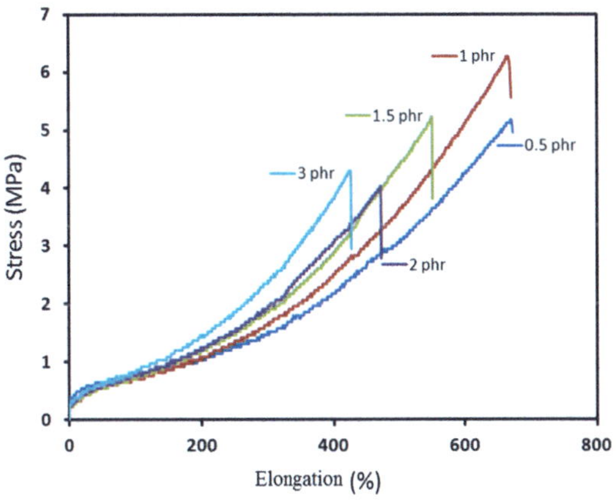

Fig. 1. Stress-strain curves of the compounds with various contents of DCP (phr)

By increasing DCP concentration further than 1 phr, an obvious decrease in tensile strength and modulus are observed. This means for having the best tensile strength and modulus, there is a critical value for the DCP content. It may be explained that by increasing the DCP, free radicals produced from its thermal degradation are increased and may attack to the SR chains and cause their scission [6]. This results the decrease in mechanical properties. The maximum value for tensile strength (6.26 MPa) and modulus (0.94 MPa) is obtained in the compound containing 1 phr DCP.

The variation of hardness (Shore A) versus the DCP content of the compounds is shown in Fig. 2. This figure indicates that by increase in DCP content its maximum is occurred in the compound with 3 phr DCP.

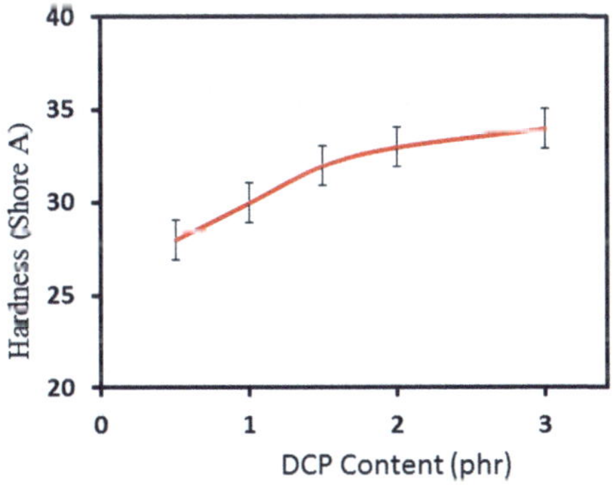

Fig. 2. Hardness of the compounds with various contents of DCP (phr)

4 Conclusion

The effect of DCP concentration on curing characteristics and mechanical properties of the vulcanized HTV-SR was investigated. The obtained results showed an increase in the difference between maximum and minimum torque of rheometry by increasing DCP amount. It was attributed to increase in crosslink density. The maximum tensile strength and elongation at break were observed for the SR with 1 phr DCP. Excessive increase in DCP leads to weaken the mechanical properties, due to production of additional radicals and creation of chain scission. Due to increase in crosslink density, the hardness was enhanced by increasing the DCP content.

References

1. Zhao, F., Bi, W., Zhao, S.: J. Macromol. Sci. Part B Phys. **50**, 1460–1469 (2011)
2. James, E., Burak, E., Frederick, R.E.: Science and Technology of Rubber. Elsevier Academic Press, San Diego (2005). Chapter 7
3. Rodgers, B.: Rubber Compounding—Chemistry and Applications. Marcel Dekker, New York, Basel (2004). Chapter 11
4. Wang, L.H., Ding, T.H., Wang, P.: Carbon **47**, 3151–3157 (2009)
5. Lipomi, D.J., Vosgueritchian, M., Tee, B.C., Hellstrom, S.L., Lee, J.A., Fox, C.H., et al.: Nat. Nanotechnol. **6**, 788–792 (2011)
6. Camino, G., Lomakin, S.M., Lazzari, M.: Polymer **42**, 2395–2402 (2001)

Porous Scaffolds of Polyvinyl Pyrrolidone/Polyethylene Glycol/Nano-Diamond Using Gamma Irradiation

Ali Jabbar[1,2], Susan Dadbin[1,2], and Masoud Frounchi[1,2](✉)

[1] Department of Chemical and Petroleum Engineering, Sharif University of Technology, Tehran, Iran
frounchi@sharif.edu

[2] Radiation Applications Research School, Nuclear Science and Technology Research Institute, Tehran, Iran

Abstract. Blends of polyvinyl pyrrolidone/polyethylene glycol (PVP/PEG) along with 1–3% nano-diamond particles (ND) were irradiated by gamma ray at various doses and then subjected to freeze-drying method to render porous scaffolds. At 35 kGy irradiation dose, the gel content was higher than 90%, though the presence of nanoparticles reduced slightly the gel content. It was found that for 10 kGy increase in irradiation dose, there was 15% reduction in pore size. On the other hand, upon the addition of 1% nano-diamond to PVP/PEG blends, the water swelling content of the scaffolds was increased by 10%. Also, the SEM images showed, at 1% nano-diamond level, the pore size was larger than pore size of neat PVP/PEG blends. The scaffolds were also tested for bio-compatibility, bio-degradability and mechanical properties. The results verified the suitability of these scaffolds for bio-medical applications.

Keywords: Polyvinyl pyrrolidone · Polyethylene glycol · Nano-diamond · Scaffolds · Irradiation

1 Introduction

Porous polymeric scaffolds are used for growth and proliferation of cells. An ideal scaffold should have interconnected-pore structure in order to facilitate cell migration along with surface biocompatibility to enhance cellular attachment and proliferation. Also suitable mechanical properties are essential during and after implantation process [1–4]. Both synthetic and natural polymers have been used to make biomedical scaffolds. In this work we have synthetized porous scaffolds of PVP and PEG and nano-diamond particles. Gamma irradiation was used to make PVP/PEG hydrogels followed by freeze-drying process to produce porous scaffolds. In particular, effect of nano-diamond particles on the properties of hydrogels and the resultant scaffolds were investigated.

H. Mirzadeh and A. A. Katbab (Eds.): ISPST 2018, *Eco-friendly and Smart Polymer Systems*, pp. 7–10, 2020.
https://doi.org/10.1007/978-3-030-45085-4_2

2 Experimental

PVP of 60000 g/mol and PEG of 4000 g/mol molecular weights were used. Nano-diamond particles were surface-treated by heating at 425 °C for 1.5 h in air atmosphere in a furnace before adding to polymers. PVP/PEG blends were prepared by dissolving both polymers in deionized water at a concentration of 10 wt%. Nano-diamond particles were added to the prepared solutions at 1, 2, and 3 wt% and sonicated for 45 min for uniform dispersion of nano-particles within the solution. Then the solutions were irradiated by gamma ray at doses of 15, 25, and 35 kGy. The formed hydrogels were freeze-dried to produce porous scaffolds.

The scaffolds were tested for gel content, water swelling ratio, cell growth and bio-compatibility, bio-degradation, mechanical properties and SEM images of pores size and micro-structure.

3 Results and Discussion

Gel content increased with gamma irradiation dose and reached 89% at 35 kGy for PVP80/20PEG blend (Table 1). In contrast, addition of ND decreased gel content showing that ND particles can absorb almost significantly irradiated gamma rays. By increase in PEG above 20%, the gel content was small because PEG is effectively insensitive to gamma irradiation. Similar results were obtained with water swelling ratio tests. Higher water swelling ratios obtained at higher ND contents (Fig. 1) as ND particles reduced the crosslinking effect of gamma irradiation and rendered less tight network for PVP/PEG hydrogels. Accordingly, at higher irradiation doses water swelling was lower as a result of tighter network gels. Also water swelling increased significantly at higher content of PEG. Indeed, PEG was removed during freeze-drying process and left behind large pores and voids within hydrogel body. SEM images of freeze-dried PVP80/20PEG/ND showed pores as large as 200 μm. Bio-degradability of PVP80/20PEG/ND at various ND contents is shown in Fig. 2. It appears that ND reduced bio-degradability of PVP80/20PEG after about two weeks. The scaffolds also showed good bio-compatibility in MTS test.

Table 1. Gel content of PVP80/20PEG/ND at different ND contents and irradiation doses

	15 kGy	35 kGy
PVP80/20PEG/ND 80/20/0%	83.3%	89.1%
PVP/PEG/ND 80/20/1%	82.1%	87.6%
PVP/PEG/ND 80/20/2%	80.5%	85.5%
PVP/PEG/ND 80/20/3%	78.3%	82.6%

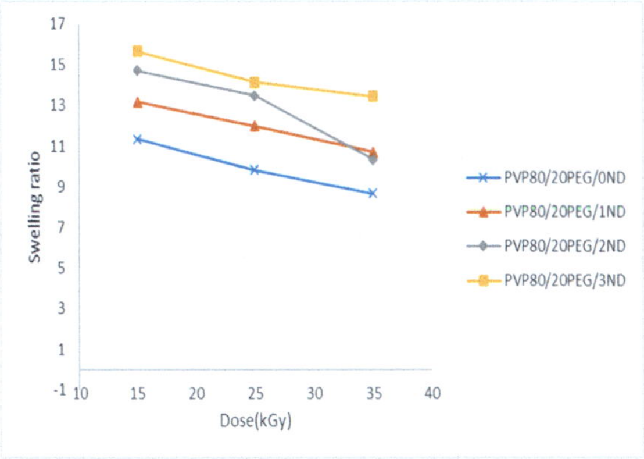

Fig. 1. Effect of nano-diamond concentration and gamma irradiation dose on swelling ratio of PVP80/20PEG/ND scaffolds

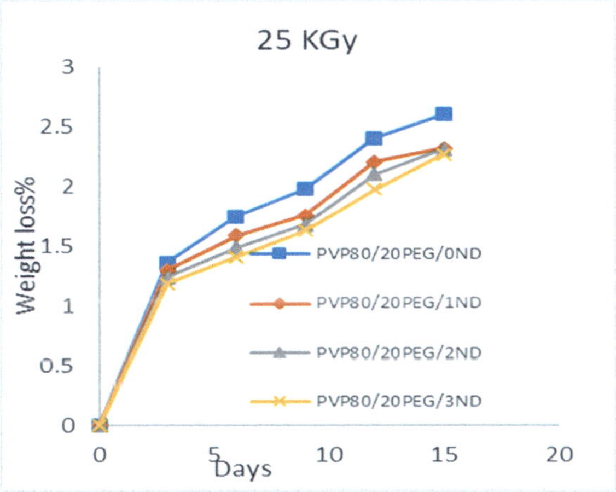

Fig. 2. Biodegradability of PVP80/20PEG/ND at 25 kGy gamma irradiation dose for different ND contents

4 Conclusion

PVP/PEG/ND hydrogels and scaffolds can be synthetized at moderate gamma irradiation doses that possess good water swelling and bio-compatibility. ND nano-particles could be used as an adjusting factor to obtain optimum properties of the scaffolds.

References

1. Zhang, F., He, C., Cao, L., Feng, W., Wang, H., Mo, X., Wang, J.: Int. J. Biol. Macromol. **48**(3), 474–481 (2011)
2. Lin, H.Y., Chen, H.H., Chang, S.H., Ni, T.S.: J. Biomater. Sci. Polym. Ed. **24**(4), 470–484 (2013)
3. Han, F., Dong, Y., Su, Z., Yin, R., Song, A., Li, S.: Int. J. Pharm. **476**(1), 124–133 (2014)
4. Zhong, S.P., Zhang, Y.Z., Lim, C.T.: Wiley Interdiscip. Rev. Nanomed. Nanobiotechnol. **2**(5), 510–525 (2010)

Chitosan and Polyethylene Glycol Nanocomposites Designed to Deliver Anticancer Drug: Molecular Dynamics Simulations

Zahra Shariatinia(✉) and Azin Mazloom Jalali

Department of Chemistry, Amirkabir University of Technology (Tehran Polytechnic), P.O. Box 15875-4413, Tehran, Iran
shariati@aut.ac.ir

Abstract. Chitosan (CS) and polyethylene glycol (PEG) nanocomposite systems containing SiO_2 nanoparticles and water molecules were designed by molecular dynamics (MD) simulations to deliver pipobromane (PIP) anticancer drug in order to discover more appropriate drug delivery system (DDS) in aqueous medium which was analogous to the human body. The CS system revealed greater fractional free volume (FFV) of 77.232%. The radial distribution function (RDF) data displayed that the PIP molecules had stronger H-bond interactions with the CS chains which reflected the drug molecules would diffuse slower inside the CS nanocomposite. Higher diffusion coefficient (0.0163×10^{-4} cm^2/s) was measured for the PEG but lower one was achieved for the CS (0.0154×10^{-4} cm^2/s) approving the drug diffusion in the CS was slower, thus providing more controlled and sustained drug delivery in this system.

Keywords: Molecular dynamics simulation · Drug delivery systems · Polymeric nanocomposites · Pipobromane anticancer drug · Diffusion coefficient

1 Introduction

Currently, chemotherapy has become a common cancer treatment by means of cytotoxic drugs [1]. However, traditional anticancer molecules may lead to severe systemic toxicity by distributing in the whole body through the blood circulation system [2]. This causes several adverse side effects including despairing the immune system, muscle pain and organ damage. Thus, chemotherapy treatment can be painful which may result in its non popularity [3]. Furthermore, most of anticancer drugs are hydrophobic with poor water solubility and this can hinder their widespread usage [4]. Consequently, it is crucial to develop targeted and controlled drug delivery systems (DDSs) for various drugs [5] and especially for the hydrophobic therapeutic anticancer

H. Mirzadeh and A. A. Katbab (Eds.): ISPST 2018, *Eco-friendly and Smart Polymer Systems*, pp. 11–14, 2020.
https://doi.org/10.1007/978-3-030-45085-4_3

drugs [6]. Nanocomposite DDSs materials composed of polymeric chains, inorganic nanoparticles and drug biomolecules have well been established as drug delivery vehicles for the anticancer drugs [7]. They have exhibited promising physicochemical properties which could be served as favorable candidates for novel DDSs.

2 Experimental

All of amorphous cells were incorporated with 500 H_2O molecules, 10 pipobromane drug molecules, 4 silica NPs (cristobalite-high) and 20 polymeric (chitosan or polyethylene glycol) chains having twenty repeating units in order to assess the influence of these polymers on the drug delivery efficacy. The three-dimensional cells were optimized by running the MD simulations at standard temperature (298.15 K) and pressure (1 atm). The computations were run using Materials Studio software (version 4.3).

3 Results and Discussion

In order to study and compare various features of CS, and PEG polymeric systems composed of silica NPs, water and PIP drug molecules, their relaxations and equilibrations were performed. Hence, the relaxation properties such as temperature along with potential and non-bond energies were projected during the MD simulations. Figure 1 exhibits the relaxed configurations of the CS and PEG polymeric composites achieved after 1 ns production run.

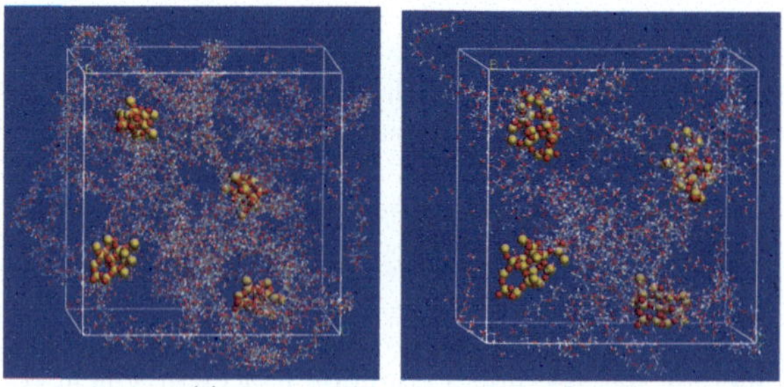

(a) **(b)**

Fig. 1. Snapshots of: (a) CS and (b) PEG polymeric drug delivery systems obtained after production MD simulation runs

It has been found that deviations in the non-bond and potential energies as well as temperature versus simulation time are very much little for all of the completely optimized CS and PEG polymeric systems indicating the systems have reached the relaxation states. Additionally, the non-bond energies of CS and PEG DDSs are smaller than their related potential energies by about 5500 and 1550 kcal/mol, respectively. Comparing the potential and non-bond energies of CS and PEG composites discloses that PEG system provides more negative energies, more positive values are measured for the CS system. The potential energies of the CS and PEG nanocomposites are about 19500 and –4600 kcal/mol while their corresponding non-bond energies are approximately 14000 and –6150 kcal/mol, respectively. Thus, it may be established that the PEG DDS with more negative energies is more apposite system. Nonetheless, in order to choose more appropriate DDS with more effectual drug delivery, other characteristics of the CS and PEG polymeric composites must be estimated and compared with each other. As a result, the relationships between polymeric nanocomposite type and PIP delivery efficacy will be analyzed in following sections to be able to state which system will be the most suitable candidate to be utilized for the delivery of PIP anticancer drug.

The FV is greater for the CS nanocomposite (260786.68 Å^3) and it is decreased to 92242.32 Å^3 in the PEG (91440.05 Å^3). The fractional free volume (FFV) is equal to the ratio of FV to the total cell volume. It is found that CS reveals greater FFV (77.232%) whereas PEG nanocomposite has a smaller FFV (70.387%). As greater FV and the FFV values belong to the CS and smaller values are obtained for the PEG system, it may be expected that more effective PIP delivery will be done using the CS system. Nonetheless, in order to determine more appropriate DDS for the PIP drug delivery, additional parameters should be assessed and compared with each other.

The PIP molecular diffusion into the CS and PEG polymeric nanocomposite systems was estimated from their MSD diagrams that are presented in Fig. 2. Also, the diffusion coefficient (D) for the PIP molecules was obtained using the slopes of these curves. The nearly linear lines in the MSD plots deduce constant diffusion of PIP drug molecules in the CS and PEG DDSs during the MD simulations. As well, it is noticeable that the CS provides smaller MSD for the PIP diffusion. Therefore, in order to have a proficient PIP delivery, a system with the smallest MSD must be nominated. This leads to the more controlled and slower passage/diffusion of PIP molecules within the CS DDS but faster transfer in the PEG nanocomposite. It is found from Fig. 2 that the CS is more effective and suitable system for the PIP drug delivery. These data are well consistent with the MSD curves so that higher diffusion coefficient (0.0163×10^{-4} cm^2/s) is obtained for the PEG whereas lower diffusion coefficient is attained for the CS (0.0154×10^{-4} cm^2/s). Thus, the diffusion of PIP in the CS matrix is slower which permits more efficient drug delivery. This result supports that the

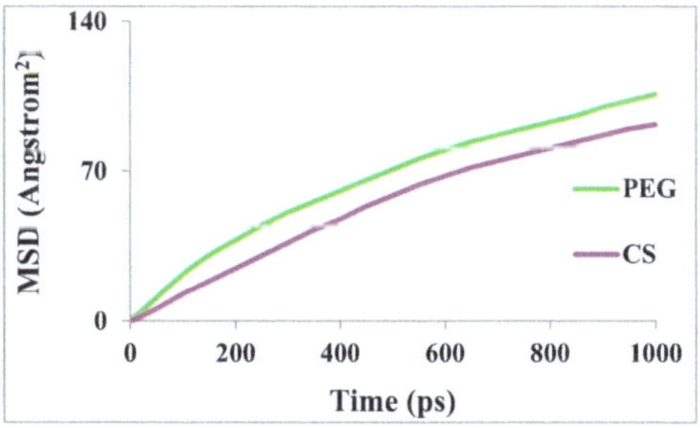

Fig. 2. MSD diagrams for the diffusion of pipobromane (PIP) in the CS and PEG drug delivery systems

delivery rate of the PIP drug using the examined systems varies in the order of CS < PEG.

4 Conclusion

MD simulations were performed on the CS and PEG systems containing silica NPs and water molecules in order to attain more suitable DDS with more proficient PIP drug delivery in the aqueous medium which is similar to the human body. It was found that CS had a greater FFV which reflected that the diffusion and transport of the PIP drug in the CS would happen more simply. A higher diffusion coefficient ($D = 0.0163 \times 10^{-4}$ cm^2/s) was obtained for the PEG but a lower D was achieved for the CS (0.0154×10^{-4} cm^2/s) justifying that the PIP diffusion in the CS was slower, thus providing more controlled and effectual drug delivery.

Acknowledgement. Authors would like to express their thanks to the Amirkabir University of Technology (Tehran Polytechnic), Tehran, Iran.

References

1. Kareva, I., Waxman, D.J., Klement, G.L.: Cancer Lett. **358**, 100–106 (2015)
2. Jain, A.K., Thanki, K., Jain, S.: Mol. Pharm. **10**, 3459–3474 (2013)
3. Xiao, H., Song, H., Zhang, Y., Qi, R., Wang, R., Xie, Z., Huang, Y., Li, Y., Wu, Y., Jing, X.: Biomaterials **33**, 8657–8669 (2012)
4. Nikfar, Z., Shariatinia, Z.: J. Mol. Graph. Model. **76**, 86–105 (2017)
5. Nikfar, Z., Shariatinia, Z.: Phys. E **91**, 41–59 (2017)
6. Li, Y., Maciel, D., Rodrigues, J., Shi, X., Tomás, H.: Chem. Rev. **115**, 8564–8608 (2015)
7. Spencer, D.S., Puranik, A.S., Peppas, N.A.: Curr. Opin. Chem. Eng. **7**, 84–92 (2015)

Molecular Dynamics Simulations on Polycaprolactone Nanocomposite Drug Delivery Systems

Zahra Shariatinia(✉) and Azin Mazloom Jalali

Department of Chemistry, Amirkabir University of Technology (Tehran Polytechnic), P.O. Box 15875-4413, Tehran, Iran
shariati@aut.ac.ir

Abstract. Molecular dynamics (MD) simulations were accomplished on polycaprolactone (PCL) nanocomposite systems composed of hydroxyapatite (HA) nanoparticles (0 and 2%) to deliver ifosfamide (IF) anticancer drug in order to achieve the most suitable drug delivery system (DDS). Smaller inter-chain distances measured for the PCL-HA2-IF established that greater inter-molecular interactions were occurred in the PCL-HA2-IF as a result of hydrogen bonds formation and electrostatic interactions among the PCL polymeric chains, HA NPs as well as IF molecules. It was found that the drug diffusion was slower/more controlled in the PCL-HA2-IF which would cause a more effective drug delivery.

Keywords: Molecular dynamics simulation · Polymeric nanocomposites · Drug delivery systems · Ifosfamide anticancer drug

1 Introduction

Cancer is second leading cause of death worldwide with an estimated increase from 14 million cases in 2012 to a projected 22 million new cases in the next two decades [1]. In 2017, more than 1.6 M new cases and more than 600,000 cancer deaths are predicted in the United States [2]. During the past decades important advances have been made to develop biodegradable polymeric materials for biomedical applications [3]. Hydroxyapatite [HA, $Ca_{10}(PO_4)_6(OH)_2$] is a bioceramic which has exceptional bio-compatibility including in vitro cell viability and osteogenic stimulatory roles of stem cells [4]. Ifosfamide (IF) is a DNA alkylating agent and one of the commonly used antineoplastic drugs [5]. The MD simulations explore the intermolecular interactions of polymeric carriers and drug molecules in polymeric nanocomposite drug delivery platforms [6].

H. Mirzadeh and A. A. Katbab (Eds.): ISPST 2018, *Eco-friendly and Smart Polymer Systems*, pp. 15–18, 2020.
https://doi.org/10.1007/978-3-030-45085-4_4

2 Experimental

All amorphous cells composed of 3 ifosfamide (IF) drug molecules, 0 and 2% HA NPs and ten PCL polymeric chains having twenty repeating units to examine the influence of nanocomposites on the drug delivery efficacy. The chemical structures of PCL, IF and HA species are displayed in Fig. 1. The 3–D cells were run using the MD simulations at standard temperature and pressure of 298.15 and 1 atm.K, respectively. All of initial amorphous cells were produced at a very low density of 0.15 g.cm^{-3} to create the cells having periodic boundary conditions through amorphous cell module existing in Materials Studio software (version 4.3).

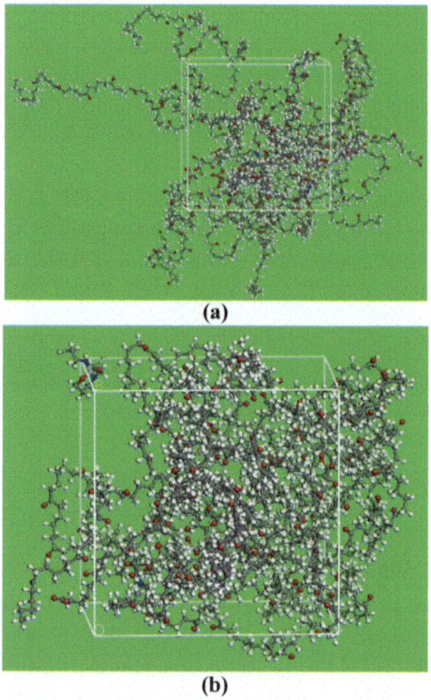

Fig. 1. Snapshots of: (a) PCL-0HA-IF and (b) PCL-2HA-IF polymeric drug delivery systems obtained by MD simulations

3 Results and Discussion

The relaxations and equilibrations of polymeric DDSs containing HA NPs (0 and 2%) were evaluated and compared with each other. Thus, some relaxation properties such as temperature in addition to the kinetic, potential and non-bond energies were acquired during the MD simulations. The relaxed configurations of the polymer composites obtained after the final 1 ns production run are shown in Fig. 1.

It was seen that the non-bond energies of all systems were more negative than their related potential energies by ~1000–1500 kcal/mol. As well, positive values were acquired for the kinetic energies of all PCL-HA-IF systems which specify the kinetic energies have the highest amounts among other energies. Comparing the potential and non-bond energies of PCL-HA-IF cells indicated that these energies were decreased by adding the HA NPs into the system. Consequently, it may be decided that the PCL-HA2-IF with the most negative energies can possibly be preferred as the most suitable system. Nevertheless, in order to find the most appropriate DDS having the most effective drug delivery, other features of the PCL-HA-IF polymeric composites must be assessed and compared with each other.

Diverse RDF values are accounted for weak, moderate and strong interactions occurred between the particles in the cell. Figure 2 demonstrates several RDF plots for the inter-atomic interactions happened within the PCL-HA-IF systems. Comparing the drug-polymer interactions in the PCL-HA-IF nanocomposites displays that the smallest drug-polymer RDFs are observed for the H(NH-IF)-O(C=O-PCL) and P(IF)-O(C=O-PCL) contacts that are H-bonding and electrostatic interactions, respectively, and change in the range of 1.95–2.99 and 2.80–2.93 Å. Further electrostatic interactions are very weaker and take place between the O, N, and Cl functional groups of IF drug with the C atom of the PCL polymer. These findings validate that the IF drug molecules have stronger interactions with the PCL matrix in the PCL-HA2-IF in these systems.

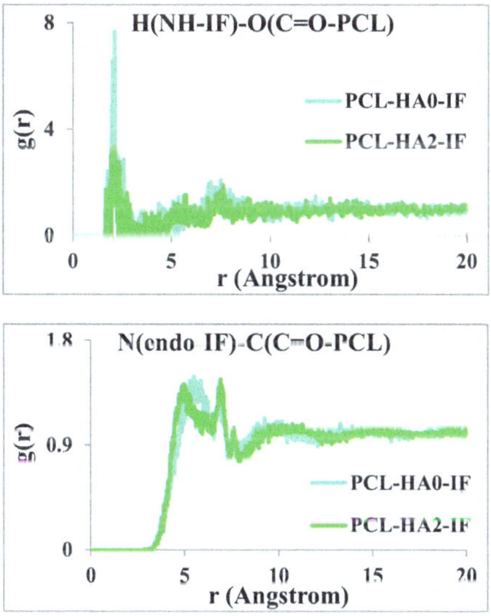

Fig. 2. RDF plots for the diffusion of ifosfamide (IF) in the PCL drug delivery systems

4 Conclusion

The MD simulations were accomplished on the PCL polymeric nanocomposite systems having HA NPs in order to achieve the most suitable DDS indicating the most efficient delivery of the IF drug. The non-bond energies of all systems were more negative by ~1000–1500 kcal/mol compared with their related potential energies. Additionally, positive values were calculated for the kinetic energies of all PCL-HA-IF cells. The RDF data illustrated that in all DDSs, the HA-polymer interactions were the strongest contacts because their RDFs were the smallest values. Also, the RDF results illustrated that the IF molecules had stronger interactions within the PCL-HA2-IF but weaker interactions inside the PCL-HA0-IF which verified the IF molecules would diffuse the most slowly/controlled in the PCL-HA2-IF system which justified the IF transport/diffusion was the most controlled/slowest in the PCL-HA2-IF that permitted the most efficient drug delivery.

Acknowledgement. Authors would like to express their thanks to the Amirkabir University of Technology (Tehran Polytechnic), Tehran, Iran.

References

1. Aftab, S., Shah, A., Nadhman, A., Kurbanoglu, S., Ozkan, S.A., Dionysiou, D.D., Shukla, S. S., Aminabhavi, T.M.: Int. J. Pharm. **540**, 132–149 (2018)
2. Gillessen, S., Attard, G., Beer, T.M., Beltran, H., Bossi, A., Bristow, R., Carver, B., Castellano, D., Omlin, A., et al.: Eur. Urol. **73**, 178–211 (2018)
3. Kohsari, I., Shariatinia, Z., Pourmortazavi, S.M.: Int. J. Biol. Macromol. **91**, 778–788 (2016)
4. Kim, H., Mondal, S., Bharathiraja, S., Manivasagan, P., Santha Moorthy, M., Oh, J.: Ceram. Int. **44**, 6062–6071 (2018)
5. Sasaki, Y., Kjellén, E., Ekblad, L., Wahlberg, P., Mineta, H., Wennerberg, J.: Oral Oncol. **48**, 61–66 (2012)
6. Shariatinia, Z., Mazloom Jalali, A.: Int. J. Biol. Macromol. **115**, 194–220 (2018)

Fabrication of Antibacterial Electrospun Chitosan-Polyethylene Oxide Nanocomposite Nanofibrous Mats

Zahra Shariatinia(✉) and Iraj Kohsari

Department of Chemistry, Amirkabir University of Technology (Tehran Polytechnic), P.O. Box 15875-4413, Tehran, Iran
shariati@aut.ac.ir

Abstract. Antimicrobial chitosan–polyethylene oxide (CS-PEO) nanofibrous mats were developed by electrospinning technique for wound dressing applications. A green method was used to achieve antibacterial mats loaded with 0.25 and 0.50 w/w% of bioactive silver nanoparticles (Ag NPs, ~70 nm diameter) reduced by *Falcaria vulgaris* herbal extract. All of the nanofibers had an average ~200 nm diameter. Interestingly, both of the CS-PEO mats containing 0.25 and 0.50% bioactive *F. vulgaris-Ag* NPs revealed 100% bactericidal activities against both *S. aureus* and *E. coli* bacteria. The improved properties of CS-PEO-0.50% *F. vulgaris-Ag* NPs relative to those of the CS-PEO 0.25% *F. vulgaris-Ag* NPs suggested that the former was superior for biomedical applications.

Keywords: Electrospinning · Chitosan · Polyethylene oxide nanofibrous mats · Silver nanoparticles · Antimicrobial activity · Falcaria vulgaris herbal extract

1 Introduction

Nowadays, polymeric nanofibers have gained much increased interest due to they are applied in a wide variety of industrial fields and present a bright future [1]. The natural polyaminosaccharide chitosan (CS) has extensively been used in the fibrous architecture for tissue scaffolds [2], wound dressing [3] and antibacterial films and coatings [4]. A number of Ag NPs embedded electrospun nanofibers with anti-microbial potential have been prepared [4]. *Falcaria vulgaris* (locally named ghazzyaghi/poghazeh) from Apiaceae family is a medicinal plant that is consumed as a vegetable in Iran, also for healing skin ulcer, stomach disorders including peptic ulcer, liver diseases and stones of kidney and bladder [5].

H. Mirzadeh and A. A. Katbab (Eds.): ISPST 2018, *Eco-friendly and Smart Polymer Systems*, pp. 19–22, 2020.
https://doi.org/10.1007/978-3-030-45085-4_5

2 Experimental

2.1 Synthesis of Silver Nanoparticles

Ag NPs were synthesized by means of simple chemical reduction of silver nitrate using *Falcaria vulgaris* herbal extract. The aqueous extract solution of the plant leaves was added to the $AgNO_3$ aqueous solution and the reaction flask was covered with aluminum foil and subjected to a constant mixing on a magnetic stirrer. The addition of the plant extract (dark green) to $AgNO_3$ aqueous solution (colorless) led to turning the solution color to dark grey indicating formation of silver nanoparticles.

2.1.1 Solution Preparation and Electrospinning Process

In a typical procedure, 50 v/v% of concentrated aqueous acetic acid solution was used as the solvent. Total concentration of the CS and PEO polymers in the solution was 4 w/w% with the CS to PEO weight ratio was 75/25. The solutions containing CS, PEO and biosynthesized Ag NPs using *F. vulgaris* extract were prepared by addition of 0.25 and 0.50 w/w% of nanobio silver aqueous solutions into the 50% acetic acid solution including CS and PEO. The CS-PEO-herbal extract-Ag NPs solution was stirred overnight in a dark brown bottle and then it was sonicated for 15 min to get Ag NPs incorporated electrospinning solution.

The electrospinning was performed in the laboratory spinning unit (ANSTCO-N/VI), which was designed in terms of a vertical working principle. The power supply (AC) was set up for a positive voltage of 21 kV. The flow rate of the solution was also determined by setting up the syringe pump at 1 mL/h. The rotational speed of the drum collector was 2500 rpm and its distance was set to 10 cm away from the nozzle. The relative humidity and temperature values ranged from 35% to 42% RH and 25 °C to 35 °C, respectively.

3 Results and Discussion

The FE-SEM images presented in Figs. 1a and 1b belong to CS-PEO nanofiber mats containing 0 and 0.25 w/w% of *Falcaria vulgaris-Ag* NPs, respectively. It is observed in both cases that very uniform fibers free of any nanobeads and droplets are formed that are about 200 nm in size. In addition, the presence of bioactive Ag NPs is observed within the nanofibers and there is not any agglomeration of Ag NPs between the fibers or on their surfaces. Similarly, according to the FE-SEM images in Figs. 1c and 1d, it is found that the mat containing 0.50 w/w% of bioactive *Falcaria vulgaris-Ag* NPs and also its related mat crosslinked with glutaraldehyde (GA) vapor are very uniform which are ~200 nm in size. Furthermore, the existence of the Ag NPs is evident in the FE-SEM image of the non-crosslinked fibers but it is not very clear for the GA crosslinked nanofibers. This is expected due to crosslinking leads in the formation of additional bonds between the polymers and crosslinker covering the Ag NPs.

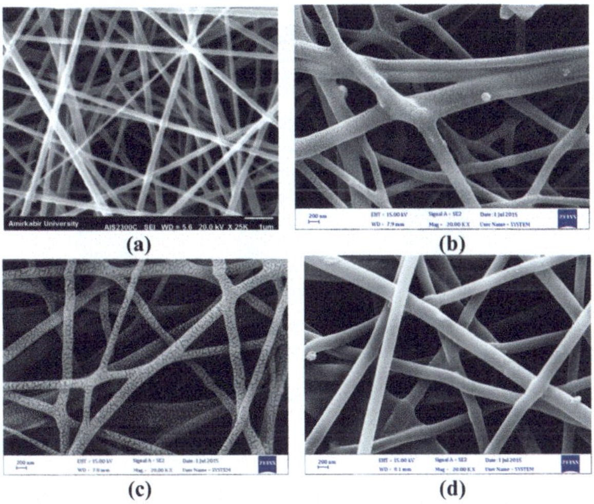

Fig. 1. SEM images of: (a) CS-PEO, (b) CS-PEO-0.25% F. vulgaris-Ag NPs, (c) CS-PEO-0.50% F. vulgaris-Ag NPs, and (d) CS-PEO-0.25% F. vulgaris-Ag NP-GA nanocomposite mats

As can be seen in Fig. 2, the contact angle of pure CS-PEO mat is slightly smaller than the mat loaded with 0.25% bioactive Ag NPs displaying the hydrophilicity is somewhat decreased in the nanocomposite mat. This may be attributed to the slightly decreased tendency of the mat containing 0.25% silver NPs to interact with water which is due to entrapment of Ag particles within the fibers so that their interactions mainly occur with the polymers. The picture observed on top of each column in Fig. 2 reveals a water droplet placed on the mat surface indicating a droplet is more dispersed (smaller angle) on a more hydrophilic fiber but it is more spherical (it is not dispersed) on a less hydrophilic fiber surface (bigger angle). The contact angle is decreased by increasing the Ag NPs amount from 0.25 w/w% to 0.50 w/w% reflecting much greater hydrophilicity for the latter. The greater tendency to interact with water is due to the increase in the amount of Ag NPs within the fibers leading to stronger interactions with water, because greater number of Ag atoms are present on the surface of the mat thus they can be accessible to the water molecules. According to these results, it can be concluded that the CS-PEO-0.25% *F. vulgaris-Ag* NPs mat is preferred over the mat containing 0.5% *F. vulgaris-Ag* NPs for biomedical applications (such as wound dressing) due to its improved hydrophilic property.

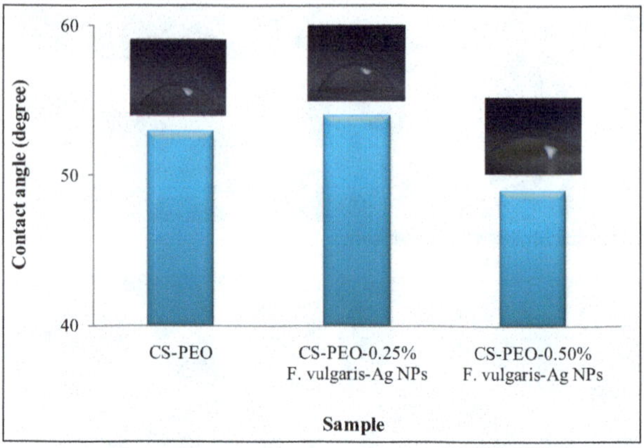

Fig. 2. (a) Contact angles of CS-PEO nanofibrous mats containing 0, 0.25, and 0.50 *F. vulgaris-Ag* NPs

4 Conclusion

Silver nanoparticles (Ag NPs) were synthesized using *Falcaria vulgaris* extract as a green reductant and applied for the fabrication of antibacterial CS-PEO nanofiber mats by electrospinning technique. The water contact angle test illustrated that the hydrophilicity of the CS-PEO-0.50% *F. vulgaris-Ag* NPs mat was the highest compared to that of pure CS-PEO and the CS-PEO-0.25% *F. vulgaris-Ag* NPs mats. Finally, the CS-PEO-0.50% *F. vulgaris-Ag* NPs mat was considered as more preferred sample for biomedical and wound dressing applications compared with the CS-PEO-0.25% *F. vulgaris-Ag* NPs.

Acknowledgement. Authors would like to express their thanks to the Amirkabir University of Technology (Tehran Polytechnic), Tehran, Iran.

References

1. Lee, K.H., Kim, H.Y., Khil, M.S., Ra, Y.M., Lee, D.R.: Polymer **44**, 1287–1294 (2003)
2. Chen, J.-P., Chen, S.-H., Lai, G.-J.: Nanoscale Res. Lett. **7**, 170 (2012)
3. Gu, B.K., Park, S.J., Kim, M.S., Kang, C.M., Kim, J.-I., Kim, C.-H.: Carbohydr. Polym. **97**, 65–73 (2013)
4. Shariatinia, Z., Fazli, M.: Food Hydrocolloid **46**, 112–124 (2015)
5. Liu, Y., Liu, Y., Liao, N., Cui, F., Park, M., Kim, H.-Y.: Int. J. Biol. Macromol. **79**, 638–643 (2015)
6. Shafaghat, A.: Nat. Prod. Res. **25**, 368–373 (2011)

Biocompatible Electrospun Nanofibrous Mats Designed for Controlled Co-release of Hydrocortisone and Imipenem/Cilastatin Drugs

Zahra Shariatinia(✉) and Yousef Fazli

Department of Chemistry, Amirkabir University of Technology
(Tehran Polytechnic), P.O. Box 15875-4413 Tehran, Iran
shariati@aut.ac.ir

Abstract. Antimicrobial chitosan–polyethylene oxide (CS-PEO) nanofibrous mats containing ZnO nanoparticles (NPs) and hydrocortisone-imipenem/cilastatin-loaded ZnO NPs were produced by electrospinning technique. The FE-SEM images displayed that the spherical ZnO NPs were ~70–200 nm in size and the CS-PEO nanofibers were very uniform and free of any beads which had average diameters within the range of ~20–130 nm. The hydrocortisone release increased to 82% within first 12 h while the release rate of imipenem/cilastatin was very much slower so that 20% of the drug was released during this period of time suggesting this nanofibrous mat is very suitable to inhibit inflammation (by hydrocortisone) and infection (using imipenem/cilastatin antibiotic and ZnO NPs) principally for the wound dressing purposes.

Keywords: Chitosan–polyethylene oxide nanofibrous mats · Electrospinning · Hydrocortisone · Imipenem/cilastatin · Zinc oxide nanoparticles

1 Introduction

Nowadays, electrospinning has become the most commonly used technique for the fabrication of nanofibers due to its simplicity, adaptability, cost-efficiency and versatility [1]. Electrospun nanofibers have outstanding properties that make them appropriate as wound dressings, i.e. high oxygen permeability, variable pore size and high surface area to volume ratio. Additionally, nanofibrous mats are morphologically similar to the extracellular matrix enhancing the cell migration and proliferation. Therefore, the electrospun mats are especially suitable for different biomedical applications such as wound dressing [2]. Chitosan is a biopolymer which is well known as being able to accelerate the wound healing in human [3] and this is due to its substantial antibacterial activity against a broad spectrum of bacteria [4]. One of the main threats in the health care is the emergence of microbial organisms which are resistant to antibiotics. As a result, researchers have tried to develop new effective antimicrobial agents such as ZnO nanoparticles which may show far lower tendency to induce microbial resistance than antibiotics [5].

H. Mirzadeh and A. A. Katbab (Eds.): ISPST 2018, *Eco-friendly and Smart Polymer Systems*, pp. 23–26, 2020.
https://doi.org/10.1007/978-3-030-45085-4_6

2 Experimental

2.1 Synthesis of ZnO NPs

An aqueous solution containing $Zn(NO_3)_2.4H_2O$ (1 g) and glycerin (5.6 mL) in 1:20 molar ratio was prepared in 20 mL of distilled water and the solution was stirred for 1 h at 90 °C to obtain a homogenueous concentrated solution. Then, it was placed in a furnace and heated up to 700 °C with a heating rate of 5 °C/min for 4 h. After calcination, the white powder of ZnO NPs were washed with acetone and dried at 100 °C.

2.1.1 Solution Preparation and Electrospinning Process

An aqueous solution of 50 v/v% concentrated acetic acid was used as the solvent for CS and PEO and total concentration of the two polymers was 4 w/w% with the CS to PEO weight ratio was 75:25. The 3% ZnO NPs (or 2% hydrocortisone-3% imipenem/cilastatin-loaded ZnO NPs) powder was added into the polymer solution and the mixture was stirred overnight in a dark brown bottle. Then, it was sonicated for 20 min to obtain the electrospinning mixture.

The electrospinning was performed in the laboratory spinning unit (ANSTCO-N/VI), which was designed in terms of a vertical working principle. The power supply (AC) was set up for a positive voltage of 21 kV. The flow rate of the solution was also determined by setting up the syringe pump at 1 mL/h. The rotational speed of the drum collector was 2500 rpm and its distance was set to 10 cm away from the nozzle. The relative humidity and temperature values ranged from 35% to 42% RH and 25 °C to 35 °C, respectively.

2.1.2 Drug Loading

The drug loading was done by dispersing the ZnO NPs in an aqueous solution containing exact amount of the imipenem/cilastatin sodium drug in distilled water. The suspension was stirred for 1 h at room temperature. After that, the mixture was centrifuged and the powder was washed with distilled water and dried at 50 °C for 5 h. The amount of imipenem/ cilastatin drug remained was determined using a UV-Vis spectrophotometer at wavelength of 300 nm.

3 Results and Discussion

The FE-SEM images presented in Fig. 1 belong to CS-PEO, CS-PEO-ZnO, and CS-PEO-hydrocortisone-imipenem/cilastatin-loaded ZnO nanofibrous mats. For the pure CS-PEO, CS-PEO-ZnO, and CS-PEO-hydrocortisone-imipenem/cilastatin-loaded ZnO nanofibrous mats, very uniform fibers free of any nanobeads have been produced and their diameters are about 90–110, 60–130, and 16–130 nm, respectively. It is notable that in case of CS-PEO-hydrocortisone-imipenem/cilastatin-loaded ZnO nanofibrous mat, all of the nanofibers have not the same average diameters but very thin nanofibers with average diameters of about 20 nm have been formed on much thicker nanofibers with average diameters of 130 nm.

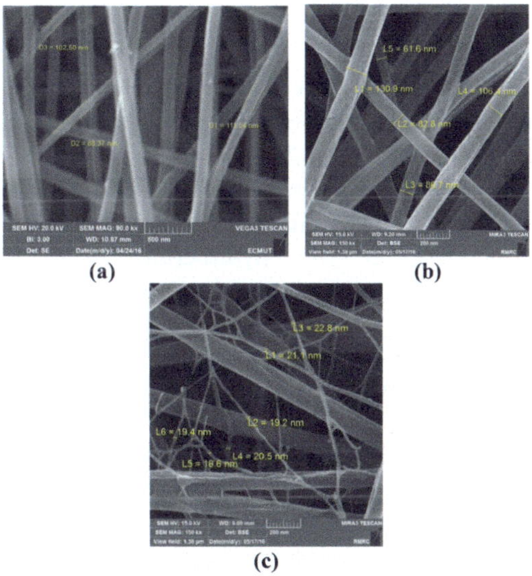

Fig. 1. FE-SEM images of: (a) CS-PEO, (b) CS-PEO-ZnO, and (c) CS-PEO-hyd-ZnO(Im) mats

To study the release behaviors of hydrocortisone and imipenem/cilastatin drugs from CS-PEO-hyd-ZnO(Im) mat, the release experiments were examined during 15 days and the results are plotted in Fig. 2. It is observed that the hydrocortisone release is sharply increased within first 12 h (burst release) so that 82 ± 4.8% of the hydrocortisone has been released during this time, it reaches 91 ± 4.7% after 24 h and after that the release rate becomes very slow (96 ± 4.6% release after 360 h). However, the release rate of imipenem/cilastatin is very much slower in comparison with that of hydrocortisone so that 20 ± 2.9% of the drug is released after 12 h, it is 28 ± 2.7% after 24 h and reaches 48 ± 2.8% after 360 h. The amount of imipenem/ cilastatin released at the same time is very much lower than that of hydrocortisone. This result may show that this nanofibrous mat is suitable to inhibit inflammation (by hydrocortisone) and infection (using imipenem/ cilastatin antibiotic and/or ZnO NPs) especially in wound dressing applications.

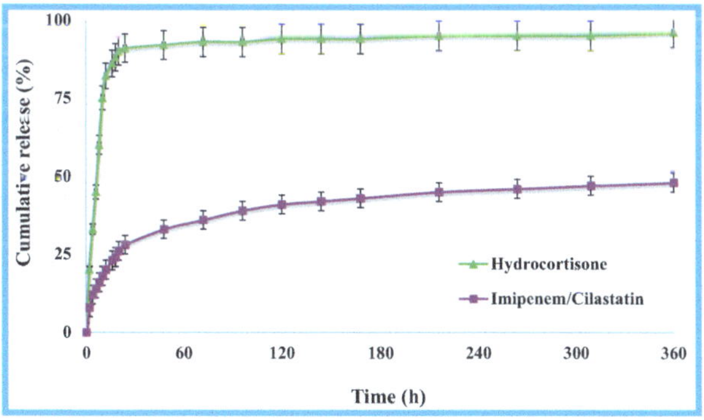

Fig. 2. Drugs release profiles (in PBS buffer) for the CS-PEO-hyd-ZnO(Im) nanofibrous mat

4 Conclusion

The CS-PEO nanocomposite nanofibrous mats containing 3% ZnO NPs and 2% hydrocortisone-3% imipenem/cilastatin-loaded ZnO NPs were fabricated. The hydrocortisone release abruptly increased to 82% within first 12 h (burst release) while the release rate of imipenem/cilastatin was very much slower during this time so that 20% of the drug was released after 12 h confirming this nanofibrous mat was very appropriate to inhibit inflammation (by hydro-cortisone) and infection (using imipenem/cilastatin antibiotic) mainly in wound dressing applications.

Acknowledgement. Authors would like to appreciatively express their thanks to the Amirkabir University of Technology (Tehran Polytechnic), Tehran, Iran.

References

1. Tamm, I., Heinämäki, J., Laidmäe, I., Rammo, L., Paaver, U., Ingebrigtsen, S.G., Škalko-Basnet, N., et al.: J. Pharm. Sci. **105**, 1239–1247 (2016)
2. Unnithan, A.R., Ghavami Nejad, A., Kurup Sasikala, A.R., Thomas, R.G., Jeong, Y.Y., Murugesan, P., Nasseri, S., et al.: Chem. Eng. J. **287**, 640–648 (2016)
3. Tian, J., Tu, H., Shi, X., Wang, X., Deng, H., Li, B., Du, Y.: Colloid Surf. B Biointerface **145**, 643–652 (2016)
4. Li, J., Wu, Y., Zhao, L.: Carbohydr. Polym. **148**, 200–205 (2016)
5. Shahmohammadi Jebel, F., Almasi, H.: Carbohydr. Polym. **149**, 8–19 (2016)

Peptide Release Behavior from *in-Situ* Forming Implants: Effect of Multi-walled CNT on Release Profile

Ehsan Avazverdi[1], Hamid Mobedi[2(✉)], and Hamid Mirzadeh[1]

[1] Polymer and Color Engineering Department, Amirkabir University of Technology, Tehran, Iran
[2] Novel Drug Delivery Group, Polymer Science Department, Iran Polymer and Petrochemical Institute, Tehran, Iran
h.mobedi@ippi.ac.ir

Abstract. The objective was to investigate the role of multi-walled carbon nano-tube MW-CNT as a nucleating agent on the phase inversion process and its effects on the morphology of PLGA in-situ forming implants and release behavior of leuprolide acetate as a model peptide. High performance liquid chromatography (HPLC) was used to measure drug release from implant and scanning electron microscopy (SEM) was conducted to study degradation and morphology of the PLGA implant. Drug release profile showed a remarkable decrease in the first and second stage bursts and a zero-order profile in the second stage which are related to phase inversion rate and implant morphology. Addition of MW-CNT led to changes in the cross sectional morphology from short and closed channels to deep and open channels. Channel depth increased 8 times from 50 μm to 400 μm by the addition of 0.8 wt% carbon nanotubes.

Keywords: PLGA · CNT · Grafting to · Esterification catalyst · Molecular weight

1 Introduction

Poly(lactide-*co*-glycolide) as random copolymer of lactide and glycolide has introduced a wide variety of potential applications in design and development of drug delivery formulations by means of controlling its physical, chemical and biological properties such as crystallinity, glass transition temperature, hydrophilicity, intrinsic viscosity and biodegradability. One of the most promising drug delivery strategies is solvent/non-solvent induced polymer phase inversion that results in formation of a drug reservoir in the body. These systems are so called the in-situ forming implants (ISFIs) [1].

Carbon nano-structures like graphene and carbon nanotube (CNT) have introduced several interesting properties such as excellent thermal, electrical and mechanical properties and have been applied in different biomedical applications such as nerve regeneration, tissue engineering and drug delivery areas [2]. Carbon nano-structures owe most of their unique properties to the high surface active area (SAA) which establishes marvelous surface and interfacial properties. Several studies have been

H. Mirzadeh and A. A. Katbab (Eds.): ISPST 2018, *Eco-friendly and Smart Polymer Systems*, pp. 27–30, 2020.
https://doi.org/10.1007/978-3-030-45085-4_7

reported the effect of graphene and CNT as heterogeneous nucleation agents on crystallinity and phase separation behavior of polymers.

To the best of our knowledge, the effect of CNT on the release of peptide from ISFIs has not been studied. The objective of this article is to study the effect of multi walled carbon nano-tubes (MW-CNT) on the morphology of ISFIs and its relationship to drug release behavior and degradation process of PLGA.

2 Experimental

Materials

−OH terminated PLGA (lactide:glycolide 50:50, M_W = 54 kDa) was provided by Evonik Industrials (Germany). MW-CNT (OD < 7 nm, 95% purity) was supplied by US Research Nanomaterials Inc. Leuprolide acetate was purchased from Bachem (Switzerland) as a model peptide.

Preparation of PLGA/CNT-*g*-PLGA Nanocomposite

Drug formulations including 32 wt%. of pure PLGA or PLGA nanocomposite containing 0.8% modified CNT, 65 wt% NMP and 3 wt% leuprolide acetate were prepared and mixed to yield a homogenous mixture and immediately injected into 5 mL phosphate buffer saline (PBS) at pH 7.4 as release medium. Release media were placed in the incubator at the temperature of 37 °C.

Instrumentation

Drug concentrations in release media were analyzed using Aliance model Waters (USA) high performance liquid chromatography (HPLC) equipped with UV-detector, online degasser. Cross section morphologies were studied using scanning electron microscopy Vegall model produced by Tescan (Czech Republic). Samples were coated with gold to improve electrical conductivity.

3 Results and Discussion

Figure 1 shows cumulative drug release profile from ISFI for formulations containing pure PLGA and PLGA/MW-CNT nanocomposite. In the first hours after injection the initial burst release for pure PLGA is more than that for PLGA-CNT nanocomposite. During phase inversion, the dissolved drug in solvent leaks into the release medium. By addition of MW-CNT the rate of phase separation increased and solvent extraction happened in higher pace. The release behavior from pure PLGA ISFIs exhibits two jumps in cumulative profile. Implant containing MW-CNT showed a linear release behavior in the second stage of the profile. However, addition of MW-CNT-g-PLGA did not change the profile trend between 48 and 336 h which is so called diffusion controlled region. Figure 2A and B displays cross-sectional micrograph of ISFIs for pure PLGA and nano-composite. Addition of the nano-tubes imposed significant changes into the formulations; whereas the channel depth increased 8 times from 50 μm dead ended to 400 μm open ended channels. Choi [3] et al. reported that carbon nano-tube increased channel depth of the polysulfun membranes by means of increasing the hydrophilicity of membrane in solvent induced phase separation process.

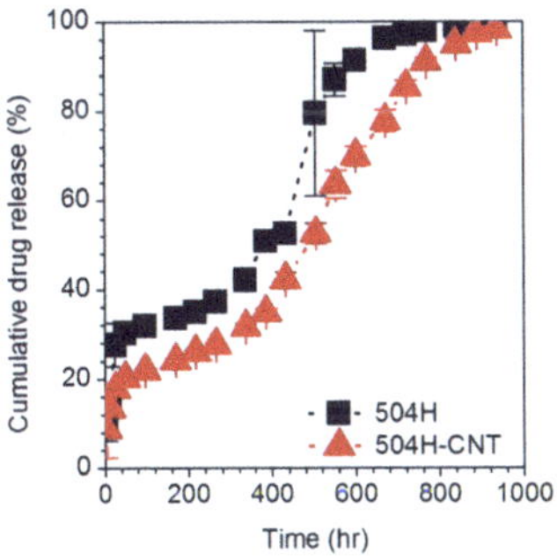

Fig. 1. Cumulative release profile of leuprolide acetate in pure PLGA and MW-CNT ISFIs

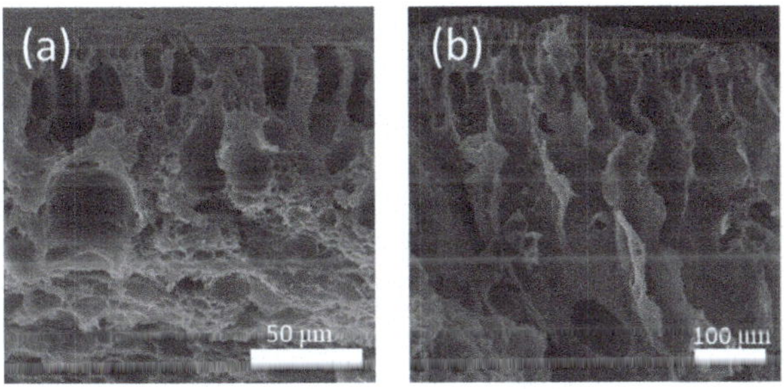

Fig. 2. Cross section SEM micrographs of pure PLGA (a) and MW-CNT loaded (b) ISFIs after 96 h injection into the release media

4 Conclusion

The effect of MW-CNT on the morphology and its relationship to drug release and degradation of PLGA ISFIs have been investigated in this study. Drug release profiles obtaining from samples containing pure and MW-CNT showed a significant change in the first and second stage burst release which were attributed to the increasing phase inversion rate and morphological changes, respectively. It leads to various degradation types from bulk to monotonic erosion and zero-order release in the second stage of the release profile as a result.

References

1. Hatefi, A., Amsden, B.: Biodegradable injectable in situ forming drug delivery systems. J. Control Release **80**(1), 9–28 (2002)
2. Fabbro, A., Prato, M., Ballerini, L.: Carbon nanotubes in neuroregeneration and repair. Adv. Drug Deliv. Rev. **65**(15), 2034–2044 (2013)
3. Choi, J.-H., Jegal, J., Kim, W.-N.: Fabrication and characterization of multi-walled carbon nanotubes/polymer blend membranes. J. Membr. Sci. **284**(1), 406–415 (2006)

Drug Release Modeling Through a 3-Stage Model on Hydrophobic Modified Carboxymethyl Chitosan Nanoparticles

Nafise Olov, Hamid Mirzadeh(✉), and Shadab Bagheri_Khoulenjani

Polymer and Color Engineering Department,
Amirkabir University of Technology, Tehran, Iran
mirzadeh@aut.ac.ir

Abstract. In this paper, the mechanism of releasing a small molecule (curcumin) from chitosan nanoparticles was investigated by a three-stage model with constant D. This model takes into account simultaneously initial burst, nanoparticle degradation–relaxation and diffusion with constant diffusion coefficient. Modified chitosan with hydrophobic functional groups is able to make self-assembled nanoparticles in aqueous media which in this study FA modified degraded carboxymethyl chitosan has been synthesized using folic acid, *N* hydroxy succinimide (NHS), *N, N*, dicyclohexylcarbodiimide (DCC) and 1-ethyl-3-(3-dimethylaminopropyl) carbodiimide EDC). The morphology of nanoparticles was observed by scanning electron microscopy which had spherical shape with an average size less than 200 nm. *In vitro* drug release studies were performed at 37 °C using dissolution medium phosphate buffered saline (pH 7/4). The release profiles exhibited three well defined phases, which were modeled as well by three-phase model with best coefficient of determination (R^2 = 0.9649).

Keywords: Carboxymethyl chitosan · Curcumin · Active targeting · Nanoparticle · Modeling

1 Introduction

Controlled release systems have several advantages, such as reducing high total dose, dosing frequency, side effects, and improving patient acceptance and compliance [1]. While there is considerable research on the systems composed of synthesis polymer such as PLGA nanoparticles due to less batch-to-batch variability as compared to degradable polymers of natural sources, much less is published on the release mechanism from natural polymer. In this research we used chitosan as a model.

Drug release from degradable polymers can be governed by (1) surface erosion of the polymer matrix, (2) cleavage of polymer bonds at the surface or within the bulk of the matrix, or (3) diffusion of the physically entrapped drug. However, drug release is often a result of a simultaneous combination of all three [2].

Complicated and more accurate mathematical models are combination of two and three phases. In 1997 Batycky et al. proposed bi-phasic release model which account the presence of initial burst release followed by diffusion via interconnected pores [3]. Another bi phasic model which was proposed by Gallagher [4] considers initial burst

H. Mirzadeh and A. A. Katbab (Eds.): ISPST 2018, *Eco-friendly and Smart Polymer Systems*, pp. 31–34, 2020.
https://doi.org/10.1007/978-3-030-45085-4_8

release and bulk degradation of the polymer in microparticles and it was developed to nanoparticles by Corrigan in 2009 [5]. Lucero (2015) by adding diffusion mechanism to Fitzgerald's model, developed a triphasic model that considers all phases of drug release involving a first order rate burst release, the bulk degradation or chain relaxation and the release by diffusion from PLGA nanoparticles [6].

$$\frac{M_t}{M_\infty} = \theta_b(1 - \exp(-k_b t)) + \theta_r\left(\frac{\exp(k_\Gamma(t - t_{\max}))}{1 + \exp(k_\Gamma(t - t_{\max}))}\right) + \theta_d\left(1 - \frac{6}{\pi^2}\sum_{n=1}^{\infty}\frac{1}{n^2}\cdot\exp(-\frac{n^2\pi^2 D_e t}{r_1^2})\right) \quad (1)$$

2 Experimental

In order to prepare carboxymethyl chitosan with 0.6° of subsitiuation, first solution of chitosan in acetic acid was pretreated by aqueous NaOH (1 M) and subsequently 2 g chloroacetic acid dissolved in isopropanol was added to reaction mixture. Reaction mixture was reacted for 8 h at room temperature. Then prepared carboxymethyl chitosan was degraded with hydrogen peroxide.

For conjugating folate to the amino groups of preactivated folate, DCC and NHS in dimethyl sulfoxide were added into a solution of 1 w/v% CMCs (10 mL) in acetate buffer (pH 4.7) and EDC (FA/EDC molar ratio = 1:1) was added to the resulting mixture and stirred for 16 h. The solution was dialyzed first against phosphate buffered saline (PBS, pH 7.4) for 1 day and then against water for 1 day. The resulting polymer was dried overnight at room temperature.

In last stage. FA-CMCs nanoparticles containing drug was prepared by using a probe type sonifier (Ultrasonic Homogenizer UH-600). MATLAB was used to fit proposed models to the experimental data of curcumin release using a non-linear least-squares algorithm.

3 Results and Discussion

The results of the Ubbelohde viscometer showed that the degraded polymer has a molecular weight of 44.3 kDa. FTIR spectrophotometer (Fig. 1) confirms synthetsis of carboxymethyl chitosan and folic acid grafted carboxymethyl chitosan.

The morphology of folic acid grafted carboxymethyl chitosan nanoparticles containing drug as shown in SEM image (Fig. 2) is spherical and mean diameter of them is about 139.44 nm. The release profile of curcumin from FA-CMCs nanoparticles in phosphate buffer at 37 °C and result of fitting of three stage model with constant D are shown in Fig. 3. As the results of fitting show, the three-stage model provides a good fitting and subsequently good R^2 (0.9649). Table 1 lists the known parameters of release.

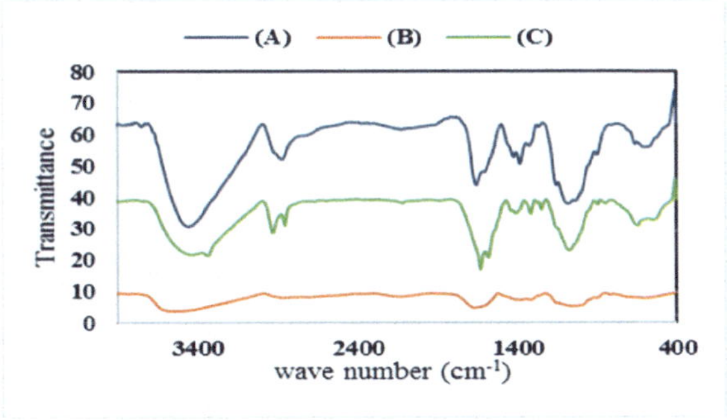

Fig. 1. FTIR spectra of (A) chitosan, (B) carboxymethyl chitosan, and (C) folic acid grafted carboxymethyl chitosan

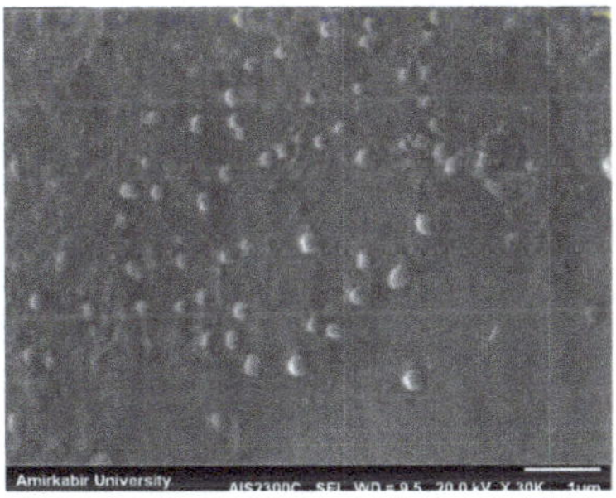

Fig. 2. SEM image of folic acid grafted carboxymethyl chitosan nanoparticles containing curcumin

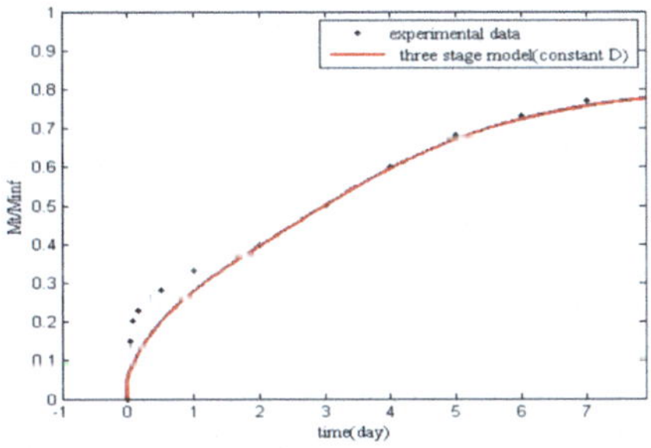

Fig. 3. Experimental and theoretical curcumin release profile from FA-CMCs nanoparticles at 37 °C

Table 1. Table of unknown parameters of three-stage model

parameters	Description	unit	Three stage model
C_S	saturation concentration in the matrix	%	-
D	diffusion coefficient	cm^2/s	9.375e-014
k_b	Burst constant	day^{-1}	1.561
k_r	Degradation constant	day^{-1}	0.6834
θ_b	Fraction of burst release	-	0.1548
θ_r	Fraction of NP degradation release	-	0.5156
θ_d	Fraction of diffusion release	-	0.3296
t_{max}	Time to achieve 50% of release	day	2.88
R^2	Coefficient of determination	-	0.9649
adjusted R^2	Coefficient of determination	-	0.9582

4 Conclusion

The SEM results show nanoparticles formation with average size of less than 150 nm. Release profile of curcumin from nanoparticles showed a three-stage release profile and result of fitting of three-stage model with constant D confirmed that it had very good coefficient of determination ($R^2 = 0.9649$) because more than one simultaneous mechanism of release to explain the three stage release profile of hydrophobic drugs from biodegradable nanoparticles is needed.

References

1. Siafaka, P.I., Barmpalexis, P., Lazaridou, M., Papageorgiou, G.Z., Koutris, E., Karavas, E., Kostoglou, M., Bikiaris, D.N.: Controlled release formulations of risperidone antipsychotic drug in novel aliphatic polyester carriers: data analysis and modelling. Eur. J. Pharm. Biopharm. **94**, 473–484 (2015). https://doi.org/10.1016/j.ejpb.2015.06.027
2. Higuchi, T.: Mechanism of sustained-action medication. Theoretical analysis of rate of release of solid drugs dispersed in solid matrices. J. Pharm. Sci. **52**(12), 1145–1149 (1963)
3. Gallagher, K., Corrigan, O.: Mechanistic aspects of the release of levamisole hydrochloride from biodegradable polymers. J. Control Release **69**(2), 261–272 (2000)
4. Corrigan, O.I., Li, X.: Quantifying drug release from PLGA nanoparticulates. Eur. J. Pharm. Sci. **37**, 477–485 (2009)
5. Lucero-Acuña, A., Guzmán, R.: Nanoparticle encapsulation and controlled release of a hydrophobic kinase inhibitor: three stage mathematical modeling and parametric analysis. Int. J. Pharm. **494**(1), 249–257 (2015)
6. Yang, S.-J., Lin, F.-H., Tsai, K.-C., Wei, M.-F., Tsai, H.-M., Wong, J.-M., Shieh, M.-J.: Folic acid-conjugated chitosan nanoparticles enhanced protoporphyrin IX accumulation in colorectal cancer cells. Bioconjug Chem. **21**(4), 679–689 (2010)

Skin Tissue Engineering Electrospun Scaffold Based on PCL/PVA/ Aloevera

Marjan Shabannejad[1], Mohammad Sadegh Nourbakhsh[1(✉)], and Amir Salati[2]

[1] Biomaterial Group, Faculty of New Sciences and Technologies, Semnan University, Semnan, Iran
s_nourbakhsh@semnan.ac.ir

[2] Tissue Engineering Group, Semnan University of Medical Sciencs, Semnan, Iran

Abstract. Artificial skins and skin substitutes are one of the most important areas of tissue engineering. In this work, three dimensional biodegradable nanofiberous scaffolds containing poly(caprolactone) (PCL), poly(vinyl alcohol) (PVA) and aloevera (AV) were successfully fabricated through two nozzles electrospinning process. For this purpose, PVA/AV blend solution was injected from one syringe and poly(caprolactone) solution from the other one. Addition of PVA and PCL improved the electrospinability of AV solution and mechanical properties of the fabricated nanofibers. Scaffolds were evaluated by SEM and FTIR analysis. Scanning electron microscopy (SEM) results showed uniform PVA/AV-PCL blend-hybrid nanofibers with the diameter average about 119.75 ± 11.75 nm. The presence of nanofibers functional groups was evaluated by Fourier transform infrared spectroscopy (FTIR). The sharp peaks observed in the nanofibers could be assigned to the asymmetric and symmetric stretching vibrations of methylene groups and carbonyl groups.

Keywords. Electrospinning · Aloevera · Poly(caprolactone) · Poly(vinyl alcohol) · Tissue engineering

1 Introduction

Wound or injury which happened due to burn, accidents or chronic ulcerations which resulted in skin lose its principal function loss. Artificial skins and skin substitutes are one of the most important areas of tissue engineering. Although much progress has been made so far, there is still no definitive treatment for second and third degree burns, and millions die annually from severe skin damage. To create a new tissue, it is necessary to have a suitable substrate as scaffold for the placement of cells. Electrospinning is the most widely used method for scaffolds fabrication. It is also a suitable method for use in skin tissue engineering [1–5]. The purpose of this study was to fabricate and evaluate the

H. Mirzadeh and A. A. Katbab (Eds.): ISPST 2018, *Eco-friendly and Smart Polymer Systems*, pp. 35–38, 2020.
https://doi.org/10.1007/978-3-030-45085-4_9

electrospun scaffold based on polycaprolactone-aloevera and polyvinyl alcohol for skin tissue engineering. The morphology of nanofibers and the presence of functional groups are evaluated using SEM and FTIR analysis.

2 Experimental

Aloevera powder was purchased from Teb Jooyan Tabiat Bonyan Co. PCL with molar mass of 80 kDa, and PVA with molar mass of 85–124 kDa was supplied from Sigma-Aldrich, Germany. All solvents, including dimethylformamid (DMF) and dichloromethane (DCM), were supplied by Merck Co. In order to obtain morphology and optimized diameter, DOE was used to investigate the effect of electrospinning parameters such as flow rate of polymer solutions, distance between syringe tip/collector and applied voltage on the morphology and diameter of blend-hybrid nanofibers. To prepare PVA solution, 0.08 mg of PVA was added into 1 ml of water and stirred for 2 h at 40 °C temperature, Then the aloevera powder was added to the solution with AV/PVA different ratios: 7.5, 5, and 1.25% and stirred for 2 h and then the solution was cooled to ambient temperature. To prepare PCL solutions, a mixed solvent system of DMF/DCM was prepared by 2/1 volumetric ratio and 0.125 g of PCL was added into 1 mL of the mixed solvent and stirred for 3 h, AV/PVA and PCL solutions were electrospun spontaneously and resultant meshs were collected on aluminum rotating drum.

3 Results and Discussion

Nanofibrous scaffolds based on PCL and PVA containing AV were designed and fabricated using electrospinning technique. Average diameter of nanofibers was measured by SEM images, shown in Fig. 1. The FTIR spectra of PVA/AV-PCL nanofibers are shown in Fig. 2. The sharp peaks observed at 2929 and 2864 cm^{-1} in the PCL nanofibers could be assigned to asymmetric and symmetric stretching vibrations of methylene groups, respectively and the band corresponding to the carbonyl stretching vibrations in PCL could be seen at 1728 cm^{-1}. The bands appeared at 1294, 1240, and 1170 cm^{-1} are the characteristic peaks of PCL which are related to stretching vibrations of C-C and C-O, asymmetric and symmetric stretching vibrations of C-O-C, respectively. The frequencies and assignments for pure PVA nanofibers are indicated as follows: 2929 cm^{-1} for asymmetric stretching vibrations of methylene groups. The absorption band of PVA at 1047 cm^{-1} corresponds to the stretching vibrations of C-O-C. The peak at 962 cm^{-1} in AV, was due to C-O stretching associated with rhamnogalacturonan side-chain constituent of pectins, and the peak at 732 cm^{-1} represents C-H out-of-plane deformation of carbohydrate monomers [1, 2].

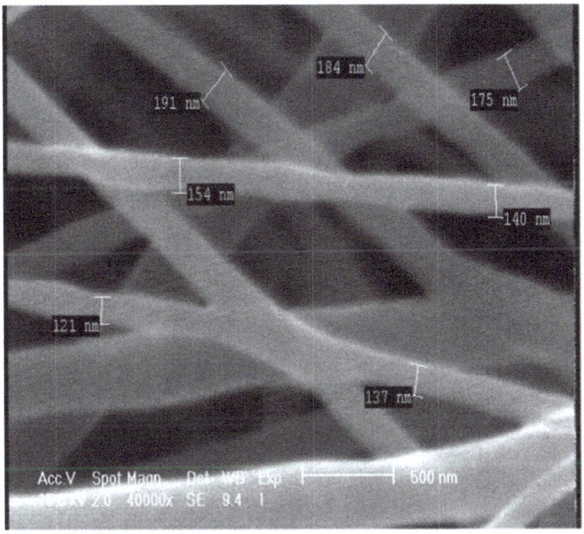

Fig. 1. SEM micrographs of electrospun PVA/AV-PCL nanofibers

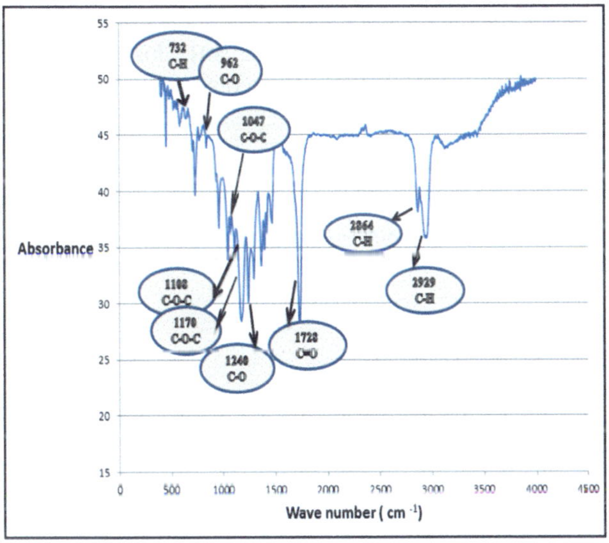

Fig. 2. FTIR spectra of PVA/AV-PCL (B-H) nanofibers

4 Conclusion

In this study, nanofibrous scaffolds containing PCL and PVA/AV (B) were successfully prepared by co-electrospinning method. The effect of electrospinning parameters on the diameter of nanofibers was evaluated by Taguchi experimental design using Minitab software. Nanofibers with the best morphology and diameter of 119.75 ± 11.78 nm were selected as optimum scaffolds for characterization.

References

1. Abdullah, N.A., Sekak, K.A., Ahmad, M.R., Effendi, T.B.: Proceedings International Colloquium in Textile Engineering, Fashion, Apparel and Design, pp. 7–11 (2014)
2. Zarekhalili, Z., Bahrami, S.H., Ranjbar-Mohammadi, M., Milan, P.B.: Fabrication and characterization of PVA/Gum tragacanth/PCL hybrid nanofibrous scaffolds for skin substitutes. Int. J. Biol. Macromol. **94**, 679–690 (2017)
3. Saeed, S.M., Mirzadeh, H., Zandi, M., Barzin, J.: Designing and fabrication of curcumin loaded PCL/PVA multi-layer nanofibrous electrospun structures as active wound dressing. Prog Biomater **6**(1–2), 39–48 (2017)
4. Abrigo, M., McArthur, S.L., Kingshott, P.: Electrospun nanofibers as dressings for chronic wound care: advances, challenges, and future prospects. Macromol. Biosci. **14**(6), 772–792 (2014)
5. Ranjbar-Mohammadi, M., Kargozar, S., Bahrami, S.H., Joghataei, M.T.: Fabrication of curcumin-loaded gum tragacanth/poly (vinyl alcohol) nanofibers with optimized electrospinning parameters. J. Ind. Text. **46**(5), 1170–11926 (2017)

Fabrication of Biocompatible PLGA/PCL/PANI Nanofibrous Scaffolds with Electrical Excitability

Houra Farkhondehnia[1], Mohammad Amani Tehran[1](✉), and Fatemeh Zamani[2]

[1] Department of Textile Engineering, Amirkabir University of Technology, 15875-4413 Tehran, Iran
amani@aut.ac.ir
[2] Hazrate Masoumeh University, Qom, Iran

Abstract. Fabrication of scaffolds with electrical properties plays a key role in neural tissue engineering. This study aimed to prepare aligned electrospun conductive nanofibers by mixing 1, 10, and 18 w/v% doped polyaniline (PANI) with polycaprolactone (PCL)/poly lactic-*co*-glycolic acid (PLGA) (25/75) solution through the electrospinning process. The fibers diameter and conductivity were measured. In addition, the shape and proliferation of the nerve cells seeded on fibers were evaluated by MTT cytotoxicity assay and scanning electron microscopy. The results revealed that the electrical stimulation enhanced neurite outgrowth compared to those PLGA/PCL/PANI scaffolds that were not subjected to electrical stimulation. As polyaniline ratio increases, electric stimulation through nanofibrous PLGA/PCL/PANI scaffolds results in cell proliferation enhancement. However, a raise more than 10% in polyaniline will result in cell toxicity. It was concluded that conductive scaffolds with appropriate ratio of PANI along with electrical stimulation have potential applications in treatment of spinal cord injuries.

Keywords: Conductive polymers · Electrospinning · Tissue engineering · Nerve cells · Electrical stimulation

1 Introduction

Peripheral nerve damage is a common condition in the clinic. The damaged neurons have the potential to redevelop. However, not only the regeneration rate is slow but also the functional performance is poor. Therefore, neural tissue regeneration approaches have received extensive attention [1–3]. Tissue engineering strategies offer alternatives to conventional transplantation methods [4]. The importance of scaffolds in tissue engineering is mainly due to their similarity in dimension with the native extracellular matrix. In addition, they provide large specific surfaces with the ability to load high amount of cells per unit of mass. Among the fabrication strategies, electrospinning is considered as the most pragmatic approach to constructing fibrous scaffolds due to cost efficiency, rapid prototyping and effectiveness [5, 6]. Recently, conductive polymers for biomedical purposes has received a rapt attention including tissue engineering. The bioelectricity in

H. Mirzadeh and A. A. Katbab (Eds.): ISPST 2018, *Eco-friendly and Smart Polymer Systems*, pp. 39–42, 2020.
https://doi.org/10.1007/978-3-030-45085-4_10

the human body plays a pivotal role in maintaining normal biological functions including signaling of the nervous system, wound healing and muscle contraction. In addition, electrical stimulation enhances several cell functions such as attachment, proliferation, migration and differentiation [7]. Electrical stimulation is an effective strategy for 2D and 3D cell cultures and in vivo studies due to its simplicity and flexibility [4]. Therefore, electrically conductive polymers were selected to electrically regulate cellular activities. This study aims at producing a biocompatible scaffold capable of electrical stimulation in order to investigate the impacts of electrical stimulation on the growth and proliferation of nerve cells. To achieve our goal, aligned scaffolds containing various levels of PANI concentrations were developed. The cells were directly seeded on the scaffolds and electrical field was applied to them. After that, they were evaluated for morphology and viability.

2 Experimental

The 5 w/v% polymer solution was prepared by dissolving PLGA/PCL (75/25) in chloroform. The solution of PANI was prepared by mixing CSA (Camphor-10-sulfonic acid) and PANI in the ratio of 1.5 and further was dissolved in chloroform/DMF (3:2) by stirring for 24 h. After stirring, the doped PANI was added to PLGA/PCL solution to make 1, 10, and 18 w/v% PLGA/PCL/PANI solutions. 1 mL of solution was fed into a plastic syringe using a needle with a diameter of 0.7 mm, using a syringe pump (KDS100, KD Scientific Inc., USA) at a feeding rate of 0.5 mL/h with applied voltages of 20 kV and the distance between the needle tip and the collector was 20 cm. Cell culture experiments were carried out with the A-172 cell line obtained from National cell bank of Pasteur Institute of Iran (IPI). The cell morphology was observed 5 days after cell seeding. The samples were naturally air-dried and coated with gold using the sputter coater and later observed under the SEM. To investigate the effect of electrical stimulation on growth of cells on nanofibrous scaffolds, assembled electrodes were used. The stainless steel electrodes (4 electrodes) were placed in each well and the electrical field of 106.25 mV/mm was applied to the cell for 1 h under the hood. The adhesion and proliferation of A-172 cells on electrospun scaffolds were determined by MTT assay.

3 Results and Discussion

Four groups of scaffolds were prepared for cell culture study. In order to apply the electrical field, stainless steel electrodes were used. It was expected that foreign objects affect cells viability. First, the negative effect of electrodes on cell viability should be calculated. To achieve this goal, MTT results of cell culture on two scaffolds with 10% PANI in different situation were investigated. One sample contained electrodes but received no electrical stimulation and no electrodes were used for the other one. It is concluded that the MTT assay difference between these two samples shows the negative effect of electrodes as foreign objects and by adding this amount to stimulated data, the negative effect of electrodes can be eliminated. For homogenization purposes, a value of 0.18 was added to all stimulated data.

Table 1 and Fig. 1 show the results of MTT assay. As indicated in Table 1, although in the stimulation mode the presence of electrodes caused a negative effect on growth of the cells, viability of the cells was improved in modified data for all the groups. Therefore, electrical stimulation showed a great positive effect on cell growth and proliferation. The highest viability was observed at 10% PANI content indicating that using conductive scaffolds containing 10% PANI along with application of electrical stimulation improves cell proliferation by 56%.

Table 1. Results of MTT assay

Data	PANI%			
	0	1	10	18
	Result MTT			
With stimulation	0.7088	0.8062	0.7802	0.5417
Stimulated modified data	0.8888	0.9862	0.9602	0.7217
Without stimulation	0.6580	0.7606	0.6134	0.4965

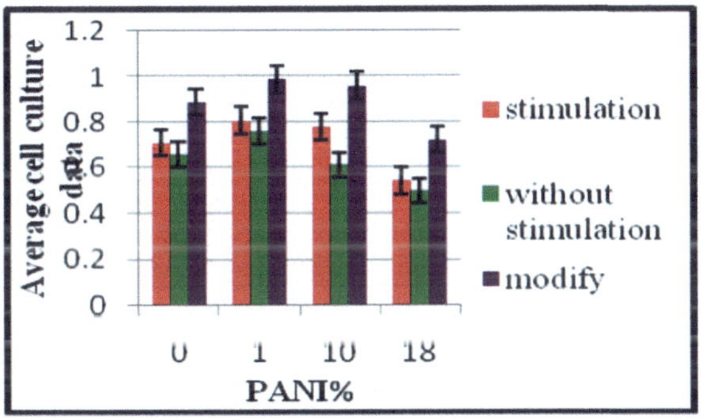

Fig. 1. Effect of electrical stimulation on the cell proliferation

4 Conclusion

This study provides an effective approach for applying electrical stimulation on nerve cells through conductive nanofibrous PLGA/PCL/PANI scaffolds. In addition, the importance of PANI concentrations and its effects on proliferation and elongation of nerve cells were identified. The study demonstrates that PLGA/PCL scaffold with 10% PANI content is effective and leads to cell proliferation and neurite outgrowth after applying electrical stimulation.

References

1. Ahn, H.-S., Hwang, J.-Y., Min, S.K., Lee, J.-Y., Kim, J.-W., Kim, H.-S., Shin, U.-S., et al.: Carbon-nanotube-interfaced glass fiber scaffold for regeneration of transected sciatic nerve. Acta Biomater. **13**, 324–334 (2015)
2. Chan, K.M., Gordon, T., Zochodne, D.W., Power, H.A.: Improving peripheral nerve regeneration: from molecular mechanisms to potential therapeutic targets. Exp. Neurol. **261**, 826–835 (2014)
3. Zamani, F., Amani-Tehran, M., Latifi, M., Shokrgozar, M.A.: The influence of surface nanoroughness of electrospun PLGA nanofibrous scaffold on nerve cell adhesion and proliferation. J. Mater. Sci. Mater. Med. **24**, 1551–1560 (2013)
4. Ghasemi-Mobarakeh, L., Prabhakaran, M., Morshed, M., Nasr-Esfahani, M.H., Ramakrishna, S.: Visual prosthesis: interfacing stimulating electrodes with retinal neurons to restore vision. Tiss. Eng. Part A **15**, 3605–3619 (2009)
5. Abadi, H.F.J., Tehran, M.A., Zamani, F., Nematollahi, M., Ghasemi-Mobarakeh, L., Nasr-Esfahani, M.H.: Int. J. Polym. Mater. Polym. Biomater. **63**, 57–64 (2014)
6. Mokhtari, F., Salehi, M., Zamani, F., Hajiani, F., Zeighami, F., Latifi, M.: Advances in electrospinning: the production and application of nanofibres and nanofibrous structures. Text Prog **48**, 119–219 (2016)
7. Pedrotty, D., Koh, J., Davis, B., Taylor, D.A., Wolf, P., Niklason, L.E.: Am. J. Phys. Heart Circ. Phys. **288**, 1620–1626 (2005)

Improved Solubility of Acetamidophenol from PEG-and Witepsol-Based Suppositories via Formation of Inclusion Complex by β-Cyclodextrin with a Controlled Release Profile

Faezeh Asgari[1(✉)], Sahar Amiri[1], and Majid Ghiass[2]

[1] Department of Polymer Engineering, Science and Research Branch, Islamic Azad University, Tehran, Iran
Asgari.Faezeh20@yahoo.com
[2] Department of Polymer Science, Faculty of Science, Iran Polymer and Petrochemical Institute, Tehran, Iran

Abstract. Acetamidophenol (ACP) is used to treat moderate pain and reduce fever but poorly dissolution of this compound has restricted its bioavailability. To enhance the solubility of ACP, inclusion complex between ACP and β-cyclodextrin (β-CD) is formed with various mass ratios of β-CD:ACP (1:1, 1:2, 1:4, 1:8) and the effect of complex formation on solubility and release of ACP from PEG (4000,400) and Witepsol (H15) based suppositories is investigated. Inclusion complexes were formed with various ratios of β-CD: ACP and loaded in the formulation of suppositories. FTIR, SEM and XRD results indicate formation of inclusion complex between β-CD and ACP. Physical characteristics and in vitro drug release of the obtained suppositories are determined by several tests such as weight variation, melting point, hardness, and release rate. Results indicate that PEG based suppositories containing β-CD:ACP 1:1 mass ratio show the best in vitro drug release characteristic which released 19.3% of ACP after 8 h.

Keywords: Suppository · PEG · Acetamidophenol · β-cyclodextrin · Controlled · Release

1 Introduction

Suppositories are medicated solid dosage forms which are melted or soften at body temperature shaped for rectal route of administration. Suppositories are prepared by using lipophilic and hydrophilic bases. Polyethylene Glycols (PEGs) are one of the most widely used water-soluble polymer suppository bases existing in a variety of molecular weights. Acetamidophenol is a white, odorless and crystalline powder with a melting point of 170 °C which has a poor solubility in suppository bases [1]. Among all the approaches implemented to enhance drug solubility, complexation with cyclodextrins is one the most promising ones. Among all the natural cyclodextrins, β-cyclodextrins (β-CDs) have the most use in pharmaceutical industry due to its cavity size, availability in pure form and low cost. Cyclodextrins have a hydrophilic exterior

H. Mirzadeh and A. A. Katbab (Eds.): ISPST 2018, *Eco-friendly and Smart Polymer Systems*, pp. 43–46, 2020.
https://doi.org/10.1007/978-3-030-45085-4_11

and a hydrophobic cavity in which the drug molecule enters and changes in its physicochemical properties occur [2]. The purpose of this study was to prepare suppositories of acetamidophenol and inclusion complex of acetamidophenol using β-CD and to evaluate its effect on the physicochemical characteristics and release parameters of suppositories.

2 Experimental

2.1 Complex Formation of β-CD/ACP

The inclusion complex of β-CD with ACP was prepared by adding ACP solution in ethanol (5.0 ml of 14.5 wt%) to β-CD solution in ethanol (5.0 mL of 5.0 wt%) at 40 °C. The suspension was stirred at room temperature (25 ± 1 °C) for 2 h, then stayed at room temperature (25 ± 1 °C) for 24 h. The obtained solid was collected and then dried at 50 °C in an oven for 24 h. The dried complex was characterized by XRD, SEM and FTIR.

2.2 Preparation of Suppositories

Suppositories were made by fusion molding method. The suppository base (PEG or witepsol) was melted and then the ACP or ACP: β-CD with various ratios were dissolved in the molten base (Table 1). The prepared suppositories were evaluated for official and unofficial parameters via visual characterization, weight variation, content uniformity, drug content, hardness, melting time, disintegration and dissolution test [3–5].

Table 1. Formulation of obtained suppositories

Formula	H1	H2	P1	P2	P3	P4	P5
PEG 4000/400 (mass ratio = 6:4)	–	–	1.2/0.8	1.2/0.8	1.2/0.8	1.2/0.8	1.2/0.8
H 15 (gr)	1.4	1.4	–	–	–	–	–
ACP (gr)	0.3	–	0.3	–	–	–	–
ACP:β-CD mass ratio	–	8:1	–	1:1	2:1	4:1	8:1

3 Results and Discussion

PEG and H15 based suppositories containing ACP and ACP/β-CD show that obtained suppositories are uniform without any fissuring and fat blooming. Weight variation studies for all suppositories show that the deviation in weight of all prepared suppositories is less than 0.5 from the average weight. All suppositories contain 85−100% indicating that drug is dispersed uniformly in suppositories. In P2, P3 and P4 by increasing the ratio of Acetamidophenol to β-CD, hardness increases due to complex formation between Acetamidophenol and β-CD molecules. P1 and P2 show the most suitable level of hardness (1.8−2 kg). This is due to formation of inclusion complex between ACP and β-CD with 1:1 mass ratio. In formulation P2 where the ACP: β-CD ratio is 1:1 the highest melting time is observed which shows that the inclusions between ACP and β-CD molecules have formed successfully. In P3, P4 and P5 amount of ACP increased compare to β-CD, so susceptible hydroxyl groups for formation of reversible hydrogen bonding are increased, so water solubility of suppositories increases significantly and disintegration time significantly decreases. The release profile from different suppository formulations is shown in Fig. 1. In P2, P3, P4 and P5 by addition of β-CD to drug, drug release rate decreases due to the formation of a complex between ACP and β-CD, which is insoluble in water, and thus this complex reduces drug diffusion through the base and prolongs the release time. Among all the formulations, P2, which is containing ACP/β-CD 1:1 mass ratio, determines the optimum ratio with the slowest release of 19.51% after 8 h [3].

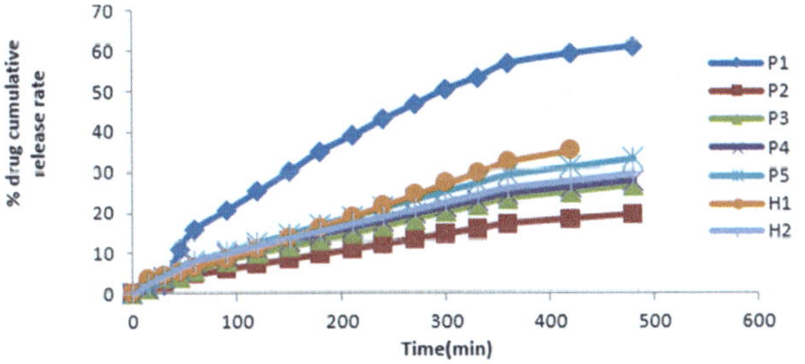

Fig. 1. In vitro release of ACP and ACP: β-CD from PEG and Witepsol based suppositories

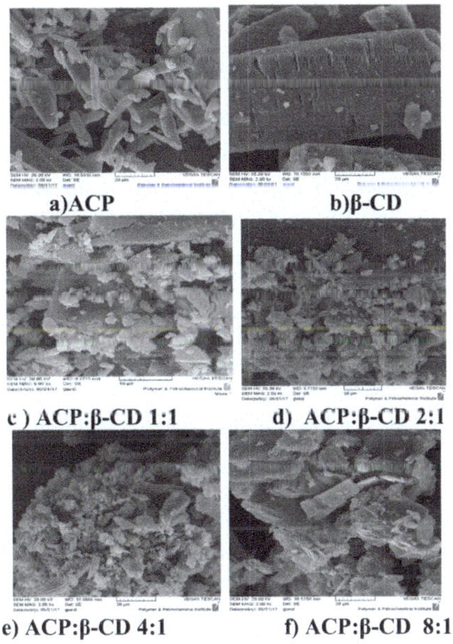

Fig. 2. SEM images of ACP and β-CD and their complexes in different ratios

The SEM images in Fig. 2 clearly show the changes in microstructure of ACP: β-CD in comparison with pure ACP and β-CD. It is found that increase in amount of ACP in P3, P4 and P5 leads to an increase in particle size [3]. X-ray diffraction spectra of ACP, β-CD, P2, P3, P4 and P5 in different conditions and their peaks have been shown in Fig. 3. X-ray diffractions clearly show the complex formation between ACP and β-CD. The results show that the complex formed between ACP and β-CD is a combination of channel and cage structure [3–5].

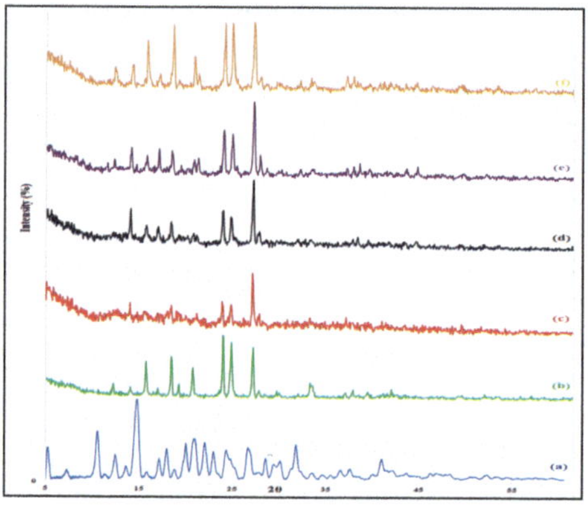

Fig. 3. XRD spectra: (a) β-CD, (b) ACP, (c) P2 containing ACP/β-CD (1:1), (d) P3 containing ACP/β-CD (2:1), (e) P4 containing ACP/β-CD (4:1), and (f) P5 containing ACP/β-CD (8:1)

4 Conclusion

It can be concluded that Acetamidophenol suppositories prepared by PEG-based suppositories offer better physical properties and higher release rate in comparison with oily bases. Incorporation of β-CD in drug formulation leads to slower drug release rate as β-CD forms a complex with ACP. The optimum ratio of ACP/β-CD for inclusion formation is 1:1 mass ratio which shows acceptable hardness and controlled drug release. Also, the results obtained from SEM, XRD and FTIR clearly indicate inclusion complex formation between β-CD and ACP at various ratio.

References

1. El-Majri, M.A., El-Basir, M.M.: Formulation and evaluation of ibuprofen suppositories. J. Res. Pharm. Pract. **7**, 87–90 (2016)
2. Bibby, D.C., Davies, N.M., Tucker, I.G.: Mechanisms by which cyclodextrins modify drug release from polymeric drug delivery systems. Int. J. Pharm. **197**, 1–11 (2000)
3. Asgari, F., Amiri, S., Ghiass, M.: J. Pharm. Innov. https://doi.org/10.1007/s12247-018-9328-y
4. Amiri, S., Rahimi, A.: Int. J. Polym. Mater. Polym. Biomater. (2018). https://doi.org/10.1080/00914037.2018.1482467
5. Amiri, S., Rahimi, A.: J. Ind. Text. (2018). https://doi.org/10.1177/1528083718764911

Rigid Semi-IPN PVC Foam Modified with Epoxidized Soybean Oil

Behnam Khaledi[1(✉)] and Farah Salehiravesh[2]

[1] Chemical Engineering Faculty, Polymer Engineering Department, Tarbiat Modares University, Tehran, Iran
khaledibehnam@gmail.com
[2] Polymer Engineering and Color Technology Faculty, Amirkabir University of Technology (Tehran Polytechnic), Tehran, Iran

Abstract. Semi-IPN rigid poly(vinyl chloride) foam had been prepared by molding and boiling foam process. The chemical structure of foam was modified by epoxidized soybean oil. The cellular structure of obtained foam was characterized. The tensile and compressive properties were also studied and compared to those of commercial foams with similar density. The obtained PVC/IPN foam showed good mechanical properties and higher compression strength than those of commercial foams.

Keywords: PVC foam · PVC/IPN · c-PVC · Sandwich structure

1 Introduction

Vehicles like boats, vessels, etc. consume energy to move. The heavier they are, the more energy consumed. Since using sandwich composites makes structural designs lighter, sandwich solutions are extremely environmentally friendly. With a sandwich solution, less material is consumed in the construction. This saves resources as well as weight in the final construction, making the construction less energy-consuming over its lifetime. Sandwich-structured composites or sandwich structures are a specific class of composites that consist of two relatively thin, stiff and strong faces separated by a relatively thick lightweight core for example, honeycomb, balsa or polymer foam cores. The core material is usually low strength material, but it provides the overall low density and high bending stiffness.

Rigid cross-linked PVC foam (c-PVC) also known as "IPN foam" and "PVC/IPN foam" is an outstanding core material for sandwich composites because of its excellent mechanical strength and stiffness to weight ratios coming from unique combination of cross-linked polyurea and PVC forming a semi-IPN structure. IPN foams also have special features, such as high chemical resistance, self-extinguishing, negligible water absorption, thermal insulation and sound absorption capabilities and low cost. Therefore, it is widely used in sandwich structures for the applications where damage tolerance and weight saving are required; e.g., wind energy, aerospace, transport and industry applications. On other hand the Fragility is the main disadvantage of these foams leading to investigations on changing the cross-linked network to enhance the toughness and flexibility. Incorporating the isocyanate groups into the chemical

H. Mirzadeh and A. A. Katbab (Eds.): ISPST 2018, *Eco-friendly and Smart Polymer Systems*, pp. 47–51, 2020.
https://doi.org/10.1007/978-3-030-45085-4_12

structure of PVC/IPN foams can impress the mechanical properties [1–3]. Shi et al. [1], replaced Maleic Anhydride (MA) with Methylhexa-hydrophtalic Anhydride (MHHPA) and named the final product "universal cross-linked PVC structural foam". They found that the majority of isocyanate groups react with hot water in the boiling stage of foam processing resulting in the formation of a polyamide network in the foam structure. In another study [4], they fabricated samples with higher heat resistance. These samples showed higher glass transition temperature (83.2 °C) and better chemical and dimensional stability compared to "universal PVC foam". It was reported that epoxy and melamine groups may be used to modify the PVC/IPN foams with isocyanate-anhydride systems, leading to higher flexibility and higher shear strain at break but lower stiffness and compression strength. Epoxy groups are able to react with both the amine groups formed during the hydrolysis of isocyanates and anhydride groups.

In this study, the PVC/IPN foam was successfully fabricated by molding and boiling process. Soybean epoxy oil was used in order to take advantage of presence of epoxy compounds in chemical structure and also to enable the biocompatibility in PVC/IPN foam.

2 Experimental

2.1 Materials

Poly (vinyl chloride) (LG-1202) was used as the matrix. Methylene-diisocyanate (MDI) was obtained from Karoon Petrochemical Company, Iran as crosslinking agent. Soybean epoxy oil was purchased from SAJO HAEPYO Corporation, South Korea as crosslinking and biocompatible agent. Azodiisobutyronitrile (AIBN) and azodicarbonamide (AC) from Merck, Germany, were used as blowing agents.

2.2 Preparation of PVC/IPN Foam

In the preparation of starting mixture, 55 g MDI and 30 g of soybean epoxy were first mixed with mechanical stirring; then, 100 g PVC, 4 g AIBN, and 3 g AC were added and mixed again. The obtained liquid starting mixture was put into a compression mold with a sealed cavity (220 mm × 220 mm × 25 mm). This mold was heated to 175 °C and kept for 10 min under a pressure of 10 MPa. In the next step, the plastic mass was cooled and solidified, and the pressure was gradually reduced to the atmospheric pressure. Then the molded sample was removed from the mold and immersed into a water bath of 95 °C to expanding to a required density.

2.3 Characterization

The cellular structure of fabricated foam was characterized by a digital microscope with 100X magnification (Dinolite). The tensile and compressive properties were conducted using a SANTAM Universal Testing Machine-STM-150 (Santam Co., Tehran, Iran) according to ISO 527 and ISO 844 standards, respectively.

3 Results and Discussion

3.1 Density

Samples with specified dimensions were cut from different areas of PVC/IPN foam and weighed. The density was calculated by dividing the weight of the samples on their geometric volume. The average density fabricated foam was 92 kg/m^3.

3.2 Cellular Structure

Figure 1, shows the cellular structure of fabricated PVC/IPN foam captured by digital microscope with 100X magnification, indicating the hexagonal closed cell structure with the diameter between 500 μm to 800 μm. Both the shape and the diameter of cells are nearly the same as the commercial PVC/IPN foams.

Fig. 1. Cellular structure of fabricated PVC/IPN foam obtained by digital microscope with 100X magnification

3.3 Tensile Properties

Figure 2 shows the tensile stress-strain plot of fabricated PVC/IPN foam and Table 1 represents the tensile properties obtained from this plot in comparison to those of some commercial grades with similar density. The tensile properties of fabricated sample are lower but comparable to commercial ones.

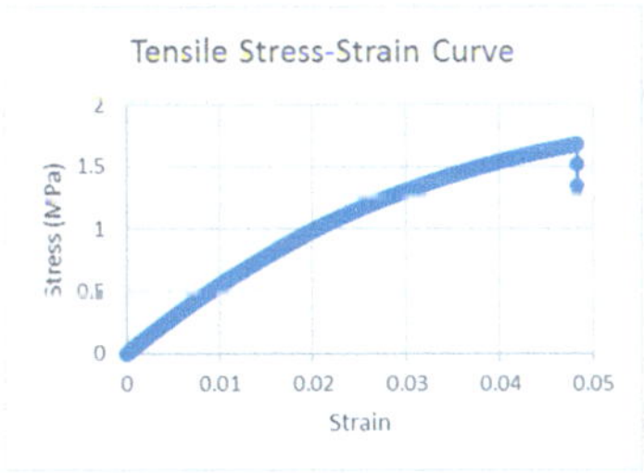

Fig. 2. Tensile stress-strain plot of fabricated PVC/IPN foam

Table 1. Tensile properties of fabricated PVC/IPN foam and commercial grades with similar density

Properties	Fabricated PVC/IPN foam	Airex C70.75	Airex C70.90	Easycell 75
Density (kg/m^3)	92	75	90	75
Tensile strength (MPa)	1.7	2	2.7	1.9
Tensile modulus (MPa)	58	66	84	75
Strain at break	0.048	–	–	–

3.4 Compression Properties

Figure 3 shows the compression stress-strain plot of fabricated PVC/IPN foam and Table 2 represents the compression properties obtained from this table in comparison to some commercial grades with similar density. The fabricated sample shows higher values of compression strength compared to all commercial grades, indicating the toughening effect of soybean epoxy on the PVC/IPN foam. The compression modulus is lower compared to that of Airex samples but higher than that of Easycell 75.

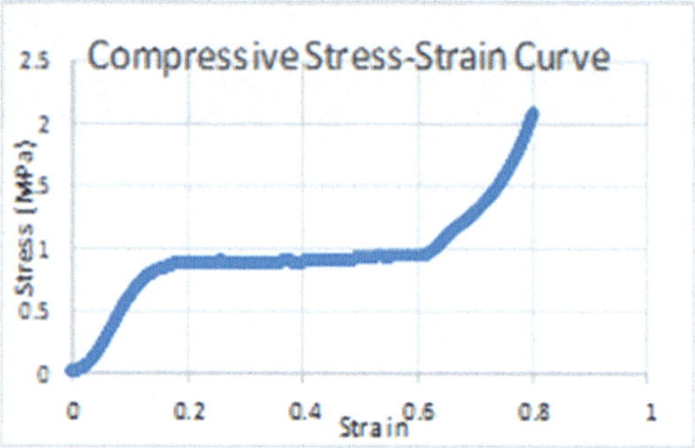

Fig. 3. Compressive stress-strain plot of fabricated PVC/IPN foam

Table 2. Compressive properties of fabricated PVC/IPN foam and commercial grades with similar density

Properties	Fabricated PVC/IPN foam	Airex C70.75	Airex C70.90	Easycell 75
Density (kg/m^3)	92	75	90	75
Compression strength (MPa)	2.1	1.45	2	1.33
Compression modulus (MPa)	87	104	130	72.5
Strain at break	0.84	–	–	–

4 Conclusion

The results show that the PVC/IPN foam with a density of 92 kg/m^3 was successfully fabricated by molding and boiling process. The shape and size of the cellular structures are similar to those of commercial grades. Their mechanical properties are comparable to those of commercial foams and in case of the compression strength, the fabricated PVC/IPN foam shows higher values due to incorporation of high amounts of soybean epoxy.

References

1. Shi, A., Zhang, G., Pan, H.: Preparation and properties of rigid cross-linked PVC foam. Adv. Mater. Res. **313**, 1056–1060 (2011)
2. Jiang, Z.W., Yao, K., Du, Z.H., Xue, J., Tang, T., Liu, W.B.: Compos. Sci. Technol. **97**, 74–80 (2014)
3. Jiang, Z., Yao, K., Du, Z., Xue, J., Tang, T., Liu, W.: J. Appl. Polym. Sci. **131**(15), 1–8 (2014)
4. Shi, A., Zhang, G., Zhao, C.: Molecules **17**(12), 14858–14869 (2012)

An Investigation on the Effect of Pressure and Temperature on Microcellular Foams of Polylactic Acid (PLA)/Acrylonitrile-Butadiene-Styrene (ABS) Binary Blends

Mohammad Reza Kamrani[1,2(✉)], Miguel Sanchez-Soto[1], Alireza Aghili[2], and Amir Ramezannejad[2]

[1] Centre Catalá del Plástico, Universitat Politècnica de Catalunya, Barcelona, Spain
mrkm6687@hotmail.com
[2] Department of Polymer Science and Technology, Azad University of Shiraz, Shiraz, Iran

Abstract. Reducing the weight of petrol-derived plastics is a common way to decrease the environmental impact of several products. Another option consists in using biopolymers instead. However, most of the times, the complete substitution is not possible and partial replacement, usually obtained by blending, appears as a solution. The foaming of these blends combines both strategies and constitutes a further step to reduce the amount of petrol-derived plastics. In this study, PLA/ABS blends were prepared by melt blending without using any type of compatibilizer and were foamed by batch using super critical carbon dioxide as foaming agent. The final morphology of the blends was studied by SEM as a function of pressure and process temperature. Results shows that PLA/ABS blends can be successfully foamed under different conditions to obtain better structures than those known for pure components.

Keywords: Biopolymer · Cellular plastic · Foaming · Super critical fluid

1 Introduction

The applications of plastic products are growing time to time. By increasing the number of polymer parts, the global concern about environmental issues is consequently growing. Therefore, many research are focused on reducing the plastic environmental impact in nature [1]. Using biopolymers could be a solution but in many cases, they cannot fulfil all the required properties. Moreover, from an economical point of view they are still not competitive with petroleum-based polymers. Therefore, blending biopolymers with commodity polymers can be an option to reduce the amount of the plastics in the nature. ABS is one of the most employed ones with a broad range of applications such as automotive and 3D printer technologies. On the other hand, due to its bio-based and bio-degradable nature and its good balance of properties, PLA has

H. Mirzadeh and A. A. Katbab (Eds.): ISPST 2018, *Eco-friendly and Smart Polymer Systems*, pp. 52–55, 2020.
https://doi.org/10.1007/978-3-030-45085-4_13

been considered an appropriate candidate for substituting some high consumption polymers [2, 3]. By blending the ABS with a bio-based and biodegradable polymer like PLA the amount of ABS will be reduced while properties would be kept in an appropriate level.

In addition, light weighting structures contributing to a reduction of polymer consumption can be further obtained by foaming. In this study, binary PLA/ABS blends having a 70 wt% of PLA and 30 wt% of ABS were prepared in an internal mixer and introduced in the batch foaming reactor in order to produce the microfoams. During this process the effect of pressure and temperature on the blends was evaluated. The scanning electron microscopy (SEM) technique was used to observe the cells morphology and the Image J software was employed to the numerical evaluation of the obtained foams.

2 Experimental

2.1 Materials

ABS under trade name Terluran® GP-22 injection grade was purchased from Styrolution Group, Germany, with a melt flow index (MFI) of 19 g/10 min. PLA (Trade name Ingeo 4032D film grade) with MFI 7 g/10 min was provided from NatureWorks LLC.

2.2 Processing

In order to remove the excess of moisture from the materials, prior the processing, PLA and ABS were dried at 55 °C for 24 h in the vacuum oven. The materials ratio in the blend was fixed (PLA/ABS-70:30). The melt processing was done with internal mixer at 200 °C for 12 min. After removing the material from the chamber, the blend was molded by compression moulding to obtain a 3 mm thick plate.

2.3 Foaming

In the final step the blends were transformed into foams using a self-designed foaming setup. It contains six parts; a pressure vessel, an electrical heater with a temperature controller, a pressure gauge; gas injection valve, safety valve and cooling system. The vessel was pressurized with carbon dioxide. Samples were put into the vessel and certain amount of carbon dioxide calculated by Wan der Waals equation of state was added into it. The pressure vessel was sealed and the heater was started. When the temperature reached the desired value, evaporation of CO_2 led to pressurized the vessel. After a period of 4 h at which temperature and pressure were kept constant, the pressure was rapidly released. Finally, the foam morphology was obtained by cooling.

2.4 Morphology Evaluation and Calculation Method

The morphology of the produced foams was evaluated using a Vega3 Tescan scanning electron microscope. Samples were prepared by brittle fracturing the foams using liquid

Nitrogen and sputter a thin layer of gold onto the fractured surface in argon atmosphere with an Q150R–ES sputter coater.

For calculating the foam density and their cell density, the produced foams were cut with laser and then weighting to calculate the specific gravity of the foams. The values of the average cell size (ϕ_{ave}) and cell density (N_f) were determined from the analysis of SEM micrographs using the ImageJ software. The N_f were calculated assuming an isotropic distribution of the spherical cells using Eq. (1):

$$N_f = \frac{6}{\pi \phi_{ave}} \left(1 - \frac{\rho_f}{\rho_s}\right) \tag{1}$$

Where, ρ_s and ρ_f are the solid and foam densities, respectively.

3 Results and Discussion

The goal of this study was to understand the influence of temperature and pressure on the cell morphology of PLA/ABS blends. As such, microcellular foams with constant composition of PLA/ABS were prepared at a saturation temperature of 120 and 165 °C and saturation pressure of 100 and 160 bar, respectively. SEM micrographs of foams are shown in Fig. 1. It can be seen that regardless of the applied conditions, micro-foams with very high densities and small cell diameters were obtained, in contrast to the usually difficulty in foaming PLA because of its low melt strength. The maximum

Fig. 1. Scanning electron microscopy of microcellular foams prepared from PLA/ABS at: (a) 100 bar and 120 °C, (b) 100 bar and 165 °C, (c) 140 bar and 120 °C, and (d) 140 bar and 165 °C

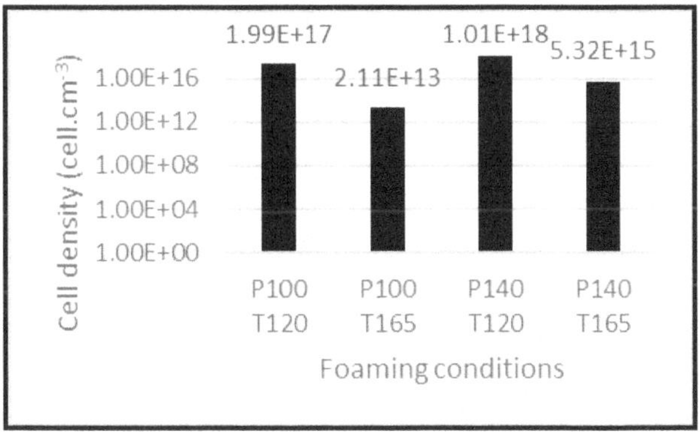

Fig. 2. Cell density values at different foaming conditions

cell density was reached at 140 bar of pressure and 120 °C of temperature, whereas minimum cell density was found at 100 bar of pressure and 165 °C (Fig. 2). Results indicate that higher pressures and temperatures lead in general to finer structures. This is attributed to the higher solubility and diffusion of CO_2 under these conditions. However, the condition at low pressure and low temperature is the optimal from the processing point of view. The limitation for the temperature was the T_g of ABS (≈100) since the one of PLA is much lower. Higher values of temperature were not used because of the risk of structure collapse. It has been seen that the presence of a small amount of ABS (30%) in the blend contributes to stabilize the morphology preventing from cell coalescence and collapse as otherwise happens in pure PLA.

4 Conclusion

PLA/ABS blends were successfully foamed under a wide range of pressures and temperatures. In general, higher values of both parameters led to finer structures but also high cell densities were obtained at low pressure and temperature, which represents the optimal condition from the processing point of view. The presence of a small amount of ABS plays an important role in the stabilization of the foamed samples since ABS is more viscous than the PLA matrix. The increase in the viscosity of the blends is thought to prevent cell coalescence and collapse stabilizing the obtained microstructures.

References

1. Scott, G.: Why degradable polymers? In: Degradable Polymers, pp. 1–15. Springer, Cham (2002)
2. Raquez, J.M., Narayan, R., Dubois, P.: Macromol. Mater. Eng. **293**(6), 447–470 (2008)
3. ABS. http://www.polypedia.ir/wiki/ArtMID/423/ArticleID/2018/Acrylonitrile-Butadiene-Styrene

Physical Properties and Antibacterial Activity of Chitosan/Carboxymethyl Cellulose/Starch Biofilms as Natural Bio-Based Polymers

Alireza Joorabloo, Zohreh Mansoori-Moghadam, Armaghan Moghaddam, and Mohammad-Taghi Khorasani(✉)

Iran Polymer and Petrochemical Institute, P.O. Box 14975-112, Tehran, Iran
m.khorasani@ippi.ac.ir

Abstract. The antibacterial property of food packaging materials plays a key role in improving the safety and shelf-life of the food. The aim of this study was to evaluate the antibacterial activity of chitosan/carboxymethyl cellulose/starch biofilms against *S. aureus* and *E. coli* microorganisms by using agar diffusion disc and optical density methods. Investigation on the properties of fabricated biofilms, including water vapor transmission rate, water solubility, tensile strength and thermal resistance was carried out. Results exhibited that the antibacterial activity was enhanced by increasing the chitosan content. Besides, increased carboxymethyl cellulose content resulted in a decline in the permeability of water vapor through samples. Furthermore, the presence of carboxymethyl cellulose improved tensile strength and thermal stability and reduced the degradation rate of biofilms within the self-life period. Also, the increased starch content in biofilms led to increased water solubility. These results suggested that chitosan/carboxymethyl cellulose/starch biofilms could be used as food packaging materials.

Keywords: Antibacterial activity · Chitosan · Food packaging · Biodegradable biofilms · Carboxymethyl cellulose

1 Introduction

Chitosan (CS) has properties like high hydrophilicity, biocompatibility, antibacterial activity, permeability against gases, providing moist environment. Antibacterial mechanism of CS is based on activity of NH_3^+ group of chitosan which can interact with the negatively charged cell membrane, leading to the cytoplasm to leak into ECM. Also, low molecular weight chitosan penetrates into bacteria cell and alters the DNA, but high molecular weight ones envelop cells and block nutritious materials to pass inside [1, 2]. Besides, cellulose derivatives, like carboxymethyl cellulose (CMC), which is a hydrophilic anionic, non-toxic and environmentally-friendly polymer, are used in medical and food industries for being accessible and cost-effective [3]. Starch (ST) is one of low cost, most commonly and biodegradable biopolymers that has been used widely in food industries [4]. In this study, the effects of CS, CMC, and ST on antibacterial activity, mechanical and thermal properties and biodegradability of fabricated biofilms were investigated.

H. Mirzadeh and A. A. Katbab (Eds.): ISPST 2018, *Eco-friendly and Smart Polymer Systems*, pp. 56–59, 2020.
https://doi.org/10.1007/978-3-030-45085-4_14

2 Experimental/Theoretical

CS/CMC/ST biofilms were prepared by casting method. Potato starch was dried at 90 °C for 4 h to remove the moisture. CS/CMC/ST solutions containing 5–10% starch, 25–40% chitosan and 50–70% carboxymethyl cellulose were prepared. The film forming solutions were casted on petri dishes and then dried overnight at 40 °C.

2.1 Microbiological Study

In agar diffusion test, 100 μL suspension of E. coli (gram-negative) and *S. aureus* (gram-positive) bacteria with a concentration of 10^8 CFU/mL was prepared and dispersed on agar plate. Then, samples were cut into circles to the diameter of 15 mm and were subjected to the bacteria in nutrient agar. The mentioned plates were incubated for 24 h at 37 °C and the inhibition zone around each sample was calculated and considered as antibacterial property of components.

In optical density method, 1000 μL suspension of each bacterium which approximately contain 10^6 CFU (colony forming unit), was poured onto the surface of biofilm samples and were kept at 37 °C with relative humidity of 90%. After 24 h, bacterial solution was collected and transferred to 96-well culture plates. Bacteria counting was performed using OD method with ELISA, ELx808 at 600 nm.

2.2 Tensile Strength

Tensile strength, Young's modulus (YM) and elongation at break point of produced biofilms were measured based on ASTM D882-02 standard using INSTRON 6025. Biofilms were cut into 4 cm × 1 cm × 0.3 cm pieces and were placed under 0.2 kN load cell, stretching speed of 20 mm/min at room temperature.

2.3 Water Vapor Transmission Rate

To measure water vapor permeability, briefly, samples were cut into circular pieces with thickness of 3 mm and diameter of 35 mm and were put on the mouth of a bottle with an inner diameter of 34 mm containing 25 mL distilled water and were tightened with Teflon tape on the bottle mouth. Then, the bottles were weighted and incubated with constant temperature and humidity (temperature of 35 °C and relative humidity of 35%). Bottles were removed after 24 h and were weighted again. Water vapor transmission rate (WVTR) was obtained from the following relation:

$$\mathrm{WVTR}\ \left(\mathrm{g/m^2\,.\,day}\right) = \left(W_i - W_f\right)/A \tag{1}$$

Where, A is the cross-section of container, and W_i and W_f are the initial and final weight of bottles, respectively.

2.4 Water Solubility

First biofilms were dried in a vacuum oven at 50 °C for 8 h. Then, the dried biofilms were weighted (W_i) and immersed in water 24 h. Afterwards, the biofilms were removed from water container and placed on aluminum foils and dried at 35 °C. The weights of dried biofilms were measured again (W_f). Water solubility (WS%) was obtained from the following equation:

$$WS\% = (W_i - W_f)/W_i \times 100 \tag{2}$$

3 Results and Discussion

Antibacterial activities could demonstrate the application of biofilms as intelligent packaging in food applications (Fig. 1). To achieve the successful development of biodegradable biofilms, and intelligent packaging, good physical and mechanical properties are necessary. Tensile strength of fabricated biofilms was in the range of 1–3 MPa (Fig. 2). With increasing CMC content, the WVTR reduced while thermal and mechanical properties increased.

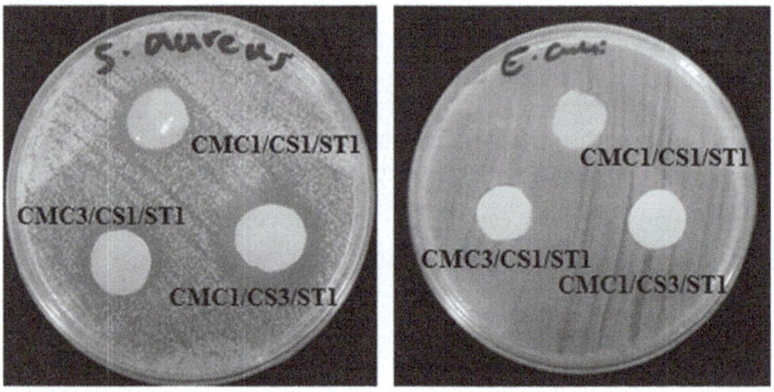

Fig. 1. Antibacterial activity of biofilms against *S. aureus* and *E. coli* bacteria

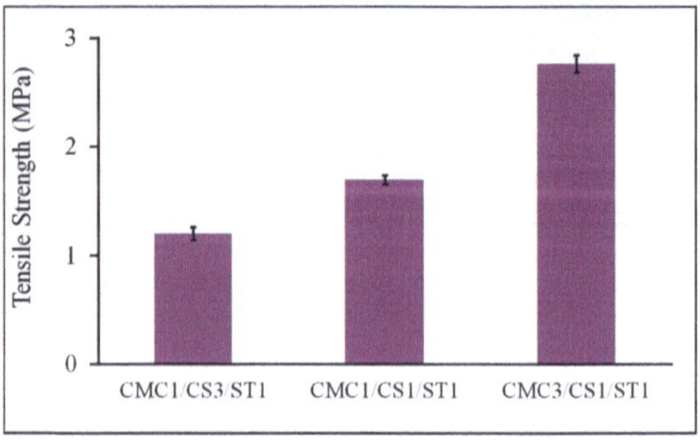

Fig. 2. Tensile strength of fabricated biofilms

4 Conclusion

Biofilms of CS/CMC/ST were prepared and characterized according to the microbiological property, mechanical and thermal analysis, water vapor permeability, water solubility depending on the composition of biofilm solution. Furthermore, the study focused on key properties of fabricated biofilms. The results showed that biofilms may find applications in food packaging systems.

Acknowledgement. The authors would like to acknowledge Iran Polymer and Petrochemical Institute (IPPI) for its technical supports.

References

1. Hu, D., Wang, H., Wang, L.: LWT-Food Sci. Technol. **65**, 398–405 (2016)
2. Khorasani, M.T., Joorabloo, A., Moghaddam, A., Shamsi, H., Mansoori-Moghadam, Z.: Int. J. Biol. Macromol. **114**, 1203–1215 (2018)
3. Sayanjali, S., Ghanbarzadeh, B., Ghiassifar, Sh: LWT-Food Sci. Technol. **44**, 1133–1138 (2011)
4. Luchese, C.L., Frick, J.M., Patzer, V.L., Spada, J.C., Tessaro, I.C.: Food Hydrocolloid **45**, 203–210 (2015)

Effect of Potassium Sorbate on the Mechanical and Thermal Behaviors of Multi-blends Based on LLDPE/LDPE/PLA with PE-g-MA as Compatiblizer

Z. Tolouei Shiviari[3] and Abdulrasoul Oromiehie[1,2(✉)]

[1] Polymer Chemistry, Azad University Tehran South Branch, Tehran, Iran
a.oromiehie@ippi.ac.ir
[2] Khatam Polymer, Tehran, Iran
[3] Polymer Engineering, Azad University Tehran South Branch, Tehran, Iran

Abstract. In the present study, the role of potassium sorbate on the mechanical and thermal properties of LLDPE/LDPE/PLA/PE-g-MA modified blends was evaluated. A series of blends of LLDPE/LDPE/PLA with the ratio of (80/20%) LLDPE/LDPE and 10, 20, and 30% PLA was prepared using co-rotating twin screw extrusion followed by extrusion blown film. In the first step, the mechanical properties of the samples were determined and the compositions of 20% PLA were optimized. In the second step, the properties were evaluated again by loading 4 phr potassium sorbate. The tensile strength of the of LLDPE/LDPE/PLA/PE-g-MA film enhanced from 7.93 Mpa to 11.71 Mpa and elongation at break from 282.57% to 551.57%, respectively. Thermal behavior of multi-blends of LLDPE/LDPE/PLA with and without potassium sorbate were studied using Differential Scanning Calorimetry (DSC). The results indicated that the presence of potassium sorbate caused the T_m and Tc of the film decrease from 122.3 °C to 111.1 °C and 110.32 °C to 98.32 °C, respectively, while the crystallinity increased from 46.7% to 54.3% because potassium sorbate acts as nucleating agent in the blends.

Keywords: Potassium sorbate · Thermal behavior · Poly-lactic acid · LLDPE · LDPE

1 Introduction

Polyethylene (LLDPE, LDPE, and HDPE) has gained special attention among different synthetic polymers in packaging industry due to its good process ability and mechanical properties, high flexibility [1]. One of the most widely used biodegradable polymer is poly-lactic acid. PLA is inherently brittle and has low elongation at break which limit its use in packaging industry [2]. So in order to enhance the biodegradability of PE-based materials and mechanical properties of PLA, blending PLA with PE-based materials has been considered as an effective solution.

Moreover, another potential problem confronts packaging manufacturer is to prolong the shelf life of products and maintain their quality as close to its fresh form as

H. Mirzadeh and A. A. Katbab (Eds.): ISPST 2018, *Eco-friendly and Smart Polymer Systems*, pp. 60–63, 2020.
https://doi.org/10.1007/978-3-030-45085-4_15

possible. While adding antimicrobial agents can play a pivotal role in changing the properties of blends, several research have carried out on evaluate the characteristics of multi-blends in order to meet an optimum point in desirable properties, biodegradability and antimicrobial property.

2 Experimental

2.1 Materials and Methods

Polymers matrix used in this study were LLDPE and LDPE (Arak Petrochemical Co.), PLA (Ö5 BIOKAS), (Chemiekas GmbH Australia), PE-g-MA (Karabond-Iran) and antimicrobial agent potassium sorbate (KS) extra pure grade (Duksan Pure Chemicals, Korea).

To prepare the initial blends firstly all materials were dried in an oven at 80 °C for 24 h in order to remove moisture. Brabender twin screw extrusion with L/D = 40, L = 80 Cm, speed ratio = 80 rpm and heating zones from 160 °C to 195 °C was used for compounding and blow molding films. The formulation of the blends were as:

Also, tensile tests were carried out in a Cometech QC-515B-LA Khatam Polymer according to ASTM D882 at tensile speed of 50 mm/min at 23 °C. The results are the average of 5 samples. Thermal properties were tested using a Differential Scanning calorimetre in Iran Polymer Petrochemical Institute from 25–200 °C at a heating rate of 10 °C/min. The percentage of crystallinity was calculated using following equation:

$$\frac{M_t}{M_\infty} = \theta_b(1 - exp(-k_b t)) + \theta_r\left(\frac{exp(k_\Gamma(t - t_{max}))}{1 + exp(k_\Gamma(t - t_{max}))}\right) + \theta_d\left(1 - \frac{6}{\pi^2}\sum_{n-1}^{\infty}\frac{1}{n^2}\cdot exp\left(-\frac{n^2\pi^2 D_e t}{r_1^2}\right)\right) \quad (1)$$

3 Results and Discussion

Figure 1 illustrates change in tensile properties of the blends after adding 4 phr potassium sorbate. The tensile strength and elongation @ break of the blends show increasing trends. The presence of PE-g-MA as compatibliser improved the compatibility between two phases along with the interaction between polar groups in potassium sorbate and PE-g-MA which caused the dispersion of potassium sorbate in the matrix phase. As a result of good dispersion of potassium sorbate, it acted as nucleating agent and improved tensile strength.

With addition of potassium sorbate, elongation @ break of the blend, on the other hand, increased dramatically. This can be due to the effect of presence of potassium sorbate which lay among poly-ethylene chains and leads to the movements of chains much easier than the blends without potassium sorbate, so the elongation @ break increases.

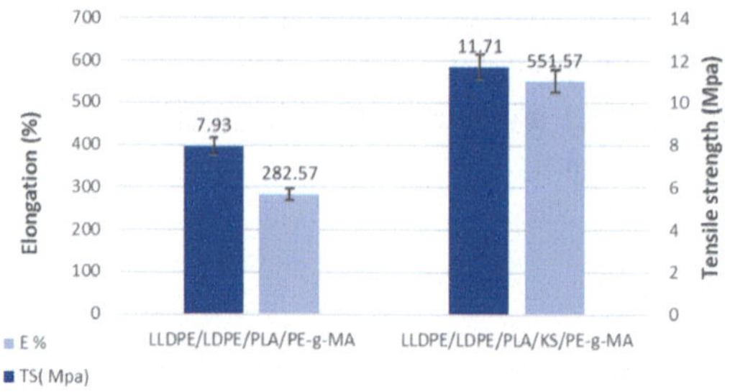

Fig. 1. Tensile strength and elongation @ break of blends with and without potassium sorbate

3.1 Thermal Properties

The role of potassium sorbate on the thermal properties and crystallinity of the blends showed in Table 2. According to the data from Table 2, it is observed that the T_c value has slightly decreased with the addition of potassium sorbate. This may have been due to the heterogeneous nucleation of potassium sorbate in the matrix of LLDPE/LDPE/PLA. Potassium sorbate acts as nucleating agent and has caused heterogeneous crystals to form at lower temperature (Table 1).

Table 1. Blends composition

KS (PHR)	PE-MA (PHR)	PLA %	LLDPE/LDPE (80/20)%	COD No.
0	0	0	100	Ref
0	3	10	90	1
0	3	20	80	2
0	3	30	70	3
4	3	20	80	4

Moreover, melting temperature (T_m) of the blend with potassium sorbate decreased from 122.3 °C to 111.1 °C. As it was mentioned above, the presence of potassium sorbate particles facilitate the flow of polymer chains, as a result, T_m decreased. On the other hand heterogeneous crystals have different sizes and melting temperatures so, they may melt in lower temperatures.

Table 2. Thermal data of blends with and without potassium sorbate

Sample	T_M (°C)	T_C (°C)	X_C (%)
LLDPE/LDPE/PLA/PE-MA (80%/20%/20%/3phr)	122.3	110.32	46.07
LLDPE/LDPE/PLA/KS/PE–MA (80%/20%/20%/4phr/3phr)	111.1	98.32	54.03

4 Conclusion

From the above mentioned results it is concluded that: The presence of potassium sorbate as antimicrobial agent, improves tensile strength and elongation at break of the blends due to its nucleation effect in LLDPE/LDPE/PLA matrix which are very favorable for packaging applications. Thermal properties of blends were improved by adding potassium sorbate.

References

1. Sigh, O., Upadhyaya, P., Chand, N., Mishra, T.K.: Int. J. Mech. Ind. Eng. **2**(2231–6477), 11–14 (2012)
2. Lu, X., Tang, L., Wang, L., Zhao, J., Li, D., Wu, Z., Xiao, P.: J. Polym. Test, 90–97 (2016)

Investigation of In Situ Systems of Progesterone in Biphasic Media

Maryam Armiun, Seyed Mojtaba Taghizadeh(✉), and Hamid Mobedi

Novel Drug Delivery Systems Department,
Iran Polymer and Petrochemical Institute, P.O. Box 14975-112, Tehran, Iran
s.m.taghizadeh@ippi.ac.ir

Abstract. Livestock breeding is one of the main concerns for animal owners. Progesterone plays a crucial role in regulating the activity of the reproductive system. In the present study, in situ formed systems that include poly(lactic-*co*-glycolic) acid (PLGA) as a biodegradable polymer, N-methyl-2-pyrrolidone (NMP) as a bioavailable solvent, and sodium chloride (NaCl) as an additive were provided to treat abortion. The release media was selected as a biphasic that contain 1,2-dichloroethane and buffer phosphate (87:13%) at 37 °C. The high-performance liquid chromatography system was also used to determine the amount of drug release. Degradation of systems was determined by examining the process of pH change. In general, it can be concluded that in situ systems based on PLGA containing NMP with 30 wt% sodium chloride salts showed a linear release profile.

Keywords: Abortion · Progesterone · *In situ* forming systems · Veterinary · Biphasic media

1 Introduction

Despite significant advances in veterinary science, preterm delivery in animals is still a challenge [1]. Abortion has a lot of financial losses for animal and livestock owners. So, veterinarians are looking for a way to prevent it. Progesterone is an effective factor in reducing the risk of abortion [2].

Finding an optimal method for prescribing progesterone that keeps pregnancy well, reduces costs, limits side effects, and easy to administer was too important for animal owners [3]. In this way, the long-term controlled systems for regulating pregnancy of cows and sheep are studied by combining progesterone into biocompatible polymer matrices [4].

2 Experimental

PLGA and NMP was purchased from Evonik and Merk Complex. PLGA (Resomer 502 H) was dissolved in NMP (Formulation 1), and NMP with 30 wt% (Formulation 2) in polymeric syringe separately, then the progesterone in drug syringe was mixed with them. PLGA matrices were prepared by injecting into release media that included

H. Mirzadeh and A. A. Katbab (Eds.): ISPST 2018, *Eco-friendly and Smart Polymer Systems*, pp. 64–66, 2020.
https://doi.org/10.1007/978-3-030-45085-4_16

1,2-dichloroethane and buffer phosphate (87:13%). Formulations were placed in buffer phosphate in the incubator at 37 °C. Progesterone was insoluble in water, so biphasic media like body made phase inversion to release the drug. In predetermined time intervals, a certain amount of the release in 1,2-dichloroethane was withdrawn, and progesterone concentrations were evaluated by HPLC.

3 Results and Discussion

Figure 1 shows that the release profile of formulation 1 containing NMP is a three-phase profile. In the first phase, it does not have an initial burst release and release of progesterone in the first 10 days is insignificant, due to the low adsorption of progesterone. Then, from the eleventh day to the thirty-fifth day, the highest amount of drug was released. In the third phase, the release profile has been steady in this interval and the release rate is negligible and around 1%. Release from 11th is starting after the destruction of polymers, continues until the 35th day and reaches from 0.5% to 49%. pH test also shows that polymer degradation started on the 9th day (Fig. 2), because of the penetration of water into the system. The system is swollen and hydrolysis phenomenon occurs.

Releasing profile of in situ forming system based-PLGA containing NMP and 30% sodium chloride (F2) is obtained in two-phase and in a linear form (Fig. 1). The first phase of F2 is from the first to the fifth day. During these 5 days, no explosion occurred, and its release began earlier than F1. Using 30% NaCl in the formulation causes faster phase inversion, more water penetrate into the system, and increases release rate. The release continues until the 50th day and ultimately total drug release was about 54%.

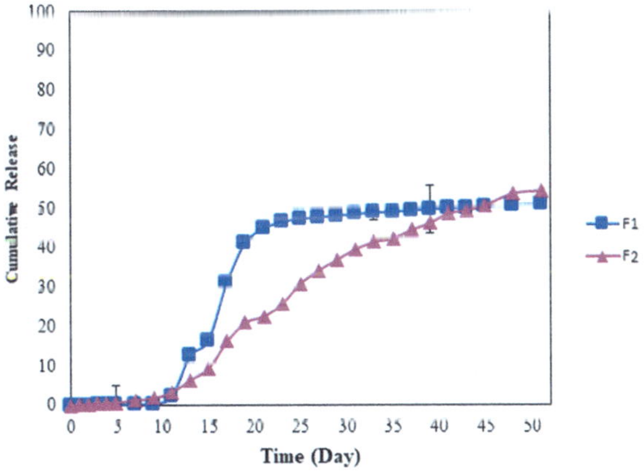

Fig. 1. Release profile of formulations contain NMP (F1) and NMP with 30% NaCl (F2)

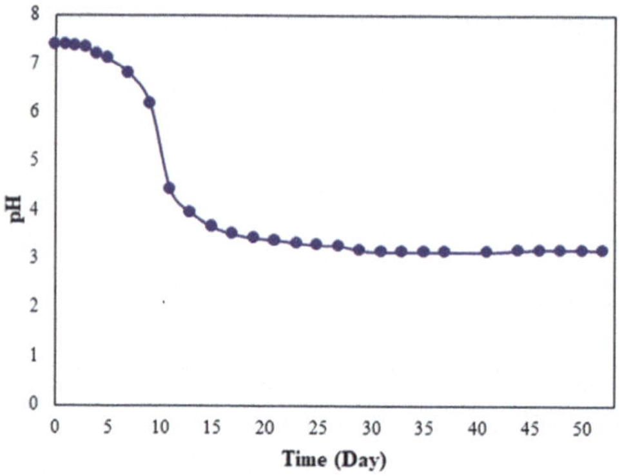

Fig. 2. pH changes in release media of the formulation based PLGA and NMP (F1)

4 Conclusion

Based on the other studies, NMP was first selected for formulation. In the study of drug release from the system, it was observed that release was delayed, began on the eleventh day, and release profile becomes three-phasic. The total drug release rate during the 50-day period was 49%. In the system containing 30% NaCl, the 10-day delay of release reduced to 5 days and release profile became two-phase and linear mode. The total amount of drug released compared to NMP-free formulation increased to about 54%.

References

1. Rode, L., Langhoff-Roos, J., Andersson, C.: Acta Obstet. Gynecol. Scand. **88**, 1180–1189 (2009)
2. Ruana, X., Mueck, A.O.: Maturitas, 248–255 (2014)
3. Haddad, G., Saguan, D.A., Maxwell, R., Thomas, M.A.: J. Assist. Reprod. Genet. **24**, 467–470 (2007)
4. Turino, L.N., Mariano, R.N., Boimvaser, S., Luna, J.A.: J. Pharm. Innov. **9**, 132–140 (2014)

Fractionation of Poly1-Decene Based Oil Synthesized with $AlCl_3$/Ethanol Cationic Catalyst in the Presence of Toluene Solvent

Amene Rahbar, Naeimeh Bahri-Laleh, Sevda Dehghani, and Mehdi Nekoomanesh(✉)

Department of Polymerization Engineering,
Iran Polymer and Petrochemical Institute, P.O. Box 14965/115, Tehran, Iran
m.nekoomanesh@ippi.ac.ir

Abstract. Polyα-olefin oil (PAO) was synthesized via cationic oligomerization of 1-decene as monomer, aluminum chloride as catalyst, ethanol as coactivator and toluene as solvent. Oligomer product was fractionated by distillation in vacuum and each fraction was characterized using viscometry, HNMR and GPC techniques. Analyses results showed that the fractions with higher boiling point and molecular weights had lower aromatic contents and higher viscosity index (VI).

Keywords: 1-decene · Cationic oligomerization · Polyalphaolefin oil · Fractionation · Microstructure

1 Introduction

Polyα-olefin oils (PAOs) are saturated hydrocarbons used as synthesized base stock oils. These materials in comparison with the mineral oils have some advantages including high viscosity index, low pour point, good thermal and oxidative stability and low toxicity. In addition, in low molecular weights they are biodegradable. They have wide applications, for example, in transmission fluids, gear, engine, and hydraulic oils as well as greases [1, 2]. In the synthesis of PAOs, linear α-olefins such as 1-decene, 1-octene, and 1-hexene have been used and generally, the best results were obtained with 1-decene monomer. Cationic catalysts are used in the industrial synthesis of PAOs. A remarkable feature of PAOs obtained with this process is the huge variety of structural isomers that the oligomer fractions consist of [1]. In this study, we prepared a poly1-decene based oil via cationic oligomerization and fractionalized by distillation technique under high vacuum condition. After successful fractionation, different fractions were analyzed by GPC, viscosity and HNMR methods to find a relationship between PAO microstructure and viscosity index.

H. Mirzadeh and A. A. Katbab (Eds.): ISPST 2018, *Eco-friendly and Smart Polymer Systems*, pp. 67–70, 2020.
https://doi.org/10.1007/978-3-030-45085-4_17

2 Experimental

Catalytic oligomerization reaction of 1-decene was performed in a 2 L stainless-steel reactor with a stirrer and an external heating jacket. The reactor was dried at 100 °C for 30 min and then cooled to 80 °C. After that, the certain amount of $AlCl_3$ as the catalyst and ethanol as a coactivator were added into the reactor. 1-Decene monomer mixed with a determined amount of toluene as the solvent was injected to the reactor in stepwise manner. During the reaction, the reactor temperature was kept in 110 °C. After 1 h, the reaction was stopped and the product was collected. Then it was filtered, washed with deionized water and NaOH solution, and distilled to remove unreacted monomer and solvent. The oligomers were fractionated by distillation in vacuum, based on boiling point into five fractions.

3 Results and Discussion

The results of the synthesized poly1-decene fractionation are tabulated in Table 1. Kinematic viscosity at 100 and 40 °C (KV^{100} and KV^{40}, respectively) and viscosity index (VI) according to ASTM D445 and D2270 were measured. VI is one of the most important properties that determine quality grade of the lube base oil and it strongly depends on the molecular structure of the hydrocarbon constitute of. The higher VI value is a sign of the better performance of oil as this index is related to the temperature-viscosity relationship [3]. As can be seen, two large weight fractions belong to the parts C and D with 29.60 and 45.10 wt% share, respectively, confirmed with the GPC curve of this oil (Fig. 1) by two big peaks.

Table 1. Characteristic of different fractions of the poly1-decene synthesized by $AlCl_3$/ethanol catalytic system in the presence of toluene diluent

Fraction label	A	B	C	D	E
Boiling range of fraction (°C)	<230	230–320	320–400	400–500	>500
Weight percent	5.32	6.30	29.60	45.10	13.68
KV^{100}(cSt)	0.44	1.34	3.30	5.88	–[a]
KV^{40}(cSt)	0.95	3.69	13.95	35.12	–[a]
VI	–[a]	79	104.5	109.9	–[a]

[a] not determined

The fraction A is related to the unreacted monomer and remaining solvent, and the fraction B can be associated to the dimmer oligomers (M_n = 280 g/mol) and 1-decene monomer bounded to toluene molecule with boiling point between 230–320 °C, (the first peak between 200–300 g/mol in GPC curve of the fraction B in Fig. 2). The GPC curve of the fraction D is also displayed in Fig. 2.

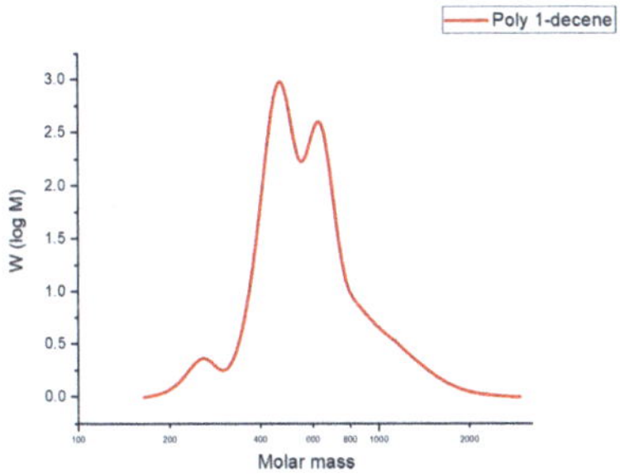

Fig. 1. GPC curve of the synthesized poly1-decene

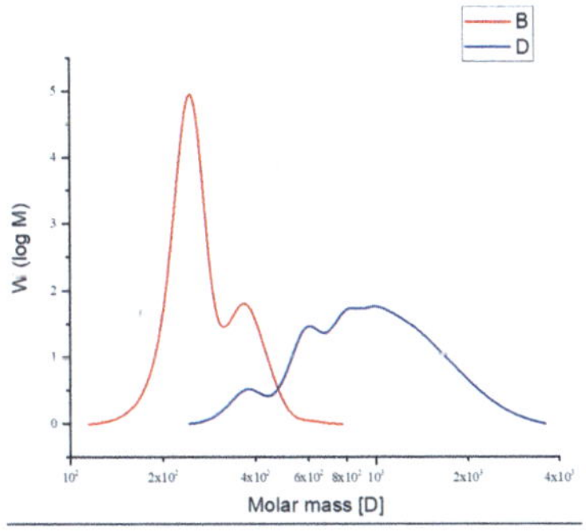

Fig. 2. GPC curves of factions B and D of the synthesized poly1-decene

To better comparison between fractions B and D, proton nuclear magnetic resonance was used to give an insight about their structures. HNMR spectra of two samples are shown in Fig. 3. The chemical shift ranges for aliphatics, olefins, and aromatics are well established in the literature [2]. The aliphatic, olefinic and aromatic regions appear at 0.8–3.0, 4.5–6.5, and 6.5–7.5 ppm, respectively. Both samples contain primarily aliphatic protons originated from 1-decene monomer. The peaks at aromatic region corresponded to the reaction between toluene and hydrocarbon chains via Friedel-Crafts mechanism [4]. The area under peak of aromatic C-H protons in sample D is far less than sample B, compare II with I, which demonstrates that chains with higher boiling points, i.e. containing more monomer units, showed less reactivity toward reaction with toluene. It maybe because of aromatic ring hindrance that does not allow to propagate chain to higher degree. Another characteristic peak as an evidence for the presence of toluene in the oligomers structures, is the sharp peak at 2.3 ppm associated to the methyl group in toluene, which is clearly decreased in the sample D.

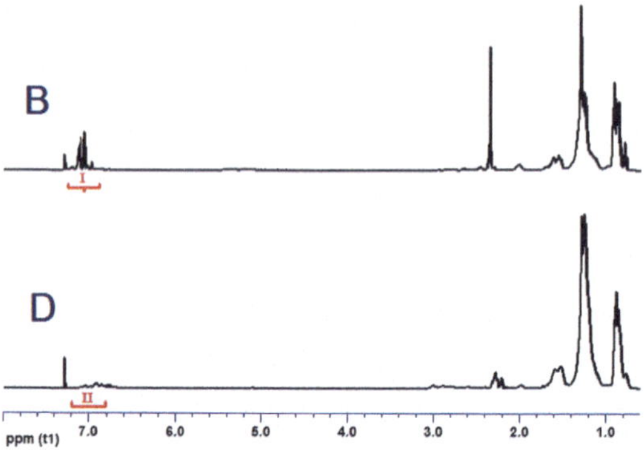

Fig. 3. HNMR spectra of fractions B and D of the synthesized poly1-decene

4 Conclusion

A poly1-decene type PAO sample was synthesized and then fractionated based on the boiling point to study the structures of various oligomers present in the fractionated product. The HNMR spectra of the fractionated samples showed a trace of aromatic C-H protons which is associated to the reaction of oligomer chains with toluene samples. Also, it is deduced that by increasing boiling point of the oligomer chain (or molecular weight), the amount of aromatic ring decreases.

Acknowledgement. The authors gratefully acknowledge Iran Polymer and Petrochemical Institute for financial support of this work under the grant number of 43791107.

References

1. Scheuermann, S.S., Eibland, S., Bartl, P.: Lub. Sci. **23**, 221–232 (2011)
2. Nicholas, C.P., Laipert, L., PrabhakarInd, S.: Ind. Eng. Chem. Res. **55**, 9140–9146 (2016)
3. Noh, K., Shin, J., Lee, J.H.: Ind. Eng. Chem. Res. (2017). https://doi.org/10.1021/acs.iecr.7b00967
4. Malgas-Enus, R., Mapolie, S.F.: Inorg. Chim. Acta **409**, 96–105 (2014)

Study on Rheology, Crystallinity and Electrical Resistance of Poly(Lactic Acid)/Graphene Oxide Nanocomposites

Fatemeh Farhanmoghaddam and Azizeh Javadi(✉)

Department of Polymer Engineering and Color Technology, Amirkabir University of Technology (Tehran Polytechnic), Tehran, Iran
ajavadi@aut.ac.ir

Abstract. Poly(lactic acid) (PLA)/graphene oxide (GO) nanocomposites containing 0.5 wt% to 2 wt% GO were prepared via solution mixing and the crystallinity, rheological and electrical properties of them were studied. It was found that the incorporation of GO leads to improve the storage modulus, loss modulus, and complex viscosity of the nanocomposite samples significantly. Also differential scanning calorimetry (DSC) results indicated that GO is an effective nucleating agent for PLA and could increase its crystallinity and decrease the crystallinity temperature. Also it was resulted that by introducing the GO in PLA matrix, the electrical resistance of samples are decreased, especially at low frequencies.

Keywords: Nanocomposite · Poly(lactic acid) · Graphene oxide · Biodegradable · Electrical resistance

1 Introduction

Poly(lactic acid) (PLA), a biodegradable thermoplastic aliphatic polyester, has been recognized as the most commonly used biomaterial due to its outstanding attributes, such as process ability, biocompatibility, high strength and modulus, etc. The excellent biocompatibility and absorbability of PLA have made it suitable for medical applications such as drug delivery systems, sutures and blood vessels. However, in practice, PLA needs to be modified for improving properties, such as melt strength, stiffness, crystallization ability, and heat deflection temperature. It has been widely proven that the addition of nanofillers could act as nucleation sites in polymer to improve their properties in recent years [1–3]. In this work, Graphen oxide is used to improve the crystallinity, rheological and electrical properties of PLA.

2 Experimental

2.1 Materials

Poly(lactide) (PLA, 4032D) with a density of 1.24 g/cm^3 and MFI of 7 g/10 min (210 °C/2.16 kg) was purchased from Nature Works USA. Anhydrous N,N-dimethyl

H. Mirzadeh and A. A. Katbab (Eds.): ISPST 2018, *Eco-friendly and Smart Polymer Systems*, pp. 71–74, 2020.
https://doi.org/10.1007/978-3-030-45085-4_18

formamide (DMF) was purchased from Mina Tajhiz Arya Iran. Graphene Oxide (GO) suspended into DMF of 1 mg/mL was provided from Graphene-X Iran.

2.2 Preparation of PLA/GO Nanocomposites

The PLA/GO nanocomposites were prepared via solution mixing using DMF in 100 °C. The resulting sample solution was immediately added into stirred water and the coagulated material precipitated continuously. Thereafter, the coagulates formed sample was isolated via filtration and washed with water. Then the samples were put in a vacuum oven at 60 °C overnight to evaporate any residual solvent. The PLA/GO nanocomposites with various GO contents were labeled as PLA/GO-0.5, PLA/GO-1, and PLA/GO-2 according to the weight concentration of GO in the composites. Neat PLA was also processed with the same procedure as PLA/GO composites for better comparison. Finally, the samples were shaped into films by compression molding at 200 °C for 3 min with a pressure of 10 MPa after preheating for 5 min. And then all the films were cooled to room temperature by cold compression molding at a pressure of 10 MPa for 5 min.

3 Results and Discussion

3.1 Thermal Properties

DSC curves (heating cycle) of neat PLA and PLA/GO nanocomposites are shown in Fig. 1. As it is observed, the glass transition (T_g) and melting temperature (T_m) of samples do not show any changes in comparison to neat PLA. In other hand, while the crystallization peak for PLA is not quite obvious, this peak is quite strong in all PLA/GO samples and the amount of crystallinity is increased by increasing the GO content, while the crystallization temperature is decreased. These phenomena indicate the heterogeneous nucleation effect of GO for this system, which leads to higher crystallinity of PLA.

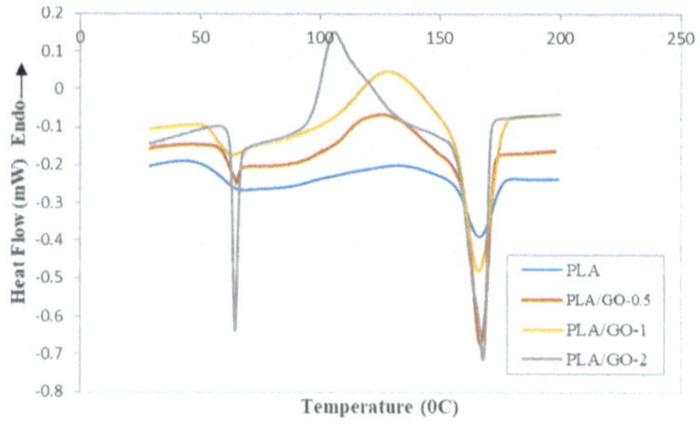

Fig. 1. DSC thermograms for neat PLA and PLA/GO nanocomposites

3.2 Rheological Properties

The rheological properties of all samples are shown in Fig. 2. It can be seen that the G′, G″, and η* of PLA/GO nanocomposites are higher than neat PLA and are increased with increasing the GO content. The differences are more pronounced at low angular frequency. It is due to the formation of an interconnected network between polymer matrix and nanoparticles, resulting in higher elasticity of PLA/GO nanocomposite samples. The analogous Cole–Cole plots of log G′ versus log G″ are shown in Fig. 2c. It is shown that the value of Cole–Cole plot's slope, approximately indicates the state of homogeneity in a sample, and changing the slope of theses curves indicates a significant microstructure change in the samples [4]. One can observe that the slope of log G′ versus log G″ is decreased with increasing GO content indicating a significant microstructure changes in the PLA/GO nanocomposites.

Figure 2d shows the variation of complex viscosity (η*) of neat PLA and PLA/GO nanocomposites. As it is observed, the complex viscosity of all composite samples are higher than neat PLA in all frequency range. Especially in samples containing 1 and 2 wt% of GO, a strong viscosity upturn is happened. As reported previously [3], this η* improvement may be attributed to the nanofiller–nanofiller interaction through hydrogen bonding and also the nanofiller–polymer matrix interfacial interaction.

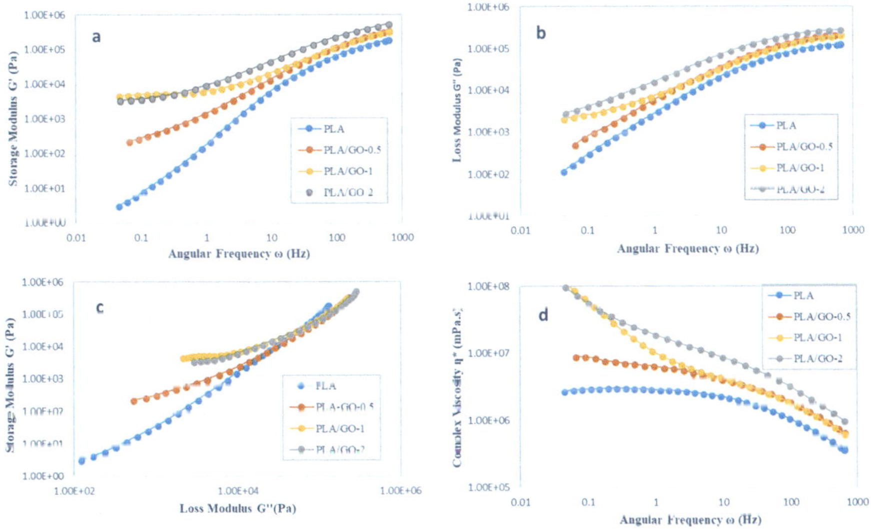

Fig. 2. Rheological properties of neat PLA and PLA/GO nanocomposites: (a) storage modulus, (b) loss modulus, (c) Cole–Cole plots of storage modulus versus loss modulus, and (d) complex viscosity

3.3 Electrical Resistance

The electrical resistance of the samples were measured using Electrochemical Impedance Spectroscopy (EIS) method in a frequency range of 0.1 kHz to 10 kHz in Alternative Current (AC) mode and the results are shown in Fig. 3. As it is seen, the maximum reduction in resistance of all samples is achieved at low frequency range (0.1–1 Hz). Also as it was expected, the resistance of samples are decreased by increasing the GO concentration.

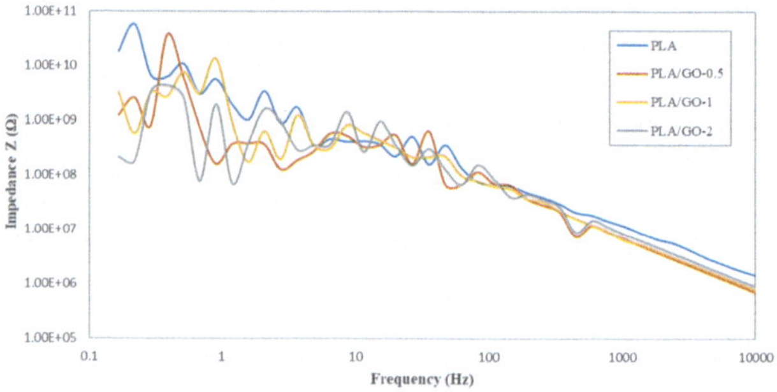

Fig. 3. Electrical resistance of neat PLA and PLA/GO nanocomposites

4 Conclusion

The results showed that the incorporation of GO improved the crystallinity of PLA matrix, indicating that GO is an effective nucleating agent for PLA. The addition of GO enhanced the storage modulus, loss modulus, and complex viscosity of PLA as well, and this improvement was increased with increasing the amount of GO. Furthermore, the electrical resistance of samples were reduced with increasing the GO content.

References

1. Gao, Y., Picot, O.T., Bilottz, E., Peijs, T.: Eur. Polym. **86**, 117–131 (2017)
2. Pinto, A.M., Gonçalves, C., Gonçalves, I.C., Magalhaes, F.D.: Eur. Polym. **85**, 431–444 (2016)
3. Tai, R.K., Hao, Y.M., Da, J.F., Xin, J., Bin, Y.C., Wen, J.M.: Ind. Eng. Chem. Res. **54**, 758–768 (2015)
4. Corre, Y.M., Maazouz, A., Duchet, J., Reignier, J.: J. Supercrit. Fluid **58**(1), 177–188 (2011)

Fabrication of NO-Releasing Silk Fibroin Nanofiber for Wound Dressing Application

Mina Vaez Ghasemi(✉) and Jhamak Nourmohammadi

Department of Life Science Engineering,
Faculty of New Sciences and Technologies, University of Tehran, Tehran, Iran
vaezmina@yahoo.com

Abstract. The aim of this study was to build a modern system with controlled release to improve the wound healing because sometimes the body alone is not capable of repair and wound closure. Due to evidence for beneficial effects of nitric oxide at any stage of wound healing (for example, induction of inflammation, angiogenesis, cell proliferation, matrix deposition, and reconstruction), treatments based on nitric oxide has been proved to be suitable for wound healing. Different amounts of sodium nitrite (2 and 4 wt%) were added to silk fibroin solution and its effect on the morphology, release profile and fibroblast viability fibers were investigated and compared to those of pure silk fibroin.

Keywords: Silk fibroin · Nitric oxide · Wound healing · Nano fiber · Sodium nitrite

1 Introduction

Wound healing is a series of complex biochemical and cellular processes including inflammation, cell proliferation, matrix deposition and tissue remodeling [1]. In recent years, electrospun nanofibrous dressings have gained much attention due to their similarity to natural extracellular matrix (ECM) structure, high specific surface area, and supporting cellular proliferation. Moreover, small pores in electrospun nanofibrous mats prevent the penetration of microbes and enhance oxygen permeability to the wound site. So far, various natural biodegradable polymers have been studied for wound healing application. Silk fibroin (SF), which is extracted from *Bombyx mori* cocoons, has long been used as a suture for treating wound injuries. Todays, SF has been widely studied for skin regeneration applications due to its good mechanical properties, oxygen permeability, controllable biodegradability, minimal inflammatory response, and promoting cellular attachment and proliferation [4].

The further requirement for nanofibrous wound dressing materials to accelerate the healing process is target delivery of various drugs such as antibiotics, anti-inflammatory drugs, and growth factors at a wound site. Recently, Nitric oxide (NO) releasing compounds have gained much popularity because of their low cost, bactericidal activity, regulating collagen formation, cell proliferation and wound contraction. It has been reported that NO-releasing formulations can improve incisional and excisional wound healing in rats [5]. Despite the beneficial effects of NO, its clinical application is limited

H. Mirzadeh and A. A. Katbab (Eds.): ISPST 2018, *Eco-friendly and Smart Polymer Systems*, pp. 75–78, 2020.
https://doi.org/10.1007/978-3-030-45085-4_19

by its gaseous property and short half-life. Therefore, designing a suitable matrix for the controlled release of NO is an indispensable property in NO delivery systems.

Thus, it seems that NO releasing compounds such as sodium nitrite is a highly potent component to be incorporated within silk fibroin nanofibers as a therapeutic agent to promote cell response and wound healing. Accordingly, different amounts of sodium nitrite (as a NO providing agent) were added to SF nanofibers and their effects on the morphology, release profile and fibroblast viability were investigated.

2 Experimental/Theoretical

SF was extracted from *B. mori* cocoons. The prepared SF sponge was then dissolved in formic acid under stirring for 3 h at room temperature to yield 13 wt% SF solution. Electrospinning (Fanavaran Nano-Meghyas, Iran) was carried out at 20 kV with a constant flow rate of 0.25 $mL \cdot h^{-1}$. Different amounts of sodium nitrite (2 and 4 wt%) were added to the solution. The samples were coded as SF (without NO), SF-2NO (containing 2 wt% $NaNO_2$), and SF-4NO (containing 4 wt% $NaNO_2$).

The morphology of the electrospun fibers was observed using scanning electron microscopy (SEM; AIS2300C, South Korea). The mean diameter of the fibers was measured by Image J software (National Institute of Health, Bethesda, Maryland, USA).

Pieces of dressings were immersed in 1 mL of PBS buffer (pH = 7.4) at 37 °C. The NO releasing profile was measured by using Griess kit assay. The absorbance was measured by using Spectrophotometer Uv-Vis (Bio-Rad, USA) at a wavelength of 548 nm. A calibration curve was prepared using different $NaNO_2$ solution ranging from 0 μg/mL to 100 μg/mL.

The fibroblast cells (L929) obtained from the National Cell Bank of Iran (NCBI; Pasteur Institute). Briefly, the fibroblasts cells (1×10^4) were first seeded on each disinfected sample and then placed in a humidified incubator with 5% CO2 for 1, 3, and 5 days. At each predetermined time, the culture medium was replaced by 10 μL MTT stock solution (12 mM) and then kept in an incubator for 4 h. Afterward, the formed purple formazan crystals were dissolved in DMSO (50 μL/well) and after 10 min incubation, the absorbance was read at 545 nm using an ELISA reader.

3 Results and Discussion

SEM images of the electrospun fibers are shown in Fig. 1. As expected, pure SF fibers exhibited continuous and bead-free fibers with the mean diameter of 643 ± 25 nm. It can be observed the mean fiber diameter of the nanofibers reduced to 385 ± 14 nm in SF-4NO and 136 ± 3 nm in SF-4NO. Such a decrease in the mean fiber diameter could be related to the increasing the electrical conductivity of electrospun solution as the $NaNO_2$ content rose. With increasing the electrical conductivity of electrospun solution the surface charge density of the electrospun jet enhanced, resulting in smaller fiber formation.

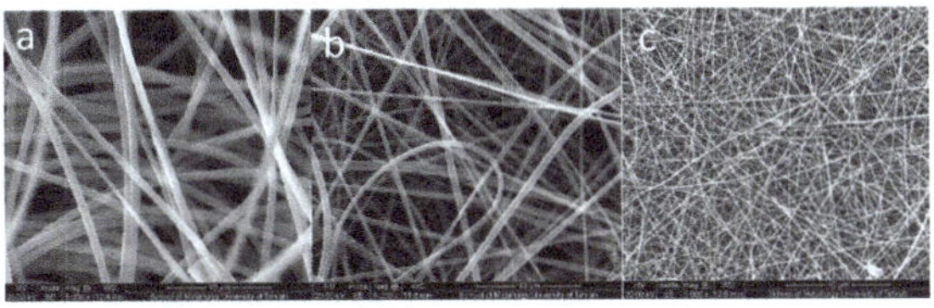

Fig. 1. SEM images of different electrospun mats (a) SF, (b) SF-2 wt% $NaNO_2$, and (c) SF-4 wt% $NaNO_2$

NO release profiles from NO-loaded electrospun fiber (with different loading doses of $NaNO_2$) are presented in Fig. 2. The release rate increased dramatically over time, and increasing $NaNO_2$ contents enhanced the NO released. This could be due to the difference between the fiber diameters in such samples. As shown in Fig. 1, the smaller size of the fibers in SF-4NO fibers increased their specific surface area, which led to the release of more NO.

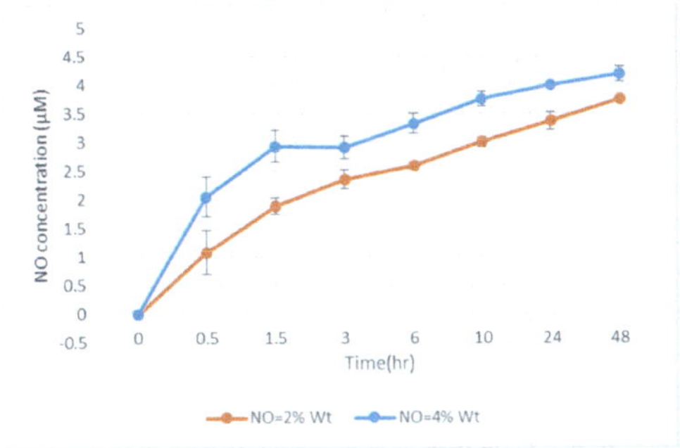

Fig. 2. NO release profiles of SF nanofibers with different $NaNO_2$ loading doses

Figure 3 shows the MTT results of the different samples as a function of culturing times. As expected, there is no significant difference between cellular viability of all electrospun mats on day 1. However, with increasing the culturing time the cellular viability of SF-4NO significantly decreased, this could be due to the release of higher amount of NO from this sample. Previous studies have also confirmed that high amount of NO can create cytotoxicity [4].

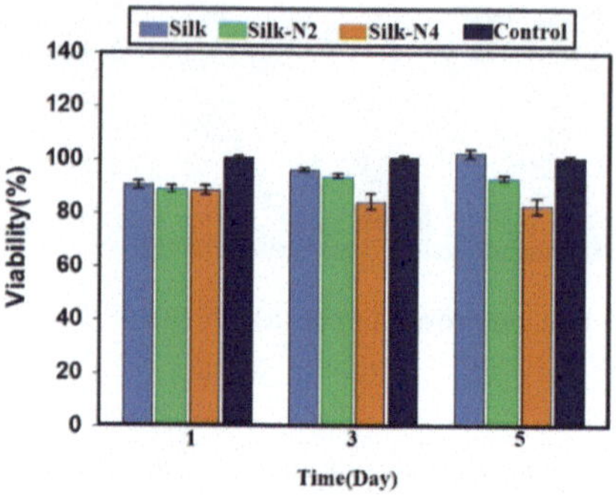

Fig. 3. The Indirect MTT assay of different mats over 5 days of culture

4 Conclusion

In this study, sodium nitrite was successfully loaded into SF nanofibers. Increasing the sodium nitrite decreased the fiber diameter which led to higher NO release. Based on the fibroblast viability, SF-2NO could be chosen as a suitable matrix for further.

References

1. Kim, J.O., et al.: Int. J. Biol. Macromol. **79**, 217–225 (2015)
2. Kasoju, N., Bora, U.: Adv. Healthc. Mater. **1**(4), 393–412 (2012)
3. Lowe, A., Bills, J., Verma, R., Lavery, L., Davis, K., Balkus Jr., K.J.: Acta Biomater. **13**, 121–130 (2015)
4. Li, C.Q., Pang, B., Kiziltepe, T., Trudel, L.J.: Chem. Res. Toxicol. **19**(3), 399–406 (2006)

Fabrication and Characterization of PCl/Gelatin/Forsterite Nanofibrous Scaffolds Used for Modification of the Implants

Fatemeh Khosravi, Saied Nouri Khorasani(✉), Hamid Zilouei, and Rasoul Esmaeely Neisiany

Department of Chemical Engineering, Isfahan University of Technology, 84156-83111 Isfahan, Iran
saied@ct.iut.ac.ir

Abstract. Bicomponent nanofibrous scaffold incorporated forsterite nanoparticles were fabricated via facile and cost-efficient electrospinning method for biomedical applications. At the first step, polycaprolactone (PCL) nanofibers were fabricated, concerning to optimize the electrospinning parameters. Afterward, gelatin was added to PCL to provide the nanofibers to be more biocompatible. Subsequently, 1 wt% of forsterite nanoparticles were incorporated in both neat PCL and PCL/gelatin nanofibers. Scanning electron microscopy was employed to investigate the morphology and diameter of the nanofibers and optimize the electrospinning parameters. SEM results showed that incorporation of gelatin to PCL nanofibers led to a significant reduction of the nanofiber's diameter. In addition, incorporation of forsterite nanoparticles led to an increase in the nanofiber's diameter for both PCL and PCL/gelatin nanofibrous scaffolds.

Keywords: Nanofiber · Electrospinning · Polycaprolactone · Gelatin · Forsterite nanoparticles

1 Introduction

Metallic implants have been widely used in the medical applications due to their low cost as well as superior mechanical strength. However, their drawbacks consisting of poor corrosion resistance leading to release of toxic metal ions such as nickel, chromate, and molybdenum which limits their applications [1]. Several surface modifications have been proposed in order to improve corrosion resistance and biocompatibility of these types of materials. Among the methods, polymeric nanocomposite coatings were investigated more than other methods [2, 3]. It is worthy to mention, from all of the ways to fabricate coatings, electrospinning is a more versatile technique used to form uniform nanofibers. Electrospinning has been widely employed for the preparation of tissue engineering scaffolds and drug delivery systems because of its relatively facility of using and adaptability. The intrinsically high surface to volume ratio of electrospun polymer fibers can increase cell adhesion, rise drug loading and controlled local drug delivery [4].

H. Mirzadeh and A. A. Katbab (Eds.): ISPST 2018, *Eco-friendly and Smart Polymer Systems*, pp. 79–82, 2020.
https://doi.org/10.1007/978-3-030-45085-4_20

Therefore, in the present research, PCL/gelatin/forsterite nanofibrous scaffolds were prepared and optimized for coating application of metallic materials, in implant applications, to overcome the aforementioned drawbacks.

2 Experimental

Before the electrospinning process, the solutions consisting of Gelatin/PCl and 1 wt% of forsterite nanoparticle were prepared. At first, PCl (100 and 60 wt%) and gelatin (0 and 40 wt%) were dissolved in a solvent including 1 mL folic acid and 3 mL of acetic acid without forsterite nanoparticle. Secondly, the above-mentioned solutions were prepared to contain 1 wt% of forsterite nanoparticle which dispersed in the solvents by 20 min sonication. The prepared solutions were fed to 1 mL syringes and the nanofibrous scaffolds were prepared by adjusting and optimizing the electrospinning parameters including, allied voltage, the distance between needle tip and collector, solution feed rates and so on.

The morphology of PCl/Gelatin as well as PCl/Gelatin/forsterite nanofibers was investigated by scanning electron microscopy (SEM, Philips, XL 30). In addition, the images were analyzed using Image J. software averaged over 50 measurements of the nanofibers diameter.

3 Results and Discussion

Figure 1 shows the SEM image of fabricated neat PCL nanofibers and the nanofibers' diameter distribution. It can be observed at the feed rate of 0.1 mL · h^{-1}, an applied voltage of 20 kV, and 15 cm distance between the needle tip and collector, the fabricated nanofibers were smooth without any signs of beads. As mentioned before, 50 measurements were performed to obtain the average and standard deviation of the nanofiber's diameter. The mean value of the neat PCL nanofiber's diameter was determined to be 203 nm, considering the ±75 nm of standard deviation.

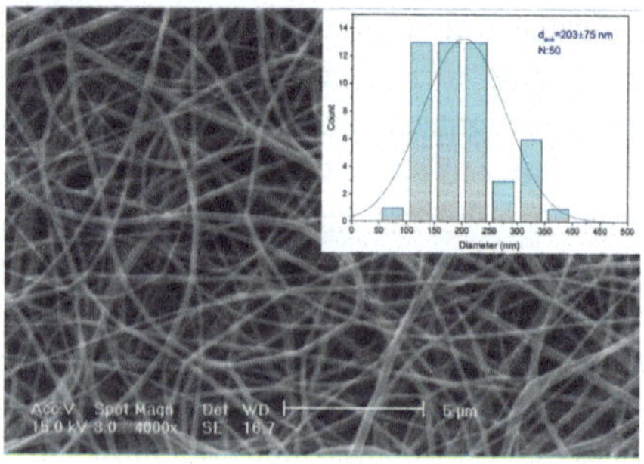

Fig. 1. SEM image of neat PCL nanofibers

Figure 2 displays the SEM micrograph of PCL/gelatin (60/40 wt%) nanofibers. For this nanofibers, the electrospinning parameters were also adjusted to prepare the nanofibers without beads with the less nanofiber's diameter. Form Fig. 2, it is clear that the diameter of the nanofibers was determined to be 168 ± 82 nm for electrospinning at following conditions: the feed rate of 0.2 mL · h^{-1}, applied voltage of 20 kV and 15 cm of distance between the needle tip and collector. It is elucidated that incorporation of gelatin led to a significant reduction in the nanofiber's diameter in comparison with the neat PCL nanofibers, which can be attributed to increasing the electrical conductivity of the solution after incorporation of the gelatin [5]. Figure 3 presents the SEM micrograph of the PCL nanofibers containing 1 wt% of forsterite nanoparticles. The electrospinning parameters for this solution were adjusted at 0.2 mL · h^{-1} of feed rate, 21 kV of the applied voltage and 15 cm of the distance. According to the SEM image, the nanofibers are smooth without beads with the nanofiber's diameter of 250 ± 72 nm. Comparing the neat PCL nanofibers and PCL nanofibers indicated that the nanofiber's diameter increased after incorporation of forsterite nanoparticles which can be attributed to increasing the viscosity of the solution [5].

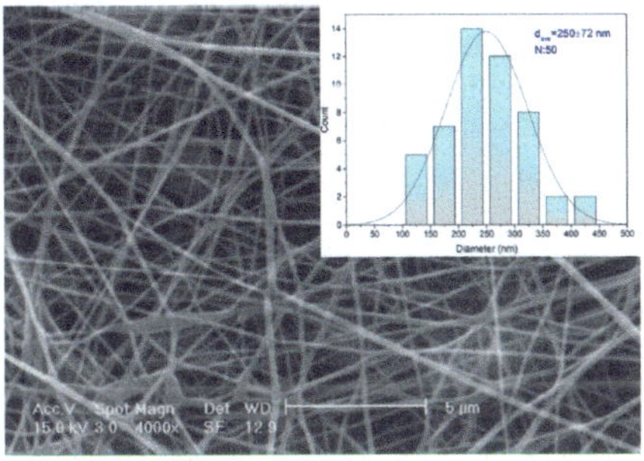

Fig. 2. SEM image of PCL nanofibers containing 1 wt% of forsterite nanoparticles

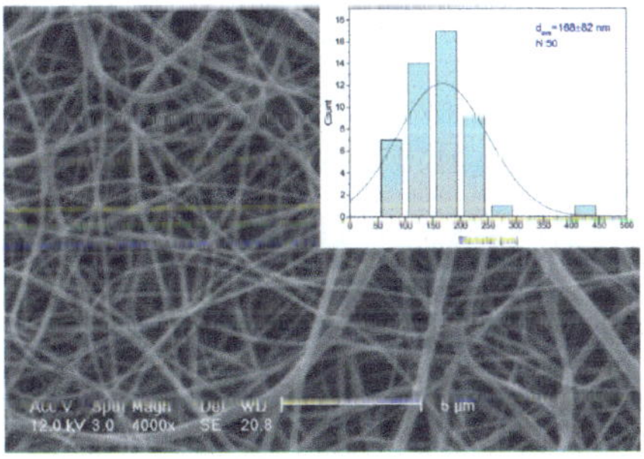

Fig. 3. SEM image of PCL/gelatin nanofibers

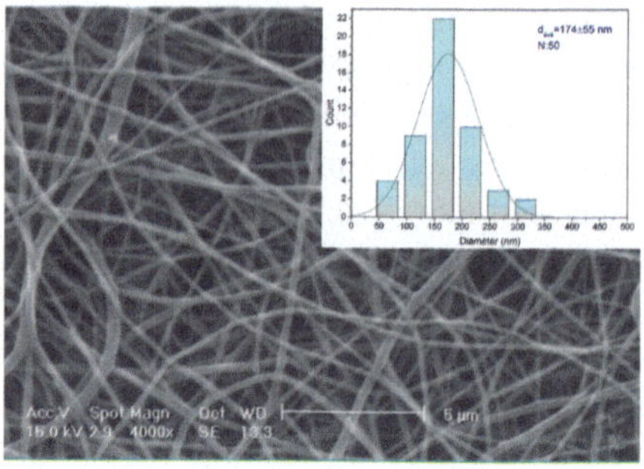

Fig. 4. SEM image of PCL/gelatin nanofibers containing 1 wt% of forsterite nanoparticles

Finally, Fig. 4 shows the SEM micrograph of the PCL/gelatin (60/40 wt%) containing 1 wt% of forsterite nanoparticles. From the assessment of nanofiber's diameter, it can be discerned that for this nanofibers the average diameter was obtained at 174 ± 55 nm. Comparing these nanofibers with other nanofibers reveals that the nanofiber's diameter increased by incorporation of forsterite nanoparticles and decreased by the addition of gelatin. The electrospinning parameters for the nanofibers were sat at feed rate of 0.1 $mL \cdot h^{-1}$ and applied voltage of 22 kV.

4 Conclusion

The novel nanofibers of PCl/gelatin containing forsterite nanoparticles were successfully fabricated using the electrospinning technique. The morphology studies of the nanofibers revealed that by optimizing electrospinning parameters the uniform nanofibers without any bead can be obtained. In addition, the results revealed that the incorporation of gelatin reduced the nanofiber's diameter whereas incorporation of nanoparticles increased the nanofiber's diameter.

References

1. Torkaman, R., Darvishi, S., Jokar, M., Kharaziha, M., Karbasi, M.: Prog. Org. Coating **103**, 40 (2017)
2. Xiao, X., Liu, R., Tang, X.: J. Mater. Sci. Mater. Med. **20**, 691 (2009)
3. Pang, X., Casagrande, T., Zhitomirsky, I.: J. Colloid. Interf. Sci. **330**, 323 (2009)
4. Xue, J., He, M., Liu, H., Niu, Y., Crawford, A., Coates, P.D., Chen, D., Shi, R., Zhang, L.: Biomater **35**, 9395 (2014)
5. Enayati, M.S., Behzad, T., Sajkiewicz, P., Bagheri, R., Ghasemi-Mobarakeh, L., Kuśnieruk, S., Rogowska-Tylman, J., et al.: Int. J. Polym. Mater. Polym. Biomater. **65**, 660 (2016)

Biodegradable 3D Printed Scaffolds Based on PLA for Bone Tissue Engineering

Shaghayegh Kaviani-Samani[1], Shadab Bagheri-Khoulenjani[1(✉)], Hamid Mirzadeh[1(✉)], and Soheil Dariushi[2]

[1] Department of Polymer Engineering, Amirkabir University of Technology, P.O. Box 15875-4413, Tehran, Iran
{s.bagheri,mirzadeh}@aut.ac.ir

[2] Iran Polymer and Petrochemical Institute, P.O. Box 14975-112, Tehran, Iran

Abstract. Fused deposition modelling (FDM) is one of the additive manufacturing techniques that offers the potential to fabricate natural tissue constructs. The aim of this study is to fabricate biocompatible and biodegradable scaffold with good mechanical properties for tissue engineering. FDM allows to design and fabricate highly porous 3D scaffolds with interconnected network. At first, filament with the use of poly lactic acid (PLA) was fabricated. Utilizing computer aided design software the scaffolds structures were designed then using fused deposition modeling, scaffolds with controlled pore size and pore structure were printed. To determine the optimum structure, the effect of several parameters, such as nozzle size, raster angle and different interior fill of software were investigated. Specimens with different nozzle sizes of 0.3, 0.4, and 0.5 mm in various raster angles of 0°/90°, ±45° and 0°/60°/120° and in various interior fill percentages were printed. Specimens were characterized with SEM and compression testing. Pore size and porosity were calculated.

Keywords: Additive manufacturing · Fused deposition modeling · Tissue engineering · 3D printed scaffold · Porosity

1 Introduction

Tissue engineering is a multidisciplinary science that includes materials science and engineering, cell biology and regenerative medicine. That is a technique for imitating natural tissue. Scaffolds are critically important for tissue engineering [1]. Scaffolds should be highly porouse and have well-interconnected pore network and consistent and adequated pore size to facilitate cell migration, infiltration and fluid diffusion [2]. The various techniques have been developed to fabricate porous scaffolds for tissue engineering such as: additive manufacturing, solvent casting, freeze-drying, sol gel technique, porogen leaching, electron spinning and etc. FDM has been widely used in additive manufacturing technology, due to its easy operation, low cost, reliability and was used to produce scaffolds with fully interconnected channel network and controllable porosity and pore size [1]. The mechanical properties, pore size and porosity of a 3D-printed scaffold can be influenced by the FDM processing parameters and

H. Mirzadeh and A. A. Katbab (Eds.): ISPST 2018, *Eco-friendly and Smart Polymer Systems*, pp. 83–86, 2020.
https://doi.org/10.1007/978-3-030-45085-4_21

setting of software such as raster angle, layer thickness, nozzle size, and interior infill of software. In this paper, the influence of interior infill of software, raster angle and nozzle size on the properties of FDM manufactured scaffold was studied.

2 Experimental

Pellets of poly lactic acid (PLA) were dried at 80 °C for 4 h to prevent hydrolysis, then were fed into a noztek extruder to prepare filaments. The temperature of extrusion was set on 185 °C. Filaments of PLA 1.7 ± 0.2 mm in diameter were obtained. Utilizing computer aided design (CAD) software we design the scaffold structure, and then specimens with various nozzle size and different interior fill percentage of software with three raster angles of 0°/90°, ±45 and 0°/60°/120° were manufactured by Sizan3, 3D printer (sizan Co., Iran) at a nozzle temperature of 210 °C. Nozzles with diameters of 0.3, 0.4, and 0.5 mm were used. A schematic of the process is shown in Fig. 1.

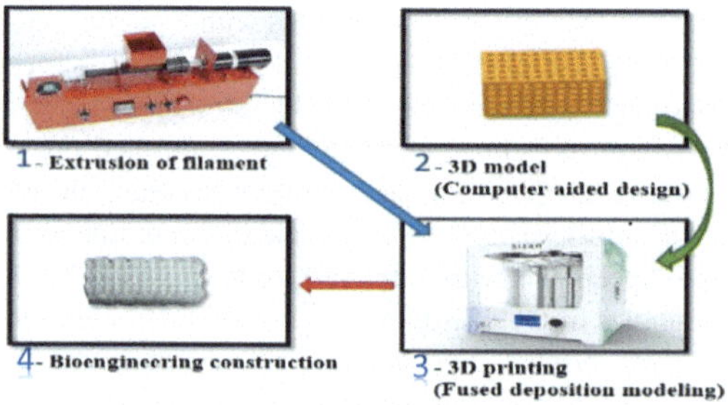

Fig. 1. A schematic of the process

3 Results and Discussion

Architectural structures of scaffolds printed with various raster angles are shown in Fig. 2. Studies of 3D porous scaffolds based on PLA were carried out by scanning electron microscopy (SEM). The SEM image of scaffolds with 40% fill printed at the different raster angles of 0°/90°, ±45°, and 0°/60°/120° and the cross section of scaffold are shown in Fig. 3. In Fig. 4 as it can be seen, by reduction of interior fill percentage of software the pore size of scaffolds increases, and as a result, the porosity also increases. The pore size of scaffolds was 300 μm to 1200 μm. With decrease in interior fill of software, porosity of scaffold increased up to 80% vol. If the interior infill of software is lower than 35% infill, the walls of the samples will not be porous. Mechanical testing was performed by compression testing of scaffold in parallel with stack. Yield strength and elastic modulus were determined (Table 1 and Fig. 5). The scaffolds printed at the angles of 0°/90° and 0°/60°/120° show the highest and the lowest yield strength and elastic modulus, respectively. The optimum scaffold had good mechanical properties and adequate pore size and porosity.

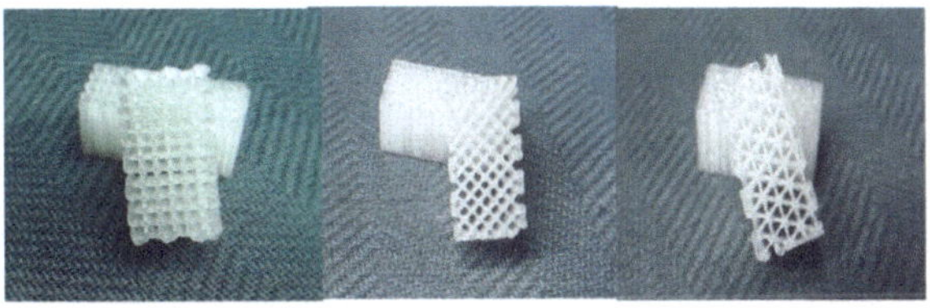

Fig. 2. Images of the specimens with 40% fill in different raster

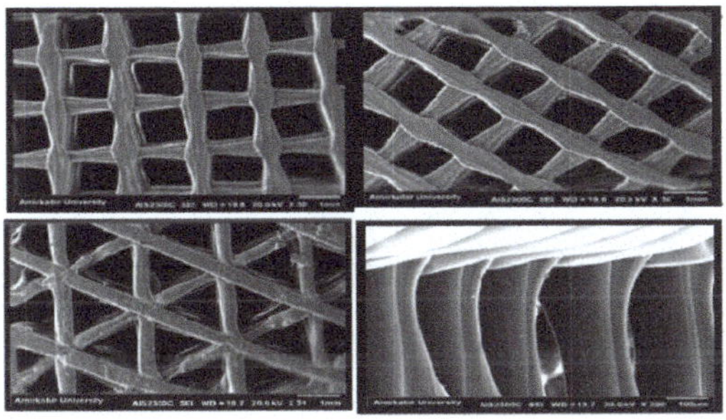

Fig. 3. SEM images of specimens with 40% fill printed at different raster angle of (a) 0°/90°, (b) ±45°, (c) 0°/60°/120°, and (d) cross section of specimens

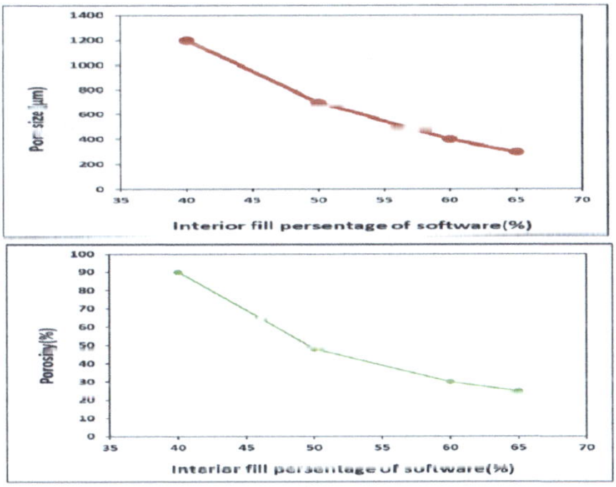

Fig. 4. (a) Pore size of scaffolds with an angle of 0°/90° and (b) porosity of scaffolds with an angle of 0°/90°

Table 1. Elastic modulus and yield strength of scaffolds (40%)

Raster angle	0°/90°	±45°	0°/60°/120°
Elastic modulus (GPa)	0.59	0.45	0.41
Yield strength (MPa)	36.15	35	28

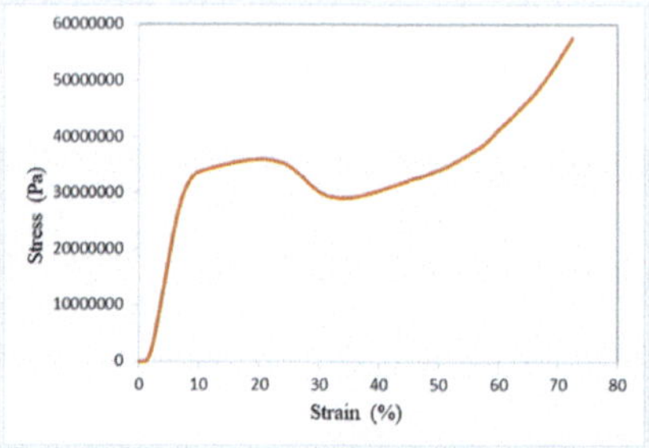

Fig. 5. Compression test of scaffold with 40% fill and 0°/90°

4 Conclusion

Porous scaffolds based on PLA for bone tissue engineering were prepared by fused deposition modelling. Properties of the scaffolds manufactured by FDM depended on several parameters. It was found Fig. 3. SEM images of specimens with 40% fill printed at different raster angles of a) 0°/90°, b) ±45°, c) 0°/60°/120°, and d) cross section of specimens that the pore size and porosity decreased with the increase of the interior fill of software also the scaffold printed at the angle of 0°/90° showed the highest yield strength and elastic modulus in comparison with the other two raster angles studied.

References

1. Thavornyutikarn, B., Chantarapanich, N., Sitthiseripratip, K., Thouas, A., Chen, Q.: Prog. Biomater. J. **3**, 61–102 (2014)
2. Senatov, F.S., Niaza, K.V., Zadorozhnyy, M., Maksimkin, A.V., Kaloshkin, S.D., Estrin, Y. Z.: Mech. Beh. Biomed. Mater. J. **57**, 139–148 (2016)

Synthesis and Characterization of Biodegradable and Nontoxic Water Borne Copoly(Ether-Urethane-Urea)s Based on Amino Acid Derivatives

Shayesteh Tafazoli and Fatemeh Rafiemanzelat(✉)

Organic Polymer Chemistry Research Laboratory, Department of Chemistry, University of Isfahan, 81746-73441 Isfahan, Iran
frafiemanzelat@sci.ui.ac.ir

Abstract. Biodegradable and biocompatible polyurethanes are a class of polymers that are useful in medical industry. In this study, new biodegradable and nontoxic water dispersed co-poly(ether-urethane-urea)s (PEUU)s were synthesized based on amino acid and peptide moiety. These polymers were prepared via the reaction of L-leucine anhydride cyclopeptid (LAC), poly-ethylene glycol (PEG), hexamethylene diisocyante (HDI) and serine (S) as dispersing agent. These polymers were characterized by FT-IR, NMR, TGA, DSC, AFM, and DLS. PEUU polymers were synthetized with three different structural architectures and block sequence. The polymer's particle size dispersions are around 300 nm. The degradation test was carried out in PBS at 37 °C and evaluated by weight loss, viscosity and particle size decrement as well as by AFM. In order to determine cytotoxicity of the polymers, in vitro toxicity of final polymer was assessed using L929 mouse fibroblast cell line. The results showed no cytotoxicity of these polymers.

Keywords: Copoly(ether-urethane-urea) · Biodegradable · Nontoxic · Water borne · Amino acids

1 Introduction

Degradable polymers have been used in biomedical applications, or as environmental friendly polymers. The biodegradability of PUs is generally achieved by incorporating labile and hydrolysable functional groups into the polymer backbones [1]. It has been shown that copolymers consisting of PEG and peptide sequences are highly water-soluble, degradable, non-immunogenic and relatively non-toxic [2].

Recently, it has been shown that the presence of amino acid functionality can lead to polymers with improved solubility, biocompatibility and biodegradability [3].

In addition during last years, preparation of waterborne polyurethanes (WBPU) has been the most interesting filed of polymerization, due to the control of volatile organic compound emissions (environmental friendship) and economical point of view [4]. In this work, L-leucine anhydride cyclo-peptide (LAC), as an amino acid derivate monomer together with PEG-1000, which have suitable properties for preparation of

H. Mirzadeh and A. A. Katbab (Eds.): ISPST 2018, *Eco-friendly and Smart Polymer Systems*, pp. 87–91, 2020.
https://doi.org/10.1007/978-3-030-45085-4_22

non-immunogenic and biodegradable polymers, were utilized to synthesize nontoxic and biodegradable polymers. Also, it is anticipated that the application of serine as the ionic dispersing agent, can play a vital role in the bio-functionalization of chains for dispersing them in water.

2 Experimental

2.1 Materials

1,6-Hexamethylene diisocyanate (HDI) (Merck, Germany) was used as received. Poly (ethylene glycol) (PEG) (Merck, Germany, Mn = 1000 $g.mol^{-1}$) was dried under vacuum at 80 °C for 6 h. L-leucine and serine (S) (Merck, Germany) were dried under vacuum at 80 °C for 3 h. Ethylene glycol (EG) and N-methyl-2-pyrrolidone (NMP) (Merck, Germany) were distilled under reduced pressure over dried BaO (Sigma–Aldrich, Germany). Dimethylformamide (DMF) (Merck, Germany) was used as received.

2.2 Synthesis of PEUUs

Three different PEUUs were synthesized which their mol fraction and block structure are shown in Tables 1 and 2, respectively. For example, PA1 was synthesized as follows: In a dried 25 mL two necked flask equipped with a N_2 balloon, serine (0.078 g, 7.47×10^{-4} mol) and NMP (0.60 mL) were stirred at 90 °C for 20 min to obtain a homogeneous solution; then temperature was reduced to 40 °C, and HDI (0.12 mL, 7.47×10^{-4} mol) was added. The solution was heated at 90 °C. Afterwards, HDI (0.06 mL, 3.734×10^{-4} mol) was added and heated according to above temperature program. LAC (0.042 g, 1.868×10^{-4} mol) which was synthesized by the same procedure as reported earlier [5] and 0.8 mL NMP were heated up to 120 °C and added to the reaction mixture. The temperature was kept at 120 °C for 10 min to dissolve LAC completely. Then the temperature was reduced down to 90 °C for 9 h. After that, the temperature was reduced to 60 °C, and PEG (0.746 g, 7.47×10^{-4} mol) was added to the reaction mixture. Subsequently, the reaction mixture was heated further at 90 °C for 8 h. Then the viscous solution was poured into 5 mL distilled water after FTIR monitoring of NCO consumption at 2270 cm^{-1}. After shaking and stirring for 0.5 h, the precipitated polymer was isolated by filtration, dried at 50 °C for 2 h under vacuum to give 0.376 g (21.49%) of the polymer. The inherent viscosity of this polymer was 0.313 dL/g in DMF, and we evaluated its chemical structure by FT-IR.

3 Results and Discussion

3.1 Characterization of the Polymers

The structures of PEUUs were characterized using FTIR. For instance, the FTIR spectrum of PA1 shown in Fig. 1 confirms the successful synthesis of PA1.

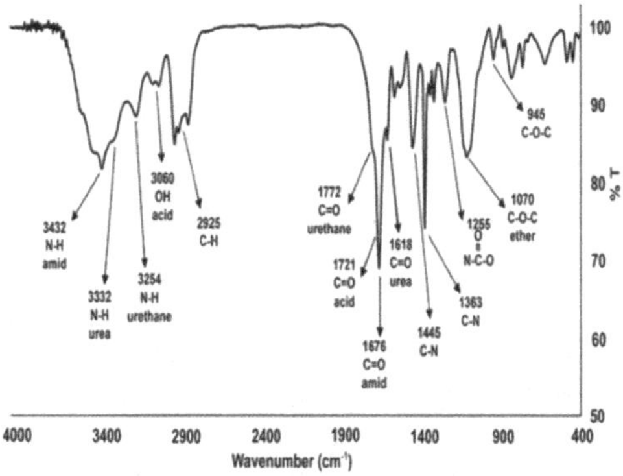

Fig. 1. FT-IR spectrum of PA1

3.2 Degradation Study of PEUs

Table 2 represents the results of degradation studies of PA1, PA2, and PA3 in PBS at 37 °C. Indeed, these degradation values are due to the extensive amount of biodegradable bonds such as those presented in peptide, urethane, amide and urea groups. The degradation rate of PA2 is more than that of PA1 and PA3. It can be attributed to the higher amount of hydrophilic PEG participation along the PA2 chains which was demonstrated in Table 2.

Table 1. Results of degradation study of PEUUs in PBS

Measured item	PA1	PA2	PA3
Hydrophilic part (%) PEG	26.66	28.57	15.38
Hydrophilic part (%) serine	26.66	30.77	28.57
Hydrophobic part (%) LAC	6.66	7.69	0
Weight loss (%) after 10 days	45.87	47.10	26.25
Primary viscosity	0.31	0.34	0.30
Viscosity decrement (%)	30.41	32.54	7.28
Primary particle size	313.9	481.3	315.1
Particle size decrement (%)	97.07	98.01	80.48

Table 2. Different structures of PEUUs

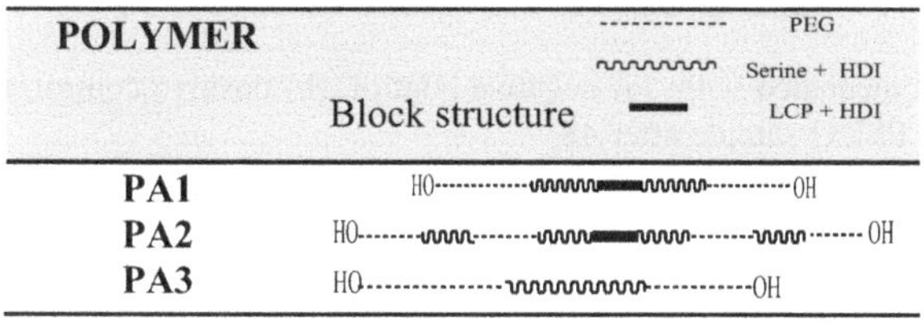

AFM was used to study the hydrolytic surface degradation of PEUUs within PBS. From phase imaging, heterogeneity is evidenced by the two distinct regions, bright regions (hydrophilic, rich in PEG) and dark regions (hydrophobic, rich in LAC). After the degradation in PBS, the increase of the bright regions and surface roughness are visible. This behavior can be due to the surface changes of the polymer as a result of water uptake and degradation (Fig. 2).

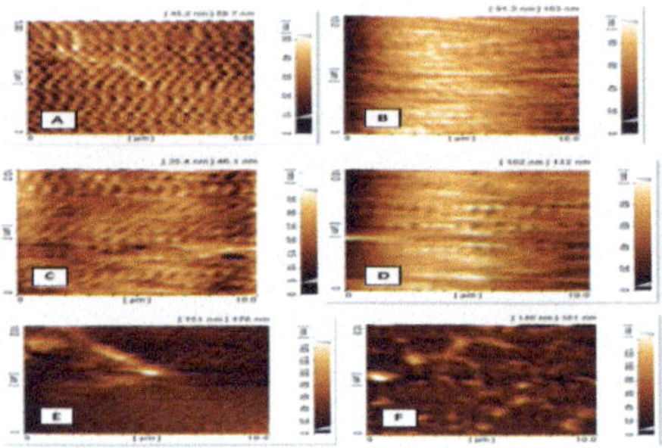

Fig. 2. AFM images of PA1(A), PA2(C), and PA3(E) before degradation in PBS and PA1(B), PA2(D), and PA3(F) after degradation in PBS

3.3 Cell Cytotoxicity Studies via Direct Contact Test

The PEUU sample was incubated with monolayer cells at 37 °C for 24–48 h. The PEUU surface was studied using the microscope for the cellular response after the specific time of incubation. Cell cytotoxicity of the PEUU sample was compared with a negative control (fibroblast cell culture) and a positive control (a toxic substance: stabilized PVC disk). The spindle-shaped L929 cells are found to adhere and expand on PEUU (Fig. 3c). The cell adhesion studies after 48 h, show that the cells have excellent adherence to the surface of PEUU and the cell proliferation on polymer surface continued (Fig. 3d).

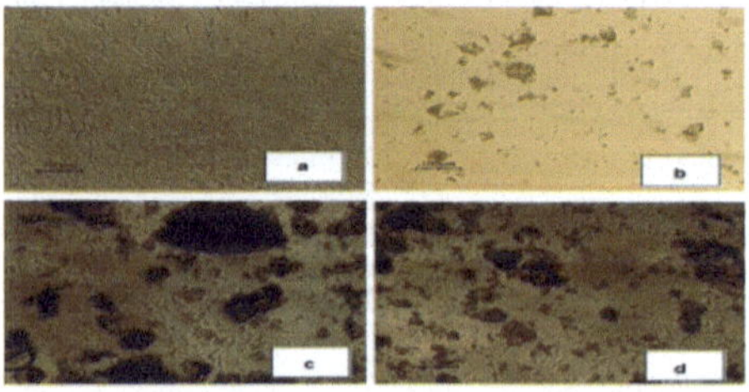

Fig. 3. L929 cells incubated with: (a) negative control, (b) positive control, (c) PEUU sample after 24 h, and (d) PEUU sample after 48

4 Conclusion

In this work, biodegradable and nontoxic water dispersed PEUUs were successfully synthesized and characterized. These polymers have high potential for using in biomedical industry.

References

1. Zhang, C., Wen, X., Vyavahare, N.R., Boland, T.: Biomaterials **29**, 3781–3791 (2008)
2. Rafiemanzelat, F., Fathollahi Zonouz, A., Emtiazi, G.: Amino Acids **44**, 449–459 (2013)
3. Torma, V., Gyenes, T., Szakács, Z., Talsma, H., Hennink, W.E.: Polym. Bull. **59**, 311–318 (2007)
4. Xia, J., et al.: Eur. Polym. J. **43**, 1838–1846 (2007)
5. Rafiemanzelat, F., Abdollahi, E.: Polym. Degrad. Stab. **95**, 901–911 (2010)

In Situ Forming Hydrogels Based on Clickable Star-PEG for Biomedical Applications

Mohammad Hossein Ghanian[1,2], Hamid Mirzadeh[1(✉)], and Hossein Baharvand[3(✉)]

[1] Department of Polymer Engineering and Color Technology, Amirkabir University of Technology, Tehran, Iran
mirzadeh@aut.ac.ir
[2] Department of Cell Engineering, Cell Science Research Center, Royan Institute for Stem Cell Biology and Technology, ACECR, Tehran, Iran
[3] Department of Stem Cells and Developmental Biology, Cell Science Research Center, Royan Institute for Stem Cell Biology and Technology, ACECR, Tehran, Iran

Abstract. Four-arm polyethylene glycol (star-PEG) macromeres are known as building blocks of homogenous hydrogels, known as tetra-PEG gels, which show excellent elasticity for load-bearing applications. However, biomedical application of the convectional tetra-PEG gels has been limited by weak bioorthogonality of the crosslinking reaction, non-biodegradability of the crosslinks, or complexity of the synthetic methods. Michael addition between simple groups of thiol and maleimide is a click reaction which is fast under physiological conditions, cytocompatible, and bioorthogonal. However, thiol-maleimide crosslinks are not biodegradable. Herein, we hypothesized that biodegradable, yet clickable tetra-PEG derivatives might be achieved by introduction of a hydrolyzable ester bond before thiol end groups. Upon mixing the degradable tetra-PEG-SH and tetra-PEG-Mal in less than 10 s, a biodegradable PEG hydrogel was produced that showed potential for loading and sustained release of proteins and encapsulation of live cells.

Keywords: Polyethylene glycol · Click chemistry · Injectable hydrogel · Biocompatible · Biodegradable

1 Introduction

Polyethylene glycol (PEG) is a synthetic hydrophilic polymer which has been widely used in biomedical researches and its use in products has been approved by the FDA due to its biocompatibility, low protein adsorption, non-immunogenicity, and potential for end-group functionalization [1]. Reactive end-groups allow for covalent coupling of PEG macromeres to produce a network. Conventional chemical reactions for production of PEG hydrogels mostly work under activation by enzymes, catalysts, initiators or photoradiation. In contrast, Michael addition is a (photo) catalyst-free click reaction which operates efficiently and quickly under physiological conditions (at 37 °C with pH 7.4) with no byproducts. The regioselectivity, fast kinetics, high efficiency, and

H. Mirzadeh and A. A. Katbab (Eds.): ISPST 2018, *Eco-friendly and Smart Polymer Systems*, pp. 92–95, 2020.
https://doi.org/10.1007/978-3-030-45085-4_23

mild reaction conditions of MA minimize the risk of cytotoxicity and cross-reactivity with the biological components [2]. In addition, compared to the conventional PEG hydrogels that are produced by random photopolymerization of PEG-diacrylate, covalent coupling of similar tetra-PEG macromeres via complementary end-groups produces more homogenous networks [3]. However, thiol-maleimide (succinimide thioether) crosslinks are not biodegradable. Herein, we hypothesized that biodegradable, yet clickable tetra-PEG derivatives might be achieved by introduction of a hydrolyzable ester bond before thiol end groups. The hydrolytic degradable thiol derivative was synthesized via esterification of PEG with a thio-acid. Upon mixing the clickable PEG solutions, sol-gel transition occurred in less than 10 s. The thiol-maleimide PEG hydrogels could degrade exclusively when an ester bond was introduced before thiol groups. The biodegradable PEG gel was successfully used as a controlled release carrier for BSA or as an in situ forming scaffold for stem cell encapsulation.

2 Experimental

The building blocks of hydrogel were synthesized by end-functionalization of four-arm PEG macromeres via coupling reactions with thiol or maleimide-bearing linkers, as previously described [4]. Conversion of hydroxyl end groups was confirmed by ^{1}H NMR. Tetra-PEG hydrogels were prepared by volumetric mixing of 10 w/v% solutions of tetra-PEG-SH and tetra-PEG-Mal. The gels were incubated in PBS at 37°C for gravimetric analysis of swelling and biodegradation during time. The biological potentials of the gel was shown by sustained release of a protein model (BSA) and viability assays of encapsulated mesenchymal stem cells. The BSA concentration in PBS was measured by BCA kit according to Pierce's protocol. Cell viability was analyzed by MTS (Promega) and Live/Dead staining (Thermo) assays following the manufacturer's instructions.

3 Results and Discussion

The tetra-PEG-SH was synthesized through formation of ester bonds between the hydroxyl end group of 4arm-PEG and carboxyl of a thioacid. The ^{1}H NMR spectroscopy confirmed that the end-groups were converted quantitatively to free thiols via ester bonds (Fig. 1A). To synthesize maleimide derivatives, hydroxyl end-groups were first converted to amine to enable covalent coupling with an amine-reactive linker. The NHS-activated carboxyl of the linker was coupled to amine end-groups of PEG-NH_2 via stable amide bonds to produce non-degradable macromeres of PEG-maleimide. As calculated based on ^{1}H NMR integration, more than 90% of 4arm-PEG end-groups were converted to maleimide (Fig. 1B). The ester bonds in tetra-PEG-SH are prone to hydrolytic degradation in aqueous media (assigned by a scissor) which is favorable for biodegradation of the click hydrogels. To address the hypothesis, the degradation of tetra-PEG hydrogels prepared from Michael addition between tetra-PEG-Mal and tetra-PEG-SH (degradable crosslinker) or DTT (a non-degradable dithiol crosslinker) was compared (Fig. 2). The results showed that the ester bonds are responsible for biodegradation of the PEG hydrogel as the tetra-PEG hydrogel was degraded after ~50 days in contrast to the stable control hydrogel. To assess the biocompatibility of the gel formation process, the PEG precursors were mixed with cell suspension to allow for gel formation in the presence of live cells. Cytocompatibility was analyzed by live/dead and MTS assays (Fig. 3). The results showed no sign of cytotoxicity as no significant difference was observed between the cells encapsulated within hydrogel and those cultured onto control tissue culture plate. Michael addition has been known as a bioorthogonal chemistry which operates under mild conditions without any deleterious effects on the encapsulated cells or other biological payloads.

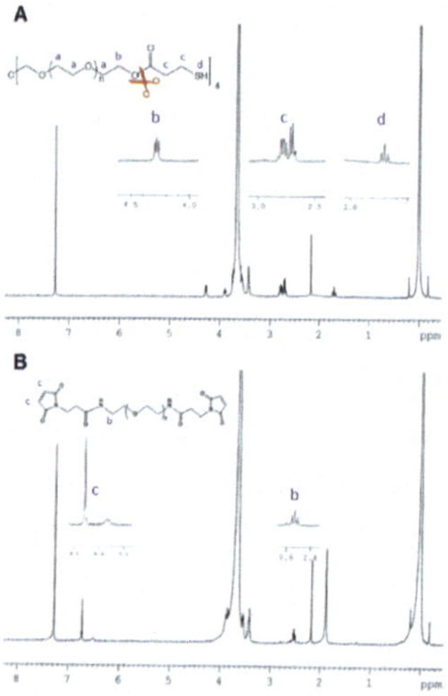

Fig. 1. ^{1}H NMR spectra of hydrogel building blocks: (A) Degradable PEG-4SH (the hydrolysable ester bond assigned by scissor) and (B) Non-degradable PEG-4Mal

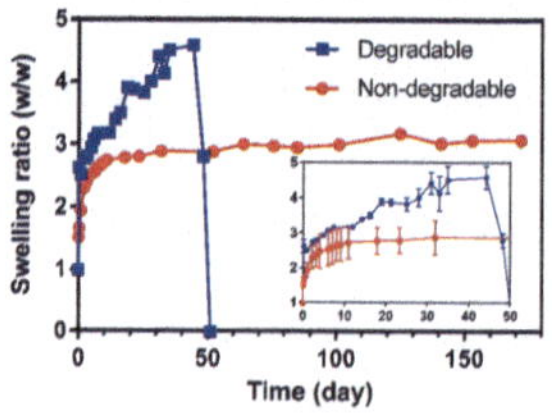

Fig. 2. Swelling over time for degradable and non-degradable (using DTT) tetra-PEG hydrogels (n = 3)

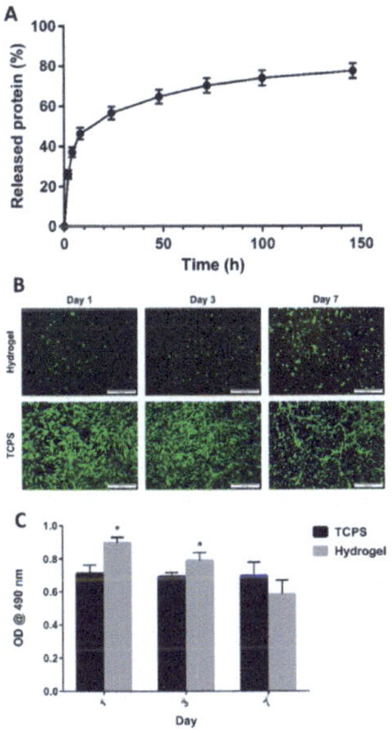

Fig. 3. Biological potentials of tetra-PEG hydrogel (A), Sustained release of protein (BSA), (B) Live cell encapsulation, and (C) Biocompatibility of hydrogel and its degradation products

4 Conclusion

In summary, a click tetra-PEG hydrogel was developed to address the urgent need for a biodegradable and biocompatible injectable hydrogel for tissue engineering and drug delivery applications.

References

1. Bakaic, E., Smeets, N.M., Hoare, T.: RSC Adv. **5**, 35469–35486 (2015)
2. Madl, C.M., Heilshorn, S.C.: Adv. Funct. Mater. **28**, 1706–1746 (2018)
3. Malkoch, M., Vestberg, R., Gupta, N., Mespouille, L., Dubois, P., Mason, A.F., Hedrick, J.L., Liao, Q., Frank, C.W., Kingsbury, K.: Chem. Commun. **26**, 2774–2776 (2006)
4. Ghanian, M.H., Mirzadeh, H., Baharvand, H.: Biomacromol. **19**, 1646–1662 (2018)

Influence of Polyolefin Elastomer Content on Morphology, Rheology and Thermal Behavior of Toughened Polylactic Acid/Polyolefin Elastomer Blends

Nadia Jalalifar[2] and Babak Kaffashi[1](✉)

[1] Department of Polymer Engineering, School of Chemical Engineering, College of Engineering, University of Tehran, Tehran, Iran
kaffashi@ut.ac.ir

[2] Department of Polymer Engineering, Kish International Campus, University of Tehran, Kish Island, Iran
nadiajalalifar@ut.ac.ir

Abstract. Improving the processability and melt strength of sustainable bio-based polymers using elastomers is essential to preserve their biodegradability making them suitable for many applications. Hence, different blends based on polylactic acid (PLA)/polyolefin elastomer (POE) were prepared by the melt mixing. The droplet-matrix morphology of blends was observed by the Field Emission Scanning Electron Microscopy. Structural development of PLA/POE blend was examined depending on the POE content (10–30 wt%). The droplet size of the dispersed POE phase initially decreased and then increased as the POE content increased. The rheological test results indicated that the incorporation of polyolefin elastomer could enhance the storage modulus and the melt complex viscosity of PLA/POE blends. Within all the compositions prepared, the blend containing 20 wt% of POE showed significant increase in the melt strength compared to neat PLA. The thermal behavior showed two different PLA crystal structures after the cooling cycle. The thermograms analysis also exhibited that the glass transition temperature was slightly decreased with the increased content of POE in PLA/POE blends. The results of structural development and the rheological properties of the blends showed that the blend performance was strongly dependent on the composition of POE in the blend.

Keywords: Polylactic acid · Polyolefin elastomer · Microstructure · Rheology · Thermal properties

1 Introduction

Polylactic acid (PLA) is an eco-friendly material that is made of renewable resources. PLA due to its fine biodegradability as compared to the petroleum-based plastics is gaining attention in the plastic market. However, several drawbacks such as its brittle behavior and low impact resistance have limited its commercial applications. The improvement of weak mechanical properties of polylactic acid using different procedures include blending with plasticizers and other polymers or mixing with nano additives.

H. Mirzadeh and A. A. Katbab (Eds.): ISPST 2018, *Eco-friendly and Smart Polymer Systems*, pp. 96–99, 2020.
https://doi.org/10.1007/978-3-030-45085-4_24

One of the main problems associated with PLA toughening through plasticizers is the reduction in strength and processability. Therefore, toughening PLA has attracted more and more attention and various materials have been used to blend with PLA for toughening [1, 2]. Polyolefin elastomers (POEs) are a relatively new class of polymers that emerged with recent advances in metallocene polymerization catalysts. POEs are often chosen over other alternatives because they are available with properties ranging from amorphous to crystalline, and low to very high molecular weight. Polyolefin elastomers are an excellent impact modifier for plastics, and offer unique performance capabilities for compounded products. Generally, saturated polymers such as polyolefin elastomers providing superior elasticity, excellent thermal stability and recyclable [3].

In current study, we demonstrate blending PLA and POE as a practical strategy for PLA toughening. The properties of such blends strongly depend on the microscopic morphology which in turn determines the micromechanical deformation mechanisms. Therefore, we investigate the correlation between rheological properties and morphology.

2 Experimental

Polylactic acid (2003D grade) is a semicrystalline extrusion material that was provided by Nature Works. Polyolefin elastomer LC565 is an ethylene-1-octene copolymer that was purchased from LG Chem. The samples, which are summarized in Table 1, were prepared in a Brabender internal mixer (WTH55, Germany) at 200 °C with mixing duration of 10 min and rotor rotation of 40 rpm. Prior to melt mixing, PLA was dried in an oven at 80 °C for 2 h. The dynamic rheological properties were measured with a parallel plate geometry using a rheometer Physica MCR 102 (Anton Paar, Austria). The morphology was observed with FESEM with an acceleration voltage of 5 kV (FEI, NOVA-450, USA). The melting and crystallization behaviors were studied using a Differential Scanning Calorimeter (METTLER TOLEDO, Switzerland) under nitrogen atmosphere.

Table 1. Sample codes and compositions

Blend sample	PLA/POE
P0	100/0
P10	90/10
P20	80/20
P30	70/30
P100	0/100

3 Results and Discussion

Figure 1a to c shows the FESEM images of three representative blends (P10, P20, and P30). The images obviously show that the POE droplets are uniformly distributed within the PLA matrix. The size of the dispersed POE phase is significantly decreased with increase in POE composition (a, b). As the POE content increases, the droplet size of the dispersed phase increases (b, c). A change in the POE composition results in various structures of the blend.

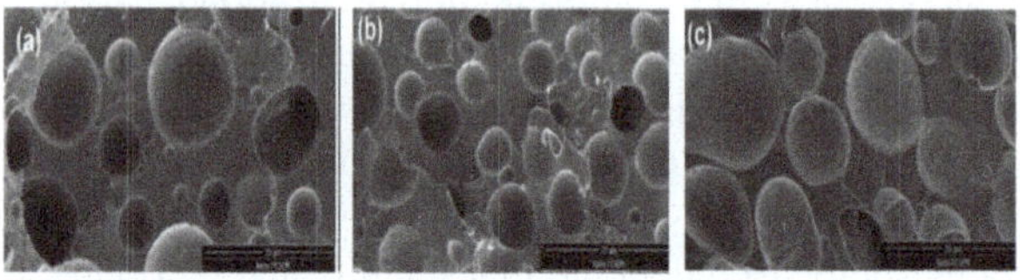

Fig. 1. FESEM micrographs at 8000x magnification of PLA/POE (a) blends 90/10, (b) 80/20, and (c) 70/30

Figure 2a and b shows the storage and loss moduli of the pure samples and PLA/POE blends. These figures show a unique frequency dependence with a general terminal behavior at low frequencies. While it is observed for the high frequency domain, the addition of POE to PLA can almost lead to decrease in the slope of the diagrams. As is demonstrated in Fig. 2c, the complex viscosity of PLA increases with POE content while the frequency dependence of viscosity is more pronounced with added POE. For the blend containing 20 wt% of POE compared with other compositions and the neat PLA, the storage and loss moduli and also the viscosity are significantly increased. One of the reasons for this increase is the reduction of droplet size in the dispersed phase, as shown in Fig. 1b.

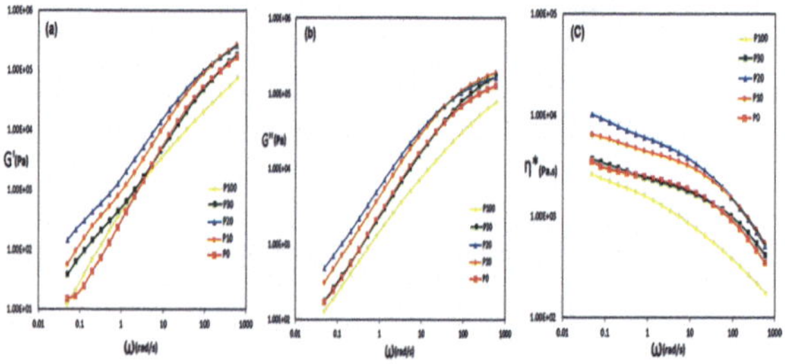

Fig. 2. (a) Storage modulus (G′), (b) loss modulus (G″), and (c) complex viscosity (η*) as a function of frequency at 150 °C

The DSC thermograms for the second heating cycle of PLA, POE and PLA/POE blends are shown in Fig. 3. As seen in the thermograms, two sharp peaks of melting temperature of PLA/POE blends caused by the two different PLA crystal structures after the cooling cycle. In other words, the PLA/POE blends exhibit the presence of a cold crystallization exotherm peak. The DSC analysis also shows that the glass transition temperature slightly decreased with increase of POE in the PLA/POE blends.

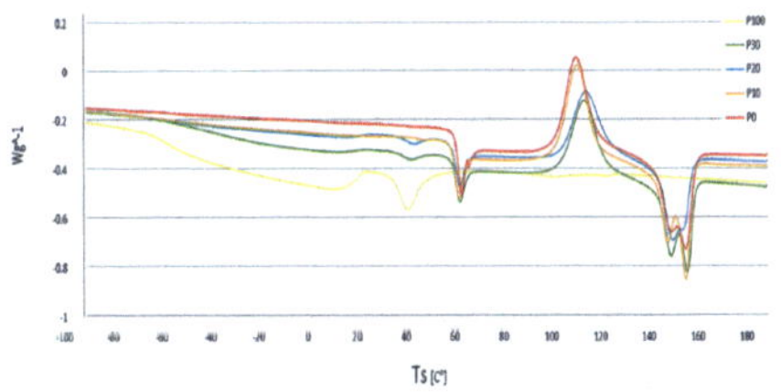

Fig. 3. DSC thermograms of second heating run of pure materials and PLA/POE blends

4 Conclusion

The effects of polyolefin elastomer (POE) on the morphological, rheological and thermal properties of polylactic acid (PLA)/POE blends were investigated. The blends exhibited a different morphological and rheological behavior depending on the contents of the POE. The FESEM analysis revealed the homogeneous dispersion of POE particles in the PLA matrix. Based on the microscopic images of the blends, it was observed that the droplet size shows clear concentration dependence. So the 20 wt% of POE reduces the interfacial tension and decreases the drop size of the dispersed phase. The incorporation of POE into PLA significantly improves the storage modulus and melts complex viscosity of the blends. The PLA/POE blend containing 20 wt% of POE exhibit significant increase in the rheological properties compared to neat PLA. The thermograms of the differential scanning calorimeters for the PLA/POE blends exhibit that the glass transition temperature slightly decreased with the increased concentration of POE. In addition, the PLA/POE blends exhibit the presence of a cold crystallization exothermic peak. In conclusion, we showed that it is possible to improve the performance of a polylactic acid by blending polyolefin elastomer by a suitable composition of the two polymers.

References

1. Mallegni, N., Phuong, T.V., Coltelli, M.B.: Materials **11**, 148–163 (2018)
2. Lai, X.L., Yang, W., Whang, Z.: Inc. J. Appl. Polym. Sci. **135**, 45675–45681 (2018)
3. Zhou, Y., Wang, J., Cai, S.H.Y.: Polym. Bull. **1**, 1–18 (2018)

Electrospinning of Polycaprolactone (PCL)/Poly(Lactic Acid) (PLA)/*Ferula Gummosa* Boiss Extract (Galbanum)

Mohaddeseh Sharifi, Melika Sharahi, Sara Labbafi,
S. Hajir Bahrami(✉), and Nahid Hemmati Nejad

Department of Textile Engineering, Amirkabir University of Technology,
P.O. Box 15875-4413, Tehran, Iran
hajirb@aut.ac.ir

Abstract. Electrospinning is one of the most efficient, cost effective and straightforward methods for fabricating nanofibers. Nanofibers with special characteristics such as high surface to mass (or volume) ratio and high porosity are best candidate for simulating the extracellular matrix structure. In this research, polycaprolactone (PCL), poly(lactic acid) (PLA), and *F. gummosa* extract were electrospun. Electrospinning was performed using solutions containing 10% PCL and 8% PLA and varying amount of *F. gummosa* extract. Extracts of *F. gummosa* with ratios of 15, 20, and 25% were added to the polymer solution and stirred for 1 h. The morphology changes of nanofibers were studied by Scanning Electron Microscopy (SEM). The antibacterial properties of nanofibers were examined against a gram positive and gram negative bacteria. SEM results showed that average diameters of nanofibers without extract of *F. gummosa* and with extract were 497 ± 30 nm and around 642 ± 35 nm, respectively. Fourier Transform Infrared Spectroscopy (FTIR) analysis revealed that extract of *F. gummosa* was loaded in nanofibers structure properly. Hydrophilicity of nanofibers was also examined by water contact angle measurements.

Keywords: Electrospinning · Nanofibers · PLA · PCL · *Ferula gummosa* boiss (galbanum)

1 Introduction

Nano scale materials pose their novel and significant chemical and physical properties [1]. Scaffolds made from nanofibers are used for applications such as tissue engineering, drug delivery, wound healing, etc. [2]. PCL and PLA are synthetic polymers with biodegradable, biocompatible, high mechanical strength and low hydrophilicity properties [3, 4]. Extract of *F. gummosa* is reported to have historical medical applications. It is said that this extract shows antibacterial properties and had been used as a wound healing and the gum of *Ferula gummosa* Boiss(Apiaceae) included a rich source of terpenoid compounds and this extract have been used in medicine for

H. Mirzadeh and A. A. Katbab (Eds.): ISPST 2018, *Eco-friendly and Smart Polymer Systems*, pp. 100–104, 2020.
https://doi.org/10.1007/978-3-030-45085-4_25

stomach pain, chorea, wound healing and epilepsy [5]. In this work, two synthesis polymers (PCL and PLA) were used along with *F. gummosa* extract to produce nanofibers. These two polymers have low degradation rate which may enable the healing patch to sustain it properties for a longer time, which in turn may help in release of the herb for a longer period of time.

2 Experimental

2.1 Materials

PCL with molecular weight 80 kDa was obtained from Sigma-Aldrich and PLA (Mw 4060 Da) was bought from Nature work LLC and DMF and DCM solvents were purchased from Merck. Extract from root of *F. gummosa* was prepared using acetone.

Polymer solutions of PCL (10 v/v%) and PLA (8 v/v%) were prepared by dissolving the polymers in DMF/DCM (1:3) for electrospinning. Electrospinning was performed using an electrospun setup. Solutions were transferred into two 5 ml syringes. The conditions of electrospinning included flow rate 0.8 mL/h, distance from the needle tip to collector 15 cm and voltages 20 and 15 kV, respectively. In this work, extracts of *F. gummosa* with ratios 15, 20, and 25% were added, respectively, to polymer solutions.

3 Results and Discussion

3.1 Morphology of Nanofibers

Morphology of the nanofibers was examined by SEM. Average diameter of nanofibers was measured at 50 points by digimizer software. The SEM images show that the nanofibers with different diameters are formed. It appears that nanofibers produced from PLA solution have higher diameter (490 $\pm$ 30 nm) whereas; diameter of nanofibers produced from PCL is in the range of 334 $\pm$ 40 nm. The difference between diameters of hybrid nanofibers produced from different solutions kept in different syringes is quite noticeable. The average fiber diameter of PLA/PCL nanofibers was in the range of 497 $\pm$ 30 nm. The loading of *F. gummosa* extract with ratios of 15 and 20% in polymer solution significantly increased the fiber diameter to 540 $\pm$ 40 and 642 $\pm$ 35 nm respectively. This is due to increase in solution viscosity and reduction of solution conductivity. But addition of 25% *F. gummosa* extract into polymer solutions reduced diameter nanofiber to 410 $\pm$ 30 nm. With increase in extract concentration, conductivity was increased. Figure 1 shows that with increasing extract of *F. gummosa* from 15% to 25% nanofibers PLA/PCL became uniform and free of bead. The produced nanofibers are suitable for antibacterial applications.

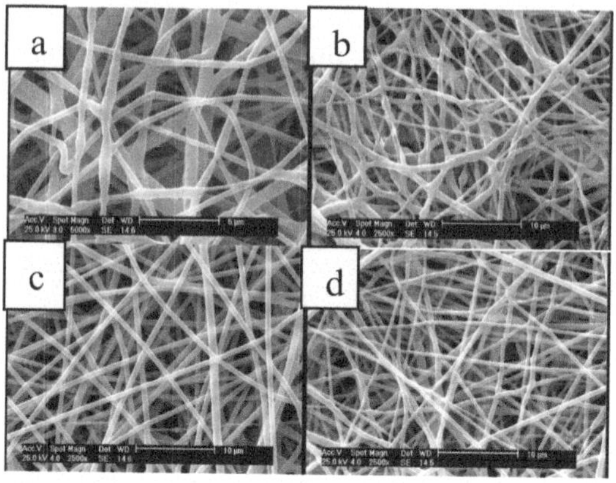

Fig. 1. SEM images of (a) PLA/PCL (b) PLA/PCL/15% F.g (c) PLA/PCL/20% F.g (d) PLA/PCL/25% F.g

3.2 Hydrophilicity of Nanofibers

Wettability of nanofibers was assessed using contact angle measuring. PLA and PCL are very hydrophobic polymers. Moreover, the behavior of PLA/PCL hybrid is similar to that of PLA and PCL alone and their hybrid has a contact angle about 103°. The loading extract *of F. gummosa* in nanofiber PLA/PCL reduced contact angle. Our results showed that the values of contact angle for PLA/PCL/15% *F. gummosa* extract, PLA/PCL/20% *F. gummosa* extract and PLA/PCL/25% *F. gummosa* extract were 100°, 105°, and 30°, respectively (Fig. 2).

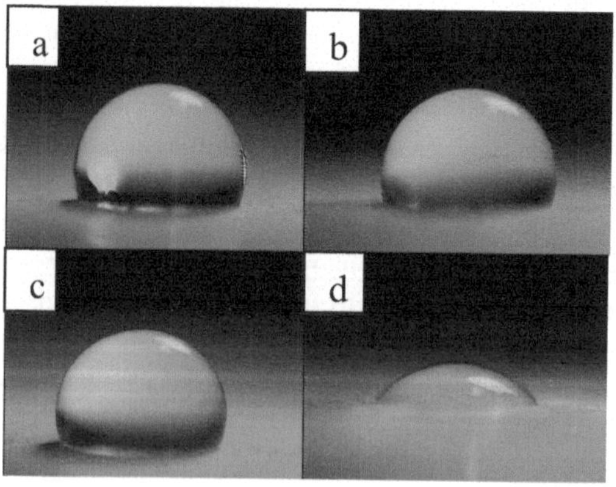

Fig. 2. Contact angles of (a) PLA/PCL (b) PLA/PCL/15% F.g (c) PLA/PCL/20% F.g (d) PLA/PCL/25% F.g

3.3 FTIR Analysis

FTIR spectra (Fig. 3) show functional groups present in the raw materials and nanofibers. In this work, FTIR spectra of the electrospun nanofibers PLA/PCL, PLA/PCL/*F. gummosa* extract and extract *F. gummosa* were analyzed. The main peaks of *F. gummosa* extract included 1021–1272 cm^{-1} C-O groups ether, 1612 cm^{-1} C=C, and 1712 cm^{-1} C=O carbonyl. The absorption bands of PLA/PCL nanofibers demonstrated the presence of aliphatic ethers (C-O) in range of 1182–1300 cm^{-1} and the presence of the esters groups (C=O) at 1730 cm^{-1}. Finally, stretching peaks for C-H appeared at 2867–2944 cm^{-1}. Additionally, the characteristic peaks of PLA/PCL/*F. gummosa* extract nanofibers are ester groups (C=O) at 1730 cm^{-1}, and 1615 cm^{-1} C=C, and ether groups at 1182 cm^{-1}. The difference between spectra of PLA/PCL/*F. gummosa* and PLA/PCL is due to C=C group that belongs to *F. gummosa*. The antibacterial activity of the nanofibers with and without the extract was examined and it showed that the extract improved the antibacterial properties significantly. The web produced without *F. gummosa* extract did not show any antibacterial property and the bacteria could grow on the web. However, with incorporation of the extract in the nanofibrous web it showed 98% inhibition of bacterial growth.

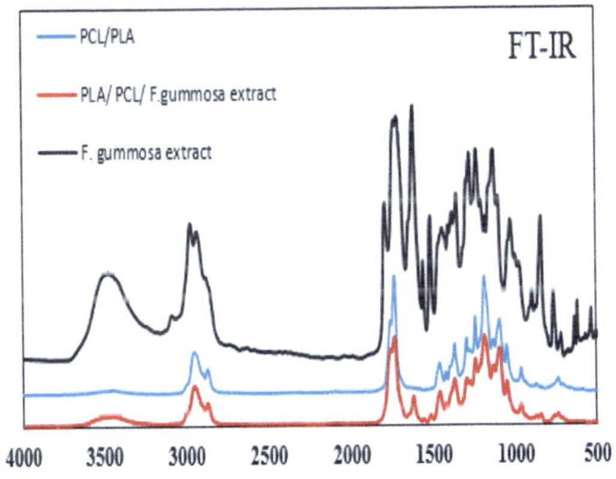

Fig. 3. FTIR spectra

4 Conclusion

In this paper, loading *F. gummosa* as herbal extract in PLA/PCL nanofibers increased average diameter and in nanofibers containing 25% extract, average diameter reduced. Also, PLA/PCL/*F. gummosa* extract nanofibers showed smooth and bead free surface along with antibacterial properties. The hydrophilicity of the webs were measured.

References

1. Feng, C., Khulbe, K.C., et al.: Sep. Purif. Technol. **102**, 118–135 (2013)
2. Rsnjbar-Mohammadi, M., Ind. Text, 1–14 (2017)
3. Bhattarai, N., Gunn, L., et al.: Adv. Mater. **21**, 2792–2797 (2009)
4. Santoro, M., Shah, S.R., et al.: Advanced Drug Deliver Rev., 1–7 (2016)
5. Mandegary Sayyah, A., et al.: Daru, **12**(2) (2004)

Synthesis of Biodegradable Unsaturated Polyester Resin Using Soybean Fatty Acid as an Acidic Monomer

Nika Parsi[1,2], Masoud Jamshidi[1(✉)], and Mohsen Sarrafi[2]

[1] Polymers and Constructional Composites Research Laboratory, School of Chemical Engineering, Iran University of Science and Technology (IUST), Tehran, Iran
mjamshidi@iust.ac.ir

[2] Research and Development Center (R&D), Resitan Company, Tehran, Iran

Abstract. In recent years, many researches have been performed on issues that affect human life. Polymers, despite of huge applications in different industries, have been considered for their long-term degradation and their negative effects on environment. Many efforts have been performed to produce environmentally friendly resins. One of the newest types of these materials are biodegradable resins. In this research, biodegradable unsaturated polyester resin was synthesized. For this purpose, soybean oil, a natural fatty acid, was selected to be used as partial replacement of acidic monomer (i.e. maleic and/or phthalic anhydride). The replacing content changed from 0 wt% to 20 wt% and its effect on the mechanical properties of the cured resin was assessed. FTIR and NMR analysis were used to characterize chemical structure of the produced biodegradable resins. For better understanding, a general purpose unsaturated polyester resin as control sample was produced and its mechanical properties compared to the biodegradable samples.

Keywords: Biodegradable resins · Soybean fatty acid · Unsaturated polyester resin · Mechancial properties

1 Introduction

Unsaturated polyester resins (UPRs) are one of the most important classes of thermoset polymers in the industry. Due to the excellent balance between their mechanical, electrical and chemical properties and their low cost, they are widely used in coatings and composite materials for construction, marine and automotive industries. However, in the last years, due to environmental and sustainability issues, the replacement of fossil-based monomers by their renewable counterparts have been the focus of several studies [1]. In the last decades, Biodegradable polymers attracted many considerations. However, the most important issue about these polymers is their lower mechanical properties which eliminate their industrial applications [2–4]. In this research, authors produced an unsaturated polyester resin with normal acidic monomers (i.e. maleic and phthalic anhydrides). Then, the monomer was replaced by 0 wt% to 20 wt% by soybean

H. Mirzadeh and A. A. Katbab (Eds.): ISPST 2018, *Eco-friendly and Smart Polymer Systems*, pp. 105–107, 2020.
https://doi.org/10.1007/978-3-030-45085-4_26

oil fatty acid as biodegradable monomer. The mechanical properties (i.e. tensile and flexural properties and impact strength) was evaluated and compared to control samples. To assess biodegradation of the samples QUV test was conducted.

2 Experimental

2.1 Materials

Soybean oil fatty acid (RFA 901) (SOFA), diethylene glycol, maleic anhydride, phthalic anhydride, styrene (St), hydroquinone (HQ), methyl ethyl ketone peroxide (MEKP), cobalt octoate (CoO, 1 w/v% styrene) were supplied from Sigma Aldrich Company.

2.2 Synthesis of Bio-Based Unsaturated Polyesters

The SOFA based unsaturated polyesters with the different percentages of soybean fatty acid were prepared by bulk polycondensation. Briefly, the monomers and a radical inhibitor (HQ, 0.01 wt% relative to the monomers in the feed) were charged into a four-necked glass reactor. The polycondensation reactions at a maximum temperature of 240 °C. Water was distilled during reaction. The end of reaction was set when the acid value (AV) reached a value around of ca. 30 mg KOH/g (according to ASTM 109-01).

2.3 UV Degradation

The specimens were degraded in QUV weathering machine, operated at 45 °C black panel temperature. Relative humidity was 65% ± 5% and the air temperature was 30 °C [4].

3 Results and Discussion

3.1 FTIR Spectra

Figure 1 shows the FTIR spectra of the different resins that synthesis. As it shows, the peaks of fatty acid that it was around 2000–1500 cm^{-1} were decreased. It means that the fatty acid was reacted and there was no free monomer of fatty acid in the resin.

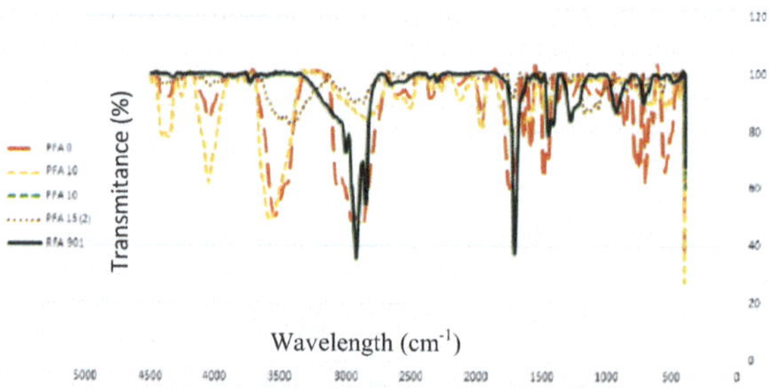

Fig. 1. FTIR spectra of the resins

3.2 Tensile and Flexural Properties

The used fatty acid, due to its long molecules, made the resin tougher and reduced the value of the tensile and bending strengths. The use of fatty acids in the resin structure between range of 0% to 20% may not be significant, but these amounts of fatty acids had considerable effects on the mechanical properties and reduced the amount of bending and tensile strengths. Table 1 shows the results of the mechanical properties.

Table 1. Mechanical properties of the resins

Code	Tensile (MPa)	Bending (MPa)
PFA 0	58.65	98.93
PFA 5	62.93	89.78
PFA 10	50.65	74.89
PFA 15	47.89	62.07

4 Conclusion

It was found that by increasing soybean fatty acid content in the unsaturated polyester resin up to 15% the mechanical properties decreased considerably. At 20 wt% of fatty acid the resin was gelled in the reactor. The best properties was achieved at 5 wt% of fatty acid.

References

1. Gobin, M., Loulergue, P., Audic, J.-L., Lemiègre, L.: Ind. Crops Prod. **70**, 213–220 (2015)
2. Costa, C.S.M.F., Fonseca, A.C., Moniz, J., Godinho, M., Serra, A.C., Coelho, J.F.J.: Ind. Crops Prod. **85**, 403–411 (2016)
3. Liu, K., Madbouly, S.A., Kessler, M.R.: Eur. Polym. J. **69**, 16–28 (2015)
4. Schoolenberg, G.E., Vink, P.: Polymer **32**, 432e7 (1991)

Fabrication of Gelatin 3-D Nanofibers via Wet Electrospinning Method

Sarah Karimi[1], Mohamad Pezeshki-Modaress[2], and Najmeh Najmoddin[1(✉)]

[1] Department of Biomedical Engineering, Science and Research Branch, Islamic Azad University, P.O. Box 14515/775, Tehran, Iran
najmoddin63@yahoo.com, najmoddin@srbiau.ac.ir
[2] Burn Research Center, Iran University of Medical Science, Tehran, Iran

Abstract. In this study, ultrafine gelatin fibers were successfully produced with the use of the wet spinning technique. Gelatin is an exceptional and versatile biopolymer with applications in various *fields* of medicine and biochemistry. Distilled water and acetic acid was used as solvents. The morphology of the gelatin fibers was found to be dependent on the voltage of wet spinning process. The SEM results showed that 20 kV is an inappropriate voltage because it formed beaded nanofibers but the 26 kV was the best voltage for successful wet spinning of gelatin and 3D nanofibrous structure formation.

Keywords: Wet spinning · Gelatin · Nanofiber · Biopolymer · Voltage

1 Introduction

Gelatin is a biopolymer derived from native collagen, the most abundant structural protein which is available from animal by-products in enormous quantities. In the conversion of collagen to gelatin, the triple-helical structure of the water-insoluble collagen is cleaved into single protein chains, yielding gelatin as a water-soluble hydrolysis product [1]. Due to a wealth of merits such as biological origin, nonimmunogenicity, biodegradability, biocompatibility, and commercial availability at relatively low cost, gelatin has been widely used in the pharmaceutical and medical fields as sealants for vascular prostheses, carrier for drug delivery, dressings for wound healing and so forth [2].

Gelatin can be produced as nanofibers. Nanofibers are very promising candidates in a variety of applications, especially in the situations where surface area is highly concerned.

In the biomedical field, gelatin nanofibers can be potentially useful in developing biomimicking artificial extracellular matrix (ECM) for engineering tissues, dressings for wound healing, and drug releases [3]. Fiber fabrication techniques include 3D printing and various spinning methods such as wet spinning, dry spinning, melt spinning, gel spinning, jet spinning, and electrospinning. Among these methods, wet spinning is an easily applicable, reproducible, and cost-efficient method that allows the production of microfibers under mild conditions. In tissue-engineering applications, wet-spinning technique has been employed successfully with both synthetic and natural polymers to fabricate three-dimensional (3-D) scaffolds with defined and controlled

H. Mirzadeh and A. A. Katbab (Eds.): ISPST 2018, *Eco-friendly and Smart Polymer Systems*, pp. 108–111, 2020.
https://doi.org/10.1007/978-3-030-45085-4_27

fibrous structure. So far, some of these approaches such as electrospinning have been carried out on two-dimensional (2-D) surfaces, which do not necessarily correlate to the response in (3-D) environments. In the body, cells are exposed to a 3-D environments which signals the cells about whether it should proliferate, differentiate, migrate or remain quiescent. Therefore, conventional 2-D patterns have limitations in mimicking the in vivo environments for tissue culture or drug screening. Manycomponents of the natural extracellular matrix (ECM) have a fibrous structure. In the present study, we expanded the research on gelatin nanofibers by introducing a wet spinning process. Wet spinning implies rapid fiber coagulation and dehydration in a solvent bath which is beneficial for large-scale processes [4].

2 Experimental

2.1 Materials

Gelatin type B from bovine skin, acetic acid, ethanol (99.9%) were all obtained from Sigma–Aldrich (St.Louis, MO).

2.2 Preparation of Wet Spun Gelatin Nanofibers

The polymer solution with concentration of 20 wt% was prepared by dissolving gelatin in acetic acid and distilled water with a volume ratio of 50:50 and stirred for 2 h. The solution was wet spun from a 5 mL syringe with a needle diameter of 0.4 mm. Polymer solution was extruded into the coagulation bath (ethanol) at a flow rate of 0.5 mL/h. A high voltage, 20 and 26 kV, was applied to the tip of the needle attached to the syringe when a fluid jet was ejected.

2.3 Characterization of Scaffolds

The morphology of nanofibers was studied by scanning electron microscopy (SEM) (JSM 5600, JEOL, Japan) Before observation, the scaffolds were dried in vacuum oven and then were coated with gold using a sputter coater (JEOL JFC-1200 fine coater, Japan). The diameter of the fiber was measured from the SEM micrographs using image analysis software (Image J, National Institutes of Health, USA).

3 Results and Discussion

As can be seen in Fig. 1, 26 kV is the best voltage for forming gelatin nanofibers without bead during wet spinning and also nanofiber diameter was estimated to be 200 nm.

As can be seen in Fig. 2, by decreasing the voltage from 26 kV to 20 kV, beaded nanofibers formed during wet spinning.

The formation of the beaded nanofibers can be considered as the capillary breakup of the wet spinning jets by surface tension, altered by the presence of electrical forces. Increasing net charge density favors formation of small diameter fibers. Decreasing the surface tension coefficient of the solvent favors the formation of larger diameter fibers. Details of ionic currents, where the jet moves through regions of different electrical potential near the orifice are also important [5].

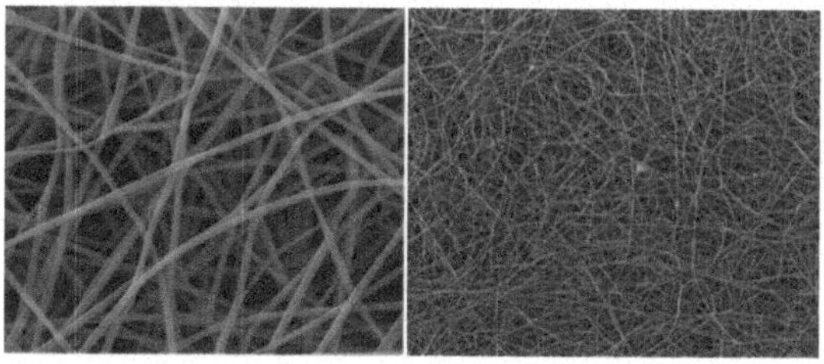

(a) (b)

Fig. 1. SEM micrographs of wet spun gelatin nanofibers with 26 kV: (a) SEM MAG: 1.50 kx and (b) SEM MAG: 10.00 kx

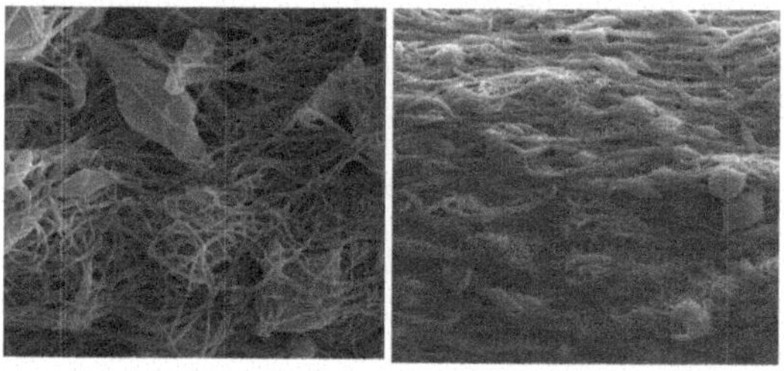

(A) (B)

Fig. 2. SEM micrographs of wet spun gelatin nanofibers with 20 kV: (A) SEM MAG: 2.00 kx and (B) SEM MAG: 5.00 kx

4 Conclusion

In this study, we produced gelatin nanofibrers with high voltages of 20 and 26 kV. The results reported in this study demonstrated that the diameter of nanofibrous scaffolds were strongly influenced by the high voltage of wet spinning. The results showed that high voltage of 26 kV is the best for wet spinning gelatin.

References

1. Stoessel, P.R., Raso, R.A., Kaufmann, T., Grass, R.N., Stark, W.J.: Macromol. Mater. Eng. **300**, 234–241 (2015)
2. Zhang, Y.Z., Venugopal, J., Huang, Z., Lim, C.T., Ramakrishna, S.: Polymer **47**, 2911–2917 (2006)
3. Ucar, S., Yilgor, P., Hasirci, V., Hasirci, N.: J. Appl. Polym. Sci. **130**, 3759–3769 (2013)
4. Hwang, C.M., Park, Y., Park, J.Y., Lee, K., Sun, K.: Biomed. Microdevices **11**, 739–746 (2009)
5. Fong, H., Chun, I., Reneker, D.H.: Polymer **40**, 4585–4592 (1999)

Fabrication and Characterization of Hybrid Scaffolds Based on β-TCP/PCL-HAp/Gel Nanofibers for Bone Regeneration

Melika Sharahi, Mohaddeseh Sharifi, Sara Labbafi, S. Hajir Bahrami, and Nahid Hemmati-Nejad(✉)

Textile Engineering Department, Amirkabir University of Technology, Tehran, Iran
hemmati@aut.ac.ir

Abstract. Design and production of nanofibrous scaffolds for medical application especially tissue engineering have attracted considerable interest in last 2 decades. Recently, 3D bioactive scaffolds have been made to repair human bone tissue. They should contain a mineral phase such as bioceramics with synthetic or natural polymers. Here, we report fibrous scaffolds made of a combination of 20% PCL and 15% gelatin with acetic acid as a co-solvent. In order to produce fibrous scaffolds, β-tricalcium phosphate loaded PCL and hydroxy apatite-Gel solutions were electrospun. For this, HAp with different weight percentages (5, 10%) were added to gelatin polymeric solution and the HAp/Gel blend was injected from one syringe. The other syringe was included 5% and 10% of β-TCP in PCL solution. After producing nanofibrous webs, their surface morphology was examined by SEM images. These results showed that PCL/β-TCP-Gel/HAp nanofibers have some features such as random arrangement with agglomeration of particles and proper porosity. Also, hydrophilicity and mechanical properties of hybrid webs were investigated. These scaffolds not only possess properties such as biocompatibility and tensile modulus, but also sufficient water absorption that will improve the adhesion process and proliferation of osteoblasts cells. Besides, cell culture studies showed good adhesion and proliferation.

Keywords: Electrospinning · Gelatin · Polycaprolactone · B-tricalcium phosphate · HAp

1 Introduction

With the advent of nanotechnology, the interest of researchers has increased in the study of the unique properties of nanofibers. Electrospinning is known as an efficient method for the fabrication of polymer nanofibers. Because of high surface area and proper porosity of nanofibers, they can be used in many fields like tissue engineering, drug delivery and so on.

The development of nanofibers has improved fabricating of scaffolds that can potentially mimic the structure of ECM (extracellular matrix) of the body tissues at the nanoscale [1]. Biomaterials play a critical role for the replacement of lost or malfunctioning tissues. Various type of biomaterials such as natural polymer, synthetic

H. Mirzadeh and A. A. Katbab (Eds.): ISPST 2018, *Eco-friendly and Smart Polymer Systems*, pp. 112–116, 2020.
https://doi.org/10.1007/978-3-030-45085-4_28

polymers, ceramics and their composites can be used in producing of different scaffold with suitable capabilities. However, natural polymers have problems due to their mechanical properties and degradability. Thus combination of natural and synthetic polymers will improve their properties. Poly caprolactone (PCL) is a hydrophobic and synthetic polymer that degrades inside the human body slowly. So, it has been investigated mainly for implant materials that use for a long time in bone remodeling. Gelatin, which is derived from collagen, has been used for biomaterial applications such as wound dressing and bone regeneration because of its excellent biocompatibility. According to previous studies, combination of PCL and Gel cannot be used for making an idea scaffold as it is not an osteoconductive or osteoinductive material [2]. Bioceramics such as HAp (hydroxy apatite) and (β-TCP) β-tricalcium phosphate are biomaterials for bone tissue engineering because of their good osteoconductivity. However, they suffer from brittleness and sometimes load bearing conditions limit their abilities especially when they are used alone [3]. To overcome this problem, hybrid systems have been introduced.

2 Experimental

2.1 Materials

HAp and β-TCP powders were purchased from Sigma-Aldrich. Gelatin, PCL and acetic acid were obtained from Merck Germany.

2.2 Hybrid Scaffold Fabrication

Electrospinning setup is consisted of a needle (gauge 19) and a syringe pump, a high voltage power supply, a cylinder collector with high rotating speed.

5% HAp and 5% β-TCP were added in polymer solution Gel (15% w/v) and PCL (20% w/v) that were separately dissolved in acetic acid (90% v/v) sooner. Then electrospinning simultaneously was performed using 10 kV voltage, 0.5 mL/h feeding rate and 15 cm distance. Similarly, other scaffolds were provided in the same way which are as follows:

a) PCL-5%TCP/Gel-5%HAp, b) PCL-10%TCP/Gel-5%HAp, c) PCL-5%TCP/Gel-10% HAp, and d) PCL-10%TCP/Gel-10% HAp.

3 Results and Discussion

3.1 Morphology of Nanofibers

SEM was used to investigate morphology and diameter of hybrid nanofibers. Figure 1 shows PCL-Gel hybrid scaffold which composed of bead-free and randomly arranged fibers with average diameter of 258 ± 57 nm. According to Fig. 2, at the same amount of HAp, fiber diameters increased significantly from 350 ± 25 nm to 485 ± 41 nm for (a–b) and from 452 ± 28 nm to 780 ± 43 nm for (c–d) scaffolds (n = 50 fibers). These may be due to presence of calcium phosphate particles in nanofibers. On the other hand, at higher percentages of particles, more agglomeration in nanofibres has been observed. The difference in the average diameters of (b) and (c) scaffolds can be expressed by the polyhedral particles of β-TCP which have larger dimensions than uniform spherical HAp particles.

Fig. 1. SEM and contact angle images of PCL/Gel hybrid nanofiber

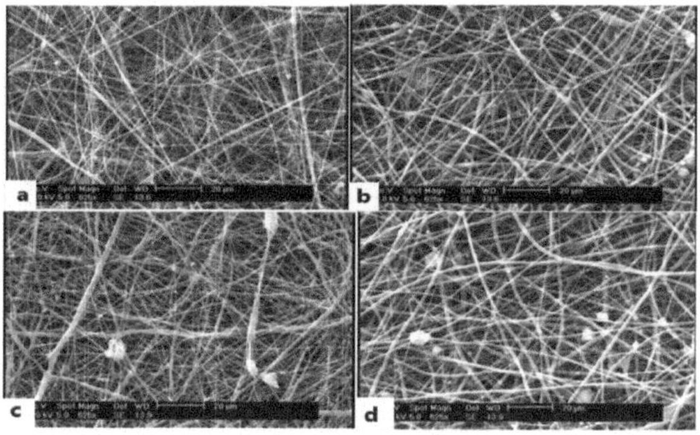

Fig. 2. SEM images of: a) PCL-5%TCP/Gel-5%HAp, b) PCL-10%TCP/Gel-5%HAp, c) PCL-5%TCP/Gel-10% HAp, and d) PCL-10%TCP/Gel-10% HAp

In order to determine the wettability of hybrid scaffolds, the contact angle of nanofibers was measured by a video contact angle system. The results showed that the contact angle of PCL-Gel scaffold was about 32.1 °C. In contrast, contact angle measurements revealed wettability of scaffolds with HAp and TCP were improved. Gel/HAp nanofibers are hydrophilic and can be considered as a nutrient in the scaffold to improve features such as bone conduction, adhesion and proliferation of cells. In addition, due to the nature of the TCP, the presence of these three components causes the wettability of PCL increases. According to Fig. 3 scaffolds can immediately absorb water. In fact, when the particle percentages increase, the scaffolds have little time to water absorption.

To investigate mechanical properties of the electrospun scaffolds, tensile stress test was performed with an Instron 5566 mechanical tester (Fig. 4). Tensile modulus was calculated in 0–10% strain region. According to the results of previous studies, the tensile strength of scaffolds was expected to increase at higher percentages of HAp,

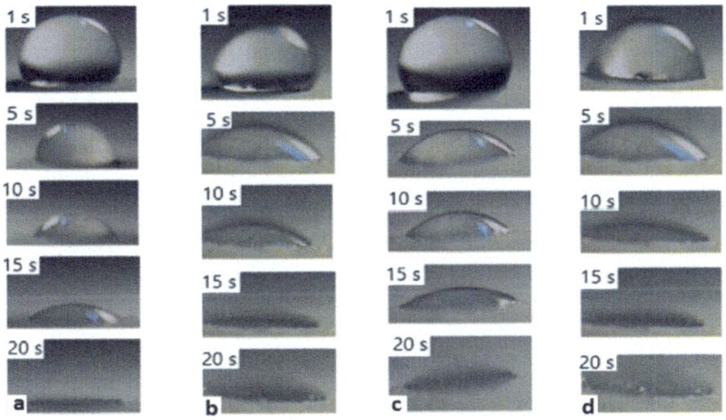

Fig. 3. Transient water absorption of one droplet on the surface for (a–d) hybrid scaffolds

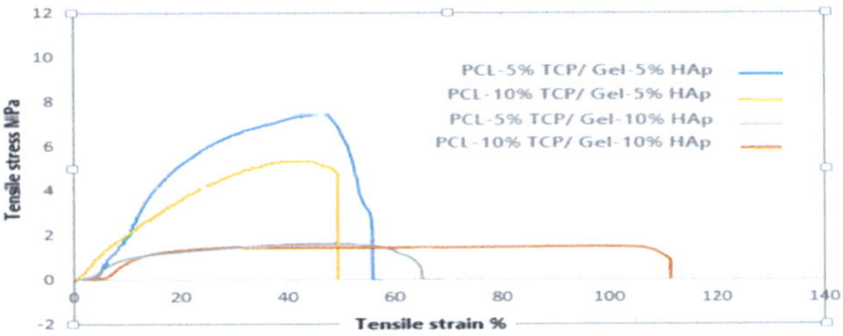

Fig. 4. Tensile stress-strain (σ-ε) curves for the scaffolds

but decreased. Besides, in scaffolds (b) and (d), with increasing percentages of TCP, the same results were achieved. It can be argued that due to the hydrophilic nature of TCP, the adhesion of particles has increased and the strength has decreased accordingly. In scaffold (a), the mechanical properties were the highest value because the agglomerations in the other scaffolds act as stress concentration points and cause rupturing of the webs. In fact, the fabricated scaffold (a) exhibited tensile strength of 7.89 MPa and tensile modulus of 3 MPa. Although in previous studies have been shown that the increase in modulus and tensile strength can be attributed to the increasing in stiffness of mineral content, however here there were a decreasing trend. Human Mesanchymal stem cells were used for cell culture studies. Results show that the fabricated nanofibers were supportive to the long term culture of hMSCS and the nanofibers will higher amount of HAp and TCP likely enhanced proliferation.

4 Conclusion

In this work, we successfully fabricated fibrous scaffolds by blending HAp and β-TCP in Gel and PCL solutions, respectively. β-TCP is a bioresorbable ceramic material and has a much faster degradation rate than biodegradable polymers such as PCL. Gel/HAp nanofibers have been used to improve features such as bone conduction and adhesion of osteoblasts cells. Contact angle measurements showed that the addition of HAp and β-TCP decrease contact angle very much and create a hydrophile surface. Among the scaffolds, the scaffold containing 5% of each minerals was selected as optimal, because it has the least agglomeration and the highest strength.

References

1. Subbiah, T., Bhat, G., Tock, R.W., et al.: J. Appl. Polym. Sci. **96**, 557–569 (2005)
2. Rajzer, I., Menaszek, E., Kwiatkowski, R., et al.: Mater. Sci. Eng. C **44**, 183–190 (2014)
3. Heather, D.: Biomed Eng, 1st edn (2014)

Surface Modification of Polyurethane Films with Chitosan and Collagen Biomolecules, Using Layer by Layer Self-Assembly Method to Improve Biocompatibility Properties

Narges Bahrami[1(✉)], Saied Nouri Khorasani[1], and Hamid Mahdavi[2]

[1] Department of Chemical Engineering, Isfahan University of Technology, 84156-83111 Isfahan, Iran
nastaran_bm@yahoo.com

[2] Department of Novel Drug Delivery Systems, Iran Polymer and Petrochemical Institute, P.O. Box 14975-112, Tehran, Iran

Abstract. Recently biocompatible polymer films have received the great attention in tissue engineering applications. One of the material surface modification methods is use of the layer-by-layer self-assembly process (LbL), especially the electrostatic self-assembly process controlled by the adsorption and desorption equilibrium in anionic and cationic solutions. In this research, the polyurethane (PU) film obtained from the reaction of hexamethylene diisocyanate and castor oil was initially functionalized by the low-pressure nitrogen plasma treatment and then grafted with acrylic acid. Acrylic acid grafted PU film was modified in three LbL ways (A, B, and C) with three layers and five layers. Each layer first consisted of collagen containing positive electrical charge, anionic sodium polystyrene sulfonate (PSS) and cationic chitosan, which was stained with quantum particles. The modified three-layer and five-layer films obtained from all three methods were characterized and analyzed with results from field emission scanning electron microscopy (FESEM) images, confocal microscopy and HGF fibroblast cell culture. The results confirmed that the method (C), firstly contained positive charge by hexamethylenediamine, was the best method through the LbL techniques.

Keywords: Surface modification · Polyurethane · Chitosan · Collagen · Biocompatibility

1 Introduction

Recently, there is a great interest in modification of biomedical polyurethane with natural polymers making them more biocompatible. To achieve intercellular cell-positive interaction, functional groups or biomacro-molecules like collagen and chitosan have been used with different surface modification methods [1–3]. One of the promising approaches in this area is electrostatic layer-by-layer self-assembly process that this method was first carried out by Ller at DuPont Company and expanded by Professor Decher exploring the application of this method in polyelectrolytes [4]. Zhu (2003)

H. Mirzadeh and A. A. Katbab (Eds.): ISPST 2018, *Eco-friendly and Smart Polymer Systems*, pp. 117–121, 2020.
https://doi.org/10.1007/978-3-030-45085-4_29

developed a new LbL method to introduce free amine groups on polyester scaffold by aminolysis of ester groups with a diamine. Chitosan containing positive charge was deposited on the surface of poly L-lactic acid membrane by layer-to-layer self-assembly and also using PSS as a negative charge polyelectrolyte [5]. Gong et al. (2007), deposited cell compatible chondroitin-sulfate with type I collagen LbL method on a poly L-lactic acid scaffold to increase the interaction of this material with the cell [6]. Antunes et al. (2011) deposited chitosan with polyglotamic acid, a type of poly-anion, on thin silicon films with a maximum of six layers by LbL method [7]. Mauquoy (2016) with the aim of studying the LbL method for two types of proteins covered polystyrene surface with two types of extracellular matrix protein, collagen and fibronectin. He received LbL method, widely used for polyelectrolytes, which is not easily transferable to proteins [8]. In this research, work type I collagen and chitosan with positive charge by using PSS as polyanionic were deposited on the polyurethane film with three different LbL methods.

2 Experimental

The acrylic acid (AAc > 99%, 147230), *N*-(3-dimethylaminopropyl)-*N*′-ethyl carbodiimide hydro-chloride (EDC, E6383), *N*-hydroxysuccinimide (NHS, 98%, 130672), phosphate buffered saline (PBS, pH 7.4, P5368), acetic acid (99%, A6283), collagen (from calf skin, C9791), chitosan (medium molecular weight, 448877), hexamethylenediamine (98%, 11696) and poly(4-styrenesulfonic acid) sodium salt (PSS, average Mw ~1,000,000, 434574) were all purchased from Sigma Aldrich. After preparation stained collagen, chitosan and PSS solutions, acrylic acid grafted PU films were modified with LbL method, presented in Fig. 1. To confirm collagen and chitosan deposition with explained LbL techniques, TESCAN-XMU MIRA3 field emission scanning electron microscopy (FESEM) and Leica TCS SPE confocal microscopy were used. To evaluate in vitro cytocompatibility of the samples, HGF fibroblast cell culture according to ISO-10993-5:2009 was performed.

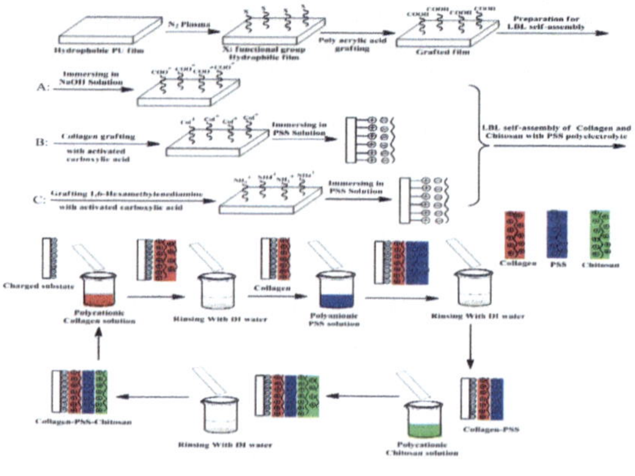

Fig. 1. Imaginal presentation PU film modified with three different LbL methods

3 Results and Discussion

With respect to the SEM obtained images in Fig. 2, the layer deposition is clearly seen as the collapsed wrinkled horizontal layers. These layers start from the region of the grooves from plasma treatment in the PU film bulk and come to the surface of the film. Among the three methods mentioned above, the highest level of the coating is observed in method (C) and then (B). Generally, the amount of these layers and folds in five-layer deposited film is more than triple-layers. In all of images, the top rows belong to triple-layers and bottom rows belong to five- layers. The cross section of the quantum particles stained samples was imaged with confocal microscopy. According to Fig. 3, it is understood clearly that the best LbL method is related to (C) technique and the maximum thickness measured of the deposited layers of the five-layer mode is 53 μm and in the second order 37 μm is related to the three-layer mode. According to the pictures, it is remarkably seen that the layer has not been made on the entire film surface, and only the points that contain the functional group have attracted poly-electrolate layers under electrostatic gravity. The amount of deposition of collagen and chitosan biochemicals on the polyurethane film influenced cell adhesion, morphology, and growth and proliferation. In other words, as the LbL self-assembly process is performed better, cell proliferation and cellular survival increases. Even the images from the cell morphology show that in method (C), the cells are more pike-drawn and have a spindle shape. It is noteworthy that in all three methods, the cell proliferation in the three-layer LbL is more than five-layers. This phenomenon can be related to the reduction of the level of gravity and height, because in the process of laying, as seen in the FESEM images, the polyelectrolyte layers fill the cavities to reach the surface. With decreasing levels of ripeness, levels of adhesion and cell growth also decrease (Fig. 4 and Table 1).

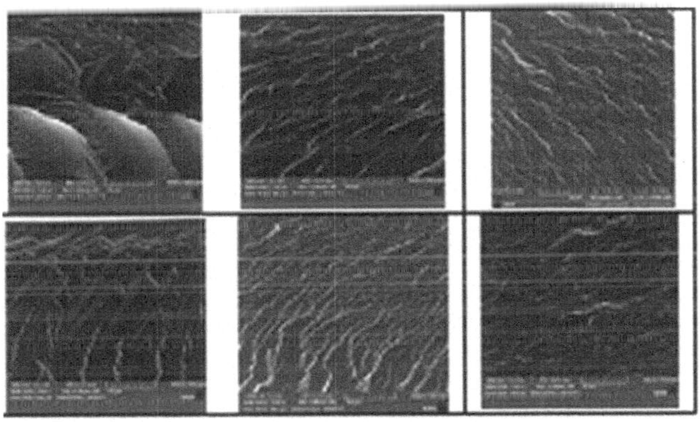

Fig. 2. FESEM images from PU films modified with (A, B, and C) LbL methods respectively with 1 Kx magnify

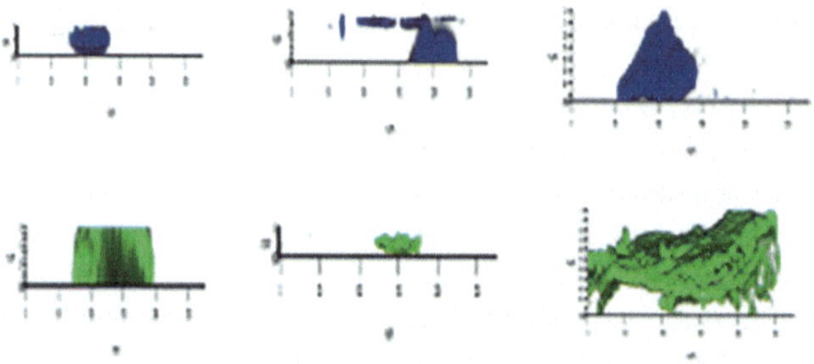

Fig. 3. Confocal images from cross sectional PU films modified with (A, B, and C) LbL methods, respectively

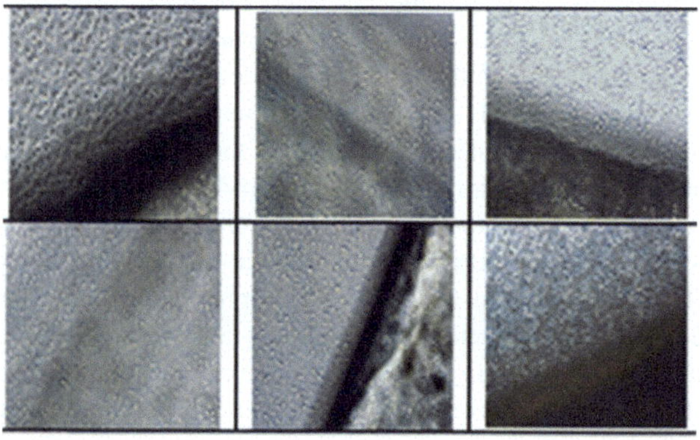

Fig. 4. Optical microscopy pictures of HGF fibroblast cell cultures grown on PU films modified with (A, B, and C) LbL methods, respectively

Table 1. MTT assay results of PU films modified with (A, B, and C) LbL methods, respectively

Mode of PU film modified with LbL	Average of abs	Cell viability
Triple layers with A method	0.521	0.87026
Five layers with A method	0.505667	0.84465
Triple layers with B method	0.524333	0.8758
Five layers with B method	0.515333	0.8608
Triple layers with C method	0.536	0.8953
Five layers with C method	0.527667	0.8814
Control	0.598667	

4 Conclusion

To PU surface modification with collagen and chitosan, LbL self-assembly method is very successful Provided that a strong polyanionic like PSS is used. Images obtained from SEM and confocal confirms the layer deposition. If the surface initially positively charged by aminolysis with a diamine, the amount of layer deposition was better. Increasing the amount of layer deposition improves fibroblast cell adhesion, morphology and proliferation.

References

1. Justyna, K.-L., Gubańska, I., Janik, H.: Polimery **59**(3), 2014
2. Deepak, R., Agrawal, Y.K.: Nanobiomed Godište **1**, 1–8 (2014)
3. Richardson, J.J., et al.: Chem Rev **116**(23), 14828–14867 (2016)
4. Kirkland, J.J.: Anal Chem **37**(12), 1146–1458 (1965)
5. Yabin, Z., et al.: Biomacromolecules **4**(2), 446–452 (2003)
6. Gong, Y., et al.: Acta Biomater **3**(5), 677–685 (2007)
7. Antunes, J.C., et al.: Biomacromolecules **12**(12), 4183–4195 (2011)
8. Mauquoy, S., Christine, D.-G.: Colloid Surface B **147**, 54–64 (2016)

Green Coatings and Engineering in Polymer Science

Performance of Room Temperature Cross-Linkable Acrylic Latexes Synthesized Using DAAM and AAEMA Functional Monomers

Saba Goharshenas Moghadam, Shahla Pazokifard(✉), and Hamidreza Parsimehr

Color and Surface Coatings Department, Iran Polymer and Petrochemical Institute, P.O. Box 14975-112, Tehran, Iran
s.pazokifard@ippi.ac.ir

Abstract. Different acrylic latexes were synthesized using acetoacetoxy ethyl methacrylate (AAEM) and diacetone acrylamide (DAAM) functional monomers (in concentration of 0.5 wt% of monomers) through seeded semi-continuous emulsion polymerization. Adipic acid dihydrazide (ADH) and 1,3-*diamino-propane* (DAP) were added as crosslink agents to the synthesized DAAM-acrylic and AAEM-acrylic, respectively to prepare two components in one-pot latexes. Crosslink reactions were carried out at room temperature during the process of latex film formation. Mechanical properties of the crosslink films were evaluated via pendulum hardness measurements and dynamic mechanical thermal analysis (DMTA). Cross-linking was proved through increase in hardness, storage modulus at rubbery plateau, and T_g values. In addition, the results showed improved mechanical properties in latex films formed from DAAM-acrylics in comparison to those formed from AAEM–acrylics.

Keywords: Hardness · Storage modulus · Crosslink density · DAAM · AAEMA

1 Introduction

Production and consumption of waterborne coatings show a growing trend in various industries due to the environmental aspects [1]. However, these coatings commonly suffer from low durability and weak mechanical properties in comparison to their solvent based counterparts [2]. This weakness can arise from the nature of common film formation process of latex films [3]. Crosslink reactions used in polymer structures and coatings are effective techniques to overcome the problem. Mechanical properties of the coatings depend strongly on the T_g of the crosslink polymer, degree and rate of crosslinking. In this study, the effect of functional monomers with different crosslink reactions has been studied on mechanical properties of waterborne acrylic coatings.

H. Mirzadeh and A. A. Katbab (Eds.): ISPST 2018, *Eco-friendly and Smart Polymer Systems*, pp. 125–128, 2020.
https://doi.org/10.1007/978-3-030-45085-4_30

2 Experimental

Methyl methacrylate (MMA), butyl acrylate (BA), acrylic acid (AA), styrene (St) monomers, sodium dodecyl sulphate (SDS) and NP-15 emulsifiers, ammonium persulphate (APS) as an initiator and $NaHCO_3$ buffer were purchased from Merck. Diacetone acrylamide (DAAM) and acetoacetoxy ethyl methacrylate (AAEM) as two functional monomers, adipic acid dihydrazide (ADH) and 1,3-diaminopropane (DAP) as crosslink agents were purchased from Sigma-Aldrich. Cross-linkable acrylic emulsion polymers with a defined theoretical T_g (15 °C, calculated by Fox equation) were synthesized through seeded semi-continuous emulsion polymerization. Copolymerization was carried out in a 500 mL, five-necked flask, equipped with a reflux condenser, a thermometer, two addition funnels and a stirrer at mild stirring (200 rpm) during the whole process. A pre-emulsified monomer mixture was prepared by mixing common monomers (i.e. MMA, BA, St, AA) and a functional monomer (DDAM or AAEM), 2/3 of mixed emulsifiers and half of the de-ionized water at room temperature for 30 min and then added to the reactor containing the rest of water and emulsifiers at 80 °C during 3 h. Initiator aqueous solution was added drop-wise to the mixture. Mixing continued for an extra hour. Ammonia solution used to adjust pH of the product at 9.0. The recipe is presented in Table 1.

Table 1. Monomer contents % (g) in polymerization

Sample code	Functional monomer	AA	St	BA	MMA
A-0	–	**2.00** (2.92)	**20.00** (29.18)	**45.20** (65.95)	**32.80** (47.86)
A-0.5D	**0.5** (0.72)	**2.00** (2.92)	**20.00** (29.18)	**45.09** (65.80)	**32.41** (47.28)
A-0.5A	**0.5** (0.73)	**2.00** (2.92)	**20.00** (29.18)	**45.20** (65.96)	**32.30** (47.14)

ADH and DAP crosslink agents were added to DAAM-acrylic and AAEM-acrylic emulsions in a stoichiometry ratio - named A-0.5D-ADH and A-0.5A-DAP - respectively. Different films were applied on a substrate in 100 μm WFT and leave to dry in ambient temperature. Pendulum hardness test was performed according to ASTM D4366 using BYK pendulum hardness tester. DMTA were carried out using Triton Tritec-2000 DMA-USA in tension mode, 1 Hz frequency, and 2 $°C.min^{-1}$ heating rate.

3 Results and Discussion

3.1 Hardness Measurements

Coating hardness is the ability to resist permanent indentation, scratching, cutting, and penetration by a hard object. As it is evidence in Fig. 1, crosslinking leads to increase in hardness of the dry latex films (A-0.5D-ADH and A-0.5A-DAP *vs* A-0, A-0.5D, and

A-0.5A). This can be assigned to imide and enamine ester formation which is shown in Fig. 2. Also, DAAM monomer in 0.5 wt% improves the hardness of the acrylic films more than that of AAEM monomer.

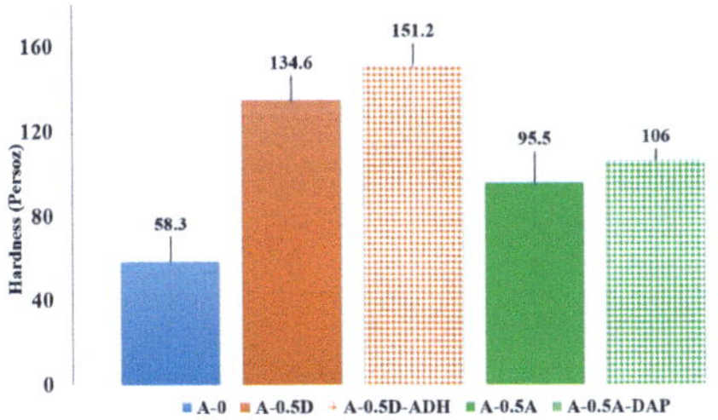

Fig. 1. Hardness of common and functional acrylic films (theoretical T_g = 15 °C)

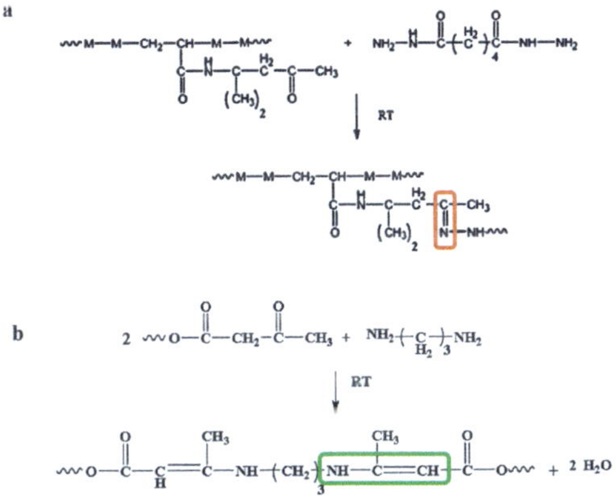

Fig. 2. Schematic reactions between acrylic polymers functionalized with: (a) DAAM and formation of imides and (b) AAEM and formation of enamine esters

3.2 Dynamic Mechanical Thermal Analysis (DMTA)

DMTA was utilized to study the effect of different functional monomers on curing behavior and crosslink density of the acrylic films. According to Fig. 3, storage modulus of acrylic latexes A-0.5D-ADH in glassy state was higher than that of the other samples. Furthermore, the modulus in rubbery plateau shifted to higher values, indicating increase in crosslink density by using functional monomers. In Fig. 4, the maximum Tan δ in the A-0.5D-ADH and A-0.5A-DAP peaks as an indication of the T_g of the films have shifted to higher temperatures. Also, the height of those peaks decreased, and the width of the peaks became boarder. Height and width of Tan δ reflect structural homogeneity and crosslink-density. Introduction of cross-link invariably reduces Tan δ and increases peak width which indicates the broadness of the distribution of length of chains between junction points in the network [4]. Also, it means damping takes place in a wider temperature rang. Crosslink density values of A-0.5D-ADH (1.86×10^{-4} mol.cm^{-3}) than A-0.5A-DAP (1.69×10^{-4} mol.cm^{-3}) and A-0.5A-DAP samples confirmed the claims.

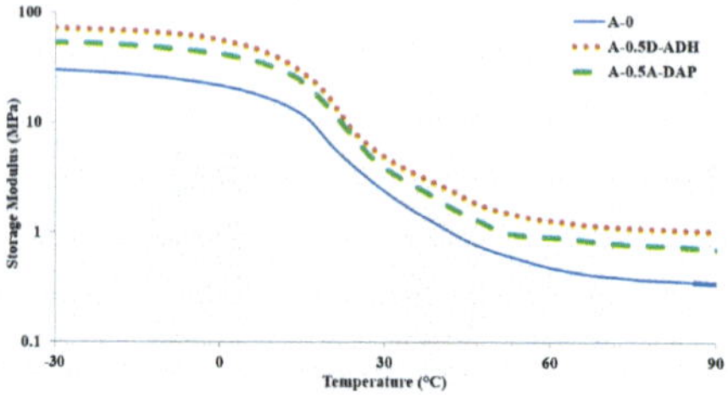

Fig. 3. Storage modulus as a function of temperature for acrylics emulsion films containing DAAM and AAEMA monomers

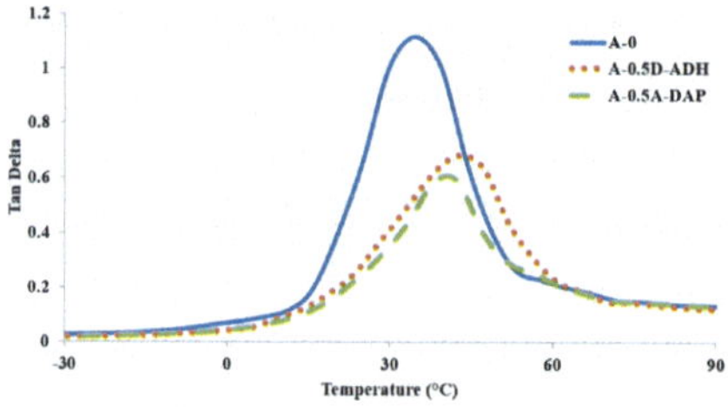

Fig. 4. Tan δ as a function of temperature for acrylics emulsion films containing DAAM and AAEMA monomers

4 Conclusion

The films of DAAM-functionalized acrylic emulsions cured with ADH at room temperature indicated improved mechanical properties in compared to AAEM-functionalized acrylic emulsions cured with DAP owing to higher hardness values, storage modulus, and T_g of the films.

References

1. Nakayama, Y.: Prog. Org. Coat. **33**(2), 108–116 (1998)
2. Hellgren, A.-C., Weissenborne, P., Holmberg, K.: Prog. Org. Coat. **35**(1), 95–99 (1999)
3. Koukiotis, C., Karabela, M., Sideridou, I.: Prog. Org. Coat. **75**, 106–115 (2012)
4. Koleske, J.V.: Paint and Coating Testing Manual, 15th edn, p. 630. ASTM International, Conshohocken (2015)

Effect of Curing Conditions on the Surface and Bulk Properties of High Solid Silicone-Epoxy Coatings

Mohammad Yasin Mollajavadi[1], Shahla Pazokifard[2](✉), and Seyed Mojtaba Mirabedini[2]

[1] Chemical and Petroleum Engineering Department, Sharif University of Technology, Tehran, Iran
[2] Color and Surface Coatings Department, Iran Polymer and Petrochemical Institute, Tehran, Iran
s.pazokifard@ippi.ac.ir

Abstract. This work reports the effect of curing at ambient conditions and elevated temperature on the properties of epoxy-silicone coatings cured by amino functional hardeners in two different weight ratios of 100:40 and 100:60. FTIR spectroscopy was used to evaluate curing process and crosslinking reaction in the coating samples. Hardness and water contact angle measurements were carried out on the coatings as the indications for bulk and surface properties of the coatings, respectively. The results revealed that increasing the curing temperature leads to higher water contact angle and lower hardness. It reason can be assigned to surface roughness of the coating caused by shrinkage or thermal stress occurred during heating the coating.

Keywords: High solid coatings · Silicone/epoxy hybrid · Curing conditions · Surface and bulk properties · Stress

1 Introduction

Nowadays, silicone-epoxy systems are considered as novel technology in coatings industry. Attractive characteristics of these coatings arise from low glass transition temperature, ability to form a very smooth surface and hydrophobic nature of their silicone part, which provide high elasticity even at low temperatures, and high crosslinking density associated with epoxy functionality reactions [1]. In general, silicone-epoxy resins provide antifouling and anti-icing properties and exhibit excellent dirt pickup and anti-graffiti properties as well. Also, they can be used in various coatings with durability under harsh operational conditions [2].

In this research, the effect of curing condition (time and temperature) was evaluated on the surface properties of a silicone-epoxy coating by measuring water contact angle and hardness of the cured coatings.

H. Mirzadeh and A. A. Katbab (Eds.): ISPST 2018, *Eco-friendly and Smart Polymer Systems*, pp. 129–132, 2020.
https://doi.org/10.1007/978-3-030-45085-4_31

2 Experimental

2.1 Materials and Coating's Preparation

SILIKOPON® EF as a hybrid of an aliphatic epoxy resin (with epoxy equivalent weight: 450 $g.mol^{-1}$) and an alkoxy silicone resin was prepared from Evonik Industries AG. SILRES® HP 2000 an amino functional, methyl phenyl silicone resin with H-amine equivalent weight: 230–255 $g.mol^{-1}$ provided from Wacker Chemie AG and used as a hardener.

Two different ratios of resin to hardener (i.e. 100:60 and 100:40 w/w) were chosen to prepare sample coatings. The coatings were applied on glass substrates with 200 μm wet film thickness, and left to dry at two different conditions: (a) room temperature for a couple of days, and (b) 1.5 h at 85 °C. S, W, O, and R in sample coding stand for silicon based resin, Wacker hardner, oven, and room temperature curing condition, respectively. Also Digits show the weight percentage of binder and hardener in coating's formulation.

2.2 FTIR Spectroscopy

FTIR is an analytical technique for analyzing chemical structure of the compounds. FTIR model Equinox X-55 was used to confirm the curing reactions between amino groups presented in SILRES® HP 2000 and epoxy groups of SILIKOPON® EF.

2.3 Hardness Measurement

Hardness of each sample was measured during 7 days and the test was repeated 3 times for each sample.

2.4 Contact Angle Measurment

Contact angle instruments (Kruss K14 tensiometer) were used and the coatings' sample surfaces were analyzed to evaluate their wettability and therefore hydrophobicity.

3 Results and Discussion

3.1 FTIR Spectroscopy

The FTIR spectra of SILIKOPON® EF as a binder, SILRES® HP 2000 as a hardener and the cured coatings (S100.W60) are illustrated in Fig. 1. The curing reaction of epoxy functional groups in silicone-epoxy hybrids with amino-group of the hardener were confirmed by IR spectral analyses. The disappearance of the peaks at 3,056 and 915 cm^{-1} indicates the opening of epoxy rings. The appearance of the band at 1,109 cm^{-1} is characteristic for C–N stretching vibrations. The absence of the absorption of epoxy ring and the presence of OH group and C–N group confirm the conversion of epoxy group into the corresponding polymer, as well as crosslinking process [4].

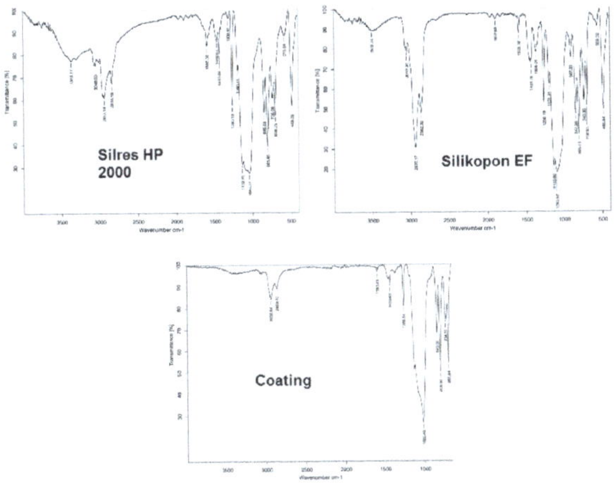

Fig. 1. FTIR spectra of silicone/epoxy binder (Silikopon EF), hardener (Silres HP 2000), and the cured coating

3.2 Hardness Measurements

High coating hardness is a measure for resistance against penetration or cutting by a hard object. Hardness changes during seven days are presented in Fig. 2. It can be considered as a bulk property of a coating. As it is evidence in this figure, hardness of all samples have increased during the time and become constant after 4 days. In the meantime, the hardness of the samples have increased by adding more hardener to the coating. Another result that is evidence in this figure is that the samples which were cured at room temperature has showed higher hardness than the samples with the same ratio of hardener but cured at high temperature.

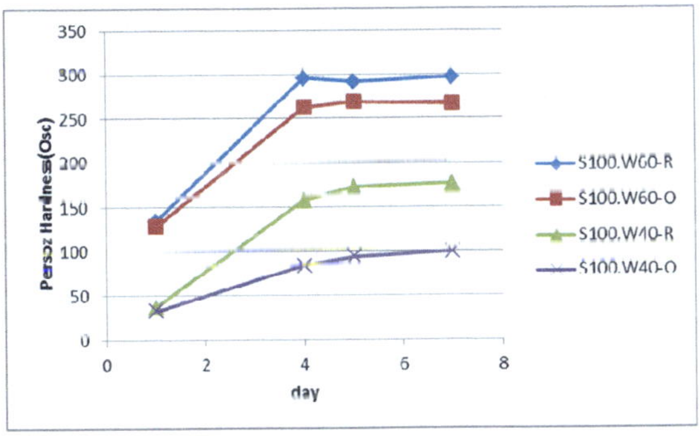

Fig. 2. Hardness changes during 7 days

3.3 Contact Angle Measurements

The results of water contact angle measurements are summarized in Table 1. As it is clear, curing in oven provides higher contact angle which means more hydrophobicity than the sample with the same composition but cured at room temperature. This can be assigned to the coating surface roughness caused by thermal stress during the high temperature curing process of the coating [3].

Table 1. Contact angle measurement results 10 days after application at room temperature

Sample	Contact angle (°)
S100.W40-O	82.9 ± 0.57
S100.W40-R	79.1 ± 0.96
S100.W60-O	85.2 ± 0.10
S100.W60-R	77.4 ± 0.98

4 Conclusion

Curing silicone-epoxy high solid coatings using amino functional hardener at temperatures higher than room temperature led to low hardness, but high water contact angle on the surface of the coatings. This can be related to the thermal stress during crosslinking reactions occurred between binder and hardener in the coatings applied on a substrate.

References

1. Mironova, G.A., Il'darkhanova, F.I., Kopteva, V.V., Bogoslovskii, K.G.: Russ. Coat. J. **7**, 26–29 (2017)
2. Bayer, I.S.: Coatings **7**, 12 (2017)
3. Rabea, A.M.: J. Coat. **8**, 497–503 (2011)
4. Zlatkovic, S., Lacnjevac, C., Stamenkovic, J., Cakic, S., Nikolic, G.S.: Study and correlation of new organic solvent free three-component waterproof epoxy/polyamine systems. In: Proceedings of the 1st International Congress of the Engineering, Materials and Management in the Processing Industry, Jahorina, Republic of Srpska, pp. 192–214 (2009)

Synthesis of UV-Curable Urethane Acrylate Zirconium Oxide Nanocomposites as Thermal Insulator

Abbas Madhi[1(✉)], Behzad Shirkavand Hadavand[2], and Ali Amoozadeh[1]

[1] Faculty of Chemistry, Semnan University, Semnan, Iran
abbas.madhi@semnan.ac.ir
[2] Department of Resin and Additives, Institute for Color Science and Technology, Tehran, Iran

Abstract. Synthesis of polyurethanes through UV-curing process contributes to the environmental protection by removing the hazardous organic solvents. At first, UV-curable urethane acrylate (UA) ZrO_2 nanocomposites with different weight percentages of ZrO_2 were synthesized and then thermal conductivity of the nanocomposites was measured by infrared thermography. Infrared thermography results showed that adding ZrO_2 nanoparticles to the urethane acrylate matrix increased surface reflection coefficient and decreased thermal conductivity in the urethane nanocomposites. Consequently, urethane acrylate ZrO_2 nanocomposites (3 and 5 wt%) acted as strong thermal insulators.

Keywords: Urethane acrylate · ZrO_2 · Nanocomposite · Thermal insulator

1 Introduction

In recent decades, use of UV-curable polymers has drawn lots of attention due to the benefits such as protection of environment because of disuse of dangerous organic solvents, saving of energy and time in curing process of nanocomposites at ambient temperature, and high chemical and thermal resistance [1]. Nowadays, with regard to the decrease in energy resources and necessity of saving them, usage of kinds of thermal insulator coatings has drawn a great deal of interest for construction and different industries [2]. Polymer thermal barrier coatings as a group of insulator coatings include some features like low heat conductivity and high thermal resistance due to the applying of proper reinforcers in polymer matrix [3]. Polyurethanes, because of flexibility, could lessen thermal pressure caused by thermal shock in electronic packages [4]. The purposes of this research are to produce urethane acrylate ZrO_2 nanocomposites, and to enhance thermal barrier feature of these polymers.

H. Mirzadeh and A. A. Katbab (Eds.): ISPST 2018, *Eco-friendly and Smart Polymer Systems*, pp. 133–136, 2020.
https://doi.org/10.1007/978-3-030-45085-4_32

2 Experimental

2.1 Materials and Instruments

Nano zirconium oxide (ZrO_2) with a particle size of 20 nm was purchased from US Research Nanomaterials Company (USA). Trimethylol-propane triacrylate (TMPTA) as a diluent was supplied from Sigma-Aldrich Company (USA). The rest of materials including 1,4-butanediol (BD), isophorone diisocyanate (IPDI), dibutyltin dilaurate (DBTDL), 2-hydroxyethyl methacrylate (HEMA), triethanolamine (TEA), benzophenone and acetone were all supplied from Merck Company (Germany).

UV radiation in a UV curing device (1 kW Hg lamp, 80 W/cm) was used for the curing process at ambient temperature. There was a 10 cm distance between the UV lamp and the samples. For measuring the thermal conductivity of samples, TEMP spectroscopy (2000 A, USA) was used by means of the Testo infrared camera (875-2, Germany); the ASTM E903-96 standard was the base of this analysis.

2.2 Synthesis of UA-ZrO_2 Nanocomposites

First, nano ZrO_2 was dispersed in acetone by means of sonication for 20 min and then added to UA resin according to the reference [1]. Then, 0, 1, 3, and 5 wt% UA-ZrO_2 nanocomposites tablets in a circular shape (3 mm thickness and 25 mm diameter) were produced by casting on glass molds. The tablets were cured for 5 min at room temperature. After UV radiation curing, the prepared tablets features were studied.

3 Results and Discussion

One method to study the thermal conductivity in a coating with a low thermal conductivity coefficient is infrared thermography that investigates the amount of thermal dispersion on the polymers and nanocomposites surfaces. The thermal conductivity coefficient of the UA-ZrO_2 nanocomposites (1, 3, and 5 wt%) was studied compared to pure polymer (urethane acrylate). The results of the infrared thermography and surface temperature of the samples are shown respectively in Table 1 and Fig. 1. The findings indicate that in the samples containing modified ZrO_2 nanoparticles (3 and 5 wt%), surface temperature decreases (40.4 and 40.2 °C, respectively) that is an indication of reflection coefficient of the samples, consequently surface temperature decreases. The least surface temperature firstly belonged to the 5% ZrO_2 nanoparticles (Z3) and secondly to the 3 wt% sample (Z2). In the 1 wt% ZrO_2 nanoparticles (Z1) sample an increased surface temperature (42.3 °C) is observed, this increment is more than that of the blank sample (42.2 °C), the reason behind this is the low amount of nanoparticles in the polymeric matrix. According to the results, as amount of ZrO_2 nanoparticles increases, the surface reflection coefficient increases too, and consequently, the thermal conductivity coefficient decreases, Therefor, one could point out that the increase of the amount of ZrO_2 nanoparticles (3 and 5 wt%) in the polymeric matrix and the proper dissipation of nanoparticles are correlated with the increase of thermal barrier feature of these nanocomposites. According to this, the higher surface reflection coefficient leads to the decrease in surface temperature; and the decreased surface temperature causes lowering the thermal conductivity and consequently reducing energy consumption.

Table 1. Surface temperature of UA and UA-ZrO_2 nanocomposites

Samples	Temperature (°C)
UA	42.2
UA/1 wt% ZrO_2	42.3
UA/3 wt% ZrO_2	40.4
UA/5 wt% ZrO_2	40.2

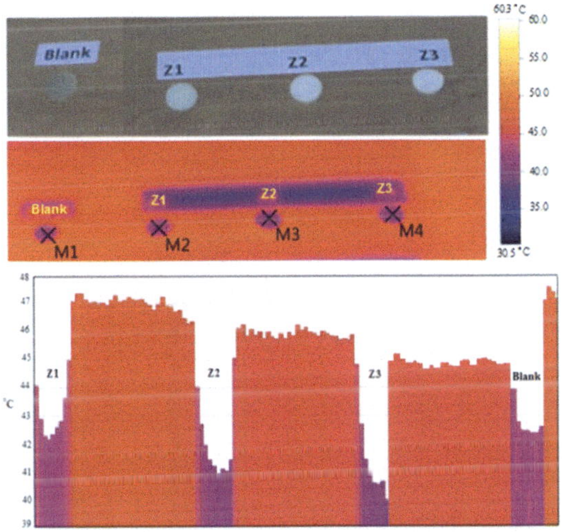

Fig. 1. Infrared thermography results for UA and UA-ZrO_2 nanocomposites

4 Conclusion

In this research, synthesis of UV curable urethane acrylate ZrO_2 nanocomposites (0, 1, 3, and 5 wt%) was conducted. According to the infrared thermography results, adding ZrO_2 nanoparticles (3 and 5 wt%) to the UA matrix caused the reflection coefficient of the surface to enhance and the amount of heat transfer of the samples to decrease. It is worth pointing out, as a conclusion, UA-ZrO_2 (3 and 5 wt%) samples are counted as optimal thermal barriers.

References

1. Madhi, A., Shirkavand Hadavand, B., Amoozadeh, A.: J. Compos. Mater. **52**, 2973–2982 (2018)
2. Moreira, D.C., Telles, M.C.O., Nunes, L.C.S., Sphaier, L.A.: ENCIT, Brazil, 18–22 November, Rio de Janeiro (2012)
3. Madhi, A., Shirkavand Hadavand, B., Amoozadeh, A.: J. Appl. Chem. **45**, 91–98 (2018)
4. Madhi, A., Shirkavand Hadavand, B., Amoozadeh, A.: Prog. Color Colorants Coat. **10**, 193–204 (2017)

Antibacterial Performance of Biomimetic Modification of Polymeric Surfaces: Rice-like and Lotus Pattern

Masoumeh Ayazi and Nadereh Golshan Ebrahimi(✉)

Department of Polymer Engineering, Faculty of Chemical Engineering, Tarbiat Modares University, P.O. Box 14115-114, Tehran, Iran
nebrahimi@modares.ac.ir

Abstract. Significant research effort has been devoted to understanding surface modification methods to achieve antibacterial surface due to the growing prevalence of antimicrobial resistant strains. In this way, the biomimetic patterning approach has considered as modification approach because of improving surface to prohibit bacteria growth in the absence of antibiotic agents. In this study, we present two biomimetic microstructures from the rice-like pattern and a lotus structure to consider their ability in reducing bacterial adhesion. Two inspired structures have developed on the polydimethyle siloxane and polypropylene surface. The results of the attachment behaviors have examined on two bacteria strains of gram-negative Escherichia coli (*E. coli*) bacteria and gram-positive Staphylococcus aureus (*S. aureus*). Bacteria adhesion function on these roughness surfaces is demonstrated by the prevention of growth on our structured plastic surfaces. The bacteria attached to the surfaces have also offered the important role of bacterial species and material type to prevent bacterial adhesion.

Keywords: Patterning · Rice pattern · Lotus leaf · Roughness surface

1 Introduction

Patterning as a known approach of surface modification has become a highly interesting approach to fight bacterial infection. The significant anti bacterial efficiency inspired from natural objects such as cicada wing [1], gecko skin [2], dragonfly [3], and etc. that motivates researches to focus on reproducing the bio-templated replica of various substrates. The roughness morphology of natural objects which known as "lotus effect" emphasized relationship between geometrical structure and superhydrophobicity [4, 5]. Micro roughness [6–8] and combined of micro and nano surface have demonstrated interrupt the cell attachments and cause death [9–11]. The patterned surface could repel bacterial contamination through drop slid away. Dirt particles and bacteria prefer to attach the water droplets on the surface, rather to adhere the surface itself. So, contaminants roll off with water droplet, simultaneously self-cleaning phenomenon. Assessment and comparison of the antibacterial efficacy of different researches have indicated that more studying need to understand a bio-complex system which can be impacted by surface energy, hydrophobicty, structural geometry, and

H. Mirzadeh and A. A. Katbab (Eds.): ISPST 2018, *Eco-friendly and Smart Polymer Systems*, pp. 137–140, 2020.
https://doi.org/10.1007/978-3-030-45085-4_33

bacteria strains. For example the gram-positive bacteria which known the higher resistance due to larger cell size, stiffness and thickness of the cell wall, so bactericidal function may be shown no similarity with negative gram strains.

So, understanding of the bacterial attachment efficacy on the roughness surface is important. In order to achieve this aim the anti-adhesion ability of bio-mimetic surface has considered through bacterial counting method. Controlling bio-adhesion by hydrophobicity, bacteria gram strains (the gram-negative (*E. coli*), gram-positive (*S. aureus*) bacteria, and materials (two hydrophobic polymers) are evaluated.

2 Experimental

Preparation method based on the patterns obtained from a Lotus-pattern with hexagonal array of circles papilla and rice-like pattern included directional grooves. The geometrical structure was prepared with soft-lithography (casting with fresh rice leaf) and optical lithography (the process of printing from designed pattern of lotus leaf on the mask and photoresis. Prepared samples were submerged in bacterial solution ($\sim 10^6$ CFU/mL), next, they were tilted at 90° to roll off the bacteria solution and rinsed with sterile 1X PBS for three times during 10 s rotation. Enumeration of colonies was performed, and bacterial growth was expressed as a reduction of cell number.

3 Results and Discussion

Figure 1 illustrates the prepared samples of natural replica mold. Table 1 presents the measured contact angle and reduction rate of bacterial growth. The role of material surface to inhibit bacterial growth indicated that hierarchical structure of PDMS has great potential for sliding away *E. coli* and the rice-like pattern of PP successfully glide the *S. aureus* (Fig. 2).

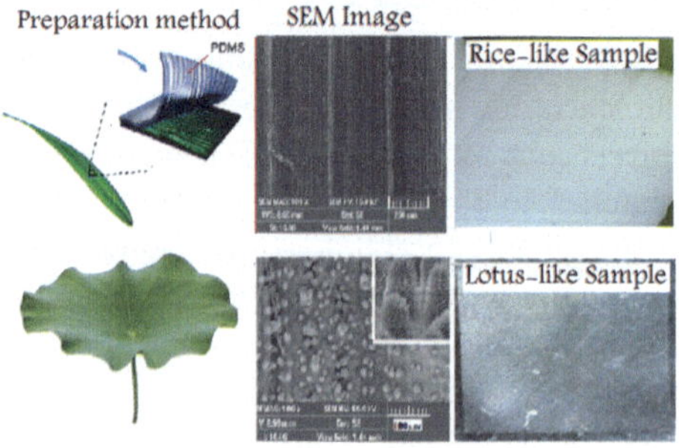

Fig. 1. Images of prepared samples

Table 1. Bacterial adhesion behavior of different prepared samples

Sample	Reduction rate (%)		Contact angle (°)
	E. coli	*S. aureus*	
Smooth PDMS			107
Micro-lotus pattern of PDMS	98	44.8	148
PDMS-micro rice-like pattern	95	54.7	131
Smooth PP			98
Micro rice-like pattern of PP	53	66	138
Micro-lotus pattern of PP	48	71	143

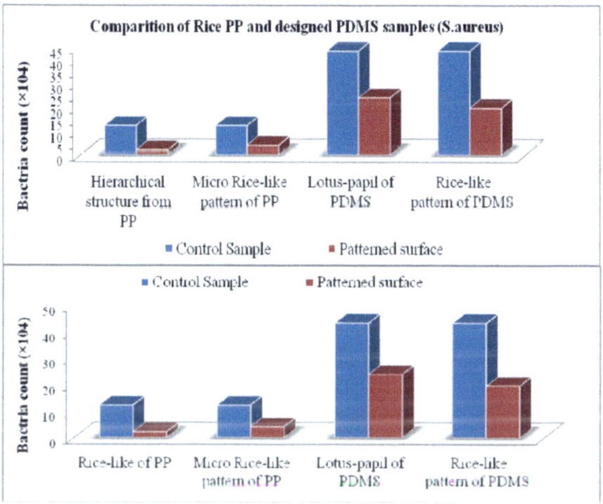

Fig. 2. Bacteria colonies reduction for different samples and two gram-stains

4 Conclusion

The bio-inspired patterned surfaces provide a meaningful reduction in the number of *E. coli* colonies up to 90% and around 50% for *S. aureus* for PDMS surface. It was also observed that the rice-like pattern of two polymers revealed different attachment ability of bacterial. Briefly, the bactericidal observations can be affected by the material and bacteria strains.

References

1. Ivanova, E.P., Hasan, J., Webb, H.K., et al.: Small **8**(16), 2489–2494 (2012)
2. Li, X., Cheung, G.S., Watson, G.S., et al.: Nanoscale **8**(45), 18860–18869 (2016)
3. Bandara, C.D., Singh, S., Afara, I., et al.: ACS Appl. Mater. Interfaces **9** (2017)
4. Dickson, M.N., Liang, E.I., Rodriguez, L.A., et al.: Biointerphases **10**(2), 021010 (2015)
5. Jaggessar, A., Shahali, H., Mathew, A., et al.: J. Nanobiotechnol. **15**(1), 64 (2017)
6. Vasudevan, R., Kennedy, A.J., Merritt, M., et al.: Colloids Surf. B **117**, 225–232 (2014)
7. Nikkhah, M., Edalat, F., Manoucheri, S., et al.: Biomaterials **33**(21), 5230–5246 (2012)
8. Elbourne, A., Crawford, R.J., Ivanova, E.P.: J. Colloid Interface Sci. **508**, 603–616 (2017)
9. Zhang, X., Wang, L., Levanen, E.: RSC Adv. **3**(30), 12003–12020 (2013)
10. Falde, E.J., Yohe, S.T., Colson, Y.L., et al.: Biomaterials **104**, 87–103 (2016)
11. Freschauf, L.R., McLane, J., Sharma, H., et al.: PLoS ONE **7**(8), e40987 (2012)

Fabrication of Ag/TiO_2-Coated Textile Fabric via a New Photoreduction Method as an Eco-Friendly Technique

Zahar Moridi Mahdieh[1], Shahla Shekarriz[2](✉), and Faramarz Afshar Taromi[1]

[1] Department of Polymer Engineering and Color Technology, Amirkabir University of Technology, Tehran, Iran
[2] Color and Polymer Research Center, Amirkabir University of Technology, Tehran, Iran
shahlashekarriz@aut.ac.ir

Abstract. A new facile *in situ* synthesis coating method was utilized to improve the photocatalytic activity of the textile fabric. An eco-friendly method of photoreduction was used to synthesize Ag/TiO_2 nanoparticles. The synthesis and coating approaches were performed simultaneously. The elemental analysis of the X-ray photoelectron spectroscopy showed a pair of separated peaks with a binding energy of 6 eV between two states which confirmed the creation of silver metallic on the fabric surface. The field emission scanning electron microscope images and map analysis showed an even distribution of coating with nanoparticle size of 27 ± 5 nm. The self-cleaning properties were estimated by measuring the discoloration of reactive blue stain under UVA, UVC, and daylight irradiation and showed a significant increase in photocatalytic activity. The anti-bacterial property showed an excellent activity after Ag-doping in TiO_2.

Keywords: Eco-friendly technology · Photoreduction method · Ag-TiO_2 nanoparticles · Anti-bacterial activity · Self-cleaning

1 Introduction

Photoreduction method is an eco-friendly technique due to reducing the volume of chemical agents, energy, and wastewater. In *ex situ* synthesis coating methods, there are two steps for applying the nanoparticles on the fabric surface: the first step is the synthesis of Ag/TiO_2 NPs by different methods, and the second step is loading the prepared nanoparticles on the fabric surface by conventional coating methods [1–3]. The *in situ* synthesis by photoreduction methods was applied in an aqueous solution with more steps and applying chemical agents that leads to producing wastewater [4, 5].

We focused on the reduction of chemical agent and wastewater of the Ag-TiO_2 NPs coating process. The Ag-TiO_2 NPs coated fabric was created by an *in situ* synthesis in a wet environment under UVA irradiation without using any solution and chemical compounds. The chemical and morphological characteristics of the Ag-TiO_2 NPs

H. Mirzadeh and A. A. Katbab (Eds.): ISPST 2018, *Eco-friendly and Smart Polymer Systems*, pp. 141–144, 2020.
https://doi.org/10.1007/978-3-030-45085-4_34

coated fabric were estimated by XPS, FE-SEM, and map analysis. Also, the self-cleaning properties and antibacterial activity were assessed.

2 Experimental

$AgNO_3$ and TiO_2 Degussa P25 were purchased from Merck and Evonik Co., respectively. The Polyester/Cellulose (P/C) (80/20) was provided by Yazd Baf Co., (Yazd, Iran).

The P/C fabric was washed by 2 g/L nonionic detergent (L:G of 40:1) for 30 min at 60 °C. 1 g TiO_2 NPs powder was added to 100 mL $AgNO_3$ solution (0.005 w/v%) and sonicated for 30 min. The P/C fabric was immersed in the suspension and padded. The wet fabric was immediately irradiated by an UVA 100 W lamp for 1 h and cured at 130 °C for 5 min. to make an *ex situ* coating, the $AgNO_3/TiO_2$ suspension was exposed under UVA for 1 h and then the fabric was padded and cured according to the above-mentioned explanation.

XRD (ESCA-3200 Bestec electron spectrometer) and FESEM (Mira 3-XMU) were used for characterization. The self-cleaning property was assessed by the measurement of variation of reactive blue 2 (0.001 w/v%) stain after 24 h under sunlight, UVA, and UVC irradiation. The color change was investigated by a DRS (X-Rite Co.; model: Color-Eye 7000 A). The antibacterial activity was evaluated according to standard of AATCC 100 against *E. coli* and *S. aureus* bacteria.

3 Results and Discussion

Figures 1a and 1b depicts the FESEM images of the untreated and Ag-TiO_2 coated fabrics. There is a smooth surface before the in situ loading process on P/C while a uniform distribution was observed for coated fabric. Therefore, the Ag-TiO_2 NPs with a size of 27 ± 5 nm and appropriate distribution was loaded on the fabric by the new *in situ* synthesis method. Also, an even distribution of Ag and Ti elements on the map image was observed for coated fabric (Figs. 1c and 1d).

The distinguished valence states of C, Ag elements were investigated by XPS analysis in Fig. 2. The C1s and Ag3d peaks were observed at binding energies of 285 and 368 eV, respectively. In Fig. 2a, the distinguished peaks of carbon can be indicated as C-C/C-H, C-O/C-O-C, C = O/O-C-O linkages at 285, 286.8, and 288.3 eV peaks, respectively, revealed the cellulose structure. Also, the C1s component spectra peaks of the polyester structure were indicated at 284.6, 286.1, and 288.7 eV for C-C/C-H, C-O, and O = C-O, respectively. The high-resolution spectra of Ag3d present a pair of separated peaks at 368.4 and 374.4 eV that revealed Ag 3d5/2 and Ag 3d3/2, respectively (Fig. 2b). The separated Ag3d doublet peaks with 6 eV between two states confirm the presence of silver metallic. Therefore, the existence of Ag metallic on the fabric surface endorses the synthesis of Ag-TiO_2 NPs on the fabric surface.

The EDS analysis showed the comparative data of *ex situ* and *in situ* coated fabrics (Fig. 3). The *in situ* coated fabric shows more content of Ag and Ti elements in comparison to *ex situ* coated fabric. The *in situ* coating method was performed using a direct photoreduction method on the surface of the fabric. The TiO_2 NPs on the fabric surface present (for *in situ*) more stability in comparison to the suspension (for *ex situ*). Therefore, It can be suggested more degree of nucleation to grow the Ag NPs. On the other hand, the secondary hydroxyl functional groups of cellulosic fibers create an ability for a gradual reduction of the silver precursor.

Table 1 presents the color changes (ΔE) to estimate the self-cleaning properties of finished fabrics. The blank fabric showed a negligible discoloration. After TiO_2, it was observed a significant increase in self-cleaning. However, the discoloration of the TiO_2 coated fabrics increased after Ag doping. The *in situ* coated fabrics showed more self-cleaning properties in comparison to *ex situ* coated fabric due to increasing the content of Ag-TiO_2 NPs.

TiO_2 coated fabric showed a 60% antibacterial activity while the blank fabric had no antibacterial activity (Fig. 4). However, both of the Ag-TiO_2 NPs coated fabric showed an excellent antibacterial activity.

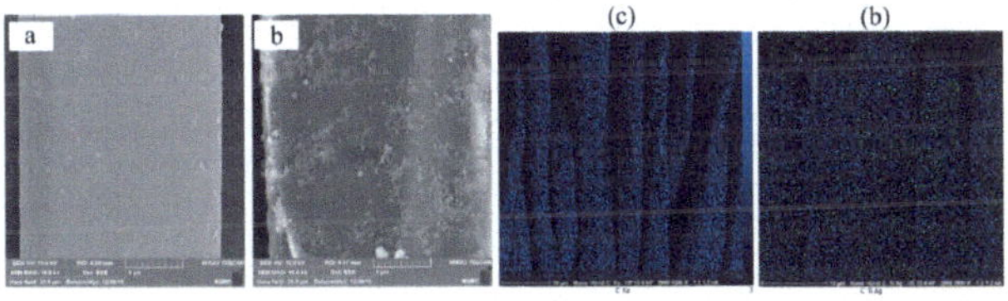

Fig. 1. FESEM images of: (a) blank fabric, (b) Ag-TiO_2 NPs *in situ* coated fabric, map analysis of (c) blank fabric, and (d) Ag-TiO_2 NPs *in situ* coated fabric

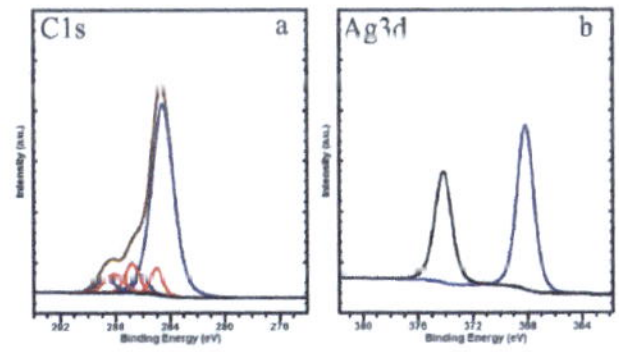

Fig. 2. XPS analysis of Ag-TiO_2 NPs *in situ* coated fabric: (a) C1s and (b) Ag3d

Table 1. Self-cleaning properties of fabrics

NPs coated fabric	ΔE after 24 h irradiation		
	UVA	UVC	Daylight
–	6.4	7.5	8.1
TiO_2	36.8	41.5	45.3
Ag-TiO_2 (Ex situ)	48.3	54.0	56.1
Ag-TiO_2 (In situ)	52.2	57.8	60.3

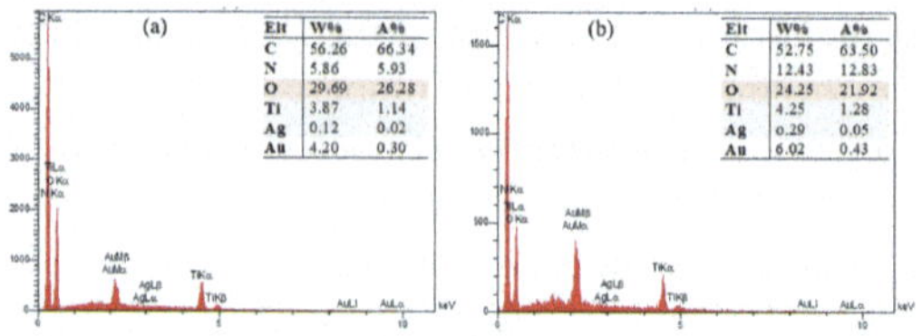

Fig. 3. EDS analysis of Ag-TiO_2 coated fabric by (a) *in situ* and (b) *ex situ*

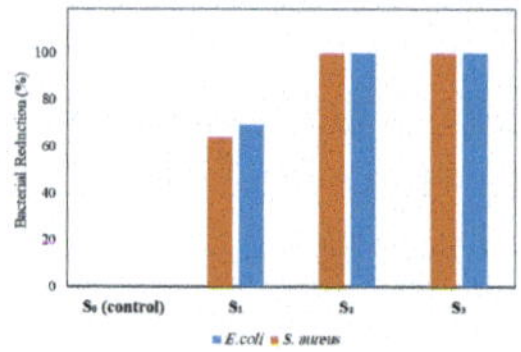

Fig. 4. Antibacterial activity of (S_0) blank fabric, (S_1) TiO_2 coated fabric, (S_2) Ag-TiO_2 NPs In-situ coated fabric, and (S_3) Ag-TiO_2 NPs In situ coated fabric

4 Conclusion

Ag/TiO_2-coated fabric was successfully fabricated by a new eco-friendly *in situ* synthesis coating which showed excellent self-cleaning and antibacterial properties. The FESEM and map analysis depicted an even distribution of the NPs. XPS analysis showed the synthesis of metallic silver on the P/C fabric. The EDS analysis showed more content of Ag and Ti elements of *in situ* coated fabric in comparison to *ex situ* coated fabrics.

References

1. Messaoud, M., Chadeau, E., Brunon, C., et al.: J. Photochem. Photobio. A Chem. **215**, 147–156 (2010)
2. Montazer, M., Behzadnia, A., Moghadam, M.B.: J. Appl. Polym. Sci. **125**, E356–E363 (2012)
3. Shih, K.Y.: US Patent 6, 686, 306 (2004)
4. Ye, W., Xin, J.H., Li, P., et al.: J. Appl. Polym. Sci. **102**, 1787–1793 (2006)
5. Milošević, M., Radoičić, M., Šaponjić, Z., et al.: Cellulose **21**, 3781–3795 (2014)
6. Milošević, M., Radoičić, M., Šaponjić, Z., et al.: J. Mater. Sci. **48**, 5447–5455 (2013)

Fabrication of Super-Hydrophobic Fabric by Eco-Friendly Plasma Technique

Zahar Moridi Mahdieh[1], Shahla Shekarriz[2(✉)], and Faramarz Afshar Taromi[1]

[1] Department of Polymer Engineering and Color Technology, Amirkabir University of Technology, Tehran, Iran
[2] Color and Polymer Research Center, Amirkabir University of Technology, Tehran, Iran
shahlashekarriz@aut.ac.ir

Abstract. A facile method was used to create a super-hydrophilic polyester/cellulose blended fabric. The fabric was exposed to air plasma treatment under different input powers and exposure times without any extra chemical modification. The wettability test showed the maximum hydrophobic properties for an input power of 300 W and exposure time of 10 min. The static contact angle of 167° was indicated for the super-hydrophobic sample. The decrease of intensity of active functional groups and the increase of roughness with a nano-scale pattern presented a super-hydrophobic surface that confirmed by ATR-FTIR and FESEM analysis.

Keywords: Eco-friendly technology · Corona discharge · Super-hydrophobic · Polyester/cellulose fabric

1 Introduction

Plasma Technology can change the polymeric surface characteristics via reactive species [1]. An atmospheric air plasma treatment is an eco-friendly technology that modifies the morphological and chemical characteristics of surfaces [2]. There are two factors that generate a super-hydrophobic surface including the low surface energy and the appropriate micro-nanoscale roughness. For textile fabrics, the super-hydrophobic property has usually been applied by pre-roughening with a non-polymerization plasma at the beginning and then post coating with fluorine components [3, 4]. Also, the use of plasma polymerization with a hydrophobic combination is a usual method for applying the hydrophobic properties on the surfaces of fabric [5].

In this research, we purpose creating a durable super-hydrophobic polyester/cellulose fabric. For this purpose, the polyester/cellulose blended fabric was directly prepared by exposing air corona discharge treatment with different input powers and exposure times without adding any hydrophobic chemical materials. The changes in the morphological and surface roughness properties after corona discharge treatment were investigated by FESEM. The wettability properties were estimated by static contact angle, water drop absorption time. The chemical properties of textile fabric were assessed by ATR-FTIR analysis.

H. Mirzadeh and A. A. Katbab (Eds.): ISPST 2018, *Eco-friendly and Smart Polymer Systems*, pp. 145–148, 2020.
https://doi.org/10.1007/978-3-030-45085-4_35

2 Experimental

The Polyester/Cellulose (P/C) (80/20) was provided by Yazd Baf Co. (Yazd, Iran). A commercial non-ionic detergent (RUCO 6670) based on fatty alcohol ethoxylated was provided from Rudolf Co. (Germany).

For the preparation of the fabrics, the P/C fabrics were washed with a non-ionic detergent at 60 °C for 30 min (L:G was 40:1). A commercial device (Naaj Corona, Naa Plastic Co., Tehran, Iran) with air gas and atmospheric pressure was used for plasma treatment. The fabrics were cut to dimensions of 10 × 10 cm^2 in size. The distance between a backing roll and the aluminum electrode nozzle was adjusted to 2 mm. The effect of corona discharge was assessed using various corona input powers (300 and 500 W) and corona exposure times (5, 10, and 15 min) according to Table 1.

Table 1. Corona discharge treatment conditions of the P/C fabrics along with the wettability results

Samples	Input power (W)	Exposure time (min)	Adsorption time (s)		
			After 0 h	After 24 h	After 12 d
S_0	–	–	710	712	709
S_1	300	5	0	0	1293
S_2	300	10	0	0	0
S_3	300	15	0	0	0
S_4	500	5	7.3	67.9	985
S_5	500	10	5.1	51.8	864
S_6	500	15	4.2	46.2	422

A field emission-scanning electron microscope (FESEM; Mira 3-XMU) and attenuated total reflection-Fourier transform infrared spectra (ATR-FTIR; Nicolet 560) were used for characterization. The static contact angle was measured with the static sessile drop method. A distilled water droplet of 6 μL was placed on the surface of the fabric by a syringe and then the video was recorded by a camera (Model: SONY SSC-DC80P). The wettability was investigated by the water drop absorption time of P/C blended fabric according to the British standard of BS 4554:1970.

3 Results and Discussion

The wettability properties of untreated and corona discharge treated samples were estimated according to Table 1. The blank fabric shows hydrophobic characteristics such that the water drop was adsorbed after 12 min. A super-hydrophobic fabric was created with an input power of 300 W. However, the increase of the aging leads to decrease the super-hydrophobic characteristics. The 300 W-10 min corona treated fabric showed the maximum durability of the super-hydrophobicity. However, the increase of the input power to 500 W leads to decrease the hydrophobicity properties.

The static contact angle is an excellent experimental data that confirms the wettability properties of surfaces. Figure 1 shows the images of water contact angle for the untreated and S_2 samples. The water contact angle was about 138° for the untreated P/C blended fabric. After corona discharge with 300 W input power, the water contact angle increased to 167° that showed a super-hydrophobic P/C blended fabric.

There is a noticeable change in the surface morphological properties of the corona discharge treated fabrics that are depicted by FESEM in Fig. 2. A relatively smooth surface was observed for the blank fabric. After the 300 W-10 min corona discharge process, a contaminant layer of the P/C surface degraded and several new nano-sized voids and cavities with an average diameter of 20 nm were created. Therefore, the roughness of the fabric surface was clearly increased after corona discharge treatment.

The ATR-FTIR spectroscopy was performed on the blank and corona discharge treated P/C blended fabrics with an input power of 300 and 500 W for the exposure time of 10 min (Fig. 3). The polyester characteristic peaks including C = O stretching vibration in the ester group, asymmetric -C-O stretching vibration of the ester group, and aromatic C-H out of plane vibration, are observed at 1711, 1240, and 722 cm^{-1} spectra, respectively. It is found that the ATR-FTIR spectra of S_0 and S_2 spectra indicate a clear difference with S_5 sample at 2322–2360 cm^{-1} that is attributed to the creation of -NH groups due to the presence of nitrogen gas in the air. There are several peaks at 3500–4000 cm^{-1} that are attributed to O-H and N-H stretching groups. The nitrogen-containing and oxygen-containing groups of the P/C fabric increased for S_5 sample due to modifying the surface by reactive species of corona discharge process.

According to the above-explanation, the functional groups minimized for S_2 sample, that can be attributed to the creation of a cross-linked layer on the P/C fabric that forms a stiffness layer on the surface of the super-hydrophobic fabric that impedes the reconstruction of the outermost layer. Consequently, the cross-linking can lead to creating a durable super-hydrophobic fabric. Also, the morphological properties showed an increase in the roughness of the S_2 sample that lead to increase the contact surface of the sample. According to Wenzel's model, the enhancement of contact surface of a hydrophobic surface leads to increase the hydrophobicity properties.

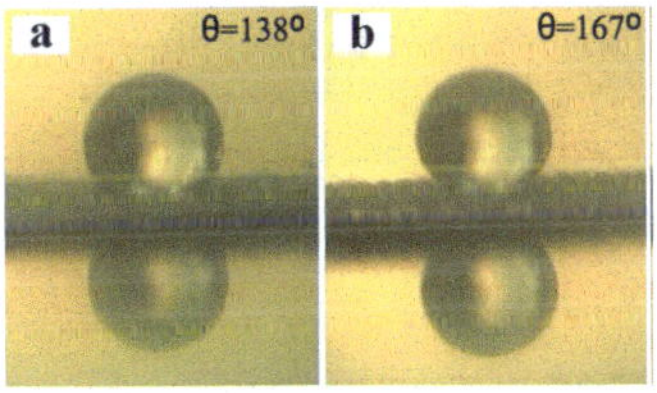

Fig. 1. Static contact angle of: (a) blank fabric and (b) S_2 sample

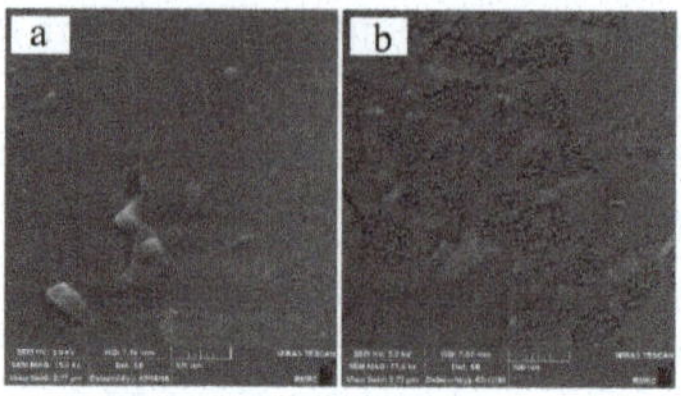

Fig. 2. FESEM images of: (a) blank fabric and (b) S_2 sample

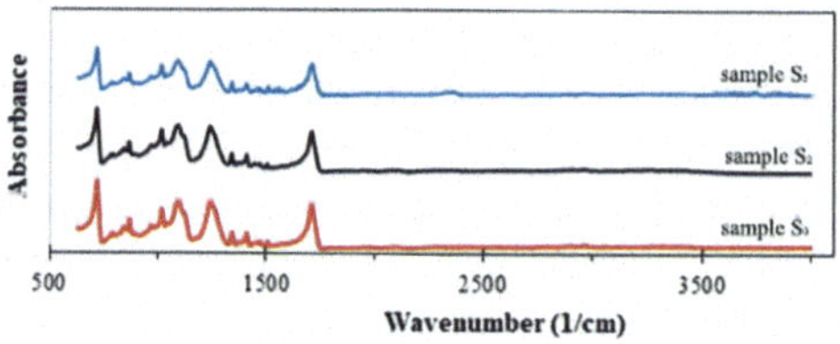

Fig. 3. ATR-FTIR analysis of S_0, S_2, and S_5 samples

4 Conclusion

The wettability test showed no water adsorption for the 300 W and 10 min corona treated fabric and the contact angle confirmed the contact angle of 167° for the S_2 sample. The morphological characteristic showed an increase in roughness with a nano-scale pattern for cavities and pores. Also, the chemical characteristic showed a decrease in oxygen and nitrogen-containing groups of the S_2 sample.

References

1. Tajima, S., Komvopoulos, K.: Appl. Phys. Lett. **89**, 124102 (2006)
2. Liguori, A., Bigi, A., Colombo, V., et al.: Sci. Rep. **6**, 1–11 (2016)
3. Nithya, E., Radhai, R., Rajendran, R., et al.: Carbohyd. Polym. **83**, 1652–1658 (2011)
4. Pandiyaraj, K.N., Selvarajan, V.: J. Mater. Process. Technol. **199**, 130–139 (2008)
5. Caschera, D., Cortese, B., Mezzi, A., et al.: Langmuir **29**, 2775–2783 (2013)

Nano-Silica Extracted from Rice Husk Ash: Effect on the Thermal Properties of Hydroxyethyl Cellulose

Niloofar Bakhtiari-Nasr and Ahmad Reza Bahramian(✉)

Polymer Engineering Department, Faculty of Chemical Engineering, Tarbiat Modares University, P.O. Box 14115-114, Tehran, Iran
abahramian@modares.ac.ir

Abstract. Thermal resistance and thermal stability of cellulose film are of particular importance in different applications. In this work, the effect of nanosilica extracted from rice husk ash on the thermal properties of hydroxyethyl cellulose has been investigated using thermal analysis test (Differential Scanning Calorimeter, DSC). Morphology and nanosilica dispersion and distribution were investigated using Scanning Electron Microscopy (SEM). The results of this work show that by increasing the amount of nanosilica from 0 wt% to 8 wt% to the hydroxyethyl cellulose, the heat of degradation increases to about 127%. The thermal degradation temperature also increases by about 33 K. Also, the burning time increased from 84 s to 300 s.

Keywords: Hydroxyethyl cellulose · Nanosilica · Thermal properties

1 Introduction

The thermal stability of cellulose nanofibrils (CNFs) can be improved by converting cellulose crystalline structure to cellulose II using an alkaline treatment method [1].

Improvement in the thermal stability of wood-based cellulose, needle-leaf, bleached, krafp pulp (NBKP) and the wood cellulose nanofibers (WCNF) obtained from the NBKP, was achieved by esterification [2].

Zinc ion as the only flame retardant of cellulose fibers was successfully grafted onto cellulose fibers [3].

The current study presents about the effect of cellulose nanofibers (CNFs) filler on the thermal and dynamic mechanical analysis (DMA) of epoxy composites as a function of temperature [4].

In this work, silica nanoparticles were used to improve thermal properties. This silica was achieved from waste of natural rice without any surface chemical compatibilization.

H. Mirzadeh and A. A. Katbab (Eds.): ISPST 2018, *Eco-friendly and Smart Polymer Systems*, pp. 149–152, 2020.
https://doi.org/10.1007/978-3-030-45085-4_36

2 Experimental

2.1 Materials

Nanosilica was obtained by pyrolysis and chemical extraction from rice husk in Tarbiat Modares University Lab. Hydroxyethyl cellulose is from Merck of Germany Company Catalog No. 8220680. Also dimethyl sulfoxide solvent was purchased from Merck Catalog No. of 102952.

2.2 Sample Preparation

Films of samples were prepared using solution casting of hydroxyethyl cellulose in dimethyl sulfoxide solvent. Nanosilica with weight percentages of 0, 4, and 8 wt% were used in the nanocomposite films preparation. Thickness of all three films was measured about 0.06 μm.

2.3 Characterization

Dynamic differential scanning calorimetry (DSC, Netzch Co. Germany) was performed to analyze the thermal properties of samples. All of the measurements were done at a heating rate of 10 °C/min in air atmosphere.

The morphology of the nanocomposite samples were investigated using a scanning electron microscope (SEM, xl30, Philips, Japan).

3 Results and Discussion

Onset, peak, endset and heat of degradation were measured (Table 1) from thermal analysis test. The results show an increase of 127% in heat of degradation and 33 K's degree on thermal degradation temperature by adding 8 wt% of nanosilica.

Table 1. Thermal specifications of samples

Samples	N.0	N.1	N.2
Nanosilica (wt%)	0.0	4.0	8.0
Onset of degradation (K)	469	489	502.4
Peak of degradation (K)	471.6	493.3	503.4
Endset of degradation (K)	483.3	522.7	551.9
Heat of degradation (J/g)	43.63	69.40	99.18

By increasing the weight percentage of nano silica, the degree of thermal degradation slope has decreased, which means increasing thermal resistance (Fig. 1). In the higher weight percentages of the nanosilica, the degradation peak is broader, which indicating a lower degradation rate than compared to the previous ones (Fig. 2).

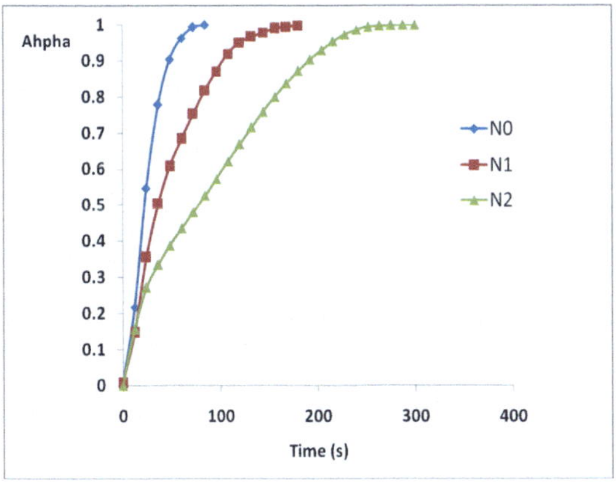

Fig. 1. Degree of thermal degradation of samples

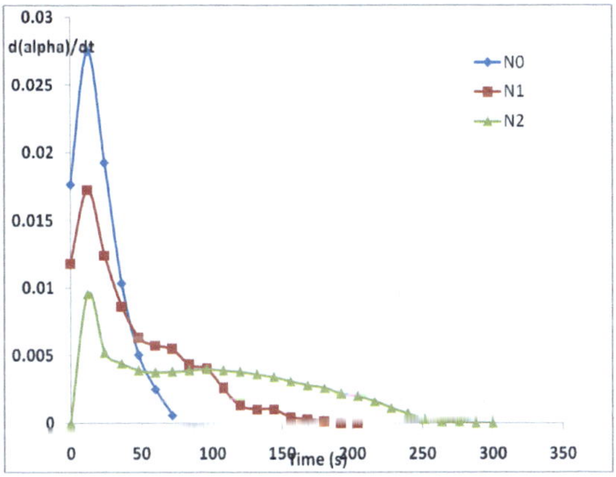

Fig. 2. Rate of thermal degradation samples

Figure 3a shows the nanosilica particles of less than 100 nm in size. The morphological investigations show a good dispersion and distribution of nanosilica in film of samples (Figs. 3b, 3c, and 3d).

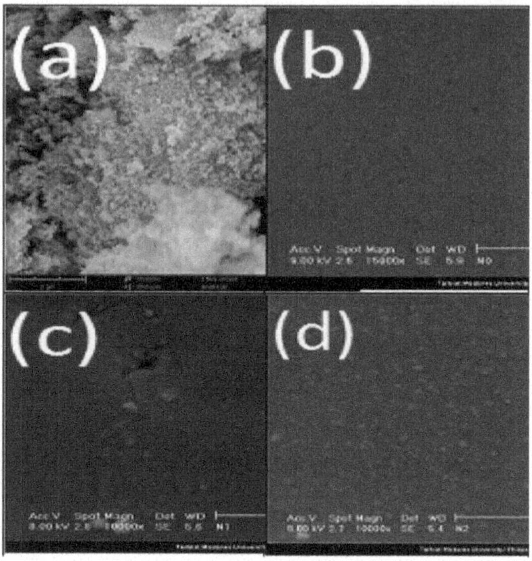

Fig. 3. SEM images of: (a) Nanosilica, (b) N.0, (c) N.1, and (d) N.2 sample

References

1. Lee, H., Sundaram, J., Zhu, L., Zhao, Y., Mani, S.: Carbohyd. Polym. **181**, 506–513 (2018)
2. Agustin, M.B., Nakatsubo, F., Yano, H.: Carbohyd. Polym. **192**, 28–36 (2018)
3. Zhang, K., Zong, L., Tan, Y., Ji, Q., Yun, W., Shi, R., Xia, Y.: Carbohyd. Polym. **136**, 121–127 (2016)
4. Saba, N., Safwan, A., Sanyang, M.L., Mohammad, F., Pervaiz, M.: Int. J. Biol. Macromol. **102**, 822–828 (2017)

Spectroscopic Study of Waterborne Polyurethane/Acrylate Hybrids Based on Acrylate Monomers Contents

Ghazal Minoofar, Hengameh Honarkar(✉), and Mehdi Barikani

Polyurethane and Advanced Polymers Department, Faculty of Science,
Iran Polymer and Petrochemical Institute (IPPI),
P.O. Box 14975-112, Tehran, Iran
h.honarkar@ippi.ac.ir

Abstract. A series of waterborne polyurethane/acrylate hybrids were synthesized using different values of acrylic monomers (butyl acrylate, methyl methacrylate, and acrylic acid). The samples were synthesized via prepolymer mixing method and followed by radical polymerization. The crucial parameters were temperature, viscosity, and rate of stirring for obtaining the stable dispersions. Here, the IR spectroscopy of waterborne polyurethane and synthesized waterborne polyurethane/acrylate hybrids is investigated.

Keywords: Spectroscopic · Waterborne polyurethane · Acrylate · Hybrid

1 Introduction

Waterborne polyurethane (WPU) has been developed and widely used because of its low toxicity, odorless, safe to use and being less harmful environmentally. However, WPU has some weaknesses such as poor water resistance, limited outdoor durability and low pH stability [1]. To obtain better cost/performance benefits, polyurethane has been modified with different kinds of polymers. Among them acrylates are the most favored because of their unique properties. Acrylate resins show excellent properties such as hardness, water and weather resistance, good pigment ability, low cost, transparency and gloss. However, they show poor toughness, abrasion resistance and adhesion. The combination of PU and acrylates leads to an improved hybrid system that includes the advantages of both components [2, 3].

This paper studies the synthesis method of waterborne polyurethane/acrylate hybrids based on different ratios of acrylic monomers. Also, the IR spectroscopic investigation is considered.

2 Experimental

2.1 Preparation of WPU

A 250 mL round-bottom, four-necked flask equipped with a mechanical stirrer and a reflux condenser, thermometer, a dropping funnel, and nitrogen inlet was used as a

H. Mirzadeh and A. A. Katbab (Eds.): ISPST 2018, *Eco-friendly and Smart Polymer Systems*, pp. 153–156, 2020.
https://doi.org/10.1007/978-3-030-45085-4_37

reactor. The reaction was performed in a constant temperature oil bath. At first, the dried PCL was charged into the dried flask. While stirring, the system was heated up to 80 °C. Then, $H_{12}MDI$ and a little of the catalyst (two droplets) were added slowly into the reactor with a dropper funnel. NCO-terminated prepolymer was prepared, after heating mixture for 4 h at 80 °C. After that, the mixture was cooled to 60 °C then dimethylol propionic acid (DMPA) and 1,4-butane diol (BDO) were added to system. After 45 min, Hydroxyethyl methacrylate (HEMA) was added dropwise to flask and reaction was continued for 2 h. The reactor was cooled to 35 °C and neutralized with triethylamine (TEA). After 30 min, water was added drop by drop into the reactor at stirring rate of 900 rpm for 30 min. It must be mentioned that the solid content is 30%.

2.2 Preparation of WPU/AC Hybrids

Proper amounts of WPU, SDS and KPS as emulsifier and initiator respectively, which dissolved in deionized water, were charged into 250 mL three-necked balloon with a flux condenser, N_2 purge and a dropping funnel, in an oil bath with 70 °C. After 15 min according to Table 1 different weight ratios of acrylate monomers were added to reaction dropwise in 30 min then the system heated up to 80 °C and after 3 h the hybrids of WPU/acrylates were obtained. It is important to point that, changes in the amounts of acrylate monomers were performed on 90/10 weight ratio of WPU/AC.

2.3 Preparation of Films

Films were prepared by casting the aqueous dispersions onto a Teflon plate at room temperature.

After 3 days, the residual water was removed by placing the films into a vacuum oven at 80 °C for 3 h.

3 Results and Discussion

The FT-IR spectra of the pure WPU and WPUA hybrids have been shown in Fig. 1. The characteristic absorption peaks in WPU spectrum are: 2927 and 2861 cm^{-1} which are due to alkane CH asymmetric and symmetric stretching vibration of CH_2 groups. Absorption at 1529 cm^{-1} is due to CHN vibration of associated secondary urethane groups (-NH bending). Also, the related peak of C-N stretching appeared at 1456 cm^{-1}. The C-O-C stretching absorption band corresponds to the ether oxygen of soft segment at 1228 cm^{-1}. The FT-IR spectrum of the mentioned polyurethane shows the characteristic bands of urethane NH groups at 3356 cm^{-1} (NH stretching) and 1725 cm^{-1} due to carbonyl (C = O) stretching urethane and ester group. This is a strong reason for PU formation. The spectral analysis is mainly used to check the completion of polymerization reaction in waterborne polyurethane, in terms of disappearance of –NCO band at 2270 cm^{-1} indicating the NCO groups have reacted completely during the chain extension step. In the hybrid samples, it can be seen that the band at about 1735 cm^{-1} is assigned to C = O stretching vibration of AC monomers, and the hydrogen –bonded C = O groups at 1725 cm^{-1} has been disappeared in the hybrid latexes. It indicates that the intra and intermolecular hydrogen bonds of the PU have been destroyed. It is due to grafting of the acrylic monomers onto PU chains and entanglements between them leading to good miscibility of components of mixture.

Also, the FT-IR spectra of hybrids show that N-H peak (3250–3400 cm^{-1}) in hybrids moves to the left into more upper wavenumbers. This conclusion approves the grafting too. Also, the disappearance of band at 1638 cm^{-1} to -C=C– confirms the copolymerization between WPU and AC components in WPUA latexes [4].

Table 1. Weight ratio of acrylate monomers

Sample	BA (wt%)	MMA (wt%)	AA (wt%)
1	100	0	0
2	50%	50%	0
3	50%	45%	5%

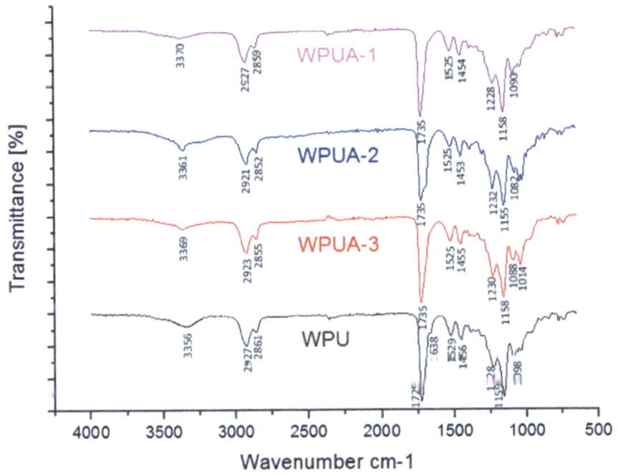

Fig. 1. FT-IR spectra of waterborne polyurethane and hybrids with 90 wt% of different acrylic monomers

4 Conclusion

In this research, waterborne polyurethane acrylate hybrids were synthesized based on various amounts of different acrylate monomers. Micro structure of them were investigated by Fourier transform infrared spectroscopy. According to the analysis, the absorbing intensity of characteristic peaks, especially the carbonyl peak at about 1735 cm^{-1} and N-H peak of films obtained from hybrids, is higher than that of film from WPU dispersion. This is due to grafting of acrylate monomers onto PU chains and compatibility between them.

The results indicate that the WPU/BA (WPUA-1) hybrid have better film structure than WPU/BA/MMA, WPU/BA/MMA/AA hybrids and pure WPU.

This result indicates that there is higher compatibility between the AC and the WPU components in the WPU/BA hybrid.

References

1. Ch, H.-T., Ya, H.-M., Li, C.-S., Hsiao, H.-Y.: Polym. Plast. Technol. Eng. **51**, 945–953 (2012)
2. Zh, G., Lu, M., Rui, X., Sun, Y.: J. Appl. Polym. Sci. **135**, 45783 (2017)
3. Al, G.A., Fu, M., Or, V.H., Gi, L.F., Martín-M, J.M.: Prog. Org. Coat. **118**, 19–30 (2018)
4. Xu, H., Yang, D., Guo, Q., Wang, Y., Wu, W., Qiu, F.: Polym. Plast. Technol. Eng. **51**, 50–57 (2012)

Enhancing Corrosion Resistance of Alkyd Coatings Using Cerium Nitrate Hexahydrate Treated Ceria Nanoparticles

Sepideh Parchami, Seyed Mojtaba Mirabedini(✉), Shahla Pazokifard, and Morteza Behzadnasab

Color and Surface Coatings Department,
Iran Polymer and Petrochemical Institute (IPPI),
P.O. Box 14975-112, Tehran, Iran
sm.mirabedini@ippi.ac.ir

Abstract. Cerium-based compounds have been considered as effective inhibitors in corrosion resistance coatings for many years. In this study, corrosion inhibition behavior of alkyd coatings containing either cerium nitrate or cerium oxide was evaluated. In order to use cerium nitrate in the coatings, cerium oxide nanoparticles were first treated with cerium nitrate hexahydrate. Then treated and untreated nanoparticles were loaded into the alkyd coating. Surface treatment of nanoparticles was studied using FTIR spectroscopy. Dispersion of nanoparticles in polymeric matrix was assessed via EDAX analysis. Corrosion performance of the coatings applied on steel substrates was evaluated using EIS technique. The results showed that presence of cerium nitrate on cerium oxide nanoparticles leads to improve corrosion resistance of alkyd coatings.

Keywords: Corrosion · Cerium oxide · Cerium nitrate · Surface treatment · Nanoparticles

1 Introduction

Corrosion prevention by organic coatings is well known in surface coatings industries. Inhibitors such as hexavalent chromium compounds as an ingredient in a protective coating can prevent or delay the corrosion process. Due to the carcinogenicity, chromates are phased out as corrosion inhibitors in organic coatings and replaced by less toxic alternatives. Cerium-based compounds are appropriate corrosion inhibitors in an aqueous environment. However, their effectivity as a corrosion inhibitor in an organic coating depends on their interaction with the coating matrix [1, 2].

Cerium compounds have different anti-corrosion abilities in various oxidation states. Cerium (III) compounds such as cerium (III) nitrate hexahydrate ($Ce(NO_3)_3.6H_2O$), named CNH, tend to oxide in presence of OH^- ions to form cerium oxide (CeO_2) and by doing so delay the corrosion of underneath surface via providing a physical barrier. To enhance the availability of Ce^{3+} ions throughout the coating while minimizing the amount of the needed inhibitor, CeO_2 nanoparticles were treated by CNH to act as a nano-sized reservoir and loaded into an alkyd-based coating. Corrosion resistance of the coatings containing untreated (CeO_2) and treated nanoparticles ($TCeO_2$) was compared.

H. Mirzadeh and A. A. Katbab (Eds.): ISPST 2018, *Eco-friendly and Smart Polymer Systems*, pp. 157–161, 2020.
https://doi.org/10.1007/978-3-030-45085-4_38

2 Experimental

2.1 Materials

CeO_2 nanoparticles with an average particle size of 15–30 nm were supplied by Nanostructured & Amorphous Materials Inc., cerium (III) nitrate hexahydrate was purchased from Sigma-Aldrich, and long oil alkyd resin (B1260) was provided by Ronass Chemical Industries.

2.2 Treatment of CeO_2 Nanoparticles

Treatment of CeO_2 nanoparticles was carried out according to the following procedure: CeO_2 particles (0.1 g) were sonicated in distilled water (40 mL) for 30 min, and then CNH (0.05 g) was added to the suspension and sonicated for additional 1 h. Solid particles were then centrifuged, washed with distilled water, and dried using a freeze dryer at −40 °C for 14 h. FTIR spectroscopy was performed in the range of 400–4000 cm^{-1} to investigate the surface treatment of cerium oxide nanoparticles. Untreated and treated nanoparticles were dispersed (3.5 wt%) into the alkyd resin under mechanical agitation for 1 h. The coating samples were then applied with an average wet film thickness of 60 ± 5 μm on degreased mild steel plates and left to dry at the ambient temperature for about two weeks. In order to characterize the nanoparticles distribution into alkyd matrix, the EDXA analysis was carried out. EIS test was performed by Ivium Compactstat in a conventional three-electrode cell arrangement consisting platinum electrode (auxiliary electrode), saturated calomel electrode (reference electrode) and the coated samples (working electrode). The selected signal amplitude was ±5 mV, 10^5–10^{-2} Hz frequency range and about 1 cm^2 testing area were employed. It should be noted that all measurements were performed in a 3.5 wt% NaCl electrolyte solution at ambient temperature.

3 Results and Discussion

3.1 FTIR Spectroscopy

FTIR spectra of CeO_2 and $TCeO_2$ nanoparticles in the range of 400–4000 cm^{-1} are illustrated in Fig. 1. The absorption peaks observed at 1041, 1350, and 1465 cm^{-1} confirm successful treatment of ceria nanoparticles with cerium III ions [3].

3.2 Energy Dispersive X-Ray Analyses

The EDXA-mapping images of CeO_2 and $TCeO_2$ nanoparticles dispersion depicted in Fig. 2, show proper dispersion of nanoparticles in alkyd matrix.

3.3 Electrochemical Impedance Spectroscopy

Figure 3 shows the Nyquist curves for neat alkyd coating and the coating containing untreated and treated CeO_2 nanoparticle. Corrosion performance of the alkyd coating containing untreated nanoparticles decreases sharply with the time, while the samples containing treated cerium oxide nanoparticles resists against corrosion. As it is clear, after 1000 h immersion the coatings show the same corrosion performance as the neat alkyd coating before immersion. In the sample containing treated ceria nanoparticle however, not only up to 4 times greater resistance against the samples containing nanoceria was observed, even a change in protection progress from decreasing to increasing of protection was detected. Cerium oxide nanoparticles can act as a barrier against corrosive media and decrease the corrosion process. By simultaneous sonication of Ce^{+} ions and ceria nanoparticles, ceria nanoparticles can act as a nano-sized reservoir for Ce^{3+} ions [4] which act as corrosion inhibitor with the following reactions [5].

$$Ce(NO_3)_3.6H_2O + 3NaOH \rightarrow Ce(OH)_3.6H_2O + 3NaNO_3 \quad (1)$$

$$4Ce(OH)_3.6H_2O + 2O_2 \rightarrow 4CeO_2 + 4H_3O^+ + O^{2-} \quad (2)$$

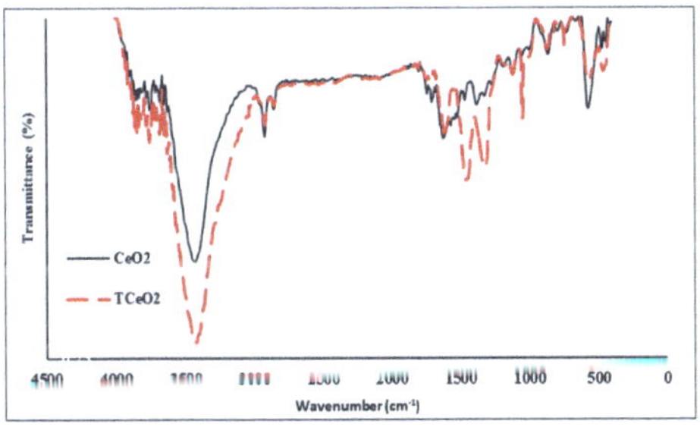

Fig. 1. FTIR spectra of: (a) CeO_2 and (b) $TCeO_2$ nanoparticle

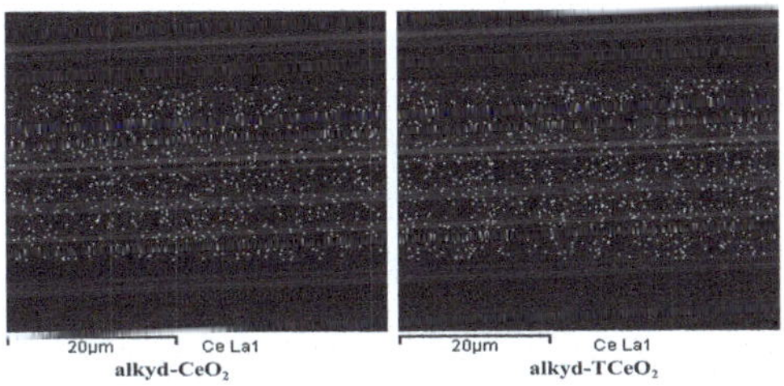

Fig. 2. Ce-mapping through EDAX in: (a) CeO_2 and (b) neat and $TCeO_2$ nanoparticle in alkyd matrix

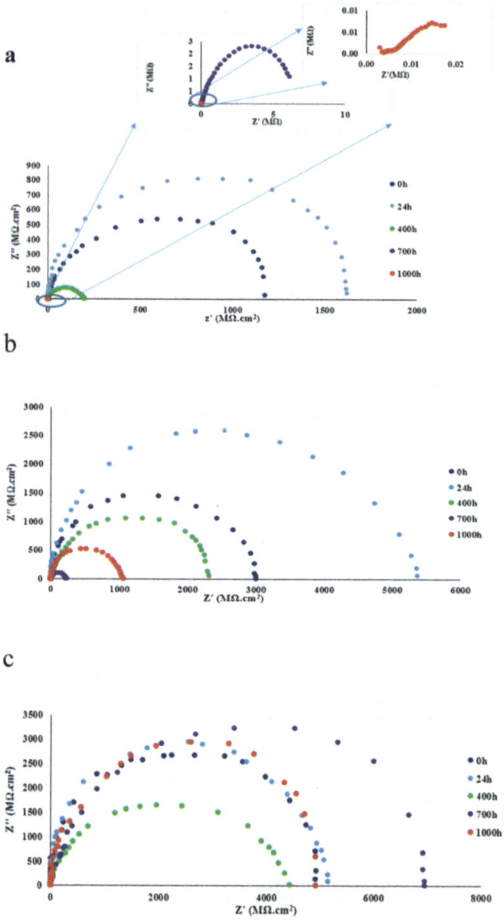

Fig. 3. Nyquist plots of: (a) neat alkyd, (b) CeO_2 embedded alkyd, and (C) alkyd containinig with TCO_2 nanoparticles

4 Conclusion

The possibility of the presence of Ce (III) ions on cerium oxide nanoparticles via simultaneous sonification was confirmed. EIS test showed that Ce (III) nitrate hexahydrate treated cerium oxide nanoparticles improved corrosion performance of an alkyd coating compared to pristine cerium oxide nanoparticles. This can be related to interfering corrosion reactions caused by the reaction of Cerium (III) ions and OH^- ions.

References

1. Jing, L., Luiz, E., Anwar, A., Jinshan, F.: J. Electr. Soc. **162**, C610–C618 (2015)
2. Ecco, L.G., Fedel, M., Ahniyaz, A.: Prog. Org. Coat. **77**, 2031–2038 (2014)
3. U Vicentini, G.: J. Inorg. Nucl. Chem. 669–672 (1972)
4. Montemor, M., Pinto, R.M.: Electrochim. Acta **54**, 5179–5189 (2009)
5. Pinjari, D.V., Pandit, A.B.: Ultrason. Sonochem. **18**, 1118–1123 (2011)

Surface Modification of BOPP Film Using a Gliding Arc Plasma and Low Pressure Atmospheric Plasma

Fahime Darvish[1] and Mohammadreza Khani[2(✉)]

[1] Department of Physics, Iran University of Science and Technology, Tehran, Iran
[2] Laser and Plasma Research Institute, University of Shahid Beheshti, Tehran, Iran
m_khani@sbu.ac.ir

Abstract. Increasing the surface energy of polymer is one of the most essential needs of industries such as battery industry, printing, textiles, etc. In this article, we present a study into the mechanisms underpinning surface energy increase of BOPP and hydrophilic using gliding arc plasma (GAP) and low-pressure atmospheric plasma (LPP). Measurements of contact angle indicate an increase in surface energy to 63 mNm^{-1}. The effect of the plasma was studied at different durations of treatment 10–120 s. The OES analysis confirms the presence of air molecules in the plasma. Atomic force microscopy (AFM) showed that plasma etching occurs in stages and may be related to the reach of the species. To analyze the chemistry on the surface, attenuated total reflectance–Fourier transforms infrared spectroscopy (ATR-FTIR) was used. Plasma is an eco-friendly process because it does not produce any pollution and is safe.

Keywords: Surface energy · BOPP film · Surface treatment · Gliding arc plasma · Contact angle

1 Introduction

Plasma surface treatments eliminate many safety risks associated with worker exposure to dangerous chemicals. The EPA has classified most plasma processes as eco-friendly. They are commonly referred to as Green Tech Plasma generally have a better consistency, and they are more repeatable than traditional chemical and mechanical processes. LPP processing is contained within a vacuum chamber, with little or no direct worker exposure. Plasma processing operates at near-ambient temperatures with no risk of heat exposure [1]. BOPP film requires surface modification due to low surface energy. In this context, plasma sources represent a fast and environmentally friendly method to modify polymeric surfaces without affecting bulk properties.

H. Mirzadeh and A. A. Katbab (Eds.): ISPST 2018, *Eco-friendly and Smart Polymer Systems*, pp. 162–165, 2020.
https://doi.org/10.1007/978-3-030-45085-4_39

2 Experimental

The surface energy of polypropylene was investigated by two plasma reactors at different times. In the APG, the nozzle gap to a polymeric surface of 15 mm, a power supply of 300 W and a voltage of 15 kV is considered. In the LPP, when the power is about 25 W, and its pressure is 230 mTorr, the plasma of air is formed. Before performing the measurements, the samples were subjected to an ultrasonic cleaning process in distilled water and acetone for 2 min in 35 °C and were further dried using air. Then, with 3 μL of distilled water and diiodomethane, contact angle test was performed. Data were collected one hour after the plasma irradiation and analyzed by using ImageJs software received their contact angle (Fig. 1). Finally, the geometric mean method of surface energy was calculated [2]. The analysis of AFM, ATR-FTIR and OES was performed by Easysacn2, Tensor27, and Avantes ULS364 devices, respectively. Surface roughness also decreased by more than 50%. In the analysis of FTIR, carboxylic groups and amino acids were observed on the surface.

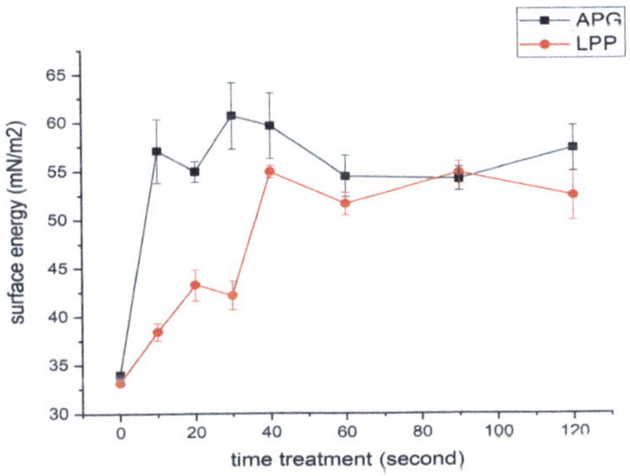

Fig. 1. Contact angle of BOPP on exposure to air plasma

3 Results and Discussion

The analysis of FTIR spectra showed that the carboxyl, alkene and carbonyl groups were formed on the polymer surface.

The spectrum of BOPP film (Fig. 2) showed that the intensity of the peaks at 840.9164 (C-H), 1379.028 (N-O), 2968.28 (O-H), 1458.103 (P—C) was increased. In these specimens, due to increased etching, the thickness of the surface is reduced. According to Eq. (1), that is the Beer-Lambert Law, at a constant concentration, adsorption decreases with decreasing thickness. In Beer-Lambert Law, A is the measured absorbance, a is a wavelength-dependent absorptivity coefficient, b is the path length, and c is the sample concentration.

$$A = a(\lambda) * b * c \tag{1}$$

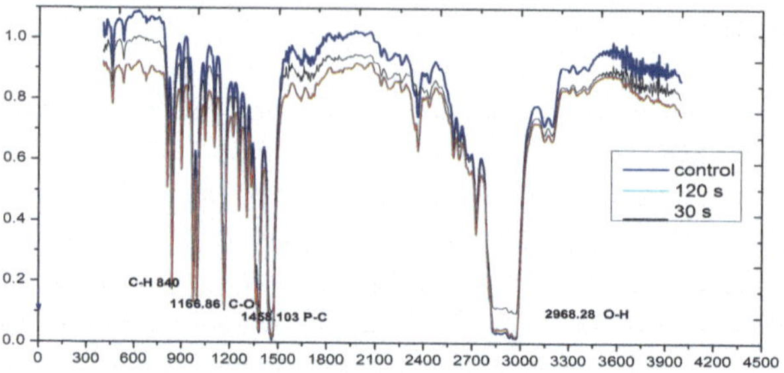

Fig. 2. FTIR spectra of BOPP exposed for different times

In Fig. 3, the results of the AFM analysis show a greater decrease in roughness in the LPP reactor than the APG reactor, and the APG surface energy is greater than the LPP, which is caused by surface roughness. The roughness in APG and APG is respectively 10.719 and 6.671 nm after 120 s of processing. The roughness of the control sample is 15.719 nm. A more probable explanation may be that polymer layer degradation had occurred during plasma exposure (Fig. 4). Active species in the OES analysis indicate that there are oxygen species and nitrogen species that will modify the surface of the film [4].

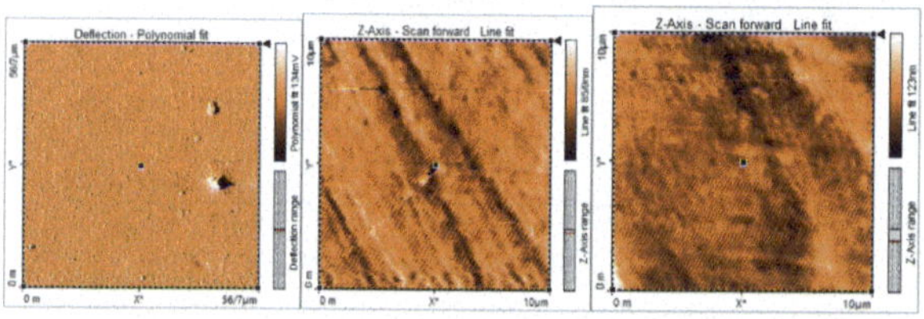

Fig. 3. Left to right roughness of control sample, APG and LPP, respectively

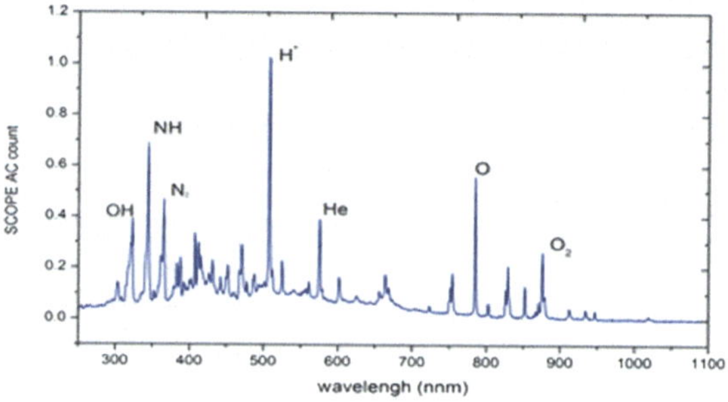

Fig. 4. Plasma spectrum of the APG. The species in the air are evident

4 Conclusion

In this study, the effect of plasma on the film was demonstrated with the aim of hydrophilicity with increasing surface energy. Plasma etching reduces the thickness of the surface and reduces the roughness and thickness. APG reactor has a higher efficiency ratio than LPP. OES analysis showed that there are groups O-H and N-O in the plasma. These active species of plasma, by altering the surface chemical structure, improve the surface properties without changing the polymer bulk structure.

References

1. Lock, E.H., Walton, S.G., et al.: Langmuir **26**(11), 8857–8868 (2010)
2. Hejda, F., Solar, P., et al.: Proceedings of WDS 2010, Part III, pp. 25–30 (2010)
3. Khosravi, Z., Kotula, S., et al.: Plasma Proc. Polym. **15**(1), 1700066 (2018)
4. Sato, T., Akiyama, H., et al.: Surf. Sci. **677**, 93–98 (2018)

Effect of Low-Functionalized Graphene Oxide on the Rheological and Electrical Properties of Water-Based Epoxy Coatings

Amir Rezvani Moghaddam[1], Milad Kamkar[2], Zahra Ranjbar[3,4](✉), Uttandaraman Sundararaj[2], and Ali Jannesari[3]

[1] Department of Polymer Engineering, Sahand University of Technology, Tabriz, Iran
[2] Department of Chemical and Petroleum Engineering, University of Calgary, Calgary, Canada
[3] Department of Surface Coatings and Novel Technologies, Institute for Color Science and Technology, Tehran, Iran
ranjbar@icrc.ac.ir
[4] Center of Excellence for Color Science and Technology, Tehran, Iran

Abstract. In the present work, rheological and electrical properties of epoxy-based nanocomposite coatings containing low-functionalized graphene nanosheets are studied. The low-functionalized, highly concentrated graphene nano-sheets in water were synthesized using a non-dispersion mechanism, which made them to be stable in water. The rheological investigations show a transition from a viscous response to a solid-like response at low frequency region, which indicates formation of nanofillers network structure in these samples at low filler contents. In addition, electrical conductivity as high as 1.34×10^{-1} S/m is achieved for the epoxy based nanocomposite coatings, which indicates a good electrical network formation through the epoxy matrix.

Keywords: Conductive coatings · Rheology · Network formation · Graphene · Epoxy nanocomposites

1 Introduction

High-performance functional polymer nano composite materials have attracted increasing attention in many industrial and high technology fields, e.g. defense, marine, aerospace, automotive, and etc. [1]. Meanwhile, electrically conductive polymer-based composites with high conductivity and lightweight are in great need for specific applications such as anti-static, electrically conductive and electromagnetic interference shielding [2]. In this regard, thermosetting polymeric coatings consisting of carbonaceous nanofillers are mainly used due to their physical and chemical features since unique properties are required such as good chemical and corrosion resistance [3]. In this work, we report the rheological and electrical properties of a novel epoxy-based coating containing low functionalized water-dispersible graphene sheets.

H. Mirzadeh and A. A. Katbab (Eds.): ISPST 2018, *Eco-friendly and Smart Polymer Systems*, pp. 166–169, 2020.
https://doi.org/10.1007/978-3-030-45085-4_40

2 Experimental

Low-functionalized, highly concentrated graphene nano-sheets in water were synthesized using non-dispersion mechanism and were added to a mixture of water based epoxy and curing agent at different graphene nano-sheets loadings ranging from 0.2 to 3.0 wt.%. The dried thickness of the coatings were adjusted to be 60 μm. Electrical properties of the coatings were characterized using a 4-point probe electrical conductivity meter.

The rheological behavior was evaluated using a rheometer (MCR 302, Anton Paar) equipped with a 25 mm cone and plate (cone angle of 2° and a truncation of 47 μm) at 25 °C. Strain sweep dynamic tests, in which the sample was subjected to increasing oscillatory strain amplitudes at controlled frequency, were used to determine the limit of linear viscoelasticity. Based on these results, strain amplitude of 1% was low enough for the epoxy/graphene nanocomposites to be in linear viscoelastic region. Thus, viscoelastic behavior of samples was studied using frequency-sweep test in linear framework (at strain amplitude of 1%). In this work the frequency range was from 0.1 Hz to 100 Hz.

3 Results and Discussion

The results of small amplitude ($\gamma = 1\%$) frequency-sweep dynamic tests on the graphene based polymer nanocomposites are shown in Fig. 1. As it can be seen, elastic modulus G' is greater than viscous modulus G'' at low frequencies and a distinct nonterminal frequency independent behavior is observable for epoxy nanocomposites containing graphene greater than 1.5 wt%. This is a sign of a transition from a viscous response to a solid-like response at low frequency region (larger relaxation times). This solid-like behavior for epoxy/graphene nanocomposites shows formation of nanofillers network structure in these samples. The higher the graphene content in the nanocomposites, the more area of nanofillers-polymer interfaces and higher the number of load bearing junctions [4] which restrain long-range molecular chain motions and reduce the mobility of the epoxy chains. This results in increase of G' in the low frequency region with increase in concentration of graphene nanosheets.

To obtain further insight into the network structure of the samples, electrical conductivity of the samples was studied. In agreement with rheological data, electrical conductivity of 1.34×10^{-1} S/m was achieved for the epoxy films with 1.5 wt% graphene nano-sheets. This also demonstrates formation of percolated electrical network of well dispersed graphene nanosheets in epoxy matrix.

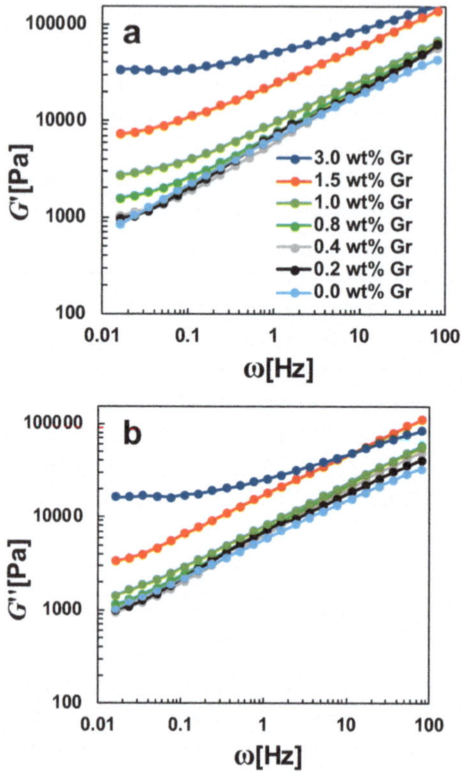

Fig. 1. (a) Storage modulus (G') and (b) loss modulus (G'') of epoxy/graphene nanocomposites containing various loadings of graphene nano-sheets for strain amplitudes of 1% using cone-plate geometry at 25 °C.

4 Conclusion

Based on the results of this study, we came to the conclusion that epoxy/graphene nanocomposites can form 3D percolated network. This was verified by apparent solid-like response observed at long time scales in the rheological response of the samples. Furthermore, high electrical conductivity was achieved at a low loading level of 1.5 wt % graphene nano-sheets, demonstrating a good electrical network formation by dispersed graphene nano-sheets in epoxy resin.

References

1. Mohan, V.B., Lau, K., Hui, D., Bhattacharyya, D.: Comp. Part B Eng. **142**, 200–220 (2018)
2. Rezvani Moghaddam, A., Ranjbar, Z., Sundararaj, U., Jannesari, A.,Kamkar, M.: A novel electrically conductive water borne epoxynanocomposite coating based on graphene: facile method and highefficient graphene dispersion. Prog. Org. Coat. **136** (2019). No. 105223
3. Montazeri, S., Ranjbar, Z., Rastegar, S.: Prog. Org. Coat. **111**, 248–257 (2017)
4. Hsiao, L.C., Newman, R.S., Glotzer, S.C., Solomon, M.J.: Proceedings of the National Academy of Sciences (2012)

Environmentally Friendly Hybrid Polysilsesquioxane-Poly(Butyl Acrylate) Nanocomposite: Film Properties

Ali Ahmadi Dehnoei and Somayeh Ghasemirad(✉)

Polymer Engineering Department, Faculty of Chemical Engineering, Tarbiat Modares University, P.O. Box 14115-143, Tehran, Iran
ghasemirad@modares.ac.ir

Abstract. Inorganic-organic core-shell polysilsesquioxane-poly(butyl acrylate) hybrid latex nanoparticles were synthesized via emulsion polymerization. FTIR spectroscopy and dynamic light scattering (DLS) confirmed core-shell structure of the resultant hybrid latex nanoparticles. Hybrid core-shell nanocomposite film underwent drying in a shorter period of time in comparison with the polysilsesquioxane latex, due to the lower Brownian diffusivity of its latex nanoparticles. It also revealed less hydrophilicity and higher resistance to water whitening, compared to the seed latex film.

Keywords: Polysilsesquioxane · Core-shell nanoparticles · Hybrid acrylic nanocomposite · Latex film drying · Water whitening

1 Introduction

Emulsion polymerization is an environmentally friendly alternative to the traditional solvent-based one [1]. Acrylate latexes prepared by the emulsion polymerization have many specific properties, such as good film-forming ability, transparency, gloss, and good cohesiveness. These properties could be synergistically improved by designing and synthesizing inorganic-organic hybrid materials, applicable in various areas such as coatings, adhesives, electronics, catalysis, diagnostics, etc. [2]. Nanosilica is used as a common inorganic component in the hybrid systems owing to its numerous resources and low price. However, in order to strengthen the interaction and promote the compatibility between silica and organic polymers, silica particles are usually modified or pretreated with a silane coupling agent [3]. To avoid the additional modification steps on silica particles in the synthesis of silica-polymers hybrid particles, polysilsesquioxanes compounds can be used [2]. Moreover, by tailoring various functional groups on their outer surface, these compounds could be compatible with most polymeric matrices. In the current study, polysilsesquioxane-poly(butyl acrylate) nanoparticles were synthesized using emulsion polymerization and its latex film formation, hydrophilicity and water whitening were compared to those of seed, polysilsesquioxane.

H. Mirzadeh and A. A. Katbab (Eds.): ISPST 2018, *Eco-friendly and Smart Polymer Systems*, pp. 170–173, 2020.
https://doi.org/10.1007/978-3-030-45085-4_41

2 Experimental

2.1 Materials

Butyl acrylate (BA) was purchased from Samchun Pure Chemical Co., Korea. 3-(Methacryloxypropyl) trimethoxysilane (MAPTS) was obtained from Sigma Aldrich. Triton X-100, sodium hydroxide (NaOH), sodium dodecyl sulfate (SDS), and potassium persulfate (KPS) were purchased from Merck Co. and used without further purification. The water used in the experiment was distilled.

2.2 Preparation of Latex Nanoparticles

Polymethacryloxypropylsilsesquioxane (PSQ) latex nanoparticles were prepared by hydrolysis and polycondensation of MAPTS in the presence of a mixture of ionic and nonionic surfactants, following the procedure proposed by Bai et al. [2]. Diluted PSQ emulsion with water, 1% KPS solution, and BA with the amount of 6.25 times PSQ were poured into a flask and emulsion polymerization was conducted at 80 °C under starved-feed conditions.

2.3 Characterization

Particle size of the diluted PSQ and core-shell PSQ-PBA latex particles were measured by dynamic light scattering (DLS) using Horiba SZ-100. FTIR analysis was carried on a Bruker Equinox 55 FTIR spectrometer. Contact angles (CA) with distilled water were measured at room temperature using the sessile drop method. Films of certain amounts of PSQ and PSQ-PBA latexes were dried separately at the same thermodynamic conditions for FTIR and CA measurements.

3 Results and Discussion

FTIR spectra of PSQ and PSQ-PBA dried films represented strong bands at 1014 and 1134 cm^{-1} corresponding to Si-O-Si, confirming synthesis of PSQ nanoparticles as a result of hydrolysis and polycondensation of MAPTS, Fig. 1. Weakening of C = C peak at 1636 cm^{-1} with respect to Si-O-Si peak was an evidence for reaction of BA on the surface of the PSQ seed. Z-average particles size of the PSQ latex nanoparticles and the core-shell PSQ-PBA hybrid latex nanoparticles were 180 and 250 nm, respectively. The difference in particle sizes proved the shell monomer polymerization onto the PSQ seed nanoparticles. The shorter drying duration of the core-shell PSQ-PBA compared to the PSQ film at the early stage of latex film formation, Fig. 2, could be attributed to the lower magnitude of its nanoparticles Brownian diffusivity, calculated from the DLS autocorrelation function data using Eq. (1) [4], Fig. 3.

$$G^2(\tau) = A\exp(-2D_m q^2 \tau) + B \quad (1)$$

Where, A, B, D_m, q, and τ are the amplitude of the correlation function, baseline, particle diffusion coefficient, magnitude of the scattering vector, and delay time, respectively. Moreover, water CA measurements revealed more hydrophilic nature of PSQ, 13°, than PSQ-PBA, 75°. PSQ-PBA dried film showed higher resistance to water whitening than PSQ, which could be used as a criterion for concentration and distribution of water in the film [5].

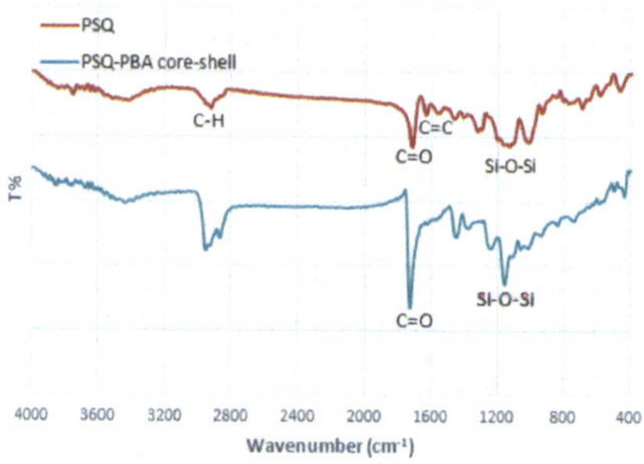

Fig. 1. FTIR spectra of PSQ and PSQ-PBA

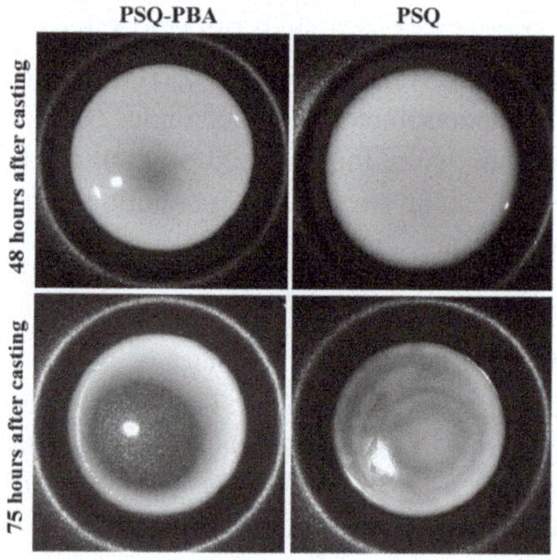

Fig. 2. PSQ-PBA and PSQ latexes film drying

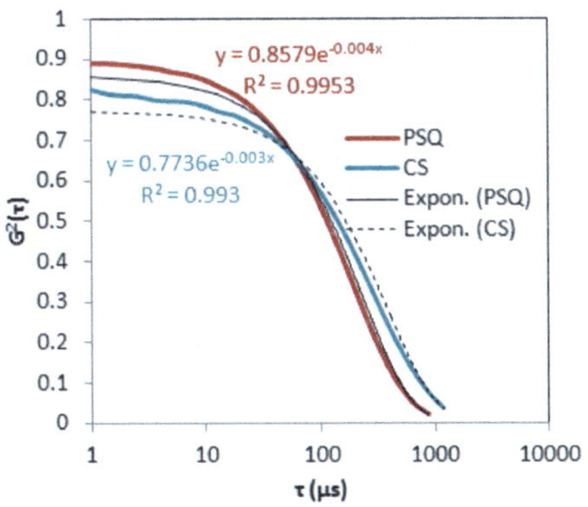

Fig. 3. Normalized autocorrelation function of the scattered intensity versus delay time

4 Conclusion

FTIR spectroscopy and dynamic light scattering (DLS) revealed successful formation of core-shell structure of the synthesized hybrid latex nanoparticles. Their lower magnitude of Brownian diffusivity expedited the latex film formation process. Besides, its film showed less hydrophilicity and higher resistance to water whitening in comparison to the PSQ film.

References

1. Tiwari, A., Syväjärvi, M. (eds.): Massachusetts (2014)
2. Bai, R., Qui, T., Han, F., He, L., Li, X.: Appl. Surf. Sci. **258**, 7683–7688 (2012)
3. Zhang, K., Chen, H., Chen, X., Cui, Z., Yang, B.: Macromol. Mater. Eng. **288**, 380–385 (2003)
4. Hassan, P.A., Rana, S., Verma, G.: Langmuir **31**, 3–12 (2015)
5. Liu, Y., Gajewicz, A.M., Rodin, V., Soer, W.J., Scheerder, J., Satgurunathan, G., McDonald, P., et al.: Polym. Sci. Part B: Polym. Phys. **54**, 1658–1674 (2016)

Protection of Structural Buildings Against Dirt Adsorption: Photoactive Cements vs. Photoactive Coating

Fatemeh Mohammadi, Seyed Mojtaba Mirabedini(✉), Khosro Khodabakhshi, and Shahla Pazokifard

Color and Surface Coatings Department, Iran Polymer and Petrochemical Institute (IPPI), P.O. Box 14975-112, Tehran, Iran
sm.mirabedini@ippi.ac.ir

Abstract. In this study, TiO_2 nanoparticles (Aeroxide P25) and synthesized TiO_2/SiO_2 nanocomposite were used to provide a photoactive concrete and water based acrylic façade coatings. For this purpose, TiO_2 nanoparticles (1 wt%) were loaded into a mixture of cement and sand (weight ratio 5:12) compared to acrylic coatings containing either 1 wt% Aeroxide P25 or synthesized TiO_2/SiO_2 nanoparticles. Photo-catalyst activity of the nanoparticles embedded samples was evaluated by color data measurements of photo-degradation of rhodamine B dyestuff implemented on the sample's surface, during exposure to the sun light for 18 days. The results showed that in spite of improving photo-activity of acrylic coatings using TiO_2/SiO_2 nano-composites, the highest photo-activity arised from concrete loaded by commercial TiO_2 nanoparticles.

Keywords: Synthesis of nanocomposite · Ti/si nanocomposite · Water based acrylic · Photocatalytic materials

1 Introduction

Photo-catalyst materials can aid and accelerate chemical reactions which convert deposited pollutants, on the surface of the coating, to more harmless ones [1]. TiO_2 nanoparticles in anatase crystalline form, is the most functional photo-catalysts among the other semiconductor materials [2]. Despite the high photo-catalyst activity, and due to the extremely high ratio of surface-area/particle-size, TiO2 nanoparticles illustrate poor dispersion stability in the aqueous solutions and their particles tend to strongly agglomerate [3]. Several reports are available concerning preparation and photo-activity of TiO2/SiO2 compounds. Khadem Hosseini et al. showed that synthesized TiO_2/SiO_2 nanocomposite has appropriate dispersion in aqueous media [4, 5]. In practical applications, nano-sized TiO_2 in the form of powder should be protected by support materials/substrates including: cementitious materials, plastics, paint and so on. Among these support materials, concrete has better performance because of inorganic and porous nature. In this study, photo-catalyst properties of acrylic/TiO_2/SiO_2, cement/TiO_2 and acrylic/TiO_2 nanocomposites using Rhodamine B (Rh.B) as a pollutant model were studied to find out whether TiO_2/SiO_2 in acrylic matrix can be used instead of photocatalyst cement or not.

H. Mirzadeh and A. A. Katbab (Eds.): ISPST 2018, *Eco-friendly and Smart Polymer Systems*, pp. 174–177, 2020.
https://doi.org/10.1007/978-3-030-45085-4_42

2 Experimental

Acrylic resin, 4410, was obtained from Simab Resin Company (Iran). TiO_2 nanoparticles, Aeroxide P25, were supplied by Evonik Industries GmbH. Cement and sand were prepared from Khuzestan and Fars Company (Iran). Tetraethylorthosilicate (TEOS, 98 wt%) and titanium tetraisopropoxide (TTIP, 98 wt%) were purchased from Sigma-Aldrich and DaeJung, respectively.

TiO_2/SiO_2 nano-composite was synthesized using Khadem Hosseini method.

TiO_2 and TiO_2/SiO_2 nanoparticles were firstly dispersed in distilled water under sonication process for 30 min. Pre-dispersed nanoparticles (1 wt% were then added to the acrylic resin, under mechanical mixing at speed of 300 rpm for 30 min. The samples with a wet film thickness of 100 μm were applied on degreased glass plates using an Elcometer Doctor Blade film applicator. Samples were then left to dry in a dark place at 30 °C for 24 h.

For preparing cementitious samples, weight ratio of cement: sand to water (containing TiO_2 nanoparticles) was adjusted as 5:12:2.5, respectively. Mixing was performed at two different rates, a low rate of 140 ± 5 rpm for 60 and 30 s at a high mixing speed of 285 ± 10 rpm. Then samples with a wet film thickness of 1000 μm were applied on cementitious substrates using Elcometer Doctor Blade film applicator. Samples aging were taken place in the wet plastic bags with 100% relative humidity and in atmospheric dark place for a further 7 days.

Rh.B solution with 0.5 M concentration was molded on each sample and kept at 50 °C for 4 h. Various Rh.B impregnated specimens were exposed to sun light for 18 days. Colorimetric measurements were performed on samples using Gretag Macbeth Color-Eye 7000A device at different time intervals. For the colorimetric measurement, a CIE L*a*b* colorimeter was used, by considering only the coordinate a* (the reference parameter for the red color). For measurement of photo-catalyst activity Δa* was calculated using Eq. (1):

$$\text{normalized}\,\Delta a^* = \frac{a^*(0) - a^*(t)}{a^*(0)} \times 100 \quad (1)$$

3 Results and Discussion

Figure 1 shows normalized Δa* of photo-catalyst acrylic containing nano-TiO_2 (Co-TiO_2), synthesis TiO_2/SiO_2 nano-composite (Co-TiO_2/SiO_2) and nano-TiO_2/cement composite (Ce-TiO_2). TiO_2 embedded acrylic (red color curve) shows maximum slope of Δa* in the first 48 h of irradiation (Δa* = 54%). Slope of the curve then decreased and after 18 days, red color paled (Δa* = 94%). For composite of acrylic and TiO_2/SiO_2 (green color curve) maximum of Δa*, (62%) was observed in 48 h and after decreasing of curve slope red color disappeared after 6 days. Reducing the degradation of red shade may indicate an increase in the acrylic self-cleaning properties corrected by synthetic TiO_2/SiO_2 nano-composite compared with the commercial type of TiO_2 (P25). The best photo-catalytic activity belongs to Ce-TiO_2 (blue color curve). At first 4 h of exposure time, normalized Δa* achieves 70% and just after 48 h achieves more than 95%. After that, the slop of the curve decreased significantly and after 120 h the red color paled. Table 1 shows visual observation of RhB dyestuff (0.5 M) molded on different samples before exposing sun light and after 120, 240, and 432 h. When normalized Δa* became 100%, red color paled.

Figure 2 shows SEM/EDXA images of samples. Cement sample has porous structure comparing to acrylic coatings. Porous structure provides more effective surface of photo-active nano particle. T_i mapping in EDAX images shows good dispersion of nano particles.

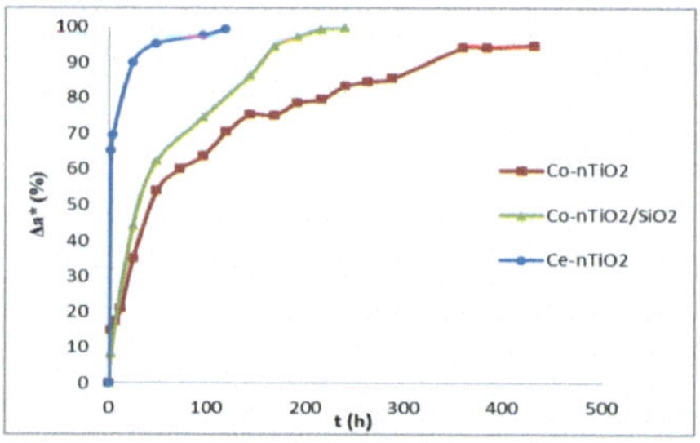

Fig. 1. Normalized Δa* of acrylic coating including nano TiO_2 (Co-nTiO_2), acrylic coating including TiO_2/SiO_2 (Co-nTiO_2/SiO_2), and cement containing nano TiO_2 (Ce-nTiO_2) vs. time

Table 1. Visual observation of RhB dyestuff (0.5 M) molded on (a) Co-nTiO_2, (b) Co-nTiO_2/SiO_2, and (c) Ce-nTiO_2

	t= 0	t= 120 h	t= 240 h	t= 432 h
(a) Co-nTiO_2	Δa*=0	Δa*=70%	Δa*=83%	Δa*=95%
(b) Co-nTiO_2/SiO_2	Δa*=0	Δa*=80%	Δa*=100%	Δa*=100%
(c) Ce-nTiO_2	Δa*=0	Δa*=100%	Δa*=100%	Δa*=100%

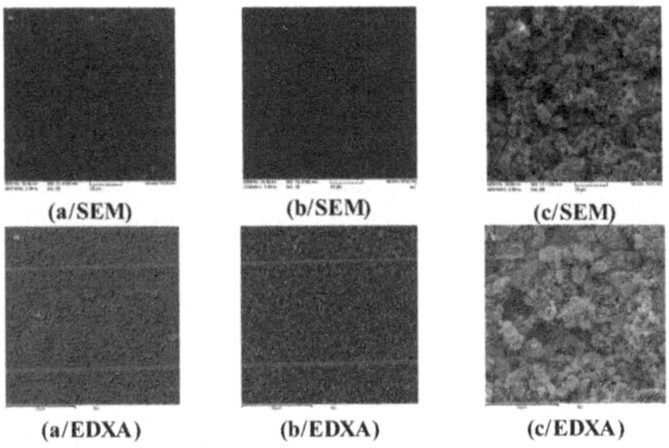

Fig. 2. SEM/EDXA images of samples: (a) Co-nTiO_2, (b) Co-nTiO_2/SiO_2, and (c) Ce-nTiO_2

4 Conclusion

Synthesized SiO_2/TiO_2 nano-composite provided better photo-catalytic properties than commercial TiO_2 nanoparticles in acrylic coatings. It increased degradation of rhodamine B red color on acrylic and enhanced photo-catalyst properties. But among acrylic and cement matrixes, cement-based system showed better photo-catalyst activity due to its porous structures and inorganic nature.

References

1. Folli, A.E., Macphee, D.: 8th fib PhD Symposium in Kgs. Lyngby, Denmark, 20–23 June 2010
2. Chen, J., Kou, S.h., Poon, C.h.: Build Environ. **46**, 1827–1833 (2011)
3. Sabzi, M., Mirabedini, S.M., Zohuriaan-Mehr, J., Atai, M.: Prog. Org. Coat **65**, 222–228 (2009)
4. Khadem Hosseini, A., Mirabedini, S.M., Pazokifard, S.: J. Mater. **51**, 3219–3230 (2016)
5. Park, H.K., Kim, D.K., Kim, C.H.: J. Am. Ceram. Soc. **80**, 743–749 (1998)

Nanoassemblies for Sustainable Energy and Environment Applications

Improved Conductivity and Film Forming of Transparent PEDOT:PSS Electrodes Using Non-ionic Surfactants

Milad Kianpisheh, Bahareh Rezaei, and Faramarz Afshar-Taromi(✉)

Department of Polymer Engineering, Amirkabir University of Technology, Tehran, Iran
afshar@aut.ac.ir

Abstract. We could successfully design and prepare transparent conductive electrodes for optoelectronic applications. The optimized electrode was formulated based on PEDOT:PSS and PVA as non-ionic surfactant. Conductivity and film forming were improved after the addition of optimum amounts of PVA to PEDOT:PSS to reduce the ionic interaction between PEDOT and PSS. UV-Vis spectroscopy, four-point probe and profilometer were used to monitor transparency, conductivity and film thickness, respectively. We could reach the conductivity of 10.12 S/cm after adding 0.625 g/L PVA to PEDOT:PSS solution, which is 15 times higher than that of bare PEDOT:PSS film. Transparency remained more than 85% after addition of less than 1 g/L PVA. The proposed electrode is a promising candidate for use in high conductive optoelectronic devices.

Keywords: Transparent electrodes · PEDOT:PSS · Solar cell · Non-ionic surfactants · Polyvinyl alcohol

1 Introduction

Optoelectronic devices such as light emitting diodes (LEDs), solar cells, and touch panel displays, have attracted significant attention. In all of these devices, at least one transparent electrode is used. The most commonly used electrode in these devices is indium tin oxide (ITO). Nevertheless, ITO has several drawbacks. Indium is a scarce element on earth; its fabrication cost is high, and it is a brittle material [1]. Therefore, finding an appropriate alternative to metal oxides is essential. The most important alternatives for metal oxides are conducting polymers [2], carbon nanotubes [3], graphene [4], and metal nanowires [5]. One promising candidate is poly(3,4-ethylenedioxy-thiophene):poly(styrenesulfonate) (PEDOT:PSS).

PEDOT:PSS has high transparency in the visible range, and it is mechanically flexible [6]. However, film forming of PEDOT:PSS on hydrophobic surfaces such as glass and PET is weak and exhibits low conductivity of lower than 1 S/cm. This conductivity is strikingly lower than that of ITO [7]. To enhance the electrical conductivity of PEDOT:PSS films, various approaches have been suggested. These methods are mostly based on the addition of solvents such as dimethylsulfoxide

H. Mirzadeh and A. A. Katbab (Eds.): ISPST 2018, *Eco-friendly and Smart Polymer Systems*, pp. 181–183, 2020.
https://doi.org/10.1007/978-3-030-45085-4_43

(DMSO) [7], polyethylene glycol (PEG) [8], etc. These additives are believed to reduce the ionic interaction between PEDOT and PSS, which facilitates the phase separation and the conformational change of PEDOT from benzoid to quinoid. A similar effect has been reported for non-ionic surfactants. The amphiphilic nature of surfactant molecules is effective to weaken the ionic stitching between PEDOT and PSS [9]. In this study, we used PVA as a surfactant to improve the PEDOT:PSS conductivity. By adding different concentrations of PVA, we reported the optimal concentration of PVA considering two factors of transparency and conductivity.

2 Experimental

PEDOT:PSS aqueous dispersion (Clevios PH 1000) was purchased from Heraeus and poly(vinyl alcohol) (MW = 13,000–23,000) was purchased from Sigma Aldrich. All reagents were used without further purification.

Conductivity of films was measured by four-point probe (SanatNamaJavan Company). Film thickness was measured on a Veeco Dektak 6 M profilometer and transmittances of films were measured by UV-Vis spectroscopy (analytikjena, spectrod 210).

3 Results and Discussion

PEDOT:PSS dispersions were mixed with desired amounts of PVA (0.1, 0.3, 0.6, 1.2, 2.5, 5, and 10 g/L of PEDOT:PSS) and stirred continuously for 24 h at room temperature. The glass substrates (2 × 1 cm) were cleaned with ultrasonic treatment in both acetone and isopropyl alcohol, before deposition of PEDOT:PSS. The formulations were spin-coated on glass. Spin speeds were varied between 2000 and 3000 rpm to control the film thickness about 40 nm After coating, each electrode was heated at 120 °C for 15 min to dry completely.

In order to find the optimal PVA concentration, conductivity and transparency of all samples were measured. As shown in Fig. 1, at concentrations less than 1 g/L PVA, the transparency of electrodes is higher than 85%, which is competitive with that of ITO.

As shown in Fig. 1, by increasing the concentration of PVA up to 0.625 g/L, the conductivity increases up to more than 15 times of the initial value and reach 10.12 S/cm. The reason for this increased conductivity is due to the generation of PEDOT nano-fibrils in the bulk PEDOT:PSS [10]. With more increase in PVA concentration, the trend is reversed and the conductivity decreases sharply, because the conductive PEDOT:PSS matrix changes to the insulator PVA matrix.

In all concentrations, the addition of PVA could greatly improve film forming in hydrophobic substrates such as glass and PET. This can make the process much easier and cost effective for electrode fabrication.

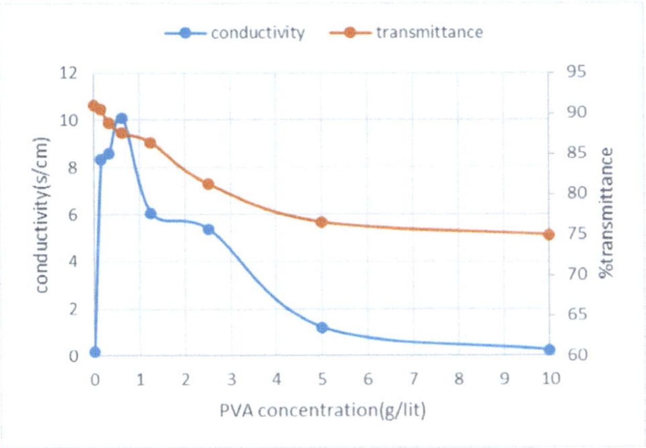

Fig. 1. Conductivity diagram (left side) and transmittance diagram (right side)

4 Conclusion

In this study, we were able to increase the polymer conductivity up to 10.12 S/cm by adding polyvinyl alcohol by maintaining more than 85% transparency. In addition to conductivity improvement, the addition of polyvinyl alcohol significantly improved the process of film forming on the hydrophobic glass, which could eliminate the surface modification step in the process of making a transparent electrode with a hydrophobic substrate.

References

1. Inganäs, O.: Nat. Photon. **5**(4), 201–202 (2011)
2. Kim, W.H., Mäkinen, A.J., Nikolov, N., Shashidhar, R., Kim, H., Kafafi, Z.H.: Appl. Phys. Lett. **80**(20), 3844–3846 (2002)
3. Gruner, G.: J. Mater. Chem. **16**(35), 3533 (2006)
4. Kim, K.S., et al.: Nature **457**(7230), 706–710 (2009)
5. Yu, Z., Li, L., Zhang, Q., Hu, W., Pei, Q.: Adv. Mater. **23**(38), 4453–4457 (2011)
6. Pei, Q., Zuccarello, G., Ahlskog, M., Inganäs, O.: Polymer (Guildf) **35**(7) 1347–1351 (1994)
7. Kim, J.Y., Jung, J.H., Lee, D.E., Joo, J.: Synth. Met. **126**(2–3), 311–316 (2002)
8. Wang, T., Qi, Y., Xu, J., Hu, X., Chen, P.: Appl. Surf. Sci. **250**(1–4), 188–194 (2005)
9. Yoon, S.-S., Khang, D.-Y.: J. Phys. Chem. C **120**(51), 29525–29532 (2016)
10. Kim, S., Lee, S.J., Cho, S., Shin, S., Jeong, U., Myoung, J.-M.: Chem. Commun. **53**(59), 8292–8295 (2017)

Sol–Gel Preparation and Characterization of Antibacterial and Self-cleaning Hybrid Nanocomposite Coatings

Sahar Amiri and Seyed Armin Zare Estekhraji(✉)

Department of Polymer Engineering, Science and Research Branch, Islamic Azad University, Tehran, Iran
armin.zare@srbiau.ac.ir

Abstract. ZnO-TiO_2, SiO_2-TiO_2, and SiO_2-TiO_2-ZnO hybrid nanocomposite coatings based on sol-gel precursors using tetra methoxysilane (TMOS), 3-glycidoxypropyl trimethoxysilane (GPTMS), tetra-n-butyl orthotitanate (TBT) and zinc acetate dihydrate were synthesized and characterized. Hybrid network was characterized by FTIR, FESEM and EDAX. The results indicated that inorganic particles sizes were in nano size range (20–30 nm) with a very uniform distribution and dispersion. Photocatalytic and self-cleaning activities of these coatings were investigated by degradation of 2-methylene blue in aqueous solution (20 ppm) at visible light irradiation which indicated photocatalytic performance of coating containing ZnO and TiO_2. The antibacterial effect of coating was investigated for inhibition and inactivation of cell growth, and the results showed that the antibacterial activity of ZnO-TiO_2 and SiO_2-TiO_2-ZnO coating against *E. coli* and *S. aureus* was the same and higher than that of SiO_2-TiO_2 hybrid nanocomposite coating.

Keywords: Sol-gel process · SiO_2 · TiO_2 · ZnO · Nanocomposite coatings · Antibacterial · Self-cleaning

1 Introduction

Nowadays, self-cleaning and antibacterial hybrid nano-composite coatings have miscellaneous applications in textile fabrics, food processing, environmental and medical applications and agricultural industry. Self-cleaning coatings have a wide variety of applications such as cements, window glasses, textiles and paints, and can reduce cleaning costs and increase their stability [1, 2]. Self-cleaning coatings can be divided into hydrophobic and hydrophilic coatings and prevent dirt from forming an intimate contact with the surface. The hydrophilic self-cleaning coatings are based on photocatalysis which in the presence of light, they are able to break down impurities. TiO_2 and ZnO are the most important photocatalysts which are non-toxic, inexpensive, well-known in household chemicals, suitable to be used as outdoor purposes and easy to handle [1]. Also, coatings based on SiO_2 that show hydrophobic and antistatic properties are developed, especially for the purpose of textile finishing. Anti-microbial finishing is a cost-effective and commercial process for cotton textiles which was performed by treating cotton textile with silica sols by sol-gel process and then with

H. Mirzadeh and A. A. Katbab (Eds.): ISPST 2018, *Eco-friendly and Smart Polymer Systems*, pp. 184–188, 2020.
https://doi.org/10.1007/978-3-030-45085-4_44

silver nitrate solution [6, 7]. Recently, investigations have been conducted to improve properties of textile fabrics and other surfaces by incorporation of various finishes into sol-gel coating. Coating deposition is done by various methods including physical vapor deposition (PVD), chemical vapor deposition (CVD), electrochemical deposition, plasma spraying and sol–gel process [3]. Sol-gel presents an extremely easy process with mild temperature and pressure to fabricate nanocomposite coatings via hydrolysis and condensation of sol-gel precursors and to form a homogeneous hybrid nanocomposite coating with unique physical and mechanical properties such as flexibility, toughness, formability, low density, hardness, stiffness, heat and chemical resistance [2, 4]. Compatibility of hybrid materials is due to formation of covalent bond between organic-inorganic polymers and inorganic components. Sol–gel processes are mainly based on hydrolysis and condensation reactions of metal alkoxides and can be used for designing of new materials with interesting properties for many applications in the form of glasses, fibers, ceramic powders and thin films. The inorganic portion imparts enhanced mechanical properties while the organic portion leads to better flexibility and functional compatibility with organic paint systems [8]. In this research, we report the preparation of ZnO-TiO_2, SiO_2-TiO_2, and SiO_2-TiO_2-ZnO hybrid nanocomposite coatings via sol-gel process and their characterization by Fourier transform infrared spectro-scopy (FTIR), energy-dispersive X-ray spectroscopy (EDAX) and field emission scanning electron microscopy (FESEM). Abrasion resistance of coating was determined by Taber abrasion test which showed ZnO and TiO_2 improved abrasion resistance of coating. Then, antimicrobial and self-cleaning properties of the coatings were investigated using UV spectroscopy and bacterial growth, respectively. The antibacterial activities of coatings were tested against *Escherichia coli* (*E. coli*) and *Staphylococcus aureus* (*S. aureus*) representing gram-negative and gram-positive bacteria, respectively. The results showed that coating containing ZnO and TiO_2 inhibited bacteria growth because of their photocatalytic property.

2 Experimental

2.1 Materials

Tetramethylorthosilicate (TMOS, >98%), 3-glycidoxy-propyltrimethoxysilane (GPTMS, >98%), hydrochloric acid (HCl, 37%), ethanol (>98%), tetra (*n*-butyl orthotitanate, >98%) (TBT), acetyl acetone (AcAc, >98%), nitric acid (HNO_3, 65%), isopropyl alcohol (>98%), zinc acetate dihydrate (>98%), monoethanol-amine (MEA, 99.8%) were purchased from Merck and used as received [2–4].

2.2 Synthesis of SiO_2 Nanoparticles

The hybrid sol was prepared by admixing silica precursors, GPTMS and TMOS. GPTMS and TMOS were mixed (2:1 molar ratio) in a beaker with ethanol (10 mL) and stirred using a magnetic stirrer (200 rpm) for 15 min at ambient temperature. The resultant two-phase solution was vigorously stirred at a rate of 240 rpm for 1 h. Water (pH 2) was used as acidic catalyst. Hydrolysis and condensation of the silane were conducted at stoichiometric and 50% substoichiometric of water (pH 2)/ethanol ratio in the presence of an acid catalyst. After hydrolysis and condensation, sol was stirred at room temperature for two hours [2–4].

2.3 Synthesis of TiO_2 Nanoparticles

TBT and isopropyl alcohol (1:10 molar ratio) were mixed and stirred by a magnetic stirrer (200 rpm) for 1 h at room temperature. Then, 1 mol AcAc (so as to control the rate of hydrolysis reaction) was added. The mixture was stirred for 3 h at the same rate in order to create a transparent solution. Next, 4 mol distilled water (pH 2, from HNO_3) was added drop by drop to above solution and stirred for 3 h at ambient temperature [2–4].

2.4 Synthesis of ZnO Nanoparticles

Zinc acetate dihydrate was dissolved in dehydrated isopropyl alcohol and stirred thoroughly using a magnetic stirrer for 30 min. Then, monoethanolamine was added to the solution as stabilizer drop by drop until it became transparent. The solution was mixed by a magnetic stirrer at 60 °C for 2 h and kept at room temperature for about 24 h for formation of a clear and homogeneous sol [2–4].

2.5 Preparation of Hybrid Coatings

SiO_2/TiO_2, ZnO/TiO_2, and $SiO_2/TiO_2/ZnO$ hybrid networks were obtained by mixing of the specific amounts of sols ($SiO_2:TiO_2$ = 2:1), ($ZnO:TiO_2$ = 1:1), ($SiO_2:TiO_2:ZnO$ = 2:1:1) and stirring by a magnetic stirrer for 2 h at room temperature at a rate of 1000 rpm. Then the obtained sols were coated on glass or Al substrates [2–4].

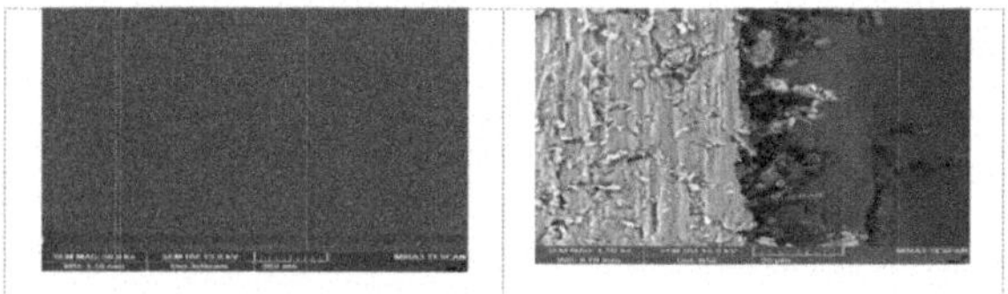

Fig. 1. FESEM of $SiO_2/TiO_2/ZnO$ hybrid coating

3 Results and Discussion

FESEM of $SiO_2/TiO_2/ZnO$ hybrid coating also indicated crack-free and smooth morphology with uniform dispersion of nanoparticles throughout the surface of hybrid coating without any aggregation. No cracks and fracture were observed on the coating surface. As it is noticed in Fig. 1 the thickness of the coating is about 10 nm and size of ZnO, TiO_2, and SiO_2 nanoparticles is between 20–30 nm. In M_1 sample, the ratio of $SiO_2:TiO_2$ is 2:1, so the low amount of TiO_2 cannot inhibit gram-negative bacterial. M_3 and M_5 show the same activity which is related to presence of ZnO nanoparticles.

MB degradation rate increased in the presence of ZnO/TiO_2 hybrid coating, which is due to excitation of TiO_2 surface by ZnO nanoparticles. In addition, a colorless solution was formed. The photocatalytic activity of $SiO_2/TiO_2/ZnO$ hybrid coating was the same as that of ZnO/TiO_2 hybrid coating, indicating that the presence of SiO_2 was not effective (Fig. 2).

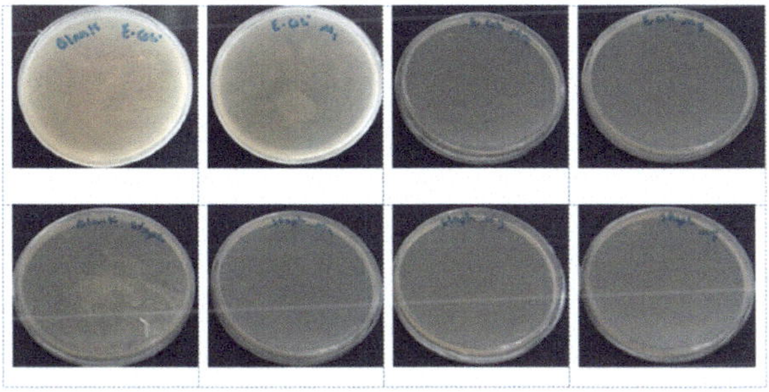

Fig. 2. The representative photographs of antimicrobial test plates *E. coli* and *S. aureus* growth n the agar medium. M_1: SiO_2/TiO_2 coating, M_3: $ZnO\text{-}TiO_2$ coating, and M_5: $SiO_2/TiO_2/ZnO$ hybrid coating

4 Conclusion

$ZnO\text{-}TiO_2$, $SiO_2\text{-}TiO_2$, and $SiO_2\text{-}TiO_2\text{-}ZnO$ hybrid nanocomposite coatings were prepared via sol-gel process. The properties of obtained coatings were investigated by FTIR, EDAX and FESEM techniques. Abrasion resistance was reported by a Taber apparatus. Self-cleaning and antimicrobial activity of coatings were investigated using UV-visible and bacterial growth on the agar medium, respectively. The FTIR results indicated formation of bonds between the organic and inorganic phases. SEM and FESEM along with elemental mapping results indicated the uniform distribution and dispersion of the inorganic nanoparticles within the polymer matrix without any agglomeration and thickness of about 10 nm for coatings. Taber abrasion experiment showed that the introduction of TiO_2 and ZnO nanoparticles increased the abrasion resistance of the hybrid coating due to the enhancement of surface hardness and elastic modulus. Also, introduction of TiO_2 and ZnO nanoparticles into the coatings increased antimicrobial and self-cleaning properties of hybrid coatings. Higher specific surface area of ZnO and TiO_2 nanoparticles enhanced particle surface reactivity and improved their antimicrobial activity. By increasing the concentration of photocatalyst, the activity of photocatalyst increased and bacteria growth was inhibited.

References

1. Dubey, R.S., Rajesh, Y.B.R.D., More, M.A.: Mater. Today Proc. **2**, 3575–3579 (2015)
2. Amiri, S., Rahimi, A.: J. Polym. Res. **21**, 566–573 (2014)
3. Kaleji, B., Mousaeia, M., Halakouie, H., Ahmadi, A.: J. Nano Struct. **5**(3), 219–225 (2015)
4. Rahimi, A., Amiri, S.: J. Coat. Technol. Res. **12**(3), 587–593 (2015)

Synthesis of Au@GO Core-Shell Structure for Solar Cell Application

Bahareh Rezaei(✉), Milad Kian Pisheh, Zahra Babaei, Bahareh Nasrollahi, and Faramarz Afshar Taromi

Amirkabir University of Technology, Tehran, Iran
baharehrezaei.chem@gmail.com

Abstract. Stable Au NPs were synthesized with Turkevich method and graphene oxide was prepared using improved Hummer method for better functionalization of graphite sheets. Core shell structure of Au@GO was synthesized by changing zeta potential of Au NPs. Band gap of Au NPs was tunable at different concentrations of GO. The Au_2@GO UV-Vis absorption peak of Au NPs shifted to a higher wavelength and the energy of band gap was determined by optical absorption of Au NPs. Tauc analysis was used for calculation of optical band gap of Au NPs. The results revealed high applicability of this structure for solar cell fabrication.

Keywords: Gold nanoparticle · Graphene · Core/shell · Solar cell

1 Introduction

As time goes by pollutions coming from fossil fuels are making the environment a non-residential place for creatures. Using a clean energy supplier is an ultimate desire of the present era. Solar cells are one of the best strategies applied for terminating damaging the world [1]. Au NPs are noteworthy for having numerous and fascinating properties like non-linear absorption of light, near-infra red region optical resonance and plasmon excitation dependent on size and shape [2]. When light reaches the surface of these metal nanoparticles, a strong interaction occurs between electromagnetic radiation and free electrons in the outer layer of nanoparticles. This phenomenon can enhance light absorption and facilitate exciton (electron and hole) generation in polymer solar cells. Besides, Au and Ag NPs LSPR effect, their shape, size and also the dielectric environment of NPs are the dominate factors in light sputtering of solar cells. Graphene oxide (GO) is a 2D material with superior properties like, ease of fabrication, low cost, very good processibility, ease of functionalization by having oxygen functional groups, which makes it a suitable candidate for encapsulating NPs [3].

2 Experimental

For the synthesis of GO, improved Hummers method was used [4]. Briefly, graphite powder (0.5 g) was mixed with sulfuric acid (60 mL) and phosphoric acid (6.6 mL) then potassium permanganate (3 g) was added to the mixture in an hour. The color

H. Mirzadeh and A. A. Katbab (Eds.): ISPST 2018, *Eco-friendly and Smart Polymer Systems*, pp. 189–192, 2020.
https://doi.org/10.1007/978-3-030-45085-4_45

changed from dark black to green and after 24 h a brown solution was produced. Hydrogen peroxide (0.5 mL) was added and was thoroughly washed with ethanol and water.

For the synthesis of Au NPs, we used Turkevich method. In this procedure, tri-sodium citrate (TSC) was used as reducing and pH controlling agent. Briefly, we used $HAuCl_4$ (0.6 mM) and 10 mL water and heated it to 90 °C, and under stirring, TSC (34 mM) was added dropwise to the solution. After 3 min, the solution was dark and slowly the color changed to wine red after 30 min.

For the preparation of Au_x@GO structure, cetrimonium bromide (CTAB), was added to the nanoparticle solution so the zeta potential of Au NPs got positive due to the micelle structure of CTAB around NPs. GO (1 mg/mL) was added to Au NPs and stirred for 24 h. The negative charge of GO interacts with positive Au NPs and forms a shell on the outer surface of gold NPs. Four ratios of Au_x:GO were prepared to study the influence of Au concentration on GO electrical properties and solar cell efficiency.

3 Results and Discussion

The UV-Vis absorption spectra of Au_x@GO for four ratios of Au:GO are shown in Fig. 1. The peaks at 520 nm are originated from the surface plasmon resonance of Au NPs, so these peaks confirm the formation of Au NPs. GO has characteristic peaks at 230 and 305 nm which corresponds to π-π^* (C = C) and n-π^* (C = O) transitions [5]. After formation of Au_x@GO structure, absorption peaks red shift to higher wavelength and we assume that it is due to the morphological changes of Au NP surface, enlargement of Au_x@GO size and also electrical changes of Au Nps in contact to GO.

GO addition to the Au NPs, can change the optical band gap of nanoparticles, in this regard the band gap energy was examined by Tauc analysis:

$$(\alpha hv) = (hv - E_g)^n \tag{1}$$

α is the optical absorption coefficient, h is planck constant and is equal to 4.136×10^{-15}eV.s, $\nu = C/\lambda$ ($C = 3 \times 10^8$ m/s) and λ is wavelength which can be extracted from UV-Vis spectra. *n* is power factor and has discrete values such as 1/2, 2, 3/2, and 3 or higher depending on the mode of transition. Figure 1 shows the E_g calculation in core/shell structure of Au_x@GO. The band gap energy of the fabricated solar cells were obtained and for the Au_2@rGO we have the lowest band gap energy (E_g = 2.02 eV) and the band gap expanded after adding more nanoparticles ($Au_{2.5}$@rGO, E_g = 2.11 eV). The larger band gap requires higher energy for exciton generation and is due to the lower interaction of Au film with GO (Fig. 2).

Figures 3a and 3b show the FESEM image of Au@Gr core shell structure after reduction with hydrazine vapor. We can see that Au NPs are separately oriented.

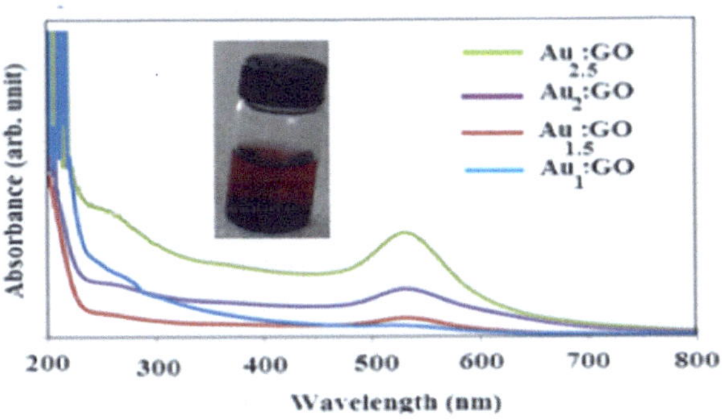

Fig. 1. UV-Vis absorption of Au NPs@GO core/shell structure in four ratios of Au:GO

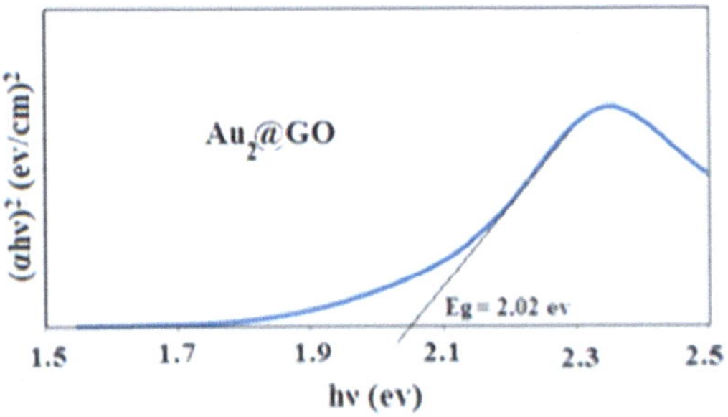

Fig. 2. Band gap calculation of Au NPs in Au_2@GO

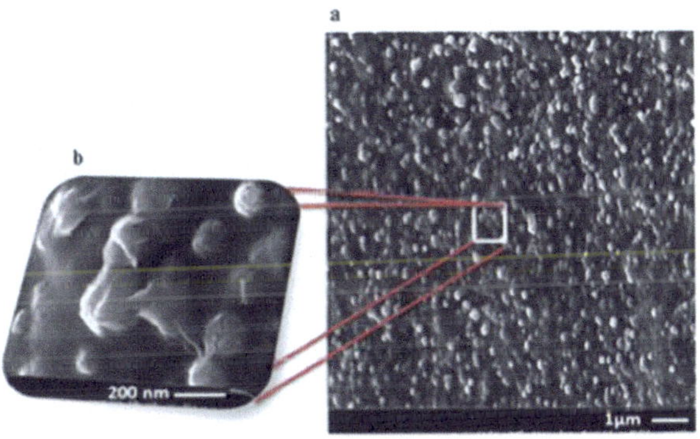

Fig. 3. (a) Close image of the same nanocomposite and (b) FESEM image of Au_2@GO structure

4 Conclusion

We could successfully synthesize Au@Gr core shell structure which were separately oriented. After addition of different concentrations of Au NPs, absorbance peak of Au NPs shifted to higher wavelength in Au_2@GO and it showed band gap modulation of Au NPs in contact with Gr. The synthesized structure is applicable for solar cell application for enhancing solar absorbance and efficiency.

References

1. Helgesen, M., Søndergaard, R., Krebs, F.C.: J. Mater. Chem. **20**(1), 36–60 (2010)
2. Tabor, C., Van Haute, D., El-Sayed, M.A.: ACS Nano **3**(11), 3670–3678 (2009)
3. Liu, F., Nunzi, J.M.: Appl. Phys. Lett. **99**(12), 203 (2011)
4. Kim, T., Kang, H., Jeong, S., Kang, D.J., Lee, C., Lee, C.H., Seo, M.K., et al.: ACS Appl. Mater. Inter. **6**(19), 16956–16965 (2014)
5. Jankovic, V., Yang, Y., You, J., Dou, L., Liu, Y., Cheung, P., Chang, J.P., et al.: ACS Nano **7**(5), 3815–3822 (2013)

Effect of Monomer Concentration in Primary Sol on the Colloid like Particle Size of Novolac Aerogel

Golnoosh Abdeali, Ahmad Reza Bahramian(✉), and Mahdi Abdollahi

Polymer Engineering Department, Faculty of Chemical Engineering, Tarbiat Modares University, P.O. Box 14115-114, Tehran, Iran
abahramian@modares.ac.ir

Abstract. Three various novolac aerogels with different monomer concentrations of 10, 15, and 20 wt% were prepared by sol-gel polymerization method. In order to investigate structure-properties relation, morphology and physical measurements were done. By increasing monomer concentration from 10 wt% to 20 wt% in initial sol, size distribution peak of aerogel colloids was moved from 120 nm to 60 nm. In addition, porosity was declined by as much as 10.8%, whereas density of ultimate aerogel was increased.

Keywords: Novolac aerogel · Physical properties · Sol/gel polymerization · Structure/properties relation · Monomer concentration

1 Introduction

Aerogels are particular type of open cell foams with low density, continuous porosity, high percentage of open nano size cavities and a matrix consisting of colloid-like particles dimensioned on a nanometer scale [1, 2]. High porosity and microstructure of particles and porous parts on a nanometer scale are the source of unique properties in these materials, among which low thermal conductivity, and high surface area are highlighted. Remarkable point about these materials is that monomer concentration has a significant effect on structure and physical properties of final gel [3]. Therefore, intention of this research is to investigate effect of novolac concentration on the colloids size and physical properties of novolac aerogel.

2 Experimental

Commercial novolac resin (IP 502 Resitan Co.) and 2-propanol (Dr. Mojallali Co.) were used to prepare novolac gel. At first step, different solutions with 10, 15, and 20 wt% of novolac resin in 2-propanol were prepared by using a mechanical stirrer. After that, all samples were placed in an autoclave at saturated vapor of solvent atmosphere at 120 °C and 15 bar for 5 h. Subsequent to this step, gelation process was completed and all samples were dried at environment condition for 3 days. Eventually, all samples went under drying process at 90 and 120 °C for 48 h.

H. Mirzadeh and A. A. Katbab (Eds.): ISPST 2018, *Eco-friendly and Smart Polymer Systems*, pp. 193–196, 2020.
https://doi.org/10.1007/978-3-030-45085-4_46

3 Results and Discussion

Physical properties of novolac aerogels with different monomer content are reported in Table 1. According to this table, by increasing novolac concentration in sol solution from 10 wt% to 20 wt%, density of samples was increased, which is due to solid mass increase in a constant volume of solution. On the other hand, this increase in density indicates a decrease in porosity of porous samples.

Table 1. Physical properties of novolac aerogels

Sample	Density (g/cm^3)/porosity (%)	Volume shrinkage (%)
10%	0.08/93.62	7.77
15%	0.13/89.12	11.48
20%	0.20/83.44	14.81

4 Morphology Analysis

To investigate microstructure of prepared aerogels, FESEM analysis has been done (Fig. 1). As can be seen in this image, by increasing monomer concentration from 10 wt% to 20 wt%, colloid size of microstructure was decreased and interconnection of all particles was increased. The source of this microstructure changing in aerogel samples is phase separation phenomenon (Fig. 2) that controls gel formation in the system. By dominance of phase separation mechanism in gelation process, particle correlation was increased and coherent structure with smaller colloidal particles was obtained [4].

Fig. 1. FESEM images of novolac aerogel with: (a) 10 wt%, (b) 15 wt%, and (c) 20 wt% monomer concentration

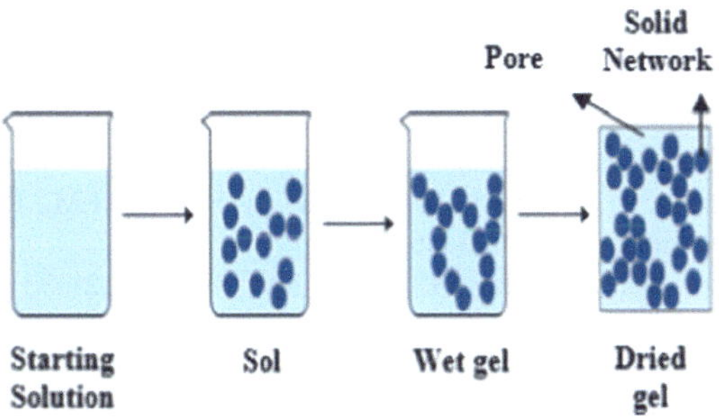

Fig. 2. Schematic of condensation-based network formation (with dominance of phase separation mechanism) starting with novolac-2 propanol solution [4]

Figure 3 shows particle size distribution images of novolac aerogels in three different monomer concentrations. As shown in Fig. 1 and proved in Fig. 3, by increasing novolac concentration, particle size was decreased. Another considerable change is type of distribution that was changed from unimodal to bimodal after increasing monomer concentration.

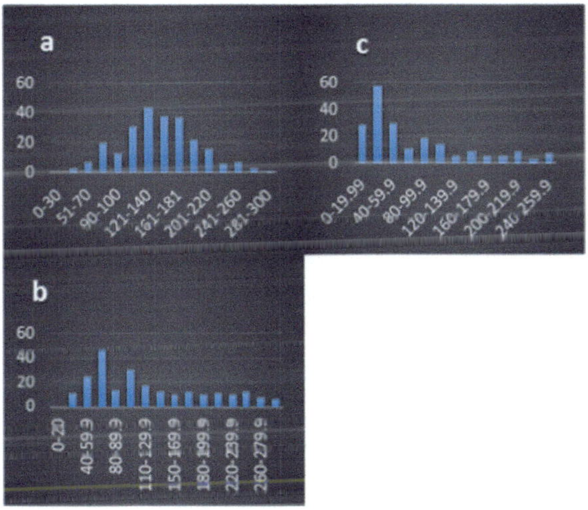

Fig. 3. Particle size distribution images of prepared aerogel with: (a) 10 wt%, (b) 15 wt%, and (c) 20 wt% monomer concentration

5 Conclusion

Rise in amount of novolac monomer concentration resulted in porosity and colloid-like particles size reduction as well as density increase in ultimate aerogel. This observation demonstrates significant effect of sol amount on phase separation, microstructure and physical properties of novolac aerogel.

References

1. Pan, Y., He, S., Gong, L., Li, C., Li, Z., Zhang, H.: Mater. Des. **113**(2), 246–253 (2017)
2. Li, C., Cheng, X., Li, Z., Pan, Y., Huang, Y., Gong, L.: Non-Cryst. Solids **457**, 52–59 (2017)
3. Biener, J., Stadermann, M., Suss, M., Worsley, M.A., Biener, M.M., Rose, K.A., Baumann, T.F.: Energy. Environ. Sci. **4**(3), 656–667 (2011)
4. Feinle, A., Elsaesser, M.S., Huesing, N.: Chem. Soc. Rev. **45**(12), 3377–3399 (2016)

Synthesis of Paraffin Wax Phase Change Material Nanoencapsulated by Polyurethane via Interfacial Polymerization

Hediyeh Nikpourian, Ahmad Reza Bahramian(✉), and Mahdi Abdollahi

Polymer Engineering Department, Faculty of Chemical Engineering, Tarbiat Modares University, P.O. Box 14115-114, Tehran, Iran
abahramian@modares.ac.ir

Abstract. This work deals with the synthesis and physico-chemical characterization of nanoencapsulated paraffin wax with polyurethane (PU) via interfacial polymerization method. In synthesized nanocapsules, the paraffin wax has a function of phase change material, while PU acts as a shell material. The nanoencapsulated PCMs were characterized chemically and morphologically by using Fourier transform infrared (FTIR) spectroscopy and scanning electron microscopy (SEM). The results of morphological study showed that the particle size distribution of the nanocapsules were 80–400 nm by digimizer analysis.

Keywords: Nanoencapsulation · Phase change material · Interfacial polymerization · Paraffin wax polyurethane

1 Introduction

The application of the heat storage system with use of phase change material is an effective way to save energy [1]. Phase change materials (PCMs) as a kind of crystalline materials, can store and release latent heat during their phase transition at a defined temperature range, which can be used in thermal energy conservation [2]. Paraffin wax is especially promising due to their characteristics such as nontoxic, chemically inert, low cost, and tunable phase change temperature and high storage energy capacity [3]. In order to prevent melt flowing during solid-liquid phase transition, encapsulation of PCMs with proper materials as shell is required [2]. This paper considers the synthesis of nanoencapsulation of phase change material, paraffin wax as a core and polyurethane shell material by in situ inter-phase polymerization.

2 Experimental

2.1 Materials

Commercial novolac resin (IP 502 Resitan Co. Iran) as a polyol and toluene diisocyanate (Merck Co. Germany) were used as precursors to form the shell.

H. Mirzadeh and A. A. Katbab (Eds.): ISPST 2018, *Eco-friendly and Smart Polymer Systems*, pp. 197–200, 2020.
https://doi.org/10.1007/978-3-030-45085-4_47

Acetone (Dr. Mojallali Co. Iran) was used as solvent. Paraffin wax was purchased from Dr. Mojallali Co. Iran and used as phase change material.

2.2 Synthesis of Nanocapsules

The encapsulation of the paraffin wax was carried out by interfacial polymerization method. For this propose; First, 0.1 g of nano particles of paraffin wax, which synthesized using solvent-anti solvent method in Tarbiat Modares University Lab, have smeared to a dozen of drops of toluene diisocyanate for 18 h. Second, 40 mL of acetone was added to the previous step and sonicated for 10 min. Third, 0.5 g of novolac were dissolved in 10 mL of acetone and added to the second step. At the end, the polycondensation polymerization reaction was done at 30 °C for 20 h.

2.3 Characterization

FTIR spectroscopy (FTIR, PerkinElmer, Frontier, USA) was applied to characterize the chemical structures of the samples using a KBr sampling sheet.

The morphology of the nanocapsule samples were investigated by using a scanning electron microscopy (SEM, xl30, Philips, Japan).

3 Results and Discussion

3.1 Morphology Analysis

The SEM images of the nanoencapsulated PCM are shown in Fig. 1. It can be seen that, most of the nanocapsules are smooth spherical shape, compact surface and regular. Also, most nanocapsules fell in the range of 80–400 nm, this was consistent with the result from the digimizer analysis (Fig. 3).

3.2 Chemical Structure Analysis

The FT-IR spectral analysis was used to confirm the formation of PU (polyurethane) shell around the paraffin wax. The FTIR spectral results of pure paraffin wax, PU and nanocapsules are shown in Fig. 2. The presence of the peak 1740 cm^{-1} (urethane group) in the FTIR of PU and nanocapsules was indicated that consisted the urethane groups. It is proved that, the paraffin wax was successfully encapsulated by PU shell material.

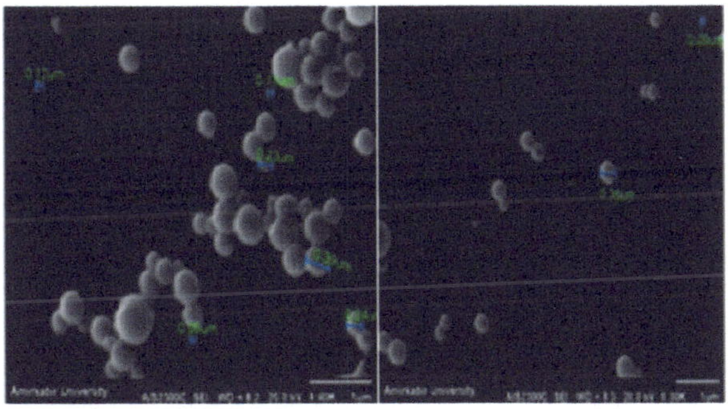

Fig. 1. SEM images of the synthesized nanoencapsulation of paraffin wax as a core and PU shell material

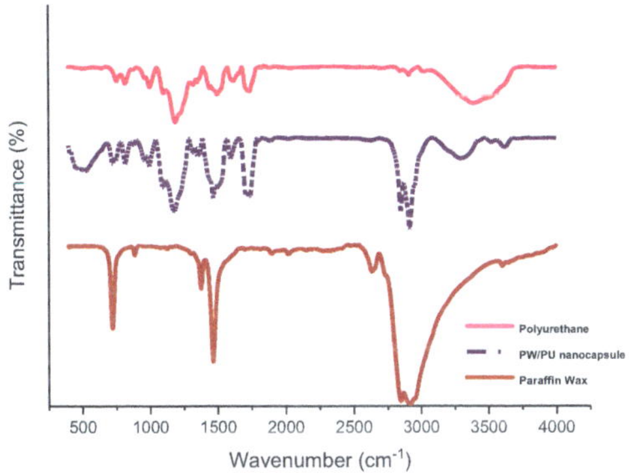

Fig. 2. FTIR of paraffin wax polyurethane and nanocapsules containing paraffin wax-polyurethane

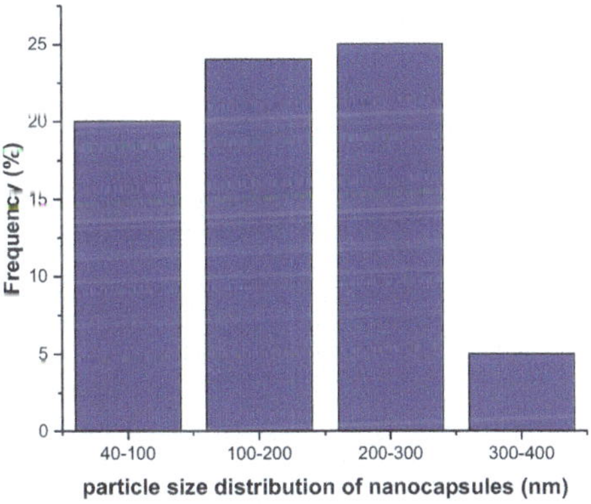

Fig. 3. Digimizer of particle size distribution of nanocapsules containing paraffin wax-polyurethane

4 Conclusion

Nanoencapsulation containing paraffin wax as a core material and PU shell material were successfully synthesized by interfacial polymerization method. The polymerization reactions regarding with the formation of PU shell around the paraffin wax were confirmed via the FTIR spectroscopy technique. The SEM analysis showed that the prepared capsules had virtually a smooth and compact surface and spherical shape.

References

1. Zalba, B., Marin, J.M., Cabeza, L.F., Mehling, H.: Appl. Therm. Eng. J. **23**, 251–283 (2003)
2. Himran, S., Suwono, A., Mansoori, G.A.: Energy Sour. J. **18**, 16–117 (2004)
3. Shchukina, E.M., Graham, M., Zheng, Z., Shchukin, D.G.: R. Soc. Chem. J. **47**, 4156–4175 (2018)

Adsorption of Acid Red 1 onto Functionalized Silica by Amine-Terminated Dendritic Polymers: Isotherms and Kinetic Models

Samane Goftari(✉) and Somaye Akbari

Textile Engineering Department, Amirkabir University of Technology, Tehran, Iran
smanegoftari@aut.ac.ir

Abstract. Pristine silica was functionalized by amine-terminated dendritic polymers via divergent synthetic routes in first generation. The chemical structure analyzed by Fourier Transform Infrared Spectroscopy (FTIR) confirmed the presence of amine terminated group on silica. The pristine and functionalized silica were investigated for adsorption capacity of acid red 1 (AR1). The results showed that the removal was enhanced from 0% to 94.8% by functionalization. It seems that this modification improved significantly the adsorption capacity of this adsorbent. Moreover, their adsorption isotherms and kinetics were further investigated to explore the application in the industrial operating system. The adsorption kinetics fitted pseudo-second order model well. Furthermore, the adsorption isotherms fitted the Freundlich model well and with the modification of pristine silica, R^2 of this model was 0/98.

Keywords: Silica · Amine terminated group · Adsorption · Isotherm · Kinetic

1 Introduction

In recent years, silica has interested, due to its large pore volume, high surface area, controllable pore size, stable aqueous dispersion, thermal and mechanical stability [1]. However, lots of investigators work on modification of silica nanoparticle to enhance behavior for various applications.

The bond between silica and the amine group prepares silica to selectively adsorb environmental pollutants in order to control and accelerate the process [2]. Dendritic polymers have been used to increase adsorption capacity. The functional groups of the materials are more accessible and can be achieved with the purpose of removing pollutants in the wastewater. In this method, the dendritic generations are located on the silica surface to enhance adsorption capacity [3].

Adsorption isotherm models explain the reaction between the adsorbent and adsorbate. Isotherm analysis of different models such as Langmuir, Temkin, and Freundlich is explained [4].

In this study, silica particles were fictionalized by amine-terminated dendritic polymers for adsorption of AR1. The influence of amine groups on the adsorption properties was also investigated by kinetic and isotherm measurements.

H. Mirzadeh and A. A. Katbab (Eds.): ISPST 2018, *Eco-friendly and Smart Polymer Systems*, pp. 201–204, 2020.
https://doi.org/10.1007/978-3-030-45085-4_48

2 Experimental

Silica was purchased from LOBA Chemie Co., India. Acid red 1 was obtained from Ciba Ltd. Aminopropyl triethoxysilane (APTES), methyl acrylate (MA), ethanol, and ethylene diamine (EDA) were of laboratory reagent grade and they were supplied from Aldrich.

2.1 Modification of Silica Nanoparticle

Amine terminated groups on silica surface have been reached by modification of silica by divergent method. First, silica, and APTES were mixed in toluene. After that solution was filtered and washed and then was dried in oven (G0 was obtained).

Under a nitrogen atmosphere, a mixture of G0 and MA was stirred in a reactor containing methanol as solvent. The resulting solution was then filtered and washed with ethanol and tetrahydrofuran and dried in oven (G0.5 was produced).

Under a nitrogen atmosphere, G0.5 and EDA and methanol as a solvent was stirred at room temperature in a reactor. The solution was then filtered and washed with ethanol and tetrahydrofuran and dried in oven. The final product of this step noted by G1.0.

2.2 Adsorption Isotherms and Kinetics Experiments

The 20-50-75-100 mg/L concentration solutions were prepared by adding 0.3 g/L adsorbent at room temperature and were stirred for 60 min at different sampling times. The adsorption of AR1 was measured by a spectrophotometer (UNICO 2100) at a maximum wavelength of AR1 that is 506 nm. These kinetic of AR1 was performed by contacting 0.3 g/L adsorbent with dye solution. Finally, adsorption was measured by a spectrophotometer.

3 Results and Discussion

3.1 FTIR

The FTIR spectra of G and G1.0 are shown in Fig. 1. The intensity of all peaks was normalized to reference peak at 1103 cm^{-1}, which is Si-O-Si peak [5]. The peaks of G1.0 at 1550 cm^{-1} were related to primary amine group and 1640 was assigned to bending primary and secondary amine groups. According to these results, the characteristic peaks of these vibration bands indicate that the amine-terminated dendritic polymers were synthesized on the silica surface successfully.

3.2 Isotherms and Kinetics

Comparison R^2 values of three models according to Table 1 and Fig. 2, shows that R^2 values of Temkin and Freundlich are highest. Due to the abundant amine terminal groups at the exterior and hollow dendritic box between cavities in exterior of branches, Freundlich model is better to follow.

The kinetic models fitted by the experimental data demonstrated that the adsorption process followed the pseudo-second-order kinetic model that shown in Fig. 3.

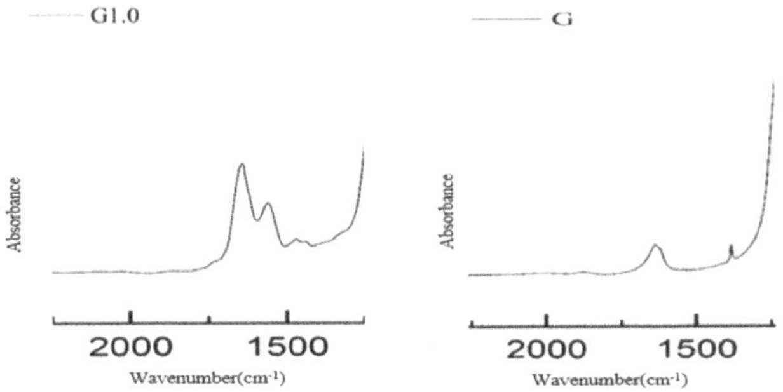

Fig. 1. FTIR spectra of G and G1.0

Table 1. Isotherm parameters for the adsorption of AR1

Langmuir				Temkin			Freundlich		
Q_m	K_L	R_L	R^2	B_1	K_T	R^2	K_F	NN	R^2
833/3	91/74	0/00033	0/93	84/3	4/49	0/99	81/85	1/22	0/98

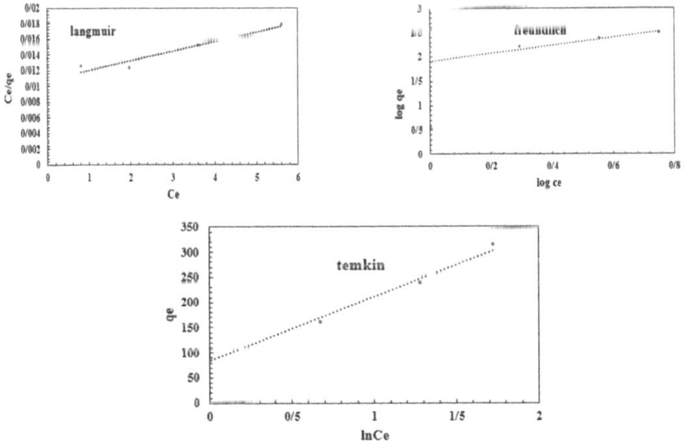

Fig. 2. Adsorption isotherm of AR1

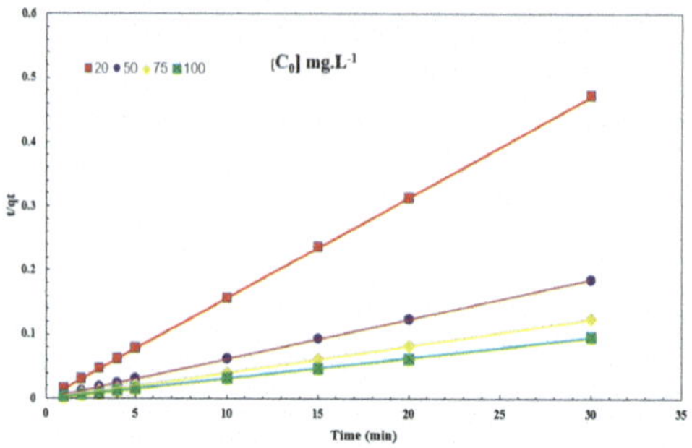

Fig. 3. Pseudo-second-order adsorption kinetic of AR1

4 Conclusion

An adsorbent with an amine terminated group was prepared. The influence of amine group on the surface of silica and adsorption behavior was investigated. The results showed that amine terminated group increased adsorption capacity of AR1. The adsorption kinetics fitted pseudo second order model well. The adsorption isotherms fitted Freundlich model well.

References

1. Zhang, Z., Li, H., Liu, H.: Environ. Sci. J. **65**, 171–178 (2018)
2. Ralph, L., Morrison, A.: Silicon material preparation and economical watering methods (1984)
3. Qu, R., Niu, Y., Sun, C., Ji, C., Wang, C., Cheng, G.: Microporous Mesoporous Mater. J. **97**, 58–65 (2006)
4. Shahamati Fard, F., Akbari, S., Pajootan, E., Arami, M.: Desalin. Water Treat. J. **57**, 26222–26239 (2016)
5. Zhang, Y.: Appl. Surf. Sci. J. **255**, 5818–5826 (2009)

Effect of Waste Paper Extracted Cellulose on the Structure and Properties of Resorcinol Formaldehyde Aerogel

Parisa Hakimi and Ahmad Reza Bahramian(✉)

Polymer Engineering Department, Faculty of Chemical Engineering, Tarbiat Modares University, P.O. Box 14115-114, Tehran, Iran
abahramian@modares.ac.ir

Abstract. Hydrophilicity properties of polymeric aerogels have an effective role in water filtration applications. In this work, the effect of cellulose extracted from the waste paper on the structure and properties of resorcinol formaldehyde aerogel was investigated. The density and porosity of resorcinol formaldehyde aerogels were decreased and increased with the addition of cellulose, respectively. Fragility and shrinkage also decreased with increasing cellulose content. The results showed an increase of 10% in porosity by adding 13 wt% of cellulose extracted from waste paper.

Keywords: Resorcinol · Formaldehyde aerogel · Structural properties · Hydrophobic and hydrophobic · Cellulose

1 Introduction

Expanding industrial oily waste water and frequent oil spill accidents have caused increasing environmental issue. The demand for developing effective and eco-friendly approaches for cleaning up oil and water mixture has been becoming critical important. It is relatively simple, efficient and economic technique to directly filtrate oil from oil/water mixture. Thus exploitation and fabrication of functionalized materials with super-hydrophilic property under water to allow water penetration was developed, strongly [1, 2].

In this work, the effect of cellulose extracted from waste paper was investigated. These samples can be used for oil/water filtration applications.

2 Experimental

2.1 Materials

Cellulose extracted from waste paper was synthesized in Tarbiat Modares Lab. Resorcinol, formaldehyde; nitric acid and sodium carbonate to form the resorcinol formaldehyde aerogel (laboratory grade) were purchased from Merck, Germany.

H. Mirzadeh and A. A. Katbab (Eds.): ISPST 2018, *Eco-friendly and Smart Polymer Systems*, pp. 205–207, 2020.
https://doi.org/10.1007/978-3-030-45085-4_49

2.2 Synthesis of Aerogels

To synthesized Aerogels; at 20–22 °C (room temperature) 25.0 g (0.227 mol) of resorcinol (R) was dissolved in 487.6 g water (W), resorcinol to water molar ratio is 0.008. Then 36.9 g of aqueous solution of formaldehyde (F, 0.454 mol), R:F molar ratio 0.5 and 0.481 g sodium carbonate (C, 0.00454 mol), R:C = 50, were added under stirring. After 5 min stirring the pH was adjusted to 5.4–5.6 using 2 N nitric acid (ca.4 mL). The stirring was continued for 60 min at room temperature [2]. Then in the same 5 container, we pour 10 g of solution. To each container; 0, 0.25, 0.5, 1, and 1.5 g of cellulose was added and mix well and put three layers of polyester in each container and place in oven at 80 °C For 2 days.

2.3 Results and Discussion

Porosity and density of samples were measured by ratio of mass/volume (Table 1), experimentally. The results show an increase of 10% in porosity by adding 13 wt% of cellulose. The density and porosity of resorcinol aerogels were decreased and increased with the addition of cellulose, respectively

Table 1. Porosity and density of samples; P, C, and R are polyester, cellulose, and resorcinol formaldehyde, respectively

Samples	P (wt%)	C (wt%)	R (wt%)
PRC0.00	0.03560	0.00000	0.9644
PRC0.25	0.03079	0.02389	0.9453
PRC0.50	0.03034	0.04652	0.9231
PRC1.00	0.02855	0.08838	0.8831
PRC1.50	0.02943	0.12638	0.8442

Samples	Density (g/mm^3)	Prosity (%)
PRC0.00	0.236	81.14
PRC0.25	0.230	81.63
PRC0.50	0.200	83.97
PRC1.00	0.186	85.14
PRC1.50	0.137	89.06

According to Fig. 1, with increasing cellulose content in aerogel sample, the contact angle of water was increased. From the point where the level becomes uniform, the contact angle decreases, this means that hydrophilicity is increased (Fig. 1). The change in hydrophilicity of sample with increasing of cellulose content from 1 g to 1.5 g is noticeable. Generally, the addition of cellulose changes a hydrophobic surface to a hydrophilic surface. The tuning of hydrophilicity properties is very important in water filtration applications.

As you can see in the Fig. 1, fragility and shrinkage of aerogel sample also decrease with increasing cellulose content.

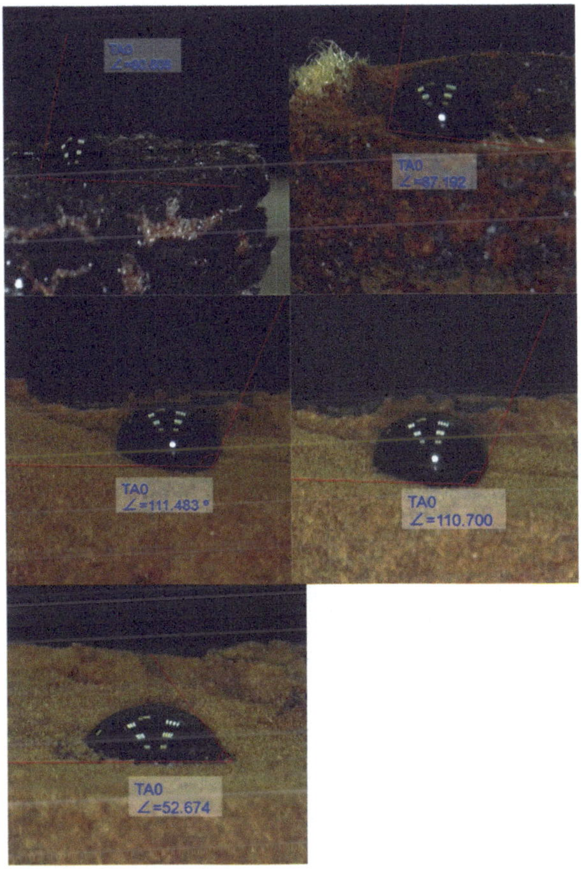

Fig. 1. Contact angle of water in aerogel sample, increasing hydrophilicity by cellulose content

3 Conclusion

An increase of 10% porosity of sample was shown by adding 1.5 g of cellulose to 10 g of resorcinol formaldehyde. In high percentage of cellulose, hydrophilic property of surface was increased. By examining the samples, we observed the improvement of properties in resorcinol aerogels with increasing cellulose content.

References

1. Sun, F., Liu, W., Dong, Z., Deng, U.: Chem. Eng. J. **330**, 774–782 (2017)
2. Schwan, M., Tannert, R., Ratke, L.: J. Supercrit. Fluids **107**, 201–208 (2016)

PVA Electrospun Hollow Nanofibers via One-Pot Coaxial Electrospinning

Niloufar Ghaedi Dehaghi, Sobhan Sharafkhani, and Mehrdad Kokabi(✉)

Department of Polymer Engineering, Faculty of Chemical Engineering, Tarbiat Modares University, P.O. Box 14115-114, Tehran, Iran
mehrir@modares.ac.ir

Abstract. Uniform and beadless PVA hollow nanofibers with a mean shell thickness of less than 100 nm were successfully prepared by one-pot coaxial electrospinning technique. The morphology and diameter of nanofibers were characterized by scanning electron microscopy (SEM). The micrographs indicated that the air flow rate could control the morphology of electrospun hollow nanofibers. Also, they showed that nanofiber diameter increased with increasing air flow rate. To obtain PVA hollow nanofibers, the ratio of air flow rate to the solution feed rate was adjusted at 75.

Keywords: PVA hollow nanofibers · Coaxial electrospinning · Morphology

1 Introduction

Electrospinning is an electrostatically driven process which has been used to fabricate nanofibers. Advances in the methodologies and strategies for synthesizing hollow nanofibers have encouraged their adoption and application in many fields over the past few years.

In many investigations, hollow fibers are fabricated by core/shell electrospinning methodology. First, core/shell fiber was prepared in coaxial electrospinning. Then, the core was removed by post treatments to produce tubular structures. Wong and coworkers showed that coaxial electrospinning could be used to produce hollow nanofibers [1]. In this method, two incompatible polymer solutions were required as core and shell solutions.

Recently, a novel technique, named as one-pot coaxial electrospinning, is applied to prepare hollow nanofibers [2]. The authors concluded that hollow structure could be obtained when the ejected air from the inner syringe was able to carry the Taylor cone. One-pot coaxial electrospinning possesses the potential for the production of hollow nanofibers by the one-step method compared to the traditional two-step method [2]. The resulting fibers were used in drug delivery and tissue engineering applications because they allowed the controlled release of drugs and growth factors [3].

PVA is a hydroxy biocompatible, biodegradable, nontoxic and water-soluble polymer which attracts attention because of promising fiber formation ability in recent years. Nanocomposites based on PVA have found various applications in medical and

H. Mirzadeh and A. A. Katbab (Eds.): ISPST 2018, *Eco-friendly and Smart Polymer Systems*, pp. 208–211, 2020.
https://doi.org/10.1007/978-3-030-45085-4_50

engineering technology due to several properties including electron transport, mechanical and optical properties [4].

In this work, PVA hollow nanofibers with a shell mean thickness of less than 100 nm were electrospun by adjusting the process conditions, containing solution concentration, feed rate and tip to target distance.

2 Experimental

2.1 Materials

Polyvinyl alcohol (degree of polymerization = 1700) was obtained from NIPPON GOHSEI Chemical Co., Japan.

2.2 PVA Solutions Preparation

An aqueous solution of 6 wt% PVA was prepared by dissolving PVA powders in distilled water under constant stirring at 75 °C for 2 h.

2.3 Coaxial Electrospinning

A high-voltage (19–20 kV) was applied to form a high electric field between the needle and the collector device. The gap between the needle and the collector was set at 190 mm. The inner diameter of the shell and core steel needles were 1.07 and 0.63 mm, respectively. Nanofibers were collected on a sheet of aluminum foil.

PVA solution was injected into the outer needle at a constant rate of 0.2 $mL.h^{-1}$. However, the air was flowed into the inner needle at different rates of 0, 0.2, 3, and 15 $mL.h^{-1}$.

2.4 Characterization of PVA Hollow Nanofibers

The morphology and diameter of the electrospun PVA hollow nanofibers were determined using (AIS2100, Seron Technology) SEM. A small section of the nanofiber mat was placed on the SEM sample holder and sputter coated with gold (BAL TEC). Accelerating voltages of 15 and 20 kV were applied to prepare the SEM micrographs. Image J was employed to calculate the average diameter of electrospun nanofibers.

3 Results and Discussion

The SEM micrographs of solid and hollow PVA electrospun nanofibers are shown in Fig. 1. It can be observed that solid nanofibers have smooth and regular morphology (Fig. 1a). Also, the frequency distribution of solid nanofibers diameter is presented in the inset of the micrograph. Solid nanofibers diameter varied in the range of 60 nm to 120 nm.

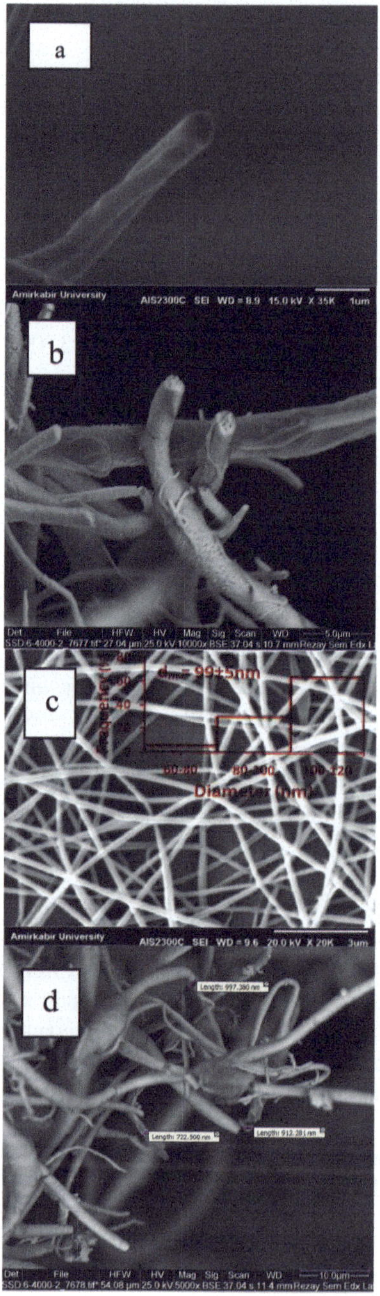

Fig. 1. SEM graphs of as-spun nanofiber in top view for solid nanofibers: (a) and cross-section view of hollow nanofiber (b, c, d)

In the presence of airflow, air trapped in the fibers and the average diameter of nanofibers increased. When the flow rate of air was set at the low value of 0.2 $mL.h^{-1}$, the cross section of PVA nanofibers was solid (Fig. 1b). By increasing the air flow rate to 3 mL/h, the nanofibers cross-section changed from solid to an irregular porous structure (Fig. 1c). With a noticeable increase in the air flow rate to 15 mL/h, uniform hollow PVA nanofibers could be prepared (Fig. 1d). The mean shell thickness of PVA hollow nanofibers was less than 100 nm. The results confirmed that the air flow rate had to be controlled accurately to ensure that the shell could be formed during electrospinning. Furthermore, it was necessary to set an appropriate ratio of the air flow rate to the PVA solution feed rate to form a stable Taylor cone. If the air flow rate is too low or too high, the electrospinning process would fail [3]. In this work, the optimum ratio of the air flow rate to solution feed rate was 75 in order to obtain uniform hollow PVA nanofibers.

4 Conclusion

The electrospinning of PVA hollow nanofibers with a mean shell thickness of less than 100 nm was successfully processed. The optimum flow rate of the PVA solution and air was set at 0.2 and 15 $mL.h^{-1}$, respectively.

References

1. Chen, P., Wong, S.-C., Hague, S., et al.: Polymer **53**, 2736–2743 (2012)
2. Xia, Z., WeiWu, H., Wong, S.-C., et al.: Mater. Manuf. Process. **33**, 1–9 (2017)
3. Pan, C.-T., Yen, C.-K., Wang, S.-Y., et al.: RSC Adv. **5**, 85073–85081 (2015)
4. Roy, A.S., Gupta, S., Sindhu, S., et al.: Compos. Part B Eng. **47**, 314–319 (2013)

A Molecular Dynamics Approach to Study Polymer/Nano-Filler Interactions

Hesam Makki(✉), Farhad Sharif, and Morteza Ebrahim

Department of Polymer and Color Engineering,
Amirkabir University of Technology, Tehran, Iran
hmakki@aut.ac.ir

Abstract. There is a common knowledge that the dynamic and structural behavior of polymer chains in the vicinity of nano-fillers is rather different from the bulk. Despite the extensive literature on this topic, this phenomenon is not yet fully understood. Apart from several experimental methods to study it, molecular simulations, in particular molecular dynamics (MD) methods, are good candidates for such study. In other words, MD simulation techniques give us a remarkable ability for mechanistic studies of dynamic and structural properties of polymer chains in nano-scales and they are proved to be highly reliable methods for estimating such properties. Therefore, in this study, we developed a time and spatially-resolved molecular dynamic method by which the structural and dynamic properties of two polymers, i.e., polypropylene (PP) and polyvinyl alcohol (PVA), are studied as a function of distance from graphene (G) and graphene oxide (GO) nano-layers.

Keywords: Polymer dynamics · Polymer orientation · Polyvinyl alcohol · Polypropylene · Graphene

1 Introduction

Understanding the structural and dynamic properties of polymers, in particular, at the interface of polymer/nano-filles is of a great importance. This paves the way of predicting different properties of nano-composites. Graphene and graphene oxides are very well known nano-fillers for polymer nano-composites due to exceptional mechanical, electrical and barrier properties they introduce to nano-composites. Moreover, due to the planar shape of these nano-layers, they tend to be in form of stacked structures, that is graphite. Oxidation of graphite is the most relevant method to exfoliate these layers and form graphene nano-layers. However, since the formation of oxygen-containing groups changes the surface properties of the G layers, one intrigues that how these modifications influence the behavior of polymer chains in the vicinity of GO layers and how the interactions between polymer and nano-fillers are altered by such modifications.

In the following section, we introduce the atomistic models which have been used to build up our polymer NC and the MD methods which have been employed to study the structural and dynamic properties of polymer chains. Next, we show the results of MD simulations and discuss our analyses.

H. Mirzadeh and A. A. Katbab (Eds.): ISPST 2018, *Eco-friendly and Smart Polymer Systems*, pp. 212–215, 2020.
https://doi.org/10.1007/978-3-030-45085-4_51

2 Experimental

2.1 Methods

Six different nano-composites, i.e., two bulk polymers (PP and PVA), two G-based polymer nano-composites (PP/G and PVA/G) and two GO-based polymer nano-composites (PP/GO and PVA/GO), have been made for our atomistic MD simulations. The sizes of both G and GO are approximately 4 × 3 nm, see Fig. 1(a, b). Based on various experimental studies, the composition of GO can vary with the synthesis method and environment [1]. We performed MD simulations on GO molecules with oxygen to carbon ratio of 20% in such a way that only hydroxyl and epoxy groups on the basal plane exist. Note that the ratio of hydroxyl to epoxy groups is retained at 3/2.

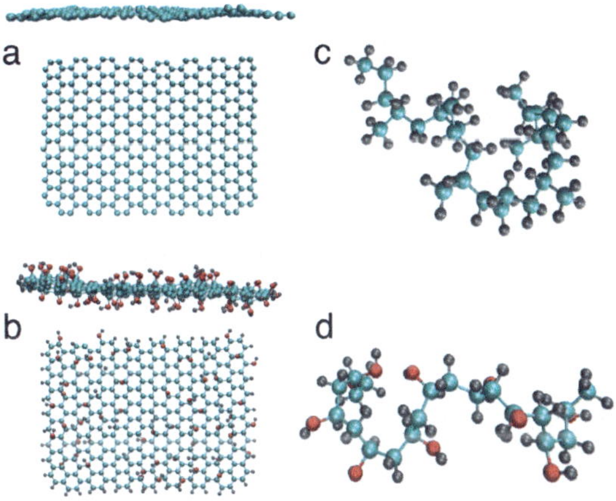

Fig. 1. (a), (b), (c), and (d) show G, GO, PP, and PVA atomistic structures. (c) and (d) images are shown in a higher level of magnification as compared to (a) and (b) [2]

Atomistic MD simulations were performed using the open-source MD simulation package Gromacs [3]. For G and GO, we applied a force field previously introduced [4]. Both polymers are represented by all-atom OPLS-AA force field. Using this force field, the bulk densities of PP at 300 K and PVA at 450 K are calculated as 783.4 and 1154.07 kg/m^3, respectively. These values are in good agreement with experimental and reported simulation data.

In this paper, we study the overall and local structural and dynamic behavior of the polymers. For local properties, we divided the whole simulation box into several domains (layers) based on the minimum distance of the center of mass of short polymer chains to G or GO molecules, see Fig. 2.

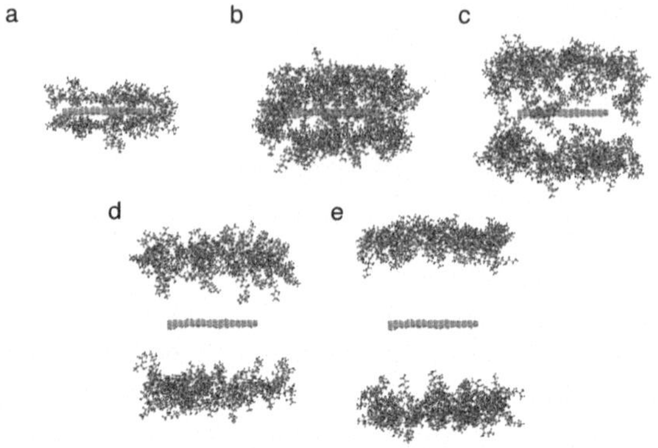

Fig. 2. (a) to (e) show the 1st to 5th layer of polymer chains, respectively. The nano-filler position (carbon atoms of G and GO molecules) is restraint in the middle of the box [2]

3 Results and Discussion

We studied the density and conformational changes of polymer chains as a function of distance from nano-filler. By analyzing such properties for different polymers at different layers, one can obtain a better insight into the effect of G/GO on the structural behavior of polymer chains. In addition, we estimated the translational and orientational relaxation times for polymer chains as a function of distance to the nano-fillers. By analyzing these quantities for all combinations of polymer/nano-filler, valuable information is gained regarding the origin of structural and dynamic behaviors of polymers used in nano-composites.

Our analyses for PP/G, PP/GO, PVA/G, and PVA/GO hybrid systems and comparing them with pure PP and PVA systems showed that significant heterogeneities in polymer chains orientation and dynamics are introduced to the systems, in the presence of G and GO. Moreover, our analyses indicate that there are different origins for polymer orientation and dynamics at the polymer/filler interface. Based on our observations, for abovementioned polymer NC, it is shown that polymer chains orientation, regardless of the type of polymer-filler interaction, are more pronounced for G-based NC as compared to GO ones. However, the mobility of the chains very much depends on the combination of polymer and nano-filler in such a way that the longer estimated relaxation times at the interface for our polar polymer (PVA) has been obtained in the presence of GO-based NC while our apolar polymer (PP) exhibits a longer relaxation time in G-based NC. Therefore, we believe that immobilization of polymer chains at the interface in G-based NC very much depends on the interactions between polymer and filler, however, the polymer chains orientation is also influenced by other factors. This has been reconfirmed by calculating the interaction and bonding energies of different systems. In addition, since GO has oxygen-containing functional groups on the basal plane, i.e., hydroxyl and epoxy groups, the increased roughness and the heterogeneous shape of the nano-filler could be a possible reason for lower amount of polymer orientation in the vicinity of GO as compared to G.

4 Conclusion

In this work, we successfully developed a MD method to study the dynamics and structural properties of polymer chains in the presence of nano-fillers. Our method is time- and spatially resolved and we believe it is capable of being used for any polymer/nano filler combinations.

References

1. Bagri, A., Mattevi, C., Acik, M., Chabal, Y.J., Chhowalla, M., Shenoy, V.B.: Nat. Chem. **2**, 581 (2010)
2. Azimi, M., Mirjavadi, S.S., Hamouda, A.M.S., Makki, H.: Macromol. Theory Simul. **26**, 1600086 (2017)
3. Berendsen, H.J.C., van der Spoel, D., van Drunen, R.: Comput. Phys. Commun. **91**, 43 (1995)
4. Tang, H., Liu, D., Zhao, Y., Yang, X., Lu, J., Cui, F.: J. Phys. Chem. C **119**, 26712 (2015)

Curing Characteristics of Phenolic Syntactic Foams Based on Oxygen Plasma Treated Hollow Glass Microspheres

Milad Karbalaei-Bagher[1(✉)], Rasool Amirkhani[2], and Mohammad-Hassan Omidi[3]

[1] Department of Polymer Engineering and Color Technology, Amirkabir University of Technology, Tehran, Iran
m.karbalaei@aut.ac.ir
[2] Chemistry and Chemical Engineering Research Center of Iran, Tehran, Iran
[3] Department of Chemistry, Faculty of Science, University of Guilan, Guilan, Iran

Abstract. Curing behavior of syntactic foams based on phenolic resins and hollow glass microspheres treated by oxygen plasma was investigated through differential scanning calorimetry (DSC) analysis. Plasma treatment was carried out and the existence of carbonyl group was approved by Fourier-transform infrared spectroscopy (FT-IR) via the appearance of an extra peak at 1730 cm^{-1} wavenumber. The composites were prepared by solution mixing and the results revealed that by using treated particles within phenolic resins, the curing temperature shifted to lower temperatures from 160 °C to 152 °C. This accelerated effect was endorsed by Flynn-Wall-Ozawa model by which the extent of reaction was traced with variations of activation energy.

Keywords: Curing · Syntactic foam · Plasma treatment · Hollow glass microsphere

1 Introduction

As an important part of composites materials, syntactic foams have abundant applications in a wide variety of fields such as deep-sea vehicles, space vehicles, aircraft, snow skis, soccer balls, and thermal insulation in underwater pipelines owing to their thermal and dimensional stability, low moisture absorption as well as low thermal conductivity and high compressive strength. Compared to other thermosetting resins, phenolic types are more difficult to burn. Even after burning, they have the least smokes and toxic gases generated [1]. It is well documented that the interaction between glass microspheres and thermosetting resin plays a critical role in determining the mechanical properties of composites. Otherwise, the microspheres may act as flaws if they are not adequately secured in the matrix [2]. In order to have a better dispersion of glass microspheres within a polar matrix various techniques are thoroughly investigated and suggested. Among them, plasma treatment has the advantages of short treatment time, an environmentally friendly manufacturing process, and the generation of numerous active species [3].

H. Mirzadeh and A. A. Katbab (Eds.): ISPST 2018, *Eco-friendly and Smart Polymer Systems*, pp. 216–219, 2020.
https://doi.org/10.1007/978-3-030-45085-4_52

A large number of charged particles can be generated by the plasma discharge, and these high-energy particles can affect the surface of materials. During this process, the clusters of atom in the treated material are exploded and distributed randomly, forming uniform nanostructures.

2 Experimental

2.1 Materials Used

The phenolic resin was a commercial novolac provided as a powder containing hexamethylenetetramine as a hardener from Resitan Co. The hollow glass microsphere particles were purchased from Potters Company, USA. The average diameter and bulk density of particles were 60 μm and 0.5 g . cm^{-3}, respectively.

2.2 Samples Preparation

The glass particles were dispersed in acetone and poured uniformly in a petri dish. After vacuuming for an hour in 60 °C to evaporate the solvent, the particles were treated by O_2 as the plasma gas at the flow of 20 cm^3/min in a chamber pressure of 10 Torr and a microwave power of 100 W for 60 s. Thereafter, both treated and untreated particles were again dispersed in acetone along with phenolic resin at 10 wt %. Finally, composite samples were vacuumed oven at 80 °C well beyond the curing temperature of resin and prepared for DSC analysis.

3 Results and Discussion

FT IR spectra of both treated and untreated samples are displayed in Fig. 1. As it is shown, for the sample treated with oxygen plasma, there is a band in the vicinity of 1730 cm^{-1} which is due to the stretching vibration of the carbonyl group induced at the surface of glass microspheres.

The curing reaction of samples has been widely studied using multiple heating rates (only the 10 °C/min is shown for the comparison) because this method allows obtaining trustworthy kinetic parameters. A careful look at the DSC thermograms in Fig. 2 reveals that with the presence of carboxyl group in the surface of the particles, the curing temperature has shifted to lower temperatures from 160 °C to 152 °C. Also, the total heat of reaction has decreased from 23 J/g to 18 J/g for the composites with treated particles. This catalytic effect of carboxylic group is due to its accelerating performance in the curing reaction. Having an accelerated reaction, the degree of conversion is increased extremely at the onset of curing transition and due to low thermal conductivity of particles as well as large 3D networks of resins, sufficient heat could not be available for other small molecules to be cured. This effect can be traced easily when considering the activation energy (Ea) from Flynn-Wall-Ozawa (FWO) model since it assumes that the changes in the extent of cure are related with the variation in activation energy. The model can be expressed as Eqs. (1) and (2):

$$Ln\beta = Ln(AE_a/R) - \ln f(a) - 5.331 - 1.052(E_a/RT) \quad (1)$$

$$f(a) = \int_0^a \frac{da}{g(a)} \quad (2)$$

Where, $g(\alpha)$ is the integral conversion function, $\beta = dT/dt$ is a constant heating rate, R is the universal gas constant, T is the absolute temperature and A is the pre-exponential factor for the reaction. A straight line should be obtained by plotting ln β vs. $1/T$ at a particular conversion α and Ea can be determined from the slope. While for the untreated samples, rising of Ea starts at the conversion of 0.8, the catalytic reaction causes the rapid formation of crosslinked network in treated samples proceeding at the conversion of 0.5 (Fig. 3).

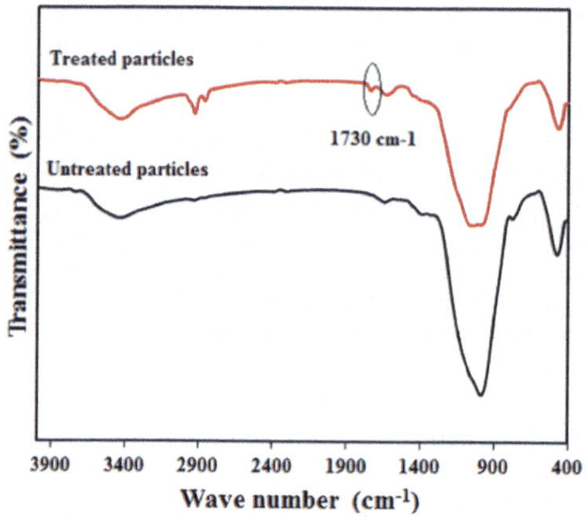

Fig. 1. FT-IR Spectra of treated and untreated particles

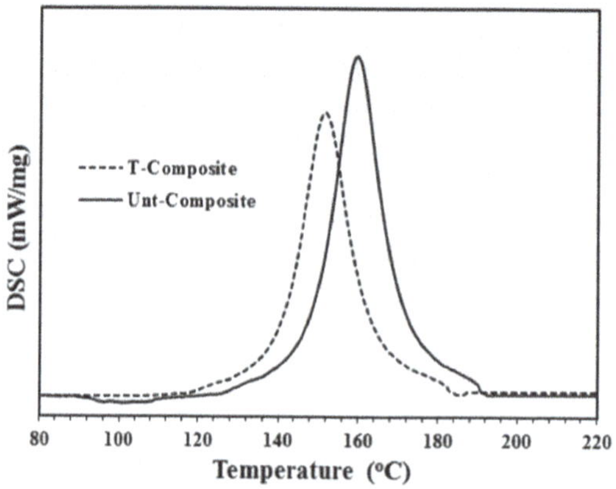

Fig. 2. DSC thermograms of composites containing treated and untreated particles at a rate of 10 min/°C

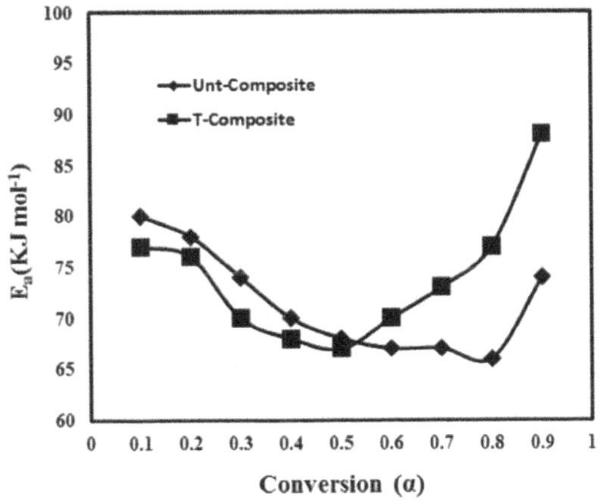

Fig. 3. Ea versus degree of conversion for composites containing treated and untreated particles

4 Conclusion

The scope of this work is oriented to evaluate the effect of oxygen plasma treated hollow glass microspheres on the curing behavior of phenolic resins. Results showed that the use of treated particles has led to an increment in the curing temperature and activation energy from 60 °C to 52 °C and from 0.8 to 0.5 conversion, respectively.

References

1. Yang, H., Jiang, Y., Liu, H., Xie, D., Wan, C., Pan, H., Jiang, S.: J. Colloid Interface Sci. **530**, 163–170 (2018)
2. Zhang, X., Wang, P., Zhou, Y., Li, X., Yang, E.H., Yu, T.X., Yang, J.: Compos. Part B Eng. **101**, 53–63 (2016)
3. Stloukal, P., Novák, I., Mičušík, M., Procházka, M., Kucharczyk, P., Chodák, I., Lehocký, M., et al.: Int. J. Polym. Mater. Polym. Biomater. **67**(3), 161–173 (2018)

Optimization of Methylene Blue Dye Batch Adsorption by Carboxylic Acid Functionalized Halloysite Nanotubes

Mohammad Hassan Kanani-Jazi and Somaye Akbari(✉)

Textile Engineering Department, Amirkabir University of Technology, Tehran, Iran
akbari_s@aut.ac.ir

Abstract. The adsorption behavior of methylene blue from aqueous solution was investigated to identify the ability of functionalized halloysite nanotubes (HNTs) to remove colored textile dye from simulated wastewater. For this purpose, carboxylic acid groups were synthesized into the halloysite as the mineral adsorbent. The synthesis process was investigated by X-ray photoelectron spectroscopy (XPS) which confirm functionalization. The statistics parameter was used to optimize the methylene blue dye batch adsorption by carboxylic acid functionalized HNTs. The optimum adsorbent dosage and pH for dye removal were obtained to be 0.3 and 9 g/L, respectively

Keywords: Halloysite · Carboxylic acid group · Optimization · Methylene blue

1 Introduction

Dyes create the attractiveness of the world, but some dyes are dangerous to human and the environment. Organic dyes are widely used in several industries such as textile, paper and pulp manufacturing, dyeing and cloth, food, cosmetic pharmaceutical, etc. [1]. Therefore, the wastewater of these industries which containing dye must be carefully purified before discharge, because it may cause serious dangerous problems such as a disorder in photosynthesis and catastrophic effect in humans being [2]. There are several methods including adsorption, ion-exchange, filtration, and membrane separation, for removing dyes. Among these methods adsorption by natural clay mineral is a suitable choice and most effective and economical method for removal dyes and pigments [3]. Despite these novel characteristics, functionalization of HNTs enhance the adsorption properties.

Halloysite nanotubes (HNTs) are natural clay minerals. HNts are made of octahedral gibbsite of Al-OH groups on the inner surface and tetrahedral sheets of Si-O-Si groups on the outer surface and consist of hollow tubular by multiple rolled layers. HNT has many hydroxyl groups at its surface and the gap space between its layers has also a significant tendency to adsorb some pollutants [4].

H. Mirzadeh and A. A. Katbab (Eds.): ISPST 2018, *Eco-friendly and Smart Polymer Systems*, pp. 220–223, 2020.
https://doi.org/10.1007/978-3-030-45085-4_53

In this study, HNTs were functionalized with carboxylic acid groups to improve the removal capacity toward cationic dye molecules. The statistic parameter was used to optimize the methylene blue dye batch adsorption.

2 Experimental

Halloysite nanotube, (diameter of 80 nm) was purchased from Delta-Dolsk Co., Poland. All chemicals were of analytical grade.

First, HNTs were purified with HCl for 24 h under constant stirring. Then, 5 g of dried HNTs-OH was added to toluene (100 mL) containing 10 mL of APTES. The suspension was refluxed at 80 °C for 12 h under constant stirring. In the next step, the solid phase was separated and washed 3 times with ethanol and methanol using Buchner funnel to remove the excess organosilans. 1 g of dried HNTs-NH_2 was mixed with 50 mL of DMF and sonicated for 5 min. Then 1.2 g of succinic anhydride was added to the solution which was then refluxed and stirred at 90 °C for 24 h. The solid phase was separated and washed 3 times with DMF using a Buchner funnel.

3 Results and Discussion

3.1 XPS Analysis

Figure 1 shows C1S spectra of HNTs-COOH. Spectra of HNTs-COOH could be curve fitted with 5 peaks at the binding energy of 284/6 eV for C-C, 286/9 eV for C=O, 287.2 eV for C–OH, 288.3 eV for O-C=O, and 282.3 eV for C–Si. The presence of O-C=O in C1S spectra confirmed the correctness of modification.

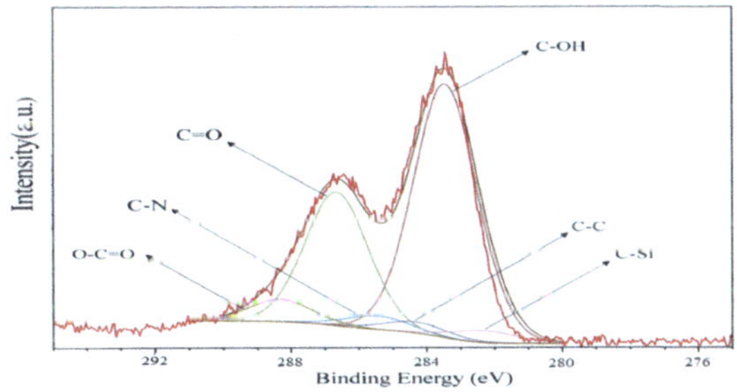

Fig. 1. XPS spectra of HNTs-COOH

3.2 Statistical Analysis and Optimization

There are various parameters which can affect the efficiency of the modification process. In our opinion, the dye concentration, adsorbent dosage, and pH of the process might have significant effect on the final properties of modified HNTs. Carboxylic acid functionalization of HNTs has improved the removal efficiency about 30%.

27 experiments were designed randomly using Box-Behnken design. Dye removal efficiency from wastewater was measured for carboxylic acid functionalized HNTs at different conditions according to the mentioned method. Results were analyzed according to P values which indicate the significant level of the model coefficients. In this study, the significant level was 95% so the P value should be lower than 0.05. The obtained model for dye removal is shown in Eq. (1):

$$Yijkm = \mu + Ai + Bj + ABij + Ck + ACik + BCjk + ABCijk + \varepsilon q(ijkm) \tag{1}$$

Where A, B, and C are the concentrations of dye, adsorbent dosage, and pH of the process in coded units, respectively. The coefficients of the response function and estimated P values of the parameters for methylene blue removal (%) has been shown in Table 1. The results showed that P value (≤ 0.05) for all independent parameters, except time, indicated that the three selected factors had a significant effect on dye removal. 3D plots for methylene blue removal are shown in Fig. 2. It can be seen that the efficiency of methylene blue dye removal went up by increasing adsorbent dosage and pH and then fell down with increasing dye concentration. The results show that the number of the active sites (carboxylic acid) on the HNTs appear to affect the amount of dye removal. It was observed that with rising the pH process and adsorbent dosage up to 0.3 g/L, the active sites increased and in turn, the dye adsorption increased due to the attraction forces between the negative carboxylic acid groups and cationic methylene blue. Also, by increasing the adsorbent dosage up to 0.3 g/L the dye removal increased up to 95%.

Table 1. ANOVA for methylene blue removal model

Source	Degree of freedom	Sum of squares	Mean of squares	F value	P value Prob > F
Model	26	26604.33	1023.24	46.16	<0.0001
A-conc	2	975.00	487.50	21.99	<0.0001
B-adsor	2	18488.78	9244.39	417.04	<0.0001
C-pH	2	4900.00	2450.00	110.53	<0.0001
AB	4	867.56	216.89	9.78	<0.0001
AC	4	305.67	76.42	3.45	0.0212
BC	4	288.22	72.06	3.25	0.0267
ABC	8	779.11	97.39	4.39	0.0017
Residual	27	598.50	22.17		
Cor Total	53	27202.83			

The model was optimized using response optimizer tool. The highest dye removal from wastewater was the aim of the present work. The optimal values of the process parameters as presented in Table 2 reached dye removal of 95%. In order to examine the accuracy, three experiments were carried out in optimal conditions and the average value is reported.

Table 2. Optimum condition for removal methylene blue by HNTs-COOH

pH	Adsorbent dosage (g/L)	Dye concentration (mg/L)	Dye removal efficiency (%)	
			Predicted	Experimental
9	0.3	20	98	95

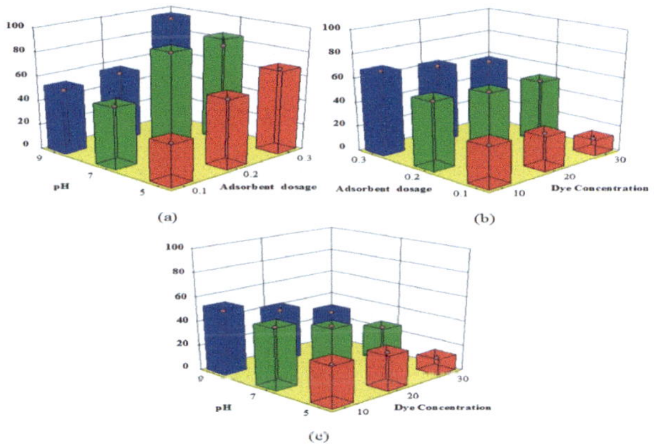

Fig. 2. Surface plots as a function of: (a) pH and adsorbent dosage for dye removal, (b) adsorbent dosage and dye concentration for dye removal, and (c) pH and dye concentration for dye removal

4 Conclusion

This study demonstrated that the addition of carboxylic acid groups via succinic anhydride can increase the interaction of the negatively charged sorbent and positive methylene blue dye molecules. The optimum parameter such as adsorbent dosage, dye concentration, and pH was determined by statistics parameter.

References

1. Hayati, B., Mahmoodi, N.M., Arami, M., Mazaheri, F.: Clean-Soil Air Water **39**(7), 673–679 (2011)
2. Yu, L., Wang, H., Zhang, Y., Zhang, B., Liu, J.: Environ. Sci. Nano **3**(1), 28–44 (2016)
3. Shahamat Fard, F., Akbari, S., Pajootan, E., Arami, M.: Desalin. Water Treat. **57**(54), 26222–26239 (2016)
4. Carli, L.N., Daitx, T.S., Soares, G.V., Crespo, J.S.: Appl. Clay Sci. **87**, 311–319 (2014)

Natural Bio-based Polymers and Composites

Effect of Crosslinking Agent Concentration on Swelling Behavior of Acrylic-Nanocellulose Hydrogels

Mahnaz Shahzamani(✉), Navid Naseri, and Fariba Safaei

Chemistry and Chemical Engineering Technical Centre,
Academic Centre for Education, Culture and Research (ACECR),
Isfahan University of Technology, Isfahan, Iran
m.shahzamani@ce.iut.ac.ir

Abstract. In this study, composite hydrogels with interpenetrated polymer networks (IPNs) based on cellulose nanofibrils (CNFs) and poly(acrylic acid) comprising low amount of CNFs were synthesized through radical polymerization for agricultural applications. N,N′-Methylene-bis-acrylamide (MBA) was used as crosslinking agent. The formation of this IPN structure was studied with thermogravimetric analysis (TGA). The onset of PAA dehydroxylation began at higher temperature in the presence of cellulose nanofibrils. The MBA concentrations of 0.05, 0.1, 0.2, 0.3, and 0.4 wt% were used to investigate its influence on the swelling degree and strength of the samples. The absorption capacities measured in distilled, drinking and saline (0.9%) water showed an increase upon increasing crosslinking agent concentration. Absorptions under load (AUL) data presented higher strength (ranged from 17 g/g to 21 g/g) with increasing the network crosslink density up to 0.3 wt% MBA concentration. The absorption rate data also followed the same trend as AUL data.

Keywords: Hydrogel · Superabsorbent · Acrylics · Cellulose nanofibers

1 Introduction

Hydrogels are hydrophilic polymers which can absorb significant amounts of water or aqueous solutions in their networks [1, 2]. These materials have been greatly suggested for horticultural purposes over the last decades to take advantage of the swelling and water release properties. Most conventional hydrogels on the market are acrylic-based products with high cost and poor biodegradability considered as potential disadvantages for agricultural uses in short cultivations [3]. A method to improve the properties and overcome the disadvantages of hydrogels is combining synthetic polymers with natural materials [1, 4]. Cellulose as one of the most abundant resource with plentiful hydrophilic functional groups attracted much attention for hydrogels preparation [5, 6].

There are many researches on the cellulose-acrylic hydrogels [6–10]. However, cellulose nanofibrilated hydrogels were rarely studied [1, 7, 8]. Nano-fibrilated cellulose (NFC) or cellulose nanofibers due to their high aspect ratio indicated high mechanical properties in nanocomposites compared to other cellulose derivatives [4]. Spagnol et al. studied on NFCs as a filler to reinforce acrylic hydrogel [7]. A few years

H. Mirzadeh and A. A. Katbab (Eds.): ISPST 2018, *Eco-friendly and Smart Polymer Systems*, pp. 227–230, 2020.
https://doi.org/10.1007/978-3-030-45085-4_54

ago, Bitis et al. patented superabsorbent composites comprising NFCs and claimed that nanofibrils may act as crosslinker in these nanocomposites [11] not reported in detail. Hence, it needs to be study more the variation of chemical crosslinker in the presence of CNFs in these networks.

The present study describes the synthesis of nanocomposite hydrogels based on wood cellulose nanofibers and polyacrylic acid via in-situ radical polymerization. Furthermore, the effect of MBA concentration as crosslinking agent on the swelling degree and strength of these kinds of interpenetrating networks (IPNs) hydrogels was investigated.

2 Experimental

Acrylic acid (AA, LG Chem Co.), N,N′-methylene-bis-acrylamide (MBA), potassium hydroxide (KOH), potassium persulphate (KPS, 99%) were all supplied from Merck Co. Cellulose nanofibrils (CNF) (2.5 wt%, with mean diameter of 32 nm) used in this work was obtained from Nano Novin Polymer Co., The hydrogels were synthesized through radical copolymerization in deionized water as solvent. In order to check the influence of MBA concentration upon the water absorption capacity of hydrogels, MBA concentration was changed from 0.05 wt% to 0.4 wt% while all other components remained constant.

To prepare the hydrogels with CNF content of 2 wt%, nanocellulose dispersion was first diluted uniformly in KOH solution in an ice bath. Then, the predetermined amount of AA was dissolved in the mixture until the neutralization was completed. Then, the crosslinking agent (MBA) and initiator (KPS) solutions were added to the monomer mixture. After mixing properly, the mixture was charged into the reactor equipped with a mechanical mixer, an efficient reflux condenser and thermometer. The polymerization carried out at 70 °C for 70 min. Finally, the obtained hydrogel was dried in a conventional oven.

3 Results and Discussion

In our previous work, it was reported that acrylic hydrogels indicated higher swelling degree and absorption under load in the presence of nanofibrilated cellulose. During hybrid hydrogel preparation, it was observed that at low MBA concentration (lower than 0.05%) gel did not form completely while in pure polyacrylic hydrogels occurred. It can be indicated that CNFs disrupted the gelation process by influencing chemical reactions. Thus, the hybrid hydrogels need higher MBA rather than pure acrylic hydrogels.

Figure 1 shows the swelling degree of synthesized hydrogel upon MBA concentration in different media. Considering the constant amounts of CNFs, it can be seen that swelling degree of the hybrid hydrogels follows a similar trend to that of pure acrylic hydrogels. It is indicated that high amount of MBA increases crosslink density leading to low swelling degrees. This is not wondering at all because additional crosslink points are formed upon IPNs formation [1]. However, reducing the absorption capacity can be useful in controlled release fertilisers for agricultural applications. The network with low swelling degree may be desired due to slower diffusion rate of fertiliser in water [1].

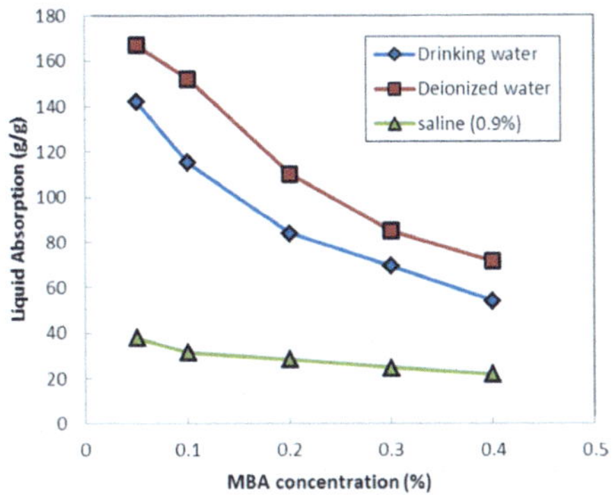

Fig. 1. Influence of MBA concentration on the hydrogel equilibrium swelling properties in distilled water (■), drinking water (♦), and saline water (0.9 wt%) (▲)

AUL can be considered as a measure of the hydrogel strength. The effects of MBA concentration on absorption under load (0.3 psi) are shown in Fig. 2. The AUL changed from 17.76 g/g to 21.24 g/g. Unexpectedly, AUL increases with increasing MBA concentration up to 0.3 wt% and beyond that the absorption does not change considerably. It may be interpreted that at higher MBA concentration, the cyclic structure was formed rather than being incorporated into the 3D network of hydrogel [12].

According to these experimental results, the reinforcement of these IPN hydrogels can be explained by both mechanical and chemical terms: the high aspect ratio of CNFs along with their ability to make bond with MBA [11].

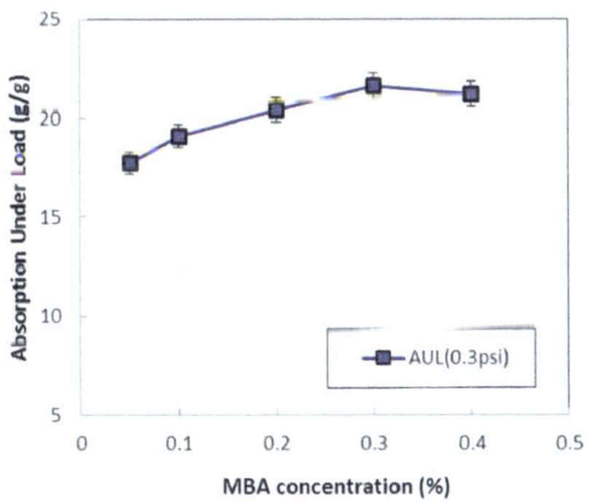

Fig. 2. Influence of MBA concentration upon the absorption under load (AUL) properties of acrylic-nanocellulose hydrogels

4 Conclusion

Superabsorbent acrylic polymers comprising nanofibrillated cellulose (NFCs) were synthesized through radical polymerization to make interpenetrating networks (IPNs). The effects of crosslinker concentration were investigated on their swelling properties. The experimental results showed that reinforcement mechanism of NFCs in these IPN networks can be proposed in terms of mechanical (physical entanglement) and chemical (chemical bond attachment with crosslinker).

References

1. Zaharia, A., et al.: RSC Adv. **8**(32), 17635–17644 (2018)
2. Zhang, J., et al.: J. Appl. Polym. Sci. **132**(34), n/a–n/a (2015)
3. Cannazza, G., et al.: Water **6**(7), 2056–2069 (2014)
4. Ma, J., Li, X., Bao, Y.: RSC Adv. **5**(73), 59745–59757 (2015)
5. Shen, X., et al.: Green Chem. **18**(1), 53–75 (2016)
6. Fekete, T.: Ph.D. thesis, Dep. Phys. Chem. Mater. Sci. (2017)
7. Spagnol, C., et al.: Carbohyd. Polym. **87**(3), 2038–2045 (2012)
8. Yuan, N., et al.: Mater. Sci. Eng. C **67**, 221–230 (2016)
9. Helmiyati, G.H.A., Kurniawan, S.: IOP Conf. Ser. Mater. Sci. Eng. **188**, 012024 (2017)
10. Kevin, J., France, T.H.D., Cranston, E.D.: Chem. Mater. **29**, 4609–4631 (2017)
11. Choudhary, M.S.: Macromol. Symp. **277**, 171–176 (2009)

Swelling Behavior of Acrylic-Cellulose Hydrogels for Agricultural Applications: Effect of Reactants Concentration

Mahnaz Shahzamani(✉) and Navid Naseri

Chemistry and Chemical Engineering Technical Centre,
Academic Centre for Education, Culture and Research (ACECR),
Isfahan University of Technology, Isfahan, Iran
m.shahzamani@ce.iut.ac.ir

Abstract. Effect of reactants concentration on swelling properties of cellulose-acrylic hydrogels was investigated in this study. The composite hydrogels with interpenetrated polymer networks (IPNs) comprising low amount of cellulose nanofibrils (CNFs) were synthesized through radical polymerization. Cellulose concentration was changed from 0.5 wt% to 2.5 wt% by using different amounts of water. The swelling behavior was measured in different aqueous media and fertilizers such as urea and ammonium sulfate solution. The absorption ranges of hydrogels in saline water (0.9 wt%) and ammonium sulfate (0.12 mol/L) are significantly lower than those in other absorption media (<25 g/g). The equilibrium absorption of the prepared hydrogels in urea solution, drinking water and distilled water changed from 62 g/g to 64 g/g, 54 g/g to 68 g/g, and 71 g/g to 79 g/g, respectively. Absorptions under load (AUL) data presented a slight decrease in strength at low cellulose concentration (0.5 wt%).

Keywords: Hydrogel · Superabsorbent · Acrylics · Fertilizer · Cellulose nanofibers

1 Introduction

The hydrogels with high water absorption capacities have been greatly suggested for horticultural applications [1, 2] especially for effective utilization of water in dry regions [3]

Cellulose-acrylic hydrogels have attracted much attention due to their biodegradability and high mechanical properties [4, 5]. Although there are many researches on cellulose-acrylic hydrogels [5–9], those comprising cellulose nanofibrilated have been rarely studied [1, 6, 7]. Nano-fibrilated cellulose (NFC) or cellulose nanofibers due to their high aspect ratio indicated high mechanical properties in nanocomposites compared to other cellulose derivatives [10]. Spagnol et al. studied on NFCs as a filler to reinforce acrylic hydrogel [6]. A few years ago, Bitis et al. patented superabsorbent composites comprising NFCs [11]. However, there are some unknown issues associated with these composite hydrogels.

H. Mirzadeh and A. A. Katbab (Eds.): ISPST 2018, *Eco-friendly and Smart Polymer Systems*, pp. 231–234, 2020.
https://doi.org/10.1007/978-3-030-45085-4_55

The synthesis of nanocomposite hydrogels based on wood cellulose nanofibers and polyacrylic acid via in-situ radical polymerization has been explained in our unpublished work. In the present work, the effects of cellulose concentration as reactant concentration on the swelling degree and strength of these kinds of interpenetrating networks (IPNs) hydrogels have been investigated. As the optimized combination of superabsorbent and conventional fertilizers may improve nutrition of plants, the swelling behavior was also studied in aqueous nitrogen-rich fertilizer solutions such as urea and ammonium sulfate.

2 Experimental

Acrylic acid (AA, LG Chem Co., industrial grade), N,N′-methylene-bis-acrylamide (MBA), potassium hydroxide (KOH), urea and potassium persulphate (KPS, 99%) were all supplied from Merck Co.

Cellulose nanofibrils (CNF) (2.5 wt%, with mean diameter of 32 nm) used in this work was obtained from Nano Novin Polymer Co.

The hydrogels were synthesized through radical copolymerization in deionized water as solvent. In order to check the influence of reactant concentration upon the water absorption capacity of hydrogels, cellulose concentration was changed from 0.5 wt% to 2.5 wt% by using different amounts of water, while all other components remained constant.

To prepare the hydrogels with CNF content of 2 wt%, nonocellulose dispersion was first diluted uniformly in KOH solution in an ice bath. Then, the predetermined amount of AA was dissolved in the mixture until the neutralization was completed. Then, the crosslinking agent (MBA) and initiator (KPS) solutions were added to the monomer mixture. After mixing properly, the mixture was charged into the reactor equipped with a mechanical mixer, an efficient reflux condenser and thermometer. The polymerization carried out at 70 °C for 70 min. Finally, the obtained hydrogel were dried in a conventional oven.

3 Results and Discussion

Various reactants concentration can be obtained by using different amounts of water. The dilution of reactants can enhance their mobility, leading to increase reaction time or polymer chain length.

Absorption capacities in distilled, drinking and saline water are presented in Fig. 1. As seen in Fig. 1, water absorbency of the samples increased slightly as media diluted. The swelling of hydrogels in saline solution was considerably decreased comparing to the measured values in distilled and drinking water. This is a well-known phenomenon in ionic hydrogels attributed to a charge screening effects of additional cations causing anion-anion electrostatic repulsion [12].

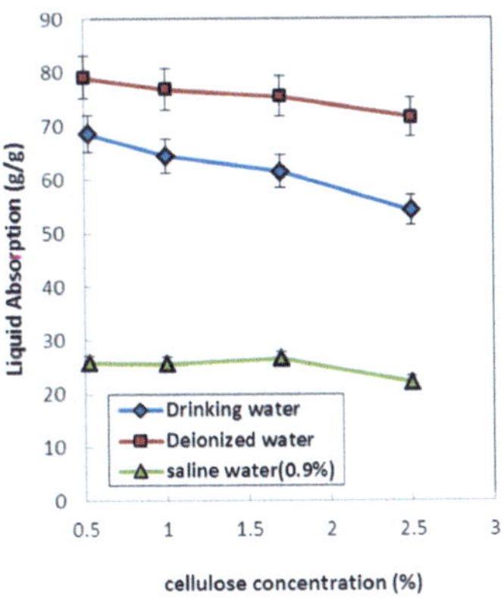

Fig. 1. Influence of cellulose concentration on the hydrogel equilibrium swelling properties in distilled water (■), drinking water (◆) and saline water (0.9 wt%) (▲)

The prepared hybrid hydrogels are anionic, thus, exhibit swelling changes as a function of pH and ionic strength. The swelling behavior of absorbents also measured in some aqueous fertilizers is shown in Table 1. Ammonium sulfate solution shows acidic behavior in nature while dissolved urea is neutral. As seen, the urea solution indicated higher absorbency compared to ammonium sulfate (AS). The experimental results also indicated that AS absorbencies changed with concentration and increased up to 32 g/g at lower concentration (0.04 mol/L). However, it does not show any significant change as a function of reactants concentration.

Table 1. Absorption behavior of the hydrogels in two aqueous fertilized solution

Absorption capacities (g/g)	Absorption capacities (g/g)	
	Urea solution (0.12 mol/L)	Ammonium sulphate (0.12 mol/L)
0.5	64.12 ± 1	20.48 ± 1
1.0	62.10 ± 1	21.38 ± 1
1.7	63.44 ± 1	18.02 ± 1
2.5	62.90 ± 1	18.50 ± 1

The absorption under load changed from 18.43 g/g to 21.64 g/g with an increase in cellulose concentration. The slight decrease in AUL at low CNF concentration (0.5 wt%) can be due to the light crosslink density and lower strength of the sample (Fig. 2).

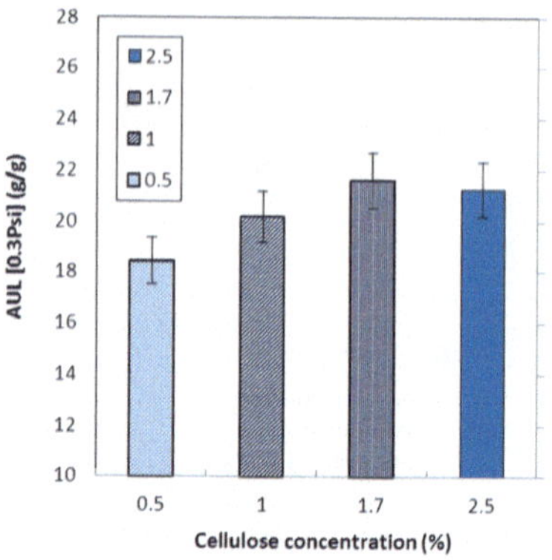

Fig. 2. Influence of cellulose concentration upon the absorption under load (AUL) properties of acrylic-nanocellulose hydrogels

4 Conclusion

In this work, the behavior of synthesized hybrid hydrogels as a function of reactant concentration in different aqueous media and fertilizers was studied. The results did not indicate any significant dependence on cellulose concentration in this range. However, it was found that these hydrogels can be good candidates for use in slow/control released fertilizers especially for urea compared to fertilizers such as ammonium sulfate. On the other hand, these hydrogels preferred to be used in combination with urea fertilizer rather than ammonium sulfate for nitrogen enrichment.

References

1. Zaharia, A., et al.: RSC Adv. **8**(32), 17635–17644 (2018)
2. Zhang, J., et al.: J. Appl. Polym. Sci. **132**(34) (2015)
3. Liu, M., et al.: Polym. Adv. Technol. **17**(6), 430–438 (2006)
4. Shen, X., et al.: Green Chem. **18**(1), 53–75 (2016)
5. Fekete, T.: Ph.D. thesis, Dept. Phys. Chem. Mater. Sci. (2017)
6. Spagnol, C., et al.: Carbohyd. Polym. **87**(3), 2038–2045 (2012)
7. Yuan, N., et al.: Mater. Sci. Eng. C **67**, 221–230 (2016)
8. Helmiyati, G.H.A., Kurniawan, S.: IOP Conf. Ser. Mater. Sci. Eng. **188**, 012024 (2017)
9. Kevin, J., France, T.H.D., Cranston, E.D.: Chem. Mater. **29**, 4609–4631 (2017)
10. Ma, J., Li, X., Bao, Y.: RSC Adv. **5**(73), 59745–59757 (2015)
11. Bitis, R., et al.: Us patent, Patent No. 2,162,007 B2
12. Bao, Y.J.M., Li, N.: Carbohyd. Polym. **84**, 76–82 (2011)

Preparation of a New Antioxidant for Polypropylene Based on Quercetin

Afarin Rahmati, Reza Jahanmardi(✉), and Sahar Amiri

Department of Polymer Engineering, Science and Research Branch, Islamic Azad University, Tehran, Iran
r.jahanmardi@srbiau.ac.ir

Abstract. Quercetin, a naturally occurring antioxidant, was converted to quercetin stearate through an esterification reaction with stearic acid. The occurrence of the esterification reaction was examined by FT-IR spectroscopy. The synthesized ester was then incorporated into polypropylene via melt mixing. Determination of oxidative induction time (OIT) and oxidation onset temperature (OOT) of the samples with the aid of differential scanning calorimetry (DSC), proved that the modified quercetin enhances the stability of the polymer outstandingly in the melt state, so that, the stabilization efficiency of quercetin stearate is higher than that of Songnox 1010, a commercially used antioxidant for polypropylene.

Keywords: Antioxidant · Degradation · Polypropylene · Stabilization · Quercetin

1 Introduction

Polypropylene (PP), due to its susceptibility to oxidative degradation, requires addition of proper antioxidants for every stage of its lifecycle [1]. On the other hand, due to the fact that oil resources are dwindling, there is a demand for production of required chemical products, such as antioxidants, from natural resources and particularly plants.

Quercetin is a natural compound found in many fruits, vegetables, leaves, and grains [2]. Its extraordinary antioxidant activity has been proven in many research works. However, due to the presence of strong H-bonding between its molecules, uniform distribution of quercetin in non-polar polymers, such as PP, is difficult.

Hence, the present work was devoted to modification of molecular structure of quercetin, so that, the compound be less polar and able to be blended more uniformly with non-polar polymers. Thus, we tried to assemble a non-polar bulky segment, i.e. stearate group, to the quercetin molecule, which disturbs strong H-bonding between these molecules.

H. Mirzadeh and A. A. Katbab (Eds.): ISPST 2018, *Eco-friendly and Smart Polymer Systems*, pp. 235–238, 2020.
https://doi.org/10.1007/978-3-030-45085-4_56

2 Experimental

2.1 Materials

PP (HP510 grade) with MFI of 9.0 g/10 min and melting point of 167 °C was obtained from Jam Petrochemical Company (Iran). Quercetin with purity of more than 95% as a yellow powder was purchased from Sigma-Aldrich. SONGNOX 1010, as a commercially used phenolic antioxidant, was supplied by Songwon Industrial Company (South Korea). All the chemicals used for the grafting reaction, i.e., stearic acid, thionyl chloride, DMF, pyridine and benzene were obtained from Merck Co. (Germany).

The esterification reaction between quercetin and stearic acid was carried out at two steps with the aid of thionyl chloride according to a well-known procedure described in the literature for the similar reactions [3]. Stoichiometric molar ratio of quercetin and acid was used in the reaction. The occurrence of the esterification reaction was examined by FT-IR spectroscopy using a Thermo Nicolet FT-IR spectrophotometer (USA).

PP was mixed with certain amounts of either of the quercetin, the chemically treated quercetin or SONGNOX 1010 at 190 °C in a Brabender internal mixer during 6 min, at 60 rpm. Formulation of the prepared samples are shown in Table 1.

Table 1. Formulation of the prepared samples

Sample code	PP	Q1	QS1	QS2	S1
PP (wt%)	100	99.9	99.9	99.8	99.9
Quercetin (wt%)	–	0.1	–	–	–
Quercetin stearate (wt%)	–	–	0.1	0.2	–
SONGNOX 1010 (wt%)	–	–	–	–	0.1

Oxidative stability of the prepared samples was examined by measuring their oxidative induction time (OIT) and oxidation onset temperature (OOT), using a METTLER TOLEDO thermal analysis system (Switzerland). In the isothermal test (OIT), samples were heated at 20 °C/min from 25 °C to 200 °C under a nitrogen stream. The gas atmosphere was then changed to oxygen immediately after the temperature was reached and was maintained at 200 °C. The gases flow rates were 50 cm^3/min according to ASTM D 3895-80.

The value of OIT was deduced from DSC curves as the time from oxygen introduction until the onset of exothermal reaction.

In the heat-ramp test, the temperature of each sample was raised from 25 °C to 250 °C at a constant rate of 10 °C/min. The samples were kept in oxygen atmosphere with a flow rate of 50 cm^3/min, right from the start of the test. The value of OOT for each sample was determined from its DSC curve as the temperature corresponding to the deviation from the baseline (onset of exothermal reaction).

3 Results and Discussion

Quercetin has five hydroxyl groups in its molecule, which are not similar with respect to reactivity toward carboxylic acids. However, it was not important for us that which hydroxyl group has the higher reactivity. The goal was only to attach a non-polar group to each quercetin molecule via an esterification reaction. The occurrence of the esterification reaction was examined by FT-IR spectroscopy. Figure 1 shows the infrared spectra of quercetin, stearic acid and the reaction product, respectively. Major characteristic absorption bands of both quercetin and stearic acid are seen in the spectrum of the reaction product. However, absorption band of carboxylic carbonyl group at 1703 cm^{-1} in the spectra of stearic acid has been disappeared. Instead, a new absorption band at 1737 cm^{-1} has been appeared in the spectrum of the reaction product, which is due to stretching vibration of esteric carbonyl group. Hence, it can be deduced that the desired esterification reaction between stearic acid and hydroxyl groups of quercetin has been occurred.

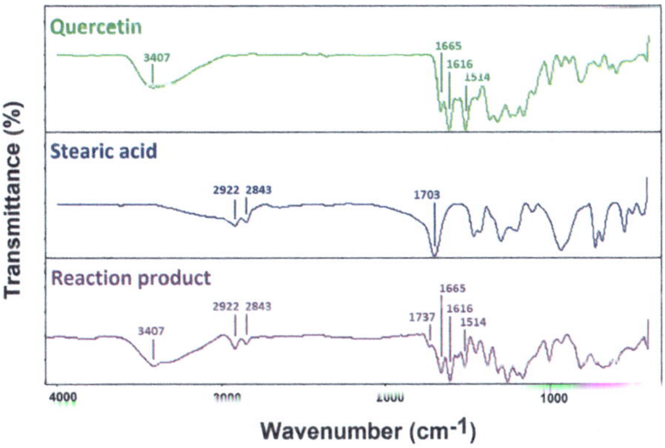

Fig. 1. FTIR spectra of quercetin, stearic acid and the reaction product, respectively

Table 2 shows the obtained OOT and OIT values. As it is seen, addition of either of the additives enhances both OOT and OIT of the polymer. However, effects of quercetin and its stearate ester in improving thermo-oxidative stability of the polymer are higher than that of the commercially used antioxidant, SONGNOX 1010. Moreover, through comparison of OIT values of QS1 and QS2 samples, or their OOT values, it can be inferred that the amount of improvement in oxidative stability of the polymer increases with elevation of the concentration of the quercetin stearate. Hence, it can be concluded that the prepared quercetin stearate acts as an effective antioxidant for PP in melt state, efficiency of which is much higher than that of the commercially used antioxidant, SONGNOX 1010 (Scheme 1).

Table 2. Measured OOT values of the prepared samples

Sample code	PP	Q1	QS1	QS2	S1
OOT (°C)	212	244	240	253	226
OIT (min)	3.9	25.3	33.5	46.8	16.6

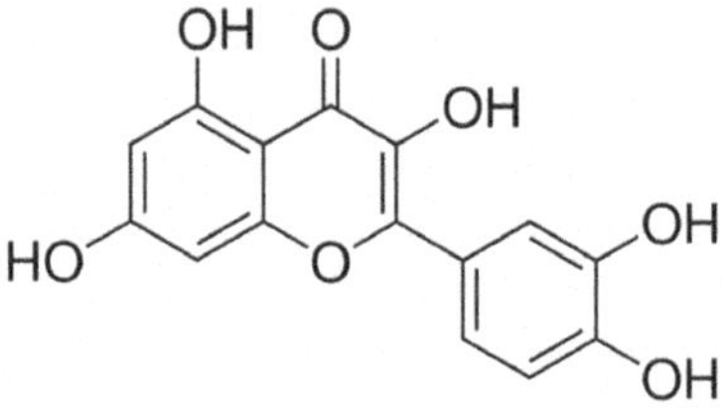

Scheme 1. Molecular structure of quercetin [4]

4 Conclusion

Successful esterification of quercetin through reaction with stearic acid was reported. The prepared quercetin stearate was shown to enhance thermo-oxidative stability of PP in melt state. More interesting achievement was that the efficiency of the new antioxidant is far higher than that of the commercially used phenolic antioxidant, SONGNOX 1010.

References

1. Allen, N.S., Edge, M.: Fundamentals of Polymer Degradation and Stabilization, 3rd edn, pp. 22–51. Elsevier, Amsterdam (1992)
2. Tátraaljai, D., Földes, F., Pukánszky, B.: Polym. Degrad. Stab. **102**, 41 (2014)
3. Bagheripour-Asl, M., Jahanmardi, R., Tahermansouri, H., Forghani, E.: Carbon Lett. **25**, 60 (2018)
4. Nabavi, S.M., Nabavi, S.F., Eslami, S., Moghaddam, A.H.: Food Chem. **132**, 931 (2012)

Controlling the Adsorption of Scleroglucan Polymer on the Surface of Carbonate Particles by Grafting Acyl Chlorides with Different Chain Lengths

Behnam Davoodi[1], Alireza Sharif[1(✉)], and Jamal Aalaie[2(✉)]

[1] Department of Polymer Reaction Engineering, Faculty of Chemical Engineering, Tarbiat Modares University, 14155/143, Tehran, Iran
asharif@modares.ac.ir

[2] Chemicals, Polymers and Petrochemicals Technology Development Division, Research Institute of Petroleum Industries (RIPI), 14665-137, Tehran, Iran
aalaiej@ripi.ir

Abstract. Scleroglucan (Sclg) polymer possesses unique properties making it especially attractive for enhanced oil recovery (EOR) applications. However, the hydrophilic nature of Sclg is a major constraint in EOR applications. Fortunately, its hydroxyl groups render it prone to different chemical reactions. In this study, Sclg is grafted by three modifying molecules with different hydrophobic chain lengths: stearoyl chloride (C18), decanoyl chloride (C10) or hexanoyl chloride (C6) and the effects of the alkyl chain length of the modifier on the adsorption of Sclg, from its saline solutions, on carbonate rock surfaces are studied. The pristine Sclg showed the lowest adsorption amount on the rock surfaces and the adsorption amounts of the modified Sclgs were increased by increasing the hydrophobic chain length of the modifier. Actually, by increasing the hydrophobic chain length in the modified samples, from hexanoyl chloride to stearoyl chloride, the possibility and strength of hydrophobic interactions among the hydrophobic sections were increased that, eventually resulted in higher adsorption amounts. Finally, the experimental results can be described by the well-known Langmuir model.

Keywords: Scleroglucan natural polymer · Hydrophobic modification · Chain length · Enhanced oil recovery · Adsorption

1 Introduction

Among the existing enhanced oil recovery (EOR) methods, recycling the residual oil by viscous fluids containing polymers has attracted much attention. To achieve suitable and optimal amounts of polymer adsorption, from the injected fluid to the surface of the reservoir rock is very important in the EOR process [1]. Scleroglucan (Sclg) is a natural polysaccharide produced by fungi of the genus sclerotium. Because of its peculiar rheological properties and its resistance to hydrolysis, temperature and electrolytes, scleroglucan has found applications in oil industries [2]. However, hydrophilic nature

H. Mirzadeh and A. A. Katbab (Eds.): ISPST 2018, *Eco-friendly and Smart Polymer Systems*, pp. 239–242, 2020.
https://doi.org/10.1007/978-3-030-45085-4_57

of Sclg limits its applications in EOR. Fortunately, the presence of many hydroxyl groups has made it prone to various hydrophobic modifications. In the present study, the possibility of controlling Sclg adsorption, from its saline solution on carbonate rock surfaces, is investigated through grafting hydrophobic molecules, with different lengths.

2 Experimental

To graft acyl chlorides with different hydrophobic chain lengths: stearoyl chloride (StCl, Sigma Aldrich), decanoyl chloride (DecCl, Sigma Aldrich) or hexanoyl chloride (HexCl, Sigma Aldrich) onto the Sclg, at first, 0.15 g of Sclg (Mw: 8×10^5 g/mol Actigum CS11, Cargill, Germany) was dissolved in distilled water (20 mL) under stirring for 2 h at 120 °C.

Then, 0.08 mL of StCl, 0.05 mL of DecCl or 0.03 mL of HexCl was added to the above mixture under stirring at ambient temperature for 1 h. To precipitate the modified polymers (StCl-Sclg, DecCl-Sclg and HexCl-Sclg), 100 mL of methanol was added to the reaction mixture.

Dilute solution viscometry (DSV) was used to measure the adsorption amount of pristine or modified Sclgs from their brine solutions on carbonate particles, obtained from one of Iran's oil-reservoir fields. The ln η_r/C calibration curves were derived for all brine solutions in the absence of carbonate rocks. Then, the reservoir rock was added to the solutions and the mixtures were stirred for 24 h at 25 °C followed by centrifugation for 15 min to precipitate the carbonate particles along with the adsorbed molecules. By performing DSV on the supernatant of the above mixtures followed by comparing the viscosity data with the above-mentioned calibration curves, the concentrations of the Sclgs in the supernatant (C_e) were obtained. Finally, the equilibrium adsorption amount, q_e, was calculated by the following equation:

$$q_e = \frac{v}{w}(C_0 - C_e) \tag{1}$$

Where, C_0, v and w are the initial concentration of the polymer (mg/mL), solution volume (mL) and amount of reservoir rock (g), respectively.

3 Results and Discussion

Figure 1 shows that at the same initial concentration (C_i) of polymers, q_e values vary in the order of StCl-Sclg > DecCl-Sclg > HexCl-Sclg > Sclg. By grafting the hydrophobic groups on the Sclg, its conformation changes from a triple helix structure to a random coil one with higher possibility to contact with the rock surfaces [3]. Moreover, adsorption nature of hydrophobically modified Sclgs can be multi-layer, while that of the pristine Sclg is mainly single-layer. This is another reason for more adsorption of modified polymers on the carbonate rock. By increasing the hydrophobic chain length in the modified samples, from hexanoyl chloride to stearoyl chloride, the amount and strength of hydrophobic interactions among the hydrophobic sections increase that eventually result in the higher adsorption amount of the StCl-Sclg sample as compared with the DecCl-Sclg and HexCl-Sclg ones (Fig. 1).

The Langmuir isotherm model (Eq. (2)) was utilized to explain the adsorption behavior of the samples.

$$\frac{1}{q_e} = \frac{1}{K_{ad} \times q_0 \times C_e} + \frac{1}{q_0} \tag{2}$$

where, q_e (mg/g), q_0 (mg/g), C_e (mg/L), and K_{ad} (L/mg) are the equilibrium adsorption amount, maximum adsorption capacity, equilibrium polymer concentration and Langmuir constant, respectively. For all the samples, the plot of $1/q_e$ against $1/c_e$ was linear ($r^2 = 0.99$). The values of K_{ad} and q_0 were obtained from the slope and intercept of the plots, respectively (Table 1).

Therefore, one can conclude that the Langmuir model can predict the experimental results very well. In addition, the StCl-Sclg sample showed the highest q_0 value indicating its higher adsorption capacity on the carbonate rock surfaces, in accordance with the results of Fig. 1.

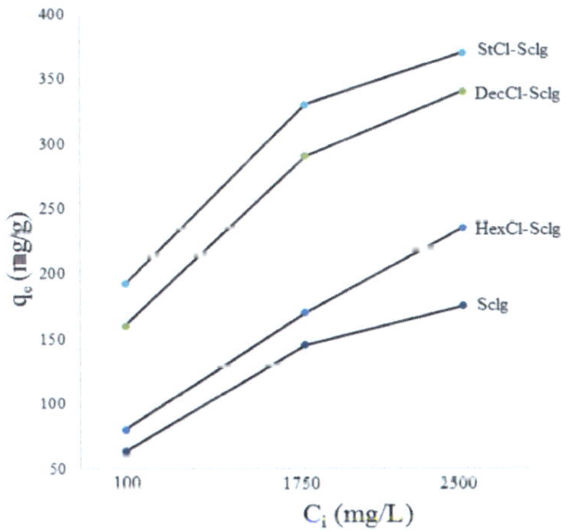

Fig. 1. q_e vs. C_i for pristine and modified Sclgs

Table 1. Langmuir model parameters

Sample	q_0 (mg/g)	K_{ad} (dL/g)
Sclg	177	730
HexCl-Sclg	236.5	800
DecCl-Sclg	342	900
StCl-Sclg	371.6	980

4 Conclusion

It was shown, in the present research, that the adsorption amount of scleroglucan polymer from its brine solution on carbonate particles can be controlled by grafting acyl chlorides with different hydrophobic chain length onto the polymer. The Sclg sample grated by StCl, with the highest chain length, showed the highest adsorption. The Langmuir isothermal model predicted the experimental results very well.

References

1. Wang, S., Li, G., Li, Y., Guo, J.: J. Appl. Polym. Sci. **134**, 45314 (2017)
2. Feeney, M., Casadei, M.A., Matricardi, P.: J. Mater. Sci. Mater. Med. **20**, 1081 (2009)
3. Bakhshi, M., Ozeiri, M., Sharif, A., Aalaie, J.: Korean J. Chem. Eng. **34**, 903 (2017)

Chemical Modification of Polyfurfuryl Alcohol Resin Using Maleic Anhydride: FTIR Study of a Diels-Alder Reaction

Roghayyeh Marefat Seyedlar[1], Mohammad Imani[1,2], and Seyed Mojtaba Mirabedini[1,3(✉)]

[1] Iran Polymer and Petrochemical Institute (IPPI), PO Box 14965-115, Tehran, Iran
sm.mirabedini@ippi.ac.ir

[2] Biobased Monomers and Polymers Division (Biobased Division), Iran Polymer and Petrochemical Institute (IPPI), PO Box 14965-115, Tehran, Iran

[3] Color and Surface Coatings Department, Iran Polymer and Petrochemical Institute (IPPI), PO Box 14965-115, Tehran, Iran

Abstract. In this study, furan bio-resin (PFA) was modified with maleic anhydride (MA) via Deils-Alder reaction. With this reaction, the number of carboxylic groups of the resin is increased and, as a result, its reactivity with polyamine-based curing agent is facilitated. The modification of PFA was performed using various wt% of MA (1, 2, 3, 5, and 15) at 40 °C. Reaction kinetic of different samples was studied using FTIR spectroscopy. The results showed that PFA modified with 1, 2, and 3 wt% of MA can be cured using 30 wt% polyamine curing agent. Higher amount of MA (5 and 15 wt%) remains intact and unreacted in the systems which is not suitable for modification.

Keywords: Furan resin · Carbonyl content · Air drying · Ambient curing · FTIR spectroscopy

1 Introduction

Furfuryl alcohol monomer can undergo a cationic polymerization reaction catalyzed by Lewis or Brønsted acid catalysts to yield poly(furfuryl alcohol) (PFA) resin. PFA is cured in the presence of acid catalysts ending to networks having excellent heat stability and resistance against acids, alkalis, and solvents. However, this such a curing reaction takes place at elevated temperature *ca.* 90–120 °C, which may limited its practical applications such as coating [1] where curing at ambient conditions is strongly preferred. As an alternative approach for curing of PFA resin at ambient conditions, curing of PFA in the presence of amine curing agents can be supposed. However, the number of carbonyl functional group is very low in PFA resin to react efficiently with an amine-based curing agent. Metil used a blend rich in terms of epoxy resin content to yield networks having acceptable mechanical properties and ambient curing [2].

Here, furan resin was chemically modified with maleic anhydride; passing through a Diels-Alder reaction mechanism, to increase its carbonyl functional group content.

H. Mirzadeh and A. A. Katbab (Eds.): ISPST 2018, *Eco-friendly and Smart Polymer Systems*, pp. 243–246, 2020.
https://doi.org/10.1007/978-3-030-45085-4_58

2 Experimental

PFA resin was polymerized from furfuryl alcohol monomer (FA, Daejung Chemicals and Metals Co., South Korea) in the presence of different wt% of maleic anhydride (MA). The details were reported elsewhere [3].

PFA resin was dissolved in THF, heated up to 40 °C for 8 h and chemically modified with 1 (PFA/1MA), 2 (PFA/2MA), 3 (PFA/3MA), 5 (PFA/5MA), and 15 (PFA/15MA) wt% of maleic anhydride. The modified PFA resin was purified by precipitation in the cold petroleum ether. The samples were characterized quantitatively using FTIR spectroscopy (Equinox 55, Bruker Optics, Germany) to determine carbonyl functional group. The extent of the resin modification was monitored considering according to Eq. (1):

$$Degree\ of\ modification = A_{3118}/A_{1712} \tag{1}$$

Where, A_{3118} and A_{1712} are the under-peak area of aromatic C-H and C=O peaks, respectively.

10, 15, 20, 25, and 30 wt% of polyamine (HY, Aradur 5052, Aradur® HY 5052, Huntsman, USA) were added to PFA/1MA, PFA/2MA and PFA/3MA samples. Different mixtures of modified resins and curing agent were applied on Teflon sheet using a film applicator at a wet film thickness of 200 μm. The samples were then left to cure at ambient temperature for a week. Curing reaction of the films was monitored by ATR-FTIR spectroscopy.

3 Results and Discussion

Chemical structure and FTIR spectra of the chemically-modified PFA resin prepared in the presence of different wt% of MA are shown in Figs. 1 and 2, respectively. Peaks positioned at 3118 and 1712 cm^{-1} are attributed to C-H stretching and C=O forming after ring opening of some furan moieties during polymerization of FA in furan resin, respectively [1]. According to the reaction mechanism, the C=O bond in the furan ring remains intact after the modification, but; the extent of furan ring conjugation reduces according to the significant reduction in furan ring aromaticity (the ratio of peak area of two peaks at 3118 cm^{-1} to 1712 cm^{-1}, shown in Table 1). However, the ratio increases with increasing wt% of MA for 5 and 15 wt%, indicating the presence of excess and unreacted MA in the system. The absorption peaks appeared at regions of 1770 and 1860 cm^{-1} are due to anhydride group, thus the amount of peak area ratio of 1860/1712 cm^{-1} increases with increasing MA concentration (Table 1).

Typical ATR-FTIR spectra of PFA/3MA samples cured with different concentrations of HY curing agent are shown in Fig. 3. The results for all samples are gathered in Fig. 4. The results indicate the 1712/1012 cm^{-1} peaks area ratio for modified resins cured with different wt% of curing agent. The peak positioned at 1012 cm^{-1} is attributed to C-O-C symmetric stretching vibrations in furan rings which remains intact during curing. HY curing agent reacts with anhydride and carbonyl group of the resin which the latter group are appeared at 1712 cm^{-1}. Thus, peak area ratio of 1712/1012 cm^{-1} for the samples reduced with HY concentration, as expected. The modified resins have shorter pot life and higher reactivity with HY curing agent in comparison with neat PFA. According to the results, a 30 wt% HY curing agent is sufficient for curing the modified PFA resin.

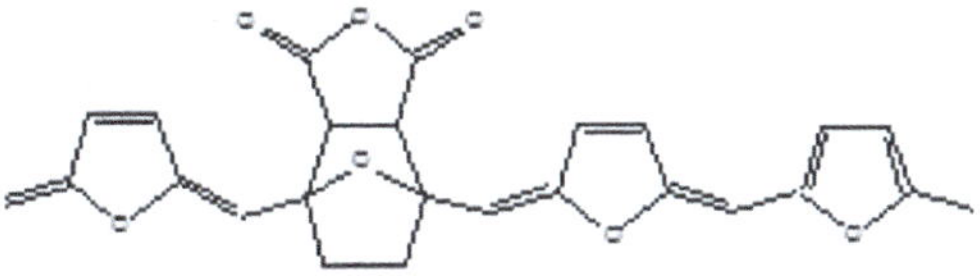

Fig. 1. Chemical structure of the MA modified PFA resin

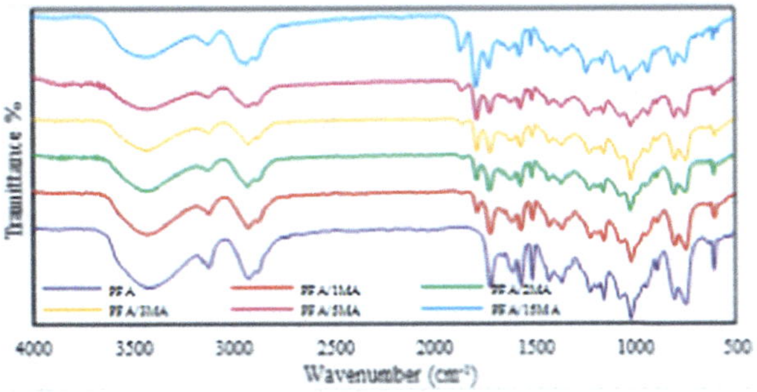

Fig. 2. FTIR spectra of PFA, PFA/1MA, PFA/2MA, PFA/3MA, PFA/5MA, and PFA/15MA resins

Table 1. Extent of area ratios for FTIR signals and aromaticity percentage

Sample	Signal area ratio		Aromaticity (%)
	3118 cm^{-1} to 1712 cm^{-1}	1860 cm^{-1} to 1712 cm^{-1}	
PFA	0.65	–	100
PFA/1MA	0.59	0.071	90.62
PFA/2MA	0.49	0.15	75.55
PFA/3MA	0.48	0.23	74.09
PFA/5MA	0.67	0.48	102.96
PFA/15MA	1.00	1.57	153.42

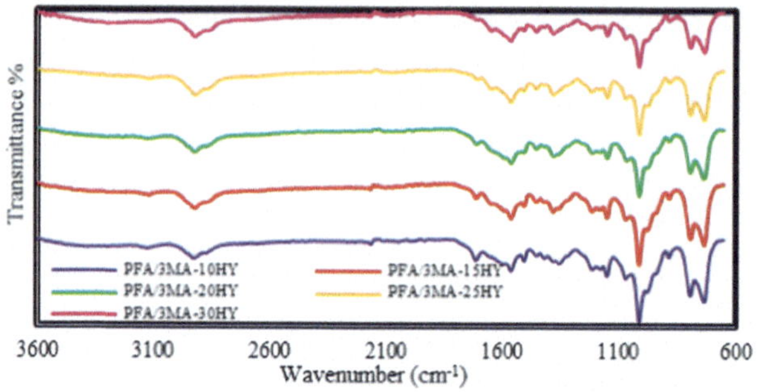

Fig. 3. ATR-FTIR spectra of cured PFA/3MA resins in the presence of different concentrations of HY curing agent

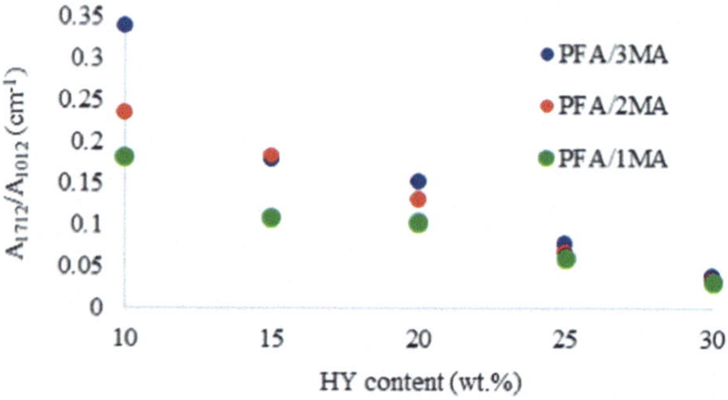

Fig. 4. 1712/1012 cm^{-1} peaks area ratio for modified resins cured at different curing agent (HY) concentrations.

4 Conclusion

Chemical modification of PFA resin with MA was performed through Diels-Alder reaction. FTIR findings support that 1–3 wt% of MA is adequate for PFA resin modification but its higher amount remains intact and unreacted. ATR-FTIR results of the specimens showed that 30 wt% of HY is appropriate for curing the resins.

References

1. Schmitt, C.R.: Polym-Plast. Technol. Eng. **3**, 121–158 (1974)
2. Metil, I.: US Patent No. 4, 267, 236 (1981)
3. Marefat Seyedlar, R., Imani, M., Mirabedini, S.M.: J. Appl. Polym. Sci. **135** (2018)

Hardness and Chemorheological Properties of Chemically-Modified Polyfurfuryl Alcohol Resin

Roghayyeh Marefat Seyedlar[1], Mohammad Imani[1,2], and Seyed Mojtaba Mirabedini[1,3](✉)

[1] Iran Polymer and Petrochemical Institute (IPPI), PO Box 14965-115, Tehran, Iran
sm.mirabedini@ippi.ac.ir

[2] Biobased Monomers and Polymers Division (BIOBASED Division), Iran Polymer and Petrochemical Institute (IPPI), PO Box 14965-115, Tehran, Iran

[3] Color and Surface Coatings Department, Iran Polymer and Petrochemical Institute (IPPI), PO Box 14965-115, Tehran, Iran

Abstract. Polyfurfuryl alcohol (PFA) is cured with amine-based curing agent via carbonyl functional groups of the resin. In this study, PFA was chemically modified via Deils-Alder reaction with different wt% of maleic anhydride (MA). Curing behavior of the modified resins was investigated in the presence of the polyamine curing agent using chemorheological and hardness techniques. According to the results, increasing MA wt% from 1 to 5, leads to increase of viscosity and storage modulus (G') of the mixture. However, the rate of viscosity increasing is higher than the increasing rate of G' for resins with higher MA content. In these systems, viscosity increases faster due to higher concentration of MA which results in reduction in chains mobility and crosslinking reaction which affects G' increasing rate. Hardness of the formulations increases with MA content and time progression up to 72 h.

Keywords: Polyfurfuryl alcohol · Chemical modification · Diels alder · Maleic anhydride · Chemorheology

1 Introduction

Polyfurfuryl alcohol (PFA) possesses a promising potential for industrial applications, especially where high thermal stability, smoke emission, residual char, chemical resistance against strong alkalis and acids are required with a relatively low cost [1]. Curing of the resin in the presence of acid catalysts occurs at high temperatures above 90 °C which limits its widespread commercial application [1]. To resolve the shortcoming, PFA is supposed to be cured using amine curing agents. However, the carbonyl functional group content of PFA resin is not adequate to react with amine curing agent.

In this study, PFA resin was modified with maleic anhydride via Diels-Alder reaction to increase the C=O functional group content of the resin. Then, curing behavior of the modified resins in the presence of amine curing agents was studied via rheology and hardness techniques.

H. Mirzadeh and A. A. Katbab (Eds.): ISPST 2018, *Eco-friendly and Smart Polymer Systems*, pp. 247–250, 2020.
https://doi.org/10.1007/978-3-030-45085-4_59

2 Experimental

Furfuryl alcohol (FA, Daejung Chemicals and Metals Co., South Korea) was polymerized to PFA resin in the presence of maleic anhydride (MA) according to a previously reported method [2].

PFA resin was modified by reacting at 40 °C for 8 h with 1, 2, 3, and 5 wt% of maleic anhydride (PFA/1MA, PFA/2MA, PFA/3MA and PFA/5MA samples, respectively) in THF solution. The modified resin was isolated by precipitation in cold petroleum ether.

The rheological time sweep measurements were performed on PFA/1MA, PFA/2MA, PFA/3MA, and PFA/5MA specimens mixed with 30 wt% of Aradur 5052 (Aradur® HY 5052, Huntsman, USA) polyamine as a curing agent using a MCR300 rheometer (Anton Paar, Graz, Austria) utilizing parallel plate measuring geometry (ϕ = 25 mm, 0.5 mm gap). Experiments were performed at room temperature, 3% strain and frequency of 1 Hz. Linear viscoelastic region was determined by performing strain sweep (0.01–100%) test.

Hardness test was performed according to ISO 1514 standard using a KONIG hardness tester (Elcometer, UK) equipped with a Persoz pendulum.

3 Results and Discussion

Changes in viscosity and modulus (G',G'') was traced during the curing reaction of modified PFA resin as a function of degree of the modification by MA in the presence of a constant concentration of the polyamine curing agent. Figures 1 and 2 show the profiles obtained for the changes in modulus and viscosity of the PFA resin modified by different concentration of MA in the range of 1–5 wt%. According to Fig. 1, viscosity increases with MA concentration but the increase for PFA/5MA specimen is much more than that for the other samples, probably due to reacting higher content of MA with PFA resin and its availability, which increase the resin molecular weight and viscosity. In addition, the excess unreacted MA present in the system can react with polyamine curing agent. The reaction is exothermic and autocatalytic, thus, curing reaction progresses rapidly. Time to reach to gel point ($t_{gel,}$ time at which G' and G'' cross over), Tanδ and modulus at 5400 s (G'_{5400}, G''_{5400}) are gathered in Table 1. For PFA/1MA specimen, t_{gel} is 315 s but it is decreased to 105 s for PFA/2MA, as expected. Because, PFA/2MA has higher carbonyl functional groups than PFA/1MA, which can react with HY curing agent. However, for PFA/3MA, the point increases to 510 s with no cross-over between the G' and G'' curves due to higher increasing rate of viscosity than that of G', which reduces chain mobility. Thus, polymerization kinetics become faster than the crosslinking reaction. In addition, Tanδ for PFA/5MA specimen is 2.31 (see Table 1) which is higher than that for the other samples.

To evaluate progress of curing reaction, hardness of the modified resins at different time intervals was measured. As illustrated in Fig. 3, hardness increases with MA concentration and time up to 72 h. However, there is no significant difference between the PFA/2MA and PFA/3MA specimens.

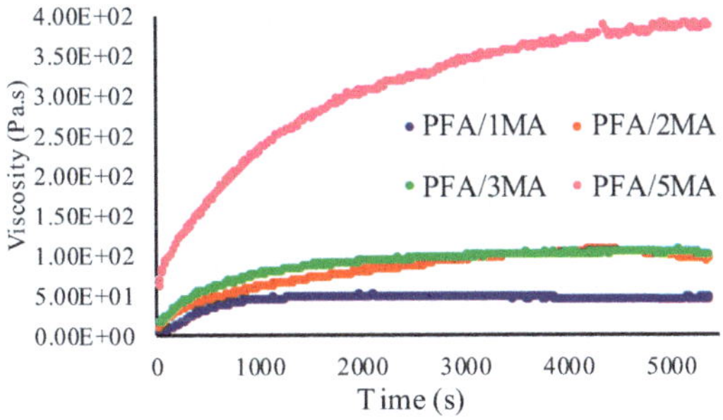

Fig. 1. Viscosity as a function of time for PFA, PFA/1MA, PFA/2MA, PFA/3MA, and PFA/5MA resins

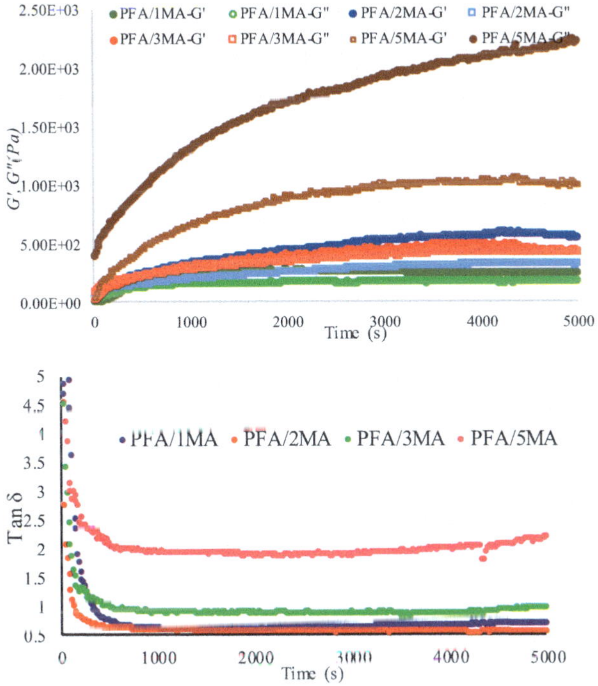

Fig. 2. Storage and loss modulus (G', G'') (a) and Tanδ (b) as a function of time for PFA/1MA, PFA/2MA, PFA/3MA, and PFA/5MA resins

Table 1. Rheological properties of resin modified with different MA contents

Samples	t_{gel} (s)	G'_{5400} (Pa)	G''_{5400} (Pa)	Tanδ
PFA/1MA	315	240	169	0.703
PFA/2MA	105	534	310	0.581
PFA/3MA	510	425	430	1.01
PFA/5MA	–	978	2260	2.31

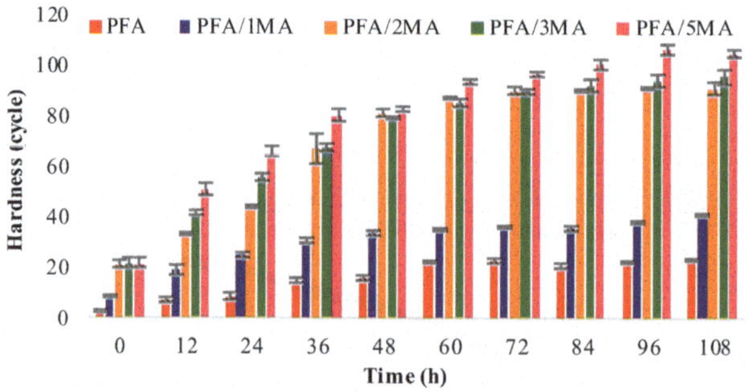

Fig. 3. Variation of hardness versus time for PFA, PFA/1MA, PFA/2MA, PFA/3MA and PFA/5MA resins

4 Conclusion

Hardness and rheological properties of the MA-modified PFA resin cured with HY curing agent were investigated. The rate of viscosity increasing is higher than G' increasing rate for PFA/3MA and especially PFA/5MA compared with PFA/1MA. This behavior is due to the excess MA remained in the composition. Hardness of the formulations increases with increasing MA concentration and time progression up to 72 h.

References

1. Schmitt, C.R.: Polym-Plast. Technol. Eng. **3**, 121–158 (1974)
2. Marefat Seyedlar, R., Imani, M., Mirabedini, S.M.: J. Appl. Polym. Sci. **133**, 1 (2016)

Influence of Cut-Out Shape on Buckling Behavior of Composite Beams

Soheil Dariushi(✉)

Composite Group, Iran Polymer and Petrochemical Institute, Tehran, Iran
s.dariushi@ippi.ac.ir

Abstract. In this study, the buckling behavior of orthotropic composite beams with circular, elliptic and square cut-out shape under compression loading has been studied. A simulation based on finite element method has been developed to evaluate critical buckling load and predict the failure around the holes with different shapes. In order to obtain empirical results, specimens were manufactured with symmetric lay-up and different hole shapes. Several key findings and behavioral characteristics are discussed. The results show that specimens without cut-out have the highest critical buckling load and specimens with circular hole have the lowest critical load. The experimental results are well matching with that of FEM solution.

Keywords: Buckling · Composite · Critical buckling load · Cut-out shape · Finite element modeling

1 Introduction

Composites offer considerable possibilities in many types of engineering application such as marine, aerospace, and automotive industries because of their high specific stiffness and specific strength. Cut-outs are unavoidable in structures mainly for practical considerations [1]. Sometimes cut-outs are made in the composites as access ports, damage inspection, altering the resonant frequency of the structures, to serve as doors and windows, to reduce weight, etc. High stresses are produced around such holes when subjected to various loads and may lead to the failure of the structure [2]. When a structure is subjected to compressive loading, buckling may occur. Buckling is a sudden sideways deflection of a member. This may happen even though when the stresses that develop in the structure are well below those needed to cause failure in the structure. Panel with cut-outs often lose stability at fairly low stress levels [3].

This article addresses the effects of a cut-out shape on the buckling behavior of composite beams using both experimental and numerical methods.

2 Experimental

For the experiments, laminated composite plates consisting of four layers of uni-directional, E-glass fabrics (300 g/m^2) with longitudinal fiber orientation angles $[0]_4$ were manufactured. Thermoset epoxy resin (Axson Epolam 2002 resin with Epolam 2002 hardener) was used as matrices.

H. Mirzadeh and A. A. Katbab (Eds.): ISPST 2018, *Eco-friendly and Smart Polymer Systems*, pp. 251–254, 2020.
https://doi.org/10.1007/978-3-030-45085-4_60

The specimens were made by hand lay-up method followed by curing process at room temperature including pressurizing (about 1.5 bar) for 24 h.

Rectangular specimens with dimensions of 150 mm * 50 mm were cut from manufactured composite plate. The material properties of composite beam are presented in Table 1.

Laser was used to create circular, elliptic and square cut-outs with equal area in the center. The diameter of circular hole is 12.25 mm. In specimens with elliptical hole, 5 and 7.5 mm are lengths of the semi-major and semi-minor axes, respectively. The length of side of square opening, is 10.85 mm.

The buckling compression tests were carried out up to ultimate fracture of specimens with STM-150 universal testing machine. The specimens' edges were clamped at the bottom and top, therefore the effective length of specimens were 100 mm.

3 Numerical

A three-dimensional finite element analysis is performed to predict the effects of cut-out shapes on the buckling behavior of composite beams. Commercially available ANSYS Workbench 16.2 was used. A rectangular beam with a central hole was created. The beam was first meshed with coarse mesh and then the meshing of the area around the cut-out was refined to enhance the accuracy of calculation. A uniform pressure in longitudinal direction was exerted on the top surface of the beam. The upper and lower edges were clamped and two other sides were free. The model consists of four layers with dimensions of 100 * 100 mm^2 and the 2 mm total thickness of the beam. The critical buckling load and maximum stress around the hole for the beams with different hole shapes were determined by static structural and linear buckling analysis.

4 Results and Discussion

The magnitude of the compressive axial stress, at which the plate becomes unstable is called the critical buckling stress. The experimental stress-strain curves of specimens are illustrated in Fig. 1. The critical stress points are shown with arrows. As it is shown in Fig. 1, the critical stress decreases in presence of cut-out. The experimental and numerical critical buckling load for each specimen were reported in Table 2. Specimen without cut-out has the highest critical load and specimen with circular hole has the lowest critical load. When the load increases beyond the critical load, the large deflection occurs in the beam. The failure strain of notched specimens is reduced.

First mode of buckling is shown in Fig. 2 for all specimens. The failure starts in the vicinity of hole where the maximum stress happened because of stress concentration around the hole in deformed beam. Figure 3 shows the good agreement achieved between the experimental and FEM results for prediction of failure point.

Table 1. Material properties of composite beam

Elasticity modulus	Shear modulus	Poisson ratio
E_1 = 29 GPa	G_{12} = 4.2 GPa	V_{12} = 0.25
E_2 = 7 GPa	G_{23} = 3.4 GPa	V_{23} = 0.35
E_3 = 7 GPa	G_{13} = 4.2 GPa	V_{12} = 0.25

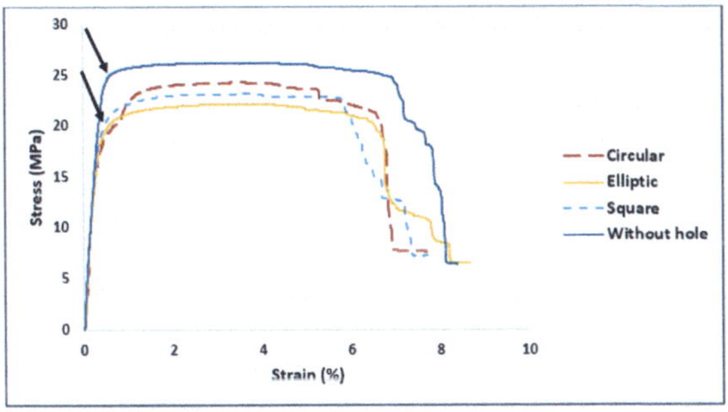

Fig. 1. Stress-strain curves of specimens

Table 2. Results of compression tests and modeling

Specimen	Experimental critical load (N)	Numerical critical stress (N)
Without hole	2230	1990
Circular hole	1600	1430
Elliptic hole	1720	1510
Square hole	1790	1620

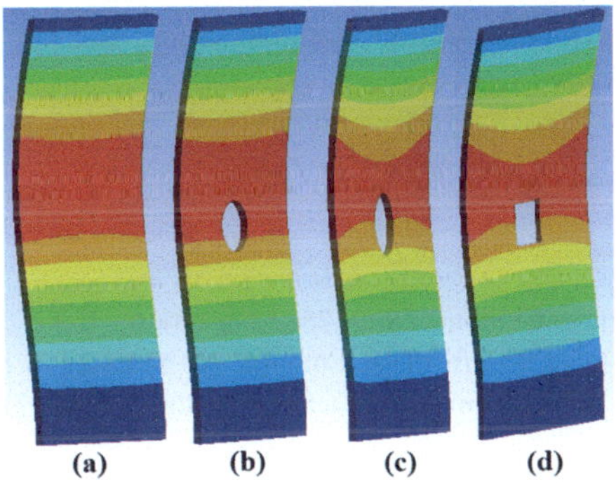

Fig. 2. First mode of Buckling: (a) without hole, (b) circular, (c) elliptic, and (d) square hole

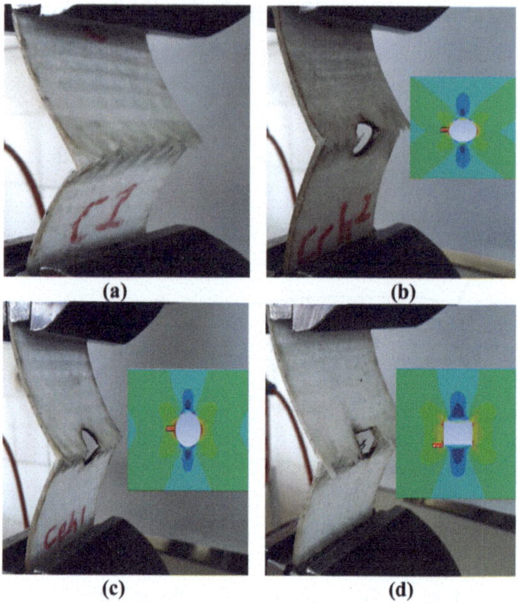

Fig. 3. Compression test on specimens: (a) without hole, (b) circular, (c) elliptic, and (d) square hole

5 Conclusion

In this study, critical buckling load for clamped composite beams under the uniaxial compressive loading with and without cut-out is investigated.

- The critical buckling load decreases 20, 16, and 14% for specimens with square, elliptic and circular loads, respectively, in compare with specimen without hole.
- Failure starts from the vicinity of cut-out toward edges.
- FEM prediction for critical buckling load and failure starting point agrees well with the experimental results.

References

1. Yazici, M.: J. Reinf. Plast. Comput. **28**, 2325–2339 (2009)
2. Akbulut, H., Sayman, O.: J. Reinf. Plast. Comput. **20**, 1112–1124 (2001)
3. Sayyad, A.S., Ghugal, Y.M.: Compos. Struct. **171**, 486–504 (2017)

Tensile Properties of Jute/Epoxy/Aluminum Hybrid Laminates

Sara Alirezaei, Soheil Dariushi(✉), and Mohammad Hosain Beheshty

Composite Group, Iran Polymer and Petrochemical Institute,
P.O. Box 14975-112, Tehran, Iran
s.dariushi@ippi.ac.ir

Abstract. In this research, tensile properties of hybrid laminates consist of layers of thin aluminum sheets and jute/epoxy composite are investigated and compared with those of jute/epoxy laminates. The main objective of this research is to study the effect of combining aluminum layers and laminated jute/epoxy composite to improve the mechanical properties. Two groups of specimens were made and tested according to ASTM D3039. The first group of specimens contained Jute/epoxy [0/90/90/0] and the second group was made up of 2024-T3 aluminum sheets and jute/epoxy laminates, the stacking sequence was set as [Al/0/90/90/0/Al]. The results show that adding aluminum layers to composites laminates caused an enhancement in modulus, maximum stress, and elongation at break point for jute/epoxy composites.

Keywords: Fiber metal laminates · Jute fiber · Aluminum · Epoxy resin · Tensile test

1 Introduction

The superior properties of the natural fibers in comparison to their synthetic counterparts make them favorable for using as the reinforcement material in the polymer composites [1]. But still the mechanical properties of these composites are lagging behind their synthetic counterparts. Jute as a natural fiber is eco-friendly, low cost, versatile in textile fields and has moderate mechanical properties, which replaced several synthetic fibers in development of many composite materials [2]. Also, aluminum is corrosion resistant, light in weight and ductile. Combination of these two materials can show extra properties over both monolithic aluminum and jute reinforced laminates [3, 4].

In this study, composites composed of epoxy resin, jute fibers, and aluminum laminates were prepared and examined experimentally.

2 Experimental

Raw materials used in the present investigation are jute fiber, aluminum 2024-T3, Epikote 828 epoxy resin, and hardener H46-A.

Preparation of jute/epoxy (JE) composites was performed as below:

H. Mirzadeh and A. A. Katbab (Eds.): ISPST 2018, *Eco-friendly and Smart Polymer Systems*, pp. 255–258, 2020.
https://doi.org/10.1007/978-3-030-45085-4_61

First, 50 g amine curing agent was added to 100 g 828 epoxy resin, and mixed. This mixture was ready to be added to fibers, laid up at 0/90/90/0 degrees. Four layers of jute fibers were placed on top of each other, impregnated using epoxy resin utilizing the hand lay-up method, and then cured under 90 bar pressure, at 50 °C for 90 min.

Preparation of aluminum/jute/epoxy (AJE) was as below:

To prepare FMLs, two aluminum laminates were used and four layers of jute fiber put between aluminums by hand lay-up. The curing was performed at 90 bar, at 50 °C for 2 h. Tensile strength and modulus of composite and FML specimens were measured using the tensile test according to ASTM D3039, with a STM-150 universal test machine, made by Santam Company, Iran. To perform this analysis, the dimensions of JE and AJE specimens were 200 × 20 × 4.6 mm^3. The analysis was carried out with a rate of 2 mm/min, at 20% relative humidity and room temperature.

3 Results and Discussion

Tensile strength and modulus of JE and AJE specimens are demonstrated at Table 1. In addition, Fig. 1 shows the stress-strain curve of JE, and AJE. Results derived by the stress-strain curve exhibited that the JE specimen showed an elastic behavior until fracture. This property is often observed in composites made up of thermoset resins. On the other hand, FML specimens had a linear elastic region until the aluminum entered the plastic phase, then, the interaction between the elastic behavior of composite layers and the plastic behavior of metal layers was observed. At maximum stress, failure in layers, especially delamination of composite and aluminum layers occurred. Then, failure and fracture of composite layer including matrix fracture, delamination, fiber pull-out from the matrix, and fiber fracture occurred. Comparing the curves of JE and AJE specimens exhibited that adding aluminum layers had led to the improvement of composite laminates properties regarding the elastic modulus and stress at fracture point. For these specimens, even the elongation at break point increased. Figure 3 demonstrates that for AJE specimen, hollow spaces were generated within the composites, indicating the fact that aluminum had entered the plastic phase and it had undergone an elongation. Although due to the entrance of aluminum to the plastic phase, it could not go back to the initial shape again, the fracture mode of the composite was elastic, and it did not undergo any elongation. Therefore, the composite layers are subjected to fracture, but the aluminum laminates entered the plastic phase and maintained their integrity. In other words, the aluminum layers delaminate from laminates. Besides, a sizable separation between the composite and aluminum layers was observed (Fig. 2).

Table 1. Tensile properties of JE and AJE

Sample	JE	AJE
Tensile strength (MPa)	63	87
Tensile modulus (GPa)	10	28

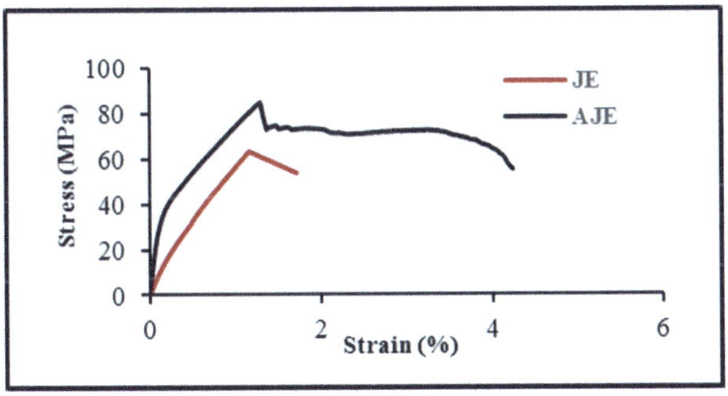

Fig. 1. Tensile stress-strain curve of JE and AJE

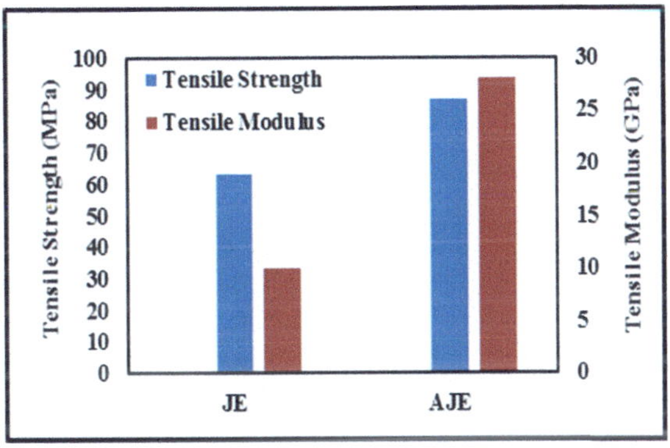

Fig. 2. Chart tensile strength and tensile modulus for JE and AJE

Fig. 3. Fracture specimens at two magnifications

4 Conclusion

The mechanical properties of jute/epoxy-aluminum laminates were studied and compared with those of jute/epoxy composite. Tensile modulus of AJE showed a significant improvement of 2.8 compared with that of JE. Also, tensile strength of AJE showed 38% improvement.

Ultimate Strain of JE was 1.71% and strain in fracture point of the AJE laminates was 4.21%.

References

1. Samanta, S., Muralidhar, M., Singh, T.J., Sarkar, S.: Mater. Today Proc. **2**, 1398–1405 (2015)
2. Hossain, M.R., Islam, M.A., Van Vuurea, A., Verpoest, I.: Procedia Eng. **56**, 782–788 (2013)
3. Sinmazcelik, T., Avcu, E., Bora, M., Coban, O.: Mater. Des. **32**, 3671–3685 (2011)
4. Kumar, G.B.V., Pramod, R.: AIP Conference Proceedings, vol. 1859 (2017)

Adsorption Behavior of Cyclodextrin-Triazole-Titanium Based Nanocomposite for Heavy Metal Ions from Aqueous Solution

Zari Fallah[1], Hossein Nasr Isfahani[1], and Mahmood Tajbakhsh[2](✉)

[1] School of Chemistry, Shahrood University of Technology, Shahrood, Iran
[2] Faculty of Chemistry, University of Mazandaran, Babolsar, Iran
mahmood.tajbakhsh32@gmail.com

Abstract. A cyclodextrin-triazole-titanium based nanocomposite (CD.COM) was successfully synthesized and fully characterized. CD.COM exhibited a good adsorption performance for removal of Zn^{+2}, Cd^{2+}, and Pb^{2+} ions from aqueous solutions by batch technique. The adsorption parameters such as pH, contact time, adsorbent dosage, temperature, metal ions concentration, coexisting ions and the regenerability of CD.COM were investigated. The Langmuir, Freundlich, Temkin and Dubinin-Radushkevich isotherm models were evaluated using adsorption equilibrium data. The Langmuir model showed the best agreement with the experimental data, resulting in the maximum adsorption capacities of 147.1, 158.7, and 200.0 (mg . g^{-1}) for Zn^{+2}, Cd^{+2}, and Pb^{+2} ions, respectively. Among the pseudo-first order, pseudo-second order, intraparticle diffusion and Elovich models, the kinetic data was well fitted with the pseudo-second order model. Thermodynamic parameters indicated a spontaneous and endothermic adsorption process. The bioadsorbent showed high selectivity for Pb(II) ions in the presence of other coexisting metal ions.

Keywords: TiO_2 · Cyclodextrin · Click reaction · Nanocomposite · Bioadsorbent

1 Introduction

Clean water is the most essential need for human's life and removing the metal ions pollution from water is one of the most vital research area which has attracted global attention due to adverse effect of pollution on the environment and human health [1]. Environmental benign nanocomposites with high surface area, more active sites and good adsorption capacity have gained a great deal of attraction [2]. In this context, CD. COM was synthesized via nucleophilic reaction of amino functionalized TiO_2 with click modified cyclodextrin and its structure was characterized by FTIR, TGA, FESEM, TEM, EDX-MAP, XRD, BET, and ICP-OES analyses. The adsorption behavior of CD.COM was investigated for removal of Zn(II), Cd(II), and Pb(II) ions from aqueous solutions.

H. Mirzadeh and A. A. Katbab (Eds.): ISPST 2018, *Eco-friendly and Smart Polymer Systems*, pp. 259–263, 2020.
https://doi.org/10.1007/978-3-030-45085-4_62

2 Experimental

CD.COM was synthesized according to the synthetic pathway shown in Scheme 1.

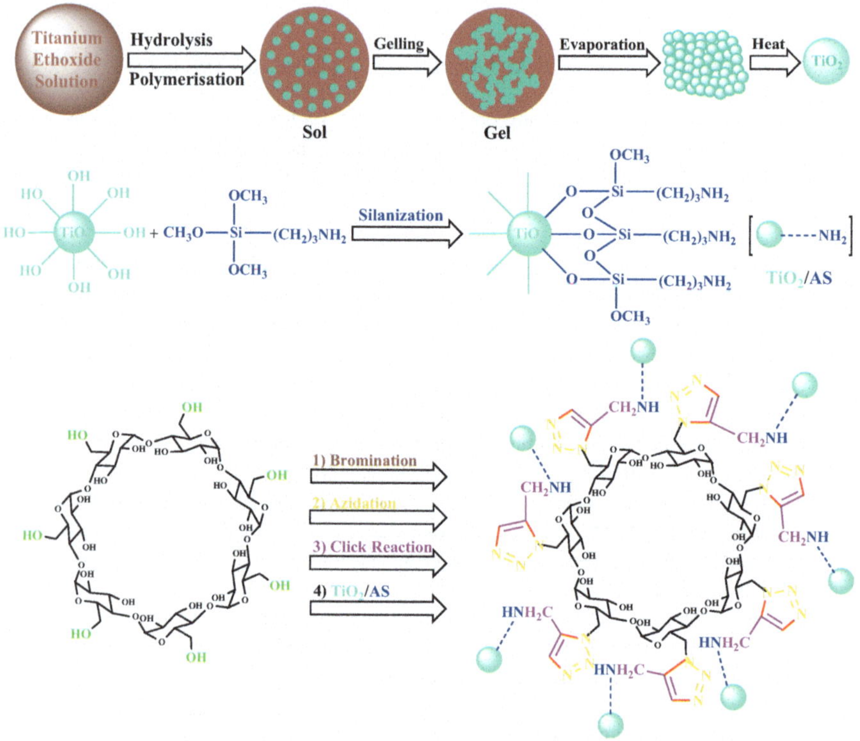

Scheme 1. Preparation of CD.COM

3 Batch Adsorption Experiments

Adsorption isotherm, kinetic and thermodynamic data were evaluated by batch adsorption experiments. The effects of pH, contact time, adsorbent dosage, temperature and initial metal ions concentration on the adsorption efficiency of CD.COM were investigated. The removal percentage ($R\%$) and adsorption capacity (Q) of CD.COM were calculated according to Eqs. (1) and (2):

$$R(\%) = \left[\frac{C_0 - C_e}{C_0}\right] \times 100 \quad (1)$$

$$Q = \left[\frac{C_0 - C_e}{M}\right] \times V \quad (2)$$

Where, C_0 and C_e (mg . L^{-1}) are the initial and final concentrations of metal ions, Q (mg . g^{-1}) is the adsorption capacity, V (L) is the volume of the metal ion solution and M (g) is the weight of adsorbent.

4 Results and Discussion

In order to gain more detail of the adsorption mechanism, different isotherms (Langmuir, Freundlich, Temkin and Dubinin-Radushkevich) and kinetic models (pseudo-first order, pseudo-second order, intraparticle diffusion and Elovich) were examined to fit the experimental data. The adsorption data were best fitted with the Langmuir isotherm and pseudo-second order kinetic models (Table 1), suggesting that the adsorption mechanism could be a monolayer chemical adsorption. CD.COM showed a preferential Q_{max} (mg . g^{-1}) for Pb^{2+} (200.0) compared with Zn^{2+} (147.1) and Cd^{2+} (158.7) ions. The maximum adsorption capacities of CD.COM for these metal ions are higher than those for many cyclodextrin-based adsorbents reported in the literature, due to the synergistic effects of cyclodextrin, triazole and TiO_2 moieties. The calculated thermodynamic parameters (ΔG°, ΔH°, and ΔS°) together with E_a demonstrated an endothermic, spontaneous and chemical adsorption process. Competitive adsorption efficiencies of CD.COM for Zn^{+2}, Cd^{+2}, and Pb^{+2} ions in single and ternary metal ion-containing solutions with equal concentration were compared to evaluate the selectivity of CD. COM towards Pb^{+2} ions, due to the larger ionic radius and lower hydration energy of Pb^{+2} ions (Fig. 1). CD.COM can be easily recovered and regenerated with HCl (0.5 M) and/or EDTA (0.05 M) solutions and reused at least five recycles without significant loss of adsorption efficiency (Fig. 2).

Table 1. Results of isotherm and kinetic models

Models	Pb(II) ion	
	Parameters	R^2
Langmuir isotherm	Q_{max} (mg g^{-1}) = 200.0 K_L (L g^{-1}) = 0.202 R_L = 0.198	**0.9999**
Pseudo-second order kinetic	$Q_{e[exp.]}$ (mg g^{-1}) = 57.00 $Q_{e[cal.]}$ (mg g^{-1}) = 59.88 K_2 (min^{-1}) = 0.005 h_0 (mg g^{-1} min^{-1}) = 16.75	**0.9999**

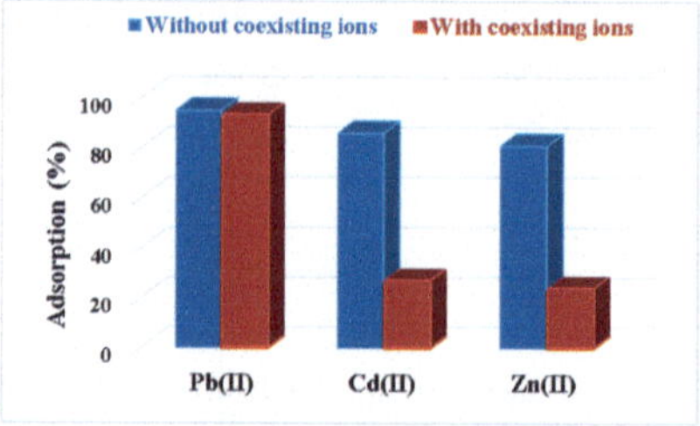

Fig. 1. Effect of competing cations on metal ions adsorption

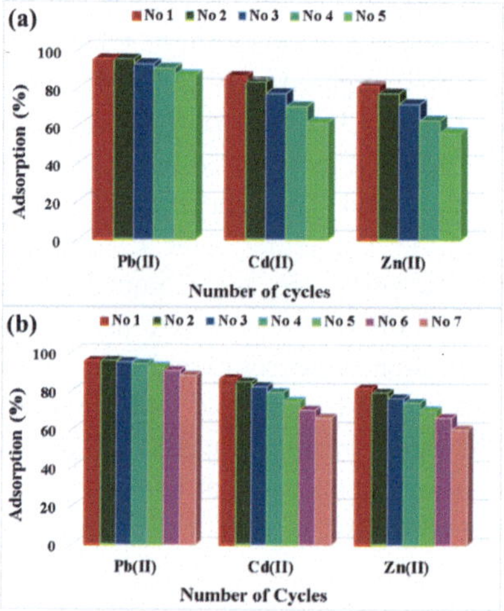

Fig. 2. Regeneration efficiency of CD.COM with: (a) HCl and (b) EDTA

5 Conclusion

CD.Com exhibited to be an efficient biosorbent for adsorption of Zn^{+2}, Cd^{+2}, and Pb^{+2} ions from aqueous solutions. The presence of multiple –OH, –NH functional groups, nitrogen atoms of 1, 2, 3-triazole ring as active binding sites and capability of cyclodextrin moiety for host-guest inclusion complex formation, makes the CD.COM as an effective nanoadsorbent for removal of metal ions. CD.COM with merits such as high adsorption capacity, good selectivity, high reusability and environmentally friendly has great potentials for removal of pollutants from water and waste water.

Acknowledgement. We wish to express our gratitude to the University of Mazandaran and research council of Shahrood University of technology for their supports.

References

1. Yidong, Z., et al.: J. Mater. Chem. A **4**, 14170–14179 (2016)
2. Fallah, Z., Nasr Isfahani, H., Tajbakhsh, M., Tashakkorian, H., Amouei, A.: Cellulose **25**, 639–660 (2018)

Study on Tensile Properties of Nanocomposites Based on PLA Reinforced by Nanoclay and Nanocalcium Carbonate

Zahra Asadi[1] and Azizeh Javadi[1,2(✉)]

[1] Faculty of Mahshahr, Amirkabir University of Technology, Khuzestan, Iran
ajavadi@aut.ac.ir
[2] Department of Polymer Engineering and Color Technology, Amirkabir University of Technology, Tehran, Iran

Abstract. In this study, binary nanocomposites of poly lactic acid (PLA) containing 3 wt% of nanoclay (NC) and 7 wt% of nanocalcium carbonate (NCC), also hybrid nanocomposite of PLA/NC/NCC containing 3 wt% of NC and 7 wt% of NC were prepared via melt mixing using an internal mixer. The effect of each nanoparticle and also both of them on the mechanical properties of nanocomposites was investigated and compared with those of pure polymer. The results showed that the modulus and tensile strength of all nanocomposites were improved in comparison with those of pure polymer. The highest increase in tensile strength was 37% for hybrid sample (PLA-NC3-NCC7). Also it was found that the highest improvement in elongation at the break was 68% and the least reduction in impact resistance was 6% which both belonged to the binary PLA-NCC containing 7 wt% of NCC.

Keywords: Nanocomposite · Polylactic acid · Nanoclay · Nanocalcium carbonate · Mechanical properties

1 Introduction

Nowadays, biodegradable polymers have been important and are a good substitute for petro-based materials. PLA which is a thermoplastic polyester with a relatively high strength and process ability is one of the most important polymers which is derived from renewable sources such as corn, rice and tapioca [1]. Despite the advantages of PLA, its mechanical properties such as strain at break and toughness are not very favorable and using this biopolymer has some limitations. Therefore, it is very important to blend this polymer with other polymers or to add nanoparticles to it and improve its properties [2].

Nanoclay is one of the most widely used nanofillers because of its affordability and optimum reinforcing. Furthermore, NCC can also be used not only as a filler, but as an impact enhancer.

Several studies have been carried out on binary systems amplified with these nanofillers. For PLA/Clay nanocomposites, it is shown that best mechanical properties are achieved by using 3 wt% of nanoparticles [3]. Moreover, for PLA/NCC, an increase in the

H. Mirzadeh and A. A. Katbab (Eds.): ISPST 2018, *Eco-friendly and Smart Polymer Systems*, pp. 264–267, 2020.
https://doi.org/10.1007/978-3-030-45085-4_63

elongation at break and decrease in the tensile strength, by adding nanoparticles was proved [4]. In addition, there is not any conduction research on hybrid nanocomposites with these two kinds of nanofillers on PLA, so in this research, PLA/NC and PLA/NCC binary, also hybrid PLA/NC/NCC nanocomposites with different compositions, were prepared and the rheology and mechanical properties of them were studied and compared with each other and with those of neat PLA.

2 Experimental

2.1 Materials

PLA 2003D is a product of Nature Works with density of 1.24 g/cm^3. The NC used was Cloisite 30B (Nanosav), with a spacing of 18 angstroms between its plates. NCC nanoparticles was a product of Golling Austria Company with an average particle size of 80 nm and was coated by stearic acid.

2.2 Process Conditions

Prior to mixing, PLA and nanoparticles were dried in an oven for 16 h at 70 °C. Nanocomposites of PLA/NC, PLA/NCC and hybrid samples of PLA/NC/NCC were prepared in an internal mixer, at 180 °C and 150 rpm for 5 min. Binary nanocomposites contained 3 wt% of NC and 7 wt% of NCC, as well as their hybrid samples with 3 wt% of NC and 7 wt% of NCC.

3 Results and Discussion

Tensile test was performed on neat PLA and its nanocomposites and the resulted stress-strain curves are shown in Fig. 1. According to the tensile curves, the modulus and tensile strength of all binary and hybrid nanocomposite samples are higher than those of neat PLA, indicating good interaction between PLA and nanoparticles. In addition, it was seen that the elongation at break for PLA-NC nanocomposites increased as the same as that reported by Pirani et al. They observed the highest elongation at break and good improvement in tensile strength for PLA-NC sample containing 3 wt% of NC [3] Furthermore, for PLA-NCC nanocomposite, the tensile strength was lower than that for PLA-NC sample but its elongation at break was higher. Jiang et al. also obtained the same results by comparing PLA-NC and PLA-NCC as well. They observed an increase in yield stress and elongation at break for PLA-NC nanocomposite containing 2.5 wt% of NC compare to those for neat PLA and also a better improvement in elongation at break and a reduction in tensile strength for 7.5 wt% of NCC compared to PLA-NC sample. They reported that shear yielding and massive crazing are two of the toughening mechanisms which were observed for PLA-NC and PLA-NCC samples, respectively, due to the platelet structure of NC particles the nature and spherical structure of NCC [4]. So the elongation at break increased for both of them. Moreover, the maximum improvement of 37% in tensile strength was only observed for hybrid nanocomposite, indicating the synergistic effect between these two nanoparticles. Alavi et al. used the same nanoparticles along with polyethylene matrix and observed a significant improvement in tensile modulus and tensile strength. They demonstrated that the maximum tensile strength was achieved for hybrid sample containing 5% nanoclay and 8% NCC [5]. Finally, to investigate the impact strength of nanocomposites, the impact test was performed on neat polymer and its binary and hybrid nanocomposites. The impact strength of neat PLA was, while the impact strength of all nanocomposites was smaller than 32 J/m. The binary nanocomposites containing NCC nanoparticles showed a less reduction in impact strength, and the lowest reduction was seen in P-NCC7 sample (30 J/m). All information related to the physical mechanical properties of binary and hybrid nanocomposites including modulus, tensile strength and elongation at break are summarized in Table 1.

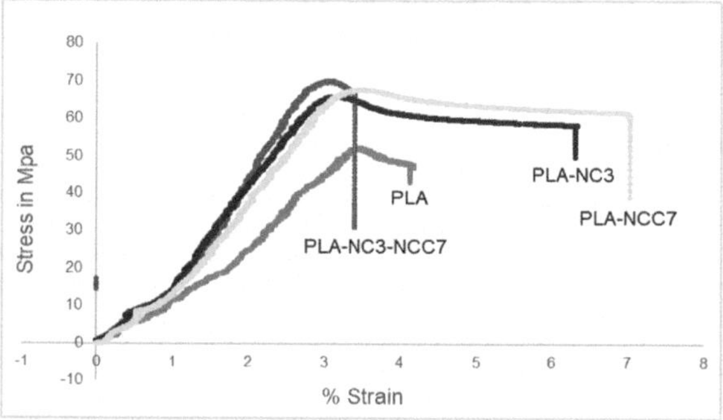

Fig. 1. Stress-Strain curves for neat PLA and its binary and hybrid nanocomposites containing 3 wt% of NC and 7 wt% of NCC

Table 1. Mechanical properties of PLA and its nanocomposites

Sample	Young's Modulus (MP)	Tensile strength (MP)	Elongation at break (%)	Impact (J/m)
P	2370 ± 20	52 ± 0.4	4.2% ± 0.2	32 ± 2
P-NC3	2650 ± 10	65.8 ± 0.28	6.3% ± 0.6	28 ± 2
P-NCC7	2570 ± 30	67.5 ± 0.10	7% ± 1	30 ± 3
P-NC3-NCC7	3640 ± 20	71.3 ± 0.11	3.1% ± 0.5	27 ± 3

4 Conclusion

In this essay, mechanical properties of binary PLA-NC and PLA-NCC, also hybrid PLA-NC-NCC nanocomposites were investigated and compared with those of neat PLA. It was found that the modulus and tensile strength of all nanocomposites were higher than those of neat polymer. As a result, the highest improvement in tensile strength was 37% compared to neat polymer, which was obtained in P-NC3-NCC7 hybrid sample. The strain at break of binary samples was higher than that of neat polymer, and its maximum, with 68% improvement was found in P-NCC7 sample. Also, the lowest reduction in impact resistance was observed for P-NCC7 as compared to neat sample.

References

1. Rafael Auras, H.T., Lim, L.-T., Sem, S.: Poly(lactic acid) **522** (2010)
2. Othman, S.H.: Agric. Agric. Sci. Procedia **2**, 296–303 (2014)
3. Jiang, L., Zhang, J., Wolcott, M.P.: Comparison of polylactide/nano-sized calcium carbonate and polylactide/montmorillonite composites: reinforcing effects and toughening mechanisms. Polymer **48**, 7632–7644 (2007)
4. Pirani, S.I., Krishnamachari, P., Hashaikeh, R.: J. Thermoplast. Compos. Mater. **27**(11), 1461–1478 (2013)
5. Garmabi, H., Emad, S., Tabari, A., Javadi, A., Behrouzi, H.: An investigation on morphology and mechanical properties of HDPE/nanoclay/nano$CaCO_3$ ternary nanocomposites, vol. 90005, pp. 1–6 (2016)

Biodegradation and Thermal Properties of Biocomposites Based on Poly(Lactic Acid)/ Thermoplasticized Rice Straw

Asa'ad Zandi, Amirhossein Zanganeh, Shayan Hosseini, Farkhondeh Hemmati(✉), and Jamshid Mohammadi-Roshandeh(✉)

Caspian Faculty of Engineering, College of Engineering, University of Tehran, Tehran, Iran
{f.hemmati, roshandeh}@ut.ac.ir

Abstract. Poly(lactic acid) as a biodegradable aliphatic polyester has some drawbacks that limit its marketability such as high cost and low biodegradation rate. In this work, these drawbacks were overcome by adding an abundant agricultural residue, rice straw, to PLA. Rice straw was thermoplasticized by using alkali pulping and benzylation processes. The benzylated rice straw pulp and pulping liquor considerably improve the biodegradation rate of PLA about 20 times faster after 128 days burying in soil. In addition, these lignocellulosic fillers can increase the PLA crystallinity degree, Vicat softening point and thermal stability comparing with untreated rice straw.

Keywords: Biodegradation · Composite · Poly(lactic acid) · Rice straw · Thermal stability

1 Introduction

One of the severe environmental concerns, especially in recent decade, is plastic waste. Many researchers have focused on novel sustainable polymers [1]. One of these polymers is poly(lactic acid) (PLA). However, PLA has some drawbacks that limit its marketability such as high cost and relatively low biodegradation rate [2]. One of the workable solutions is the PLA compounding with agricultural waste like rice straw [1]. In this work, rice straw as an abundant lignocellulosic biomass is modified by alkali pulping and benzylation reactions and employed to diminish the PLA cost and improve its biodegradability, which have not been reported yet to the best of our knowledge.

2 Experimental

PLA (Ingeo™ Biopolymer 3052D) was supplied by Nature Works Co. Rice straw was taken from a local mill (Guilan). After drying rice straw powder, it was mixed with 10 wt% NaOH solution, refluxed for 1 h at 120 °C. Then, the pulping liquor was separated from the pulp. After washing the rice straw pulp, it was mixed with benzyl chloride and 40% NaOH solution and refluxed at 110 °C for 5 h to prepare benzylated pulp (BP).

H. Mirzadeh and A. A. Katbab (Eds.): ISPST 2018, *Eco-friendly and Smart Polymer Systems*, pp. 268–271, 2020.
https://doi.org/10.1007/978-3-030-45085-4_64

Similarly, the rice straw pulping liquor was benzylated to obtain benzylated pulping liquor (BL) (refer to [3]). The photograph and microscopic images of these products are shown in Fig. 1. After drying the materials, the PLA samples containing BL and BP at 10, 20, and 30 wt% were prepared using Rondol Microlab twin-screw extruder. The aerobic degradation of virgin PLA, PLA/BL, and PLA/BP samples was investigated in an agricultural-rich soil for a period of 128 days at the temperature of 20–25 °C and relative humidity of 50–70%. Thermal analyses was performed using a Perkin Elmer Pyris STA6000. Vicat softening point of samples was also measured by using ATS-FAAR MP3 machine. The cryofractured surface of PLA/BL and PLA/BP composites was examined employing AIS-2100 SERON SEM.

3 Results and Discussion

The distribution and dispersion state of BP and BL micron-size particles are shown on the cryofractured surface of samples in Fig. 2. The biodegradation of PLA involves a two-step mechanism; first, hydrolysis of ester linkage in PLA backbone, then, metabolism of the products by microorganisms [2]. The weight retention versus time of the prepared samples in addition to the water uptake data during soil burial test are illustrated in Fig. 3. As can be seen, the weight loss of composites, especially PLA/BL ones at higher filler loadings, reaches to 20–35 wt% after 128 days positioning in soil while the neat PLA samples retain quite their whole weight. The higher biodegradation rate of PLA composites can be due to larger water uptake, as demonstrated in Fig. 3(b), and lower glass transition temperature of PLA matrix in the presence of the modified lignocellulosic fillers shown in Table 1. Both used fillers, especially BL, are more hydrophilic than PLA and have some amount of lignin, which is a natural plasticizer [3]. The DSC results of samples, which are gathered in Table 1, show that BP and BL increase the crystallinity weight fraction (X_c) of PLA owing to their effective role in nucleation and growth stage of PLA crystallization. Furthermore, BP microparticles that mostly consist of benzylated cellulose enhance the Vicat softening point of PLA at 10 wt% and make one of the main disadvantages of PLA less pronounced. Even though rice straw can be a useful filler in polymers, the presence of it diminishes the thermal stability of matrix. As can be found from Fig. 4 and Table 1, the benzylated products, BP and BL, have higher thermal stability than rice straw owing to the larger amount of components such as lignin and silica [3]. The incorporation of the lignocellulosic fillers into PLA reduces the thermal stability of matrix. However, PLABL10 sample has higher degradation temperatures among other composites. At larger filler loadings, the composites containing BP show higher thermal stability, which stem from the formation of a protective char layer from silica on the surface.

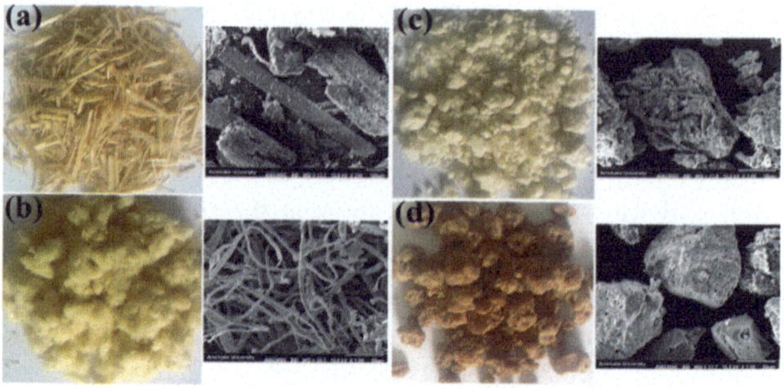

Fig. 1. Photograph and SEM micrograph of: (a) rice straw powder, (b) rice straw pulp, (c) BP, and (d) BL

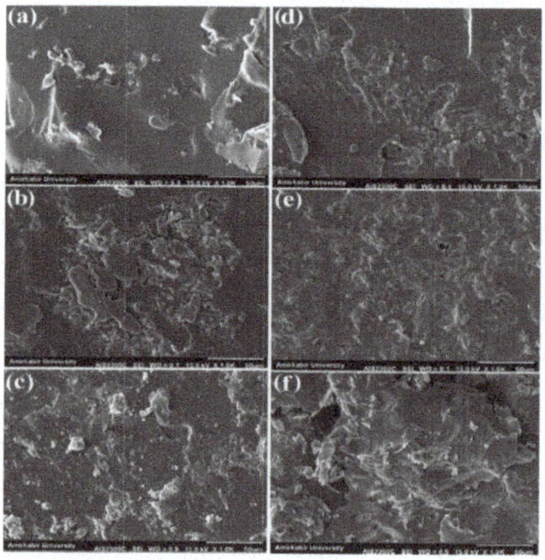

Fig. 2. SEM micrographs of the cryofractured surface for: (a) PLABL10, (b) PLABL20, (c) PLABL30, (d) PLABP10, (e) PLABP20, and (f) PLABP30

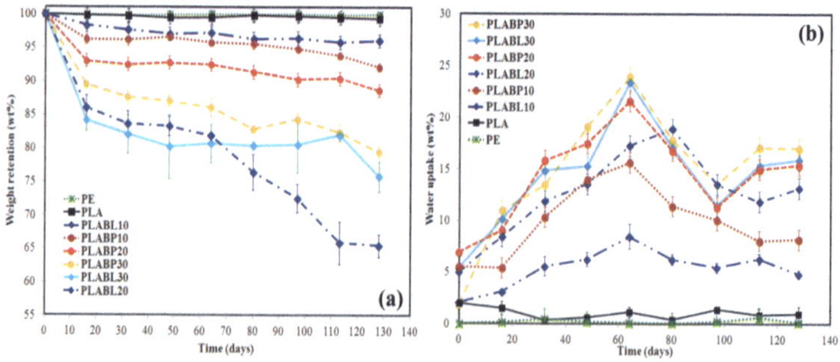

Fig. 3. (a) Weight retention and (b) Water uptake of virgin PLA, PLA/BL and PLA/BP during soil burial test for 128 days

Table 1. Vicat softening point and DSC results for virgin PLA, PLA/BL and PLA/BP samples

Sample code	Vicat softening point (°C)	T_g (°C)	T_m (°C)	X_c (wt%)
PLA	63.8 ± 0.9	55.2	155.3	9.1
PLABL10	61.0 ± 0.8	53.7	152.5	28.5
PLABL20	54.7 ± 0.7	52.2	135.0	32.8
PLABL30	50.8 ± 3.8	48.3	135.7	34.3
PLABP10	71.1 ± 0.9	52.4	151.8	27.9
PLABP20	61.8 ± 1.4	49.6	147.8	26.5
PLABP30	58.3 ± 0.6	48.1	147.5	30.0

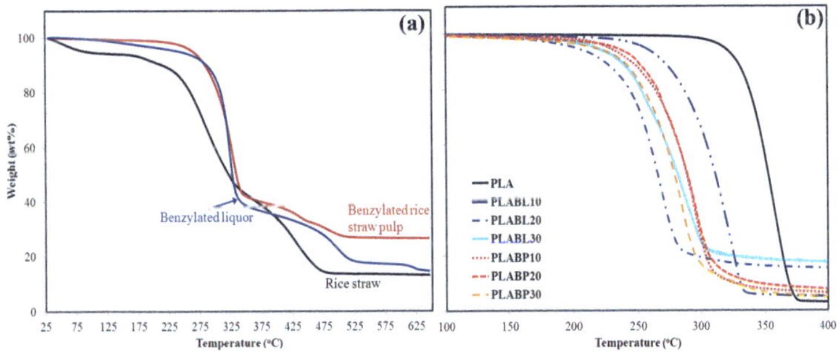

Fig. 4. TGA thermograms of: (a) rice straw powder and the used modified fillers and (b) virgin PLA, PLA/BL and PLA/BP samples

4 Conclusion

In this study, the effects of cheap fillers based on treated rice straw on the biodegradation and thermal stability of PLA were investigated. These fillers could improve the PLA biodegradation rate, crystallinity degree, Vicat softening point and thermal stability comparing with the untouched rice straw.

References

1. Wu, C.S., Liao, H.T., Jhang, J.J., Yeh, J.T., Huang, C.Y., Wang, S.L.: Polym. Bull. **70**, 3221–3239 (2013)
2. Karamanlioglu, M., Robson, G.D.: Polym. Deg. Stab. **98**, 2063–2071 (2013)
3. Mohammadi-Rovshandeh, J., Kaffashi, B., Davachi, S.M., Pouresmaeel-Selakjani, P., Chaboki, M.G., Dirini, F.: Cellul. Chem. Technol. **51**, 223–235 (2017)

Vancomycin Loaded Silk/Pectin Hydrogel for Bone Tissue Engineering

Akbar Karkhaneh(✉), Fereshteh Ahadi, and Sajedeh Khorshidi

Biomedical Engineering Faculty, Amirkabri University of Technology, Tehran, Iran
a.karkhaneh@aut.ac.ir

Abstract. Designing a biomaterial with efficient drug delivery has great importance in bone tissue engineering. In the present study, based on Schiff mechanism, a self-crosslinking hydrogel of oxidized pectin (OP) and silk fibroin (SF) was synthesized and vancomycin, a small charged molecule was loaded into hydrogel precursor by electrostatic interactions. The hydrogels with 1:1 and 3:1 ratio of SF and OP were screened with respects to swelling ratio, internal morphology and cell interactions. The hydrogels were well able to preserve cell viability and proliferation toward mesenchymal stem cells.

Keywords: Schiff mechanism · Oxidized pectin · Silk fibroin

1 Introduction

Besides clinical treatments, bone tissue engineering (BTE) gained a lot of interest in treatment of bone defects. Different forms of biomaterials have been used in BTE and among these biomaterials hydrogels are a large group of hydrophilic networks made of synthetic and natural polymers [1]. Silk fibroin as a natural polymer has unique properties such as good biocompatibility, high mechanical strength, simplicity of processing in aqueous or organic solvents and manageable rate of degradation. Remarkable mechanical properties and controlled release of drugs and biomolecules make SF a suitable choice for bone tissue engineering and drug delivery applications [2]. Pectin, an anionic polysaccharide can be functionalized by periodate oxidation to endow aldehyde groups to its macromolecules. In this study, self-crosslinking hydrogels without any external cross-linkers are formed through Schiff mechanism between aldehyde functions of OP and amino groups of SF [3].

2 Experimental

SF aqueous solution was extracted. Briefly, Bombyx mori cocoons were boiled in 0.02 M Na_2CO_3 for 30 min and washed 3 times with deionized water to remove the glue-like sericin protein. SF fibers were dried overnight and dissolved in 9.3 M LiBr for 4 h at 60 °C. Then the solution was dialyzed in deionized water for 48 h and subsequently centrifuged at 4000 RPM [4]. For periodate oxidation of pectin, 1 g of pectin was dissolved in 20 mL ethanol. Then an aqueous solution of 0.47 M sodium

H. Mirzadeh and A. A. Katbab (Eds.): ISPST 2018, *Eco-friendly and Smart Polymer Systems*, pp. 272–275, 2020.
https://doi.org/10.1007/978-3-030-45085-4_65

periodate with a 1:1 volume ratio was slowly added to the mixture, stirred in dark for 6 h and dialyzed for 72 h. To prepare hydrogels, vancomycin was blended in 8 w/v% SF solution and added to 8 w/v% OP solution with volume ratios of 1:1 and 3:1. Final weight ratio of vancomycin to total polymer was 10% in all samples. The samples were incubated at 37 °C overnight and dried afterward. Swelling behavior of samples were compared by weighting them after soaking in phosphate buffered saline (PBS) for 24 h [5]. Morphology of freeze-dried hydrogels were examined by scanning electron microscopy (SEM). To observe cellular attachment, human adipose derived mesenchymal stem cells (hAD-MSCs) were seeded onto scaffolds and cultured for 48 h.

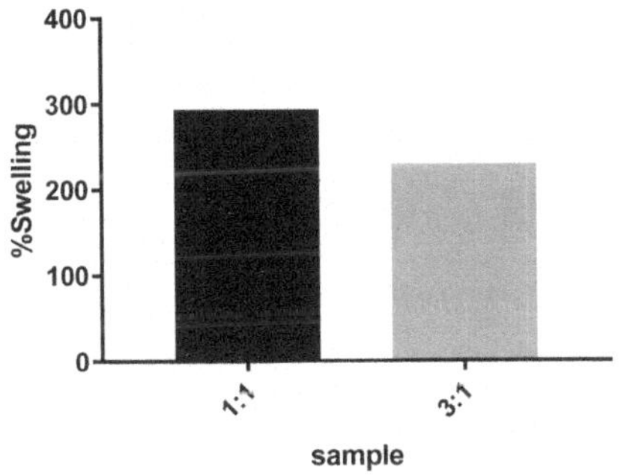

Fig. 1. Swelling behavior of samples with 1:1 and 3:1 SF to OP ratio

3 Results and Discussion

Swelling behavior of samples was studied and reported in Fig. 1. Sample with 1:1 ratio indicated more swelling and it could be attributed to less cross-linking density. As SF ratio increased, swelling ratio decreased. Cross sectional SEM images of freeze-dried samples demonstrated a highly porous structure with semi-uniform pore sizes in desirable range. Comparison between images revealed that the sample with 3:1 ratio has more pore interconnectivity and exhibits better uniformity in pore morphology as shown in Fig. 2.

After cell culture, fixation and dehydration, morphology of hAD-MSCs was observed using SEM. As evidenced from the Fig. 3, both samples showed good cell adhesion. In both cases a cellular monolayer was formed on the surface of hydrogels. Nevertheless, the cells were more elongated on scaffold with higher fibroin ratio. This occurrence can be explained by multiple reasons; first, the less water content of the hydrogel with higher fibroin ratio provided more spaces for cell adhesion. Second, fibroin has multitude amine groups that positively affect cell behavior. And finally, more crosslinking of this hydrogel created a stiffer substrate for cell residence.

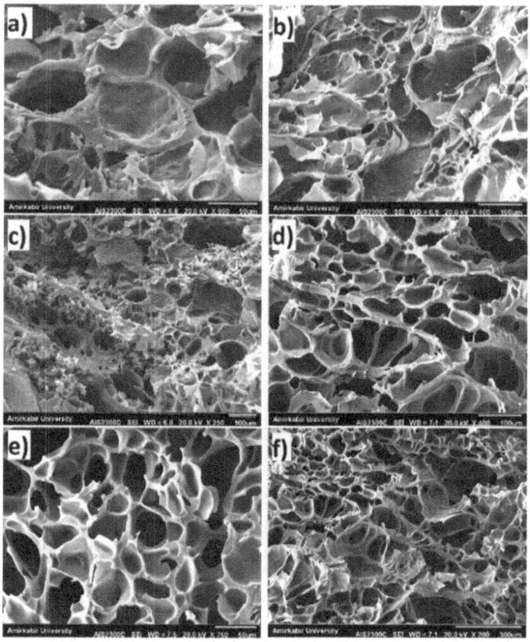

Fig. 2. Cross sectional SEM images of freeze-dried hydrogels: (a), (b), and (c) with 1:1 & (d), (e), and (f) with 3:1 SF to OP ratio

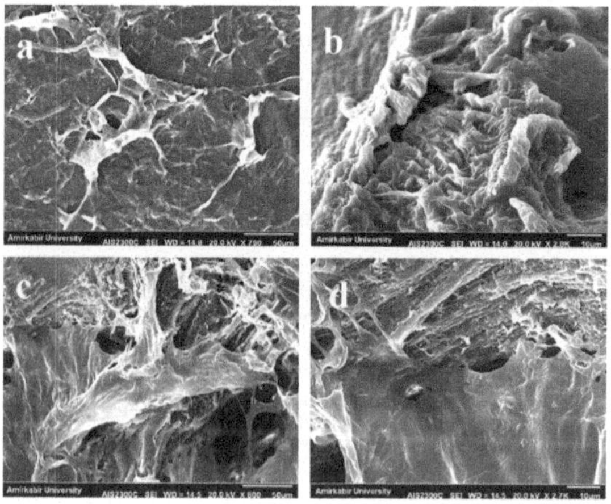

Fig. 3. SEM images of hAD-MSCs cultured on scaffolds: (a) & (b) with 1:1, and (c) & (d) with 3:1 SF to OP ratio

4 Conclusion

Silk based hydrogels can offer suitable scaffolds for bone regeneration. Their capacity to load therapeutics such as antibiotics enables them to be used for infection prevention, while providing a porous structure for cell growth. Hydrogels with 1:1 and 3:1 ratio of SF to OP were prepared. Increased in SF ratio has led to reduction in swelling and enhanced pore uniformity.

References

1. Mandal, B.B., Grinberg, A., Seok, E., Panilaitis, B., Kaplan, D.L.: Proc. Nat. Acad. Sci. **109** (20) (2012)
2. Melke, J., Midha, S., Ghosh, S., Ito, K., Hofmann, S.: ACTA Biomater, no. September (2015)
3. Khorshidi, S., Karkhaneh, A.: Mater. Lett. **164**, 468–471 (2016)
4. Mottaghitalab, F., Farokhi, M., Shokrgozar, M.A., Atyabi, F., Hosseinkhani, H.: J. Control Release (2015)
5. Numata, K., Yamazaki, S., Katashima, T., Chuah, J., Naga, N., Sakai, T.: Macromol. Biosci. **14**(6), 799–806 (2014)

Fabrication of Hybrid Suture Yarn Containing Chitosan/Poly(Vinyl Alcohole)-Poly(Lactic Acid) Micro and Nanofibers and Investigation of Their Mechanical Properties

Vida Saa'di, Marzieh Ranjbar Mohamadi(✉), Meysam Moezzi, and Reza Sghafi

Department of Textile Engineering, University of Bonab, Bonab, Iran
ranjbar.aut@gmail.com

Abstract. The low diameter of nanofibers, high specific surface, flexibility, desirable mechanical properties, and the porous nature of nanofibers, make such nanostructures as a suitable choice for many medical applications. Suture yarn as a biodegradable material is one of the mechanical methods for closing and healing of wounds. Suitable suture should have some characteristics such as making the least tissue reaction, antibacterial properties, non-sensitivity, non-carcinogenicity, low price, comfortable handling, proper tensile strength and suitable binding ability. In this paper, hybrid nanofiberous yarns containing chitosan/poly(vinyl alcohole) (PVA/CS) with the blend ratio of (75:25) and poly (lactic acid) (PLA) were fabricated by electrospinning process. For this, CS/PVA and PLA solutions were injected from two different syringes. These two different syringes had positive charge, and the collector for assembling fibers had negative charge. In the composition of produced micro and nano fibrous yarn, both PLA and PVA/CS exist with the blend ratio of 75:25. Fabricated yarns were assessed for their morphological, surface, mechanical characteristics. However, the hydrophilicity nature of produced mat was analyzed by reporting contact angle measurements. The SEM images exhibited porous and smooth fibers with parallel arrangement. The mechanical strength of produced yarn was about 79.26 MPa. The presence of chitosan/PVA component in yarn composition changed the hydrophilic/hydrophobic balance. The results of this paper showed that these yarns are a good candidate for application as wound sutures.

Keywords: Nanofiber · Suture yarn · Biocompatible · Biodegradable · Mechanical properties

1 Introduction

Sutures are categorized according to their physical and chemical properties from various aspects, such as the type of material (natural or synthetic), the shelf-life of the body and the ability to absorb it, as well as their structure (single-filament or multiple filament) [1]. Sutures are one of the most important bioactive materials that can be used in the wound infection, since infection usually extends along or near the lines and

H. Mirzadeh and A. A. Katbab (Eds.): ISPST 2018, *Eco-friendly and Smart Polymer Systems*, pp. 276–281, 2020.
https://doi.org/10.1007/978-3-030-45085-4_66

closing wound environment [1, 2]. Stitching with antibacterial sutures can support the wound for a long time and play a major role in preventing infection as well as in treating it [3]. Recently, the production of yarn from micro and nanofibres has been considered by many researchers for potential applications. Electro-technology is considered as an economic, easy and available method for the production of fibers for application in suture applications. When the fiber diameter decreases from micro to nano, outstanding characteristics such as high surface area, flexibility and mechanical properties are observed [4]. One of the challenges ahead in fabrication nanofibers for suture applications, is the choice of suitable polymer for fibers [6]. Chitosan is also a natural biopolymer with antibacterial characteristic that has non-toxic biocompatible properties [6]. The successful use of poly-lactic acid in medicine has attracted much attention to this polymer. The substance caused by the destruction of this polymer, lactic acid, is the metabolic product of all animals and micro-organisms, which ultimately turns into water and carbon dioxide [7]. PVA is a synthetic polymer with hydrophilicity nature that has been used in several papers for improving spin ability of natural based polymers such as chitosan and gum tragacanth. In this research, for first time in literature, we study the possibility of producing suture yarns from poly lactic acid (PLA) and chitosan/poly(vinyl alcohole) (CS/PVA) and investigating their mechanical properties.

2 Experimental

2.1 Materials and Methods

At the first 10 g of PVA powder was dissolved into 100 mL of distilled water to get the solution with concentration of 10 w/v%. The solution was stirred at room temperature for 8 h using electromagnetic stirrer to reach homogenous and crystal clear solution. 1 g of CS powder (at a concentration of 5 wt%) was dissolved into 20 mL of acetic acid 80% solution (v/v) at room temperature under magnetic stirring. Then 12 g of PLA was dissolved into 100 mL of dichloromethane at room temperature under magnetic stirring to form a proper solution with concentration of 10 w/v%. The solution was stirred at room temperature for 8 h using electromagnetic stirrer to reach homogenous and crystal clear solution. 1 g of CS powder (at a concentration of 5 wt%) was dissolved into 20 mL of acetic acid 80% solution (v/v) at room temperature under magnetic stirring. Then 12 g of PLA was dissolved into 100 mL of dichloromethane at room temperature under magnetic stirring to form a proper solution. The electrospinning process is exhibited in Fig. 1. Because the standard length of suture yarn is 70 cm we selected the external pickup environment of collector about 70 cm. Two different syringes for fabrication of hybrid fibers had positive charge. In this case, two polymers had the same condition to fabricate hybrid yarn and the collector for collecting fibers had negative charge. The polymer solution was taken up in a 10-mL glass syringe equipped with a 20 - gauge, stainless steel needle (diameter = 0.6 mm). The needle was connected to the emitting electrode of positive polarity of a High Voltage Research device. The electric potential was fixed at 15 kV. The nanofibers were collected as-spun on an aluminum rotating drum. The solution was electrospun at room temperature, and the collection distance was fixed at 17 cm. The solution feed rate from syringe containing PLA (12%) was 1.8 mL/h and for CS/PVA solution was about 1.8 mL/h for preparation PLA-CS/PVA fibers with ratio of 75:25. The electrospun nanofibers were firstly sputtered with thin layer of gold before SEM observation. The morphology of the fibers was observed utilizing scanning electron microscopy (SEM) (Hitachi, Model S-4160) at an accelerating voltage of 30 kV, and the average fiber diameter (AFD) was estimated using Microstructure Measurement Software. A sessile drop method was adopted in which deionized water was automatically dropped in the form of droplets on the surface of the fibrous specimens until equilibrium was applied to examine the effects of addition of PVA and CS on the hydrophilicity of the electrospun fibers at room temperature. The drop shapes on the nanofibers surfaces were imaged by Video contact analyzer, and contact angles were evaluated using Image J Software. These test was investigated for two samples, PLA and PLA-PVA/CS separately. The mechanical property of the yarns are tested by an Iranian universal tensile strength tester (Santam Co., type STM 20) with load cell 20 N. The twisted fibrous yarns were randomly cut into pieces with an equal length of 80 cm and then weighed to obtain the linear mass density (Tex) of the yarns. The sample was fixed firmly in the clamps with gage length of 50 cm and test speed was set to 10 mm/min. Sutures fabricated from synthetic polymeric materials exhibit the viscoelastic behavior in final applications. Sutures shows stress relaxation when they are hold on the fixed elongation in knots. Then, a theoretical model for the description of relaxation curves after dynamic loading is presented in this paper. According to Fig. 2 (schematic view of our model), the model is based on a nonlinear spring placed in parallel with two Maxwell element the obtained formula is presented in Eq. (1). The obtained formula is fitted on experimental stress-relaxation curves using the Marquardt algorithm and the resultant coefficients are obtained.

$$F = E_b \varepsilon^2 + \exp(-E_1 t/\eta_1) + \exp(-E_2 t/\eta_2) \quad (1)$$

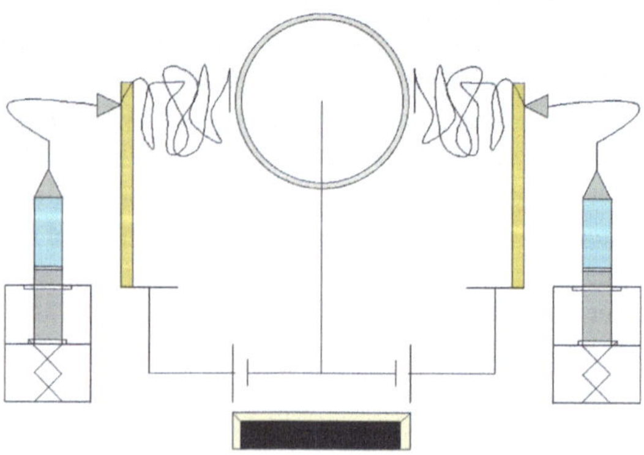

Fig. 1. Suture yarn preparation setup

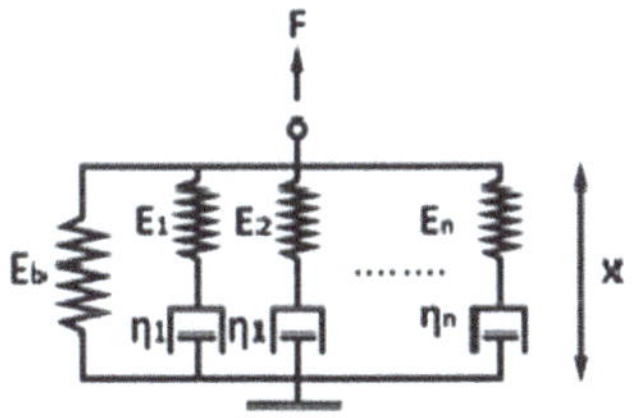

Fig. 2. Schematic view of the nonlinear extended Maxwell model

3 Results and Discussion

3.1 Morphology of Electrospun Fibers

Figure 3 shows the SEM images of micro and nanofibers formed in electrospinning. The images captured showed good alignment and porous structure of fibers. The porous structure can be created by quicker evaporation of solvent. The diameter of PVA/CS and PLA fibers was about 169 nm (with CV% = 23.92) and 1221 µm (with CV% = 27.81), respectively.

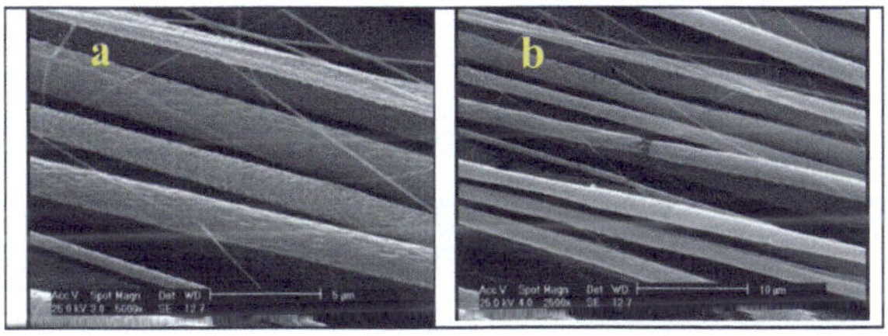

Fig. 3. SEM images of fabricated yarns: (a) PLA and (b) PLA-PVA/CS fibers

3.2 Hydrophilicity Nature of Yarn

As it has been shown in Fig. 4, the contact angle of PLA is more than that of PLA PVA/CS, that means PVA/CS increases hydrophilicity of suture yarn (Fig. 4). Chitosan and PVA have hydrophilic functional groups that help to increase hyrophilicity of fibers.

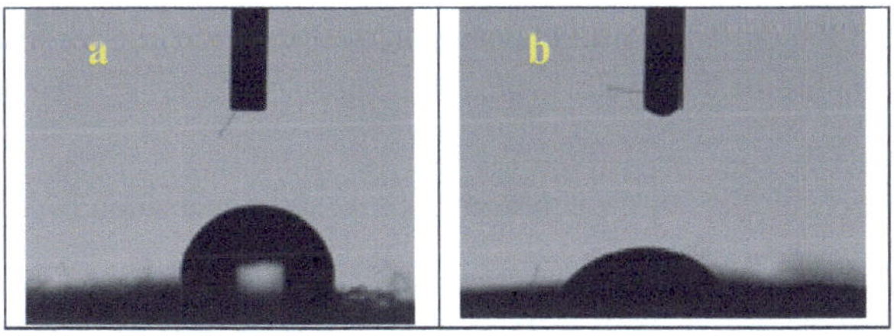

Fig. 4. Contact angle results: (a) PLA and (b) PLA-PVA/CS fibers

3.3 Mechanical Strength of Yarn

Investigation on the mechanical properties of textile products is important technically and practically. The mechanical behavior of yarns is influenced from the yarn geometry and properties of individual fibers in the yarn structure. Changes in the structure of the fiber during the electrospinning process, especially crystallinity and molecular orientation, plays an important role in the final behavior of them. The geometric arrangement of fibers and interaction between the fibers in the yarn structure (applied twist number) also have an important effect on the mechanical properties of yarns. In addition, the diameter of the fibers forming the yarn, the molecular orientation and crystal structure morphology, and structural homogeneity of fibers are important factors affecting the mechanical behavior of electrospun yarns. Force-extension curve of produced yarn was exhibited in Fig. 5. The yarn count of produced yarn was 112.5 Tex and its diameter was 0.3 mm. The maximum stress of the produced yarn at breaking point and the other mechanical results are shown the Table 1 and Fig. 5.

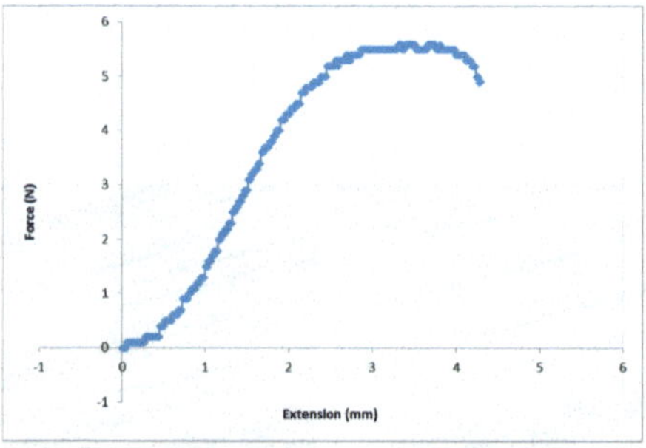

Fig. 5. Force-extention curve

Table 1. Mechanical result of the prepared yarns

Yarn type	Stress at maximum point (CN/Tex)	Stress at maximum point (MPa)	Extension at maximum point (mm)	Elastic modulus in 2% elongation (MPa)
PLA-PVA/CS	3.28	79.26	5.5	271.945

3.4 Stress Relaxation Results of Organic Suture Materials

The coefficients of the equation fitted on the stress-relaxation curves for prepared yarns are obtained and listed in Table 2.

Table 2. Resultant coefficients of the equation fitted on stress-relaxation curves of the prepared yarns

Yarn type	Eb	E1	E2	η1	η2
PLA-PVA/CS	2.093	76.57	0.1317	1.598	0.029435.5

4 Conclusion

In the present work, the feasibility of electrospinning process, as a method for producing absorbable suture, was analyzed. In this regard, the mechanical behavior of sutures made of PLA-PVA/CS fibers was reported. The results showed that the presence of CS/PVA influenced the mechanical, morphological and hydrophilic nature of yarns. CS/PVA and PLA solutions produced nano and micro sized fibers, respectively. Presence of CS/PVA in the formulation of nanofibers increased hyrophilicity of yarn. Fabricated yarn exhibited proper mechanical strength which is not enough to use for application as suture in closing wounds.

References

1. Padmakumar, S., Joseph, J., Neppall, M.H., Mathew, S.H.: ACS Appl. Mater. Interf. **1** (2016)
2. Abellán, D., Nart, J., Pascual, A., Cohen, R.E., Sanz-Moliner, J.D.: Polymers **8**, 147 (2016)
3. Mankod, H.R.: Coat. Int. J. Text. **11** (2013)
4. Chen, S.H., Ge, L., Mueller, A., Carlson, M., Matthew, J.: Nanomed. Nanotechnol. Biol. Med. **54** (2017)
5. Chellaman, K.P.I., Veerasubramanian, D., Vignesh Balaji, R.S.: South India Text. Res. Assoc. **778**, 2278–5213 (2013)
6. Kazuo, A., Izumi, R.: Funct. Biomater. **6**, 104–142 (2015)
7. Nguyen, T., Dinh, T.: Nanosci. Nanotechnol. **15**, 9585–9590 (2015)

Benzylated Products of Rice Straw as Effective Fillers for Starch Biocomposites

Maryam Shoja[1], Farkhondeh Hemmati[2(✉)], and Jamshid Mohammadi-Roshandeh[2]

[1] Fuman Faculty of Engineering, College of Engineering, University of Tehran, Tehran, Iran
[2] Caspian Faculty of Engineering, College of Engineering, University of Tehran, Tehran, Iran
f.hemmati@ut.ac.ir

Abstract. Rice straw is an abundant biomass that can be used as filler in biodegradable polymers like starch. In this study, different microfillers based on rice straw including untreated rice straw, alkali pulp of rice straw, benzylated rice straw and benzylated pulp and pulping liquor were incorporated in plasticized starch with glycerol. The findings show that the benzylated products of rice straw having stronger interaction with starch are more effective fillers than untreated rice straw in starch biocomposites.

Keywords: Starch · Biocomposites · Rice straw · Benzylation process · Alkali pulping

1 Introduction

In the last decades, eco-friendly polymers have attracted great interest due to raising awareness of environmental issues such as the depletion of non-renewable fossil resources and the contamination of soil and water by plastic waste [1]. One of these polymers is starch (ST), which has gained more marketability among others owing to low price, biodegradability, accessibility and flexibility. However, similar to most natural polymers, the starch properties have to be improved and lots of researches have focused on the subject by applying different methods such as mixing of starch with natural fillers obtained from biomass. Rice straw is one of the cheap agricultural wastes that has been used as reinforcement agent/filler in starch [2, 3]. High interfacial strength between the natural fiber and starch is an absolute prerequisite for reinforcing. Therefore, in this study, rice straw is modified by alkali pulping method and/or benzylation process to obtain thermoplastic lignocellulosic/cellulosic fillers and their effects on the mechanical performance, water absorption and solubility of starch are investigated.

H. Mirzadeh and A. A. Katbab (Eds.): ISPST 2018, *Eco-friendly and Smart Polymer Systems*, pp. 282–285, 2020.
https://doi.org/10.1007/978-3-030-45085-4_67

2 Experimental

Corn starch was supplied by Cargill Co. and rice straw (RS) was taken from a local mill in Guilan. Chemicals such as glycerol, benzyl chloride, sodium chloride, chloroform and methanol from Merck were also used. Dried rice straw was mixed with 30% NaOH solution (by weight), refluxed for 2 h at 110 °C. Then, the pulping liquor was separated from the pulp (PRS). After washing the rice straw pulp for several times with water and drying, it was mixed with benzyl chloride and 40% NaOH solution and the mixture was refluxed at 110 °C for 6 h. After a period, two phases were formed. For preparing the benzylated rice straw pulp (BP), the organic phase was separated. The benzylation reaction of rice straw and pulping liquor was performed in a similar way. For more detail refer to the previous work [3]. After drying the materials, the starch biocomposites containing RS, PRS, BP, benzylated rice straw (BR) and benzylated rice straw pulping liquor (BRPL) at fixed loading of 20 wt% and glycerol at 40 wt% were prepared using Rondol Microlab twin-screw extruder at a temperature profile of 110 °C to 160 °C and screw speed of 50 rpm. The mechanical properties of the samples were determined by using a Sanaf Tensile Machine according to ASTM D638 standard. The applied microfillers as well as the cryofractured surface of the biocomposites was examined employing AIS-2100 SERON SEM. The water solubility of the starch-based composites was measured by exposing the samples to distillated water for 4 days and weighting the samples after drying at 105 °C. The thickness swelling and water absorption of the samples were determined by soaking them in distillated water for 2 h at ambient temperature.

3 Results and Discussion

Figures 1 and 2 show the SEM micrographs of the used microfillers and the cryofractured surface of the prepared starch biocomposites, respectively. While the bundle of natural fibers is clearly evident in starch/RS composite, the other samples have much smaller filler particles especially starch/BP (cellulose-rich filler) and starch/BRPL (lignin-rich filler) composites. Moreover, the micron-size cellulosic fibers of PRS are also obvious, which are dispersed uniformly in the starch/glycerol matrix. The mechanical characteristics of plasticized starch and its biocomposites are illustrated in Fig. 3. As can be found, maximum Young modulus and tensile strength are attained for starch/PRS and starch/RS composites due to the stronger thermoset-like nature of the cellulosic and lignocellulosic fibers. In contrast, starch composites containing benzylated products of rice straw show higher elongation at break and dissipated energy, which stems from the thermoplasticization modification of rice straw, rice straw pulp and pulping liquor. Among all prepared samples, the starch composites containing BR and BP fillers have largest water solubility. It can be resulted from high water absorption and enhanced compatibility between starch chains and BP/BR particles. Contrary to them, rice straw pulp and benzylated pulping liquor decrease the water solubility and water absorption of starch (Figs. 4 and 5). The plausible reason is that the benzylated pulping liquor, which is rich in lignin, is relatively less hydrophilic. Besides, cellulosic fibers in PRS have a profound impact on the starch chain entropy and molecular dynamics.

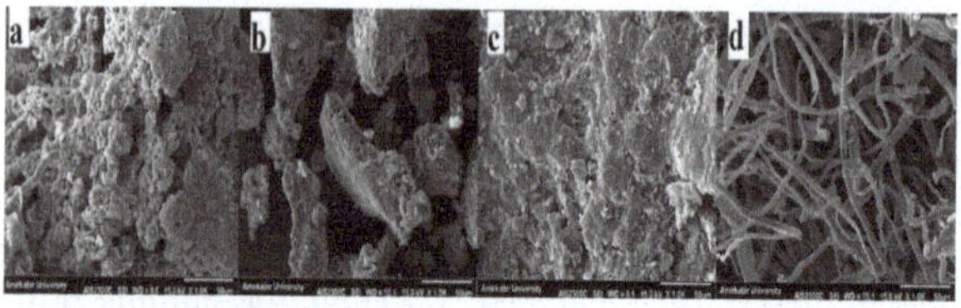

Fig. 1. SEM micrograph of: (a) BP, (b) BRPL, (c) BR, and (d) PRS

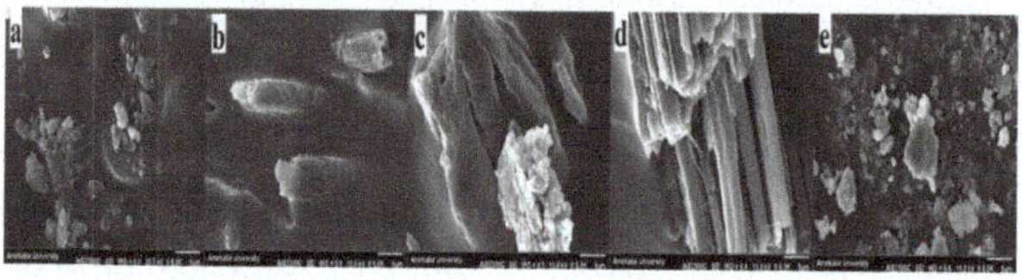

Fig. 2. SEM micrograph of starch biocomposite containing: (a) BP, (b) PRS, (c) BR, (d) BRPL, and (e) RS

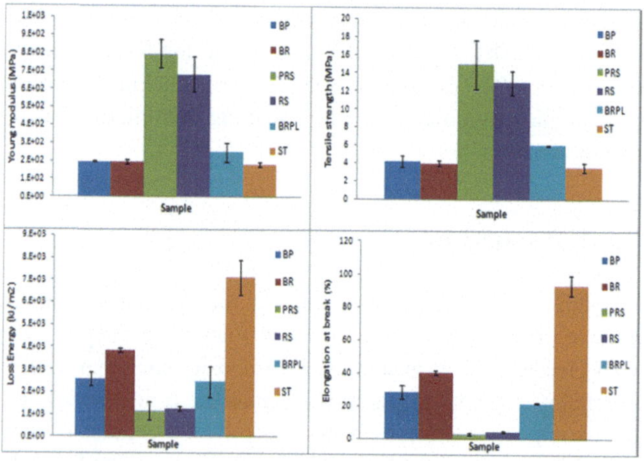

Fig. 3. Mechanical properties of pure starch/glycerol blend (ST) and the prepared biocomposites containing different microfillers, including Young modulus, tensile strength, dissipated energy and elongation at break

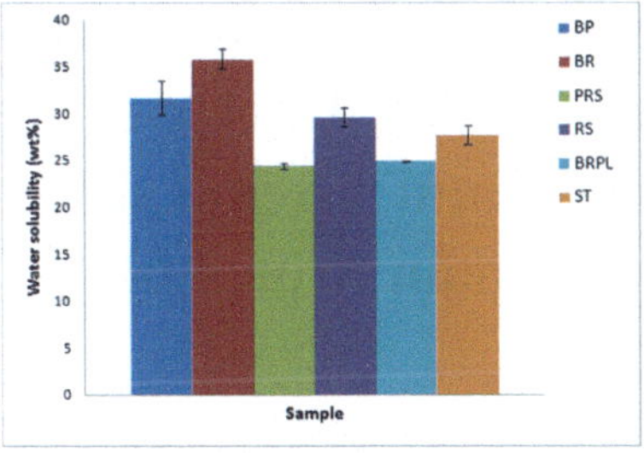

Fig. 4. Water solubility (wt%) of pure starch/glycerol blend (ST) and the prepared biocomposites containing different microfillers

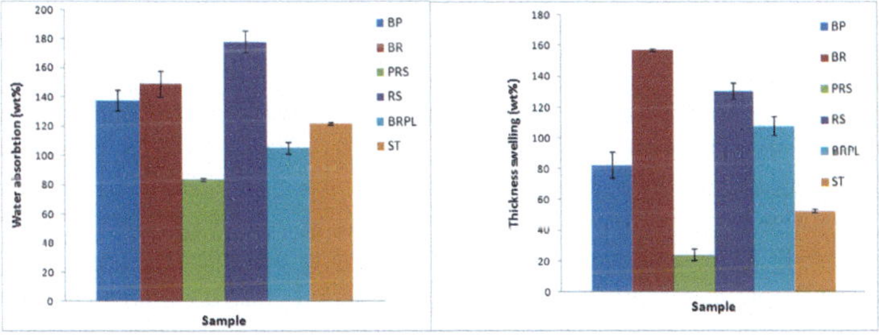

Fig. 5. Water absorption (wt%) and thickness swelling of pure starch/glycerol blend (ST) and the prepared biocomposites containing different microfillers

4 Conclusion

In this study, the effects of five different fillers based on rice straw on the mechanical performance and water absorption and solubility of starch were investigated. Benzylation process of rice straw, alkali pulp and pulping liquor of rice straw can result in effective microfillers for starch, which improve the elongation at break, dissipated energy and water solubility of starch/rice straw biocomposites.

References

1. Thakur, V., Singha, A., Thakur, M.: Int. J. Polym. Anal. Charact. **17**, 401–407 (2012)
2. Majeed, Z., Mansor, N., Man, Z.: Adv. Mater. Res. **45**, 1133 (2016)
3. Mohammadi-Rovshandeh, J., Kaffashi, B., Davachi, S.M., Pouresmaeel-Selakjani, P., Ghorbani-Chaboki, M., Dirini, F.: Cellul. Chem. Technol. **51**, 223–235 (2017)

Extrusion Foaming of PLA-Based Nanocomposites: Effects of Different Nanofillers and Chemical Reagents on Foam Morphology

Mohammad-Hasan Khademi, Sirwan Ghavami, Farkhondeh Hemmati(✉), and Mohammad-Ali Aroon

Caspian Faculty of Engineering, College of Engineering, University of Tehran, Tehran, Iran
f.hemmati@ut.ac.ir

Abstract. Poly(lactic acid) (PLA) foams have great potential for replacing synthetic polymer foams owing to their special properties such as biodegradability, biocompatibility and good mechanical performance. However, PLA has some drawbacks in common foaming processes to gain marketability. In this work, the PLA melt strength in extrusion foaming process was improved by adding different nanofillers, i.e. nanosilica, expanded graphite, pristine nanoclay and organoclay as well as chemicals including benzoyl peroxide and maleic anhydride grafted polyethylene. The morphological results and void content measurements show that the incorporation of the chemicals into PLA matrix comparatively causes better foam structural characteristics than nanofillers.

Keywords: Biodegradable foams · Poly(lactic acid) · Nanoparticles · Extrusion · Chemical foaming

1 Introduction

Recently, a lot of effort has been internationally made to obtain biodegradable polymers from renewable resources owing to the growing concern about the polymer wastes, fluctuating prices of petroleum and natural gas and severe depletion of fossil fuels [1]. One of these biopolymers that have been attracted a great interest is poly(lactic acid) (PLA). PLA foams have excellent potential to replace their non-biodegradable counterparts, polystyrene and polyethylene foams, due to the appropriate mechanical performance, accessible raw materials, comparative processing cost, biocompatibility and biodegradability [2]. However, PLA has several drawbacks in common foaming processes. Low melt strength of PLA as the main drawback can be overcome by using different methods such as applying PLA/nanoclay masterbatch in physical foaming process [3]. In this study, the melt strength of PLA in extrusion foaming process was improved by incorporating different nanofillers including nanosilica, expanded graphite, pristine nanoclay and organoclay along with different chemical reagents, a peroxide initiator and maleic anhydride grafted polymer, which has not been reported yet to the best of our knowledge.

H. Mirzadeh and A. A. Katbab (Eds.): ISPST 2018, *Eco-friendly and Smart Polymer Systems*, pp. 286–289, 2020.
https://doi.org/10.1007/978-3-030-45085-4_68

2 Experimental

PLA (Ö5 BIOKAS) with melt flow rate of 7 g/10 min (210 °C, 2.16 kg) was supplied by Chemiekas GmbH. Hydrophobic nanosilica (Si) from Evonik Co., organically-modified nanoclay (ONC, Cloisite 15A) and unmodified nanoclay (UNC, Cloisite Na^+) from Southern Clay Products Inc. and graphite powder after thermal expansion (EG, 104206) from Merck were used as nanoparticles in this work. Moreover, benzoyl peroxide (BP) from Merck and maleic anhydride-g-polyethylene (PEMA, Fusabond® E100) from Dupont as chemical reagents of PLA and azodicarbonamide (ADA) from Kum Yang Co. as chemical blowing agent were also applied. For preparing PLA chemical foams, first, PLA/nanofiller masterbatches containing each nanoparticle at 5 wt% loading were obtained using Rondol Microlab co-rotating twin-screw extruder at a temperature profile of 165 °C to 175 °C and screw speed of 150 rpm; second, the resultant masterbatches were pelletized and diluted by virgin PLA. Third, the physical mixtures of PLA and masterbatches were chemically foamed by incorporating ADA at fixed content of 4 phr through the twin-screw extruder with a slit die at a screw speed of 150 rpm and temperature profile of 165, 160, 170, 160, 150, 135, and 135 °C. The void content (V_f) of the prepared foams was determined by measuring the density of foamed and unfoamed samples with the same composition (except for ADA) and thermal history using solvent displacement method (ASTM D792). The foam morphology of the prepared films was observed by using HITACHS-4160 scanning electron microscopy (SEM) on the cryofractured and gold-sputtered surfaces. In the following parts, samples will be coded as PLAxY or PLAxYzW, where Y and W are the components incorporated into PLA and x and z are their weight fractions, respectively.

3 Results and Discussion

In the foaming extrusion process of PLA, the temperatures of slit die were chosen to be close to the PLA crystallization temperature, 135 °C, in order to improve the matrix melt strength. The SEM micrographs of the prepared PLA foams are illustrated in Fig. 1 and the foam void content along with the structural characteristics including number-average and volume-average cell diameter (D_n and D_v) and cell density are gathered in Table 1. The results show that by reducing the extruder screw speed to 75 rpm, the buddle size and void content increase and the cell density diminishes to lower values owing to higher residence time and intensified ADA decomposition and PLA hydrolysis. The addition of BP and PEMA has a considerable influence on the PLA foam morphology. While both of them results in larger cell density than virgin PLA, the BP incorporation causes smaller cells indicating the occurrence of chemical reaction between BP and PLA backbone and the resultant improvement in the PLA melt strength. Among the incorporated nanoparticles, Si and EG containing foams have higher void content, which stems from their stronger hydrophobic nature and interactions with PLA chains. On contrary, both nanoclays reduce the PLA void content, particularly organoclay, indicating their catalytic effect on PLA hydrolysis during extrusion process and correspondingly, the resultant poor melt strength [4]. However, an increment in ONC loading brings about smaller bubbles and larger cell density.

The plausible reason is the cell nucleation-assisted effect of nanoparticles that is more pronounced at higher nanofiller loadings. The simultaneous addition of nanopaticles and PEMA results in smaller cells and cell density just for PLA1EG sample.

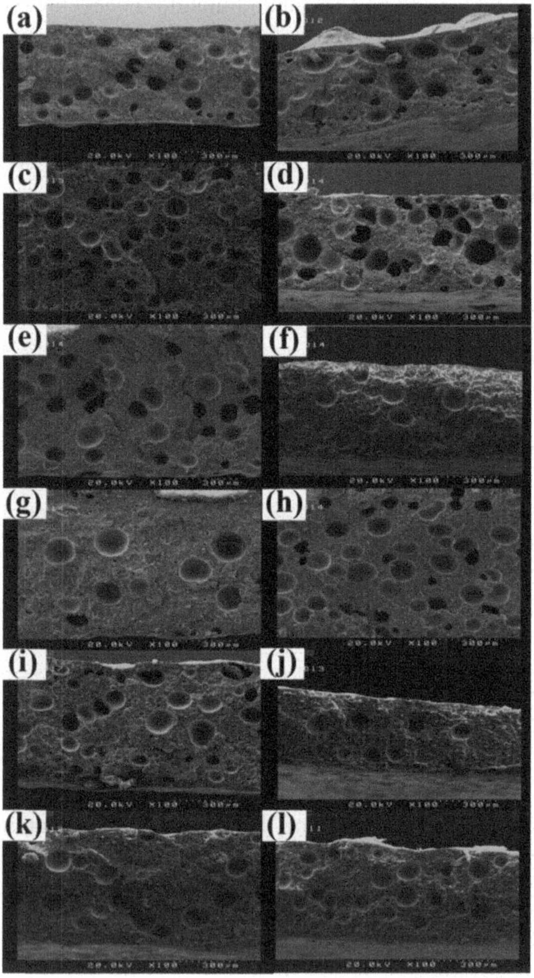

Fig. 1. SEM micrographs of the prepared foams: (a) PLA, (b) PLA (75 rpm), (c) PLA0.5BP, (d) PLA3PEMA, (e) PLA1Si, (f) PLA1Si3PEMA, (g) PLA1ONC, (h) PLA3ONC, (i) PLA1UNC, (j) PLA1UNC3PEMA, (k) PLA1EG, and (l) PLA1EG3PEMA

Table 1. Void content and structural properties of the obtained foams

Sample code	Void content (%)	D_n (μm)	D_v (μm)	Cell density (No./cm^3)
PLA (150 rpm)	16.4	67.3 ± 21.5	83.7	3.6E + 5
PLA (75 rpm)	26.1	78.79 ± 33.5	112.7	3.1E + 5
PLA0.5BP	15.0	65.8 ± 19.5	81.7	8.1E + 5
PLA3PEMA	22.0	73.5 ± 28.3	104.1	1.3E + 6
PLA1Si	15.6	76.1 ± 22.6	93.1	2.2E + 5
PLA1Si3PEMA	8.4	84.1 ± 23.1	99.8	1.1E + 1
PLA1ONC	12.5	99.6 ± 42.1	136.2	7.4E + 3
PLA3ONC	11.6	73.4 ± 23.9	93.9	7.6E + 5
PLA1UNC	10.7	70.5 ± 27.7	97.9	5.6E + 5
PLA1UNC3PEMA	9.3	83.1 ± 18.8	91.9	3.6E + 4
PLA1EG	15.6	79.8 ± 33.6	111.0	9.0E + 4
PLA1EG3PEMA	14.4	66.8 ± 22.7	83.6	1.5E + 5

4 Conclusion

In this study, efforts were made to overcome the PLA drawbacks in the extrusion foaming process by adding different nanoparticles, i.e. nanosilica, pristine nanoclay, organoclay and expanded graphite, and chemical reagents including benzoyl proxide and PEMA. The findings show that the incorporation of chemicals such as the thermal initiator has more considerable effect on the PLA foaming process and foam morphology than the used nanoparticles.

References

1. Julien, H., Quantin, C., Benezet, J., Bergeret, A., et al.: Eur. Polym. J. **67**, 40–49 (2015)
2. Nofar, M., Park, C.B.: Prog. Polym. Sci. **39**, 1721–1741 (2014)
3. Nofar, M.: Mater. Des. **101**, 24–34 (2016)
4. Fukushima, K., Abbate, C., Tabuani, D., Gennari, M.: Polym. Degrad. Stab. **94**, 1646–1655 (2009)

Effect of Microorganisms on Rheological Behavior of Polymer Fluids

Behtab Montaser Kouhsari and Nadereh Golshan Ebrahimi(✉)

Chemical Engineering Faculty, Polymer Engineering Department, Tarbiat Modares University, P.O. Box 14115-114, Tehran, Iran
ebrahimn@modares.ac.ir

Abstract. The presence of microparticles in polymeric fluids increases their viscosity. Here, we experimentally investigated the effect of the presence of the bacterium *E. coli* which has flagellates behind the cell body. We found that even a small amount of bacteria reduces viscosity in the polymeric solutions.

Keywords: Active suspensions · Microorganisms · Reynolds numbers · E. coli · Viscosity

1 Introduction

The hydrodynamics of active suspensions is at the center of many recent underlying studies [1, 2]. In nature, wide classes of living microorganisms move in fluids at very low Reynolds numbers [3]. Their motility arising from a variety of propulsive flagellar systems powered by nanomotors. For bacteria such as Escherichia coli (*E. coli*), the motive force comes from the rotation of helix-shaped flagella creating a propulsive force at the rear of the cell body [4]. Microswimmers such as *E-Coli* or Bacillus subtilis produce dipolar fluid flows through the combined action of their flagella and cell body on the fluid. In the far field, they are well described as "pusher" [5], corresponding to the case of flagella behind the cell body. The motion mechanism of pushers inside the fluid is shown in Fig. 1.

2 Experimental

2.1 Materials and Methods

The active fluids considered here are prepared out of E-coli suspended into a minimal medium where the bacteria are still motile but do not divide. E-coli is cultured overnight at 25 °C in LB medium shaken at 240 rpm [6]. Next, the culture is washed twice by centrifugation (6000 rpm for 10 min), and the cells are resuspended into a motility medium containing 10 mM potassium dihydrogen phosphate, 0.1 mM K-EDTA, 34 mM manganese (II) acetate, 20 mM sodium lactate, and 0.05 g polyvinylpyrrolidone (PVP-k60).

The rheological test was performed on a bacterial solution and a control solution (a non-bacterial solution), by the Austro-made Anton Paar-MCR 502 rheometer device.

H. Mirzadeh and A. A. Katbab (Eds.): ISPST 2018, *Eco-friendly and Smart Polymer Systems*, pp. 290–293, 2020.
https://doi.org/10.1007/978-3-030-45085-4_69

Determination of rheological properties of samples was done using bob and cup rheometry. The inner bob (radius r_i 13.332 mm, length 40.002 mm, and underside cone angle 120°) is suspended by a torsion wire into a cup (inner radius r_i 14.455 mm). The cup rotates at an angular rate Ω controlled by a computer. Viscosity and shear stress changes for each sample were evaluated at shear rates in the range of 0.01 (1/s) to 1000 (1/s).

3 Results and Discussion

Figure 2 shows the viscosity graph in terms of shear rate for a solution containing bacteria and control solution. Figure 3 shows the effect of bacteria on shear stress, which is similar to the previous one.

If the body of the bacterium and its flagellum are considered to be free from stress and torque, the Ω (the angular velocity of the bacterial body) should be as the following equation:

$$\Omega - \frac{D(A_0 + A) - B^2}{D_0(A + A_0)}\omega = \beta\omega \tag{1}$$

In which A_0, D_0, A, B, D are proportional to the solvent viscosity (η_s). Therefore, Ω can be predicted as a function of solvent viscosity [7].

$$\Omega \propto \eta^{-1} \tag{2}$$

Hydrodynamic interactions, in a fluid, lead to a significant reduction in viscosity and shear stress. The effective viscosity of 'pusher' bacteria is significantly lower than the pure solvent viscosity (activity-induced 'thinning'), as shown in Fig. 2.

The collective dynamics of swimming microorganisms exhibit a complex interplay with the surrounding fluid: the motile cells stir the fluid, which in turn can reorient and advect them. This feedback loop can result in long-range interactions between the cells, an effect whose significance remains controversial [8].

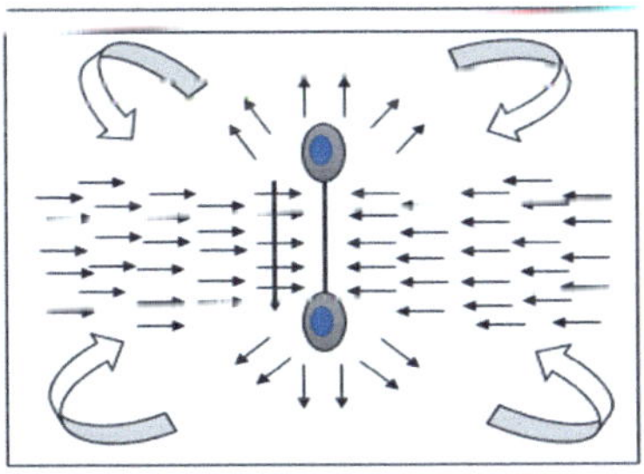

Fig. 1. Motion mechanism of pushers inside the fluid

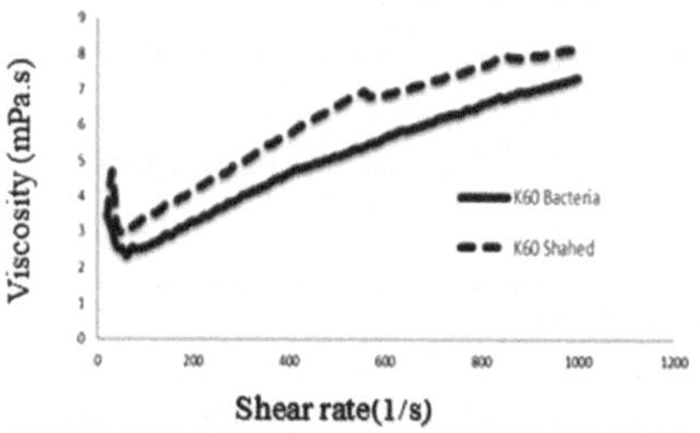

Fig. 2. Viscosity variation in terms of shear rate for control and bacterial-containing polymer solutions

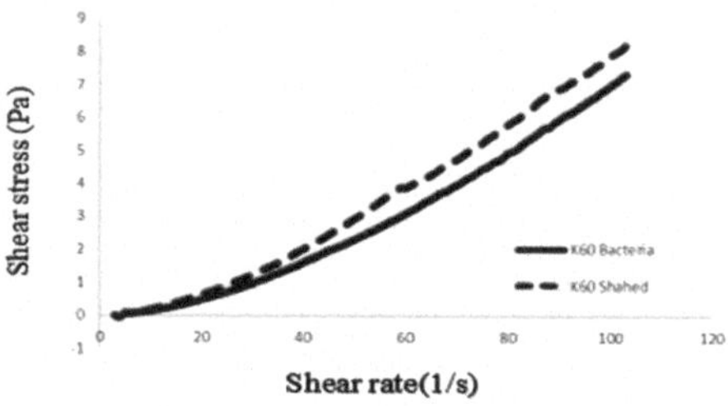

Fig. 3. Shear stress variation in terms of shear rate for control and bacterial-containing polymer

4 Conclusion

Active fluids are a new field of many fundamental studies which are not uptight by usual thermodynamic laws. The swim pressure provides a tool to analyze how activity imparts new behaviors to active matter, and how the combination of activity and confinement (e.g., in polymers) affects the nonequilibrium physics and rheology of active matter suspensions. In cases where there is a need for low fluid viscosity, a pusher microorganism can be used. Bacteria convert chemical energy into fluid motion, making fluid flow easier and diluting the solution.

References

1. Koch, D.L., Subramanian, G.: Annu. Rev. Fluid Mech. **43**, 637 (2011)
2. Marchetti, M.C., Joanny, J.F., Ramaswamy, S., Liverpool, T.B., Prost, J., Rao, M., Aditi Simha, R.: Rev. Mod. Phys. **85**, 1143 (2013)
3. Lauga, E., Powers, T.R.: Rep. Prog. Phys. **72** (2009)
4. Berg, H.C.: E. coli in Motion. Springer, New York (2004)
5. Drescher, K., Goldstein, R.E., Michel, N., Polin, M., Tuval, I.: Phys. Rev. Lett. **105**, 168101 (2010)

6. López, H.M., Gachelin, J., Douarche, C., Auradou, H., Clément, E.: Phys. Rev. Lett. **115**, 1–5 (2015)
7. Martinez, V.A., Schwarz-Linek, J., Reufer, M., Wilson, L.G., Morozov, A.N., Poon, W.C.K.: Proc. Natl. Acad. Sci. U.S.A. 1–18 (2014)
8. Lushi, E., Wioland, H., Goldstein, R.E.: Phys. Sci. Biophys. Comput. Biol. **111**, 9733–9738 (2014)

Fabrication of Antibacterial PVP/Keratin Hydrogel Embedded with Lavender Extract

Faezeh Tajik, Niloofar Eslahi(✉), and Aboosaeed Rashidi

Department of Textile Engineering, Science and Research Branch,
Islamic Azad University, Tehran, Iran
niloofar.eslahi@srbiau.ac.ir

Abstract. In this research, chicken feathers as the source of keratin and polyvinyl pyrrolidone (PVP) as the synthetic polymer were used to produce composite hydrogel by UV irradiation. The results revealed that adding hydrogen peroxide accelerated the formation of crosslinking during the irradiation process. Swelling and mechanical analysis were performed on the produced samples with different formulations. To induce antibacterial properties to the hydrogel, herbal lavender extract was embedded in the hydrogels. The antimicrobial activity was finally assessed against *Escherichia coli* (Gram-negative) and Staphylococcus aureus (Gram-positive) bacteria and the incorporated hydrogels exhibited excellent antibacterial properties.

Keywords: Hydrogel · Polyvinyl pyrrolidone · Keratin · UV radiation · Lavender

1 Introduction

Hydrogels are three-dimensional polymer networks in which hydrophilic polymer chains are connected by physical or chemical bonds [1]. There are numerous applications for hydrogels particularly in medicine owing to their biocompatibility and high water content [2].

Keratin proteins are structural fibrous proteins that are the key structural component of hair, feathers, wool, nails, and horns. These proteins are composed of cysteine-rich structural proteins, which have a considerable number of disulfide bonds with significant stability [1]. Keratin hydrogels are highly attractive for wound healing due to their inherent bioactivity, biocompatibility and physical properties [3]. Besides, PVP, a neutral water-soluble and biocompatible polymer, is used in pharmaceutical and cosmetic formulations. Lopergolo et al. [4] investigated PVP crosslinking using UV radiation from arc mercury lamps. However, the necessity of the high doses precludes the use of this process, since photodegradation competes with crosslinking.

In this study, the extracted keratin from chicken feathers was mixed with PVP in different ratios to produce composite hydrogel under UV irradiation. Then, the effect of lavender on the antibacterial properties of the hydrogels with potential application in wound dressing was studied.

H. Mirzadeh and A. A. Katbab (Eds.): ISPST 2018, *Eco-friendly and Smart Polymer Systems*, pp. 294–296, 2020.
https://doi.org/10.1007/978-3-030-45085-4_70

2 Experimental

At first, keratin protein was extracted from cleaned chopped feather by using 10 g/L sodium sulfide for 3 h at 60 °C. Then, the purified keratin solution was mixed with 30% PVP (MW 360 kDa, Sigma) at different ratios of 1:0, 2:1, 3:1, and 4:1 (PVP: keratin). Hydrogen peroxide (0.1 mL) was also added to the mixtures to have better PVP cross-linking. Moreover, 25% glycerol was employed to improve the flexibility of the hydrogels. In order to induce gelation, the prepared solutions were then put under ultraviolet radiation (254 nm) for 30 min. The produced hydrogels were evaluated by swelling and mechanical analysis. Finally, the antibacterial effect of the aqueous extract of lavender embedded in the hydrogels was examined.

3 Results and Discussion

Table 1 shows the results of swelling test in water at different times. As it can be seen, swelling was increased with prolonging time as well as PVP content in the hydrogels due to the hydrophilic nature of this polymer.

The result of tensile test was also shown in Fig. 1. According to the obtained curves, addition of keratin to PVP enhanced both stress and strain of the samples. In fact, the possible interactions such as hydrogen bonding between the employed polymers caused this improvement in mechanical properties.

Finally, the antibacterial properties of the incorporated hydrogels with 3% lavender were investigated against Escherichia coli (Gram-negative) and *Staphylococcus aureus* (Gram-positive) bacteria. Figure 2 confirms excellent antibacterial properties of the samples. The antibacterial activity of lavender extract was due to the presence of linalool [5].

Table 1. Swelling percentage for different samples

Sample (PVP:Keratin)	1 h	2 h	4 h	24 h
1:0	250%	300%	320%	340%
2:1	122%	126%	129%	130%
3:1	136%	133%	135%	135%
4:1	145%	148%	150%	156%

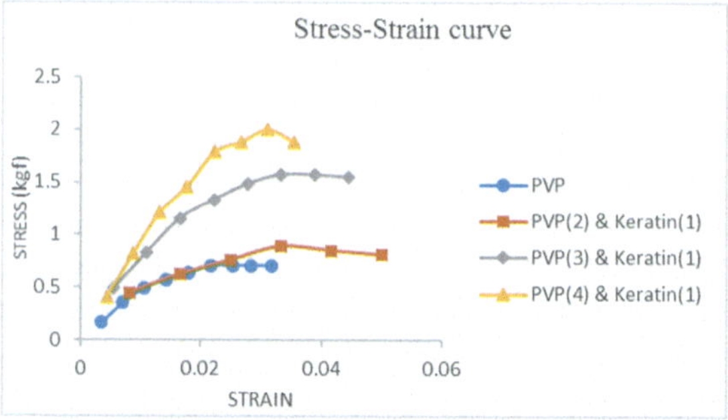

Fig. 1. Stress-strain curves for hydrogel specimens

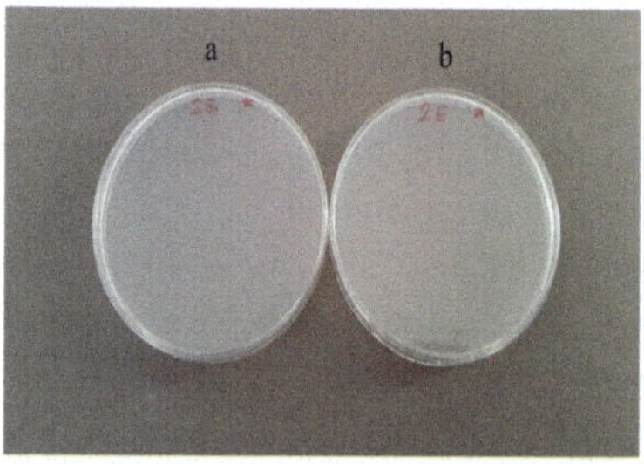

Fig. 2. Antibacterial properties of lavender incorporated hydrogel against: (a) *S. aureus* and (b) E. coli

4 Conclusion

In this study, we investigated the feasible fabrication of keratin and PVP composite hydrogels by UV irradiation. For this purpose, different ratios of polymers were mixed and their swelling and mechanical properties were evaluated. To induce antibacterial properties to the hydrogels, aqueous extract of lavender was added to the samples. The results of the antibacterial test against both gram positive and gram-negative bacteria revealed that the obtained incorporated hydrogel had excellent antibacterial activity and therefore it could be used in wound dressing applications.

References

1. Park, M., Shin, H.K., Park, S.J., Kim, H.Y.: Mater. Sci. Eng. **33**, 5051–5057 (2013)
2. Schacht, E., Dubruel, P., Vlierberghe, S.: Biomacromol **12**, 1387–1408 (2011)
3. Wang, J., Hao, S., Luo, T., Cheng, Z., Li, W., Gao, F., Guo, T., et al.: Biointerfaces **149**, 341–350 (2016)
4. Lopérgolo, L.C., Lugão, A.B., Catalani, L.H.: Polymer **44**, 6217–6222 (2003)
5. Dalilian, S., Heidari Kashl, S., Zamanian Azodi, M., Omidi, R., Roeintan, R., Hosseini, R.: J. Ilam Univ. Med. Sci. **21**, 143–149 (2013)

Soybean Oil Derived Non-isocyanate Polyurethane with In Situ Formed Nano-silver as Antimicrobial Wound Dressing

Hoshyar Gholami and Hamid Yeganeh(✉)

Iran Polymer and Petrochemical Institute, P.O. Box 14965/115, Tehran, Iran
h.yeganeh@ippi.ac.ir

Abstract. Reaction of a carbonated soybean oil (CSBO) with a multifunctional amine curing agent containing both primary and secondary amine groups led to the formation of a non-isocyanate polyurethane network (NPU) with free secondary amine groups chemically anchored to the polymer network. The resulting NPU shaped as free standing membrane was immersed into a solution containing silver salt. Silver nanoparticles (AgNPs) were simply prepared and loaded into dressing membrane through a redox reaction between silver ions and free amine groups. Evaluation of the physical, mechanical and biological properties for this nanocomposite showed its high utility for using as an antimicrobial wound dressing for low exuding wounds.

Keywords: Wound dressing · Polyurethane · Non-isocyanate chemistry · Renewable resource · Nano silver

1 Introduction

An open dermal wound can be colonized with aerobic and anaerobic bacteria, and optimal wound care is required to prevent progression to infection. Application of an antimicrobial wound dressing is a promising strategy to treat such wounds [1]. In fact, an antimicrobial wound dressing is a multifunctional medical device. It combats bacterial invasion by the action of the antimicrobial agents loaded into the dressing membrane and in the meantime, it provides optimum moist condition over damaged tissue by proper control of segregated wound exudates present at the wound site. Therefore, proper selection of antimicrobial agents and the materials used as the main matrix of wound dressing are key factors for the successful utilization of such medical device. Among the antibacterial agents, AgNPs with high antimicrobial activity against wide range of bacterial strains was selected in this work and a vegetable oil derived polyurethane prepared through a non-isocyanate synthetic methodology was used as the main body of the designed dressing membrane [2]. The problem of suitable loading and dispersion of AgNPs into the dressing membrane was solved by innovative in situ reduction of silver ions by secondary amine groups embedded in the dressing backbone.

H. Mirzadeh and A. A. Katbab (Eds.): ISPST 2018, *Eco-friendly and Smart Polymer Systems*, pp. 297–300, 2020.
https://doi.org/10.1007/978-3-030-45085-4_71

2 Experimental

2.1 Preparation of Dressing Membrane (AgNPU)

CSBO, pentaethylenehexamine (at 1:1 molar ratio) and a catalytic amount of LiCl were dissolved in minimum amount of toluene. After proper mixing, it transferred to a Teflon mold and subjected to thermal treatment according to a predetermined temperature programming schedule. The resulting light brown, transparent and flexible polyurethane membrane was immersed into a beaker containing a certain concentration of silver nitrate. It was removed and washed with plenty of distilled water. AgNPs loaded NPU was dried at room temperature under vacuum and used for subsequent studies.

3 Results and Discussion

The structure of prepared polyurethane was confirmed by examination of its FTIR spectrum as shown in Fig. 1. The almost complete disappearance of cyclic carbonate peak at 1798 cm^{-1} and generation of new bands at 1694 and 1530 cm^{-1} related to the urethane carbonyl and C-N stretching vibration confirmed ring opening reaction of cyclic carbonate groups of CSBO with primary amine functions of curing agent. The broad peak appeared at 3334 cm^{-1} was attributed to hydroxyl and secondary amine N-H groups as well as urethane N-H bond. The peak at 1740 cm^{-1} related to ester groups of CSBO was remained unchanged after curing reaction.

The prepared NPU showed gel content value of about 90%, which showed the proper progress of network formation through non-isocyanate synthetic route.

The redox reaction of secondary amine moieties with Ag^{+} ions available in the media used for swelling of dressing led to the simultaneous reduction of ions to AgNPs and trapping/distributing them into the bulk of dressing matrix. Despite the visual color change of dressing to a deep brown as a common sign for AgNPs formation, it was confirmed by UV-Vis spectroscopy. Dressing was placed in distilled water and the AgNPs released from the swollen membrane into the water was subjected to UV-vis spectroscopy analysis. The presence of a broad surface plasmon resonance band at about $\lambda = 420$ nm was a strong evidence for the formation of AgNPs (Fig. 2).

It is well-known that the biological activity of AgNPs is concentration dependent. Therefore, dressings containing different amounts of AgNPs were prepared by changing the initial concentration of Ag+ ions. These dressings were then subjected to biological analyses. Direct fibroblast cell contact and MTT assay were done to find information regarding their cytocompatibility and colony counting against Staphylococcus aureus (ATCC 6538), Pseudomonas aeruginosa (ATCC 15442), and Candida albicans (ATCC 10231) to measure their antimicrobial activity. At a proper concentration of AgNPU loaded into the dressing membrane, it can support growth and proliferation of fibroblast cells as it is shown in Fig. 3. In addition, the fibroblast cells can keep their viability in the presence of dressing up to 92%, which is another sign of proper cytocompatibility of the prepared AgNPU. In the meantime, examination of biocidal property performed according to the procedure described in E 2180–07 showed very promising activity with almost complete (100%) bacterial reduction for all of studied strains for the same dressing. In fact, high antimicrobial activity and preservation of desire cytocompatibility are attainable just by adjusting the concentration of AgNPs.

Preservation of optimum moist environment over the damaged skin achievable by the balanced water vapor transmission rate (WVTR) and water absorption capacity (EWA) is a necessary factor for fast wound healing. The recorded data presented in Table 1 and comparing them with some of the commercially available products shows that the specification of the prepared dressing is best suitable for low exuding wounds.

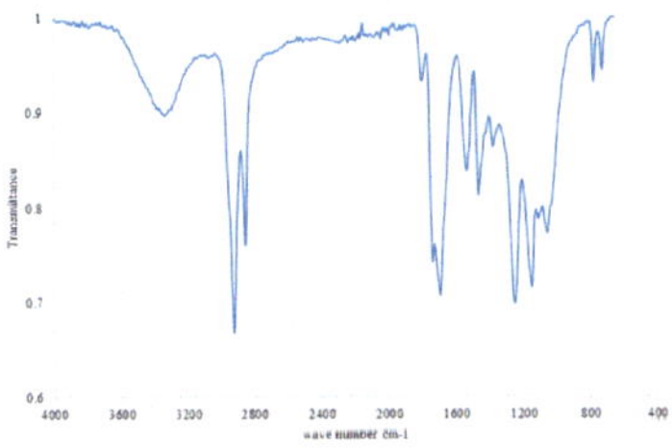

Fig. 1. FT-IR spectrum of NIPU2

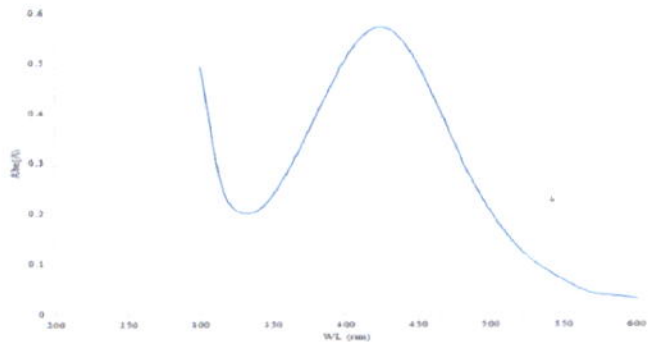

Fig. 2. UV spectrum of Ag

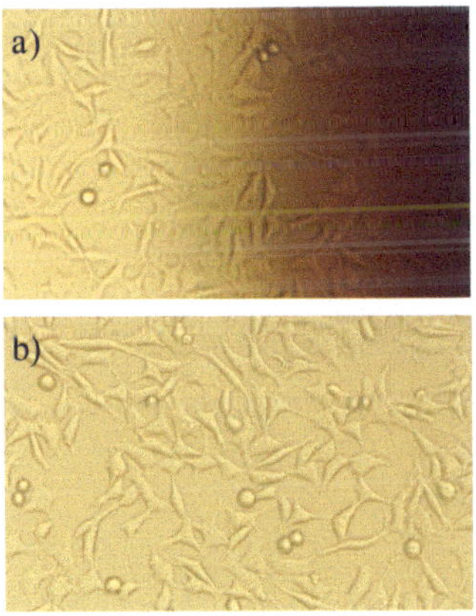

Fig. 3. Optical microscopy images of L929 mouse fibroblast cells in direct contact with: (a) AgNPU and (b) control after 48 h

This behavior is best correlated to the hydrophobic nature of soybean oil used as a main chemical building block of the AgNPU. However, the hydroxyl groups generated during the urethane bond formation through non-isocyanate mechanism is responsible for the partial ability of dressing for absorption of wound exudates.

Table 1. Hydrophilicity data of AgNPu and some commercial samples

Sample	Contact angle	EWA (%)	WVTR ($g/m^2 day^1$)
AgNPU	79	32	401 ± 13
Dermiflex	–	–	394 ± 12
Bioclusive	–	24	205 ± 10

4 Conclusion

A non-isocyanate chemistry utilizing a renewable resource-based raw material and a redox reaction between amine and silver ions were used to prepare a wound dressing loaded with AgNPs. The measured chemical and biological properties showed potential applicability of this dressing for low exuding wounds.

References

1. Gholami, H., Yeganeh, H., Burujeny, S.B., Sorayya, M., Shams, E.: J. Polym. Environ. **26**(2), 462–473 (2018)
2. Jalilian, S., Yeganeh, H.: Polym. Bull. **72**(6), (2015)

Diameter-Tuning of Electrospun Pectin-Based Nanofibers

Shahrzad Zirak Hassan Kiadeh[1], Azadeh Ghaee[1(✉)], Addie Bahi[2], Mehdi Farokhi[3], Jhamak Nourmohammadi[1], and Frank K. Ko[2]

[1] Department of Life Science Engineering, Faculty of New Sciences and Technologies, University of Tehran, P.O. Box 14395-1561, Tehran, Iran
sh.zirak.hk@gmail.com

[2] Department of Materials Engineering, The University of British Columbia, Vancouver, BC V6T 1Z4, Canada

[3] National Cell Bank of Iran, Pasteur Institute of Iran, Tehran, Iran

Abstract. There is an increasing interest to fabricate natural nanofibers with tunable structures using electrospinning. Pectin (PEC), an anionic hetero-polysaccharide present in plant cell-walls, is biodegradable and biocompatible so it can be used in biomedicine and food industry. In this study, various blends of pectin and polyethylene oxide (PEO) solutions were prepared. The viscosities of the samples containing Triton-X-100, as a nonionic surfactant, were measured prior to electrospinning. The electrospinning parameters, PEC: PEO ratio and molecular weight (Mw) of PEO were assessed using SEM, and ATIR-FTIR spectroscopy in order to improve the spinnability of pectin with appropriate morphological appearance and mechanical properties. SEM micrographs implied that higher Mw of PEO is a feasible way to apply higher ratio of PEC: PEO, leading to bead-free fiber formation. In addition, it was also found that higher ratio of PEC: PEO resulted in thinner nanofibers.

Keywords: Electrospining · Pectin · Polyethylene oxide · Tensile strength · Average diameter

1 Introduction

Natural polymers have advantages, such as biodegradation ability, and biological activities [1]. Among them, pectin is a natural heterogeneous polysaccharide found in terrestrial plant cell walls, such as sunflower head, sugar beet, soybean, citrus, and fruits [2].

The composition, molecular weight and esterification degree (i.e. methoxyl group content) of pectin vary depending on the extraction source and used method affecting their physicochemical and biological properties. Typically, pectin is composed of esterified/amidated galacturonic acid (GalA) residues forming three main domains; mainly (98%) linear homogalacturonan (HGA), rhamnogalacturonan-I (RG-I) and rhamnogalacturonan-II (RG-II).

There are smaller proportions of other segments, such as xylogalacturonans (XGA). In the more recent model, it is proposed that mentioned part of the pectin are covalently

H. Mirzadeh and A. A. Katbab (Eds.): ISPST 2018, *Eco-friendly and Smart Polymer Systems*, pp. 301–305, 2020.
https://doi.org/10.1007/978-3-030-45085-4_72

conjugate together and form block copolymer, included both HGA and RG-I in its core [3–5]. Pectin is insoluble in common organic solvent and nanofiber fabrication of aqueous pectin solution is challenging. Therefore, spinnable synthetic polymers, such as poly(ethylene oxide) (PEO) can be blended with pectin in order to facilitate the electrospinning process of aqueous pectin solution. However, these synthetic polymers may impair the integrity and mechanical strength of the nanofibers in biological environments. It is highly desirable to fabricate pectin nanofibers with least amount of carrier polymer as possible.

Herein, we assessed the effect of the PEO content and molecular weight on morphological and mechanical properties of pectin nanofibers. A nonionic surfactant was used additionally to facilitate the electrospinning of bead free pectin nanofibers.

2 Experimental

2.1 Materials and Electrospinning

4% aqueous citrus pectin (PEC, galacturonic acid $\geq$ 74.0%, Sigma-Aldrich) solutions and PEO (Mw = 600 & 900 kDa, Sigma-Aldrich) solutions were prepared separately and blended with the desirable PEC/PEO ratios. Triton-X-100 (Emd, USA) was added to the mixture at the concentration of 1 wt%. The resulted solution was stirred for 6 h and stored overnight before electrospinning. 10 mL of the final polymer solution was electrospun using a 20-gauge stainless steel needle at 17 kV. The gap distance and the relative humidity were 18 cm and 20–30%, respectively. Distilled water contained 2% acetic acid (Acros, USA). The electrospining solutions characteristics were summarized in Table 1.

2.2 Characterization

2.2.1 Morphology

Morphology of the nanofibers was studied using a scanning electron microscope (Hitachi, Japan). Diameters of the nanofibers were measured (statistical average values for 50 nanofibers in the SEM pictures) using imageJ software.

2.2.2 Composition

Attenuated Fourier transform infrared spectroscopy (ATR-FTIR, Agilent Technologies, USA) was used to determine the composition of nanofibers.

2.2.3 Mechanical Property

The tensile properties of electrospun rectangular specimens (approximately 6 mm 50 mm, W $\times$ H) were characterized using a testing machine (KES-G1, Kato tech LTD, Japan) followed standard mechanical testing methods for fabric textiles. The gauge length and rate were 40 mm and 10 mm/min, respectively. The ultimate tensile strain (UTSA), Young's modulus (E) and ultimate tensile strength (UTS) were calculated from the stress–strain curves. This test repeated six times for each sample [6, 7].

3 Results and Discussion

3.1 Morphology of PEC: PEO Nano-fibers

Inspired by previous studies [2], flexible-chain and water-soluble synthetic polymers, PEOs, with Mw of 600 and 900 kDa (PEO600 & PEO900) were introduced first into aqueous pectin solution to enhance chain entanglements. PEO is generally considered as a biocompatible polymer but its dissolution in biological aqueous environments affects the integrity and mechanical property of the materials. It was, therefore, attempted to reduce the content of PEO in electrospun pectin nanofibers. The PEC: PEO ratios of 50:50 and 75:25 were investigated. As observed using SEM micrographs (Fig. 1), PEO and Triton-X-100 addition led to improve solution property and jet stability, and subsequently, fiber formation, and droplet formation during the electrospinning of pure pectin aqueous solution. Triton X−100, as a nonionic surfactant, was reported to be able to reduce surface tension of pectin solution without affecting the conductivity, and thus enhancing jet stability [2].

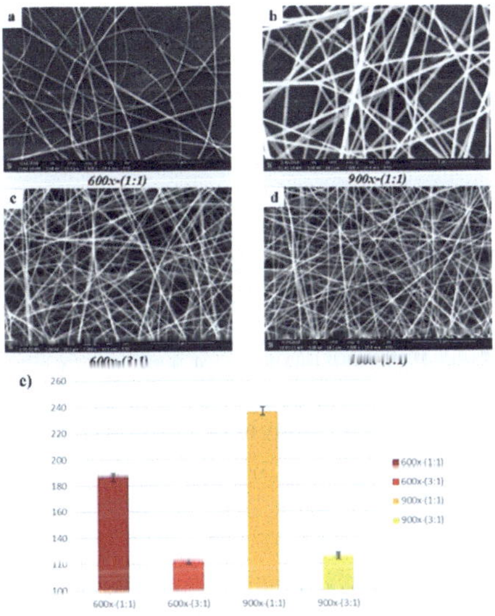

Fig. 1. SEM images of pectin-based nanofibers from the aqueous solution containing PEO of different content and Mw. (a) 600x-(1:1), (b) 900x-(1:1), (c) 600x-(3:1), (d) 900x-(3:1). Concentran of polymer and triton x-100 was fixed at 4 wt% and 1 wt%, respectively, and (e) Average fiber size of pectin elecrospun nanofibers obtained by ImageJ analysis

Figure 1 also shows two points: first fewer beads were generated if PEO of higher molecular weight was introduced into pectin solution due to better chain entanglements. The second point is that PEC:PEO ratio and PEO Mw affect the average diameter of nano fibers, reversely.

3.2 Composition of PEC: PEO Nano-fibers

PEC: PEO nanofiber composition was characterized using ATR-FTIR spectroscopy. As shown in Fig. 2, all of the pectin nano-fibers showed two characteristic peaks at 1751 and around 1620 cm^{-1}, corresponding to the C = O stretching of ester and carboxylate groups of pectin, respectively [4]. PEO and Triton X−100 have similar –CH_2–CH_2–O– repeating units in their molecular chains, and their C–H and C–O bonds resulted in the absorbance peaks at 2875 and 1104 cm^{-1}, respectively. A characteristic absorbance of Triton X−100 appeared at 1511 cm^{-1}, which is assigned to the aromatic ring stretch.

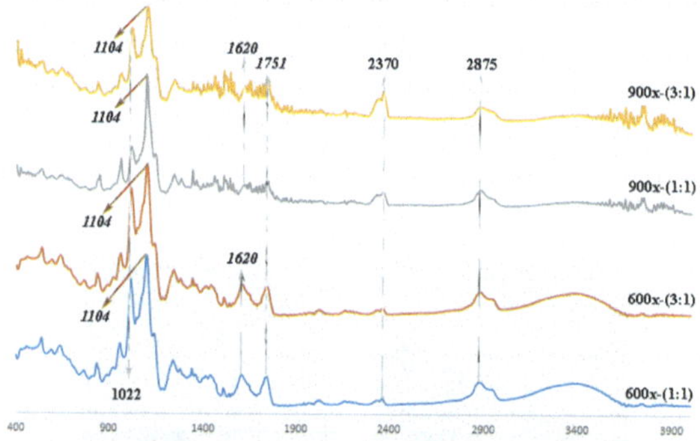

Fig. 2. ATIR spectra of pectin nano-fibers from solutions with PEC:PEO mass ratio (w/w) of 0.5 and 0.75 and different Mw of PEO.

3.3 Mechanical Property

Tensile properties of the electrospun pectin nano fibers are shown in Table 2. As shown in Table 2, there were no significant differences in the Young's modulus of 600x-(1:1), 600x-(3:1), and 900x-(3:1) samples. Young's modulus of these nanofibers was approximately between 205 to 218 MPa but in the case of 900x-(1:1) sample, dramatically increased to 492 MPa. The data confirmed the effect of Mw increasing on enhancing the mechanical properties (tensile strength and Young's modulus) of the electrospun nanofibers. In contrast, the ultimate tensile strain of the pectin nanofibers was noticeably reduced by adding higher Mw of PEO (900 kDa) from 58% to 5%, indicating their high rigidity. The results also revealed that the ultimate tensile strength of the sample of 900x-(3:1) was the most at 4.41 MPa.

Table 1. Solutions characteristics of electrospun nanofibers used in this study

Sample name	PEC:PEO	TritonX-100 (wt%)	Mw of PEO
600x-(1:1)	50:50	1%	600 kDa
600x-(3:1)	75:25	1%	600 kDa
900x-(1:1)	50:50	1%	900 kDa
900x-(3:1)	75:25	1%	900 kDa

Table 2. Mechanical properties of pectin textiles

Sample name	UTS (MPa)	Y modulus (E)	UTSA (%)
600x-(1:1)	3.4	2.09	58.23
600x-(3:1)	2.0	2.19	37.76
900x-(1:1)	3.45	4.92	9.07
900x-(3:1)	4.41	2.05	4.73

4 Conclusion

By electrospinning method, we successfully fabricated pectin nano-fibers, with different Mw and content of carrier polymer (PEO) in pectin aqueous solutions, which were assessed in terms of morpholgical and mechanical properties. To overcome the jet instability problem and enable the electrospinning of pectin nanofibers, a non-ionic surfactant was used. The incorporation of pectin in the fibers was confirmed by FTIR analysis. It is of interest to note that the mechanical properties (both tensile strength and tensile modulus) were improved significantly by incorporating of PEO of 900 kDa. In short, it is believed that these pectin nanofibers possess low cost and rich bioactivity, consequently promising potential for a wide range of biomedical applications.

References

1. Cui, S., Yao, B., Gao, M., Sun, X., Gou, D., Hu, J., Zhou, Y., et al.: Carbohyd. Polym. **157**, 766 (2017)
2. Cui, S., Yao, B., Sun, X., Hu, J., Zhou, Y., Liu, Y.: Mater. Sci. Eng. C **59**, 885 (2016)
3. Kaczmarek, H., Bajer, K., Gałka, P., Kotnowska, B.: Polym. Degrad. Stabil. **92**, 2058 (2007)
4. Cui, S.-S., Sun, X., Yao, B., Peng, X.-X., Zhang, X.-T., Zhou, Y.-F., Hu, J.-L., et al.: J. Nanosci. Nanotechnol. **17**, 681 (2017)
5. Ropers, M.-H., Meister, A., Blume, A., Ralet, M.-C.: Biomacromolecules **9**, 1306 (2008)
6. Li, M., Mondrinos, M.J., Gandhi, M.R., Ko, F.K., Weiss, A.S., Lelkes, P.I.: Biomaterials **26**, 5999 (2005)
7. Bahi, A., Shao, J., Mohseni, M., Ko, F.K.: Sep. Purif. Technol. **187**, 207 (2017)

Synthesis of Chitosan/Zeolite/Silver Nanoparticles Composites: Antibacterial Activity Against Aquatic Bacteria

Fatemeh Salar Hosseini[1], Navid Ramezanian[1(✉)], and Zoheir Shokouh Saljoghi[2]

[1] Department of Chemistry, Faculty of Science, Ferdowsi University of Mashhad, Mashhad, Iran
ramezanian@um.ac.ir

[2] Department of Fisheries, Faculty of Natural Resources, Isfahan University of Technology, Isfahan, Iran

Abstract. This study was conducted to develop novel chitosan nanocomposites and to investigate the antibacterial activity of chitosan/zeolite/silver nanoparticles composites for the aquaculture application. Novel chitosan/zeolite/silver nanoparticles composites were prepared via the method of solution blending. Streptococcus iniae bacteria was used to test antibacterial activity of the obtained chitosan/zeolite/silver nanoparticles composites by disc diffusion method. Finally, it was found that composites B, C, and D exhibited good antibacterial activity and composite C had the highest antibacterial activity. The above analysis suggested that these composites could be used as potential candidates for the aquaculture hatcheries systems.

Keywords: Aquaculture · Antibacterial activities · Chitosan · Zeolite · Silver nanoparticles

1 Introduction

The continued development of aquaculture relies on improved microbial control to prevent the proliferation and spreading of pathogens in water [1]. Water quality control in a recirculation system is achieved by combining treatment and reuse of water with water exchange affected by supply of fresh water [2]. To support this concept, biopolymer materials and inorganic materials, such as metal and metal oxides, have been considered because of their special antimicrobial properties, biocompatibility and eco-friendly nature [3].

Among many natural biopolymers, chitosan is one of the most commonly used raw materials, because of its abundance, low cost, renewability, biocompatibility and biodegradability. Recently, hybrid materials based on chitosan have been developed, including metal nanoparticles and oxide agents. Hence, chitosan has a tremendous ability to form metal composite with silver metal because of its NH_2 and OH groups [3].

H. Mirzadeh and A. A. Katbab (Eds.): ISPST 2018, *Eco-friendly and Smart Polymer Systems*, pp. 306–309, 2020.
https://doi.org/10.1007/978-3-030-45085-4_73

The antibacterial application of zeolites is currently being tapped by researchers. Zeolites incorporated with metallic ions were observed to achieve, greater significance comparing to conventional antibacterial agents. The inclusion of the metallic ions within their structure allowed controlled release and prevention of concentration-dependent toxicity since metallic ions such as Ag, Cu, and Zn can be easily oxidized and lose their antimicrobial property. The selectivity of the different types of zeolites in their ion-exchangeability is higher towards the Ag ions over Zn and Cu ions [4].

Silver nanoparticles (Ag NPs) have been used for a long time as antibacterial agent due to its strong inhibiting effect on bacteria [5]. The advantage over conventional and narrow-target antibiotics is that microbes are less likely to develop resistance against silver, as they do against antibiotics [6].

Hence, the present study describes the synthesis of chitosan/zeolite/Ag nanoparticles composites, and reports the antibacterial activity against Streptococcus iniae bacteria isolated from aquatic environment.

2 Experimental/Theoretical

Chitosan (medium molecular weight) and glacial acetic acid ($C_2H_4O_2$, 99%) were purchased from Sigma Aldrich Chemical Co. Ltd. Zeolite was locally sourced. Silver nanoparticles were obtained from Merck. Sodium hydroxide pellets (NaOH) was purchased from Chem-Lab NV. The bacterial strains used to provide the antibacterial activity was Streptococcus iniae (ATCC, 29178) from the Pasteur Institute of Iran.

1 g of chitosan was dissolved in 200 mL aqueous solution of glacial acetic acid (2 v/v%) at 25 °C under continuous stirring and then the solution was allowed to keep at room temperature for one or two days to remove the air bubbles formed during stirring so as to obtain a clear solution. Then different ratios of Ag NPs were dispersed into the chitosan solution and after stirring for 15 min, different ratios of zeolite were added into the mixed solution and stirred continuously for about 30 min (Table 1). NaOH 1 M was then added drop-wise until the solution attained pH 9.5. Then the obtained composites were washed with distilled water and dried in an oven at 60 °C for 24 h.

The antibacterial activity of chitosan/zeolite/silver nanoparticle composites was evaluated by disk diffusion method. So that a specific amount of each composite was placed under UV light to sterilize them and then added to the medium containing Streptococcus iniae bacteria. After incubation, the diameter of the inhibition zone of bacteria around each sample was measured and the amount was reported.

3 Results and Discussion

According to the results, the composite A had no antimicrobial activities, but the diameter of the zone inhibition of bacteria for composites B, C, and D was 11.54, 16.25, and 14.95 mm (Fig. 1), respectively. The result indicated that composite C showed the highest antibacterial activity against Streptococcus iniae bacteria as compared to other composites.

Table 1. Proportion of chitosan, zeolite and Ag NPs in composites

Composite code	Chitosan (g)	Zeolite (g)	Ag NPs (mg)
A	4	0.8	20
B	4	1.2	20
C	4	0.8	60
D	4	1.2	60

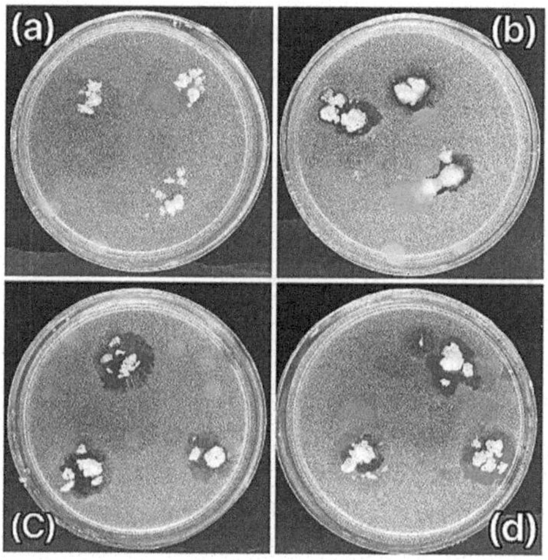

Fig. 1. Representative photographs of the antibacterial activity of chitosan/zeolite/silver nanoparticle composites against Streptococcus iniae bacteria: (a) composite A, (b) composite B, (c) composite C, and (d) composite D

4 Conclusion

The antibacterial activity of chitosan/zeolite/silver nanoparticles composites, especially composite C, recommended that the combination of these materials could be deployed for application in aquaculture industry.

Acknowledgement. The authors would like to acknowledge Department of Chemistry, Faculty of Science, Ferdowsi University of Mashhad, Mashhad, Iran and Department of Fisheries, Faculty of Natural Resources, Isfahan University of Technology, Isfahan, Iran for the support to carry out this research project (project code: 3.46671).

References

1. Olafsen, J.A.: Aquaculture **200**, 223–247 (2001)
2. Bovendeur, J., Eding, E., Henken, A.: Aquaculture **63**, 329–353 (1987)
3. Vaseeharan, B., Sivakamavalli, J., Thaya, R.: J. Compos. Mater. 1–8 (2013)
4. Taaca, K., Vasquez, M.R.: Micropor. Mesopor. Mat. **241**, 383–391 (2017)
5. Cao, X.L., Cheng, C., Ma, Y.L., Zhao, C.S.: J. Mater. Sci. **21**, 2861–2868 (2010)
6. Banerjee, M., Mallick, S., Paul, A., Chattopadhyay, A., Ghosh, S.: Langmuir **26**, 5901–5908 (2010)

Other Aspects of Polymer Science and Technology

Effect of Nanoclay Concentration on the Curing and Mechanical Behavior of Chlorobutyl Rubber Nanocomposites

Matin Khamseh-Kajoori and Azam Jalali-Arani(✉)

Department of Polymer Engineering and Color Technology,
Amirkabir University of Technology, Tehran, Iran
ajalali@aut.ac.ir

Abstract. Chlorobutyl rubber (CIIR) as a special grade of rubbers is widely used in tire industries thus, the reduction in its consumption by decreasing the products thickness/weight along with improvements in mechanical properties of the composite has economic significance. In this work, samples containing different amounts of organoclay (OC, cloisite 30B) were prepared on a two-roll mill and characterized. The rhometry test of the samples, which was performed by an oscillating disk rhometer, showed slightly increase in scorch and cure times by addition of OC and increasing its amount, which this behavior was in contrast to what was reported for the other elastomers. Measuring mechanical properties of the samples indicated an increase in tensile strength and elongation at break in the presence of OC and increasing its amount up to 3 phr. This was attributed to the dispersion of the OC, the presence of OC modifier and increase in crosslink density.

Keywords: Chlorobutyl rubber · Nanocomposite · Curing characteristics · Mechanical behavior

1 Introduction

The chlorinated copolymer of isobutylene and isoprene is called chlorobutyl rubber (CIIR). This polymer similar to butyl rubber (IIR) exhibits superior resistance to moisture, solvent and heat as well as low air tightness, while with respect to IIR, it shows a higher vulcanization rate and better compatibility with other rubbers [1]. To improve the mechanical properties of rubbers, reinforcing agents such as carbon black and silica have been used. Substitution of these conventional reinforcing fillers with nanoclay has improved some of the desired properties of rubber compounds in addition to some reductions of the final weight and price of the products [2]. In this work, nanocomposites of CIIR containing different concentration of OC were prepared by melt intercalation and their curing behavior and mechanical properties were evaluated and discussed.

H. Mirzadeh and A. A. Katbab (Eds.): ISPST 2018, *Eco-friendly and Smart Polymer Systems*, pp. 313–315, 2020.
https://doi.org/10.1007/978-3-030-45085-4_74

2 Experimental

The chlorobutyl rubber used in this study (CBK139, chlorine content 1.13%, Mooney viscosity (ML1 + 8 at 125 °C) = 38) was from Nizhnekamsk, Russia. Cloisite30B (C30B) as a modified clay with methyl, tallow, bis-2-hydroxyethyl ammonium organic modifier) was supplied by Southern Clay products, USA. The other ingredients included zinc oxide (3 phr), stearic acid (1 phr), and sulfur curing system were purchased locally. Preparation of the compounds with different amounts (1.5, 3, and 5 phr) of OC was carried out on a laboratory-size two roll-mill (Farrel Bridge, UK). Curing behavior of the samples were evaluated by using an oscillating disk rheometer (ODR, Monsanto 100S, UK) at temperature of 160 °C. Then the samples were molded in a hot press according to the optimum cure time obtained from the rheometry test.

Stress–strain measurements were carried out on dumbbell-shaped specimens by using a tensile testing machine (HIWA 200, Iran) according to ASTM D412 test method.

3 Results and Discussion

The curing characteristics of the prepared samples are compared with each other in Fig. 1. As can be seen, by adding 1.5 phr OC the scorch (t_s) and optimum cure (t_{90}) times of the compound were increased and vulcanization rate was decreased. These results indicate that the used OC, contrary to its effect on other elastomers has reduced the curing reaction of the CIIR compound. This effect of organoclay on vulcanization of CIIR compound may be attributed to the chemical structure of the CIIR and its effect on amine functional groups in the organoclay. In fact, the possible formation of a Zn complex with sulfur and alkylamine groups was delayed, which affects the curing reaction of nanocomposites. The scorch and optimum cure times of the nanocomposites gradually increased with further addition of organoclay (Fig. 1). The difference between maximum and minimum torque of rheometry (M_H–M_L) was slightly increased by the addition of the OC and increasing its amount, which can be attributed to an increase in cross-link density of the vulcanized samples [3].

The tensile properties of the samples were measured and the variation of tensile strength and elongation at break for the samples versus their OC content is shown in Table 1. Rubber–clay interactions, compounding conditions, and the clay content are three important factors that affect the mechanical strength of nanocomposites. As expected, the tensile strength of the samples was increased by adding OC and increasing its amount. This can be attributed to a high aspect ratio and platelet structure of organoclay, which may cause an improvement in the resistance to crack growth as well as the increase in crosslink density. An increase in elongation at break with the incorporation of OC and increasing its concentration up to 3 phr is observed in Table 1. In fact, the presence of OC modifier, intercalation/exfoliation and dispersion tendency of the OC causes the mobility of some rubber chains to be increased, and these are responsible for increasing the elongation at break of nanocomposites [4, 5]. In general, improvement in tensile properties is an indication of reinforcement of CIIR with this type of reinforcing filler. As a result, it can be explained that among the used concentration OC in this work, using 3 phr OC can be considered as the optimum one for having improvement in mechanical properties.

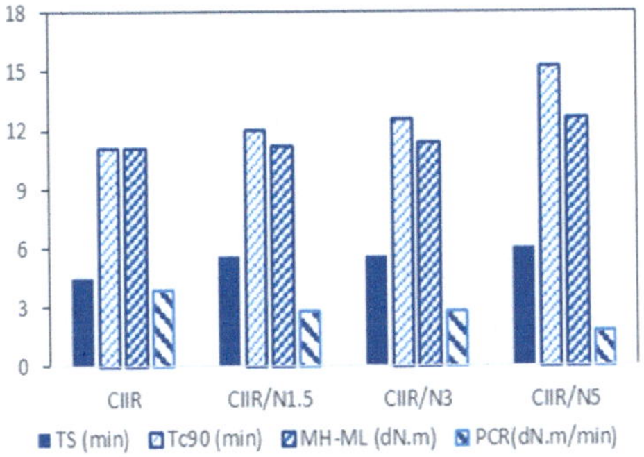

Fig. 1. Curing characteristics of CIIR and CIIR/OC nanocomposites

Table 1. Mechanical properties of CIIR and CIIR/OC nanocomposites

Test	CIIR	CIIR/N1.5	CIIR/N3	CIIR/N5
Elongation-at-break (%)	653	698	728	608
Tensile strength (MPa)	2.42	2.85	3.16	3.18

4 Conclusion

In summary, study on the effect of OC and its concentration on curing and mechanical properties of CIIR showed slightly increase in scorch and cure times which was different from other elastomers, and the toque difference of rheometry slightly increased. The tensile strength and elongation at break were increased by using OC and increasing its amount up to 3 phr.

References

1. Qu, L., Huang, G.: J. Mater. Sci. **42**, 7256–7262 (2007)
2. Maiti, M., Sadhu, S., Bhowmick, A.K.: J. Appl. Polym. Sci. **96**, 443–451 (2005)
3. Ghari, H.S., Jalali-Arani, A.: Appl. Clay Sci. **119**, 348–357 (2016)
4. Liang, Y., Cao, W., Li, Z., Wang, Y., Wu, Y., Zhang, L.: Polym. Test **27**, 270 (2008)
5. Razzaghi-Kashani, M., Samadi, A.: Plast. Rubber Compos. **44**(7), 253–258 (2015)

Effect of Hydroxyl Value of Acrylic Resin on the Anti-graffiti Properties of Acrylic-PU Nanocoating

Saba Nemati Mahand[1], Shahla Pazokifard[2](✉), and S. Mojtaba Mirabedini[2]

[1] Chemical and Petroleum Engineering Department, Sharif University of Technology, Tehran, Iran
[2] Color and Surface Coatings Department, Iran Polymer and Petrochemical Institute, Tehran, Iran
s.pazokifard@ippi.ac.ir

Abstract. In this study, the effect of hydroxyl value of acrylic resin on the anti-graffiti properties of a silica nanoparticles embedded polyurethane coating was evaluated. Hardness property of different coating samples was determined using pendulum hardness (Persoz). Hydrophobicity and graffiti resistances of the coatings were studied through water contact angle measurements and removability of permanent ink marker from the surface of the coatings, respectively. The results showed that using high hydroxyl value acrylic resin in combination with applying hydrophilic silica nanoparticles (0.7 wt%) and hydroxyl-functional silicone-containing surface additive (2.0 wt%) improves easy-cleaning and mechanical properties of resulted PU coatings. This can be related to the synergic effects of crosslinking reaction of hydroxyl acrylic resin with IPDI in presence of hydrophilic silica nanoparticles along with hydrophobic character of silicone-based additive on the surface of the coating.

Keywords: Polyurethane · Acrylic resin · Nanosilica · Contact angle · Easy-clean · Synergic effect

1 Introduction

The problem of graffiti originated, cause a lot of damages to buildings and public areas [1–3]. Also, the cost of graffiti removal is very high. In fact, sacrificial and permanent anti-graffiti coatings provide a surface with low energy and easy-clean character. Easy-cleaning property of a coating depends on the texture and hydrophobic character of its surface. Permanent coatings are more appreciated due to their cost effective and high durability. Polyurethane based coatings are widely used in this application.

In this study, easy-clean permanent polyurethane coatings were prepared using hydroxyl acrylic resins with different hydroxyl values, isophorone diisocyanate (IPDI), silica nanoparticles, and a hydroxyl-functional silicone-containing surface additive. The effect of hydrophilic silica nanoparticles and a silicone-based additive on the easy-cleaning behavior of the coatings were evaluated.

H. Mirzadeh and A. A. Katbab (Eds.): ISPST 2018, *Eco-friendly and Smart Polymer Systems*, pp. 316–319, 2020.
https://doi.org/10.1007/978-3-030-45085-4_75

2 Experimental

Acrylic resins with hydroxyl functionality were prepared from TAK Resin Kaveh (Iran) with the characteristics mentioned in Table 1. Isophorone Diisocyanate (IPDI), (37.5% isocyanate in solution form) was supplied by Bayer Corporation. The mixing molar ratio –OH: –NCO was 1:1.2 [5, 6]. BYK-Silclean 3700 –a hydroxyl-functional silicone-containing surface additive– was obtained from BYK Altana group and used in amount of 2 wt% in coatings formulations [4]. Silica nano-powder (Aerosil 200v from Evonic Industries) used in preparation of the coatings (with 0.4, 0.7, and 1.0 wt% of solid coating formulation). The ingredients used in coatings formulations are shown in Table 2. The coating samples were applied on degreased glass plates with a wet film thickness of 200 μm using Doctor Blade Film Applicator) and left to dry at ambient temperature (28 °C, and 50% RH). Hardness and contact angle measurements were performed after 4 days using Pendulum persoze hardness instrument (Elcometer Co.) and Kruss K14 tensiometer, respectively. Staedtler Lumocolor Permanent Multifunctional Green Marker, used as a graffiti model in assessment of feasibility of surface cleaning of coatings surfaces.

Table 1. Specifications of hydroxyl acrylic resins

Acrylic resin coding	Solid content (%)	OH value (%)	Solvent
B	65 ± 1	4.5	Butyl acetate
392H	55 ± 1	1.8	Xylene
392HB	55 ± 1	1.8	Butyl acetate
X765	60 ± 1	2.7	Xylene
1210 N	60 ± 1	4.5	Xylene

* X stands for Xylene, B stands for butyl acetate

Table 2. Ingredients used in polyurethane coatings

Coating coding	Acrylic resin		IPDI (g)	Nano-silica (wt%)	Silicone additive (wt%)
	Resin code	Resin (g)			
C-4.5-B	R-4.5-B	2.80	0.99	0.0	–
				0.4, 0.7, 1.0	2
C-4.5-X	R-4.5-X	2.85	1.01	0.0	–
				0.4, 0.7, 1.0	2
C-2.7-X	R-2.7-X	3.35	0.72	0.0	–
				0.4, 0.7, 1.0	2
C-1.8-B	R-1.8-X	3.60	0.50	0.0	–
				0.4, 0.7, 1.0	2
C-1.8-X	R-1.8-B	2.80	0.40	0.0	–
				0.4, 0.7, 1.0	2

3 Results and Discussion

3.1 Hardness Measurements

Hardness measurement results for different coating samples using acrylic resins with different hydroxyl values and containing different nanoparticles amounts have shown in Fig. 1. As it is clear, the coatings with different OH-values and solvents (i.e. C-4.5-B vs C-4.5-X, and C-1.8-B vs C-1.8-X) show completely different results. In fact, increase of OH values improves hardness values, so C-4.5 had the highest level of hardness. Another result that is evidence in this Figure is that the C-4.5-X sample with 0.7% nanosilica (C-4.5-X-0.7) had the highest hardness.

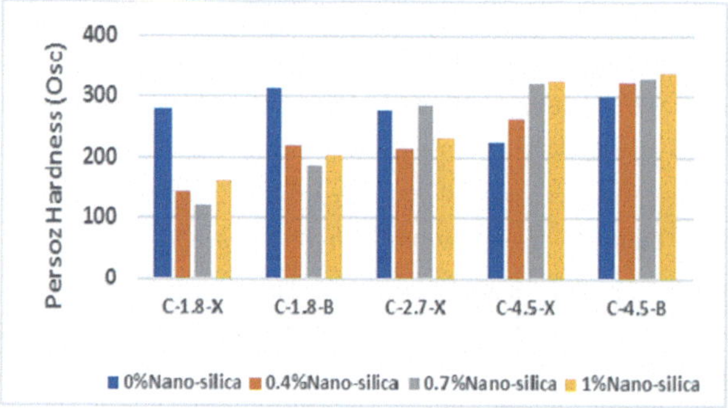

Fig. 1. Hardness of permanent anti-graffiti coatings containing hydrophilic silica nanoparticles (all samples contain BYK-Silclean 3700 (2 wt%) except the samples without any silica nanoparticles)

3.2 Contact Angle Measurement

Contact angel measurements were performed for the samples with highest hardness (C-4.5-X-0.7). The results of water contact angle measurements have illustrated in Fig. 2. The contact angel values were increased from 76.3° and 87.8 ± 0.8° to near 90°, after adding additives. This contact angle value is sign of easy-cleaning properties.

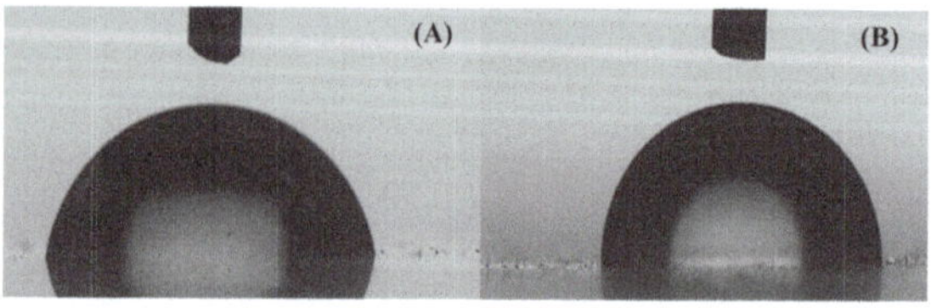

Fig. 2. Contact angle images of C-4.5-X: (A) 0% nanosilica and (B) 0.7% nanosilica

3.3 Graffiti Removal

Remarkably test of sample C-4.5-X was done before and after adding additives with Staedtler Lumocolor Permanent Multifunctional Green Marker. As it is shown in Fig. 3, the graffiti drawn on the film is remarkably removed after adding silicone-based additive. Due to a very low surface free energy, the silicone part of additives which is chemically bonded to polymer back-bone, moved toward the coating surface and therefore the surface is turned to low surface energy. Moreover, the presence of nanoparticles in the film will enhance the physical network and as a result, reduce the permeability of the coating. The decreasing coating permeability in combination with reducing coating free energy, leads to create anti-graffiti properties of the coating.

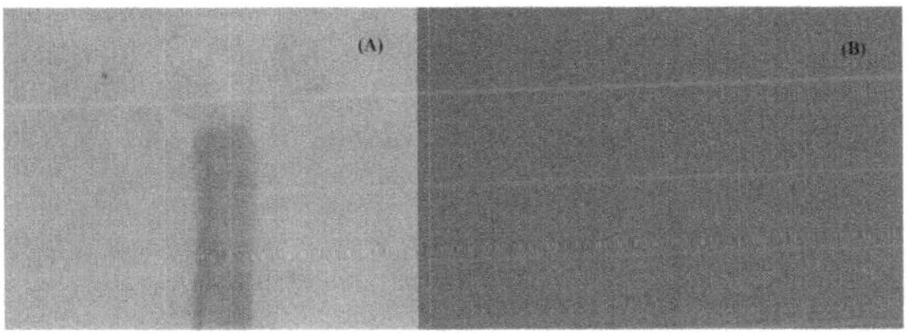

Fig. 3. Cleaning ability images of C-4.5-X marked with Staedtler Lumocolor permanent multifunctional green marker: (A) 0% nanosilica and (B) 0.7% nanosilica

4 Conclusion

As the results showed the increasing OH values in acrylic resin as a binder in polyurethane coatings leads to increase in crosslink reactions and therefore improves hardness values. Moreover, the presence of nanoparticles in the film will enhance the physical network and as a result, reduce the permeability of the coating. The decreasing coating permeability in combination with reducing coating free energy, leads to create anti-graffiti properties of the coating.

References

1. Lettieri, M., Masieri, M.: Appl. Surf. Sci. **288**, 466–477 (2014)
2. Watzlawik, M.: Cult. Psychol. **20**, 404–415 (2014)
3. Sanmartin, P., Cappitelli, F., Mitchell, R.: Constr. Build. Mater. **71**, 363–374 (2014)
4. Rabea, A.M., Mohseni, M., Mirabedini, S.M.: J. Coat. Technol. Res. **8**, 497–503 (2011)
5. Bayer: Material Science. The Chemistry of Polyurethane Coating, pdf file, p. 10. http://digitalcollections.library.cmu.edu/awweb/awarchive?type=file&item=488891
6. Koleske, J.V.: Polyurethane coatings. In: Paint and Coating Testing Manual, 15th edn., ch. 13, pp. 102–112. ASTM International

Microstructure and Mechanical Properties of Nanocomposite Based on PA6/NBR/Graphene

Mohammad Reza Nakhaei[1,2(✉)], Ghasem Naderi[2], and Mir Hamid Reza Ghoreishy[2]

[1] Mechanics and Energy Faculty, Shahid Beheshti University, Tehran, Iran
mr_nakhaei@yahoo.com
[2] Faculty of Polymer Processing, Iran Polymer and Petrochemical Institute (IPPI), P.O. Box 14975 112, Tehran, Iran

Abstract. Polyamide 6 (PA6)/nitrile butadiene rubber (NBR) thermoplastic elastomers (TPEs) reinforced with graphene oxide were compounded by a two-step melt mixing process via an internal mixer. The effect of graphene oxide on tensile and impact strength was investigated. Samples containing 30 wt% of NBR, 0.5, 1.25, and 2 wt% of graphene oxide were produced. Increasing the graphene oxide content from 0.5 wt% to 2 wt%, increased the tensile strength from 63.6 MPa to 69.3 MPa, whereas impact strength decreased from 78.1 MPa to 75.3 MPa when graphene oxide content increased. The morphology study of prepared nanocomposites with scanning electron microscopy (SEM) shows that the size of NBR droplets in TPE nanocomposites reduces by adding of graphene oxide into the PA6 phase.

Keywords: PA6/NBR/graphene · Thermoplastic elastomer · Nanocomposites · Tensile modulus · Impact strength

1 Introduction

Thermoplastic elastomeric materials (TPEs) are defined as a group of polymer blends which consist of two-phase structure of the materials with both thermoplastic and elastomeric properties. TPEs nanocomposites based on thermoplastic containing an elastomer and an organic or inorganic filler have found wide application in automotive and airplane industries, shipbuilding, household appliances and medical apparatus [1]. PA is well known as an engineering thermoplastic that is widely used in industrial applications for its remarkable mechanical and thermal properties, high resistance to crack initiation and processability. However, these advantages are accompanied by limitations such as moisture absorption, notch sensitivity, a low resistance to crack propagation and poor toughness under impact loading [2]. Blending PA with rubbers like NBR improves the toughness of this material. Although compounding rubber phase with PA improves toughness but decreases certain mechanical properties such as stiffness and hardness. These mechanical and thermal properties can be improved dramatically by

H. Mirzadeh and A. A. Katbab (Eds.): ISPST 2018, *Eco-friendly and Smart Polymer Systems*, pp. 320–323, 2020.
https://doi.org/10.1007/978-3-030-45085-4_76

adding a very small proportions of nanofillers such as nanoclay, nanotube and graphene to PA. The effect of nanoparticles extremely depends on the amount of dispersion in the matrix [1–3]. There is some reports on the study of TPEs nanocomposites based on PA6, NBR and different nanoparticles. Mahalati et al. [2] has introduced the Cloisite 30B into the PA6/NBR thermoplastic elastomers for oil resistance improvement and dynamic mechanical properties development. Paran et al. [4] investigated the effect of Halloysite nanotubes (HNTs) on the morphology and mechanical properties of the PA6 thermoplastic elastomers with various contents of NBR and found that HNTs can decrease the size of NBR droplets and improve mechanical behavior of the PA6/NBR blend. While so far, the graphene have not been used in reinforcing PA6/NBR TPE blends. Thus, in this work we have focused our attention on the effects of graphene on the microstructure and mechanical properties of PA6/NBR blends.

2 Experimental

2.1 Materials

The PA6 (KOPA, KN136) was supplied by KOLON Plastics, INC. (Korea) with density of 1.14 g/cm^3, melt flow index (MFI) of 31.4 g/10 min (2.16 kg & 230 °C) and melting point of 220 °C. Nitrile butadiene rubber (NBR), KNB 35L was purchased from KUMHO Korea with acrylonitrile content of 34 wt%, density of 0.98 g/cm^3 and Mooney viscosity of 41 (ML (1 + 4), 100 °C). Graphene oxide sheets (GOSs) were prepared through oxidation of graphene nano sheets (GNs) powder using a procedure based on modified Hummers and Offeman's method.

2.2 Preparation of PA6/NBR Nanocomposites

The TPU nanocomposites based on PA6/NBR with various graphene contents were prepared through a direct melt mixing process in a laboratory size internal mixer (Brabender Plasti-Corder, W50, Germany) at 230 °C and 80 rpm for 6 min. PA6 and Graphene were dried at 80 °C for 24 h in a vacuum oven before TPV nanocomposite preparation.

2.3 Characterization

For tensile and impact test, specimens were prepared according to standard ASTM D638 and D256, respectively. To tensile test samples tested in a Zuker tensile test machine (Zwick Co., Germany) with cross-head speed of 1 mm/min and at room temperature. Impact test was measured with notched specimens; using a Zwick with energy of 1 J. Scanning electron microscopy (SEM) images of prepared TPE nanocomposites were captured by Field emission SEM (FE-SEM) using ZEISS SUPRA 40 VP, Germany, operated at voltage of 20 kV.

3 Results and Discussion

The tensile and impact strength of PA6/NBR (70/30) blend are 60.14 and 81.72 MPa, respectively. The tensile and impact strength of prepared PA6/NBR/Graphene nanocomposites are shown in Figs. 1 and 2. It is obviously clear that, the tensile strength of the all samples increased with an increase in graphene content. The increase in the tensile strength is attributed to the presence of graphene oxide, which provides strong interaction between the graphene and polymer matrix and also adding graphene makes the NBR particles' size in PA6 matrix to decrease according to Table 1 and Fig. 3. In Fig. 3 the rubber droplets which have been removed by selectively etching in acetone, were displayed as dark holes. In agreement with the previous works, this may be related to the increase of the melt viscosity of the PA6 phase in the presence of graphene which leads to a decrease in the viscosity difference between the two phases.

As presented in Fig. 2, increasing the graphene content decreases this impact strength value. The decrease in impact strength is mainly due to the reinforcement effect. The reinforcement effect of graphene has dominated the reduction of elastomer particle size in our samples, resulting in lower impact strength by increasing graphene concentration.

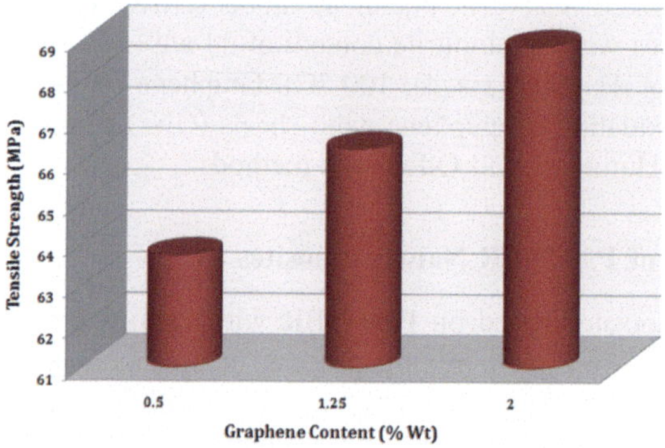

Fig. 1. Effect of graphene content to tensile strength

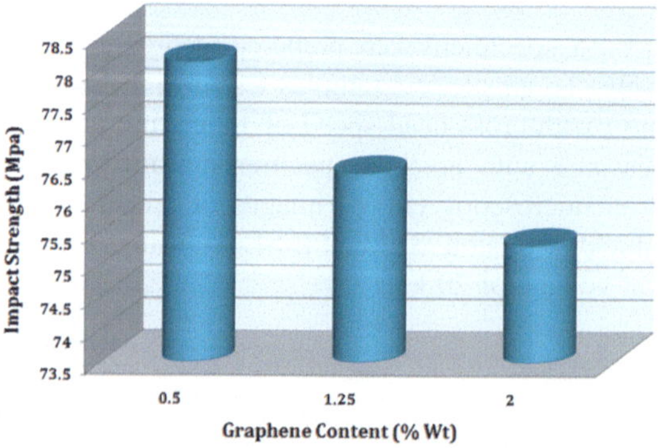

Fig. 2. Effect of graphene content to impact strength

Table 1. The effect of graphene content to NBR particle size

Graphene content (wt%)	0	0.5	1.25	2
NBR particle size (μm)	0.28	0.21	0.15	0.1

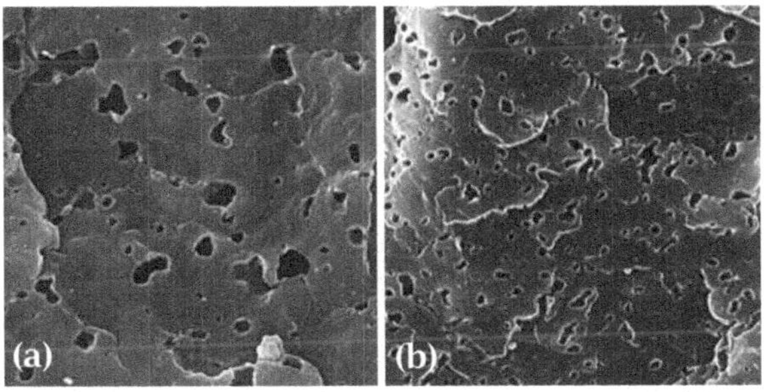

Fig. 3. Effect of graphene content to NBR particle size: (a) PA6/NBR blend and (b) PA6/NBR with 2 wt% graphene oxide

4 Conclusion

In this study, Graphene based PA6/NBR (70/30) nanocomposites were prepared through melt mixing. The effect of graphene content (wt%) on the mechanical properties of PA6/NBR (70/30) nanocomposites fabricated by melt mixing has been investigated. PA6/NBR/Graphene blend with 0.5, 1.25, and 2 wt% graphene oxide was produced to investigate the effect of graphene content on the tensile and impact strength. Increasing the graphene oxide content from 0.5 wt% to 2 wt% increased the tensile strength from 63.6 MPa to 69.3 MPa, whereas impact strength decreased from 78.1 MPa to 75.3 MPa when graphene oxide content increased.

References

1. Taghizadeh, E., Naderi, G., Razavi-Nouri, M.: Polym. Test **30**, 327 (2011)
2. Mahallati, P., Arefazar, A., Naderi, G.: Iran J. Chem. Eng. **8**, 56 (2011)
3. Ghasem, N., Lafleur, P.G., Charles, D.: Polym. Compos. **29**, 1301 (2008)
4. Paran, S.M.R., Naderi, G., Ghoreishy, M.H.R.: Polym. Compos. **38**, 451–461 (2017)

Numerical Simulation of Mixing Polymeric Fluids in Rotating Cylinder Rheometer

Ebrahim Khalaj(✉), Mohammad Hossein N. Famili, and Faezeh Ghodratpour

Tarbiat Modares University, Tehran, Iran
ebrahimkhalaj@gmail.com

Abstract. In this study, mixing of non-Newtonian polymeric fluids was simulated three dimensionally using a power-law rheological model in a rotational rheometer with coaxial cylinders. The simulation was performed by obtaining 3D components of velocity for each grid point using finite volume method and tracking the grid points in specified time intervals. To generalize the method to immiscible polymers, a modification was done by introducing interfacial tension and viscosity ratios of two fluids in solving velocity equations. The developed simulation method can be used to simulate the mixing of immiscible polymers.

Keywords: Polymer blends · Numerical simulation · Mixing · Morphology development · Non-newtonian

1 Introduction

The mixing or blending of polymers is an important and complex issue in the polymer processing industry. A better understanding of the mixing process will help to achieve optimum processing conditions and increase the quality of the final part. Several researches have been done in the regard. Much of these researches are about the mixing of Newtonian fluids.

In the field of studies on the simulation of mixing of polymer blends, a general classification of existing studies can be divided into four groups of phenomenological methods, statistical methods, surface tracking and concentration tracking.

From phenomenological methods, we can refer to tracking of immiscible interfaces using volumetric methods, generally in the case of extension of fluids and drop formation. Neural networks and genetic studies are categorized in statistical methods. Using surface tracking methods, local traction and distribution of phase particles can be obtained at predetermined boundaries after several periods of times. Surface tracking methods can be divided into two general categories of surface tracking and volume tracking. Front Tracking, Volume of Fluid, Smoothed Particle Hydrodynamics, Level Set and Phase Field are in this category. In the category of concentration tracking methods, Markers and Cells (MAC) [1] and Mapping methods have been proposed. These methods are explicit and indicate the location of each phase over a given time period. Mapping method is based on the repeatability of mixing flow and can be used for Newtonian fluids with the same properties. However, the modified mapping method has also been developed for polymer fluids [2].

H. Mirzadeh and A. A. Katbab (Eds.): ISPST 2018, *Eco-friendly and Smart Polymer Systems*, pp. 324–327, 2020.
https://doi.org/10.1007/978-3-030-45085-4_77

The method used in this study is a combination of the mapping and MAC methods and of course it can also be called a modified or simplified MAC. In this way, the velocities are obtained for each point of the network and the secondary location of each point is given according to its velocities in three directions of space in the desired time interval. Therefore, choosing the size of cells, method of solving equations, and choosing appropriate time intervals are very important in this method.

2 Experimental

2.1 Numerical Method

To trace the movement of phases, you must first obtain the velocity value in three directions for each control volume. Flow field is obtained by solving the equations of continuity and momentum. For this purpose, the flow of fluid was considered as stable, non-compact, and three dimensional. In this case, the continuity and momentum equations are written in the following form in 3 dimensions (Eqs. (1–4)):

$$\frac{\partial u}{\partial x}+\frac{\partial v}{\partial y}+\frac{\partial w}{\partial z}=0 \tag{1}$$

$$\frac{\partial}{\partial x}\left(u^2-\frac{1}{Re}\frac{\partial u}{\partial x}\right)+\frac{\partial}{\partial y}\left(uv-\frac{1}{Re}\frac{\partial u}{\partial y}\right)+\frac{\partial}{\partial z}\left(uw-\frac{1}{Re}\frac{\partial u}{\partial z}\right)=-\frac{\partial p}{\partial x} \tag{2}$$

$$\frac{\partial}{\partial x}\left(uv-\frac{1}{Re}\frac{\partial v}{\partial x}\right)+\frac{\partial}{\partial y}\left(v^2-\frac{1}{Re}\frac{\partial v}{\partial y}\right)+\frac{\partial}{\partial z}\left(vw-\frac{1}{Re}\frac{\partial v}{\partial z}\right)=-\frac{\partial p}{\partial y} \tag{3}$$

$$\frac{\partial}{\partial x}\left(uw-\frac{1}{Re}\frac{\partial w}{\partial x}\right)+\frac{\partial}{\partial y}\left(wv-\frac{1}{Re}\frac{\partial w}{\partial y}\right)+\frac{\partial}{\partial z}\left(w^2-\frac{1}{Re}\frac{\partial w}{\partial z}\right)=-\frac{\partial p}{\partial z} \tag{4}$$

By integrating the momentum equations in three dimensions on the control volume in the discretized computational domain, the finite difference form of these equations was obtained. The discretized form was obtained in the same way for the other two directions. Finally, the obtained equations were solved by a known algorithm called tridiagonal matrix algorithm (TDMA).

Semi-Implicit Method for Pressure-Linked Equations (SIMPLE) algorithm was used to calculate the pressure field. In this way, by extending continuity equation to equations in terms of pressure, which is called the correction of pressure equation. Because if the pressure is correct, the continuity equation is satisfied.

To generalize the problem to polymeric fluids, Power-law rheological model was used and flow field was obtained using modified coefficients. In order to generalize the flow equations to immiscible or partially miscible fluids and to introduce effect of hydrodynamic forces in phase movement and formation of interfaces, interfacial tension and viscosity ratio of studied two fluids were used. In this way, viscosity and surface tension in each control volume are considered to be the average viscosity of six surrounding control volumes.

3 Results and Discussion

Considering that for solving velocity equations, two fluids are considered to be non-Newtonian, it is not possible to use rheological and physical properties to distinguish between two polymers. Because these properties change with shear strain, which itself depends on the place of particles. Therefore, a hyphenated parameter (C) according to the spatial coordinates of particles of each fluid in the beginning time is needed. All particles in control volumes were traced to the end of the desired simulation time according to the assigned amount of C, and their new coordinates were obtained. The results of mixing simulation for three pairs of polymers are shown in Fig. 1 in specified time steps. Experimental images for mixing systems, at the same time steps and frame relative to results of numerical simulation, are also shown in Fig. 1. So comparison of theoretical and experimental results is simply possible.

Figure 1a shows simulation results for homogenous system of PE-PE. As the system name implies, both fluids are identical and therefore the miscibility parameters are not entered into the simulation. As seen in the Fig. 1a, two section of PE penetrate in each other in direction of moving cup. This penetration is more progressive in the lower part of rheometer at first time step, because of more adhesion of the polymer to the metal surface of the rheometer and continues to blend in angular direction of cylinder. Due to the homogeneity of physical and rheological properties of the fluid on both sides of the rheometer, all particles are under the same interfacial forces and this uniform motion is justified by flow field.

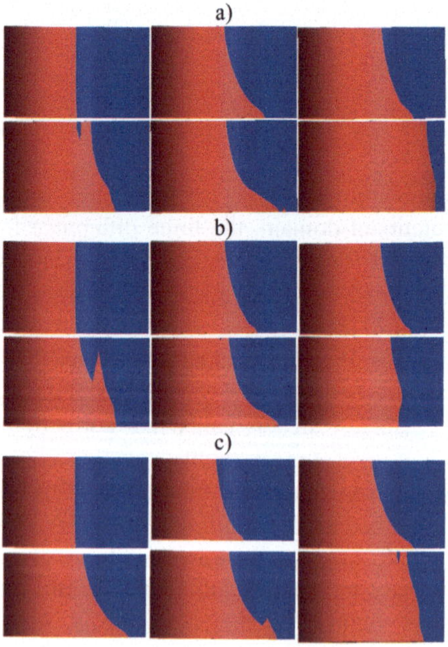

Fig. 1. Mixing simulation results in rotating cylinder rheometer at time intervals required for rotation of rheometer cup for three polymer pairs with different miscibilities; (a) homogeneous system of PE-PE, (b) partially miscible system of PE-PP, and (c) immiscible system of PE-PS

Figures 1b and 1c show simulation results for heterogeneous systems of PE-PP and PE-PS respectively in which the ratio of viscosities and interfacial tension of polymer pairs are also introduced in simulation. In the case of partially miscible system of Polyethylene-Polypropylene, penetration of one phase in another is performed at a lower rate than the homogeneous system. This progress is even slower in the case of immiscible polymer pair of Polythene-Polystyrene. Because in the case of heterogeneous systems, interfacial tension compensates for some of the effects of the current and prevents movement of two phase of polymers.

4 Conclusion

A three dimensional method based on modified mapping and MAC methods has developed for mixing simulation of polymeric fluids considering non-Newtonian rheological behavior and miscibility of different polymer pairs. This method can be a valuable tool for mixing simulation of polymeric fluids and provides useful information on the development of morphology during mixing of miscible and immiscible polymeric binary systems.

References

1. McKee, S., Tome, M.F., Ferreira, V.G., Cuminato, J.A., Castelo, A., Sousa, F.S., Mangiavacchi, N.: Comput. Fluid. **37**, 907–930 (2008)
2. SarhangiFard, A., Hulsen, M.A., Anderson, P.D.: Int. J. Numerical Method Fluid. **70**(6), 775–792 (2012)

Numerical Simulation of Mixing Polymeric Fluids in Twin Screw Extruder

Ebrahim Khalaj(✉), Mohammad Hossein N. Famili, and Faezeh Ghodratpour

Tarbiat Modares University, Tehran, Iran
ebrahimkhalaj@gmail.com

Abstract. In this study, mixing of non-Newtonian polymeric fluids was simulated three dimensionally using a power-law rheological model in a twin screw extruder. The simulation was performed by obtaining 3D components of velocity for each grid point using finite volume method and tracking the grid points in specified time intervals. To generalize the method to immiscible polymers, a modification was done by introducing interfacial tension and viscosity ratios of two fluids in solving velocity equations. The developed simulation method can be used to simulate the mixing of immiscible polymers.

Keywords: Polymer blends · Numerical simulation · Mixing · Morphology development · Non-Newtonian

1 Introduction

The mixing or blending of polymers is an important and complex issue in the polymer processing industry. A better understanding of the mixing process will help to achieve optimum processing conditions and increase the quality of the final part. Several researches have been done in the regard. Much of these researches are about the mixing of Newtonian fluids.

In the field of studies on the simulation of mixing of polymer blends, a general classification of existing studies can be divided into four groups of phenomenological methods, statistical methods, surface tracking and concentration tracking.

From phenomenological methods, we can refer to tracking of immiscible interfaces using volumetric methods, generally in the case of extension of fluids and drop formation. Neural networks and genetic studies are categorized in statistical methods. Using surface tracking methods, local traction and distribution of phase particles can be obtained at predetermined boundaries after several periods of times. Surface tracking methods can be divided into two general categories of surface tracking and volume tracking. Front Tracking, Volume of Fluid, Smoothed Particle Hydrodynamics, Level Set and Phase Field are in this category.

In the category of concentration tracking methods, Markers and Cells (MAC) [1] and Mapping methods have been proposed. These methods are explicit and indicate the location of each phase over a given time period. Mapping method is based on the

H. Mirzadeh and A. A. Katbab (Eds.): ISPST 2018, *Eco-friendly and Smart Polymer Systems*, pp. 328–331, 2020.
https://doi.org/10.1007/978-3-030-45085-4_78

repeatability of mixing flow and can be used for Newtonian fluids with the same properties. However, the modified mapping method has also been developed for polymer fluids [2].

The method used in this study is a combination of the mapping and MAC methods and of course it can also be called a modified or simplified MAC. In this way, the velocities are obtained for each point of the network and the secondary location of each point is given according to its velocities in three directions of space in the desired time interval. Therefore, choosing the size of cells, method of solving equations, and choosing appropriate time intervals is very important in this method.

2 Experimental

2.1 Numerical Method

To trace the movement of phases, you must first obtain the velocity value in three directions for each control volume. Flow field is obtained by solving the equations of continuity and momentum. For this purpose, the flow of fluid was considered as stable, non-compact, and three dimensional. In this case, the continuity and momentum equations are written in the following form in 3 dimensions (Eqs. (1–4)):

$$\frac{\partial u}{\partial x}+\frac{\partial v}{\partial y}+\frac{\partial w}{\partial z}=0 \tag{1}$$

$$\frac{\partial}{\partial x}\left(u^2-\frac{1}{Re}\frac{\partial u}{\partial x}\right)+\frac{\partial}{\partial y}\left(uv-\frac{1}{Re}\frac{\partial u}{\partial y}\right)+\frac{\partial}{\partial z}\left(uw-\frac{1}{Re}\frac{\partial u}{\partial z}\right)=-\frac{\partial p}{\partial x} \tag{2}$$

$$\frac{\partial}{\partial x}\left(uv-\frac{1}{Re}\frac{\partial v}{\partial x}\right)+\frac{\partial}{\partial y}\left(v^2-\frac{1}{Re}\frac{\partial v}{\partial y}\right)+\frac{\partial}{\partial z}\left(vw-\frac{1}{Re}\frac{\partial v}{\partial z}\right)=-\frac{\partial p}{\partial y} \tag{3}$$

$$\frac{\partial}{\partial x}\left(uw-\frac{1}{Re}\frac{\partial w}{\partial x}\right)+\frac{\partial}{\partial y}\left(wv-\frac{1}{Re}\frac{\partial w}{\partial y}\right)+\frac{\partial}{\partial z}\left(w^2-\frac{1}{Re}\frac{\partial w}{\partial z}\right)=-\frac{\partial p}{\partial z} \tag{4}$$

By integrating the momentum equations in three dimensions on the control volume in the discretized computational domain, the finite difference form of these equations was obtained.

The discretized form was obtained in the same way for the other two directions. Finally, the obtained equations were solved by a known algorithm called tridiagonal matrix algorithm (TDMA).

Semi-Implicit Method for Pressure-Linked Equations (SIMPLE) algorithm was used to calculate the pressure field. In this way, by extending continuity equation to equations in terms of pressure, which is called the correction of pressure equation. Because if the pressure is correct, the continuity equation is satisfied.

To generalize the problem to polymeric fluids, Power-law rheological model was used and flow field was obtained using modified coefficients. In order to generalize the flow equations to immiscible or partially miscible fluids and to introduce effect of hydrodynamic forces in phase movement and formation of interfaces, interfacial tension and viscosity ratio of studied two fluids were used. In this way, viscosity and surface tension in each control volume are considered to be the average viscosity of six surrounding control volumes.

3 Results and Discussion

Considering that for solving velocity equations, two fluids are non-Newtonian, it is not possible to use rheological and physical properties to distinguish between two polymers. Because these properties change with shear strain, which itself depends on the place of particles. Therefore, a hyphenated parameter (C) according to the spatial coordinates of particles of each fluid in the beginning time. All particles in control volumes were traced to the end of the desired simulation time according to the assigned amount of C, and their new coordinates were obtained. The results of mixing simulation for three pairs of polymers are shown in Fig. 1 in specified time steps. Experimental images for mixing systems, at the same time steps and frame relative to results of numerical simulation, are also shown in Fig. 1. So comparison of theoretical and experimental results is simply possible.

Figure 1a shows simulation results for homogenous system of PE-PE. As the system name implies, both fluids are identical and therefore the miscibility parameters are not entered into the simulation. As seen in the Fig. 1a, two section of PE penetrate in each other by rotating of berrel. This penetration is more progressive in the intermeshing area of two screws. Due to the homogeneity of physical and rheological properties of the fluid on both sides of the extruder, all particles are under the same interfacial forces and this uniform motion is justified by flow field.

Figures 1b and c show simulation results for heterogeneous systems of PE-PP and PE-PS respectively in which the ratio of viscosities and interfacial tension of polymer pairs are also introduced in simulation. In the case of partially miscible system of Polyethylene-Polypropylene, penetration of one phase in another is performed at a lower rate than the homogeneous system. This progress is even slower in the case of immiscible polymer pair of Polythene-Polystyrene. Because in the case of heterogeneous systems, interfacial tension compensates for some of the effects of the current and prevents movement of two phase of polymers.

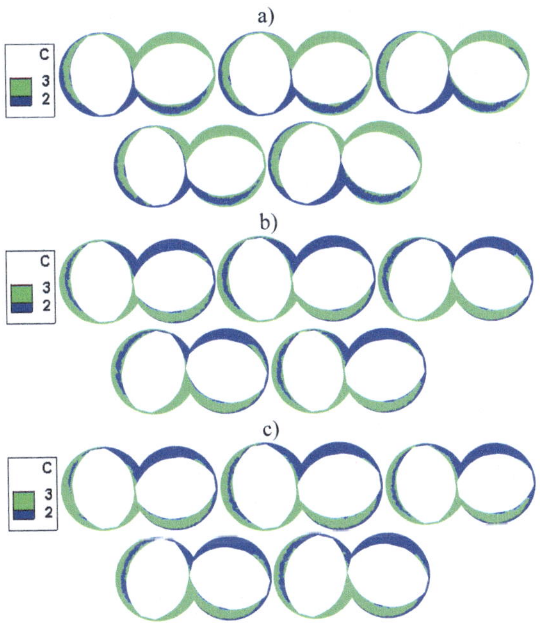

Fig. 1. Mixing simulation results in different cross sections of twin screw extruder at the same time for three polymer pairs with different miscibilities; (a) homogeneous system of PE-PE, (b) partially miscible system of PE-PP, and (c) immiscible system of PE-PS

4 Conclusion

A three dimensional method based on modified mapping and MAC methods has been developed for mixing simulation of polymeric fluids considering non-Newtonian rheological behavior and miscibility of different polymer pairs. This method can be a valuable tool for mixing simulation of polymeric fluids and provides useful information on the development of morphology during mixing of miscible and immiscible polymeric binary systems.

References

1. McKee, S., Tome, M.F., Ferreira, V.G., Cuminato, J.A., Castelo, A., Sousa, F.S., Mangiavacchi, N.: Comput. Fluid **37**, 907–930 (2008)
2. SarhangiFard, A., Hulsen, M.A., Anderson, P.D.: Int. J. Numer. Method Fluid **70**(6), 775–792 (2012)

3D Simulation of Velocity Field for Polymeric Fluids in Twin Screw Extruder by Finite Volume Method (FVM)

Ebrahim Khalaj(✉), Mohammad Hossein N. Famili, and Faezeh Ghodratpour

Tarbiat Modares University, Tehran, Iran
ebrahimkhalaj@gmail.com

Abstract. In this research, the velocity field of polymeric fluids is simulated in twin screw extruder by finite volume method (FVM). The simulation was three dimensional and the discretization performed using central difference method. For optimizing of the simulation, an Under Relaxation Factor (URF) parameter was considered. It was found that by using URF equal to 0.3, simulation had more rapid convergence. Considering a lower power law index (n) also had a similar effect.

Keywords: Simulation · Velocity field · Finite volume · Non-Newtonian fluid · Twin screw extruder

1 Introduction

In order to understand different polymer processes and their equipment and also polymer melt's transport to petrochemical processes it is necessary to predict flow field of different polymers. Simulation will lead to substantial reduction of lead times and costs of new designs. Computational fluid dynamics (CFD) codes are structured around the numerical algorithms that can tackle fluid flow problems. For this purpose, several methods were developed. The flow field of polymer melts has been studied both experimentally and computationally. The majority of the studies have focused on finite element method and 2D geometry by considering specific conditions [1]. The control volume integration distinguishes the finite volume method from all other CFD techniques. The resulting statements express the (exact) conservation of relevant properties for each finite size cell. This clear relationship between the numerical algorithm and the underlying physical conservation principle forms one of the main attractions of the finite volume method and makes its concepts much more simple to understand by engineers than the finite element and spectral methods [2]. In this research, a comprehensive computer program was developed for simulation of velocity field for polymer fluids with arbitrary consistency and power law index in the rotational cylinder rheometer.

H. Mirzadeh and A. A. Katbab (Eds.): ISPST 2018, *Eco-friendly and Smart Polymer Systems*, pp. 332–335, 2020.
https://doi.org/10.1007/978-3-030-45085-4_79

2 Experimental

2.1 Numerical Method

In order to develop a numerical method for solving Navier-Stokes equations, 3 steps should be performed: Integration of the governing equations of fluid flow over all the (finite) control volumes of the domain, discretization for conversion of the resulting integral equations into a system of algebraic equations, and finally solution of the algebraic equations by an iterative method.

The geometry is a three dimensional rotational cylinder rheometer that polymer melt placed between two cylinder.

Velocity field was obtained by solving the equations of the continuity and momentum. Flow was considered to be steady state, incompressible, isothermal and 3D. Discretization of the equations was performed using control volume (C.V.) method. Pressure correction was conducted by using semi-implicit method for pressure-linked equations (SIMPLE) assumption. Two terms of flow and penetration were written separately in order to choose type of discretization according to Peclet number. Primary 3D equations of continuity (Eq. (1)) and momentum (Eqs. (2–4)) are as follows:

$$\frac{\partial u}{\partial x} + \frac{\partial v}{\partial y} + \frac{\partial w}{\partial z} = 0 \tag{1}$$

$$\frac{\partial}{\partial x}\left(u^2 - \frac{1}{Re}\frac{\partial u}{\partial x}\right) + \frac{\partial}{\partial y}\left(uv - \frac{1}{Re}\frac{\partial u}{\partial y}\right) + \frac{\partial}{\partial z}\left(uw - \frac{1}{Re}\frac{\partial u}{\partial z}\right) = -\frac{\partial p}{\partial x} \tag{2}$$

$$\frac{\partial}{\partial x}\left(uv - \frac{1}{Re}\frac{\partial v}{\partial x}\right) + \frac{\partial}{\partial y}\left(v^2 - \frac{1}{Re}\frac{\partial v}{\partial y}\right) + \frac{\partial}{\partial z}\left(vw - \frac{1}{Re}\frac{\partial v}{\partial z}\right) = -\frac{\partial p}{\partial y} \tag{3}$$

$$\frac{\partial}{\partial x}\left(uw - \frac{1}{Re}\frac{\partial w}{\partial x}\right) + \frac{\partial}{\partial y}\left(wv - \frac{1}{Re}\frac{\partial w}{\partial y}\right) + \frac{\partial}{\partial z}\left(w^2 - \frac{1}{Re}\frac{\partial w}{\partial z}\right) = -\frac{\partial p}{\partial z} \tag{4}$$

By integrating the momentum equations in three dimensions on the control volume in the discretized computational domain, the finite difference form of these equations was obtained.

The discretized form was obtained in the same way for the other two directions. Finally, the obtained equations were solved by a known algorithm called tridiagonal matrix algorithm (TDMA).

Semi-Implicit Method for Pressure-Linked Equations (SIMPLE) algorithm was used to calculate the pressure field. In this way, by extending continuity equation to equations in terms of pressure, which is called the correction of pressure equation. Because if the pressure is correct, the continuity equation is satisfied.

To generalize the problem to polymeric fluids, Power-law rheological model was used and flow field was obtained using modified coefficients.

3 Results and Discussion

Equations obtained from the explained numerical method were solved via writing cods in MATLAB program using tridiagonal Matrix Algorithm (TDMA).

Velocity field for a typical low density polyethylene (LDPE) with consistency index ($\eta_{\circ}$) of 6×10^3, power law index of 0.39 at 160 °C, and Reynolds number (Re) of 0.01 was plotted via Tecplot 360EX 2016 R2 software. Intended geometry is shown in Fig. 1. The barrel rotates at a known speed.

Figure 2 shows 2D velocity contours for LDPE in a cross section of twin screw extruder in the intermediate slice of the extruder.

An iterative method is used to solve algebraic equations with nonlinear terms. In order to reduce the number of iterations required to achieve the correct velocity field, coefficients of velocity in three directions were divided by an Under Relaxation Factor (URF). Best URF for the considered polymer and geometry was 0.3 in all 3 directions.

Fig. 1. 3D geometry of screws obtained by the computer program

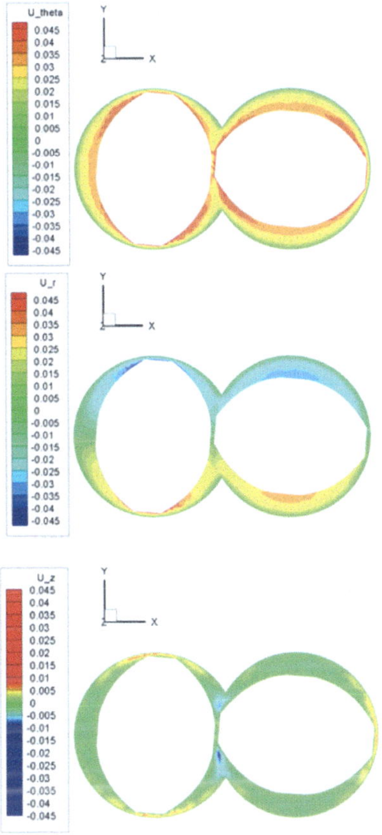

Fig. 2. 2D velocity profile for LDPE (n = 0.39, η_0 = 6 × 10^3) in a cross section of twin screw extruder in three dimensions

4 Conclusion

Velocity field was three dimensionally simulated for a typical polymer fluid with known consistency and power law index (LDPE with $\eta_0 = 6 \times 10^3$, n = 0.39 at 160 °C, and Re = 0.01) in a twin screw extruder via MATLAB codes. The number of iterations needed for convergence was lower compared with a Newtonian fluid. On the other hand, the best under relaxation factor for optimized convergence was 0.3 for all three directions.

References

1. Escudier, M.P., Oliveira, P.J., Pinho, F.T., Smith, S.: Exp. Fluid **33**, 101–111 (2002)
2. Versteeg, H.K., Malalasekera, W.: An Introduction to Computational Fluid Dynamics: The Finite Volume Method, 2nd edn. Pearson Education, London (2007)

Study on Rheology, Morphology and Mechanical Properties of PLA/Nanoclay Nanocomposites

Fatemeh Mohammadzadeh, Romina Esmailzadeh, and Azizeh Javadi(✉)

Department of Polymer Engineering and Color Technology, Amirkabir University of Technology, Tehran, Iran
ajavadi@aut.ac.ir

Abstract. In this research, nanocomposites of polylactic acid (PLA)/nanoclay (NC) containing 1–3 wt% of Cloisite 30B were prepared via melt mixing. The mechanical properties, rheology and morphology of resulted nanocomposites were studied. It was observed that the addition of nanoclays led to increase in modulus and tensile strength and decrease in the impact resistance of the neat PLA. The maximum increase in tensile strength was 33% corresponded to sample containing 3 wt% of Cloisite 30B. X-ray diffraction results showed an increase in the interlayer spacing of Cloisite 30B, which indicates the intercalation of this nanoparticle in the PLA matrix. Rheological assessments on nanocomposites containing Cloisite 30B displayed that the elastic modulus, loss modulus and complex viscosity of poly lactic acid were increased by increasing the amount of nanoclay and this increase is much higher at lower frequencies.

Keywords: Polylactic acid · Nanoclay · Nanocomposite · Properties · Melt mixing

1 Introduction

Poly lactic acid (PLA) is a thermoplastic aliphatic polyester that, because of its origin from renewable sources and also its consumption of carbon dioxide during production, is known as a preferable biopolymer. PLA is obtained from fermentation of raw material containing carbohydrates. PLA has been categorized as a harmless material by the FDA and its applications in food and liquid packaging is approved [1].

Some of the main advantages of PLA and its usage are biocompatibility, abstaining from environment pollution, reduction of the usage of petroleum-based material, good process ability, good mechanical properties and low friction coefficient [2]. On the other hand, this polymer has some disadvantages like its brittleness, gas and water vapor permeability, low glass transition temperature, low strain at break, weak flexibility, low toughness, low heat resistance and hydrophobicity too [3].

One of the known methods in order to improve the properties of PLA is the addition of nanofillers [1, 3]. Among nanofillers, montmorillonite which is a layered

H. Mirzadeh and A. A. Katbab (Eds.): ISPST 2018, *Eco-friendly and Smart Polymer Systems*, pp. 336–339, 2020.
https://doi.org/10.1007/978-3-030-45085-4_80

silicate has drawn more attention. In this study, nanoclay Closite 30B is used for enhancing the properties of PLA.

2 Experimental

PLA 2003D was supplied by NatureWorks LLC (USA). Cloisite 30B (C30B) was purchased from Southern Clay Products (USA). PLA and nanoclays were dried in a vacuum oven at 70 °C for 16 h before usage. PLA nanocomposites with 1, 2, and 3 wt% of Cloisite 30B were prepared in an internal mixer (volume 60 mL), using 150 rpm and 190 °C for 5 min. For mechanical tests, the prepared nanocomposites were then pressed at 200 °C under the pressure of 100 bar. X-ray diffraction (XRD), rheological, and mechanical tests were done.

3 Results and Discussion

3.1 XRD

By adding nanoclay to the PLA matrix, the 2θ diffraction peak is transmitted to a smaller angle due to the fact that the polymer chains diffuse between clay plates, increase the inter layer spacing and cause the formation of layered structure. In low clay loadings, the clay layers are able to have a better dispersion. So the PLA chains can diffuse between clay layers easier. By increasing the amount of nanoclay loading from 2 wt% to 3 wt%, a slight increase in the 2θ diffraction peak is observed. This observation can be due to inappropriate dispersion and the formation of agglomerates in samples containing 3 wt% of nanoclays, Fig. 1 [1, 3, 4].

3.2 Rheological Measurement

Storage modulus, loss modulus and complex viscosity of the samples were higher than of neat PLA in all frequencies. These values were increased with increasing the nanoclay loading, especially in lower frequencies. The reason for this observation could be the better dispersion of nanoclay layers in the polymer matrix and producing a network between polymer matrix and nanoparticles. By increasing the nanoclay loading, the behavior of samples went from liquid like to solid like. This can be understood from the decrease of the slopes of loss modulus and storage modulus curves. The maximum increase in storage modulus, loss modulus and complex viscosity was observed in sample containing 3 wt% of Cloisite 30B. Also the dependence of storage modulus to frequency was significantly decreased by increasing the amount of nanoclays, which could be due to the formation of silicate network, prolonging the relaxation time of the samples, Fig. 2 [4].

3.3 Mechanical Properties Testing

As it is observed in Table 1, modulus of all nanocomposite samples were increased by increasing the amount of nanoclay loading. Also in nanocomposite samples containing

1, 2, and 3 wt% of nanoclays, the tensile strength was increased by 18%, 27% and 33%, respectively. These observations indicate a good dispersion and appropriate interaction of nanoparticles with polymer matrix and more yielding of samples under tensile tension. The elongation at break of samples containing 1 and 2 wt% of nanoclay were increased compared to the neat PLA, but a great decrease in the elongation at break was observed in sample containing 3 wt% nanoclay. This could be due to the agglomeration of nanoparticles or formation of some voids in this sample during its preparation [5].

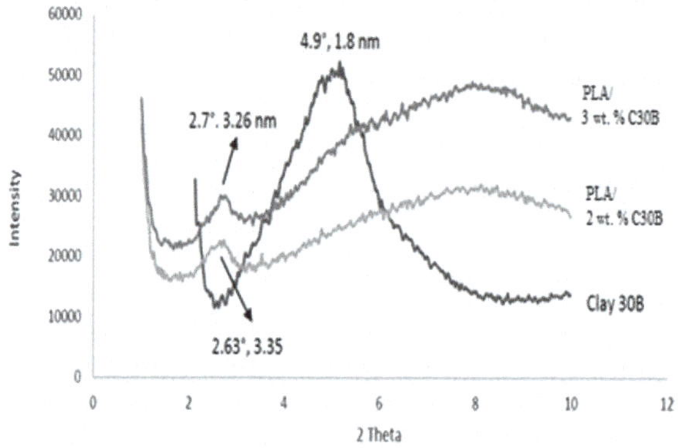

Fig. 1. XRD test for Cloisite 30B and nanocomposites containing 2 and 3 wt% of Cloisite 30B

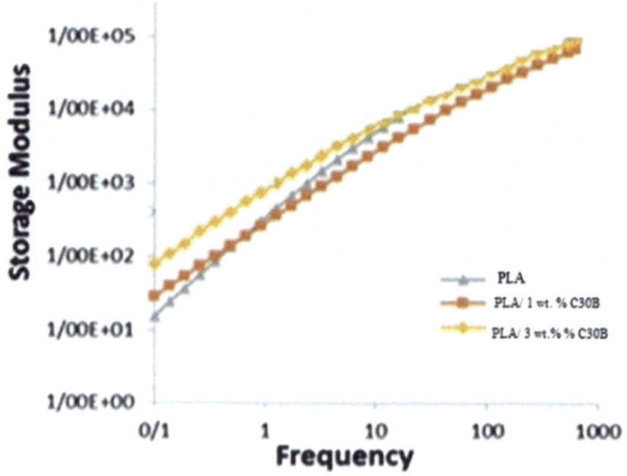

Fig. 2. Storage modulus curves of neat PLA and its nanocomposites

Table 1. Tensile and impact test results

Samples	Modulus (MPa)	Elongation at break (%)	Tensile strength (MPa)	Impact resistance (J/m)
Neat PLA	2371	4.18	52.02	32
PLA/1 wt% C30B	2543	6.77	61.26	31
PLA/2 wt% C30B	2693	7.105	66.125	25.16
PLA/3 wt% C30B	3458	2.98	69.24	19.33

4 Conclusion

The results of mechanical tests showed an increase in modulus and tensile strength with increasing the amount of nanoclays. The highest tensile strength was 33% for sample containing 3 wt% Cloisite 30B. On the other hand, the addition of nanoparticles to the PLA matrix decreased the impact resistance. The elongation at break in the samples containing 1 and 2 wt% of nanoclay were increased compared to neat PLA. The morphological studies for nanocomposites showed an increase in the distance between nanoclay layers and the formation of a layered structure. The rheological results showed that by adding Cloisite 30B to the PLA matrix, the storage modulus, loss modulus and the complex viscosity were increased, especially at low frequencies.

References

1. Molinaro, S., Cruz, M., Boaro, M., Sensidoni, A., Lagazio, C., Morris, M., Kerry, J.: J. Food Eng. **117**(1), 113–123 (2013)
2. Lai, S.-M., Wu, S.-H., Lin, G.-G.: Eur. Polym. J. **52**, 193–206 (2014)
3. Acik, E., Orbey, N., Yilmazer, U.: Appl. Polym. Sci. **42915**, 1–10 (2016)
4. La Mantia, F.P., Scaffaro, R., Ceraulo, M., Mistretta, M.C., Dintcheva, N.T., Botta, L.: Compos.-Part B Eng. **98**, 382–388 (2016)
5. Farahnaky, A., Mohammad, S., Dadfar, M., Shahbazi, M.: J. Food Eng. **122**, 78–83 (2014)

Investigation on the Effects of Nanoclay and Nano Calcium Carbonate on Properties of Binary and Hybrid Nanocomposites Based on PLA

Kosar Alavi[1], Zahra Asadi[2], Fatemeh Mohammadzadeh[1], and Azizeh Javadi[1(✉)]

[1] Department of Polymer Engineering and Color Technology, Amirkabir University of Technology, Tehran, Iran
ajavadi@aut.ac.ir

[2] Faculty of Mahshahr, Amirkabir University of Technology, Mahshahr, Iran

Abstract. Polylactic acid (PLA)/nanoclay, PLA/nano calcium carbonate (NCC) and PLA/nanoclay/NCC binary and hybrid nanocomposites were prepared via melt mixing containing 1, 2, and 3 wt% nanoclay and 5, 7 wt% NCC. It was observed via X-ray diffraction that nanoclay interlayer distance was enhanced from 1.8 nm to 3.26 nm indicating intercalation in binary samples. Interlayer distance was increased more in presence of NCC comparing to binary PLA/nanoclay samples. Rheological studies showed an increase in the storage modulus of samples at low frequencies comparing to neat polymer. Modulus was less increased in NCC containing binary and hybrid samples at higher amounts of NCC. An improvement in complex viscosity of all samples was observed and the highest increase was witnessed for hybrid sample containing 5 wt% NCC. According to Differential Scanning Calorimetry results, crystalline temperature of hybrid sample containing 7 wt% NCC was about 8 °C higher than that of neat sample.

Keywords: Nanocomposite · Poly(lactic acid) · Nanoclay · Nano-sized calcium carbonate · Rheology

1 Introduction

Synthetic biodegradable polymers were fine replacements of petroleum-based materials in many industries, and could be transformed to different shapes required for many applications. PLA was one of the biodegradable polymers which was a linear thermoplastic polyester obtained from fermentation of raw materials containing carbohydrates. This biodegradable polymer had good strength and process ability but meanwhile, it had some thermal limitations and brittleness which leads to limitation of its industrial use [1].

One way for improving PLA properties is using different shapes of nanoparticles like nanoclay (NC), carbon nanotube, nanosilica and NCC [2].

H. Mirzadeh and A. A. Katbab (Eds.): ISPST 2018, *Eco-friendly and Smart Polymer Systems*, pp. 340–343, 2020.
https://doi.org/10.1007/978-3-030-45085-4_81

Previous studies indicated that nanoclay did not modify crystallinity evolution of PLA [3]. In addition, producing hybrid nanocomposites was another approach to achieve different properties at the same time. In order to enhance reinforcement simultaneously, some had used both NC and NCC together [4]. To the best of our knowledge, PLA/NC/NCC hybrid system has not been studied among hybrid nanocomposites. In this research, the effect of simultaneous using of these two nano-fillers on rheology and crystallinity of PLA was investigated for the first time.

2 Experimental

PLA (Ingeo 2003D) with a density of 1.24 g/mL and MFI of 6 g/10 min was purchased from Nature Works. Nanoclay Cloisite 30B modified with methyl, tallow, bis-2-hydroxyethyl, quaternary ammonium was supplied by Songhan Plastic Technology (China). Nano calcium carbonate (HAKUENKA CC-R) coated with fatty acid was procured from Shiraishi-Omya GmbH (Austria).

Prior to mixing, PLA and nanoparticles were dried in an oven for 16 h at 70 °C. Nanocomposites were prepared in an internal mixer at 180 °C and 150 rpm for 5 min according to compositions mentioned in Table 1. Samples were then pressed at 200 °C under the pressure of 100 bar. In order to investigate the effect of nanoparticles on crystallinity of PLA matrix, DSC was employed. The nanocomposite structure was evidenced using X-ray diffraction. Rheological measurements were then performed.

Table 1. Composition of PLA nanocomposites

Name	Nanoclay (wt%)	Nano calcium carbonate (wt%)
P	–	–
P-NC1	1	–
P-NC2	2	–
P-NC3	3	–
P-NCC5	–	5
P-NCC7	–	7
P-NC3-NCC5	3	5
P-NC3-NCC7	3	7

3 Results and Discussion

3.1 DSC

Glass transition temperature (T_g) of nanocomposites did not change significantly comparing to neat sample and an increase of about 5 °C was observed in hybrid sample. This observation was due to reduction of free volume and slow dynamics of polymer chains resulting from interaction between nanoparticles and PLA matrix. Crystallinity of P-NC3 and P-NCC7 binary nanocomposites only slightly increased (about 2%).

In contrast to binary samples, crystallinity of P-NC3-NCC7 decreased by 10% comparing to neat PLA. This phenomenon was mainly because of chain dynamic restriction in presence of higher nanoparticle content in comparison with binary nanocomposites.

The crystallization temperature of hybrid sample was increased about 8° compared to neat polymer which indicates an increase in the speed of crystallinity, whereas the crystallinity percentage was not improved. This means that more incomplete crystals had been formed.

3.2 XRD

Addition of NCC shifted nanoclay peak to lower 2θ comparing to binary sample containing the same amount of nanoclay and interlayer distance increased from 3.26 nm to 3.31 nm, Fig. 1. Addition of NCC increased melt viscosity and as a result applied higher shear stress on nanoclay aggregates leading to better layer dispersion. On the other hand, high melt viscosity limits molecular diffusion into interlayer spacing and confines further dispersion. Thus the amount of incorporated NCC should be optimized.

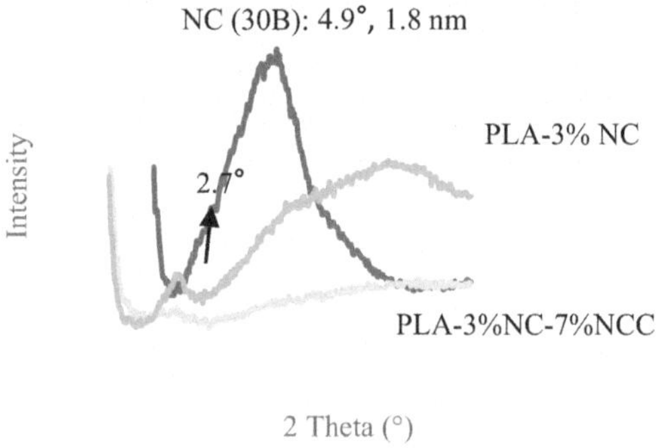

Fig. 1. XRD patterns of NC, PNC3 and PNC3NCC7 nanocomposites

3.3 Rheology

Considering oscillatory shear test results, it was observed that storage modulus (G') of samples increased in the whole frequency range with addition and increase of nanoclay loading. The same trend was observed for binary NCC containing nanocomposites which indicates the improvement in elastic behavior of these samples in comparison to neat PLA. However, G' of sample containing 5 wt% of NCC was higher than the one containing 7 wt% of this nanoparticle. This can be attributed to plasticizing effect of stearic acid treatment on NCC caused by PLA transesterification leading to thermal degradation. For hybrid samples of PLA/NC/NCC, as is seen in Fig. 2, nearly the same behavior of PLA/NCC binary samples is observed but in hybrid nanocomposites, G' of sample containing 7 wt% of NCC, is lower than the nanocomposite with 5 wt% of NCC in all frequency range. Also the amount of this decrease is higher in comparison to binary PLA/NCC binary samples (Table 2).

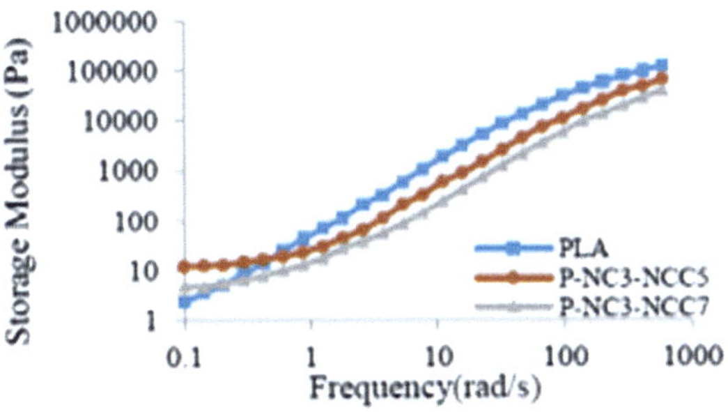

Fig. 2. Frequency dependence e of storage modulus of binary and hybrid sample

Table 2. DSC characteristics of PLA and its nanocomposites

Name	PLA	P-NC3	P-NCC7	P-NC3-NCC7
T_g (°C)	60.50	59.90	60.40	65.90
T_{cc} (°C)	130.00	129.10	119.00	138.70
ΔH_m (J/m)	27.62	29.03	28.09	17.23
α (%)	29.70	31.02	30.20	26.00

4 Conclusion

XRD results revealed better dispersion of nanoclay layers in presence of NCC. Crystallinity enhancement was negligible in binary nanocomposites and decreased about 10% in hybrid samples. In rheological measurements, at low frequencies, increasing NC content leads to higher G' in all samples comparing to neat polymer. Also, by adding NCC to binary and hybrid samples less enhancement of G' was observed.

References

1. Hamad, K., Kaseem, M., Yang, H.W., Deri, F., Ko, Y.G.: Express Polym. Lett. **9**(5), 435–455 (2015)
2. Ching, Y.C., Yahya, R.Y., Li, G.: J. Nano. Res. **23**, 7 15 (2013)
3. Dai, X., Shang, Q., Jia, Q., Li, S., Xiu, Y.: Polym. Eng. Sci. **50**, 3–8 (2010)
4. Sorrentino, L., Berardini, F., Capozzoli, M.R., Amitrano, S., Iannace, S.: Appl. Polym. Sci. **113**, 3360–3367 (2009)

Qualitative Characterization of Branches in Various Polyethylene Species via Fourier Transform Spectroscopy

Mohsen Mohammadi(✉) and Alireza Azizi

Department of Polymer Engineering, Faculty of Engineering,
Qom University of Technology, P.O. Box 37195-1519, Qom, Iran
mohammadi@qut.ac.ir

Abstract. In this paper the microstructure of different polyethylenes was characterized via FTIR spectroscopy. Two commercial low density polyethylenes (PE) (L1 and L2) and one commercial high density polyethylene were characterized. For comparison, one branched metallocene-PE and one linear metallocene-PE were also used. Test specimens were prepared in the form of film through hot compression molding of the polymers. Based on the FTIR spectra the exact position of the methyl deformation band for L1 and L2, was located at 1379.1 cm^{-1}, which has a position between the ethyl branch (1379.1 cm^{-1}) and the butyl (1377.8 cm^{-1}) and for mLCB-PE, the absorbance band is at 1377.5, which indicates that most of its branches are of the hexyl type. Absorbance bands in the 790–700 cm^{-1} region correspond to the branches of ethyl, butyl and hexyl or longer groups, indicating the presence of ethyl branches in LDPE samples, butyl but not any ethyl branch in mLCB-PE, and no ethyl and butyl branches in the HDPE. According to the absorption bands of 888 and 910 cm^{-1} unsaturated species in the microstructure of polyethylene samples, mainly vinylidene and vinyl end groups, were investigated.

Keywords: FTIR spectra analysis · Polyethylene · Short chain branches · Methyl and methylene rocking mode

1 Introduction

Polymers properties, such as melting point, density and mechanical behavior which have a close relation to crystallinity of polymer are related to branch type and its content. Moreover, data of methyl content when coupled with data of molecular weight and active end groups like vinyl results in perception of structure of end groups and polymerization mechanism. One of useful techniques for characterization of microstructure of polyethylene is FTIR spectroscopy [1, 2]. In this paper, different species of polyethylene were distinguished through FTIR.

H. Mirzadeh and A. A. Katbab (Eds.): ISPST 2018, *Eco-friendly and Smart Polymer Systems*, pp. 344–347, 2020.
https://doi.org/10.1007/978-3-030-45085-4_82

2 Experimental

2.1 Materials and Methods

High density polyethylene (Poliran HB0035), and two low density polyethylenes (Poliran LF0200 and Poliran LH0075) were prepared from Iran Petrochemical Company, respectively designated as HD, L1 and L2. For comparison and qualitative analysis of the samples, two synthesized metallocene-polyethylene, a linear one (m-HDPE) and a branched one (mLCB-PE) which is a kind of LLDPE were also used.

Films of polyethylenes in the desired thickness were molded through a hot press at 170 °C for two minutes and then quickly plunged into a bucket containing slurry of ice and water. The films were carefully removed and dried with a cloth or tissue.

Fourier Transform Infrared (FTIR) spectra were taken with a Bruker IFS 48. The test specimens were conditioned at 23 ∓ 2 °C and 50 ∓ 5% relative humidity for not less than 40 h prior to any measurements. Compensation were used with a wedge of polymethylene or a polyethylene of known low methyl content according to ASTM D2238-92, *Test Method B*. Compensation minimizes interference from methylene group absorption bands at 1368 cm^{-1} and 1352 cm^{-1} with the 1378-cm^{-1} methyl deformation band [3].

3 Results and Discussion

Thickness of the test specimens was measured 0.0264, 0.0273, 0.0305, 0.0279, and 0.0234 cm for L1, L2, HD, mLCB-PE, and m-HDPE samples, respectively. Figure 1 presents the infrared spectra of samples L1, L2, HD, and mLCB-PE in the 610 cm^{-1} to 4000 cm^{-1} region. The distinguishing factor of the three types of LDPE, LLDPE, and HDPE polyethylene is the side branches and the spectrum of these three types of polymers varies in the region of 1330 cm^{-1} to 1390 cm^{-1}. In spite of the absence of functional groups in the structure of these polymers, the infrared spectrum of polyethylene is not surprisingly simple and it exhibits various absorbencies at different frequencies (wave numbers).

The spectra of LDPE (L1, L2) and LLDPE (mLCB-PE) are very similar. The HDPE (H1) species have a much lower absorption than LDPE and LLDPE at 11378 cm^{-1}, and clearly show the absorption of methylene in the crystalline phase at 1352 and 11367 cm^{-1}. But the LDPE species have an absorption at 1378 cm^{-1}, which is broadly due to overlapping with the 1367 cm^{-1} band. To find out exactly the position of the methyl deformation band and the type of branches in the polyethylene structure, a comparison with a high density polyethylene with density of about 0.96 (m-HDPE) was done. For L1 and L2, the exact position of the band is at 1378.9 cm^{-1}, which has a position between the ethyl branch (1379.1 cm^{-1}) and the butyl (1377.8 cm^{-1}).

Blitz and McFaddin characterized type and amount of branches in polyethylenes trough methyl deformation and rocking band [4]. Based on their studies it can be assumed that the branches in L1 and L2 are mostly butyl, ethyl and hexyl. For mLCB-PE, the absorbance band is at 1377.5, which indicates that most of its branches are of the hexyl type.

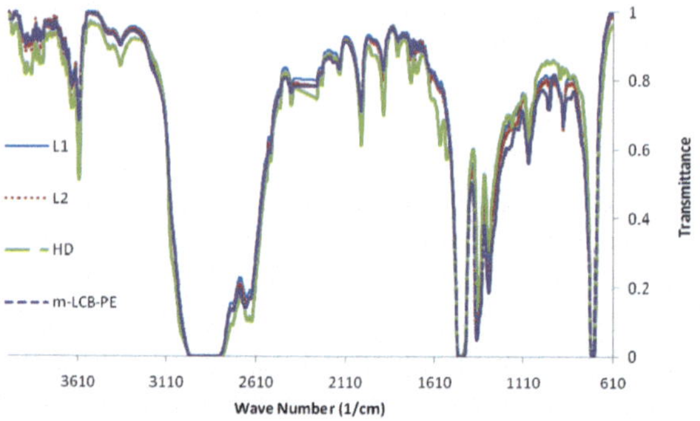

Fig. 1. Comparison of FTIR spectra of polyethylene samples

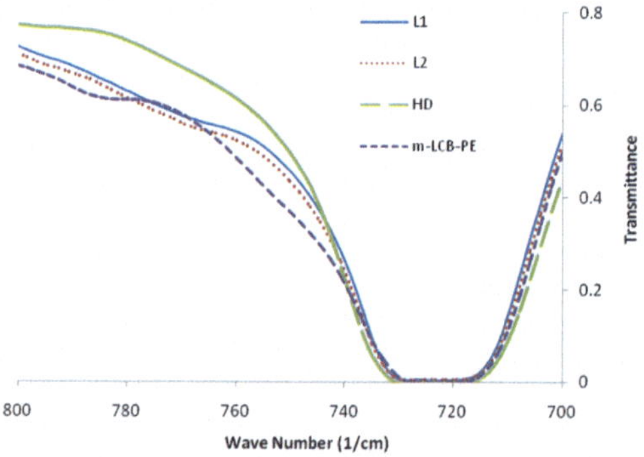

Fig. 2. FTIR spectra of the samples in the region of methylene rocking mode

The 790–700 cm^{-1} region is related to the methylene rocking mode (Fig. 2). Knowing that the absorption bands at 772, 745, and 740–720, respectively, correspond to the branches of ethyl, butyl and hexyl or longer groups, it can be qualitatively analyzed the existence of these branches in the sample [5]. For the LDPE samples a shoulder at 776 cm^{-1} demonstrates some ethyl branches in the polymer chains. For mLCB-PE, there is an apparent absorption of 755 cm^{-1}, which indicates the presence of butyl but not any ethyl branch. For HD there is no absorbance related to ethyl and butyl branches, therefore, its low amount branch could be hexyl or longer.

According to Fig. 3, the absorption bands of 888 and 910 cm^{-1} indicate unsaturated species in polyethylene chain, which are mainly vinylidene and vinyl end groups, respectively, and are related to the double bond in the structure of $RR'C = CH_2$ and $RCH = CH_2$, respectively. For L1, L2, and mLCB-PE, the amount of the vinyl type $RR'C = CH_2$ is greater than $RCH = CH_2$ because it has a larger absorption peak. But for HD it is the opposite, and also the spectrum suggests that the content of the vinyl in HD sample is much lower than that in the other samples.

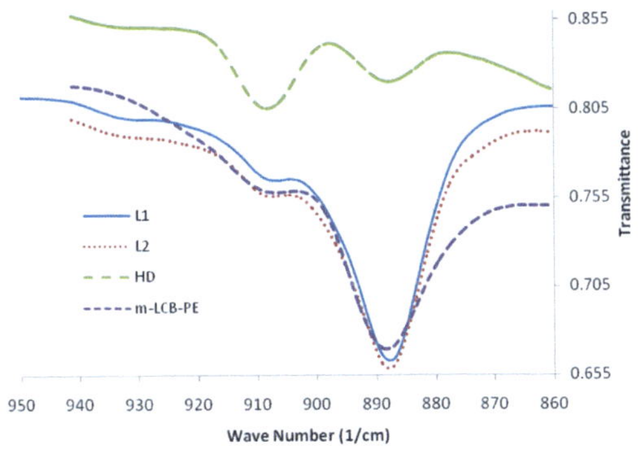

Fig. 3. FTIR spectra of the samples in methyl rocking region

4 Conclusion

Based on the infrared spectrum of polyethylene samples in the region of the methyl-deformation band and by comparing with a linear metallocene polyethylene sample in the methylene deformation region, as well as drawing the infrared spectra of samples in the methylene rocking region, the presence of branches of ethyl, butyl, hexyl, and longer in the samples was discussed. Unsaturated species in the samples were also characterized.

References

1. Caro, E., Comas, E.: Talanta **163**, 48 (2017)
2. Gulmine, J.V., Janissek, P.R., Heise, H.M., Akcelrud, L.: Polym. Test. **21**, 557 (2002)
3. ASTM D 2238–92: Standard test method for absorbance of polyethylene due to methyl groups at 1378 cm^{-1}. American Society for Testing and Materials, West Conshohocken (2004)
4. Blitz, J.P., McFaddin, D.C.: J. Appl. Polym. Sci. **51**, 13 (1994)
5. Willbourn, A.H.: J. Polym. Sci. **34**, 569 (1959)

Investigation of Thermal Stability of Polyethylene/Silica Aerogel Compounds

Mohammad Amin Abbas-Zadeh, Amir Hesam Arab Sorkhi, and Mohsen Mohammadi(✉)

Department of Polymer Engineering, Faculty of Engineering, Qom University of Technology, P.O. Box 37195-1519, Qom, Iran
mohammadi@qut.ac.ir

Abstract. To investigate the effects of adding silica aerogel (SA) nanoparticles to low density polyethylene (LDPE) on the thermal stability and thermal degradation of their compounds, SA/LDPE composites were prepared. Silica aerogel and low density polyethylene were mixed via melt mixing method using an internal mixer and then compression molded to sheets. The composites had different weight percent of silica aerogel. Thermal stability and thermal degradation of these composites were examined using thermal gravimetric analysis. It was observed that increasing the silica aerogel content, shifted T_{max} to higher temperature that shows increase of thermal stability.

Keywords: Low-density polyethylene · Silica aerogel · Nanocomposite · Thermal stability

1 Introduction

Low-density polyethylene (LDPE) is widely used because of its properties such as acceptable flexibility, transparency, low cost, easy processability and thermal stability [1, 2].

Aerogels are excellent nanoporous materials with high specific surface area, low density, low dielectric constant, low thermal conductivity and high insulation [3].

The purpose of this study is to prepare low-density polyethylene/aerogel nanocomposites used in packaging and to improve thermal stability.

2 Experimental

2.1 Materials and Methods

The components used were low density polyethylene and silica aerogel. Low density polyethylene granules, which have 2 g/10 min melt flow index (MFI) value, were provided by Bandar Emam Petrochemical (grade LF0200). A grade of LLDPE was also used. Silica aerogel (SA) nanoparticles (Aerulate) were obtained from Vakonesh Sanat Part Company. These nanoparticles were fully hydrophobic and had average pore diameter about 20 nm. Their average bulk density was about 0.07 g/cm^3.

H. Mirzadeh and A. A. Katbab (Eds.): ISPST 2018, *Eco-friendly and Smart Polymer Systems*, pp. 348–350, 2020.
https://doi.org/10.1007/978-3-030-45085-4_83

The LDPE/LLDPE blend containing 10% LLDPE were compound with silica aerogel with different weight percent (3, 9, 15, and 25 of SA) by melt mixing using an internal mixer (60 mL, Brabender, Germany). First, LDPE and LLDPE was melted for 3 min and SA was then mixed with them. The total mixing time was 12 min at 175 °C.

The obtained compounds formed as different sheets in a same shape by compression molding for 12 min at 175 °C.

Thermal gravimetric analysis (TGA) was used to characterize thermal stability and thermal degradation.

3 Results and Discussion

To examine the thermal resistance of LLDPE/LDPE/SA nanocomposites, thermal gravimetric analysis was performed. Figures 1 and 2 indicate TGA and DTG analysis of nanocomposites with different weight percent of silica aerogel. The results have been shown in Table 1.

It is observed that increasing the silica aerogel content, shifted T_{max} to higher temperature that shows increase of thermal stability. This phenomenon was due to increasing incombustible phase of silica aerogel, development of intumescence property of porous aerogel nanoparticle during combusting and making thermal insulator.

The amount of ash residue increased with increase of the aerogel nanoparticles content.

T_{onset} of these composites are lower than T_{onset} of pure polyethylenes.

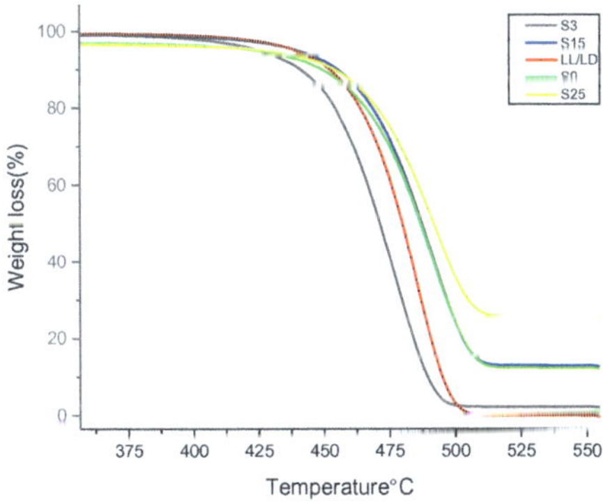

Fig. 1. TGA of LLDPE/LDPE/SA composites and pure LLDPE/LDPE

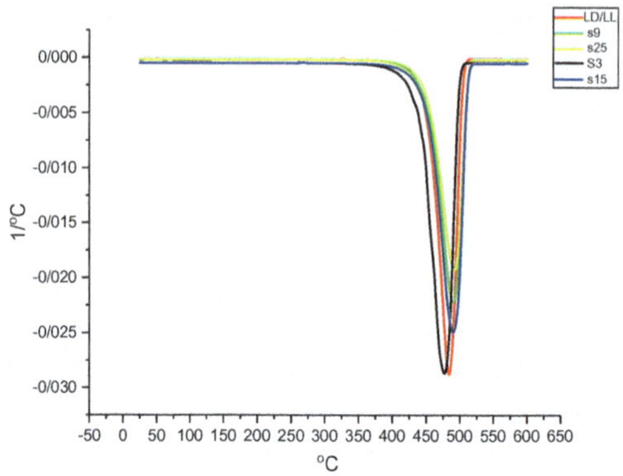

Fig. 2. DTG of LLDPE/LDPE/SA composites and pure LLDPE/LDPE

Table 1. Summary of TGA and DTG data

	LL/LD	S3	S9	S15	S25
Weight loss (%)	2.96e−3	2.11	7.06	12.69	21.54
T_{max} (°C)	484	476	489	490	492
T_{onset} (°C)	454.08	444.92	452.69	449.23	452.71

4 Conclusion

Low density polyethylene and silica aerogel composites were prepared with different silica aerogel contents. The thermal behavior and thermal stability were detected by thermal gravimetric analysis (TGA). The results showed the increase of thermal stability with increasing silica aerogel contents.

References

1. Marsh, K., Bugusu, B.: J. Food Sci. **72**, 39 (2007)
2. Del Nobile, M.A., Conte, A., Buonocore, G.G., Incoronato, A.L., Massaro, A., Panza, O.: J. Food Eng. **93**, 1 (2009)
3. Motahari, S., Motlagh, G.H., Moharramzadeh, A.: J. Macromol. Sci. Part B **54**, 1081 (2015)

Novel Polysulfone and Related Sepiolite Nanocomposites for Water Desalination

Samaneh Khodami, Samal Babanzadeh, and Shahram Mehdipour-Ataei(✉)

Iran Polymer and Petrochemical Institute, P.O. Box 14965/115, Tehran, Iran
s.mehdipour@ippi.ac.ir

Abstract. In this study new types of polysulfone, related sulfonated polysulfones, and also nanocomposites were prepared and investigated for desalination application. For this purpose, a novel monomeric diol was prepared and identified. Through its reactions with bis(4-fluorophenyl)sulfone monomer, the related polysulfones were prepared and characterized. In the second step by the aim of improving the hydrophilic nature of the polymer, the sulfonated groups were introduced into the polysulfone structure by direct sulfonation of dihalide, and in this way copolymers with 20–50 wt% sulfonation content were prepared. Another strategy for improving the hydrophilicity of the polysulfones was applied via the addition of hydrophilic nanoparticles to the polymeric matrix. In this part, by addition of 1, 3, 5, 7, and 10 wt% of sepiolite nanoparticles to the polymeric matrix, the related nanocomposites were prepared. Finally, membranes with 30 wt% of the samples were produced by phase inversion method and the operation tests of the prepared membranes were done and compared.

Keywords: Polysulfone · Water desalination · Sepiolite nanoparticles · Sulfonated polysulfone · Reverse osmosis

1 Introduction

Preparation of desalinated water from the brackish water becomes a basic challenge for the scientists. Reverse osmosis (RO) is one of the methods in this field that a well established membrane process and an efficient technique developed for separation of dissolved salts and small organic molecules. This method have the advantages of less energy consumption, low cost, and specific properties that have been widely used over the years [1, 2]. In this study, a novel polysulfone was selected as the main polymer for the preparation of the desalination membranes because of good mechanical, thermal, and chemical properties. It was synthesized via a condensation reaction between a new diol and a dihalide [3], however the structure was modified to improve the efficiency of some characteristics, such as hydrophilicity by introducing sepiolite nanoparticles into polymer matrix and addition various sulfonation content in the polymer structure.

H. Mirzadeh and A. A. Katbab (Eds.): ISPST 2018, *Eco-friendly and Smart Polymer Systems*, pp. 351–354, 2020.
https://doi.org/10.1007/978-3-030-45085-4_84

2 Experimental

A novel diol (diol-A) was obtained via reaction of 2,6-dichloropyridine with bisphenol A in presence of K_2CO_3 as a catalyst. The structure of novel diol was shown in Scheme 1. In the second step, polycondensation reaction between the synthesized diol with bis(4-fluorophenyl) sulfone as a dihalid afforded the desired polymer (PSf-A) (Scheme 2).

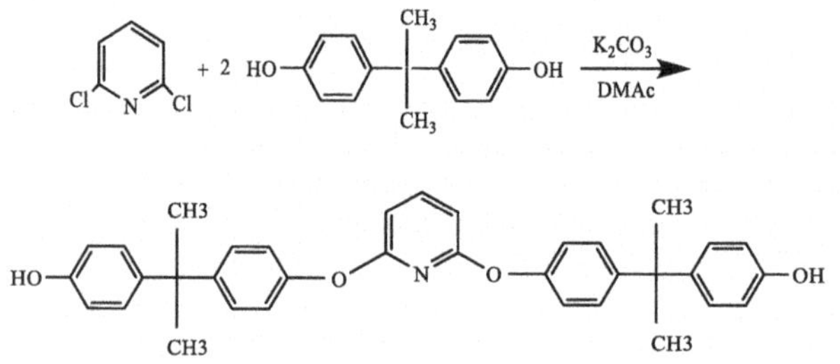

Scheme 1. Preparation of diol-A

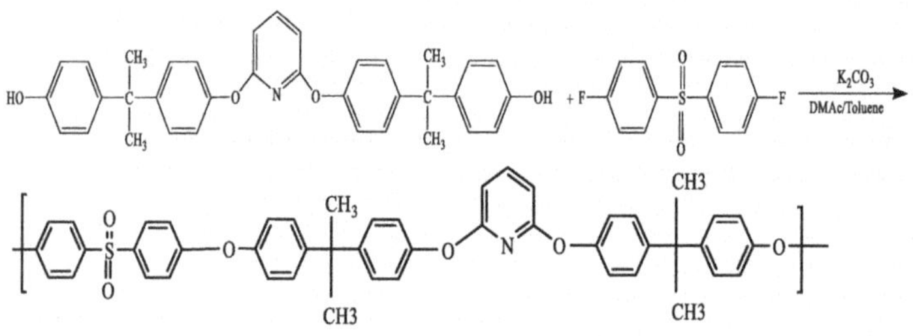

Scheme 2. Preparation of PSf-A

For improving hydrophilicity of the PSf, two pathways were employed. In the first method, sulfonated copolymers with 20, 30, 40, and 50 wt% sulfonation were synthesized via reaction of sulfonated and non-sulfonated bis(4-fluorophenyl) sulfone with diol-A. In the second method, nanocomposites were prepared by introducing 1, 3, 5, 7, and 10% of sepiolite nanoparticles into PSf matrix.

3 Results and Discussion

The performance of the prepared membranes from the pristine polysulfone-A, nanocomposites with various concentrations of sepiolite, and sulfonated copolymers with different sulfonation contents in terms of water flux (Figs. 1 and 2) and salt rejection (Figs. 3 and 4) were investigated using 500, 5000, and 30000 ppm aqueous solution of NaCl salt as feed and also, pure water permeability (PWP) by a cross-flow filtration system. The water flux (J) and NaCl rejection (R) ratio were calculated by the Eqs. (1) and (2), respectively:

$$J = V/AP\Delta t \tag{1}$$

$$R = \left(1 - \frac{C_{permeat}}{C_{feed}}\right) \times 100 \tag{2}$$

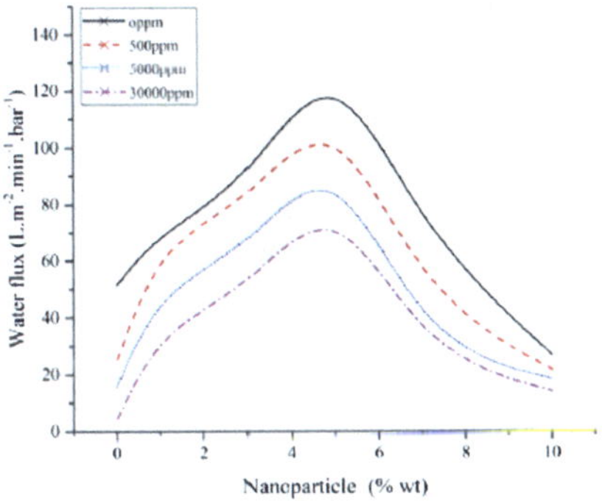

Fig. 1. Effect of nanoparticle on water flux of the prepared membranes

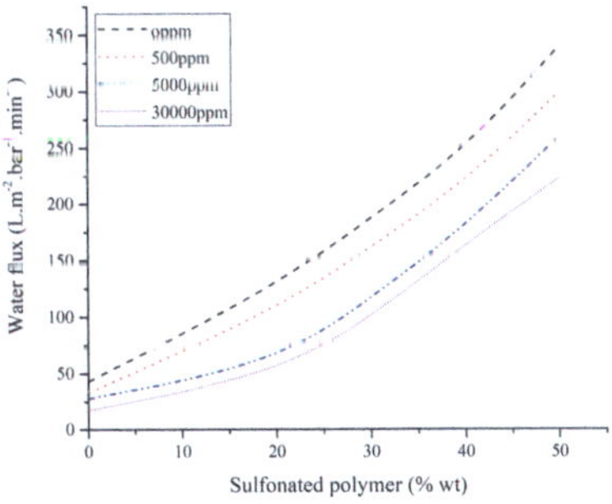

Fig. 2. Effect of sulfonation on water flux of the sulfonated membranes

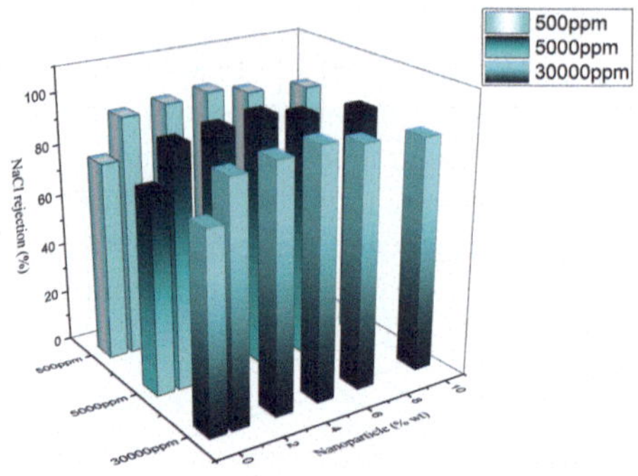

Fig. 3. Effect of nanoparticle on salt rejection of the prepared membranes

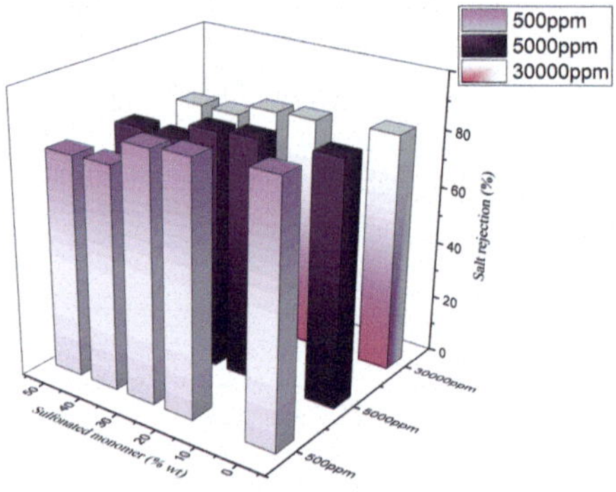

Fig. 4. Effect of sulfonation on salt rejection of the prepared membranes

4 Conclusion

A novel aromatic diol was prepared via nucleophilic reaction of bisphenol-A with 2, 6-dichloropyridine. A new polysulfone was synthesized via polycondensation reaction of the diol-A with bis(4-fluorophenyl) sulfone. The attention of the present study was to improve the hydrophilicity of the polysulfone by introducing sepiolite nanoparticles and sulfonation units into the polymer structure and to investigate their effects on the performance in the desalination process. The results indicated that the samples contained 30 wt% sulfone group and nanocomposite with 5 wt% sepiolite were successfully showed optimum performance (water flux and salt rejection) among other membranes.

References

1. Greenlee, L.F., Lawler, D.F., Freeman, B.D., Marrot, B., Moulin, P.: Water Res. **43**, 2317–2348 (2009)
2. Li, D., Wang, H.: J. Mater. Chem. **20**, 4551–4566 (2010)
3. Chakrabarty, B., Ghoshal, A.K., Purkait, M.K.: J. Membr. Sci. **325**, 427–437 (2008)

Synthesis and Characterization of Novel Poly(Arylene Ether Sulfone) for Desalination Application

Rahil Nahvi, Shahram Mehdipour-Ataei, and Samal Babanzadeh(✉)

Iran Polymer and Petrochemical Institute, P.O. Box 14965/115, Tehran, Iran
s.babanzadeh@ippi.ac.ir

Abstract. Poly(arylene ether sulfone)s are known as high performance thermoplastic polymers which are broadly used in membrane technology. In this study, a novel poly(arylene ether sulfone) was synthesized through condensation polymerization of 4,4'-thiodiphenol and dichlorodiphenyl sulfone. ^{1}H-NMR and IR spectroscopies were used to determine the chemical structure of polymer. Differential scanning calorimetry (DSC), thermogravimetric analysis (TGA), gel permeation chromatography (GPC), and scanning electron microscopy (SEM) were employed to characterize thermal properties, molecular weight, and morphologies of the polymer. The static contact angle measurement was demonstrated hydrophilicity of the polymer. Polymer films and related membranes were prepared by means of phase inversion method. The obtained results confirmed that these membranes had potential applications for water desalination process.

Keywords: Poly(aryl ether sulfone) · Water desalination · Membrane · Reverse osmosis

1 Introduction

Water scarcity is a forthcoming crisis all around the world. Reverse osmosis (RO) is a progressive common method for desalination process, because of its quite low energy consumption. There are several polymers that used in membrane preparation such as polyamides, polysulfones, and cellulose acetates. Ploy(arylene ether sulfone)s have been recognized as high performance polymers which are potential candidates for RO process, due to their high thermal and chemical stability and good mechanical strength [1]. Several researches have been examined on modifications of poly(arylene ether sulfone)s for improving their efficiency. Introduction of hydrophilic groups into poly(arylene ether sulfone) backbone is an important modification that has been investigated [2].

The objective of this research was to synthesize a new poly(arylene ether sulfone) via condensation polymerization. Thermal stability and behavior of the polymer was investigated. Membrane was cast from the polymer solution for desalination performance investigation. The morphology of prepared membrane was studied by scanning electron microscopy (SEM). The hyrophilicity of membrane was also studied based on static contact angle.

H. Mirzadeh and A. A. Katbab (Eds.): ISPST 2018, *Eco-friendly and Smart Polymer Systems*, pp. 355–358, 2020.
https://doi.org/10.1007/978-3-030-45085-4_85

2 Experimental

Poly(arylene ether sulfone) was prepared via reaction of 4,4'-thiodiphenol (TDP) and dichlorodiphenyl sulfone in DMAc as a solvent (Scheme 1).

For preparation of membrane via phase inversion induced immersion precipitation, the dried polymer was dissolved in DMAc at 25 wt% polymer concentration. The solution was cast onto glass plate and then immediately moved to the coagulation bath. A thin polymeric film separated from the glass within a few minutes.

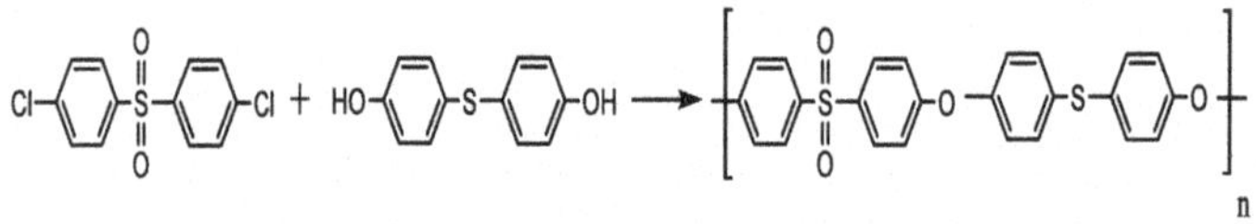

Scheme 1. Preparation of poly(arylene ether sulfone)

3 Results and Discussion

Reaction of 4,4'-thiodiphenol (TDP) and dichlorodiphenyl sulfone in DMAc solvent in the presence of K_2CO_3 resulted in preparation of a novel poly(arylene ether sulfone). The chemical structure of the polymer was confirmed by ^{1}H-NMR and FT-IR spectroscopy. The polymer was designed in such a way that presence of ether and sulfide units in the polysulfone backbone resulted in improving of the hydrophilicity of the polymer.

The molecular weight of polymer was determined by Gel permeation chromatography (Agilent 2100, THF) which showed the number average molecular weight around 57000 (g.mol^{-1}) (Fig. 1).

Polymer film with 0.04 mm thickness was cast from 25 wt% DMAc solution. Mechanical properties of the polymer were investigated. The mechanical strength of the polymer was 7.9 MPa and elongation at break was about 9% that was a reasonable result.

Thermal behavior and stability of polymer was investigated by DSC and TGA. Table 1 represents thermal properties data of poly(arylene ether sulfone) (Fig. 2).

The morphology of membrane was observed by SEM. The porous structure of membrane was shown in Fig. 3.

The static contact angle was used to evaluate hydrophilicity of the film surface. The contact angle of polymer was about 39 that showed high hydrophilicity of the polymer in comparison to the similar polymers.

Using 5000 ppm aqueous solution of NaCl salt as feed by a cross-flow filtration system, the water flux and salt rejection of the prepared membrane were in the range of 34.7 lit.m^{-2}.bar^{-1}.h^{-1} and 47.35%, respectively. Also, for pure water permeability (PWP) the water flux was 38.2 lit.m^{-2}.bar^{-1}.h^{-1}.

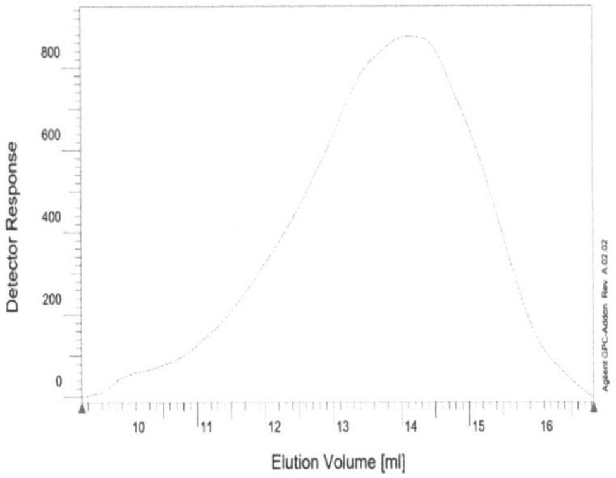

Fig. 1. GPC curve of poly(arylene ether sulfone)

Table 1. Thermal analysis data

Polymer	T_g (°C)	T_0 (°C)	$T_{10\%}$ (°C)	Char yield at 650 °C%
Pristine	166	310	506	48

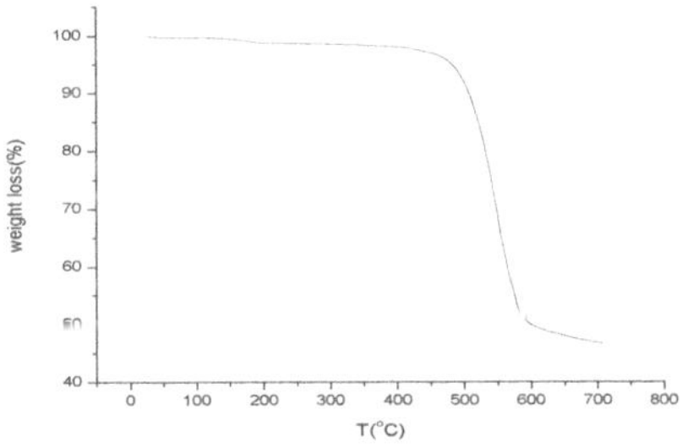

Fig. 2. Thermogeravimetric trace of the polymer

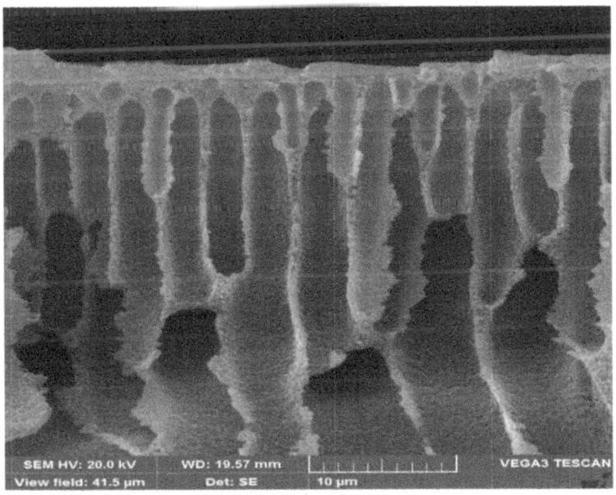

Fig. 3. SEM image of poly(arylene ether sulfone)

4 Conclusion

A novel poly(arylene ether sulfone) was prepared through condensation polymerization of DCDPS with TDP in aprotic solvent.

Obtaining high molecular weight polymer caused proper chemical and mechanical stability in the membrane. Presence of ether and sulfide hydrophilic units in the polymer led to high flux and salt rejection of the polymer and made it a suitable candidate for water desalination.

References

1. Checchia, S., Sabatini, V., Farina, H., Ortenzi, M.A.: Polym. Bull. **74**, 3939–3954 (2017)
2. Pirali-Hamedani, M., Mehdipour-Ataei, S.: Design Monom. Polym. **20**, 54–65 (2017)

Low Compression Set VMQ/XNBR Rubber Blends with Improved Aging Properties

Mustafa Azizi[1,2(✉)] and Ghodratollah Hashemi Motlagh[1]

[1] College of Engineering, School of Chemical Engineering, University of Tehran, Tehran, Iran
mu.azizi@ut.ac.ir

[2] Meshkin Halgheh Engineering Company, Zagros Industrial Town, Shahr-e-Ghods, Tehran, Iran

Abstract. The supply of consumable parts has always been a limiting factor for diverse industries. Sealings are considered as vital consumable parts in industry. Vinyl Methyl Siloxane (VMQ) is mainly used in the sealing industry, which needs improved mechanical and aging properties. Furthermore, compression set is a crucial property needed in this industry. In this research, the effect of blending VMQ with XNBR was studied to improve mechanical and aging properties especially compression set of VMQ. VMQ blends containing 0 phr to 20 phr Carboxylated Acrylonitrile Butadiene Rubber (XNBR) were prepared. Optimum cure conditions were obtained by a moving die rheometer. Samples were produced by compression molding. The results of tensile and hardness tests on the initial samples; showed a peak. Blend containing 10 phr of XNBR has showed a great improvement in compression set. Moreover, blending improved mechanical properties after aging than the non-blended sample. Thus, ultimately increases service life.

Keywords: Vinyl Methyl Siloxane (VMQ) · Carboxylated Acrylonitrile Butadiene Rubber (XNBR) · Polymer blend · Sealing

1 Introduction

Elastomers mostly play a major role in sealing industries. A sealant is used to close unwanted passages to prevent fluid loss [1]. VMQ has excellent resistance to ozone, weathering at higher temperature and remains serviceable at very low temperatures [2, 3]. Thus, this elastomer is very interesting in sealing industry. One major problem of VMQ is its high compression set which may cause leakage [3]. On the other hand, XNBR is an elastomer being used for different applications because of low compression set, high friction resistance and desirable oil and heat resistance [4]. Improvement of compression set of VMQ by blending with XNBR is the aim of this research.

H. Mirzadeh and A. A. Katbab (Eds.): ISPST 2018, *Eco-friendly and Smart Polymer Systems*, pp. 359–362, 2020.
https://doi.org/10.1007/978-3-030-45085-4_86

2 Experimental

For VMQ rubber Elastosil R401/70 S was used. Nipol NX775, used as a grade of XNBR and Perkadox BC-FF, grade of DCP was selected as a crosslinking agent. Curing properties of all blends were studied by a Moving Die Rheometer (MDR) at 180 °C. Tensile and hardness properties of volcanized blends were measured. The compression set of the samples was evaluated according to ASTM D395 method B. Aging conditions were according to, ASTM oil no.1 at 150 °C and air at 225 °C for 70 h.

In this research, five different blends containing 0 phr to 20 phr of XNBR were prepared by two roll mill though the following steps. The two roll mill was stabilized at 100 °C and the nip roll was adjusted at 1 mm. Then XNBR was milled for 5 min for mastication and reduction of viscosity. When the viscosity of XNBR was decreased, VMQ was added slowly to XNBR and mixed together for about 15 min. The blend was released from the rolls and relaxed for 10 min to release the stresses. In the meantime the temperature of the mill was decreased to 60 °C. After temperature stabilization at 60 °C, DCP as crosslinking agent was added to the blend and mixed for 3 min. The compound was then collected and used for testing and further characterization.

3 Results and Discussion

In Table 1 detailed curing data of blends have been presented. As seen, the MH values are analogous to each other which represent a very well crosslinked structure with uniform curing blends. More torque difference (MH-ML) is attributed to the blends compatibilization without phase separation [5]. As the table shows, the torque difference stays almost constant by increasing XNBR which shows very low phase separation. Scorch time (Ts5) shows the prevulcanization safety of rubber blends and optimum cure time (Tc90) presents completion of curing to an optimum level and formation of a stable network. Table 1 shows that scorch time and optimum cure time have little change by changing XNBR weight fraction. On further heating, blends torque curves show plateau behavior which represents no chain scission or crosslink breakage at 180 °C. Curing rate index is derived using the following equation:

$$CRI = \frac{100}{Tc90 - Ts5} \tag{1}$$

Table 2 presents initial tensile properties of blends and hardness changes after aging. Tensile strength, elongation at break and hardness see a maximum. It shows that phase interaction and crosslinking the blends improved by adding XNBR up to 10 phr, and decreased after that. Besides, hardness changes decrease after aging in ASTM oil because of more polarity of XNBR and its good fluid resistance which was predictable; however, air aging makes slight decrease in hardness by adding more XNBR to blends. This shows slight difference for high temperature resistance which is attributed to releasing small molecules trapped in XNBR phase. Figure 1 is another indication to prove this statement. Figure 2 shows tensile strength and elongation at break changes after aging. As it is clear, both of them show improvement in aging properties especially at 10 phr of XNBR. At last, the most critical property which was measured during this project was compression set. Figure 3 reveals the effect of blending on this property. Due to the best interaction between both phases at 10 phr, it shows the lowest compression set.

Table 1. Moving Die Rheometer results

XBNR (phr)	0	5	10	15	20
ML (N.m)	0.27	0.35	0.47	0.6	0.72
MH (N.m)	4.56	4.63	4.71	4.78	4.84
MH-ML (N.m)	4.29	4.28	4.24	4.18	4.12
Ts5 (m:s)	00:16	00:18	00:20	00:20	00:20
Tc90 (m:s)	01:25	01:34	01:32	01:34	01:30
CRI (min^{-1})	86.96	78.74	83.33	81.30	86.21
Trend	Plateau	Plateau	Plateau	Plateau	Plateau

Table 2. Initial tensile properties and hardness changes after aging in different fluids

XBNR (phr)	Tensile strength (MPa)	Elongation at break (%)	Hardness (Shore A)	Hardness changes after air aging (Shore A)	Hardness changes after ASTM oil no. 1 aging (Shore A)
0	7.78	328	68	1	9
5	8.32	332	74	−2	6
10	9.45	354	77	4	5
15	7.32	329	77	−3	6
20	6.30	291	74	−3	13

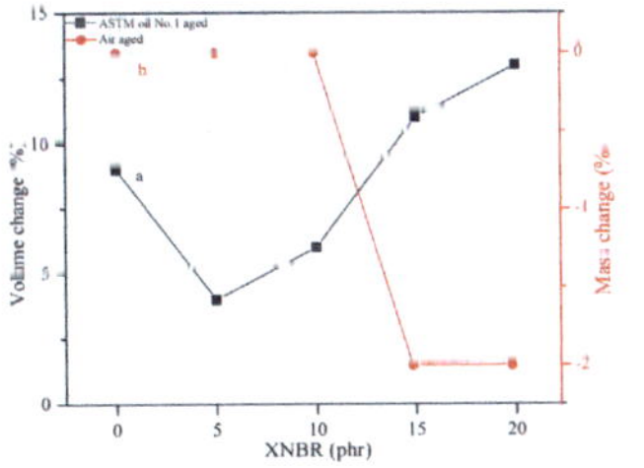

Fig. 1. Volume changes vs. XNBR phr after aging in ASTM oil no. 1 (a) Mass changes vs. XNBR phr after aging in air medium (b)

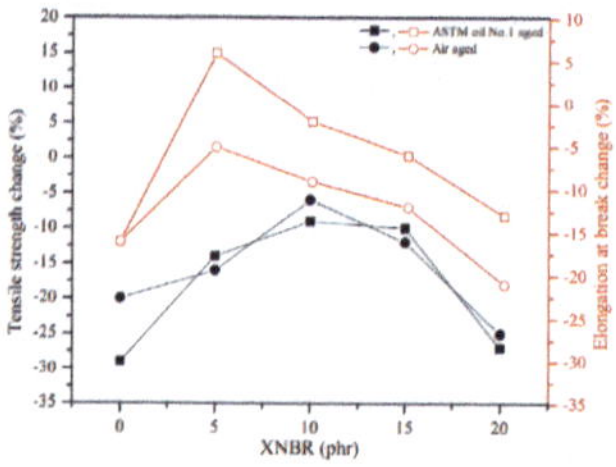

Fig. 2. Tensile strength and Elongation at break changes vs. XNBR phr after aging in different fluids

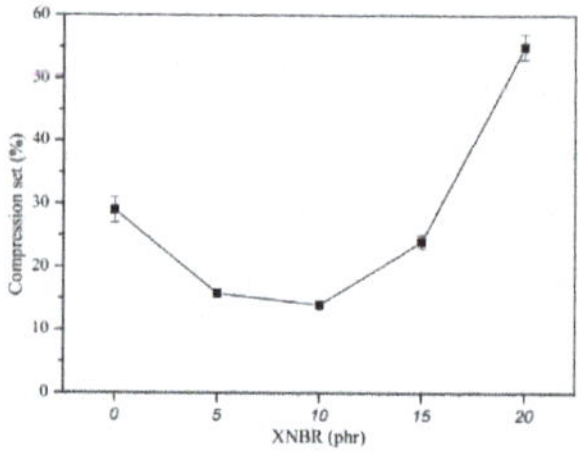

Fig. 3. Compression set changes vs. XNBR phr

As a conclusion, VMQ/XNBR blends have an optimum weight fraction to improve both mechanical and aging properties which is 10 phr of XNBR. This elastomer would help VMQ to be used in more various applications.

References

1. Seals, P.: Parker O-Ring Handbook. Park Hannifin Corp, Subang (2007)
2. Corning, D.: FSRs in Extreme Applications: Proof of the New Paradigm. A Dow Corning Publication, Midland (2004)
3. Polmanteer, K.E.: Rubber Chem. Technol. **61**(3), 470–502 (1988)
4. Przepio, A., Zaborski, M., Prochon, M.: Keratin as a Filler for Carboxylated Acrylonitrile-Butadiene Rubber XNBR (2007)
5. Bhuvaneswari, C.M., Dhanasekaran, R., Kale, S.S., Gouda, G.: Prog. Rubber Plast. Recycl. Technol. **31**(3), 207–218 (2015)

Morphology, Rheology and Impact Resistance of PS/EOC/SEBS Ternary Blends

Hasti Pourmadadkar[1], Mehdi Entezam[1(✉)], Hossein Ali Khonakdar[2], and Seyed Hassan Jafari[3]

[1] Department of Chemical and Polymer Engineering, Faculty of Engineering, Yazd University, 891581–8411 Yazd, Iran
m.entezam@yazd.ac.ir
[2] Iran Polymer and Petrochemical Institute (IPPI), P.O. Box 14965/115, Tehran, Iran
[3] School of Chemical Engineering, College of Engineering, University of Tehran, 11155-4563 Tehran, Iran

Abstract. Immiscible polystyrene (PS)/ethylene-octene copolymer (EOC) blends with different amounts of styrene-ethylene–butylene-styrene copolymer (SEBS) as compatibilizer were prepared by melt mixing method and the morphology, rheology and impact resistance of the blend samples were investigated. The scanning electron microscopy (SEM) micrographs revealed that incorporation of SEBS into PS/EOC blend leads to change the shape of EOC dispersed phase from droplet to fibrillar. From the melt linear viscoelastic rheological studies, the interaction between SEBS and each of PS matrix and EOC dispersed phases in the ternary blends was confirmed. Moreover, the micelle formation of the compatibilizer in PS matrix phase of the ternary blends with relatively high amounts of SEBS was founded by the rheological analysis. The impact resistance of PS/EOC blend enhanced with increasing the SEBS quantity.

Keywords: PS · EOC · SEBS · Ternary blend · Morphology

1 Introduction

It has been well known that compatibilization is an efficient method to improve the performance of incompatible polymer blends. But, the effectiveness of a compatibilizer is strongly depended on its effect on the morphology and the interfacial interaction between phases of the polymer blend. On the other hand, there is a relationship between the rheology and the morphology as well as the interfacial properties of polymer blends, so that reliable information about them could be obtained by the rheological studies [1, 2].

In this work, the amount of influence of SEBS as a compatibilizer on the morphology, rheology and impact resistance of incompatible PS/EOC blend is reported.

H. Mirzadeh and A. A. Katbab (Eds.): ISPST 2018, *Eco-friendly and Smart Polymer Systems*, pp. 363–366, 2020.
https://doi.org/10.1007/978-3-030-45085-4_87

2 Experimental

PS (MFI = 11 g/10 min; 200 °C, 2.16 kg) from Tabriz Petrochemical Co. (Iran), EOC (MFI = 1 g/10 min; 190 °C, 2.16 kg) with an octene content less than 30% from Dow Chemical Co. (USA) and SEBS triblock copolymer (Kraton G1652: styrene content = 32 wt%; Mw values of EB and PS blocks are 35,000 and 7500, respectively) from Shell Co. (USA) were used as received.

PS/EOC blend with a composition of 80/20 wt% was prepared by the melt blending in a laboratory batch internal mixer (Brabender W350 EHT) at a temperature of 200 °C and at a rotor speed of 60 rpm. In order to prepare the compatibilized blend samples, Various amounts, 1, 5, and 10 wt% (relative to the total weight of the blend) of SEBS were simultaneously added to the blend. The samples were compression molded into sheets with proper dimensions in a DR Collin (25 MPa) laboratory hot press at 200 °C under 10 MPa pressure.

Morphology of the blend samples were examined by SEM using a Vega Tescan model microscope. The samples were cryofractured in liquid nitrogen and the fractured surfaces were gold coated prior to microscopic observation. The melt linear viscoelastic shear rheological properties of the samples in oscillating sinusoidal mode were measured by a MCR501 rheometer equipped with parallel plate geometry (d = 25 mm, gap = 1 mm). Frequency sweep measurements were performed in a frequency range of 0.05–500 rad/s under nitrogen atmosphere at 200 °C and at a strain amplitude of 1% to get the response of samples in the linear viscoelastic regime. The impact experiments were carried out according to ASTM D256 by using a pendulum impact testing machine (B511 300, Zwick).

3 Results and Discussion

The observations of the SEM micrographs (Figs. 1a and 1b) reveal that all the blend samples have the matrix-dispersed morphology, but the shape of EOC dispersed phase changes from droplet in the uncompstibilized blend to fibrillar in the compatibilized blends. This change in the morphology of blend was intensified with increasing the amount of SEBS (as compatibilizer), due to the decrease of the interfacial tension between PS and EOC phases in the compatibilized blends and therefore more stability of EOC fibrillar structures formed during the melt processing [3]. Noteworthy is that the EOC droplet dispersed phase domains in the compatibilized blends have a lower average size in comparison to the uncompatibilized blend.

Corresponding to the results of rheological studies (not shown here), both neat PS and EOC show a pseudo-plastic rheological behavior and EOC with higher viscosity and elasticity than PS displays more shear thinning behavior. All the blends also have the pseudo-plastic rheological behavior and their rheological properties are between those of the neat components. By considering the complex viscosity and in particular the storage modulus of the blends, they are higher for the compatibilized blends compared to the uncompatibilized blend at low frequencies. This result observed at low frequency ranges, where flow induced orientation has a less effect on the interfacial interaction between phases, could be attributed to reinforcement of the interfacial interaction between PS and EOC in the presence of SEBS [4, 5]. On the other hand, for the compatibilized blends with relatively high value of SEBS (about 5 wt%), the complex viscosity and the storage modulus also are more than that of the uncompatibilized blend at high frequency ranges, indicating the micelle formation of SEBS in the PS matrix phase of these blends.

The impact resistance results of the samples are given in Table 1. Inclusion of SEBS in the blend improves its impact resistance. This improvement pronounced with the increase of SEBS amount in the blend could originate from several functions including, modification of blend morphology, reinforcement of the interfacial interaction between phases in the presence of SEBS and also high toughness of SEBS as an elastomer. Of course, the effect of SEBS toughness on the impact resistance could be more for the compatibilized blends with the micelles of SEBS in PS matrix phase.

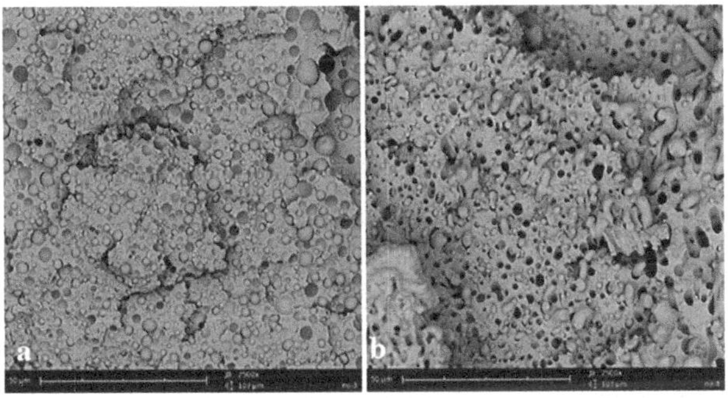

Fig. 1. SEM microghraphs of PS/EOC (80/20): (a) without SEBS and (b) with SEBS (5 wt%)

Table 1. The Impact resistance (IR) results of the blend samples

Sample	IR (kJ/m^2)
PS/EOC	2.8
PS/EOC/SEBS (1 wt%)	3.3
PS/EOC/SEBS (5 wt%)	4.2
PS/EOC/SEBS (10 wt%)	5.5

4 Conclusion

The effect of SEBS quantity as a compatibilizer on the morphology, rheology and impact resistance of immiscible PS/EOC blend was investigated. The important conclusions obtained from this work are as follows:

1- The shape of EOC dispersed phase changed from droplet to fibrillar, due to the incorporation of SEBS into the blend.

2- SEBS-induced reinforcement of the interfacial interaction between PS and EOC phases and the micelle formation of SEBS in PS matrix phase of the compatibilized blends could be founded by the linear viscoelastic rheological analysis of the blend samples in low and high frequency ranges, respectively.
3- The compatibilization of PS/EOC blend using SEBS led to improve its impact resistance and this improvement was more pronounced for the compatibilized blends with the micelles of SEBS.

References

1. Ajji, A., Utracki, L.A.: Polym. Eng. Sci. **36**, 1574–1585 (1996)
2. Faker, M., Razavi Aghjeh, M.K., Ghaffari, M., Seyyedi, S.A.: Eur. Polym. J. **44**, 1834–1842 (2008)
3. Li, J., Ma, P.L., Favis, B.D.: Macromolecules **35**, 2005–2016 (2002)
4. Razavi Aghjeh, M.K., Khodabandelou, M., Khezrefaridi, M.: Appl. Polym. Sci. **114**, 2235–2245 (2009)
5. Entezam, M., Khonakdar, H.A., Yousefi, A.A., Jafari, S.H., Wagenknecht, U., Heinrich, G.: Macromol. Mater. Eng. **297**, 312–328 (2012)

Acidic Dye Removal by Chitosan Treated Membrane Containing Spacer Fabric/Nylon6 Nanofibers/TiO_2 Nanoparticles

Ehsan Yousefi, Marziyeh Ranjbar Mohammadi(✉), and Mohammad Nouri Taba

Textile Department, University of Bonab, Bonab, Iran
m.ranjbar@bonabu.ac.ir

Abstract. The major aim of this paper was to assess the capability of polymeric nanofibers membranes containing nylon 6/titanium which were electrospun onto both sides of the spacer fabric and then was treated with chitosan for removing acidic dyes from textile waste water. The prepared compositions were characterized by SEM analysis. Chitosan treated nylon/TiO_2/spacer (CNTS) compositions showed the highest amount of discoloration for acidic dyes at pH = 6. The results of this study exhibited CNTS structures could be used as a great absorbent in removing azoic dyes of wastewater.

Keywords: Spacer fabric · Nylon6/TiO_2 composite · Methylene blue · Acid dye · Dye removal

1 Introduction

Due to the undeniable increasing of industries, industrial wastewater has become a serious challenge in the world. Chitosan, as a natural biopolymer taken from N-deacetylation of chitin, is the unique basic amylose in nature [1]. One of the most important pollutants in the sewage is azo dyes. There are azoic (-N = N-) and aromatics groups in the chemical structure of these dyes. The aromatic groups in their structure are usually toxic and largely non-degradable [2]. Nowadays, using from electrospun nanofibers for dye removal has been grown. Electrospinning is a well technique for fabricating continuous nanofiberous mat. Produced nanofibers expose excellent mechanical strength, water flux, and particle rejection. These characteristics make electrospun membranes as a great candidate for microfiltration applications [3]. In this work, nylon-6/TiO_2 nanofibers were fabricated onto both sides of spacer fabric and then prepared membrane was treated with chitosan and its efficiency in discoloration and adsorption of azoic dyes was assessed.

H. Mirzadeh and A. A. Katbab (Eds.): ISPST 2018, *Eco-friendly and Smart Polymer Systems*, pp. 367–370, 2020.
https://doi.org/10.1007/978-3-030-45085-4_88

2 Experimental/Theoretical

2.1 Materials

In this study, nylon-6 with a relative viscosity $\eta_{rel} = 2.63$ in the form of granules and spacer fabric (a surface and beneath fiber of 75 den, a monofilament connector fiber of 30 den, a weight of 230 g/m^2) were supplied by Aliaf Public Joint Stock Co. TiO_2 rutile powder with diameter of 30 nm, and formic acid were used. Chitosan (average molecular weight was about 80,000 and the degree of deacetylation was over 90%) was purchased from Sigma Co. Acidic dye.113 with Mw = 681.65 g/mol which is principle applied dyes in textile industry was chosen as a model for testing discoloration.

2.2 Preparation of Polymeric Solution

Solution containing 1.5% titania was prepared in formic acid. Then, the prepared solutions were sonicated (Endicott Corporation, Iran) for 15 min. After complete homogenization of TiO_2, nylon at a concentration of 15 (w/v%) was added to disperse and stirred for 24 h. electrospinning solution were placed in a 10 mL syringe with a metal needle of 0.4 mm in diameter. Fabricated nanofibers were assembled onto rotating collector with diameter of 5 cm.

2.3 Treating Fabricated Membrane with Chitosan

Activation of spacer fabric wrapped with nylon and TiO_2 (NTS) was done by formaldehyde. Membrane were shaken in 1 M HCl for 4 h at 40 °C. After that, washing of membranes was done several times in the water with the temperature of 40–50 °C for partial hydrolysis amide bonds. Then, the activation was done and the NTS membranes were incubated in a solution of 20 mL formaldehyde (>36.5 wt%) and 0.2 mL phosphoric acid (85 wt%) for 7 h at 60 °C. The washing process was repeated as before. Chitosan solution with the concentration of 2 wt% was prepared in 1 vol.% acetic acid solution. The activated membranes were immersed and shake in chitosan solution for 1 h at room temperature. Then the treated sample with chitosan was incubated in an oven at 80 °C for 1 h. Rinsing process with 1 vol.% acetic acid solution and deionized water was used to separate unbound chitosan polymers from the membrane surface [4].

2.4 Dye Removal Process

Photocatalytic process was carried out in a reactor with a 6 W UV lamp and wavelength of 245 nm. The schematic for doing this experiment was shown in Fig. 1. Sample were collected during at 30 time in different PH and its absorbance was calculated by spectrophotometer at the maximum wavelength of 570 nm. Dye removal percentage was calculated using Eq. (1).

$$Dye\,Removal\,(\%) = [(C0 - C)/C0] \tag{1}$$

Where, C0 and C are dye concentration before and after the treatment, respectively.

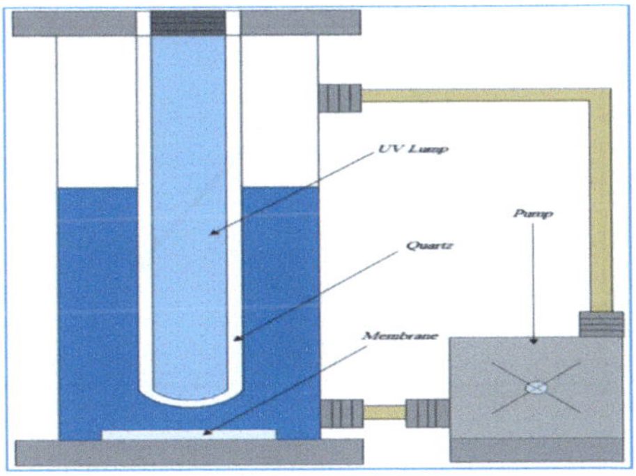

Fig. 1. Schematic representation of photocatalytic reactor

3 Results and Discussion

3.1 Morphology and Cross Section of Chitosan Treated Membranes

The morphology of chitosan treated SNT membranes was indicated in Fig. 2a. This image shows that uniform nanofibers have been formed. Figure 2b shows the cross section of the CSNT membrane. This images indicate that nylon/TiO2 nanofibers adhered well onto both surfaces of spacer fabric and wrapped all over the surface.

3.2 Effect of pH

These samples were applied for study about acid blue 113 (AB113) dye removal efficiency and the effect of pH on the amount of dye removal was investigated. Chitosan is a weak base and is insoluble in water and organic solvents. However, in acidic dilute solution, PH < 6.5, it can be converted into glucosamine units and $RNH^{+\ 3}$ ions can be formed in solution [5]. Also, at pH less than 6, the H^+ ion concentrations in solution increases, as a result, the amount of active sites for absorbing anionic dye is reduced [6]. According to Fig. 3, dye removal efficiency after passing 30 min. at pH = 3, 6, and 11 was about 46, 72, and 28%, respectively.

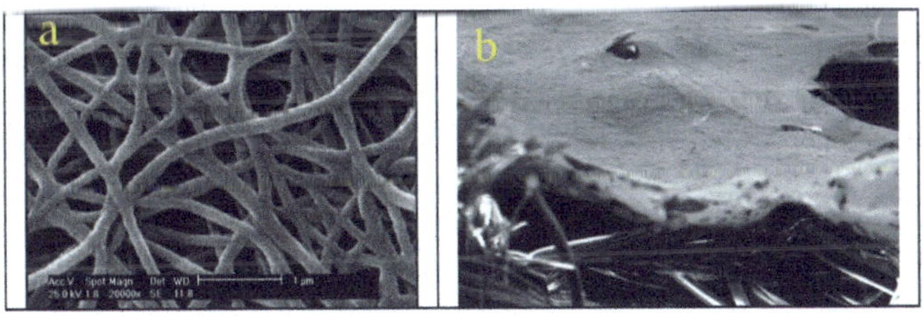

Fig. 2. SEM image: (a) Morphology and (b) Cross section of membrane

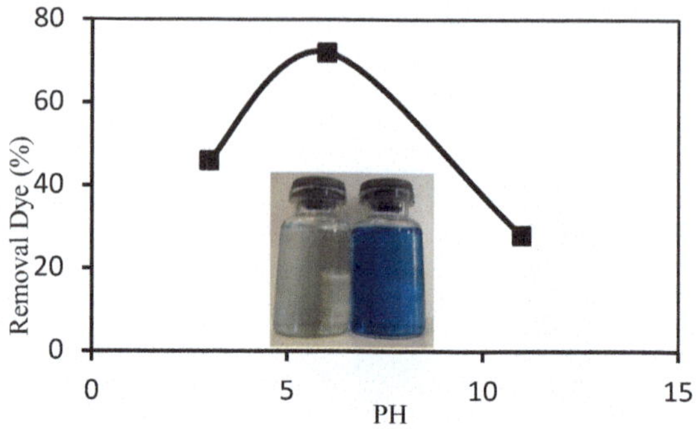

Fig. 3. Effect of pH on AB113 removal (%) (initial dye concentration: 10 mg/L, membrane size: 5 cm^2, 50 mL of dye solution)

4 Conclusion

In this paper, for the first time in literature, nylon/TiO_2 nanofibers were electrospun over and behind the polyethylene terephthalate spacer fabric and treated by chitason. After passing 30 min, CSNT membrane exhibited a dye removal about 73% at pH = 6 for AB113 dye.

References

1. Lou, T., et al.: Carbohyd. Polym. **170**, 182–189 (2017)
2. Shirzad-Siboni, M., et al.: J. Ind. Eng. Chem. **20**(4), 1432–1437 (2014)
3. Gopakumar, D.A., et al.: Acs Sustain Chem. Eng. **5**(2), 2026–2033 (2017)
4. Zhang, H.-T., et al.: Surface modification of electrospun nylon nanofiber based dye affinity membrane and its application to papain adsorption. In: 3rd International Conference on 3rd International Conference on Bioinformatics and Biomedical Engineering, ICBBE 2009. IEEE (2009)
5. Kaur, M., Verma, N.: J. Mater. Sci. Technol. **30**(4), 328–334 (2014)
6. Kumar, P.S., et al.: Desalination **261**(1–2), 52–60 (2010)
7. Lawrie, G., et al.: Biomacromolecules **8**(8), 2533–2541 (2007)

Preparation of Fe_3O_4/Polymethyl Methacrylate Composite Particles from Monolayer Oleic Acid-Modified Fe_3O_4

Roxana Moarref[1,2] and Saeed Pourmahdian[1,2(✉)]

[1] Polymer Engineering Department, Amirkabir University of Technology, Tehran, Iran
pourmahd@aut.ac.ir

[2] Laser and Plasma Institute, Shahid Beheshti University, Tehran, Iran

Abstract. Mono disperse magnetic nanocomposite particles with high saturation magnetization have wide applications in biomedical fields. In this article, we have studied the structure of Fe_3O_4 nanoparticles with polymethyl methacrylate shell. First of all Fe_3O_4 nanoparticles were synthesized by co-precipitation method and then they were modified by oleic acid in order to put the poly methyl methacrylate layer via mini emulsion method. The surface properties of OA modified magnetite nanoparticles have great effects on preparation of composite particles. The monolayer OA modified Fe_3O_4 nanoparticles are more hydrophobic than the multilayer coated ones and by multilayer OA coated Fe_3O_4, there was a large amount of free poly methyl methacrylates in the solution of product. After coating, the monolayer nanocomposite particles formed monodisperse nanoparticles as it was demonstrated by DLS tests. The PMMA/Fe_3O_4 nanoparticles have spherical core shell structure with a proper size distribution. The VSM tests have showed that the Fe_3O_4/PMMA nanoparticles have high magnetic power. High saturation magnetization demonstrates the super paramagnetic property.

Keywords: Mini-emulsion polymerization · Oleic acid · Nanocomposite particles · Monolayer oleic acid · PMMA/Fe_3O_4 nanocomposite particles

1 Introduction

Magnetic polymer coating nanoparticles have attracted a lot of attention due to their potential in biomedical fields. These nanocomposite particles have remarkable properties, they are superparamagnetic and they have high saturation magnetization, they have narrow size distribution and they are mono disperse.

In different applications, different size of nanocomposite particles is required. For most of the applications, nanocomposite particles should have no sedimentation, have nano size and narrow distribution, high super paramagnetic content and have no toxicity [1].

Magnetic Fe_3O_4 nanoparticles tend to aggregate and they are not chemically stable. They oxidize very easily and aggregate very quickly in liquid form. For making Fe_3O_4 nanoparticles stable and prevent them from aggregation scientists have studied many

H. Mirzadeh and A. A. Katbab (Eds.): ISPST 2018, *Eco-friendly and Smart Polymer Systems*, pp. 371–374, 2020.
https://doi.org/10.1007/978-3-030-45085-4_89

methods [2], in this article Fe_3O_4 nanoparticles are coated with polymethyl methacrylate shell. Mini-emulsion polymerization is the most efficient way to synthesis core shell nano spheres this is because of droplet nucleation of nano spheres [3].

First of all Fe_3O_4 nanoparticles was synthesized by co-precipitation method, in this method the size of nanoparticles is 30 nm. After that for depositing poly methyl methacrylate shell, we should first modify the surface of Fe_3O_4 by oleic acid. By this method the surface of hydrophilic nanoparticles changes to hydrophobic and it becomes compatible with the monomer. So the poly methyl methacrylate layer can be deposited on the modified core. The Fe_3O_4/PMMA composite particles with defined structure can be prepared from monolayer OA modified Fe_3O_4 nanoparticles. In the multilayer OA coated Fe_3O_4, large amounts of free polymethyl methacrylate monomer exist. The surface of monolayer OA modified Fe_3O_4, is more hydrophobic than the multilayer one [4]. The best way to deposit polymethyl methacrylate shell on the surface of nanoparticles is mini-emulsion method. In this method the ferrofluid was added to methyl methacrylate and form a micro-emulsion. Then sodium dodecyl sulfate (SDS) as emulsifier with cetyl alcohol (CA) as osmotic agent in water form another micro-emulsion. The first micro emulsion is added drop wise under ultrasonic treatment and polymerize at 80 °C with nitrogen inlet [5].

2 Experimental Materials

All of the chemicals were analytical grade. Water was deionized with double distillation and deoxygenated.

2.1 Preparation of Fe_3O_4 Nanoparticles and Surface Modification by Oleic Acid

Fe_3O_4 nanoparticles were synthesized by the chemical co-precipitation method. This synthesis was done under N_2 protection, 0.046 mol $FeCl_3$ was dissolved in 75 mL H_2O and 0.023 mol $FeCl_2$ was dissolved in 75 mL H_2O. Then the mixture was transferred to a four necked flask with vigorous stirring, 50 mL NH_3 (25%) was added drop wise to the solution. After 30 min the black Fe_3O_4 precipitated, in this step 3 mL oleic acid was added drop wise under vigorous stirring (400 rpm). The mixture was stirred until 2 phase separation. The whole synthesis last for 1 h and 30 min. After that the OA-coated Fe_3O_4 nanoparticles were separated by magnetic field and was washed 3 times by DI water and 1 time by ethanol for omitting excess oleic acid.

2.2 Preparation of PMMA/Fe_3O_4 Nanocomposite Particles

PMMA/Fe_3O_4 nanocomposite particles were prepared by mini-emulsion polymerization method. 10 mL of methyl methacrylate (MMA) was mixed with 0.5 g Divinyl benzene and 2 g of OA-coated Fe_3O_4 nanoparticles as the oil phase. SDS and CA with a molar ratio of 1:3 were dissolved in 100 mL of DI water as the water phase. Then they were mixed together with ultrasonic treatment to form an oil in water mini-emulsion and transferred to a 250 mL four necked flask equipped with a condenser, a nitrogen inlet, and a stirrer. The mixed solution was agitated for 30 min.

When the temperature increased to 80 °C, 1 g of benzoyl peroxide (BPO) was added to initiate the polymerization, which proceeded for 5 h at 80 °C. The resulting magnetic nanospheres were separated by the magnetic decantation and washed with DI water for several times.

3 Results and Discussion

3.1 OA Modified Fe_3O_4 Nanoparticles

The monodispersity of nanocomposite particles with mono layer OA is 0.1 which indicates homogeneity, but the monodispersity of nanocomposites with multilayer OA is 0.4. This result indicates that the surface properties of Fe_3O_4 nanoparticles are so important for depositing methyl methacrylate monomer on Fe_3O_4 nanoparticles and PDI index. The excess OA could be removed by purification with water [4].

The size of nanocomposite particle depends on the polymeric shell, and the surface of Fe_3O_4 nanoparticles, and it refers to the polymethyl methacrylate amount used during synthesis. As the ratio of Fe_3O_4 to PMMA was increased, the thickness of the PMMA shell w increased too [3].

3.2 Magnetic Properties

For evaluating the magnetic properties of nanocomposite particles, VSM test was used. This test demonstrated the super paramagnetism of nanocomposite spheres. From this test, it was showed that, both the Fe_3O_4 nanoparticles and Fe_3O_4/PMMA nanocomposite particles are super paramagnetic at room temperature because there is no hysteresis in the plot. And the magnetic remenance and coercivity are approximately zero. Another test was done for demonstrating magnetic sensitivity [3]. The bottle of two (monolayer and multilayer OA) brown suspension of Fe_3O_4/PMMA nanocomposite particles in aqueous solution was attached to external magnetic field. In the monolayer OA coated Fe_3O_4/PMMA suspension, the nanocomposite particles were attracted near the magnet when the external magnetic field was removed nanoparticles disperse again in water without aggregation. This test showed that the nanoparticles have super para magnetism and are sensitive to magnetic field.

4 Conclusion

The super para magnetic Fe_3O_4/PMMA nanocomposite particles were prepared from OA modified Fe_3O_4 via mini emulsion polymerization. The monolayer OA modified Fe_3O_4 nanoparticles are more hydrophobic than the multilayer coated ones and by multilayer OA coated Fe_3O_4, there was a large amount of free poly methyl methacrylates due to poor dispersion of Fe_3O_4 nanoparticles in MMA monomer. The excess OA could be removed by purification with water. After coating, the monolayer nanocomposite particles formed monodisperse nanoparticles as it was demonstrated by DLS tests.

References

1. He, L., Li, Z., Fu, J., He, N., Wang, Z., Wan, H., Shi, Z., et al.: Biomed. Nanotechnol. J. **5**, 596–601 (2009)
2. Liu, J., Qiao, S.J., Qing, H.H., Qing, G.: Small J. **4**, 425–443 (2011)
3. Lan, F., Liu, K., Jiang, W., Zeng, X., Wu, Y.: Nanotechnol. J. **22** (2011)
4. Yan, F., Li, J., Zhang, J., Liu, F., Yang, W.: Nanopart Res. J. **11**, 286–296 (2009)
5. Lei, H., Zhiyang, L., Jing, F., Fang, W., Chao, M., Yan, D., Zhiyang, S., et al.: Nanosci. Technol. J. **11**, 2256–2262 (2011)

A Silicone-Based Macro-Initiator for RAFT Polymerization

Saeed Hanifi, Farhid Farahmand(✉), and Mohammad Imani

Novel Drug Delivery Systems Department,
Iran Polymer and Petrochemical Institute, P.O. Box 14975/115, Tehran, Iran
f.farahmand@ippi.ac.ir

Abstract. Reversible addition-fragmentation chain transfer (RAFT) polymerization is an efficient technique for synthesis of polymers with well-defined architecture. Here, a macro-initiator based on dihydroxy-terminated polydimethylsiloxane (PDMS) and 4-Cyano-4-(phenylcar-bonothioylthio) pentanoic acid as a chain transfer agent (PDMS-macro CTA) were prepared via esterification reaction promoted in the presence of dicyclohexylcarbodiimide (DCC) coupling agent and 4-(dimethylamino) pyridine (DMAP) as a catalyst. PDMS-macro CTA was purified and characterized by FTIR spectroscopy. According to the FTIR findings, absence of hydroxyl functional groups, a broad peak in the region of 3340 cm^{-1} and appearance of carbonyl ester groups at 1743 cm^{-1} supported the performance of the esterification reaction.

Keywords: RAFT polymerization · PDMS · Esterification · DCC

1 Introduction

Recently, application of amphiphilic block copolymers containing of polydimethylsiloxane (PDMS) segments has attracted considerable attention for many applications, especially in biomedical and drug delivery systems. An amphiphilic copolymer consists of hydrophilic and hydrophobic segments and is capable to form nanostructures like micelles by self-assembling [1, 2]. The most researches of amphiphilic copolymers have been focused on polyethylene oxide (PEO) copolymers, while have not been more studies in silicone- based block copolymers [1]. More recently, the synthesis of amphiphilic copolymers has been extended using controlled/living radical polymerization [2–6]. Bas et al. were prepared tri-block copolymers of 2-hydroxyethyl methacrylate and PDMS by atom transfer radical polymerization (ATRP) [2]. Another amphiphilic diblock copolymer of poly(methacrylic acid-*co*-poly(ethylene oxide) methyl ether methacrylate) copolymer was successfully synthesized using RAFT polymerization by Zhang et al. [4]. Maiti et al. have been used the RAFT technique to copolymerize the fatty acid containing methacrylate monomers from bio-renewable resources [5]. Alejandra et al. were prepared PDMS and benzyl methacrylate (BzMA) amphiphilic diblock copolymers via RAFT Polymerization [6]. In this article, a Silicone-Based macro chain transfer agent (PDMS-macro CTA) as a macro-initiator was synthesized via an esterification reaction catalyzed by DCC/DMAP coupling agent

H. Mirzadeh and A. A. Katbab (Eds.): ISPST 2018, *Eco-friendly and Smart Polymer Systems*, pp. 375–378, 2020.
https://doi.org/10.1007/978-3-030-45085-4_90

for RAFT polymerization. Chemical structure of the PDMS-macro CTA and the reactants were characterized by FTIR spectroscopy.

2 Experimental Materials

Dihydroxy-terminated PDMS (TEGOMER H-Si2311, Mn = 2500 g.mol^{-1}, Evonik), N,N-dicyclohexylcarbodiimide (DCC, Aldrich), 4-(dimethylamino) pyridine (DMAP, Aldrich), and 4-Cyano-4-(phenylcar-bonothioylthio) pentanoic acid (RAFT agent, Aldrich) were used as received. Dichloromethane (DCM, Aldrich) was distilled in the presence of magnesium sulfate for drying.

PDMS macro-CTA was synthesized by adding the RAFT agent (0.557 g, 2.0 mmol) was added to DCM (50 mL) charged in a 100 mL, round-bottomed reactor and dissolved after 20 min stirring in 300 rpm at room temperature. PDMS (2.5 g, 1.0 mmol), DCC (0.619 g, 3.0 mmol), and DMAP (0.037 g, 0.30 mmol) were then added in order. Solution was purged with N_2 for 30 min, then sealed and heated up to 50 °C for 24 h under reflux condition with continuous stirring. PDMS-macro CTA was purified according to a procedure previously reported by Alejandra et al. [5]. A clear yellowish oily liquid remained after purification (yield = 89 wt%). The reaction is schematically shown in Fig. 1.

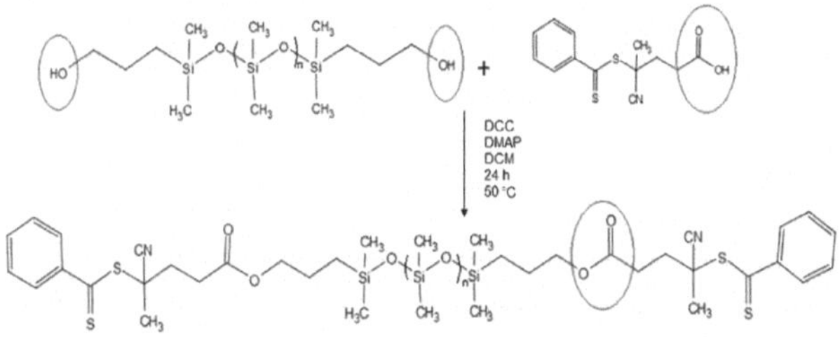

Fig. 1. Synthesis pathway of PDMS-macro CTA via DCC-activated esterification

Fourier transform infrared (FTIR) spectroscopy was used for determining of the chemical structure of the product. Spectra were obtained using a BRUKER Equinox 55 spectrometer.

3 Results and Discussion

PDMS-macro initiator was prepared from PDMS and RAFT agent by esterification reaction. Activation of carboxylic acid from RAFT agent by DCC in the presence of DMAP catalyst is a highly efficient performed when esterification reaction performed in non-aqueous solvents [5]. Figure 2 illustrates the FTIR spectra of the reactants and PDMS-macro CTA. In the spectrum of PDMS, strong peak at 1260 cm^{-1}, broad absorption at 1015-1093 cm^{-1}, strong peak at 798 cm^{-1}, broad peak in the region of

3340 cm^{-1} can be attributed to $Si\text{-}CH_3$ stretching vibrations, Si-O-Si asymmetric stretching vibrations, Si-C band stretching and O-H stretching vibrations, respectively. The strong sharp peak at 1704 cm^{-1} and broad peak in the region of 3447 cm^{-1} corresponding to carbonyl (C=O) and hydroxyl (O-H) functional groups of carboxylic acid, are shown in the RAFT agent spectrum. The spectrum of PDMS-macro CTA exhibited no peak for OH functional groups in the region of 3340 cm^{-1}, while a new peak appeared in 1743 cm^{-1} corresponding to -C=O groups. Disappearing the broad peak of hydroxyl groups and observation of carbonyl ester groups, indicates that esterification between PDMS and RAFT agent was performed successfully. Also, a characteristic peak at 1704 cm^{-1} for -C=O groups from carboxylic acid functionality in RAFT agent shifted to 1743 cm^{-1} for -C=O groups from ester functionality in PDMS-macro CTA, is another clue for confirmation of esterification reaction. Summarized the characteristic peaks for reactants and PDMS-macro CTA are shown in Table 1.

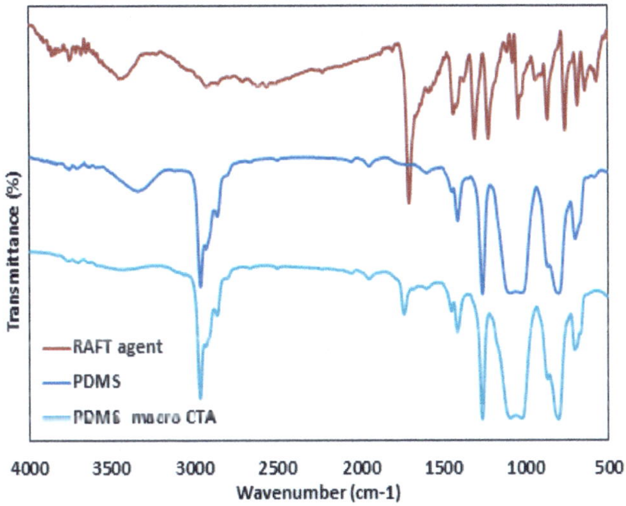

Fig. 2. FTIR spectra of the reactants and PDMS-macro CTA

Table 1. Characteristic peaks for the reactants and PDMS-macro CTA

Functionality	Wavenumber (cm^{-1})	Related to
Si-C	798	PDMS
		PDMS-macro CTA
Si-O	1015–1093	PDMS
		PDMS-macro CTA
$Si\text{-}CH_3$	1260	PDMS
		PDMS-macro CTA
C=O	1704	RAFT agent
	1743	PDMS-macro CTA
O-H	3200–3700	PDMS
		RAFT agent

4 Conclusion

In the present study, activation of PDMS with RAFT agent for RAFT polymerization has successfully been performed via DCC esterification promoting agent. Chemical structure confirmation was approved by FTIR analysis.

References

1. Quémener, D., Deratani, A., Lecommandoux, S.: Top Curr. Chem. **322**, 165–192 (2012)
2. Bas, S., Mark, D.S.: Polym. J. **44**, 1087–1097 (2012)
3. Moad, G., Rizzardo, E., Thang, S.H.: J. Chem. **58**, 379–410 (2005)
4. Zhang, X., Boisse, S., Zhang, W., Beaunier, P., Agosto, F., Rieger, J., Charlex, B.: Macromolecules **44**, 4149–4158 (2011)
5. Maiti, B., Priyadarsi, D.: RSC Adv. **3**, 24983–24990 (2013)
6. Alejandra, P., Nicholas, J., Arthi, R., Oleksandr, O.: Macromolecules **48**, 3547–3555 (2015)

Synthesis of Silicone-Based Amphiphilic Copolymer via RAFT Polymerization

Saeed Hanifi, Farhid Farahmand(✉), and Mohammad Imani

Novel Drug Delivery Systems Department, Iran Polymer and Petrochemical Institute, P.O. Box 14975/115, Tehran, Iran
f.farahmand@ippi.ac.ir

Abstract. Reversible addition-fragmentation chain transfer (RAFT) polymerization has been used to prepare amphiphilic silicone-based copolymer via copolymerization of 2-hydroxyethyl methacrylate (HEMA) at the end of activated polydimethylsiloxane (PDMS). PHEMA-b-PDMS-b-PHEMA copolymer synthesized using direct polymerization of PDMS-macro CTA (macro chain transfer agent) and HEMA monomer in the presence of AIBN as an initiator. After completion of reaction, the prepared copolymer was successfully purified. The chemical structure characterization was assessed by FTIR spectroscopy. According to the FTIR results, the appearance of new broad peak in the region of 3200–3700 cm^{-1} was attributed to the hydroxyl groups. Also presence of high-intensity sharp peak at 1743 cm^{-1} related to the carbonyl functionalities and PDMS finger print peaks in the product spectrum, showed that copolymerization reaction was successfully performed.

Keywords: Amphiphilic copolymer · RAFT polymerization · Self-assembly · PDMS · PHEMA

1 Introduction

Currently, silicone-based amphiphilic copolymers are widely used for various applications such as drug delivery systems, bio-medical applications, coatings, membranes and stabilizer for emulsions and colloids and dispersion polymerizations [1, 2]. An amphiphilic copolymer consisting of hydrophilic and hydrophobic segments, is enable to self-assembly and formation of nanoparticle micelles with spherical, worm-like and vesicles morphologies in the selective solvents [1–3]. Developments of controlled radical polymerization (CRP) techniques were used to successfully preparing of these amphiphilic copolymers. Reversible addition-fragmentation chain transfer (RAFT) polymerization, as one of CRP methods, is an efficiently technique for synthesis of well-defined architecture silicone-based copolymers [4, 5]. Only few researchers have proved that incorporation of poly(2-hydroxyethyl methacrylate) (PHEMA) to the end of polydimethylsiloxane (PDMS) chains, permits hydrophilic segments (PHEMA) in amphiphilic copolymer acts as a hydrating and a lubricating agent for biomedical applications [6]. Reported here is the copolymerization of previously prepared PDMS-macro CTA (PDMS-macro chain transfer agent) and HEMA monomer using RAFT polymerization. FTIR spectroscopy was utilized to chemical structure characterization.

H. Mirzadeh and A. A. Katbab (Eds.): ISPST 2018, *Eco-friendly and Smart Polymer Systems*, pp. 379–382, 2020.
https://doi.org/10.1007/978-3-030-45085-4_91

2 Experimental

Materials
Previously prepared PDMS-macro CTA (using DCC/DMAP catalyzed esterification of 4-Cyano-4-phenylcarbonothioylthio pentanoic acid (RAFT agent) and dihydroxy-terminated polydimethyl siloxane), 2-Hydroxyethyl methacrylate (HEMA, Sigma-Aldrich), Azobisisobutyronitrile (AIBN, Merck), 1,4-dioxane (Sigma-Aldrich), Toluene (Sigma-Aldrich), Methanol (Sigma-Aldrich), N-hexane (Sigma-Aldrich) were used in this study.

Synthesis of PHEMA-b-PDMS-b-PHEMA
PDMS triblock copolymer with PHEMA was synthesized using RAFT polymerization technique. Previously prepared PDMS-macro CTA (3.925 g, 1.283 mmol) and AIBN (0.3154 g, 1.922 mmol) were added to dried 100 mL round-bottomed reactor and dissolved in 50 mL pre-dried dioxane: toluene (1:1) after 30 min stirring at room temperature. The choice of this solvent mixture ensures the solubility of PHEMA and PDMS segments. HEMA monomer (15.685 g, 128.2 mmol) was then added gradually. The mixture was stirred continuously at 85 °C upon reflux system and N_2 purge during the polymerization. After 20 h the reaction was stopped. Solvent was removed in vacuum at 70 °C for 24 h. The obtained viscose product was washed three times, using 25 mL methanol and 25 mL n-hexane for elimination of un-reacted monomer and initiator. The product was washed three times by 25 mL distilled water to remove remaining methanol. Finally, n-hexane was removed in vacuum oven at 40 °C for 48 h. The yields of reaction were 83.2% by weight. PHEMA-b-PDMS-b-PHEMA was synthesized according to reaction scheme shown in Fig. 1.

Characterization
Fourier transform infrared (FTIR) spectroscopy (BRUKER Equinox 55 spectrometer) was used for monitoring the chemical changes.

3 Results and Discussion

Figure 2 shows the FTIR spectra of the reactants and the synthesized PHEMA-b-PDMS-b-PHEMA. In the spectrum of PDMS-macro CTA, Strong peak at 1260 cm^{-1}, broad absorption at 1015–1100 cm^{-1}, strong peak at 800 cm^{-1}, sharp peak in 1733 cm^{-1} are attributed to Si-CH_3 stretching, Si-O-Si asymmetric stretching vibrations, Si-C band stretching and-C=O stretching vibrations, respectively. The strong sharp peak at 1720 cm^{-1} and broad peak in the region of 3200–3700 cm^{-1} corresponding to carbonyl (-C=O) and hydroxyl (O-H) functional groups, are shown in the HEMA monomer spectrum. According to the PHEMA-b-PDMS-b-PHEMA spectrum, observed new broad peak in the region of 3200–3700 cm^{-1} and high intensity peak in 1743 cm^{-1} relate to the OH and -C=O functional groups, respectively. Also, other finger print peaks of PDMS are shown in copolymer spectrum. Observation of hydroxyl groups and finger print peaks of PDMS in the copolymer spectrum, indicates that copolymerization of PDMS-macro CTA and HEMA was successfully performed. The characteristic peaks for reactants and prepared copolymer are shown in Table 1.

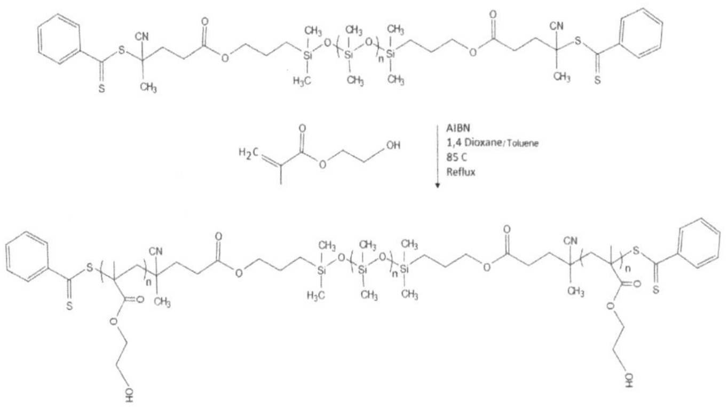

Fig. 1. Synthesis of PHEMA-b-PDMS-b-PHEMA via RAFT polymerization

Table 1. Characteristic peaks for reactants and copolymer

Functionality	Wavenumber (cm^{-1})	Related to
Si-C	800	PDMS-macro CTA
		Copolymer
Si-O	1015–1100	PDMS-macro CTA
		Copolymer
Si-CH_3	1260	PDMS-macro CTA
		Copolymer
C=O	1720	HEMA
	1733	PDMS-macro CTA
	1743	Copolymer
O-H	3200–3700	PDMS-macro CTA
		HEMA
		Copolymer

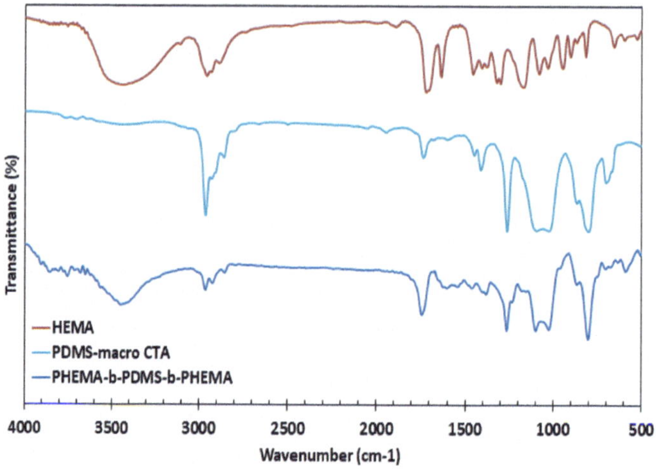

Fig. 2. FTIR spectra of the reactants and copolymer

4 Conclusion

Synthesis of PHEMA-b-PDMS-b-PHEMA triblock copolymers has been successfully achieved using RAFT polymerization technique. FTIR spectroscopy was used to analyze the chemical structure of triblock copolymer and reactants. According to the results, presence of hydroxyl broad peak, carbonyl ester sharp peak and PDMS finger print peaks in the copolymer spectrum, indicating that RAFT polymerization is a powerful technique in the preparation of siloxane-based block copolymers.

References

1. Alejandra, P., Nicholas, J., Arthi, R., Oleksandr, O.: Macromolecules **48**, 3547–3555 (2015)
2. Bas, S., Mark, D.S.: Polym. J. **44**, 1087–1097 (2012)
3. Zhang, X., Biosse, S., Zhang, W., Beaunier, P., Agosto, F., Rieger, J., Charlex, B.: Macromol. J. **44**, 4149–4158 (2011)
4. Maiti, B., Priyadarsi, D.: RSC Adv. **3**, 24983–24990 (2013)
5. Beers, K., Boo, S., Gaynor, S., Matyjasewski, K.: Macromolecules **32**, 5772–5776 (1999)
6. Vargun, E., Usanmaz, A.: Polym. Int. **59**, 1586–1597 (2010)

Toughening Mechanism of Polyetheramine-Modified Epoxy Resins

Hossein Abdollahi(✉), Ali Samadi, and Somayyeh Hosseini Rad

Department of Polymer Engineering, Faculty of Engineering, Urmia University, P.O. Box 5756151818-165, Urmia, Iran
h.abdollahi@urmia.ac.ir

Abstract. In this research polyetheramine (PEA)-modified epoxy was prepared and the toughening mechanism of the modified networks was studied. According to the mechanical test results, incorporation of PEA into epoxy networks can significantly toughen the epoxy matrix. The addition of 15 phr PEA increased the critical stress intensity factor (K_{IC}) and critical strain energy release rate (G_{IC}) of the epoxy from 0.85 MPa.m$^{1/2}$ to 1.86 MPa.m$^{1/2}$ and from 0.38 kJ.m^{-2} to 2.27 kJ.m^{-2}, respectively. The toughening and failure mechanisms were discussed based on microscopy of the crack tips. As a result, shear yielding as a possible toughening mechanism was proposed based on microscopical investigations. These in-depth studies could uncover underlying structure-property relationships in a relevant class of PEA-like modifiers, shedding light on the future design of top-performing homogeneous tough polymer networks.

Keywords: Polyetheramines · Epoxy · Toughening mechanism · Microscopy · Toughness

1 Introduction

Epoxy resins are a class of thermoset materials that have drawn much attention due to their excellent mechanical properties, high adhesiveness in addition to superior heat and chemical resistance. However, their high crosslinking density often leads to undesirable properties such as brittleness and low fracture toughness [1].

Several techniques introduced for improving toughness of the epoxies that could be classified in two categories of physical modification (*e.g.*, using liquid elastomers, etc.) and chemical modification of the epoxy structure (*e.g.*, using polyetheramines (PEAs), etc.) [2].

The PEAs have been used as modifiers and curing agents for epoxy resins and their influence on the epoxy properties has been investigated previously [3]. Despite the dependence of epoxy resin toughness on structural parameters of polyethers, a detailed study on the toughening mechanism of PEAs has not been investigated.

The aim of our study is to determine the efficiency of PEA for toughness improvement of a DGEBA-type epoxy resin and study the toughening mechanism of PEA-modified epoxies. For this propose, 15 phr of polyethylene glycol diamine (PEG-amine) was used as a PEA.

H. Mirzadeh and A. A. Katbab (Eds.): ISPST 2018, *Eco-friendly and Smart Polymer Systems*, pp. 383–386, 2020.
https://doi.org/10.1007/978-3-030-45085-4_92

2 Experimental

The DGEBA epoxy resin under the trade name EPON 828 from USA was used. The curing agent was diethylenetriamine (DETA) which was purchased from Sigma Aldrich. Polyethylene glycol (PEG) with the molecular weight 400 g/mol was obtained from Merck. The chemical reagents were benzene sulfonyl chloride (BSC, Merck), triethylamine (TEA, Merck), and aqueous ammonia (25%, Daejung). Polyetheramine (PEA) was synthesized and utilized for sample preparation according to the procedure that reported elsewhere [4].

Fracture toughness was carried out according to ASTM D5045 by a Santam STM 20 testing machine. Double-notched four-point bending (DN-4PB) technique was performed to investigate the mature plastic zone ahead of the crack tips as described in reference 2. The toughening mechanism was analyzed by a transmission optical microscope (TOM, Leica DMLS, Germany) from the area ahead of the unbroken sub-critical crack tip. The morphological properties of fractured surfaces were investigated by scanning electron microscopy (SEM, VEGA\\TESCAN) at an accelerating voltage of 15 kV.

3 Results and Discussion

The results of the fracture toughness test of the neat and PEA-modified epoxy networks are summarized in Table 1.

As seen in Table 1, the fracture toughness of the epoxy resin clearly changes with introduction of the PEA in the epoxy network. A great strategy to develop a tough material is to introduce relatively short and long polymer chains in polymer networks. Under deformation, the short-chain networks can be fractured to dissipate mechanical energy while the long-chain networks will maintain the elasticity of the networks [2]. Therefore, if PEA and DETA are considered as long and short chains, respectively, the long chains are disentangled with the application of stress and thus a higher amount of strain is exhibited. For explicit specification, a schematic illustration of the cross-linked networks of the neat and PEA-modified epoxy systems are shown in Fig. 1.

Figure 2 shows the TOM and SEM micrographs of the neat and PEA-modified epoxies after DN-4PB technique. Figures 2b and 2c attribute to the TOM images of subcritical crack tip of DN-4PB samples of the neat and modified epoxies, respectively. As observed, no plastic deformation zone at the crack tip of neat epoxy could be find, whereas the PEA-modified sample shows the yielding region in the damage zone. When PEA was incorporated in the epoxy network, the damage zone is comprised of two zones, an outer larger relatively pale zone surrounding an inner smaller dark zone closer to the crack tip. Besides, Figs. 2d and 2e illustrate the SEM micrographs of the fractured area of the mentioned samples that confirm the TOM observations. A mirror-like surface was detected for the unmodified sample (Fig. 2d) that reflect the brittle fracture mechanisms. The sample with PEA, however, shows a slightly rough surface (Fig. 2e) which is due to the plastic deformation.

On the other hand, during the initial loading step, elastic deformation displaces some epoxy segments from their original positions, causing a small local volume expansion, and decreasing resistance to further deformation. The outer larger relatively pale zone in TOM image may be attributed to this phenomenon. The strains in these fewer resistant regions increase progressively, hence producing shear bands, and producing elastic strain concentrations at the edges of the bands. Due to their high plastic strains, shear bands are strongly birefringent, and thereby easy to see in an optical microscope, which obviously seen as an inner smaller dark zone closer to the crack tip in Fig. 2c. These appearances could be attributed to shear yielding mechanism [5]. As a result, it could be said that shear yielding mechanism is a possible mechanism in both multi-phase and mono-phase toughened systems.

Table 1. Mechanical properties of the PEA-modified epoxy networks

Sample	K_{IC} (MPa.m$^{1/2}$)	G_{IC} (kJ.m^{-2})
Neat epoxy	0.85 ± 0.10	0.38 ± 0.08
PEA-modified epoxy	1.86 ± 0.08	2.27 ± 0.20

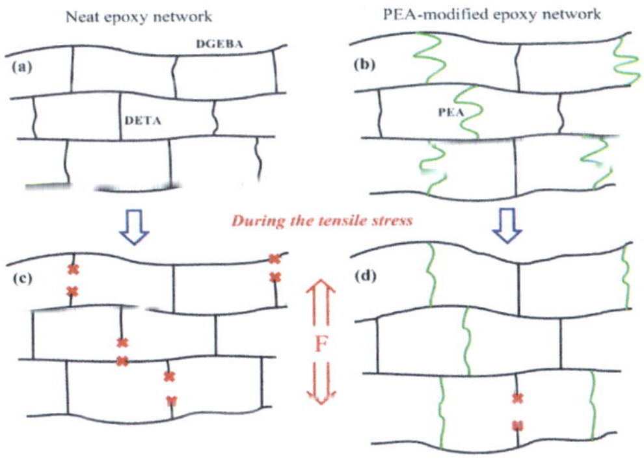

Fig. 1. Schematic representation of epoxy network structures of: (a) unmodified, (b) PEA-modified epoxy systems before load, (c) unmodified, and (d) PEA-modified epoxy systems during load

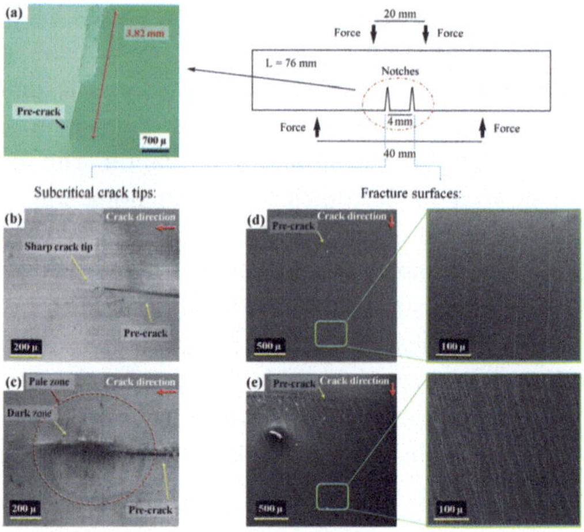

Fig. 2. Schematic representation of DN-4PB test to analysis the toughening mechanism. TOM micrographs of: (a) initial crack tip of the DN-4PB sample, (b) subcritical crack tip of unmodified, and (c) subcritical crack tip of PEA-modified epoxy systems. SEM micrographs of fractured surface for (d) unmodified, and (e) PEA-modified epoxy systems

4 Conclusion

In this work, a PEA was used to toughen the epoxy resin. The fracture toughness results show PEA is an effective approach for toughening of brittle epoxy resins. From TOM observations a shear yielding mechanism was proposed to explain the fracture results. This work opens up the possibility for the better interpretation of relationship between microstructure and properties of epoxy systems treated with polyetheramines and various structural parameters.

References

1. Abdollahi, H., Salimi, A., Barikani, M.: J. Appl. Polym. Sci. **133**, 44061 (2016)
2. Abdollahi, H., Salimi, A., Barikani, M., Samadi, A., Hosseini Rad, S., Zanjani, A.R.: J. Appl. Polym. Sci. (2018, Accepted)
3. Yang, G., Fu, S.: J. Polym. **48**, 302 (2007)
4. Abdollahi, H., Salimi, A., Barikani, M., Zeynizadeh, B.: Mater. Manuf. Process **32**, 1296 (2017)
5. Bucknall, C.B.: J. Microsc. **201**, 221 (2001)

Investigation of Encapsulation of 1-Methylimidazole Latent Curing Agent Using Solid Epoxy Resin

Sayed Morteza Mozaffari[1], Mohammad Hosain Beheshty[2(✉)], and Sayed Mojtaba Mirabedini[3]

[1] Faculty of Engineering and Technology, University of Mazandaran, Babolsar, Iran
[2] Department of Composites, Iran Polymer and Petrochemical Institute, P.O. Box 14975-112, Tehran, Iran
m.beheshty@ippi.ac.ir
[3] Department of Coatings, Iran Polymer and Petrochemical Institute, P.O. Box 14975-112, Tehran, Iran

Abstract. Encapsulation of 1-methylimidazole (1-MI) was investigated to prepare the latent 1-MI curing agent by using solid epoxy resin as the wall materials. The morphology of 1-MI/solid epoxy microcapsules was investigated by SEM. The thermal behavior of the prepared microcapsule was investigated using differential scanning calorimetry (DSC) and thermal gravimetric analysis (TGA). According to the TGA and DSC results, the weight loss of 1-MI/solid epoxy microcapsules although not showing the actual value of the encapsulated 1-MI, but it can be thought-out as the weight of 1-MI that participates in the reaction with the matrix liquid epoxy resin.

Keywords: Microcapsule · Solid epoxy · 1-methylimidazole · TGA · DSC

1 Introduction

As one of the most significant classes of high performance thermosetting polymers, epoxy resins have been widely used as adhesives, protective coatings, composite matrices, electronic-packaging mold compounds, insulating and embedding materials, and sealants. This is because of their high performance such as good mechanical and electrical properties, strong adhesion to various substrates, high chemical and corrosion resistance, low cure shrinkage and formulation variety [1].

Imidazoles are one of the generic curing agents for epoxy resins. It is also known that materials cured by imidazoles have excellent mechanical properties and high glass-transition temperatures. Curability of imidazoles can be pretty controlled by introducing exchange on the imidazole ring; however, because of their high reactivity, it is totally difficult to prepare epoxy resin–imidazole compositions with high storage stability. This inflicts a limitation for their use in one-component industrial materials [2].

To overcome this case, latent curing agents have been developed to produce imidazoles in recent years. Various kinds of latent curing agents have been reported,

H. Mirzadeh and A. A. Katbab (Eds.): ISPST 2018, *Eco-friendly and Smart Polymer Systems*, pp. 387–390, 2020.
https://doi.org/10.1007/978-3-030-45085-4_93

such as imidazole carboxylate, epoxy-imidazole adducts, metal cation–imidazole salt complexes, and imidazole reacted with acidic substances. Among the methods of forming non-reactive imidazoles, the encapsulation of the imidazole is a promising and economic method [3, 4]. The application of solid epoxy resin for encapsulation of 1-methylimidazole (1-MI) was reported in our previous study [5].

In this study, DSC and TGA for the thermal analysis of microcapsule with solid epoxy shells and 1-MI core was used and their results are discussed.

2 Experimental

Materials

1-methylimidazole (DY 070, Huntsman, Germany) and solid epoxy resin (NPES-903, Nan Ya epoxy resin, Taiwan) were used as the core and shell materials, respectively. The sodium dodecyl sulfate (SDS) surfactant and acetone solvent were obtained from Aldrich.

Microencapsulation Procedure

The aqueous phase containing the emulsifier (200 mg dissolved SDS in 200 mL distillated water) was firstly prepared. 7 g 1-methylimidazole and 3 g solid epoxy resin, with the ratio 30/70 of core/shell, were dissolved in acetone (15 mL) and at a rate of approximately one drop per second gently added to the continuous phase (1 wt% SDS in an aqueous solution) under mechanical stirring (RZR 2102, Heidolph) at 500 rpm to get an O/W emulsion. Then, the solvent was evaporated in an open vessel with continuous stirring at room temperature for 4 h and microcapsules containing the core material (1-MI) were obtained. The prepared microcapsules were then separated from the solution with a filter paper, rinsed with deionized water a few times and dried in a low pressure oven for 3 h at 25 °C.

Equipment

Thermal behavior of the microcapsules were evaluated using a DSC 1 METTLER TOLEDO differential scanning calorimeter at a heating rate of 10 °C/min, with a sample weight of about 15 mg. DSC scans of the microcapsules were monitored from 10 to 300 °C.

TGA was used to study the thermal stability of the microcapsules and to determine the core content (1-MI) of microcapsules. TGA was carried out using TGA/DSC 1 (Stare system METTLER TOLEDO equipment) at heating rate of 20 °C/min under N_2 atmosphere from 25 °C to 600 °C.

3 Results and Discussion

Figure 1 shows the DSC and TGA curves of microcapsule with a solid epoxy wall containing 1-MI. According to the figure, the DSC curve of microcapsule is divided into three sections.

In section (a), plateau region is not observed in the DSC curve. Since the wall of this microcapsule is solid epoxy and solid epoxy can react with the nucleus (1-MI), this suggests that by heating the microcapsule, the microcapsule wall melts and at the same time it participates in the reaction with capsulated 1-MI. In fact, there is a competition between the melting of solid epoxy that absorbs heat and the ring-opening polymerization of solid epoxy that generates heat; but the amount of heat absorption in the melting stage is high and lower heat is generated, at the first. On the other hand, according to the DSC tests of microcapsules it can be derived that however the solid epoxy shell was in contact with imidazole curing agent for long time, it had melting point. Maybe anionic polymerization [6] increased the chain length of solid epoxy resin.

When the melting of solid epoxy is completed; without occurring the evaporation of the 1-MI from the microcapsule (as TAG curve shows), the imidazole of core reacts with solid epoxy and an exothermic peak is observed (in section (b)). So in the presence of a layer of the solid epoxy that its chain length is increased, 1-MI cannot escape from the microcapsule. On the other hand, all the epoxy functional groups of the microcapsule's shell do not participate in the curing reaction with the core curing agent at room temperature. When the wall melts, the 1-MI curing agent crosslinks the solid epoxy.

After this exothermic peak in the DSC curve, in section (c), there is an endothermic peak that can be related to the capsulated 1-MI that is evaporated. Therefore, according to the TGA curve, the weight loss in the 1-MI/solid epoxy microcapsules can be considered as the weight of 1-MI encapsulated in microcapsule that participates in the reaction with the matrix liquid epoxy resin.

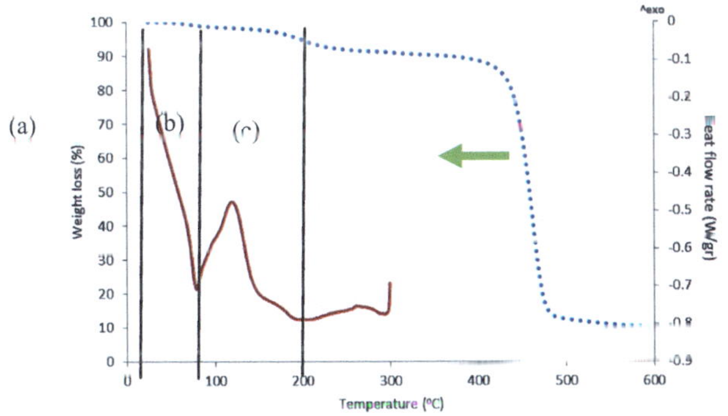

Fig. 1. DSC and TGA curves of microcapsule (solid epoxy/1-MI = 3/7)

4 Conclusion

The DSC curve of microcapsule is divided into three sections: Melting-reaction, reaction, and evaporation. At the first, there is a competition between the melting of solid epoxy and the polymerization; but the amount of heat absorption in the melting stage is high and lower heat is generated. Also, the weight loss of microcapsule obtained from the TGA test not showing the actual value of the encapsulated 1-MI, but it shows the amount of 1-MI that can react with the liquid epoxy of matrix.

References

1. Wu, F., Zhou, X., Yu, X.: RSC Adv. **8**, 8248–8258 (2018)
2. Kudo, K., Furutani, M., Arimitsu, K.: J. Polym. Sci. Part A: Polym. Chem. **56**, 471–474 (2018)
3. Xing, S., Yang, J., Huang, Y., Zheng, Q., Zeng, J.: Mater. Des. **85**, 661–670 (2015)
4. Tripathi, M., Dwivedi, R., Kumar, D., Roy, P.K.: Polym. Plast. Tech. Eng. **55**, 129–137 (2016)
5. Mozaffari, S.M., Beheshty, M.H., Mirabedini, S.M.: Iran. Polym. J. **26**, 629–637 (2017)
6. Hart, K.R., Sottos, N.R., White, S.R.: Polymer **67**, 174–184 (2015)

Using the Rheo-Dielectric Method in Analyzing the Effect of System Dynamics on the Dispersion State of CB/PS Nanocomposites

Mona Janipour and Mohammad Hossein Navid Famili(✉)

Chemical Engineering Faculty, Tarbiat Modares University, Tehran, Iran
nfamili@modares.ac.ir

Abstract. Simultaneous dielectric and dynamically mechanical investigations have a tremendous number of impressive accomplishments on microscopic properties of polymeric systems containing dipoles, specially nanocomposites. So far, a huge amount of research has been carried out to analysis such properties of nanocomposites by using dielectric and rheo-dielectric method. In this research we traced the percolation threshold and the changes in dielectric properties of nanocomposites of CB/PS through rheo-dielectric tests. We observed the remarkable differences of dielectric properties and also percolation threshold between before and after of imposing the oscillatory shears on CB/PS nanocomposites with three different volume fractions of nanoparticles.

Keywords: Dielectric · Nanocomposite · Rheo-dielectric · Percolation threshold

1 Introduction

Recently a considerable amount of research is being carried out on nanocomposites. The merit of nanocomposites in comparison of microcomposites is its higher specific area which causes in better stounding properties even at low volume fraction of nanoparticles [1–3]. Analyzing the rheological and dielectric measurement simultaneously is a technical method to survive the dynamic and viscoelastic behaviors of polymeric systems [4, 5]. This technique that is called the rheo-dielectric method is applicable to each system which has permanent or induced dipoles [6].

The aim of this research is to analysis the effect of oscillatory shear sweep tests on percolation threshold of nanocomposites of CB/PS that would be traced through dielectric properties and compared with the results of nanocomposites which have the amount of nanoparticles below the percolation threshold.

H. Mirzadeh and A. A. Katbab (Eds.): ISPST 2018, *Eco-friendly and Smart Polymer Systems*, pp. 391–393, 2020.
https://doi.org/10.1007/978-3-030-45085-4_94

2 Experimental

Materials and Sample Preparation
The nanoparticles of Carbon Black with the average size of 25 nm was purchased from Nanosany Company and the polystyrene was purchased from Tabriz Petrochemical Company. The nanocomposites at three volume fractions and alcohol as anti-solvent were mixed by a mechanical stirrer and an ultra-sonic homogenizer for distribution and dispersion, respectively.

3 Methods

Dielectric measurements have been carried out at the frequencie of 1 MHz. For dielectric measurements we were to instruct a dielectric analyzer and we have done it by combining a function generator and oscilloscope through wires and an electrical circuit. A parallel plates rheometer with the diameter of 5 cm was used. In this research we carried out the rheological measurements through oscillatory shear at the range between 64% to 450% and at the constant amount of shear rate means 1 rad/s.

4 Results and Discussion

Figure 1, shows the effect of shear domain on the CB/PS nanocomposites at three various volume fractions. Increasing in shear domain leads to reduction of dielectric constant except of the third volume fraction means 17%. By imposing the first shear on the 17% of volume fraction sample, it first faced with a jumping to a higher amount for dielectric constant. This phenomenon is because of breaking the agglomerations and increasing in the particle-matrix interactions that is called interfacial polarization and decreasing in electrical conduction that this exists because of the conductive nanoparticles of carbon black. Another imposed shear domain leads to the reduction of dielectric constant because of decreasing the nanoparticles interactions. As it is seen the 5% volume fraction sample has no obvious changes by growing the shear amplitude while in contrast a huge amount of variations could be seen at two other samples, it is concluded that the 10 and 17% volume fraction nanocomposites have amount of nanoparticles at percolation and over percolation threshold respectively. At the maximum amount of shear domain, it can be observed that dielectric constant at three percentage of volume fraction got close together and the 10 and 17% volume fraction samples changed to the samples with amount of nanoparticles under the percolation threshold.

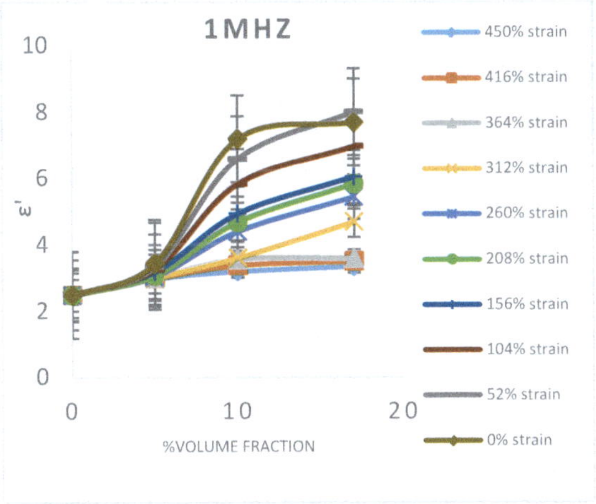

Fig. 1. Dependence of dielectric constant to percentage of volume fraction at different shear amplitudes

5 Conclusion

It has been demonstrated that agglomeration of nanocomposites could break to smaller one and reduce the percolation threshold. The samples which have amount of nanoparticles at percolation threshold or more, after the strong imposed shear domain behave like samples that are under the percolation threshold.

Acknowledgement. Financial support from Tarbiat Modares University for Scientific and Industrial Research is greatly acknowledged.

References

1. Camargo, P.H.C., Satyanarayana, K.G., Wypych, F.: Mater. Res. **12**(1), 1–39 (2009)
2. Hussain, F., et al.: J. Compos. Mater. **40**(17), 1511–1575 (2006)
3. Islam, M.S., Masoodi, R., Rostami, H.: J. Nanosci. (2013)
4. Steinhauser, D., Möwes, M., Klüppel, M.: J. Phys. Condens. Mat. **28**(49), 495103 (2016)
5. Pakula, T., et al.: Rheol. Acta **35**(6), 631–644 (1996)
6. Watanabe, H., Matsumiya, Y., Inoue, T.: J. Phys. Condens. Mat. **15**(11), S909 (2003)

Analyzing and Comparing the Effect of System Dynamics on Agglomerations of CB/PS and TiO_2/PS Nanocomposites Through Rheological Dielectric Method

Mona Janipour and Mohammad Hossein Navid Famili(✉)

Chemical Engineering Faculty, Tarbiat Modares University, Tehran, Iran
nfamili@modares.ac.ir

Abstract. Various particles containing dipoles have different dielectric properties which would be applicable in varied industries like electronics, aerospace, food and etc. Carrying out dielectric and rheological experiments called "Rheo-dielectric" would result in dramatic achievement in macro and micro structures of polymeric composites. In this study we used TiO_2 (as a ceramic material) and CB (as a conductive material) nanoparticles in PS (which is neutral) as a matrix to investigate the dispersion and agglomeration of each kind of nanocomposite and then compare them together. The dielectric experiments have been done at the frequency of 100 Hz and a sweep of shear strain has been imposed to the nanocomposites at the constant shear rate of 1 rad/s. We observed the breaking down of agglomerations, resulting in changes in dielectric data which were different in each nanocomposite.

Keywords: Dielectric · Ceramic nanoparticle · Conductive nanoparticle · Rheo-dielectric · Agglomeration · Break down

1 Introduction

In this present decade, the applications of nanocomposites filled with dipolar nanoparticles have increased at the industry remarkably and a wide range of research about their specific properties has been carried out [1, 2]. Dipoles according to their particular inherent property which could be ceramic, conductive or magnetic or the combination of each of them would produce various properties [3, 4]. Dielectric experiments are one the bests which could trace the unique feature of these materials [5]. We could use the dielectric tests to find out the filler-filler and polymer-filler interactions [6]. Carbon black as a highly conductive particle and titanium oxide as a good ceramic material are used for instructing the nanocomposites with the aims of conductivity and dielectric respectively [7].

The goal of this research is the discovering of the effect of rheological experiments on dielectric results and agglomerations in rheological dielectric tests.

H. Mirzadeh and A. A. Katbab (Eds.): ISPST 2018, *Eco-friendly and Smart Polymer Systems*, pp. 394–397, 2020.
https://doi.org/10.1007/978-3-030-45085-4_95

2 Experimental

Materials and Sample Preparation

The nanoparticles of Carbon Black and TiO_2 with the same average size of 25 nm was purchased from Nanosany Company and the polystyrene was purchased from Tabriz Petrochemical Company. The nanocomposites at three volume fractions and alcohol as anti-solvent were mixed through a mechanical stirrer and an ultra-sonic homogenizer for distribution and dispersion respectively.

3 Methods

A parallel plate rheometer with the diameter of 5 cm was used. In this study we carried out the rheological measurements through oscillatory shear at the range between 64% to 450% and at the constant shear rate means 1 rad/s. Dielectric measurements have been carried out at the frequencies of 100 Hz. For dielectric measurements we were to instruct a dielectric analyzer and we have done it by combining a function generator and oscilloscope through wires and an electrical circuit.

4 Results and Discussion

Figures 1 and 2, show the dielectric constants of CB/PS and TiO_2 nanocomposites at various strain domains in three volume fractions. It could be seen that for CB/PS samples when there is no strain domain, the dielectric constant of 10% volume fraction is much more than the dielectric constant of 17% volume fraction. By imposing the prime strain domain, we see an abrupt jump in the dielectric constant amount of 17% volume fraction sample which even gets more than the 10% volume fraction sample while in contrast, the dielectric constant of 10% volume fraction decreases. This phenomenon is because of breaking down of the agglomerations which of the 17% volume fraction sample are much bigger than the 10% volume fraction's that filler-filler interactions overweigh the polymer-filler interactions and through breaking the agglomerations of 17% the polymer-filler interactions increase and electrical conduction decreases while for the 10% volume fraction sample by fracturing the agglomerations the filler-filler interaction decrease. While for TiO_2/PS the results of this step are much different and the dielectric constant of 17 and 10% volume fraction samples decrease that is because of the reduction of filler filler interactions. The reason of growing of dielectric constant with increase in volume fraction is more polymer-filler interaction through higher amount of nanoparticles. Another shear strain domain for both CB/PS and TiO_2/PS nanocomposites leads to the reduction of dielectric constant at both volume fractions which is because of more breaking down of agglomerations and decrease in the filler-filler interactions and finally the amount of dielectric constant of both volume fraction samples arrive the amount of 5% volume fraction's which means that the nanoparticles of 17 and 10% volume fraction nanocomposite.

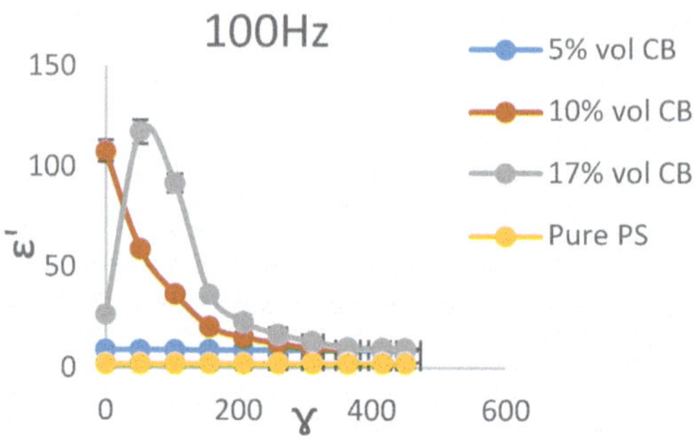

Fig. 1. Dependence of dielectric constant to shear strain amplitude at different volume fractions

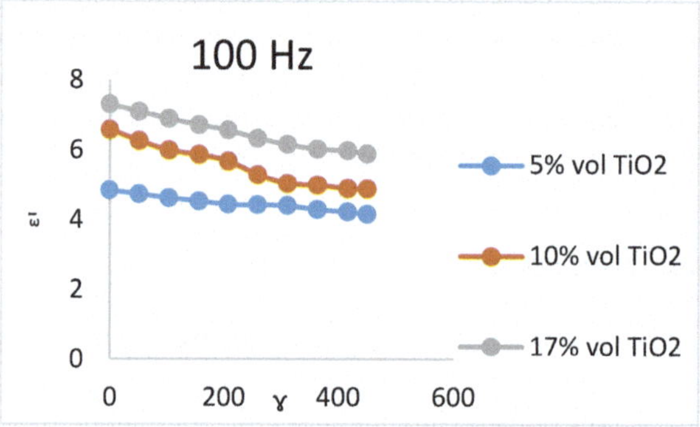

Fig. 2. Dependence of dielectric constant to shear strain amplitude at different volume fractions

5 Conclusion

It has been demonstrated that the curve of 17% volume fraction CB/PS nanocomposites was a bell curve that is because of conductivity of CB nanoparticles while in contrast, this phenomenon was not seen for TiO_2/PS nanocomposites at any volume fractions. Another reason for this phenomenon is that in the CB/PS nanocomposite in this volume fraction the filler-filler interactions are much more than the polymer-filler interactions while in the TiO_2/PS nanocomposites this does not apply.

Acknowledgement. Financial support from Tarbiat Modares University for Scientific and Industrial Research is greatly acknowledged.

References

1. Watanabe, H., Ishida, S., Matsumiya, Y.: Macromolecules **35**, 8802 (2002)
2. Yamamoto, R., Onuki, A.: Phys. Rev. E **70**, 041–801 (2004)
3. Onogi, S., Matsumoto, T.: Polym. Eng. Rev. **1**, 45 (1981)
4. Tirado-Mirande, M., Schmitt, A.: Prog. Colloid. Polym. Sci. **104**, 138 (1997)
5. Jona, F., Shirane, G.: Perg Press, New York (1962)
6. Merz, W.J.: Phys. Rev. **76**, 1221 (1949)
7. Ponomarenko, A.T., Shevchenko, V.G., Enikolopyan, N.S.: Adv. Polym. Sci. **96**, 125 (1990)

Improving Polypropylene Sediment Filter by Nano Surface Coating to Reduce Biomass Infection

Farzaneh Ghorab[1,2](✉) and Zarrin Es'haghi[1]

[1] Department of Chemistry, Payame Noor University, 19395-4697, Mashhad, Iran
farzaneh1980@yahoo.com
[2] Nuclear Science and Technology Research Institute, The Photonic and Quantum Technology Research School, Tehran, Iran

Abstract. Polypropylene sediment filter which is made from multi-layers of tissues with fine pore of polypropylene yarn and high filtering efficiency that can remove dirt, rust, dust, silt, algae and the other, could not prevent bacterial growth and remove harmful bacteria and microorganisms unless being coated by nano silver coating. So, physical vapor deposition, as a special technique to coat a thin layer of nano silver with an average of arbitrary thickness on the filter surface, was used. It improved the polypropylene sediment filter application with 40 nm thickness of silver coating in specific temperature (up to ten degree more than 25 °C), evaporation rate (0.1 nm/s) and vacuum condition (3×10^{-5} Torr). As a result, both the filter pore size reduction and anti-microbial nano silver thin layer caused harmful bacteria and biomass affection to decline.

Keywords: Polypropylene sediment filter · Biomass effects · Nano silver coating · P.V.D coating technique

1 Introduction

Polypropylene filters, which can be located in the pure water supply systems, may support surface harmful bacteria growth and the micro-organisms aggregation. Microbial communities and micro-organisms can adhere to surfaces because of their nanostructure characteristics. The cells aggregation constitutes serious problems in various processes and can cause pollution, corrosion, heat and mass transfer limitations in industrial processes. Moreover, water filters have a significant influence on the bacterial community composition in the post filtration water [1]. A huge percent of micro-organisms live in microbial communities. The nanostructure of micro-organisms such as bacteria [2], viruses and yeast are attractive scaffolds for biological molecules and cells gathering. Microbial cells or micro-organisms as the matrix of exo-polymeric substances (EPS) play important roles in various environments. Biomass growth and bio-fouling within pure water systems present potentially serious problems [3]. Protection of surfaces can be accomplished by different methods like adding a non-adhesive or antimicrobial material on the surface of various things by coating. Using physical vapor deposition (P.V.D)

H. Mirzadeh and A. A. Katbab (Eds.): ISPST 2018, *Eco-friendly and Smart Polymer Systems*, pp. 398–401, 2020.
https://doi.org/10.1007/978-3-030-45085-4_96

coating [4] method is useful to reduce the surface harmful bacteria growth and aggregation of microorganisms on the filter surface. So pure water and water distribution systems as the microbial infections sources can be safe by using this fabricated filters which is located before the water supply system. In this proposal the fabrication of nano silver [5] filter is explained by using the electron beam gun in P.V.D machine for decreasing microbial infections.

2 Experimental

The fine pore polypropylene type of filter was used for coating process. Nearly 2 grams of pure silver (99.99%), which is known as a safe, naturally anti-bacterial, and anti-viral material was put in special crucible. The BALZERS 760 as the P.V.D (Physical Vapor Deposition) coating machine with an electron beam gun ESQ 110 was used for fabrication of the nano thin films in specific temperature with 0.1 nm/s evaporation rate. Three types of coated filters with 25, 40, and 90 nm of silver thicknesses were prepared by programing the coating system. The average of coated layer thicknesses was revealed by the microprocessor unit of the Balzers machine during the coating process. The beam gun, as the P.V.D machine heat energy source, causes partially ionized vapor atoms of silver that leave the crucible to fabricate a nano thin film on the filter, in specific conditions of temperature (up to ten degree more than 25 °C), evaporation rate (0.1 nm/s) and vacuum condition (3×10^{-5} Torr). The effect of polypropylene filters (which were coated with three various thicknesses of nano silver) on pure water was compared with a raw filter effect, after staying all of them in passing pure water. Then the experiment was completed by bacterial testing on pure water samples. Finally, Scanning Electron Microscopy (SEM) technique helped to compare the polypropylene filters images in arbitrary magnitudes.

3 Results and Discussion

The results of bacterial testing on pure water samples show the nano silver coated effects successfully. The reduction of CFU was observed in three types of coated polypropylene filters with 25, 40, and 90 nm of silver thicknesses. The microbial growth results of coated filters with 25 and 90 nm thicknesses were more than 500 colonies for unit per milliliter (CFU/mL). But the result of the 40 nm thickness filter was less than 100 CFU/mL and completely acceptable (Table 1). Bacterial tests were accomplished on TSA (Tryptose Soy Agar) during 48 h in 25 °C temperature to compare the growth of colonies in treated pure water by raw filter and the others. Comparing between raw filter (a) and 40 nm silver coated filter (b) shows the microbial growth in cultured water sample (a) (Fig. 1). Scanning Electron Microscopy (SEM) technique used to characterize the filter surfaces before and after nano silver coating by comparing the polypropylene filters images (Fig. 2).

Table 1. Bacterial test on pure water samples to compare the reduction of CFU/mL according to different silver thicknesses on PP filters

Silver thickness on filter surface (nm)	Bacterial test condition in pour plate	Detection limit	Bacterial growth in pure water (CFU/mL)
Zero (raw filter)	(25 °C/48 h)	<1 cfu/mL	26850
25 and 90	(25 °C/48 h)	<1 cfu/mL	>500
40	(25 °C/48 h)	<1 cfu/mL	<100

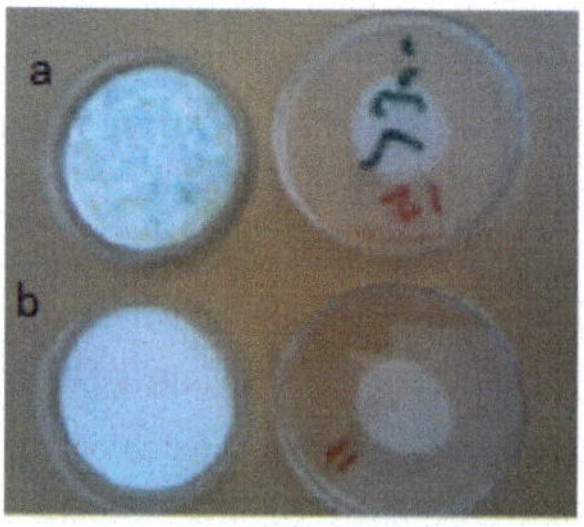

Fig. 1. Comparing the growth of colonies on raw filter: (a) and Silver 40 nm coated filter (b)

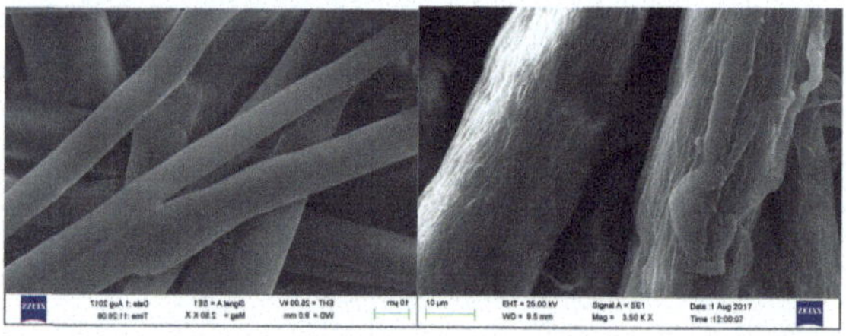

Fig. 2. Scanning electron microscopy (SEM) image of raw filter (left picture) with treated filter (right picture)

4 Conclusion

Study on coated polypropylene filter by P.V.D method show that the best thickness of nano silver to decrease the infections of nano-biomass in pure water was in 40 nm. Nano silver coated filter was useful to reduce the surface harmful bacteria growth and aggregation of microorganisms on the filter surface. So the growth of bacterial populations on filter surface, in addition of filter pore size, can be controlled by using modified polypropylene filter. It means that the biomass affection will decrease by improving the polypropylene sediment filter.

References

1. Vignola, M., Werner, D., Wade, M.J., Meynet, P., Davenport, R.: Wat. Res. **129**, 499–508 (2018)
2. Hochbaum, A.I., Aizenberg, J.: Nano Lett. **10**(9), 3717–3721 (2010)
3. Zhao, G., Chen, W.N.: Roy. Soc. Chem. (RSC) Adv. **7**, 37990 (2017)
4. Heidarpour, F., Ghorab, F., Torabian, A.: First International Conference on "Advances in Wastewater Treatment and Reuse" 10–12 November 2009. University of Tehran, Tehran, Iran (2009)
5. Varshney, R., Bhadauria, S., Gaur, M.S.: Nano Biomed. Eng. **4**(2), A Review (2012)

Microwave Curing of Epoxy Resin by Embedding Core-Shell Structure Particles

Sara Akbarpour Esfahlani[1], Jalil Morshedian[1(✉)], and Habibollah Baharvand[2]

[1] Department of Polymer Processing, Iran Polymer and Petrochemical Institute, P.O. Box 14975-112, Tehran, Iran
j.morshedian@ippi.ac.ir
[2] Department of Polymer Science, Iran Polymer and Petrochemical Institute, P.O. Box 14975-112, Tehran, Iran

Abstract. Magnetite particles with a polymeric shell were synthesized and used for heat generation and curing of an epoxy resin. The synthesized particles were characterized using several analysis methods. Samples containing particles, epoxy resin, and curing agent were cured by microwave and thermal oven and were then investigated. DSC measurements showed an increase in degree of conversion and T_g by increasing the microwave irradiation time. Using high contents of particles led to improved mechanical properties. The identified chemical groups of cured samples through FTIR analysis approved the occurrence of curing reactions.

Keywords: Magnetite particles · Epoxy resin · Microwave curing

1 Introduction

Recently, material processing by microwave has been utilized in applications like polymerization, synthesis of materials and curing of thermoset polymers and rubbers [1]. Among these applications, due to the ability to heat the sample directly through specific interactions with materials, microwave process has many advantages over the conventional heating methods. Neither proper curing nor the optimization of properties is readily feasible using conventional curing processes. Hence, the final properties of epoxy parts undergo losses. One of the reasons for inappropriate curing is non-uniformity in the curing of the resin due to the low heat transfer during the curing process that causes non-uniformity of properties in the whole part [2]. Curing the epoxy resin using microwave can be a favorable solution to get over the problem [3]. Good efficiency of microwave heating requires the components inside the material to be heated by absorbance of the microwave [4].

In this study, core-shell structure magnetite particles were used to increase the microwave absorption for curing of epoxy resin. Also, the curing behavior, and mechanical properties were investigated.

H. Mirzadeh and A. A. Katbab (Eds.): ISPST 2018, *Eco-friendly and Smart Polymer Systems*, pp. 402–405, 2020.
https://doi.org/10.1007/978-3-030-45085-4_97

2 Experimental

First, magnetite particles with polymeric shell were synthesized through the encapsulation method. Then the mixture of epoxy resin, curing agent and capsules were cured by microwave and thermal oven.

3 Results and Discussion

The synthesized particles were investigated by FTIR, TGA, and SEM. SEM micrographs of the synthesized particles are demonstrated in Fig. 1. This graph showed that the magnetite particles were surrounded by a rough spherical skin that formed due to deposition of the polymer at the surface of the magnetite particles. The identified chemical groups from FTIR spectrum and obtained weight losses in TGA thermo grams of synthesized particles confirmed the existence of polymeric phase around the magnetite particles.

DSC measurements were applied on the samples cured by a microwave oven for various irradiation times. The results of DSC analysis are illustrated in Table 1.

According to Table 1, the T_g and the degree of conversion values were increased by increasing the microwave irradiation time. Also, curing kinetics of cured samples were studied by common prediction models, and good accordance was observed between experimental data and models [5].

To investigate the mechanical properties, tensile test was applied on microwave cured specimens for different irradiation times. The results showed the enhancement of mechanical properties by increasing the microwave irradiation time. The microwave cured specimens for short irradiation times had a more ductile behavior than the long time cured samples due to the incomplete crosslinking of epoxy. Besides, mechanical properties of samples cured by both conventional and microwave methods were compared by tensile test. The stress-strain curves are shown in Fig. 2. Mechanical properties of microwave cured samples were higher than conventional cured sample. Increasing the content of particles led to improved mechanical properties. The curing behavior of microwave cured epoxy resin was studied by determining chemical groups through FTIR analysis. Appearance of ether peak and reducing of epoxy peak area with increasing the microwave irradiation time was an implication of the occurrence of curing process and crosslinking [6].

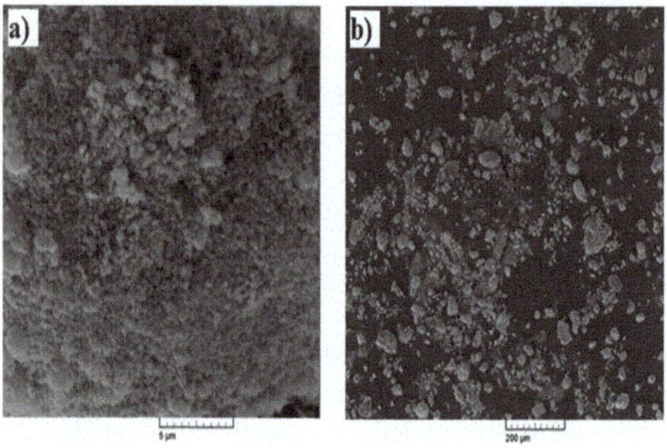

Fig. 1. SEM micrographs of synthesized particles at: (a) high magnification and (b) low magnification

Table 1. Results of DSC analysis of microwave cured samples for various irradiation times

Irradiation time (min)	T_g (°C)	Degree of conversion (%)
5	31	51
10	37	61
15	42	73
20	43	85
25	45	89
30	53	97
35	56	96
40	78	98

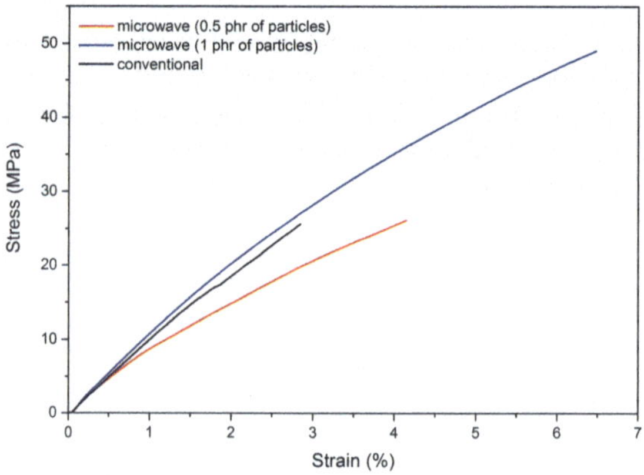

Fig. 2. Stress-strain curves of conventional and microwave cured samples with two different particle contents (0.5 and 1 phr)

4 Conclusion

Core-shell structured particles as microwave absorbers were successfully synthesized and utilized for curing of epoxy resin. Thermal and mechanical analyses showed that crosslinking epoxy resin could be achieved quickly using the microwave method and led to high degrees of conversion and T_g values accompanied with superior mechanical properties.

References

1. Jacob, J., Chia, L.H.L., Boey, F.Y.C.: J. Mater. Sci. **30**, 5321–5327 (1995)
2. Joshi, S.C., Bhudoli, S.K.: J. Compos. Mater. **48**, 3035–3048 (2014)
3. Fotiou, I., Baltopoulos, A., Vavouliotis, A., Kostopoulos, V.: J. Appl. Polym. Sci. **129**, 2754–2764 (2013)
4. Menendez, J.A., Arenillas, A., Fidalgo, B., Fernandez, Y., Zubizarreta, L., Calvo, E.G., Bermudez, J.M.: Fuel Process. Technol. **91**, 1–8 (2010)
5. Boey, F.Y.C., Qiang, W.: Polymer **41**, 2081–2094 (2000)
6. Johnston, K., Pavuluri, S.K., Leonard, M.T., Desmulliez, M.P.Y., Arrighi, V.: Thermochim. Acta **616**, 100–109 (2015)

Enhancing In-plane Mechanical Properties of Carbon/Epoxy Composite Using Poly(methyl methacrylate) Nanofibers

Rasoul Esmaeely Neisiany[1], Saied Nouri Khorasani[1], Jeremy Kong Yoong Lee[1], Alireza Safdari[1](✉), and Seeram Ramakrishna[2]

[1] Department of Chemical Engineering, Isfahan University of Technology, 84156-83111 Isfahan, Iran
a.safdari@ce.iut.ac.ir
[2] Centre for Nanofibers and Nanotechnology, Department of Mechanical Engineering, National University of Singapore, Singapore 117575, Singapore

Abstract. In this study, poly(methyl methacrylate) (PMMA) nanofibers were employed to enhance the in-plane mechanical properties of a conventional carbon/epoxy composite. The optimized PMMA nanofibers were straightly deposited on the conventional carbon fiber fabric surfaces via the electrospinning process. Vacuum assist resin transfer molding (VARTM) was used for the preparation of the carbon/epoxy composite. Evaluation of in-plane mechanical properties of the composites revealed that incorporation of PMMA nanofibers between carbon layers led to 14% and 8% improvement in tensile strength and tensile strain at break, respectively in comparison with the results obtained for the control composite. T-test analysis was also carried out and confirmed that incorporation of PMMA nanofibers significantly increased the in-plane mechanical properties of the composite. Finally, fractographical analysis of the composites cross-sections, investigated by field emission scanning electron microscope (FE-SEM), revealed that incorporating PMMA nanofibers in the hybrid composite led to a tougher fracture in resin rich-area.

Keywords: Electrospinning · Carbon/epoxy composite · Nanofibers · Mechanical properties · Poly(methyl methacrylate)

1 Introduction

Carbon fiber reinforced composites are widely employed in various applications such as aerospace and automotive [1]. Because of the laminate nature, resin rich areas, between the reinforcement layers, are weak areas leading to initiates of fractures such as: delamination and buckling. Delamination is one of the most ordinary failure modes for composite laminates structures. During the years many novel techniques have been proposed and developed to improve mechanical properties of resin-rich area and to prevent the crack propagation of the cracks. Recently, by comprehensively using the benefits of the nanofibers, innovative delamination prevention was used to enhance the

H. Mirzadeh and A. A. Katbab (Eds.): ISPST 2018, *Eco-friendly and Smart Polymer Systems*, pp. 406–409, 2020.
https://doi.org/10.1007/978-3-030-45085-4_98

resin rich layers properties [2–4]. The toughening mechanisms of using nanofibers at the interfaces of polymer matrix composites are attributed to the improvement of the interlaminar fracture toughness of polymer matrix composites and suppression of the singular stresses near laminate free-edges since the entangled nanofibers at interfaces. In addition, nanofibers breakage and nanofibers pull-out are the other mechanisms for stress dissipation and mechanical properties improvement [2].

In this research, PMMA nanofibers were utilized to enhance the in-plane properties of carbon fiber fabric-epoxy composite.

2 Experimental

Unidirectional carbon fibers 300 g/m^2 were purchased from Shanghai Horse Construction, China. PMMA (M_w = 350 kDa) and N,N-dimethyl formamide (DMF) was provided by Sigma-Aldrich. The resin employed in this study was epoxy (EPON™ Resin 828) and its hardener (EPIKURE™ Curing Agent F205) was obtained from Hexion. PMMA nanofibers were deposited on the surfaces of carbon fiber fabric which rolled on the cylindrical collector. To provide stable electrospinning, a DC voltage of 16 kV was applied. Solution flow rate was set at 1 mL/h, and the distance between needle and collector was adjusted at 15 cm. The angular velocity of the roller was set at 50 rpm.

The VARTM technique was employed to produce epoxy composites reinforced with electrospun nanofibers mats. The mass ratio of the epoxy resin and its curing agent was set at 100/58 according to manufacturer's datasheet.

The morphology of the electrospun nanofiber and fractured surfaces of the composite coupons were investigated using a Hitachi S-4300 FE-SEM. Tensile, tests were conducted according to standards ASTM D3039. The results were compared with the control composite, which published in our previous work [3].

3 Results and Discussion

Figure 1 presents the FE-SEM micrograph of the prepared PMMA nanofibers. It can be discerned that the nanofibers were smooth without any signs of beads at the electrospinning parameters. The average of nanofibers' diameter is 254 ± 42 nm.

Table 1 summarizes the tensile property results of reinforced composite with embedding PMMA nanofibers in the interlayers of carbon fibers and control composite. It can be observed that tensile strength and elongation to break of the composite increased 14 and 8%, respectively after embedding nanofibers at the interlayers of unidirectional carbon fiber layers. This phenomenon is due to the nanofibers breakage and nanofibers pull-out which lead to energy dissipation and consequently, increasing tensile properties of the composite. Furthermore, the statistical results confirmed that incorporation of PMMA nanofibers significantly increased the tensile properties of the composites.

Table 1. Tensile properties of the composites

	Tensile strength (MPa)	Strain to break (%)
Control composite [3]	15.09 ± 1.37	0.87 ± 0.08
Reinforced composite	17.19 ± 0.85	0.94 ± 0.08

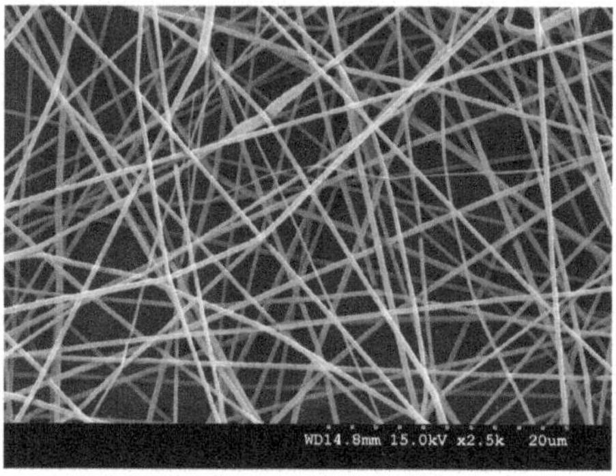

Fig. 1. FE-SEM micrograph of the PMMA nanofibers

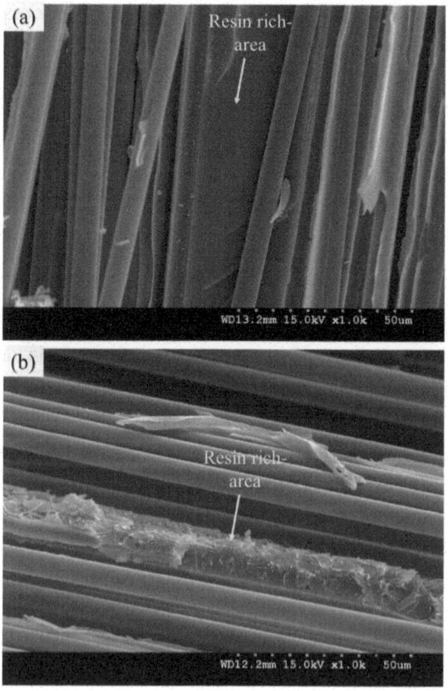

Fig. 2. FE-SEM images of fractured surfaces of: (a) control composite and (b) composite containing PMMA nanofibers

Figures 2a and 2b shows the fractured surfaces of control composite and composite containing PMMA nanofibers after tensile tests. It is obviously clear that incorporation of the PMMA nanofiber at interlayers makes resin rich layer tougher, therefore more surfaces created during the fracture.

4 Conclusion

Mechanical studies and fractographical studies of fracture surfaces confirmed that using PMMA nanofibers improved the in-plane properties of carbon/epoxy composite. In should be mentioned that the improvement did not change the fabrication method. Furthermore, this novel method does not any adverse effect such as increasing cost or weight.

References

1. Neisiany, R.E., Khorasani, S.N., Naeimirad, M., Lee, J.K.Y., Ramakrishna, S.: Macromol. Mater. Eng. **302**, 1600551 (2017)
2. Wu, X.F., Yarin, A.L.: J. Appl. Polymer Sci. **130**, 2225 (2013)
3. Neisiany, R.E., Khorasani, S.N., Lee, J.K.Y., Naeimirad, M., Ramakrishna, S.: Theor. Appl. Fract. Mech. **95**, 242 (2018)
4. Razavi, S.M.J., Neisiany, R.E., Khorasani, S.N., Ramakrishna, S., Berto, F.: Theor. Appl. Mech. Lett. **8**, 126 (2018)

Preparation and Characterization of Novel Polyureas Containing Amide Units with High Thermal Stability

Fatemeh Sadat Siadatian-Haghighi[1] and Shahram Mehdipour-Ataei[2(✉)]

[1] Department of Chemistry, Science and Research Branch, Islamic Azad University, Tehran, Iran
[2] Iran Polymer and Petrochemical Institute, P.O. Box 14965/115, Tehran, Iran
s.mehdipour@ippi.ac.ir

Abstract. A novel diamine with built-in ether and amide structure was prepared via two-step reactions. Nucleophilic reaction of 4,4′-oxydianiline with 6-chloronicotinoyl chloride led to preparation of N,N′-(oxybis(1,4-phenylene))bis (6-chloronicotinamide) (OPCN). OPCN was reacted with 4-aminophenol to provide a novel diamine (EADA). After complete characterization, the prepared diamine (EADA) was used to prepare related polyureas by reaction with different diisocyanates. The obtained poly(ether-amide- urea)s were characterized and their properties were studied. Improved thermal stability was observed for the prepared polyureas in comparison to the common ones. This modification extends application of these polyureas in various industrial purposes.

Keywords: Diamine · Polyuria · Thermal properties · Synthesis · Polycondensation

1 Introduction

Polyureas may be classified as heterochain macromolecular compounds, which contain urea groups in their structure. Polyureas are one of the most important engineering polymers that have many different practical applications in industry as elastomers, adhesives, spray foams, microcellular foams, liquid plastics, encapsulation of pharmaceuticals, inks, dyes, and also in the preparation of coating, varnishes, and lacquers. Polyureas were developed because of their good mechanical properties, hydrolytic stability, chemical resistance, and optical properties [1–3].

A number of methods have been used to prepare polyureas, however the best method is reaction of diamines with diisocyanates in polar aprotic solvents such as N-methyl-2-pyrrolidone (NMP) at room temperature. This paper describes the preparation and properties of novel poly(ether urea)s with improved thermal stability containing amide units through condensation polymerization of a new diamine with different diisocyanates.

H. Mirzadeh and A. A. Katbab (Eds.): ISPST 2018, *Eco-friendly and Smart Polymer Systems*, pp. 410–413, 2020.
https://doi.org/10.1007/978-3-030-45085-4_99

2 Experimental

2.1 Monomer Synthesis

Synthesis of N,N′-(oxybis(1,4-phenylene))bis(6-chloro- nicotinamide) (OPCN): A round-bottom flask was charged with 0.005 mol of oxydianiline. 20 mL of DMAc was added to the flask and it was maintained at 0 °C for 30 min. Then, 7 mL of triethylamine and 0.01 mol of 6-chloronicotinoyl chloride were added to the mixture and the temperature was raised to ambient temperature and kept for 24 h. Then, it was precipitated in water. The product was filtered, washed with hot water and methanol. It was dried in an oven at 60 °C.

Synthesis of diamine: N,N′-(oxybis(1,4-phenylene))bis(6-(4-aminophenoxy)nicotinamide) (EADA): A round-bottom flask was charged with 0.001 mol of OPCN, (0.004 mol) of 4-amino- phenol, 6 mL of NMP, and 4 mL of dry toluene. Then 0.008 mol of K_2CO_3 was added to the mixture and the reaction mixture was heated to 140 °C for 6 h with continuous stirring. The reaction temperature was raised to 165 °C and kept at the same temperature for 21 h. Then, it was precipitated in water. The mixture was filtered and washed with hot water and methanol.

2.2 Polymer Synthesis

A round-bottom flask equipped with an N_2 inlet tube, a condenser, and a thermometer was charged with 0.001 mol of diamine and 3 mL of dry DMAc. Then, a solution containing 0.001 mol of diisocyanate in 2 mL of dry DMAc was added in one lot to the flask contents. The solution was stirred for 1 h at room temperature and for 2 h at 60 °C, and then it was precipitated in water. The polymer was filtered, washed with hot water and methanol, and dried in a vacuum oven at 100 °C.

3 Results and Discussion

At first, 4,4′-oxydianiline was reacted with 6-chloronicotinoyl chloride. Subsequent reaction of the obtained product with 4-aminophenol led to the preparation of a novel diamine (EADA). Spectroscopic methods were applied for the characterization of the diamine (Table 1) and the diamine was polycondensed with toluene diisocyanate (TDI), diphenylmethane diisocyanate (MDI), and isophrone diisocyanate (IPDI) to obtain related polyureas. Characteristic bands in the FT-IR spectra of the polymers were observed at 3361-3401 (N-H), 2901-2972 (C-H), 1643-1655 (C=O), 1489-1490 (C=C), and 1217-1220 (C-O) cm^{-1} (Fig. 1). The inherent viscosity of the polymers at a concentration of 0.5 g/dL in DMAc at 30 °C that was in the range of 0.64-0.71 dL/g showed suitable molecular weights. Thermal stability and behavior of the polymers was evaluated using TGA and DSC methods in air at a heating rate of 10 °C/min. Polymers revealed high thermal stability according to the derived data in Table 2. A representative TGA thermogram is shown in Fig. 2. The glass transition temperature of the polymers was in the range of 161–185 °C.

Table 1. Characterization of the diamine

Substrate	IR, KBr (cm^{-1})	NMR (DMSO-d_6, δ, ppm)
$C_{36}H_{28}N_6O_5$ mp: 237 °C Yield: 97%	3284(N-H amine), 3364(N-H amide), 1650(C=O amide), 1200,1251(C-O ether)	10.31(N-H, 2H, amide) 8.70 (C-H, 2H, pyridine) 8.28 (C-H, 2H, pyridine) 7.75 (C-H, 4H, phenyl) 7.03 (C-H, 2H, pyridine) 6.96 (C-H, 4H, phenyl) 6.82 (C-H, 4H, phenyl) 6.59 (C-H, 4H, phenyl) 5.06 (N-H, 4H, amine)

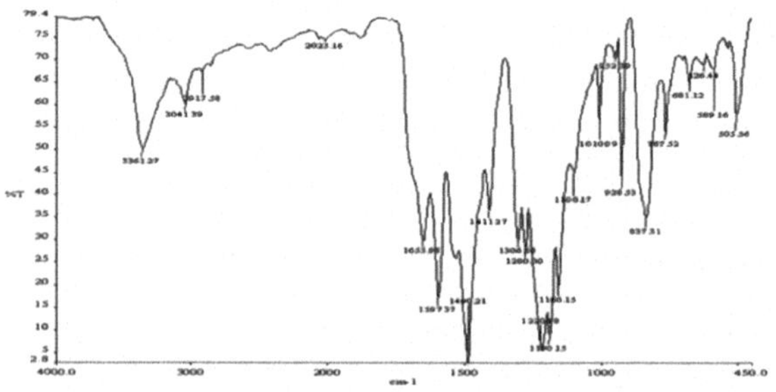

Fig. 1. FT-IR spectrum of EADA-TDI polyuria

Table 2. Thermal analysis data

Polymer	T_g (°C)	T_0 (°C)	T_{10} (°C)	T_{max} (°C)	Char yield at 600 °C (%)
EADA-MDI	185	295	370	475	48
EADA-TDI	170	285	360	460	45
EADA-IPDI	161	260	325	420	39

T_g: glass transition temperature; T_0: initial decomposition temperature; T_{max}: maximum decomposition temperature; T_{10}: temperature for 10% weight loss; char yield: weight of polymer remained

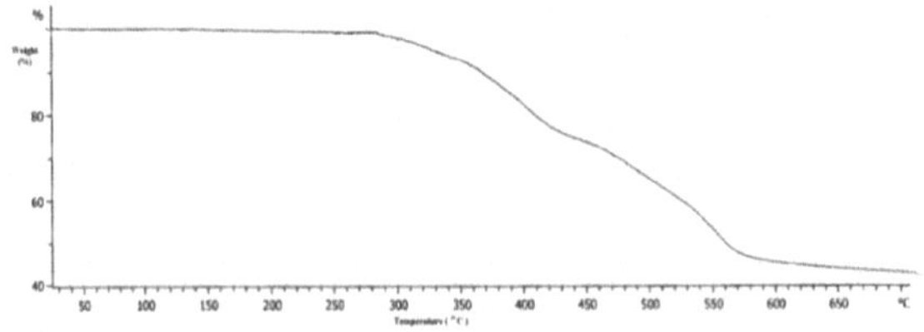

Fig. 2. TGA curve of EADA-TDI polyurea

4 Conclusion

New polyureas containing amide and ether units showed unique properties including high thermal stability and improved solubility in comparison to common polyureas. Design of the aromatic diamine with no weak linkages containing amide units induced thermal stability to the final polymers, while presence of ether, and pyridine groups improved the solubility and processability of them.

References

1. Allen, S.G., Bevington, J.C.: Comprehensive Polymer Science, 1st edn, p. 427. Pergamon Press, Oxford (1989)
2. Tamami, B., Koohmareh, G.A.: Des. Monomers Polym. **10**, 221–233 (2007)
3. Mehdipour-Ataei, S., Akbarian-Feizi, L.: E-Polymers **126**, 1–11 (2010)

Synthesis and Properties of Novel Pyridine-Based Polyamides with Improved Solubility

Milad Lotfnejad-Moghimi[1], Amir Ahmad Latifi[1], and Shahram Mehdipour-Ataei[2(✉)]

[1] Department of Chemistry, Science and Research, Islamic Azad University, Tehran, Iran
[2] Iran Polymer and Petrochemical Institute, P.O. Box 14965/115, Tehran, Iran
s.mehdipour@ippi.ac.ir

Abstract. A novel aromatic diamine containing pyridine unit was prepared via two-step reactions. Nucleophilic substitution of oxydianiline with 6-chloronicotinoyl chloride led to preparation of a dichloro-diamide compound. Subsequent reaction of this compound with 3-aminophenol resulted in synthesis of a particular diamine. After complete characterization, polycondensation reaction of the diamine with terephthaloyl chloride and isophthaloyl chloride led to preparation of related polyamides. The polymers were characterized and their physical and thermal properties were studied.

Keywords: Diamine · Polyamide · Thermal properties · Synthesis · Polycondensation

1 Introduction

Thermally stable polymers have received extensive interest due to the increasing demands for high temperature polymers as replacements for ceramics and metals in the automative, aerospace and microelectronics industries. Aromatic polyamides generally exhibit outstanding mechanical properties and have excellent thermal and oxidative stability [1, 2]. To extend the utility of these high performance materials, it has been suggested to improve their flexibility and solubility [3]. In this article we wish to describe the synthesis and characterization of a new diamine based on 6-chloronicotinoyl chloride, and then preparation of related aromatic polyamides therefrom with high solubility.

2 Experimental

2.1 Monomer Synthesis

Synthesis of Dichloro-Diamide Compound

Into a round-bottomed flask was placed 1.0012 g (0.005 mol) of oxydianiline. Then 20 mL of DMAc was added to the flask and it was maintained at 0 °C for 30 min with continuous stirring under inert nitrogen gas. Then 7 mL of triethyl amine was added to

H. Mirzadeh and A. A. Katbab (Eds.): ISPST 2018, *Eco-friendly and Smart Polymer Systems*, pp. 414–417, 2020.
https://doi.org/10.1007/978-3-030-45085-4_100

the mixture and stirred at the same temperature for a few minutes. A total of 1.814 g (0.01 mol) of 6-chloro nicotinoyl chloride was added to the mixture and the temperature was raised to ambient temperature and kept for 8 h. The precipitated product in water was washed with hot water and methanol, repeatedly.

Synthesis of Diamine-Diamide Compound

Into a round-bottomed flask equipped with a Dean-Stark trap, a condenser, a nitrogen inlet tube, a thermometer, an oil bath, and a magnetic stirrer were placed with 1.589 g of dichloro compound, 0.914 g of 3-aminophenol, 12.5 mL of NMP and 8 mL of dry toluene. Then, 2.43 g (0.017 mol) of K_2CO_3 was added to the mixture and the reaction mixture was heated to 140 °C for 6 h with continuous stirring. The reaction temperature was raised to 165 °C and kept at the same temperature for 21 h. The resulting reaction mixture was cooled and poured into water.

2.2 Polymer Synthesis

Synthesis of polyamides: Into a round-bottomed flask equipped with a nitrogen inlet tube, a drying tube, 0.5 g (1 mol) of diamine and 10 mL of DMAc were stirred for 30 min at 0 °C. Then, about 1.2 mL of pyridine was added an after a few minutes 0.1676 g (1 mol) of terephthaloyl chloride was added. This mixture was then stirred first at 0 °C for 0.5 h and then at room temperature for 6 h. The polyamide product was precipitated by pouring the contents of the flask into water. It then was filtered, and dried overnight.

3 Results and Discussion

In order to obtain a new category of polyamides with good solubility and processability while maintaining heat resistance, a new dichloro-diamide was firstly prepared. The presence of amide linkage in this structure was confirmed by existence of a band at 1663 cm^{-1} in FT-IR attributed to the amide carbonyl bonds. Also, appearance of a peak at δ 10.60 in ^{1}H-NMR was assigned to the amide N–H bonds. In the second step of diamine synthesis (Scheme 1 and 2), a band related to C–O ether group at 1228 cm^{-1} in FT-IR in addition to a peak related to amine bonds at δ 5.33 in ^{1}H-NMR was observed. This diamine was polycondensed with TPC and IPC in order to prepare pyridine-based polyamides. The polymers were fully characterized using FT-IR, 1H-NMR, and elemental analysis (Fig. 1).

The inherent viscosities of the polyamides measured at a concentration of 0.5 g/dL in DMAc at 30 °C was about 0.72–0.75 dL/g, indicating relative high molecular weight.

The thermal stabilities of the polyamides were evaluated by TGA and DSC analysis. In the TGA analysis the polyamide started to lose weight, because of thermal degradation at about 300 °C. The 10% polymer weight loss, as a criterion for the thermal stability of the polyamide, was in the range of 400–420 °C indicating that the prepared polymer had good thermal stability. The obtained results are tabulated in Table 1. DSC analysis for TPC-derived polyamide is shown in Fig. 2. The glass transition temperature of the polymer was about 240 °C. The solubility of the polymers

Table 1. Thermal characterization of the polymers

Polymer	T_0 (°C)	T_{10} (°C)	T_{max} (°C)	Char yield at 600 °C
AP-TPC	305	420	545	50
AP-IPC	295	400	510	48

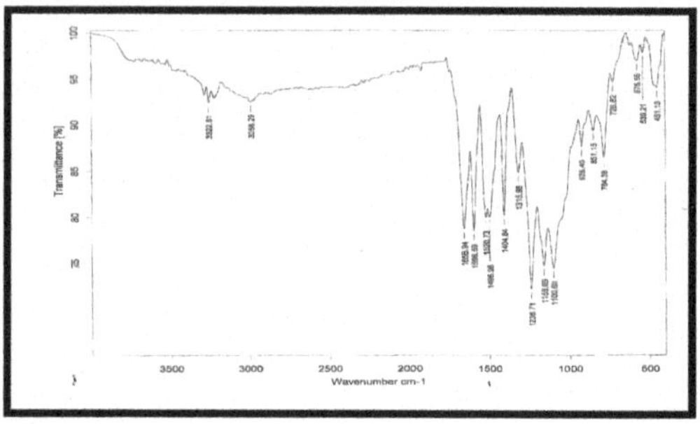

Fig. 1. FT-IR spectrum of TPC-derived polymer

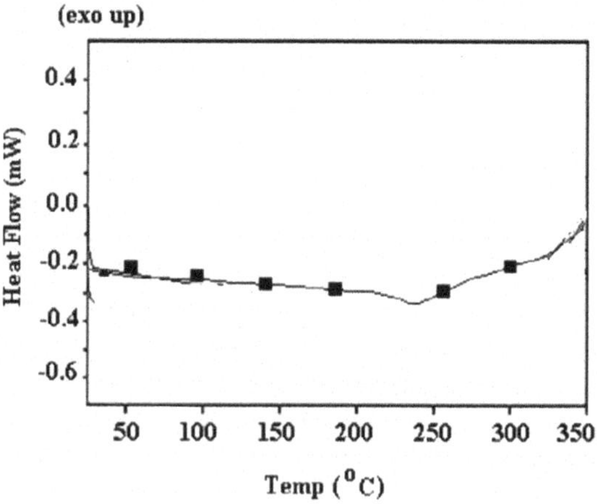

Fig. 2. DSC curve of TPC-derived polymer

was studied in different solvents. Due to the presence of ether linkages, para-catenation, and polar pyridine unit, these polymers showed improved solubility in DMAc, DMSO, NMP, DMF, m-cresol, THF, and pyridine.

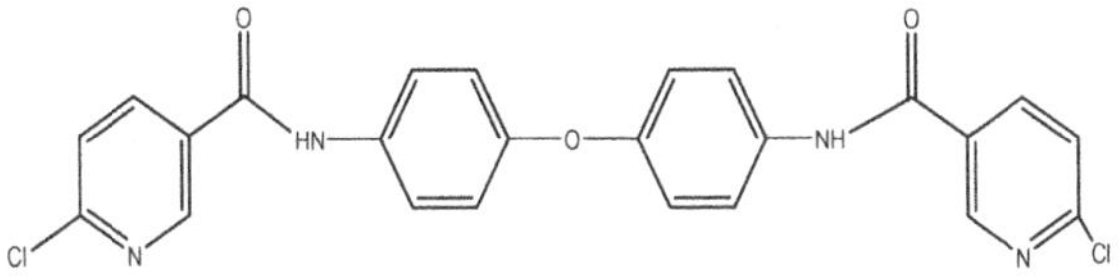

Scheme 1. Structure of dichloro-diamide

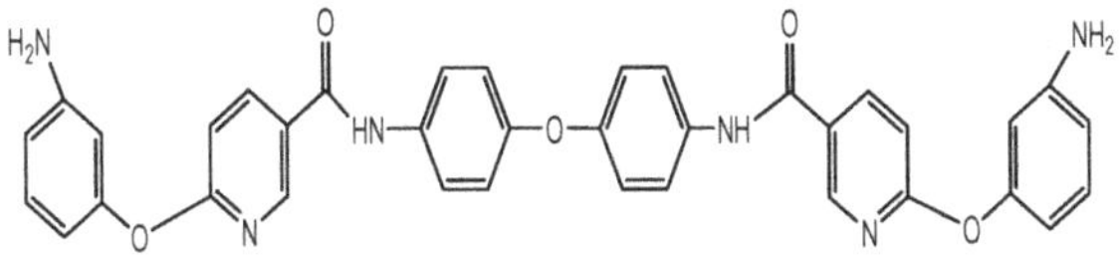

Scheme 2. Structure of damine-diamide

4 Conclusion

A new pyridine–based diamide-diamine was prepared and related polyamides were synthesized. Presence of flexible ether, bulky phenyl and polar pyridine units and the incorporation of these structures in to the backbone of the polymer led to the preparation of a new category of polyamide with favorable balance of properties including high heat resistant and improved solubility.

References

1. Mehdipour-Ataei, S., Amirshaghaghi, A.: Eur. Polym. J. **40**, 503–507 (2004)
2. Mehdipour-Ataei, S., Amirshaghaghi, A.: Polym. Int. **53**, 1186–1190 (2004)
3. Maglio, G., Palumba, R., Vignola, M.C.: Macromol. Chem. Phys. **196**, 2775–2783 (1995)

Nicotinic-Based Poly(ether-amide-imide)s: Preparation and Properties

Sattar Gholamrezaei Boshehri Nejad[1], Mehrdad Hasani[1], and Shahram Mehdipour-Ataei[2(✉)]

[1] College of Basic Sciences, Science and Research Branch, Islamic Azad University, Tehran, Iran
[2] Science Faculty, Iran Polymer and Petrochemical Institute, P.O. Box 14965/115, Tehran, Iran
s.mehdipour@ippi.ac.ir

Abstract. A novel aromatic diamine containing pyridine units, amide, and ether groups was generated via two-step reactions: nucleophilic substitution reaction of oxydianiline with 6-chloronicotinoyl chloride led to preparation of a dichloro compound. Subsequent reaction of this compound with 4-aminophenol resulted in synthesis of the particular diamine. After characterization, polycondensation reaction of the diamine with different dianhydrides led to preparation of related poly(ether-amide-imide)s. The nicotinic-based poly(ether-amide-imide)s showed high heat resistance and improved solubility in polar solvents.

Keywords: Polyimide · Heat resistant · Nicotinic · Polycondensation

1 Introduction

Aromatic polyimides (PIs) have been widely used by various industries, because of their excellent balance of thermal and mechanical properties. However, despite the excellent combined properties, their widespread applications are often limited by their poor solubility and high processing temperature, which make it impossible for most PIs to be directly processed in their imidized form [1]. In order to overcome these drawbacks, many methods have been tried, such as introduction of flexible linkages, bulky units, bulky pendent groups in the polymer backbone. Another solution for making polyimides processable is copolymerization like poly(ether-amide-imide) [2]. The incorporation of flexible units (like –O-) into the polymer backbone may have interrupted the chain packing, thus leading to improved solubility [3]. In this research, new types of nicotinic-based poly(ether-amide-imide)s with balanced properties were prepared via designing of a new diamine.

H. Mirzadeh and A. A. Katbab (Eds.): ISPST 2018, *Eco-friendly and Smart Polymer Systems*, pp. 418–421, 2020.
https://doi.org/10.1007/978-3-030-45085-4_101

2 Experimental

2.1 Synthesis of Monomer

1 g of 4,4-Oxydianiline and 13 mL of NMP was added to the flask and it was maintained at 0 °C for 30 min with continuous stirring. Then, 7 mL of triethylamine was added to the mixture. In the next step 1.81 g of 6-chloronicotinoyl chloride was added to the mixture and the temperature was raised to ambient temperature and kept for 8 h. After that, the flask content was poured into distilled water with stirring. The product was filtered and removed.

1.58 g of the above product, 0.91 g 4-aminophenol, 12.5 mL of NMP, and 8 mL of dry toluene were added to a flask. Then, 2.43 g of K_2CO_3 was added to the mixture and the reaction mixture was heated to 140 °C for 6 h with continuous stirring. The reaction temperature was raised to 165 °C and kept at the same temperature for 21 h. Then, the resulting reaction mixture was poured into water and the mixture was filtered and washed with hot water.

2.2 Synthesis of Poly(ether-amide-imide)

2.2.1 Synthesis of Ploy(Amic Acid)

Into two necked, round-bottomed flask equipped with drying tube, a nitrogen inlet tube, and an ice bath and a magnetic stirrer, 0.5 g (0.0008) of diamine, 4 mL of dried DMAc were stirred for 30 min at 0 °C. Then, 0.1801 g of pyromellitic dianhydride (PMDA) was added to mixture and it was stirred for 1 h. The ice bath was removed and the mixture was stirred for 24 h at room temperature. The mixture was poured into 200 mL of distilled water. Then, the mixture was filtered and washed with hot water.

2.2.2 Cyclization of Ploy(Amic Acid)

Thermal method was used for cyclization of poly(amic acid). The obtained poly(amic acid) was placed in an oven. First, the oven temperature was adjusted at 100 °C and kept for 1 h. Then, the temperature was raised to 150 °C for 1 h, 200 °C for 1 h, 250 °C for 1 h, and for 3 h kept 300 °C.

3 Results and Discussion

As shown in Scheme 1, the diamine monomer (OPAN) was obtained through a two-step synthetic route. The new aromatic diamine monomer is stable in atmosphere at room temperature and pure enough for polymerization with commercial aromatic dianhydride monomers to prepare PIs (Scheme 2). The formation of PIs was confirmed by FTIR spectroscopy. The thermal properties of the resulting PIs were investigated by DSC and TGA methods (Figs. 1 and 2). The TGA curve proved that poly (ether-amide-imide)s were high heat resistant (Table 1).

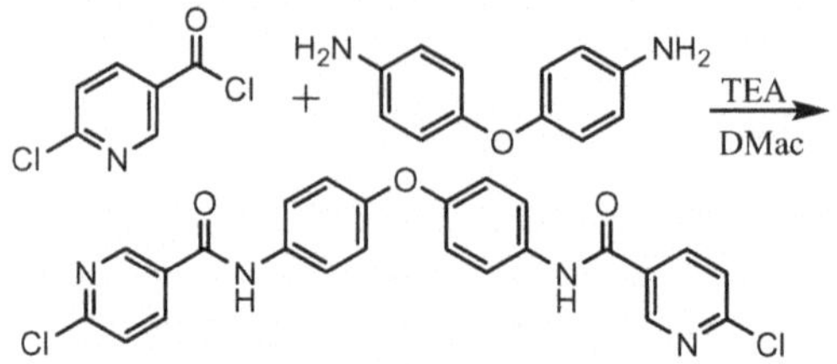

Step 1. Preparation of (OPCN)

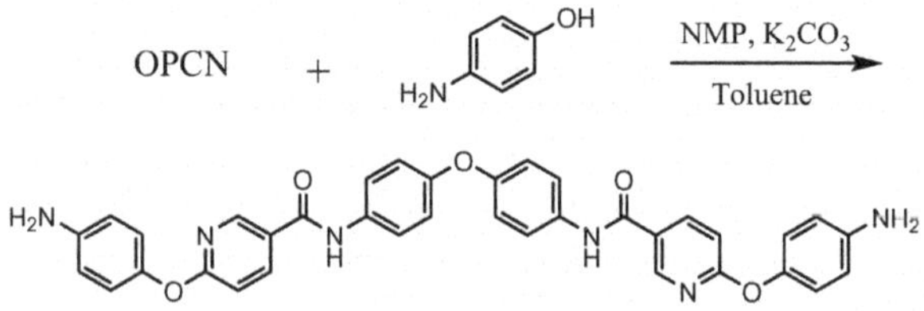

Step 2. Preparation of diamine (OPAN)

Scheme 1. The diamine preparation steps

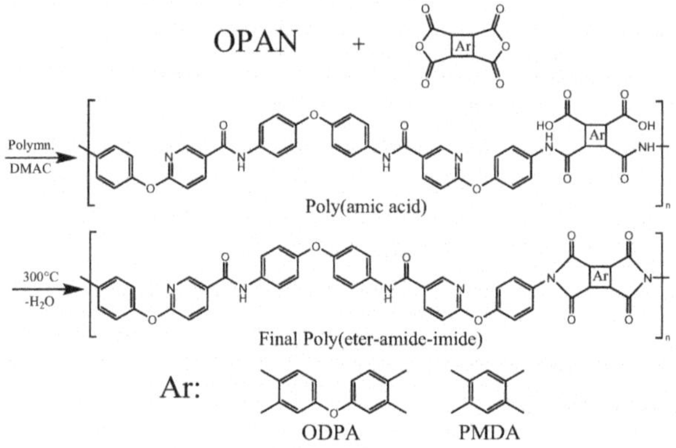

Scheme 2. Synthesis of PIs

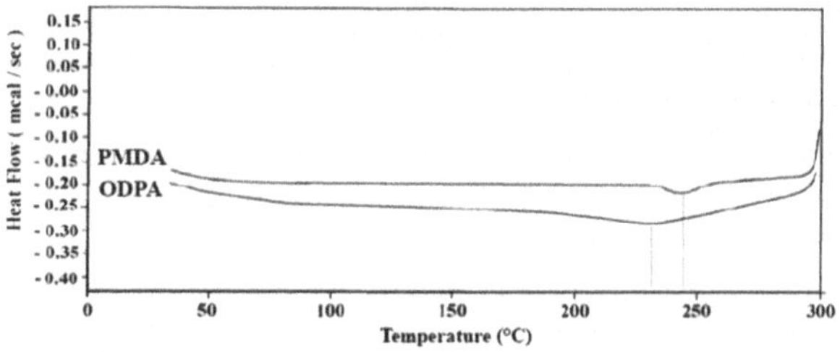

Fig. 1. DSC curve of PIs

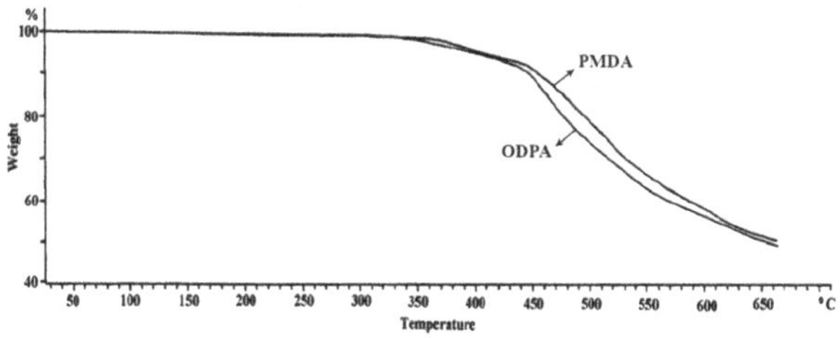

Fig. 2. TGA curve of PIs

Table 1. Thermal characteristic date

Polymer	T_g (°C)	T_d (°C)	T_5 (°C)
PMDA-PI	242	365	428
ODPA-PI	227	340	385

The presence of ether groups, pyridine units and bulky groups in addition to amide units in the final PIs chains caused higher solubility of these polymers in respect to other polymers.

4 Conclusion

A new nicotinic-based diamine containing ether, amide and pyridine units was successfully synthesized and characterized. Polycondensation reaction of this nicotinic-based diamine with dianhydrides (PMDA and ODPA) via two-step thermal imidization led to preparation of poly(ether-amide-imide). The new PIs have fairly high T_g values, excellent thermal stability and good solubility in polar solvents. All of these properties were enhanced to achieve favorite balance between high heat resistant and improved solubility.

References

1. Liaw, D.J., Hsu, P.N., Chen, W.H., Lin, S.L.: Macromolecules **35**, 4669–4676 (2002)
2. Mehdipour-Ataei, S., Amirshaghaghi, A.: Eur. Polym. J. **40**, 503–507 (2004)
3. Mehdipour-Ataei, S., Amirshaghaghi, A.: Polym. Int. **53**, 1186–1190 (2004)

Synthesis and Characterization of Core/Shell Titanium Dioxide Nanoparticle/Polyacrylamide Nanocomposite

Erfan Rezvani Ghomi[1,3(✉)], Saied Nouri Khorasani[1], Mohammad Dinari[2], Shahla Ataei[1], and Rasoul Esmaeely Neisiany[4]

[1] Department of Chemical Engineering, Isfahan University of Technology, 84156-83111 Isfahan, Iran
e.rezvani@ce.iut.ac.ir
[2] Department of Chemistry, Isfahan University of Technology, 84154-83111 Isfahan, Iran
[3] Center for Nanofibers and Nanotechnology, Department of Mechanical Engineering, Faculty of Engineering, Singapore 117576, Singapore
[4] Department of Materials and Polymer Engineering, Faculty of Engineering, Hakim Sabzevari University, 9617976487 Sabzevar, Iran

Abstract. Core/shell titanium dioxide (TiO_2) nanoparticle/poly(Acrylamide-2-Acrylamido-2-methylpropane sulfonic acid) [P(AM-AMPS)] nanocomposite was successfully synthesized via in situ emulsion polymerization. TiO_2 nanoparticles were firstly modified by silane coupling agent, 3-(Triethoxysilyl) propyl Methacrylate (MPS), to increase the dispersibility of TiO_2 nanoparticles into the polyacrylamide matrix. The core/shell TiO_2/polyacrylamide nanocomposite was then synthesized by in situ emulsion polymerization. The MPS-modified TiO_2 nanoparticles and the synthesized core/shell nanocomposite were characterized by Fourier transform infrared spectra (FTIR) and X-Ray Diffraction (XRD). The results showed that MPS coupling molecules was chemically bonded on the TiO_2 nanoparticles surface. According to XRD patterns the copolymer composite was formed.

Keywords: TiO_2 nanoparticles · In situ emulsion polymerization · Nanocomposite · Core/shell · Polyamide

1 Introduction

Over the past decades, synthesis of inorganic/polymer nanocomposites have received special attention due to their unique features with the mixture of both inorganic nanoparticles and organic polymers. The nanocomposites synthesized by inorganic nanoparticles show improvement on mechanical, electrical, optical, and other properties, which have an extended range of applications, such as optical devices, paints, catalysts, waste water treatment, and drug delivery systems [1]. However, inorganic nanoparticles tend to agglomerate in the polymer matrix, which spoils the electrical and mechanical properties of the nanocomposites. Surface modification by silane coupling agents is an effective method to improve the dispersibility of nanoparticles in the

H. Mirzadeh and A. A. Katbab (Eds.): ISPST 2018, *Eco-friendly and Smart Polymer Systems*, pp. 422–425, 2020.
https://doi.org/10.1007/978-3-030-45085-4_102

polymer matrix [2]. Incorporation of various kinds of inorganic nanoparticles, such as TiO_2, ZnO, SiO_2, Al_2O_3, and so on, have been carried out, which results in many types of high-performance nanocomposites. TiO_2 has attracted attentions in recent years because of its outstanding properties, such as photocatalytic properties, great thermal stability, and excellent UV light absorption [3, 4].

In this work, we synthesized TiO_2 nanoparticle/P(AM-AMPS) nanocomposite with core/shell morphology via in situ emulsion polymerization.

2 Experimental

TiO_2 was purchased from Notrino Co. with average particle sizes of 20 nm. AM monomer, sodium hydroxide, hydrochloric acid, ammonium persulfate (APS), N,N,N', N'-tetramethylethyleneediamine (TEMED), sodium dodecyl sulfate (SDS) stabilizer and ethanol were all provided by Merck. MPS and AMPS monomer were purchased from Sigma-Aldrich Co. The deionized water was used for all the synthesis processes.

TiO_2 nanoparticles (0.5 g) initially were added to water/ethanol solution (1:4, v/v) and sonicated with ultra-sonication horn of 400 W for 1 h. MPS was added to ethanol (10 mL) and the mixture stirred for 1 h. MPS mixture was then fed to the TiO_2 nanoparticles and heating reflux for 8 h at 70 °C. The obtained product was centrifuged and washed by ethanol for three times and dried in a vacuum oven at 50 °C for 10 h. The TiO_2 nanoparticle/P(AM-AMPS) nanocomposite was synthesized in a 250 mL four-necked glass reactor, equipped with a reflux condenser, a mechanical stirrer and a nitrogen inlet gas. Firstly, SDS was dissolved in deionized water. Then modified TiO_2 nanoparticles dispersed in deionized water and equimolar of AAm and AMPS were added to the mixture. The pH of reaction solution was adjusted to 8.5 ± 0.5 by addition of NaOH to ensure neutralization of the AMPS monomer to its sodium salt. The water soluble initiator APS and TEMED were dissolved in deionized water and dropped in to the reaction mixture. The mixture was heated at 60 °C for 5 h. At the end of the reaction, the reaction mixture was cooled at room temperature. The product was precipitated into acetone while stirring.

The FTIR spectra of samples have been recorded with a Fourier transform infrared spectrometer (Jasco-680, Japan) in the range of 1000–400 cm^{-1} to provide the proof of the grafting. The XRD pattern for the samples was obtained with Philips X' Pert-MPD system at the radiation ranging from 5° to 80°.

3 Results and Discussion

Figure 1 shows the FTIR spectra of the materials. From Fig. 1a, it can be observed that the surface of the unmodified TiO_2 nanoparticle is covered only with -OH groups (peaks at 3430 and 1640 cm^{-1}). The ethoxy groups of the MPS is hydrolyzed to form the reactive silanetriols, which are rapidly deposited on the TiO_2 surface and reacted with the -OH groups on the TiO_2 surface to form Si-O-Ti covalent bond, leading to the chemically bonded MPS on the surface of TiO_2 nanoparticles. The vinyl groups of the MPS on the TiO_2 surface can further polymerize with acrylamide monomers via free

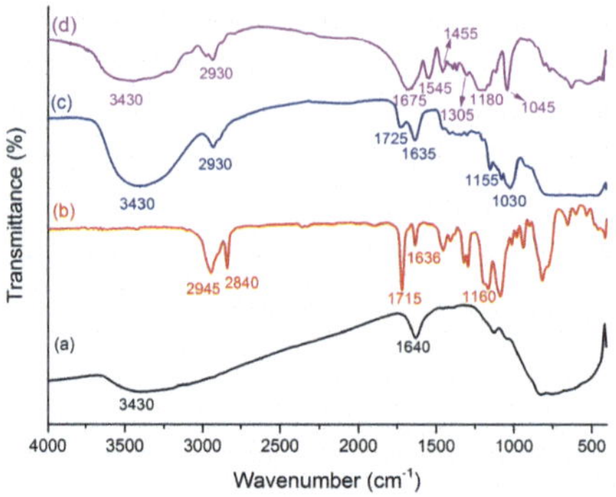

Fig. 1. FT-IR spectra of (a) TiO_2 nanoparticles, (b) MPS, and (c) MPS-modified TiO_2 nanoparticles, (d) TiO_2 nanocomposite

radical polymerization, resulting in a polymer shell bonded to the TiO_2 core. Comparing to Fig. 1a of the unmodified TiO_2 with Fig. 1c, it can be discerned that the new bands at 2930, 1725, 1635, 1155 and 1030 cm^{-1} are observed from Fig. 1c of the modified TiO_2 nanoparticles. They are corresponded to the -CH_3 and -CH_2 stretching, C = O stretching, C = C stretching vibration, = CH_2 band, and Si-O-Ti stretch vibration band, respectively. There is a strong and broad peak at 3450–3400 cm^{-1}, which is probably due to the physically adsorbed water on the surface of TiO_2 nanoparticles. Considering Fig. 1d, The peaks at 1675, 1545, 1455, 1305, and 1180 cm^1 assigned to the presence of C = O amide group, C = O carboxyl group, CH_2 amide group, C-N amide and SO_3 sulfonyl group of AMPS monomer, respectively. The FTIR results confirmed the chemical grafting of copolymer to modified TiO_2.

As shown in Fig. 2, the XRD spectra revealed the diffraction peaks of the TiO_2 nanoparticles, MPS modified TiO_2 nanoparticles and TiO_2 nanocomposite. Regarding to diffractograms, the diffraction peaks at 2θ = 25.3°, 37.9°, 48.2°, 54°, 55.1°, 62.8°, 68.8°, 70.5°, and 75°, corresponding to the (101), (004), (200), (105), (211), (204), (116), (220), and (215) crystal faces of TiO_2 nanoparticles, indicate the existence of TiO_2 in the nanogel composite. The strong diffraction peaks corresponding to 101, 004, 200, 211, for TiO_2 nanoparticles were attributed to the anatase TiO_2 phase. It can be seen that functionalizing TiO_2 nanoparticles by MPS does not change the crystalline structure of TiO_2 nanoparticles. The appearance of broad amorphous diffraction peak at 2 theta = 15–35° indicates the formation of copolymer composite. Careful inspection of XRD spectrums of the MPS modified TiO_2 nanoparticles and TiO_2 nanocomposite confirm the decreasing of peak intensity of the (101) crystal face of TiO_2 nanocomposite that is relating to the AMPS/AAm nanogel coverage at the surface of MPS modified TiO_2 nanoparticles.

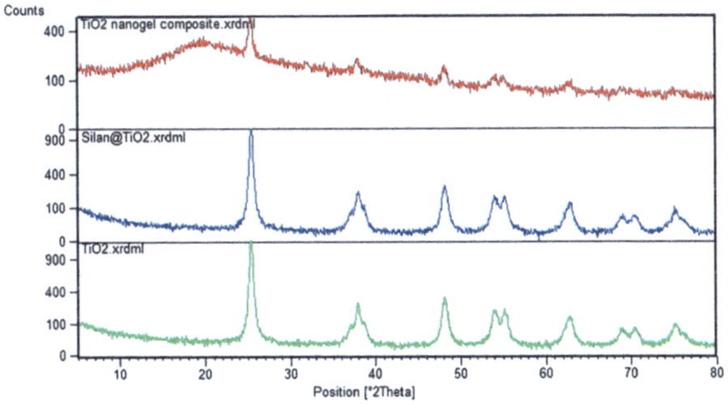

Fig. 2. XRD pattern of TiO_2 nanoparticles, MPS modified TiO_2 and TiO_2 nanocomposite

4 Conclusion

The core/shell TiO_2/P(AM-AMPS) nanocomposite was successfully prepared via in situ emulsion polymerization. FTIR and XRD results confirmed that MPS was chemically bonded on the surface of TiO_2 nanoparticles and also the copolymerization was carried successfully out on the surface of modified TiO_2 nanoparticles.

References

1. Inkyo, M., Tokunaga, Y., Tahara, T., Iwaki, T., Iskandar, F., Hogan, C.J., Okuyama, K.: Ind. Eng. Chem. Res. **47**, 2597 (2008)
2. Kango, S., Kalia, S., Celli, A., Njuguna, J., Habibi, Y., Kumar, R.: Prog. Polym. Sci. **38**, 1232 (2013)
3. Atta, A.M., Al-Lohedan, H.A., El-saeed, A.M., Al-Shafey, H.I., Wahby, M.H.: Prog. Org. Coat. **105**, 291 (2017)
4. Atta, A.M., El-Saeed, A.M., Al-Shafey, H.I., Al-Lohedan, H.A., Tawfeek, A.M., Wahbey, M.: Int. J. Elec. Sci. **12**, 1167 (2017)

Electrochemical Conversion of Heavy Residues to Bitumen and Other Light Products by Silver (II) Mediator

Shahrnaz Mokhtari, Fereidoun Mohammadi(✉), and Zahra Panahi

Iran Polymer and Petrochemical Institute, P.O. Box 14965/115, Tehran, Iran
f.mohammadi@ippi.ac.ir

Abstract. Ag (II) ions generated by Mediated Electrochemical Oxidation (MEO) was employed to convert a heavy residue feed (mazut) to a porous solid polymer and some other valuable products. At atmospheric pressure, conversion of the heavy residue was carried out at 50–60 °C for 2 h. The obtained liquid and solid products were then separated and analyzed by FTIR and simulated distillation (Sim-Dist) analysis. FTIR results specified that the reactions between Ag (II) and heavy residue resulted in a decrease in methylene groups and an increase in aromatics, branched aliphatic hydrocarbons and carbonyl groups. Also, Sim-Dist results proved that during this process the heavy hydrocarbons of the feed were cracked and changed into lighter hydrocarbons and bitumen; however a liquid product was formed which comprises gasoline, kerosene and atmospheric gasoil.

Keywords: Mazut · Electrochemical oxidation · Silver (II) · Bitumen

1 Introduction

Petroleum residues are known to contain such heavy and highly viscous hydrocarbons as asphaltenes in addition to light and highly volatile ones. Improvement of physical and chemical characteristics of heavy petroleum residues requires polymerization of asphalten molecules. In order to increase stiffness and reduce penetration during the manufacture of some petroleum products, polymerization of asphalten molecules of heavy residues is necessary [1].

Mediated Electrochemical Oxidation (MEO) process leads to a successful method due to its high efficiency, compact facilities, low operating costs and friendly environment. The main objective of this study is to investigate the electrochemical conversion of heavy residues such as mazut and asphalten by electrochemically generated silver (II) ions at the optimum condition obtained formerly [2]. As reported earlier [3], during the electro-chemical process about 10% of the mazut feed was converted to a solid polymer product. Melting point of obtained solid product was about 150 °C compared to 75–120 °C for bitumen which can be counted as a similar substance. Agglomerate and polymerized asphalten particles were also clearly observed in the SEM micrographs [3]. For more confirmation in this research characteristics of the solid polymer obtained were determined by FTIR and simulated distillation (Sim-Dis) techniques.

H. Mirzadeh and A. A. Katbab (Eds.): ISPST 2018, *Eco-friendly and Smart Polymer Systems*, pp. 426–429, 2020.
https://doi.org/10.1007/978-3-030-45085-4_103

2 Experimental

The electro-chemical polymerization of mazut was employed in an electro-chemical cell at 2.4 V constant potential. The cell was contained of anolyte and catholyte compartments which were separated by a Nafion® 115 membrane. DSA oxygen and titanium plate were employed as anode and cathode, respectively. After electrochemical oxidation of silver (I) to silver (II) at optimum conditions, mazut 230 obtained from Tehran oil refinery was charged to anolyte. During the process the temperature was maintained at 50 °C for 2 h.

3 Results and Discussion

In order to compare the chemical structure of the products with that of the feed, FT-IR analysis of the heavy residue and the polymeric solid and liquid products were carried out. The corresponding spectra are depicted in Fig. 1. As observed, the intensity of the peaks is maximum for the heavy residue feed; however, they are reduced significantly after electrochemical conversion. The peaks have a minimum intensity for the solid product. Absorption peaks of 1602 cm^{-1} of heavy residue feed, 1559 cm^{-1} of liquid product, and 1552 cm^{-1} of solid product representing an aromatic structure and ascribes to stretching vibrations of aromatics (C = C). These peaks are sharper for the solid and liquid products in comparison with the initial heavy residue suggesting formation of more aromatic products. Furthermore, absorption peaks of 1708 cm^{-1} for the liquid product and 1710 cm^{-1} for the solid product reveals the presence of carbonyl groups which is not exist in the feed [4]. These carbonyl groups are formed due to the oxidation reactions of the heavy residue during electrochemical conversion process. Finally, existing picks in the range of 3200–3650 cm^{-1} shows the hydroxyl group which can be attributed to alcohols, phenols, carboxylic acids or water. Since these peaks are only found in the solid product spectrum and not seen in the spectra of the feed and liquid product, it is more likely that 3562 and 3419 cm^{-1} peaks are due to water absorption by the solid product [5]. Figure 2 shows the boiling range for the heavy residue, the liquid and solid products. The boiling range of heavy residue feed is 206–629 °C, while the solid product's boiling range is 220–617 °C and that of the liquid product is 181–598 °C. By comparing the boiling range of the solid product with the feed, it can be inferred that only 10% of the solid product has a boiling point higher than the feed, and the boiling range of 90% of the solid product is less than the feed. Also, the boiling range of the produced liquid by mediated electrochemical process of the heavy feed is lower than the primary feed. The results revealed that, 10% of the liquid product seems to be gasoline and kerosene, and only 30% of the liquid produced contains heavy hydrocarbons, while the furnace feed contains only 0.5% kerosene and 50% of heavy hydrocarbons. Therefore, it can be concluded that during the mediated electrochemical conversion of the heavy residue feed by silver (II) ions, cracking reactions were also occurred leading to a more value-added products.

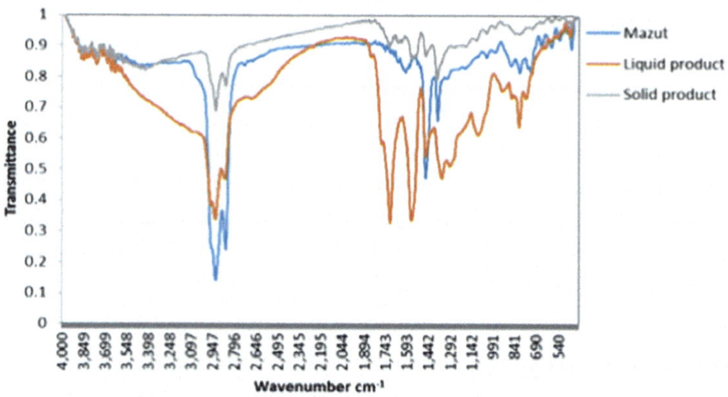

Fig. 1. FTIR spectrums of the heavy residue feed and the products obtained by MEO conversion by Ag (II)

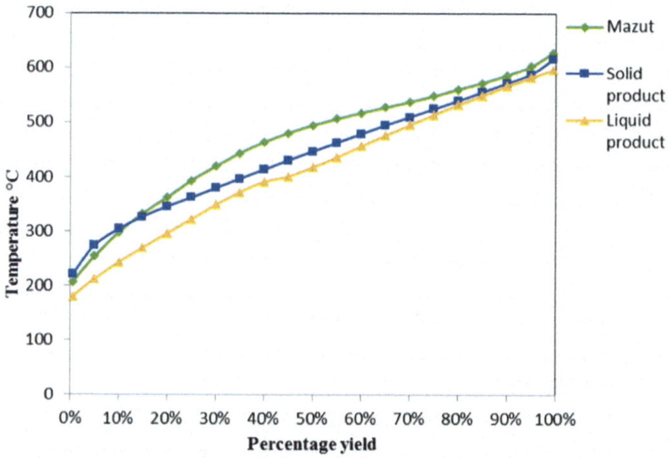

Fig. 2. Boiling range of the heavy residue feed and the products of mediated electrochemical conversion by Ag (II)

4 Conclusion

Electrochemical conversion of mazut culminated in production of some porous solid substance such as bitumen and more valuable products. By comparing the FTIR results of the initial feed with the products of the MEO process, significant changes were observed. During this process while the methylene groups were reduced, aromatic compounds, branched aliphatic and carbonyl groups increased. Sim-Dist results also confirmed that cracking reactions were performed and part of heavy residue feed converted into more valuable products.

References

1. Le Guern, M., Chailleux, E.: Fuel **89**, 3330–3339 (2010)
2. Mokhtari, Sh, Mohammadi, F., Nekoomanesh, M.: Iran. Polym. J. **25**, 309–320 (2016)
3. Panahi, Z., Mohammadi, F., Mokhtari, Sh.: 12th International Seminar on Polymer Science & Technology (ISPST 2016), Azad university, Iran, 2–5 November 2016
4. Vahabzadeh, F., Dermanaki, A.: Oil Gas Sci. Technol. **70**, 1035–1049 (2015)
5. Feng, Z., Wang, S., Bian, H., Guo, Q., Li, X.: Constr. Build. Mater. **115**, 48–53 (2015)

Synthesis of Polycarboxylate Superplasticizer: Influence of Temperature of Polymerization

Maryam Mohammadi[1(✉)], Alireza Redaei[1], Mohammad Ali Karimizarchi[2], and Mohammad Ali Mohammad Mirzaei[1]

[1] Faculty of Engineering, Department of Chemical and Polymer Engineering, Yazd University, Yazd, Iran
mohamadimaryam002@gmail.com
[2] Department of Chemistry, College of Science, Yazd University, Yazd, Iran

Abstract. In this study, Polycarboxylate (PC) superplasticizers based on acrylic acid (AA) were synthesized via free-radical copolymerization with an ethylene glycol monomer and characterized. The temperature of polymerization (ranging from 50 °C to 70 °C) appeared to be the key operating factor governing the chemical structure of the superplasticizers. The chemical structures of the products were analyzed by gel permeation chromatography, Fourier transform infrared spectroscopy and nuclear magnetic resonance. The results showed that the PCs synthesized at different polymerization temperature have similar functional groups but the temperature has affected on the molecular weight (MWs) and MW distribution. By increasing of temperature of polymerization the MW of PC has increased.

Keywords: Polycarboxylate · Superplasticizer · Acrylic acid · Polymerization temperature

1 Introduction

Wide use of polymer admixtures is an element of the development of contemporary concrete technology. Water reducing admixture and rust inhibitors are widely used cement concrete additives in the construction. Polycarboxylate series of high performance water reducer develops easterly in recent years. Polycarboxylate-type water reducer has the following advantages such as low dosage, strong dispersity, controllable molecular structure, environmental protection, and so on. Hence, it becomes the development direction and the research hotspot of concrete admixture [1]. It is obvious that the microstructure of a superplasticizer is mostly controlled by the type of materials and the synthesis parameters [2, 3]. The possibility of changing several synthesis parameters of the PCEs (molecular weight, backbone length and flexibility, side-chain length, charge density) has led to the development of products characterized by tailored properties for specific applications [4]. One of the important factor of the polymer admixtures efficiency is temperature. The direction of changes depends on the type (chemical nature) of the modifier as well as its content and the composition of the concrete mix. Considering the diversity of the polymer admixtures, the prediction of

H. Mirzadeh and A. A. Katbab (Eds.): ISPST 2018, *Eco-friendly and Smart Polymer Systems*, pp. 430–433, 2020.
https://doi.org/10.1007/978-3-030-45085-4_104

the temperature characteristics is difficult and requires detailed investigations. On the basis of the gathered information, one can conclude that in the case of the new generation superplasticizers, the growth of temperature usually leads to worsening of effectiveness, and in consequence to the increasing dosage of admixtures.

The direction of changes depends on the type (chemical nature) of the admixture, and often also on its concentration and composition of the concrete mix. Taking into consideration the big diversity of the used polymer admixtures, it is very hard to predict their temperature characteristics; this requires detailed testing. It is possible to find out, on the base of the collected data, that in the case of the new generation of the superplasticizers an increase of temperature usually leads to diminishing of the admixture efficiency. This means the necessity of increasing dosage of the superplasticizers of new generation at higher temperature. This study aims at elucidating the correlations of synthesis temperature of PC superplasticizers with their performance and to lay the experimental preclinical for the development of new superplasticizers with more efficient and robust performance.

2 Experimental

The reaction was done in two steps. Firstly, A three-necked, round glass flask reactor equipped with a magnetic stirrer and an oil-bath heater was used for the polymerization. Two necks were needed to feed the AIBN and acrylic acid (AA) and to place the thermocouple. The ratio of AA/3chloro-2methyl-1propene was fixed at 1:1. The AIBN and AA were added to the mixture drop wise during the first 4 h of the reaction. AIBN was used as the initiator of radical polymerization. The reaction time was set to 8 h, and polymerization was conducted at different temperatures of 50, 60, and 70 °C for the AA. The reaction conditions were kept noticeably unchanged, except we tried to find the most suitable temperature for synthesis and at the same time guaranteed the highest concrete performance when the super-plasticizer was introduced into the cement. After 8 h, the reactor was cooled to room temperature. In the following stage, PEG and K_2CO_3 (as catalyst) was added into the solution and reacted for 6 h on a magnetic stirrer until completion.

The superplasticizers synthesized in this manner were coded as SP–1, SP–2, and SP–3 copolymers. It is worth mentioning that the monomer concentrations were kept constant by drop wise addition during the time span of the reaction to prevent as far as possible the homo-polymerization of AA.

3 Results and Discussion

The MW of the superplasticizer increased over a temperature rise. The increase in the MW could be explained by the fact that the reactivity of PEG was quite low at 50 °C and a step rise in the polymerization temperature up to 70 °C provided sufficient activation energy for the reaction between AA and PEG. The infrared spectrum of the PCs synthesized at different polymerization temperature are given in Fig. 1. In general, the peak at 1100 cm^{-1} was assigned to the adsorption of ether linkage, that at

1150 cm^{-1} was the characteristic adsorption of sulfonic groups, and that at 1715 cm^{-1} was the stretching vibrations of the carboxylic groups. The appearance of the peak at 1630 cm^{-1} strengthened the possibility of the existence of C-C double bounds. In this way, the presence of unreacted monomers in the media in which the superplasticizers were synthesized could be proven [5].

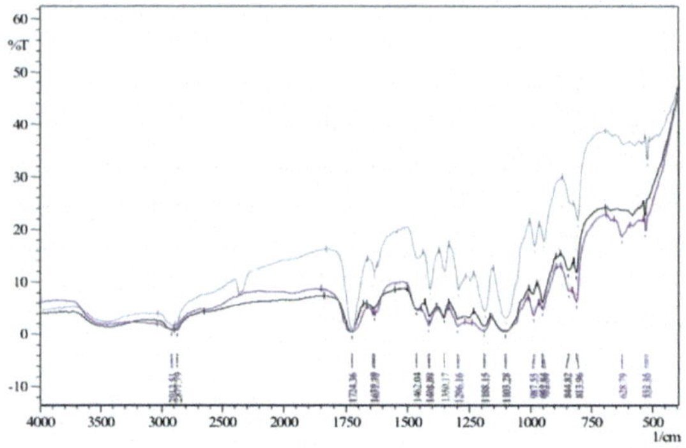

Fig. 1. IR spectrograms of SP–1, SP–2, and SP–3

The ^{1}H-NMR spectra of copolymers proved copolymerization by the presence of the PEG signal of the PEG copolymer and signals in the 0–3.5 ppm region (Fig. 2).

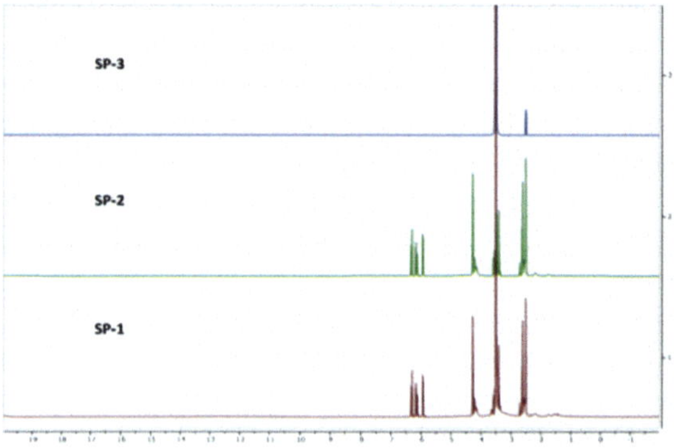

Fig. 2. H-NMR spectra (500 MHz) of SP–1, SP–2, and SP–3

The test results of ^{1}H nuclear magnetic resonance confirmed the occurrences of polymerization. There are very similar spectra curves that exhibit some common features.

4 Conclusion

Polycarboxylate superplasticizers based on acrylic acid were synthesized by using 3-chloro-2-methyl-1-propene and the main results are as follows:

- Among superplasticizers based on the AA monomer, the one synthesized at 50 °C showed the shortest backbone length and the highest side-chain density. In addition, an increase in the temperature of polymerization caused a fall in the side-chain density and a rise in the backbone length of this family of plasticizers.
- A comparative assessment revealed that the AA-based superplasticizers possessed higher MWs. This group also had considerably longer backbones.
- The optimum temperatures guaranteeing the highest performances were 60 °C for the AA-based superplasticizers.

References

1. Sakai, E., Ishida, A., Ohta, A.: Adv. Concr. Technol. J **4**, 211–223 (2006)
2. Voit, B.: Polym. Sci. Part A Polym. Chem. J. **38**(14), 2505–2525 (2000)
3. Jikei, M., Kakimoto, M.: Prog. Polym. Sci. J. **26**(8), 1233–1285 (2001)
4. Meyer, L.M., Perenchio, W.F.: Concr. Int. J. **1**, 36–43 (1979)
5. Yinwen, L., Chaolong, Y.: Constr. Build. Mater. J. **64**, 324–332 (2014)

Effect of Silane Curing Conditions on Properties of the Metallocene-Based Polyethylene-Octene Copolymers Compound

Mehri Dana[1], Gholam Hossein Zohuri[1,2](✉), and Saeid Asadi[3]

[1] Department of Chemistry, Faculty of Science, Ferdowsi University of Mashhad, P.O. Box 91775-1436, Mashhad, Iran
zohuri@um.ac.ir

[2] Environmental Chemistry Research Center, Department of Chemistry, Faculty of Science, Ferdowsi University of Mashhad, P.O. Box 91775-1436, Mashhad, Iran

[3] Baspar Sazeh Toos Co., Part Lastic Group, P.O. Box 91851-75596, Mashhad, Iran

Abstract. A compound was prepared from crosslinking of blend from two type of metallocene-based polyethylene-octene copolymers by different octene content through a two-step silane grafting method (Sioplas process) in a twin-screw extruder. The silane grafted compound was used as wire and cable coatings. The curing condition and specimen preparation method (injection molding and/or extrusion) were factors influenced the hot-set test results at 200 °C. The results of tensile and elongation studies showed a maximum value of 9 MPa for tensile strength and 397% for elongation-at-break after 6 h curing.

Keywords: Sioplas process · Blend · Silane grafting · Curing condition

1 Introduction

Polyethylene copolymers are usually crosslinked to ensure that they maintain their stability and mechanical properties above the melting point of their crystalline phase [1, 2]. In industry, the crosslinking process is generally performed by using peroxide or silane compounds, and sometimes through electron beam crosslinking processes [1, 3, 4]. The crosslinking using silane grafting reagents is one of the most common methods for improving the properties of Polyethylene copolymers [5, 6]. Easy processing, low capital investment and favorable properties of processed materials are the advantages of using silane crosslinking method [3, 4, 6, 7]. The Monosil and Sioplas processes are two of the major technologies in the manufacturing of silane crosslinked materials. At low temperatures, the rate-determining step of the cross-linking reactions is water diffusion, rather than the hydrolysis and the subsequent condensation reactions of the silyl trimethoxy groups. However, at high temperatures and high degrees of silane grafting in the samples, the chemical reactions dominate in the crosslinking process [3]. Hot set tests (duration of 15 min, at temperature of 200°C, and under definite static load) are carried out on cross-linked samples to evaluate and quantify the cross-link density [8].

H. Mirzadeh and A. A. Katbab (Eds.): ISPST 2018, *Eco-friendly and Smart Polymer Systems*, pp. 434–437, 2020.
https://doi.org/10.1007/978-3-030-45085-4_105

In this work, the crosslinking of a blend of comprising metallocene-based polyethylene-octene copolymers by different octene content was carried out using the Sioplas process, which represents a two-step silane crosslinking process.

2 Experimental/Theoretical

2.1 Materials

EOC and LLDPE were purchased from SK Global Chemical Co., Ltd, (South Korea) and Amirkabir Petrochemical Co., (Iran), respectively. Pentaerythritol tetrakis (3-(3,5-di-tert-butyl-4-hydroxyphenyl) propionate), Irganox 1010 antioxidant, was supplied by BASF Co., Germany. Vinyl trimethoxysilane (VTMS) was purchased from RUI Chemical (China). Dibutyltin dilaurate (DBTDL), the catalyst, was supplied by Merck (South Korea). Dicumyl peroxide and benzoyl peroxide, the initiators, were supplied by Coin AKZO Nobel Co., (Netherlands) and Arkema (France), respectively.

2.2 Preparation of Sample

The silane grafting of EOC/LLDPE blend was performed in a co-rotating twin-screw extruder with twelve heating/cooling zones (Model CTE 65, COPERION Co. Germany). A temperature profile of 170–210 °C were applied. A constant silane and peroxide concentration of 5 and 0.13 wt% was used, respectively. In the next step, a blend of the silane grafted compound (~95 parts) and the catalyst masterbatch (~5 parts) was shaped as wire and cable coatings. For extrusion coating of a cable using the prepared silane grafted compound, an extruder with seven heating/cooling zones was used at a production speed of 10 m/min under a temperature profile of 170–220 °C from the feed zone to the die zone. The specimens (sheets) with different thicknesses were cured under the same conditions in a water bath operated at 85 °C for 4–6 h per 1 mm sheet thickness.

3 Results and Discussion

3.1 Hot-Set Test at 200 °C

3.1.1 Effect of Curing CONDITIONS

The results of hot-set test at 200 °C at different curing conditions are shown in Table 1. As seen, the morphology of specimens is affected by the curing conditions. The hot-test was not confirmed because of possible degradation of the polymer as a result of curing.

Table 1. Effect of curing conditions on results of the hot set test at 200 °C

Curing temperature (°C)	Curing time (h)	Results	
85	8 h	Extension < 175%	Set < 15%
		−24.85	31.05
		−26.50	28.85
		−29.45	26.05
100	Shelf life at 200 °C (min)		
	2 h	3:26	
	4 h	4:28	
	6 h	3:56	

3.1.2 Effect of Processing Type

The hot-set test at 200 °C was performed on the specimens prepared by injection molding and extrusion processes (Table 2). Differences in the obtained results can be attributed to the difference in the shape of the test specimens prepared. The pressure in extrusion process typically is 5000 Psi (35 MPa) which is less than that in injection molding process [9]. Due to high pressure of injection, the test specimens prepared by injection molding process is more compressed than those prepared by extrusion process. It may state that more compressed specimens have higher stress resistance.

Table 2. Hot-set test results for specimens with different shapes

Specimen type	Gel content (%)	Hot-set test at 200 °C	
		Extention < 175%	Set < 15%
Cable	74.0	154	2.25
	78.2	146	0.25
Dumbbell-shaped	75.6	29.6	−23.9
	75.0	29.6	−27.2

3.2 Effect of Curing Time on Mechanical Properties of the Crosslinked Compound

The tensile strength-at-break and elongation-at-break versus curing time are shown in Fig. 1. As seen, the highest amount of tensile strength and elongation are obtained after 6 h of curing time. The loss in the mechanical properties is probabily due to the polymer degradation occurred after curing. The curing conditions can improve the direction of the polymer chains, thereby reducing the tensile strength [10].

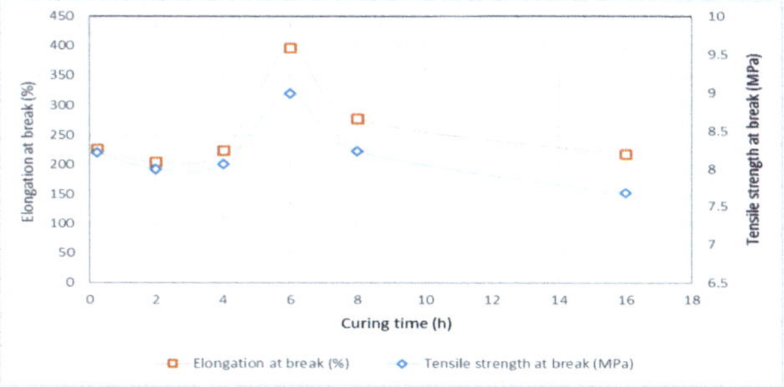

Fig. 1. Effect of curing time on the elongation-at-break and tensile strength of the crosslinked compound

4 Conclusion

Crosslinking of EOC/LLDPE blend was carried out using the two-step Sioplas process in a twin-screw extruder. The blend of the silane-grafted compound (~95 parts) and the catalyst masterbatch (~5 parts) was used to make coatings for wire and cable. The curing conditions and the type of specimen were factors influenced the hot-set test results at 200 °C. Study of tensile strength and elongation-at-break of the crosslinked compound showed a maximum value of 9 MPa and 397% after 6 h curing time, respectively.

Acknowledgment. The authors gratefully acknowledge the Ferdowsi University of Mashhad, Mashhad, Iran, for its financial support (project code: 3/43281), and Baspar Sazeh Toos Co. of Part Lastic Group Co. for providing equipment and chemicals for conducting this study.

References

1. Stephen, H.C.: Polyolefin elastomer and polysiloxane blends; US Patent 2016/0108239 A1, 21 April 2016
2. Svoboda, P., Poongavalappil, S., Theravalappil, R., Svobodova, D., Mokrejs, P.: J. Polym. Int. **62**, 184–189 (2013)
3. Sirisinha, K., Meksawat, D.: J. Polym. Int. **54**, 1014–1020 (2005)
4. Jiao, C., Wang, Z., Gui, Z., Hu, Y.: Eur. Polym. J. **41**, 1204–1211 (2005)
5. Chen, W.C., Lai, S.M., Qiu, R.Y., Tang, S.X.: J. Appl. Polym. Sci. **124**, 2669–2681 (2012)
6. Lai, S.M., Lui, J.L., Chen, Y.C., Chang, K.H.: J. Appl. Polym. Sci. **101**, 2472–2481 (2006)
7. Zhang, G., Wang, G., Zhang, J., Wei, P.: J. Appl. Polym. Sci. **102**, 5057–5061 (2006)
8. Morshedian, J., Mohammad Hoseinpour, P.: Iran. Polym. J. **18**, 103–128 (2009)
9. Handbook of Thermoplastic Elastomers: Processing Methods Applicable to Thermoplastic Elastomers, Elsevier Inc., pp. 26–141 (2014). https://doi.org/10.1016/b978-0-323-22136-8.00004-1
10. Roumeli, E.: Structure and properties of crosslinked high Ddensity Polyethylene Nanocomposite materials, Doctoral Dissertation, Aristotale University, Greece, Thessaloniki (2014)

The Relationship Between Adhesion and Electrochemical Parameters of Epoxy Resin on Mild Steel

Amir Hossein Jalali Kandeloos
and Mohammad Reza Mohammadzade Attar(✉)

Faculty of Polymer Engineering and Color Technology,
Amirkabir University of Technology, Tehran, Iran
attar@aut.ac.ir

Abstract. Electrochemical impedance spectroscopy (EIS) has been employed in the investigation of the corrosion characteristics of epoxy polymeric film both coated and attached as a free film without adhesion on mild steel panels during immersion in 3.5 wt% NaCl solution. Meanwhile gravimetric liquid sorption experiments were used to evaluate water uptake inside the coating. The results of impedance spectroscopy indicate that epoxy resin shows greater anticorrosion protection characteristics when applied on mild steel than attached as a free film. Attached Free films present higher value of water uptake compared with coated samples. Adhesion seems to play a major role in the protection of the metals. The first step of electrolyte penetration through the coating is related to water uptake and after that a different process occurs, that could be either the penetration of the water in the bulk of the polymer, or the slow diffusion of ions.

Keywords: Epoxy coatings · Adhesion · Water diffusion coefficient · EIS · Free film

1 Introduction

The most widely used protective coatings for steel substrates are epoxy-based because of their excellent chemical resistance, mechanical strength, and adhesion to metal surfaces [1]. These coatings are extensively utilized in smart coatings applications, e.g. self-healing, fluorescence for early detection of corrosion, etc. The corrosion protective action of organic coatings consists mainly of resistance inhibition and the introduction of a barrier for the transport of corrosive species [2, 3]. Defects within the coating introduce direct pathways for water, oxygen and electrolyte to reach the metal substrate, resulting in the localized loss of barrier protection [3, 4]. Water molecules at the metal/coating interface may reduce the coating adhesion, thus favoring corrosion of the metal underneath the film [5]. One of the commonly used methods to study corrosion performance of coatings is EIS [4–6]. Gravimetric transport measurements can also be used to obtain coating properties [7, 8].

H. Mirzadeh and A. A. Katbab (Eds.): ISPST 2018, *Eco-friendly and Smart Polymer Systems*, pp. 438–441, 2020.
https://doi.org/10.1007/978-3-030-45085-4_106

The aim of the investigations reported in this paper, was to study the relationship between adhesion and electrochemical parameters of epoxy resin on mild steel and mechanisms of water uptake and ion penetration which cause coating degradation.

2 Experimental

Two component commercial epoxy coatings with a thickness of 30 μm were applied on degreased mild steel (St37) and glass (in order to form free films) substrates by a film applicator. Obtained Free films were attached on steel from four sides. EIS study of epoxy polymeric films both coated and attached as a free film on steel was performed using a standard electrochemical cell including coated sample, saturated (KCl) Ag/AgCl electrode and platinum electrode as working, reference and auxiliary electrodes respectively during exposure to 3.5 wt% NaCl solution. Determination of water uptake was carried out on free films according to ASTM D-570.

3 Results and Discussion

The fitting of experimental EIS data, using the program of Nova 1.11, with a model of general equivalent electrical circuit for polymer-coated metal shows that impedance plots correspond well to the model in Fig. 1, where R_s, R_{pore}, CPE_{pore}, R_{ct}, and CPE_{dl} correspond to the solution resistance, pore resistance, pore capacitance, charge transfer resistance and double layer capacitance, respectively. The impedance of CPE is given by:

$$Z_{CPE} = 1/Q(j\omega)^n \tag{1}$$

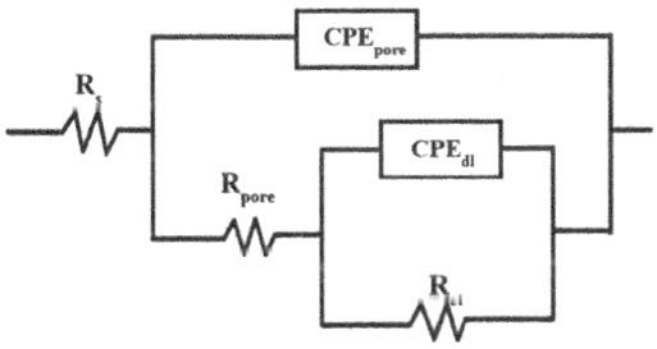

Fig. 1. An impedance model of a polymer-coated metal

Where, ‘n’ is the CPE exponent and Q (CPE) is the capacitance element. When n = 1 Q will be pure capacitance while it will be pure resistance when n = 0. Q is called as CPE when 0 < n < 1 and it prevails under conditions of surface heterogeneity. The time course of the impedance parameters used to establish the effectiveness of the anticorrosive properties. Figure 2 shows the variations of the pore resistance and pore capacitance for epoxy coating and epoxy attached free film, over time. From Fig. 2 it could be inferred that epoxy coating shows greater resistance when applied on mild steel than attached as a free film. Adhesion seems to play a very important role in the protection of steel. When the adhesion is low, anodic and cathodic ions beneath the coating can easily reach and neutralize each other so corrosion could occur in a much faster rate and lower impedance values would be expected. The rate of capacitance growth in the free film is higher than that of the coating Fig. 2. After several days of exposure, the capacitance was still higher in the free films, suggesting that the total amount of water uptake was higher in the free films. The best explanation for the different variation in capacitance seems to be the ingress of ions from the solution. Applied coatings are rigidly held on the surface of the metal and substrate act as a support for the coating reducing stresses and distortions which are a consequence of water penetration [7].

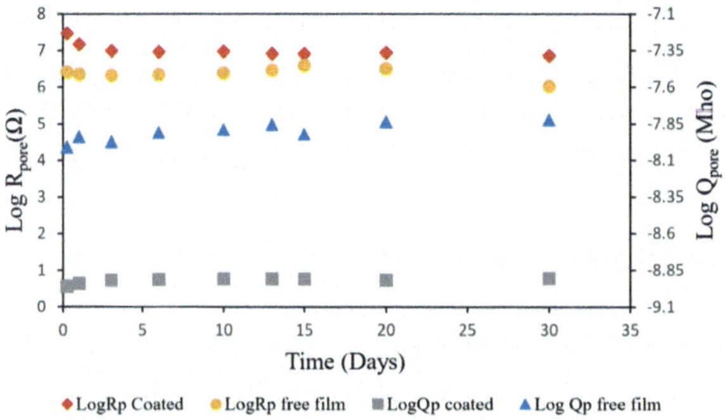

Fig. 2. The time dependence of pore capacitance and pore resistance for an epoxy resin both coated and attached as a free film on mild steel, during immersion in 3.5% NaCl solution

The gravimetric sorption liquid data are reported in Fig. 3. This initial linear behavior suggest that the sorption is controlled by fickian diffusion. For short times, where $M_t/M_\infty < 0.6$, the sorption data should obey the following Eq. (4):

$$\frac{M_t}{M_\infty} = \frac{4t^{1/2}}{\pi^{1/2} l} D^{1/2} \quad (2)$$

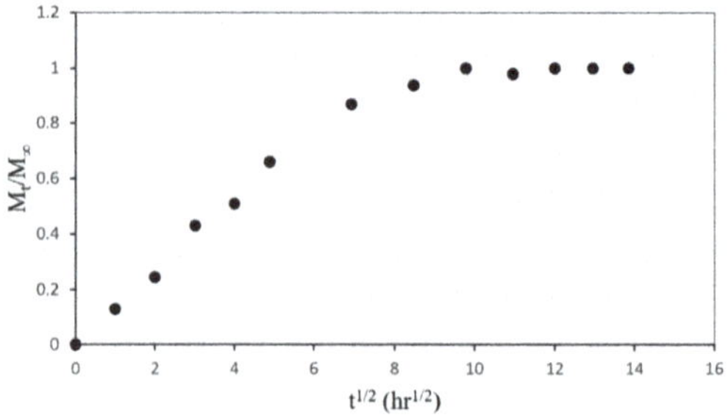

Fig. 3. The reduced sorption curve for epoxy coating during exposure to DI water

By using Eq. (2), the diffusion coefficient for water across epoxy coating was calculated to be 2.9×10^{-11} $cm^2.s^{-1}$. The diffusion coefficient of Cl^- ions across an epoxy coating has been reported as 4.7×10^{-12} $cm^2.s^{-1}$, *i.e.* one order of magnitude smaller than the diffusion coefficient of water [3]. The saturation time of coating with water, (Fig. 3), almost coincides with the initial increase in coating capacitance and initial decrease in pore resistance, (Fig. 2), denoting the entry of electrolyte into the coating. The first step of electrolyte penetration through the coating is related to water uptake, when molecules of water diffuse in the micro pores of polymer net according to Fick's law [3]. The capacitance of attached free films grows slowly even after apparent stabilization. It is possible that after the initial penetration of water in the pores of the coating, a different process occurs, that could be either the penetration of the water in the bulk of the polymer, or the slow diffusion of ions. Given the lack adhesion, it seems that these processes occur faster [8].

4 Conclusion

Epoxy resin coated on steel showed better resistance and lower water uptake compared with epoxy free films attached without adhesion to the substrate, which indicates the important role of adhesion on the resistance of coatings. Electrolyte penetration through the coating occurs in several steps. First is related to water uptake and after that a different process occurs.

References

1. Augustyniak, A., Tsavalas, J., Ming, W.: ACS Appl. Mater. Inter. **1**(11), 2618–2623 (2009)
2. Lyon, S.B., Bingham, R., Mills, D.J.: Prog. Org. Coat. **102**, 2–7 (2017)
3. Mišković-Stanković, V.B., Drazić, D.M., Kacarević-popović, Z.: Corr. Sci. **38**, 1513–1523 (1996)
4. Belluci, F., Nicodemo, L.: Corr. Sci. **49**, 235–247 (1993)
5. González-García, Y., González, S., Souto, R.M.: Corr. Sci. **49**, 3514–3526 (2007)
6. Liu, J., Zhang, L., Mu, X., Zhang, P.: Prog. Org. Coat. **111**, 315–321 (2017)
7. Walter, G.W.: Corr. Sci. **32**(10), 1085–1103 (1999)
8. Castela, A.S.L., Simões, A.M., Ferreira, M.G.S.: Prog. Org. Coat. **38**, 1–7 (2000)

Dielectric Properties of Polydimethylsiloxane (PDMS) Composites Containing Hybrid Silica-Decorated Reduced-Graphene Oxide (SiO_2@rGO)

Mohsen Sadroddini and Mehdi Razzaghi-Kashani(✉)

Department of Polymer Engineering, Faculty of Chemical Engineering, Tarbiat Modares University, P.O. Box, 14115-114 Tehran, Iran
mehdi.razzaghi@modares.ac.ir

Abstract. The aim of this study is to investigate the effect of a new hybrid graphene-based filler, namely silica-decorated reduced-graphene oxide (SiO_2@rGO) nano-plates on dielectric characteristics (dielectric constant, dielectric loss, and dielectric breakdown strength) of PDMS composites. The SiO_2@rGO hybrid was synthesized using tetraethylorthosilicate (TEOS) and chemical reduction by hydrazine solution. Dielectric results revealed that by replacing rGO with SiO_2@rGO in PDMS composites, a significant improvement in dielectric performance of the final composites was achieved. Such improvement was explained and quantified by "dielectric efficiency" concept which will introduce in this study. In comparison of rGO/PDMS with SiO_2@rGO/PDMS composites, later ones have higher dielectric efficiency and greater dielectric breakdown strength. This achievement provides a simple and efficient approach in relation to high performance dielectric smart polymer composites, especially in electric energy storage devices and dielectric elastomer actuators applications.

Keywords: Graphene Oxide (GO) · Silica-decorated reduced · Graphene oxide (SiO_2@rGO) · Polydimethylsiloxane (PDMS) · Dielectric properties · Dielectric breakdown strength

1 Introduction

Dielectric polymer composites as new smart materials have found very interesting potential applications in wide range of electronic industry from electrical energy storage devices up to sensors and actuators [1, 2]. Generally, low dielectric loss, high dielectric breakdown strength and high dielectric efficiency are three key parameters in a high performance dielectric composite for practical applications.

Incorporation of electrically conductive fillers into polymers is a known approach to increase dielectric constant of polymers. Despite this positive point, the main drawback of this approach is their high dielectric loss and low dielectric breakdown strength, which restricts their practical applications [3].

H. Mirzadeh and A. A. Katbab (Eds.): ISPST 2018, *Eco-friendly and Smart Polymer Systems*, pp. 442–445, 2020.
https://doi.org/10.1007/978-3-030-45085-4_107

With the aim of high performance dielectric composites, present study was employed a facile and effective route to achieve a high performance dielectric polymer composites. This task is accomplished by using new hybrid graphene-based filler. The dielectric results showed that using SiO_2@rGO instead of rGO could be an efficient approach toward a high performance dielectric polymer composites.

2 Experimental

GO was synthesized by the oxidation of graphite powder according to improved Hummer's method. The SiO_2@GO was synthesized in water/ethanol mixture containing specific amount of TEOS under a reflux system at room temperature for 18 h for decorating silica onto GO surface. The SiO_2@rGO hybrids was obtained by chemical reduction of SiO_2@GO using hydrazine hydrate solution. Finally, rGO/PDMS and SiO_2@rGO/PDMS composites were prepared by solution mixing method using THF as solvent. Composites were cured and post-cured at 55 °C for 24 h and at 120 °C for 4 h, respectively, to form thin composite films.

3 Results and Discussion

Figure 1 shows dielectric constant and dielectric loss of the composites at room temperature and frequency of 1 kHz. All SiO_2@rGO/PDMS composites have lower dielectric constant and dielectric loss compared to their corresponding rGO/PDMS composites. An important point to mention here is that by using SiO_2@rGO the reduction in dielectric loss is much higher than the reduction in dielectric constant. For example, dielectric constant decreases from 19.84 (for rGO/PDMS, 3 wt%) to 6.31 (for SiO_2@rGO/PDMS, 3 wt%), while dielectric loss value decreases, respectively, from 4.24 to 0.58, in other words, we almost see a seven-fold decrease in dielectric loss and three-fold decrease in dielectric constant. This concept is "dielectric efficiency" which is the ratio of "increase in dielectric constant" to "increase in dielectric loss" due to effect of filler addition.

To achieve a high performance dielectric composite, a high dielectric efficiency is required. Another important factor in dielectric composites for practical applications is the dielectric breakdown strength, thus it has also been investigated in this study.

Based on the above mentioned results, Fig. 2 shows dielectric breakdown strength and dielectric efficiency of the composites. Dielectric breakdown strength and dielectric efficiency were found to decrease with increasing filler content. This may be due to dominant effect of the dielectric loss with increasing filler content.

The most important result is that the SiO_2@rGO/PDMS composites have much higher dielectric breakdown strength and dielectric efficiency compared to their corresponding rGO/PDMS composites which prove the major effect of silica decoration on dielectric performance of the final composites.

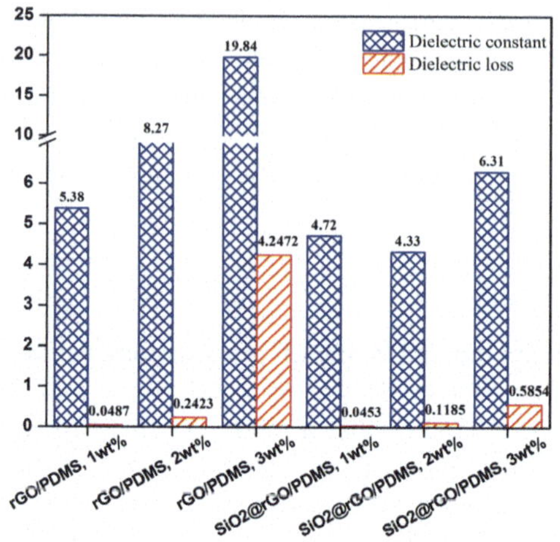

Fig. 1. Dielectric constant and loss of the composites

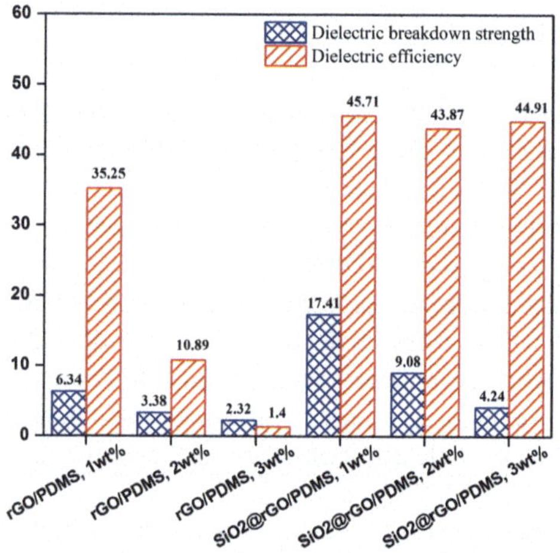

Fig. 2. Dielectric breakdown strength and dielectric efficiency of the composites

In summary, the existence of an insulating silica shell in SiO_2@rGO acts as a barrier layer to suppress dielectric loss through blocking free electron transfer across the adjacent rGO through the matrix [3]. This phenomenon leads to lower dielectric loss, higher dielectric breakdown strength, and higher dielectric efficiency of the final composites.

4 Conclusion

An efficient approach toward high performance dielectric PDMS/graphene-based composites was utilized. New hybrid filler with rGO core and an insulating silica shell was synthesized and named SiO_2@rGO. Dielectric results obviously revealed fundamental role of the silica decorating onto GO surface and its positive impact on dielectric properties of the composites. In detail, it was shown that SiO_2@rGO/PDMS composites have much higher "dielectric efficiency" and dielectric breakdown strength compared to their corresponding rGO/PDMS composites. This facilitates wide applications of dielectric polymer composites, especially in smart applications of polymers such as dielectric elastomer actuators (DEAs) and electric energy storage devices.

References

1. Tian, M., Wei, Z., Zan, X., Zhang, L., Zhang, J., Ma, Q., Ning, N., Nishi, T.: Compos. Sci. Technol. **99**, 37–44 (2014)
2. Wang, Q., Zhu, L.: J. Polym. Sci. Part B: Polym. Phys. **49**, 1421–1429 (2011)
3. Sadroddini, M., Razzaghi-Kashani, M., Miranzadeh, M., Kassaee, M.Z.: Polym. Compos. (2015, in-Press). https://doi.org/10.1002/pc.24068

Effect of Ethylene Monomer Pressure on BCE Ziegler-Nata Catalyst Fragmentation in the Early Stage of Polymerization

Mohammad Mansouri, Gholam Hossein Zohuri(✉), and Navid Ramezanian

Department of Chemistry, Faculty of Science, Ferdowsi University of Mashhad, P.O. Box 91775-1436, Mashhad, Iran
zohuri@um.ac.ir

Abstract. In order to investigate ethylene monomer pressure on fragmentation and in the morphology of the BCE Ziegler-Nata catalyst particles ($TiCl_4/MgCl_2$) in the early stages of polymerization reaction, polymerization was carried out at a 3.0 bar to 4.5 bar of the ethylene monomer pressure. Scanning electron microscope (SEM) has been used to investigate the effect of ethylene monomer pressure on morphology of the catalyst particles. SEM images show that with increasing pressure size of pores and cracks of the catalyst increases. Meanwhile activity of the catalyst was linearly increased from 16.0 to 18.6 ($gPE.mmolTi^{-1}.h^{-1}$) with increasing of the monomer pressure.

Keywords: Ziegler-Natta catalyst · Ethylene polymerization · Fragmentation · Morphology · SEM

1 Instruction

Undaptedly one of remarkable discoveries in the field of polyolefins was the discovery of the Ziegler-Nata catalyst [1]. In the catalytic polymerization of polyolefins, the final polymer morphology produced is under influence of the changes which may happen in the morphology of the catalyst and its fragmentation in the early stages of the polymerization [2]. The catalyst fragmentation requires a high porosity in the catalyst to lead the monomer into the catalyst particles and carry out the polymerization on the internal active sites. The hydraulic pressure of the polymer produced inside the catalytic pore causes the catalyst fragmentation [3, 4].

2 Experimental/Theoretical

Polymerization reaction of ethylene was carried out in a buchi metal reactor of 1 L in the presence of triethylaluminum (TEA) as a cocatalyst. In order to investigate the effect of pressure on BCE Ziegler-Natta catalyst ($TiCl_4/MgCl_2$) fragmentation, the polymerization reaction was carried out at to 3.0 bar to 4.5 bar of ethylene monomer pressures, 84 °C, reaction time of 5 min and 25:1 M ratio of cocatalyst to catalyst.

H. Mirzadeh and A. A. Katbab (Eds.): ISPST 2018, *Eco-friendly and Smart Polymer Systems*, pp. 446–448, 2020.
https://doi.org/10.1007/978-3-030-45085-4_108

In all of the reactions, slurry polymerization was carried out in hexane. At the end of the polymerization catalyst particles was washed using hexane containing anhydrous THF (THF/Ti: 200/1 mol ratio) [5] and dried under vacuum.

3 Results and Discussion

SEM images of the original catalyst are shown in Fig. 1, with an average catalyst particle diameter of about 12.0 μm. In order to investigate the effect of ethylene pressure on morphology and the catalyst activity in the early stages of the polymerization, polymerisation was performed at different pressures of ethylene monomer. The results of the activity analysis indicate that by increasing of the pressure of ethylene monomer, activity of the catalyst was linearly increased (Fig. 2). SEM images show that with increasing of the monomer pressure, size of the porosity and cracks in the catalyst was increased. The behavior could be due to the greater penetration of the monomer into the catalyst particle and the polymerization in the active sites, which produce higher hydrolytic pressure into the pores of the catalyst (Fig. 3).

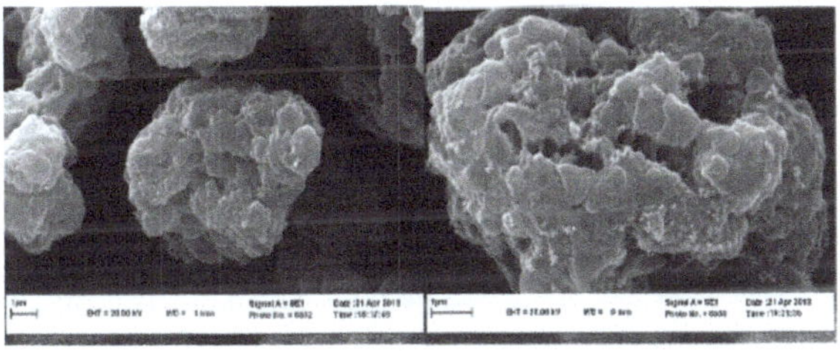

Fig. 1. The original SEM image of the catalyst

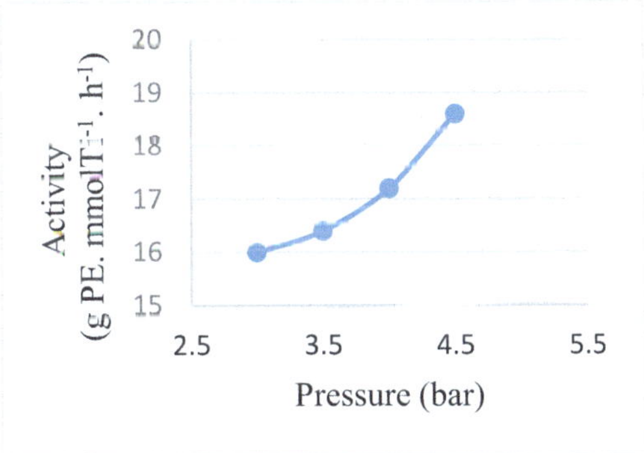

Fig. 2. Effect of monomer pressure on activity of the catalyst

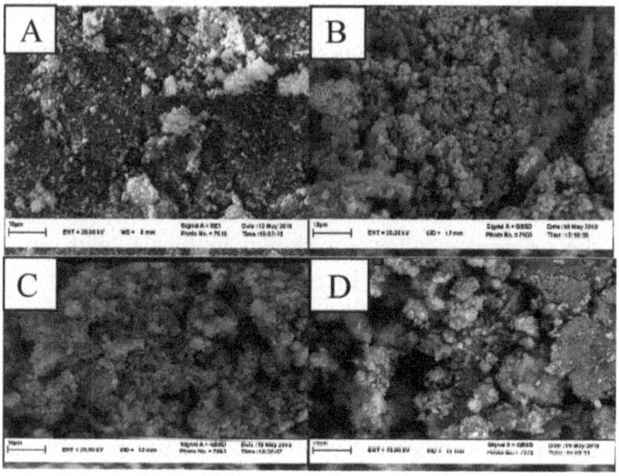

Fig. 3. SEM images of monomer pressure on the catalyst fragmentation: (A) 3.0 bar, (B) 3.5 bar, (C) 4.0 bar, and (D) 4.5 bar

4 Conclusion

The polymerization results was shown that with increasing pressure of the monomer, the catalyst activity was linearly increased. The SEM images was shown that the size of pore and catalyst cracks tends to increase with increasing of the monomer pressure too.

Acknowledgement. This work was supported by Department of Chemistry, Faculty of Science, Ferdowsi University of Mashhad (Project code: 3/46093), Mashhad, Iran which is appreciated.

References

1. Dwivedi, S., Taniike, T., Terano, M.: Macromol. Chem. Phys. **215**, 1698–1706 (2014)
2. Abboud, M., Denifl, P., Reichert, P.K.H.: Macromol. Mater. Eng. **290**, 558–564 (2005)
3. Mayrhofer, L., Paulik, C.: Macromol. React. Eng. **8**, 194–200 (2014)
4. Ronkko, H.L., Korpela, T., Knuuttila, H., Pakkanen, T.T., Denifl, P., Leinonen, T., Kemell, M., et al.: J. Mol. Catal. A **309**, 40–49 (2009)
5. Thang, V.Q., Taniike, T., Umemori, M., Ikeya, M., Hiraoka, Y., Nguyen, N.D., Terano, M.: Macromol. React. Eng. **3**, 467–472 (2009)

Evaluation of Mechanical Properties of Polylactic Acid/Nanoclay Nanofibers Using Taguchi Method

M. Aloushi, Marziyeh Ranjbar-Mohammadi, Ali Vahidifar, Elnaz Esmizadeh(✉), and Golshan Hajizadeh

Faculty of Engineering, Department of Polymer Science and Engineering, University of Bonab, P.O. Box 5551761467, Bonab, Iran
esmizadeh@bonabu.ac.ir

Abstract. In this study, nanofibers based on polylactic acid (PLA)/nanoclay were produced using electrospinning method. The mechanical properties of the nanofibers were studied by ANOVA method. The effects of formulation parameters as nanoclay contents 0 wt% to 6 wt%, and processing parameters such as spinning voltage, solution concentration and feeding rate on modulus and strength of PLA/nanoclay nanofibers were studied.

Keywords: Nanofiber · Electrospinning · Taguchi · Nanoclay · PLA · Morphology

1 Introduction

Electrospinning has been known as an efficient technique to produce nanofibers through imposes of an external electric field on a polymer solution or melt [1, 2]. In a typical process, upon the applied electric field overcomes the surface tension of the droplet, a charge jet of polymer grows longer and thinner due to bending instability until it solidifies on the collector [3]. The main feature of the electro-spinning process is to fabricate a non-woven fabrics composed of electrospun fibers have a large specific surface area and small pore size compared to commercial textiles, making them excellent candidates for use in filtration, biomedical materials and membrane applications [4]. Taguchi's approach provides a systematic methodology for conducting experiments utilizing orthogonal arrays to study a large number of variables using a small number of experiments [5]. Dr. Taguchi advocates the use of orthogonal array designs to allocate selected factors. The most common designs of orthogonal arrays are L_{18}, L_{16}, and L_9 [6]. This method uses statistical methods in engineering processes.

In this research, we tried to evaluate the mechanical properties of polylactic acid nanofibers reinforced by nanoclay using Taguchi method.

H. Mirzadeh and A. A. Katbab (Eds.): ISPST 2018, *Eco-friendly and Smart Polymer Systems*, pp. 449–452, 2020.
https://doi.org/10.1007/978-3-030-45085-4_109

2 Materials and Methods

- Organo-modified clay (Cloisite®15A, Southern Clay Products, USA),
- Polylactic acid (PLA 2003D, NatureWorks LLC, USA),
- Dichloromethane (DCM, Merck, Germany).

The prepared nanoclay solution in DCM was mixed for 30 min and homogenized by ultrasonic homogenizer. Then it was added into the polymer solution and mixed overnight. A horizontal electrospinning apparatus was used consisting of a 10 ml syringe with a needle tip (gauge 19 mm), a syringe pump, and a high-voltage power supply of 40 kV. Applied voltage, solution feed rate, nanoclay content and PLA concentration were chosen as the most influential electrospinning parameters according to Table 1 based on L_9 orthogonal array designed by Taguchi.

Table 1. L_9 Orthogonal array of Taguchi method

Sample no.	Nano clay	PLA concentration	Feed rate	Voltage
1	3	3	2	1
2	1	3	3	3
3	3	2	1	3
4	1	2	2	2
5	3	1	3	2
6	2	1	2	3
7	2	3	1	2
8	2	2	3	1
9	1	1	1	1

For studying the mechanical properties, PLA and PLA/nanoclay films were mechanically made using pulling device made by the company. The films were rectangular with dimensions of 1×3 in.

3 Results and Discussion

The relative contribution percentage (P) of each factor obtained by the ANOVA method is given in Table 2. It can be concluded that the significance of factors prevails in the following order of importance (1) PLA concentration, (2) nanoclay content, (3) feed rate, and (4) voltage.

Table 2. ANOVA table

Col/factor	Dof (f)	Sum sqrs	Variance (v)	Pure sum	Percent (%)
Nanoclay (wt%)	2	0.001	0	0.001	16.086
PLA (wt%)	2	0.008	0.004	0.008	74.601
Feed rate (mL/h)	2	0	0	0	6.614
Voltage (kV)	2	0	0	0	1.813

Before study on the mechanical properties of nanofibers, it should be mentioned that the thickness of pure PLA nanofibers was about 0.3, which it was increased to 0.4 by adding 6 wt% nanoclay. It can be said that nanoparticles presence increases the viscosity of PLA solution and causes increment in nanofiber diameter.

In order to study the mechanical properties, effects of nanoclay content on modulus and tensile strength are shown in Table 3. By increasing nanoclay from 0 phr to 6 phr, modulus and strength increased from 0.04 to 0.12 GPa and from 0.47 MPa to 1.25 MPa, respectively. The results obtained for nanoclay films show that nanoparticles have a significant effect on the mechanical properties of the PLA/nanoclay nanofibers. Effect of processing parameters such as feeding rate, voltage and PLA concentration in solution on the modulus of nanofibers is illustrated in Fig. 1. As it can be seen, in order to produce fibers with high modulus, it is useful to use solution having higher nanoclay content with low PLA concentration. While, higher spinning voltage of 20 V, can produce high modulus PLA/nanoclay nanofibers.

Table 3. Mechanical properties of poly-lactic acid and nanocomposites

Sample	Strength (MPa)	Modulus (GPa)
PLA pure	0.47	0.04205
PLA/6% clay	1.25	0.11908
PLA/4% clay	0.73	0.11417
PLA/2% clay	0.57	0.06368

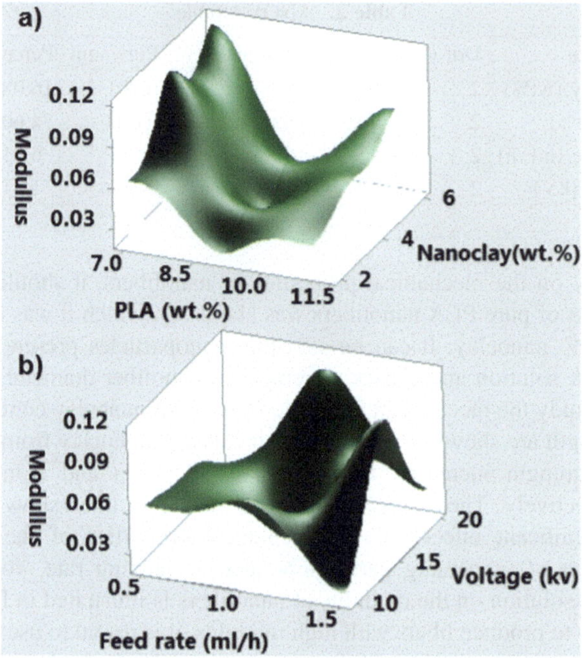

Fig. 1. (a) Surface plot of modulus vs nanoclay (wt%) and PLA (wt%) and (b) surface plot of modulus vs voltage (kV) and feed rate

4 Conclusion

The Taguchi method was used for PLA/nanoclay nanofibers with nanoclay contents of 2, 4, and 6 wt%. The results showed that PLA concentration, nanoclay content, feed rate and voltage had highest effect on the PLA/nanoclay nanofiber properties, respectively. Nanoclay content had linear effects on modulus, strength and physical properties including thickness.

References

1. Deitzel, J.M., Kleinmeyer, J., Harris, D., et al.: Polymer **42**, 261–272 (2001)
2. Shih, K.Y.: US Patent 6,686,306 (2004)
3. Reneker, D.H., Yarin, A.L., Fong, H., et al.: J. Appl. Phys. **87**, 4531–4547 (2000)
4. Chuang, W.Y., Young, T.H., Yao, C.H., et al.: Biomaterials **20**, 1479–1487 (1999)
5. Phadke, M.S.: Quality Engineering Using Robust Design. Prentice Hall, Upper Saddle River (1989)
6. Antony, J., Antony, J.F.: Teaching the Taguchi method to industrial engineers. University Press **50**(4), 141–149 (2001)

Creating Gas Lenses Using Polymer in a Minimum Pressure Range of 0.01 to 0.1 Bar

Soheila Shojaei[1(✉)], Mohammad Bohloul[2], and Zhila Shah Ali[3]

[1] Department of Physics, Islamic Azad University, Qom, Iran
shoja_l@yahoo.com
[2] Department of Physics, Payame Noor University, Tehran, Iran
[3] Department of Physics, Almahdi University, Isfahan, Iran

Abstract. Given the increasing importance of lenses, this study has focused on designing and introducing gas lenses by putting forward the question "Is it possible to compress gas using pressure at the ambient temperature so that it forms a gas-like lens with a defined focal length?" Poly methyl Methacrylate (PMMA) and carbon dioxide were used to create the lenses. In this study, the model was simulated using ANSYS (Ver.11). This work has been done for the first time in the world. Due to the specifications of the polymer and the gas used, the focal length of the lenses was calculated at various pressures of 0.01 bar to 0.1 bar.

Keywords: Gas lenses · PMMA · Pressures · Carbon dioxide · Focal length

1 Introduction

Due to the wide and diverse use of lenses in the optical and medical industries and new technologies, further research is needed in this regard. This study has focused on designing and introducing gas lenses. According to the present study, this is a new method to use gas was chosen as the lens structure from the gases studied.

To date, a large number of lenses have been made with solids, liquids and gases for different applications such as optics, industry, medicine, and so on.

Marcus and Miller [1] made use of gas to focus light in order to use in the welding industry. Mitchell et al. [2]. used a gas lens to build a telescope.

Considering that a gas lens with applied pressure variation at the pressure of 64 bar (point of gas to liquid transition) has already been done by researcher Shojaei [3], in this study, carbon dioxide was used in the gas lens to produce energy.

S. Shojaei—Researcher and Inventor in Physics.

H. Mirzadeh and A. A. Katbab (Eds.): ISPST 2018, *Eco-friendly and Smart Polymer Systems*, pp. 453–456, 2020.
https://doi.org/10.1007/978-3-030-45085-4_110

2 Assumptions and Specifications

Between the nitrogen, hydrogen, and carbon dioxide studied, carbon dioxide was chosen for a variety of reasons, including its ease of accessibility, insulation, less environmental pollution, reasonable price, as well as its molecular properties in thermodynamic processes when applying different pressures.

In this study, the model was simulated using ANSYS (Ver.11). CO_2 has very high permeability and has been used for the flexibility of a transparent polymer due to its limited pressure variations. It was assumed that the thermal stresses did not affect the structural calculations and the materials at the time of applying pressure and after were in the elastic range. In order to cover a wide range of flexible materials, the elasticity modulus of this material was considered to be between 100 and 500 kNm^{-2} (Table 1).

Table 1. Specifications of the polymer [4]

E (kN/m^2)	C_{01} (kN/m^2)	C_{10} (kN/m^2)	N
10.39	6.2766	10.39	0.49967
200	12.5532	20.78	0.49967
300	18.8298	31.17	0.49967
400	25.1064	41.56	0.49967
500	31.3829	51.95	0.49967

3 Modeling

The material of selective behavior was in fact the equivalent of the flexible part of the empirical model defined in the simulation with the following specifications:

$$C_{01}.C_{10} : (Mooney\ Rivilin\ factors, \frac{C_{01}}{C_{10}} \approx 0.6, C_{01} + C_{10} \approx E/6$$

The flexible section was simulated with the CONTAC52 element. Modeling and loading was done in ANSYS environment. In the following figure, a three-dimensional lens model is presented.

The refractive index of carbon dioxide as a function of gas pressure was calculated according to the following equation [5] (Fig. 1).

$$n_{co2} = 1 + \tfrac{\alpha P}{RT} \times P, \quad \alpha P = 9/935 \left(\frac{\mathrm{Cm}^3}{\mathrm{mol}}\right)$$

$$n_{co2=1+4/01\times} 10^{-9}\ \mathrm{p}\left(\frac{\mathrm{N}}{\mathrm{m}^2}\right)$$

$$n_{co2=1+4/01\times10^{-9}\times10^4}, \quad n_{co2=1/0000401}$$

$$R^{/} = \frac{44/5533}{cos\ cos\left(\frac{44/5533}{\delta x} - \frac{\delta x}{44/5533}\right)}$$

$$R^{/} = 44/55486314$$

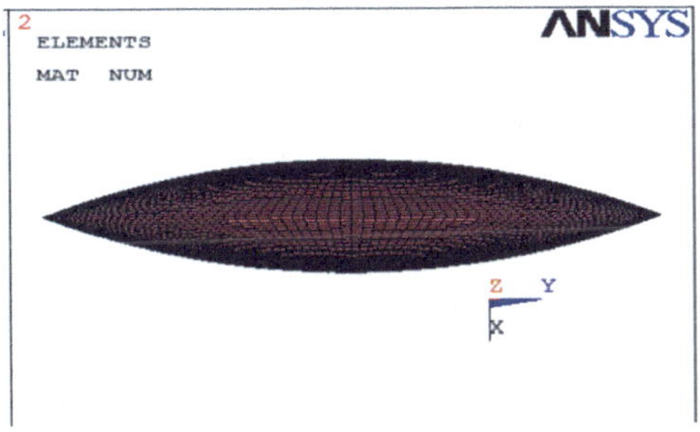

Fig. 1. The lens wall thickness was considered to be 1 cm and the most amount of lens thickness was in the middle with 20 cm. The radius of the lens curvature was 104.25 cm and the lens height was 90 cm

The focal length of the gas lens is calculated according to the following equation [5]:

$$(f_{co2})^{-1} = (n_{co2-1}) * [2 * (R_1)^{-1} + (n_{co2-1}) * D/((R_2 * R_1) * n_{co2}$$

$R_1.R_2$: Radius curvature of the sides of the lens
n_{co2}: The refractive index of carbon dioxide
f_{co2}: Focal length of the gas lens, D = 20 cm

As a result, the focal length of the flexible parts and the gas parts are calculated from the following equation:

$$(f)^{-1} = \frac{1}{(f_{co2})} + \frac{2}{(f_{flex})} + \frac{1}{f_{Glass}} \quad R_{Glass=\infty} \rightarrow (f)^{1} = \frac{1}{(f_{co2})} + \frac{2}{f_{(flex)}}$$

As a result, the Mooney-Rivilin model was used [5]:

$$d = \frac{1 - 2v}{c_{10} - c_{01}} = -3.96 * 10^{-5}$$

And the maximum displacement of the lens radius was calculated using simulation data. The three gas pressures equaled to 0.1, 0.05 and 0.01 respectively, the changes in the polymer radius are as follows:

$$p = 10\ \frac{\text{kN}}{\text{m}^2}(0.1\ \text{bar}) \quad \delta_{max} = 36\ \text{cm}$$
$$p = 5\ \text{kN/m}^2(0.05\ \text{bar}) \quad \delta_{max} = 18\ \text{cm}$$
$$p = 1\ \text{kN/m}^2(0.01\ \text{bar}) \quad \delta_{max} = 5.5\ \text{cm}$$

Additionally, the simulation of the focal length variation with gas pressure according to the elastic coefficient of the flexible material showed that E = 500 kNm^{-2} with the difference that it was tested at 0.1 bar.

The black curve shows the lens surface before applying the pressure. Due to the laboratory conditions, the pressure applied was from 1 kNm^{-2} to 10 kNm^{-2} to make the results more accurate. In the radial variation diagram, the higher the pressure was applied, the bigger the radius became. The elasticity modulus of polymer from 100 to 500 was investigated, as shown below (Diagram 1).

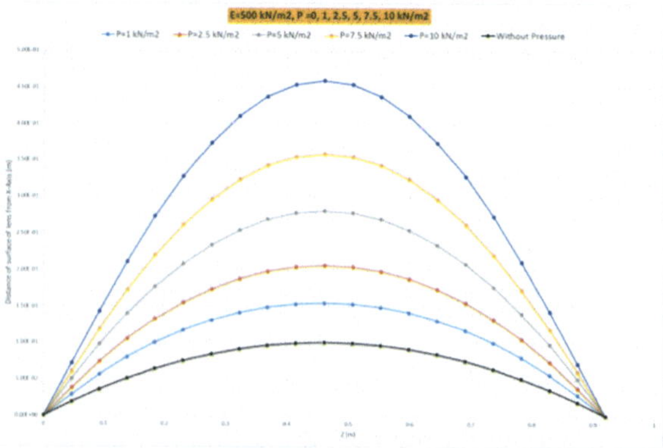

Diagram 1. Results of the polymer with an elasticity modulus of 500 and applying pressure from 1 kNm^{-2} to 10 kNm^{-2}

The elasticity modulus of polymer was considered to be from 100 to 500 and, in ratio to each elastic modulus, pressure from 1 kNm^{-2} to 10 kNm^{-2} was calculated in the simulation to consider the laboratory conditions and factors to increase accuracy.

4 Conclusion

As a result, a lens with a defined focus can be obtained from a gaseous environment. This lens can be used in industrial applications for energy storage and creating optical properties of gases when charged.

References

1. Marcuse, D., Miller, S.D.: Analysis of a tubular gas-lens. Bell Syst. Tech. J. **43**, 1759–1782 (1964)
2. Michaelis, M.M., et al.: Nature **353**, 547–548 (1991)
3. Shojaei, S.: US Patent 76940, 26 September 2012
4. Derouet, D., Tran, Q.N.: J. Appl. Polym. Sci. **112**, 788–799 (2009)
5. Marchetti, M., Simili, R.: Infrared Phys. Technol. **47**, 263–266 (2006)

Effect of Polymers on the Damping Capacity of Automotive Bitumen Anti-vibration Insulators

Haniyeh Miranmousavi[1], Gholamhossien Zohuri[2(✉)], and Mohammad Nourmohammadi[1]

[1] Department of Research and Development, Ayegh Khodro Toos (AKT) Co. of Part Lastic Group, Mashhad, Iran

[2] Department of Chemistry, Faculty of Science, Ferdowsi University of Mashhad, Mashhad, Iran

zohuri@um.ac.ir

Abstract. Sound and vibration damping in automobile industry cause manufacturers to use insulators in automobile parts. High performance polymers are the best choice for noise reduction application. The addition of commercial polymers to bitumen base insulators of Ayegh Khodro Toos (AKT) Co. improved their performance. Effect of polymers such as styrene-butadiene-styrene (SBS), styrene butadiene rubber (SBR) and butyl rubber with different weight percentages (2, 4, 6, and 8 w%) was studied in laboratory scale (3.5 kg product). Butyl rubber showed the highest vibration absorption (about 27%) among the polymers studied. Generally, the higher the concentration of the polymer used, the lower tensile strength, elongation at break and also the higher of damping was obtained.

Keywords: Insulator · Polymer insulator · Bitumen · Vibration absorption · Damping

1 Introduction

Bitumen insulator products are widely used in the automotive and commercial vehicle market due to their performance and low cost [1]. Anti-vibration products are utilized to minimize noise and vibration for vehicle occupants and operators, as well as improving longevity and reliability of electrical and mechanical components [2]. Sheets are manufactured from bitumen, synthetic polymers, and minerals to create flexible and viscoelastic materials. Polymers such as Low density polyethylene (LDPE) [3], styrene-butadiene-styrene copolymer (SBS) and ethylene-vinyl-acetate (EVA) were used to provide anti vibration properties for bitumen insulators [3]. In a study to reduce noise and vibration using a polyurethane foam insulator, good results suitable for automobile industry were obtained [4]. Also, in another study, the sound absorption properties of nonwoven polyester/polyethylene were investigated and improved noise absorption properties were revealed [5]. Researchers investigated low weight, high elastic thermal and sound insulators as well as used polyether as main material

H. Mirzadeh and A. A. Katbab (Eds.): ISPST 2018, *Eco-friendly and Smart Polymer Systems*, pp. 457–460, 2020.
https://doi.org/10.1007/978-3-030-45085-4_111

of insulators [6]. Furthermore, there are another polymers used as sound absorption and damping like polyurethane/epoxy interpenetrating polymer networks [7] and polyurethane/poly(methyl methacrylate) [8]. Besides, use of fillers to increase mechanical properties and durability, for example, calcium carbonate [9] and palm slog [9] also is common.

In this work, polymers such as SBS, SBR and butyl rubber were used to improve anti-vibration and sound absorption of bitumen insulator.

2 Experimental

Bitumen (21%), carpet fiber, polyethylene (PE) granule and styrene butadiene rubber (2, 4, 6, and 8 w%) were used, and in some cases, SBS and Butyl rubber were used instead of SBR with the same weight percentage. The other chemicals were obtained from AKT Co. Calcium carbonate with mesh 400 was supplied by Omnia Pars Co. In this process, bitumen, PE, calcium carbonate, SBR and the other chemicals were added and mixed in one stage to obtain the insulating compound. The mixing process was carried out at the temperature of 110 °C for 35–40 min. The mixture was completely mixed to obtain a relatively soft homogenous compound with high viscosity.

3 Results and Discussion

Study on the effect of SBS, SBR and butyl on the damping properties showed that the highest damping capacity belonged to butyl polymer. The highest damping (about 27%) was obtained at 4 w% (Fig. 1). This behavior is due to isobutylene pendant group of the butyl rubber which can lose lots of energy [11].

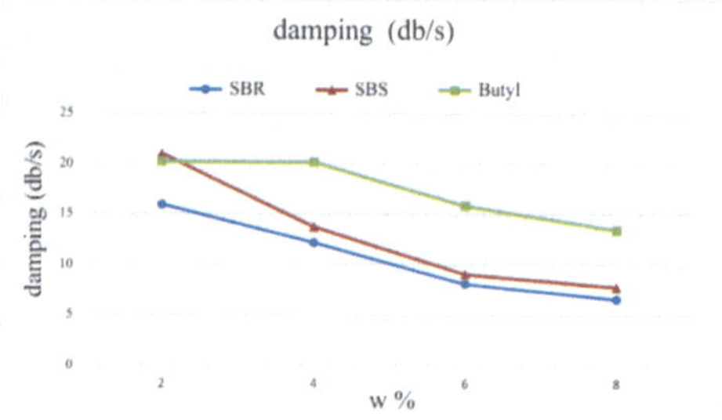

Fig. 1. Effect of polymer concentration on noise damping

However, using SBS (2 w%) provided the highest loss factor (about 0.1260) (Fig. 2). The benzene ring of the polymer is responsible for adsorption of heat, and consequently increasing loss factor. No shrinkage was observed in the compounds containing SBS (Table 1).

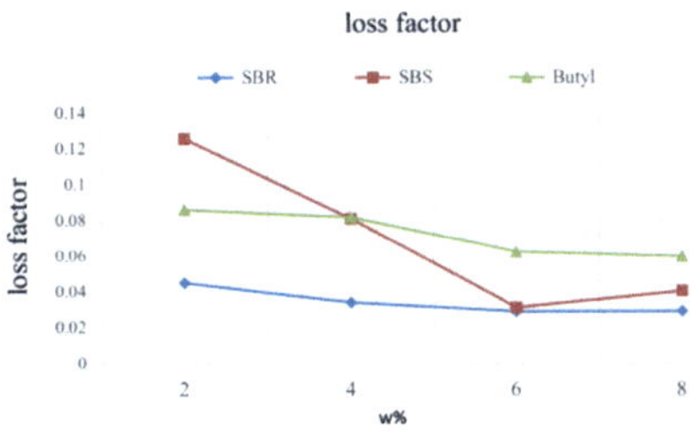

Fig. 2. Effect of polymer concentration on loos factor

No regular behavior was obtained for tensile strength using all of the studied polymers in the compound. SBS caused to increase tensile strength to a limited value (about 0.2 kg/cm^2) (Fig. 3). However, butyl rubber decreased the tensile strength from 0.197 kg/cm^2 to 0.103 kg/cm^2 (Fig. 3). The higher the concentration of the polymers, the higher elongation at break was obtained (Table 1).

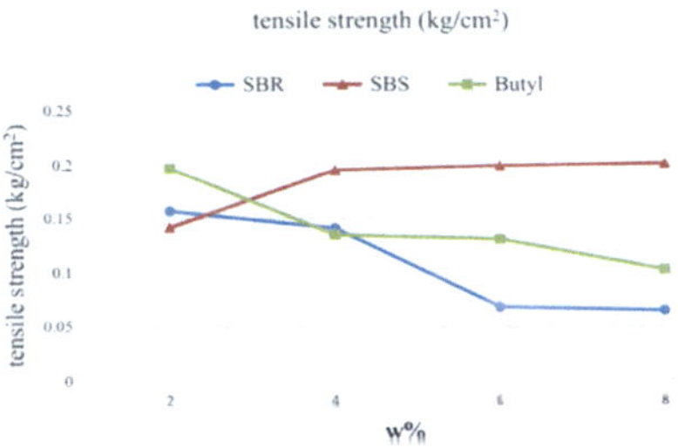

Fig. 3. Effect of polymer concentration on tensile strength

Table 1. Results for insulator samples prepared in this study

No.	Polymers	w%	Elongation (%)	Damping (db./s)	Shrinkage (mm)
1	SBR	2	31.9	15.91	1
2		4	35.16	12.15	1
3		6	69.4	8.06	1
4		8	73.1	6.56	1
5	SBS	2	22.7	20.99	0
6		4	25.3	13.74	0
7		6	26.6	9.07	0
8		8	40.9	7.77	0
9	Butyl	2	12.5	20.21	0.5
10		4	22.5	20.15	0
11		6	28.7	15.85	0
12		8	31.9	13.43	0

4 Conclusion

Using of the polymers (SBS, SBR, butyl) up to 4 w% in bitumen insulating compound reasonably affected the damping capacity, which is the main characteristic of insulators. Among the polymers used, butyl rubber showed the highest damping capacity.

Acknowledgement. This work was supported by AKT Co. Mashhad, Iran which is strongly appreciated.

References

1. Chatterjee, D., Nayak, S.M., Saha, P.: Sound Vibr. **48**, 14–15 (2014)
2. Federico, C., Valeria, C., Fabbrocino, F., Nanni, F.: Open Mater. Sci. J. **12**, 14–28 (2018)
3. Xia, L., Wu, H., Guo, S., Sun, X., Liang, W.: Appl. Sci. Manuf. **81**, 225–233 (2016)
4. Chen, S., Jiang, Y.: Polym. Compos. **39**, 1370–1381 (2018)
5. Tiuc, A.E., Vermeşan, H., Gabor, T., Vasile, O.: Energy Proc. **85**, 559–565 (2016)
6. Yang, T., Xiong, X., Venkataraman, M., Mishra, R., Novák, J., Militký, J.: J. Text. Inst. 1–6 (2018)
7. Wang, G., Zhao, G., Dong, G., Mu, Y., Park, C.B., Wang, G.: Eur. Polym. J. **103**, 68–79 (2018)
8. Zhou, J., Li, H., Lu, X.: Polym. Adv. Technol. **29**, 2308–2316 (2018)
9. Moradi, G., Nassiri, P., Ershad-Langroudi, A., Monazzam, M.R.: Plast. Rubber Compos. **47**, 221–231 (2018)
10. Bonilla-Cruz, J., Hernández-Mireles, B., Mendoza-Carrizales, R., Ramírez-Leal, L.A., Torres-Lubián, R., RamosdeValle, L.F., Paul, D.R., Saldívar-Guerra, E.: Polymers **9** (2017). https://doi.org/10.3390/polym9020063

Effect of Nanoclay on the Morphological Properties of Ternary PA6/POE-g-MAH/LDPE Blend

Tahere Keramati, Shervin Ahmadi(✉), and Ali Salimi

Polymer Processing Department, Iran Polymer and Petrochemical Institute, P.O. Box 14975-112, Tehran, Iran
sh.ahmadi@ippi.ac.ir

Abstract. In recent years, polymer blends have gained more attention because of their applications and extra ordinary properties. Blending in polymers has been divided in two groups: miscible and immiscible blends. Immiscible blends are usually used to improve the properties of blends because of taking advantage of diverse polymers with different characters. It is more probably to achieve specific goals. In this type of polymers, it is necessary to use compatibilizers. Nowadays, nanoparticles play a crucial role as compatibilizer in immiscible polymers because they are able to improve the properties of blends. In this case, nano particles should be located in interface. The effect of nanoclay Cloisite15A on the ternary blend PA6/POE-g-MA/LDPE was investigated. The XRD results demonstrated that nano particles were exfoliated completely. The SEM results showed that average diameters of particles firstly decreased and after reaching of the minimum of average diameters, improved with increasing nano particles.

Keywords: Ternary blend · Compatibilizing · Morphology · Polyamide · Nanoclay

1 Introduction

One of the most successful methods to produce advanced materials with specific properties and practical goals is blending. One of the main reasons for blending polyolefin with polyamide6 is associated with increasing properties of polyamide namely ductility, water absorption and processability and enhancing polyolefins functions related to their thermal stability, hardness and impermeability for oxygen or solvents. In recent years, modified nano particles and making polymer nanocomposites have been gained the attention for achieving better properties. As a result, with dispersing nanoparticles homogeneously into a polymer matrix, different properties of matrix are greatly enhanced. Some optimum properties were demonstrated for the ternary blend system of PA6/LLDPE/POE-g-nanoclay. Moreover, playing a role as a compatibilizer depends on the location of nano particles in polymer blend. They can increase the properties of blends [1].

H. Mirzadeh and A. A. Katbab (Eds.): ISPST 2018, *Eco-friendly and Smart Polymer Systems*, pp. 461–464, 2020.
https://doi.org/10.1007/978-3-030-45085-4_112

2 Experimental

PA6 (NB40NLE grade) produced by Eurotec Company from Turkey with a density 1.13 g/cm, and a melting point 223 °C was purchased. LLDPE (density 0.92 g/cm^3 and MFI 2 g/10 min) was purchased from Imam Petrochemical Complex. Polyolefin elastomer (density 0.92 g/cm^3) was prepared from Karangin. Nanoclay, modified by the ammonium alkyl ion, manufactured by Southern Clay Product (USA) was used in this study. First, a masterbatch containing 2% nanoclay and polyolefin elastomer was prepared using an internal mixer brabender at 140 °C temperature and at maximum rotor speed. Then, mixing was done by a ZSK extruder in the temperature range from 230 °C to 250 °C and at the rotor speed of 250 rpm. Test samples were prepared by injection molding (Table 1).

Table 1. Composition of samples

Composition	Sample code
70/20/10/0	M15A0
70/20/10/0.5	M15A0.5
70/20/10/1	M15A1
70/20/10/2	M15A2

3 Results and Discussion

SAXS test was performed for all samples containing 2% nanoclay. As shown in Fig. 1, for pure Cloisite15A a peak appears at 2θ = 2.8°. When the nanoclay was added to the POE-g-MA, the characteristic peak of clay disappeared, indicating an exfoliated structure. It seems that greater viscosity of the elastomeric phase and the high torque applied to the system lead to locate the chains between the layers of nanoparticles and better distribution of clay in the masterbatch [2].

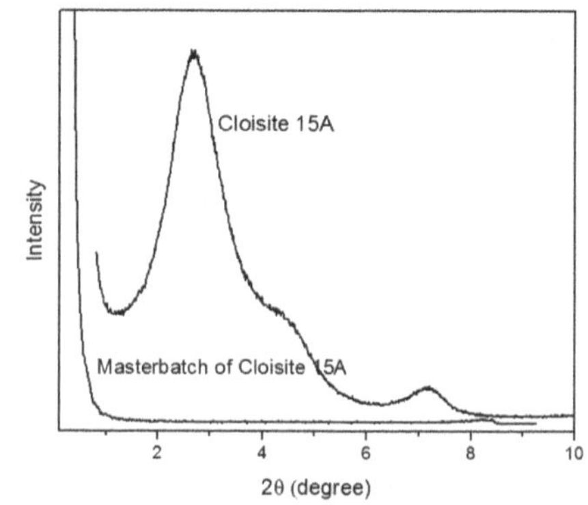

Fig. 1. XRD patterns of samples

On the other hand, polyolefin elastomer grafted with maleic anhydride has maleic anhydride as an active agent. The polar nature of polyamide and nanoparticles lead to migration of nanoclay to polyamide phase. When the nanoclay inclination to both phases becomes balanced, it is expected that nanoclay presumably locate in interphase and curb the coalescence of dispersed particles [3]. As a result, morphology and compatibility were improved greatly. So, locating nanoparticles in the interface leads to a decrease in interfacial tension and compatibility [4]. Figure 2 shows the SEM images of the samples, and Table 2 illustrates the results obtained from ImageJ software. The average diameter of dispersed phase was diminished by adding 0.5 phr masterbatch. The average diameter was enhanced with increasing nanoclay. It was found that the polyolefin elastomer grafted with maleic anhydride was saturated, and migration did not happen. In fact, nanoclay could act as a compatibilizer when used in a certain content [5].

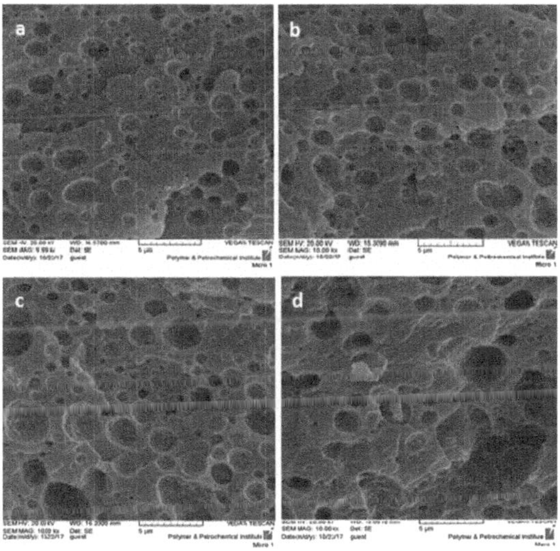

Fig. 2. Micrographs of PA6/POE-g-MAH/LDPE/Nanoclay (a) M15A0, (b) M15A0.5, (c) M15A1, and (d) M15A2

Table 2. Average diameter of dispersed particles

Dn (μm)	Sample
1.36	M15A0
1.20	M15A0.5
1.49	M15A1
1.66	M15A2

4 Conclusion

The results showed that nanoclay Cloisite15A was exfoliated in all compositions. Findings of morphology indicated that nanoclay could act as a compatibilizer when located in the interface. In addition, it was concluded that nanoclay changed the morphology and improved the compatibility. The results depicted that, the size of dispersed phase was improved because of nano particle agglomeration.

References

1. Khoshkava, V., Dini, M., Nazockdast, H.: Appl. Polym. Sci. **125**, E197–E203 (2012)
2. Babaei, A., Arefazar, A.: Appl. Polym. Sci. **132**(20), 1–8 (2015)
3. Chow, W.S., Ishak, Z.A.M., Karger-Kocsis, J.: Macromol. Mater. Eng. **290**(2), 122–127 (2005)
4. Elias, L., Fenouillot, F., Majeste, J.C., Cassagnau, P.: Polymer **48**, 6029–6040 (2007)
5. Dorigato, A., Pegoretti, A., Penati, A.: Express Polym. Lett. **4**(2), 115–129 (2010)

Optical Properties of PMMA Nanofibers with Different Fiber Diameters

Fatemeh Ahmadkhani Khari[1], Saeideh Gorji Kandi[1(✉)], Maryam Yousefzadeh[2], and Farhad Panahi[3]

[1] Polymer Engineering and Color Technology Department, Amirkabir University of Technology, Tehran, Iran
s.gorji@aut.ac.ir
[2] Textile Engineering Department, Amirkabir University of Technology, Tehran, Iran
[3] Mahshahr Campus, Amirkabir University of Technology, Mahshahr, Iran

Abstract. Electrospinning is an effective method for fabrication of nanofibers. Polymethyl methacrylate (PMMA) was electrospun to prepare nanofiber. For investigation of effect of diameter on the optical properties of fibers, fibers were electrospun with different diameters by changing in polymer concentration, needle inner diameter, and flow rate. Fibers with the same morphology were selected. It was observed that in the studied range of diameter by increasing fiber diameter, the reflectance of fibers increased.

Keywords: Electrospinning · PMMA · Nanofibers · Polymer concentration · Fiber diameter

1 Introduction

Nanofibers have attracted much attention due to their special properties and applications [1]. Electrospinning of polymers is a superior method for fabrication of high-aspect-ratio fibers with controllable diameters that range from microns to several tens nanometers [2, 3]. A variety of factors affects the morphology and diameter of produced fibers, such as polymer solution, type of solvent, tip to collector distance, applied voltage, flow rate, needle diameter, and etc. [4].

In this work, PMMA nanofibers were prepared by electrospinning and fibers with similar morphologies and different diameters were produced. Then, the effect of fiber diameter on the optical properties of fibers was investigated.

2 Experimental

2.1 Materials and Methods

Polymethylmethacrylate (PMMA) (M_w = 120000) was purchased from Sigma Aldrich. N,N-Dimethylformamide was purchased from Fluka. These chemicals were used as received, without further purification.

H. Mirzadeh and A. A. Katbab (Eds.): ISPST 2018, *Eco-friendly and Smart Polymer Systems*, pp. 465–468, 2020.
https://doi.org/10.1007/978-3-030-45085-4_113

2.2 Preparation of PMMA Nanofibers

PMMA solutions were prepared with different concentrations. For preparation of polymer solution, PMMA was dissolved in DMF and magnetically stirred overnight at ambient temperature in order to obtain homogeneous solutions. The PMMA concentrations were 2, 3, and 4 wt%.

In order to conduct electrospinning process, the PMMA solution was placed in a syringe with 22 and 25-gauge stainless-steel needle (needles inner diameter was 0.41 and 0.26 mm, respectively). A self-assembled electrospinning apparatus consists of a high voltage power supply and a digitally controlled syringe pump were used. The polymer solution was forced through the syringe needle, at 0.5 and 1 mL/h via a syringe pump. The electric voltage applied by a high voltage power supply was 17 kV. The electrospun fibers were collected on a piece of aluminum foil placed at a distance of 20 cm. The nanofibers used for reflectance measurements were collected on a piece of glass paper that was put on an aluminum foil to facility the separation of nanofibers from the substrate.

2.3 Characterization

The morphology of the fibers was investigated by scanning Electron Microscopy (SEM). The average diameter of the fibers was determined by analysing the SEM image by Image J software. The reflectance of nanofibers was measured by an X-Rite Gretag Macbeth ColorEye 7000A spectrophotometer in specular component exclude mode.

3 Results and Discussion

In this work, for obtaining nanofibers with different diameters, solutions with different concentrations were used. For PMMA solution of 2 wt% concentration, and with 22-gauge needle at the flow rate of 0.5 mL/h, the average of fibers diameter was 166 nm, and when the concentration of PMMA solution increased, in addition to diameter, the morphology of the fibers was changed in the same condition of electrospinning. As the morphology of fibers affects the optical properties, the samples with the same morphology were prepared. For overcoming this problem, needles with different gauges were used. 3 and 4 wt% PMMA solutions were electrospun using 25-gauge needle. The nanofiber diameter was 312 and 483 nm for PMMA concentrations of 3 and 4 wt%, respectively, and flow rate of 0.5 mL/h. To achieve another fiber diameter, the 4 wt% solution was electrospun at 1 mL/h flow rate, and average diameter of fiber was 531 nm. These results are summarized in Table 1 and are shown in Fig. 1.

For measurement of reflectance of nanofibers and investigation of the effect of fiber diameter on the optical properties, fibers with the same morphologies were electrospun on glass paper and then the nanofibers were separated from substrate and the reflectance of the samples was measured by a spectrophotometer. Figure 2 shows the spectral reflectance of the prepared PMMA nanofibers with different diameters. In this range of diameter, it can be observed that the reflectance increases by increasing fiber diameter.

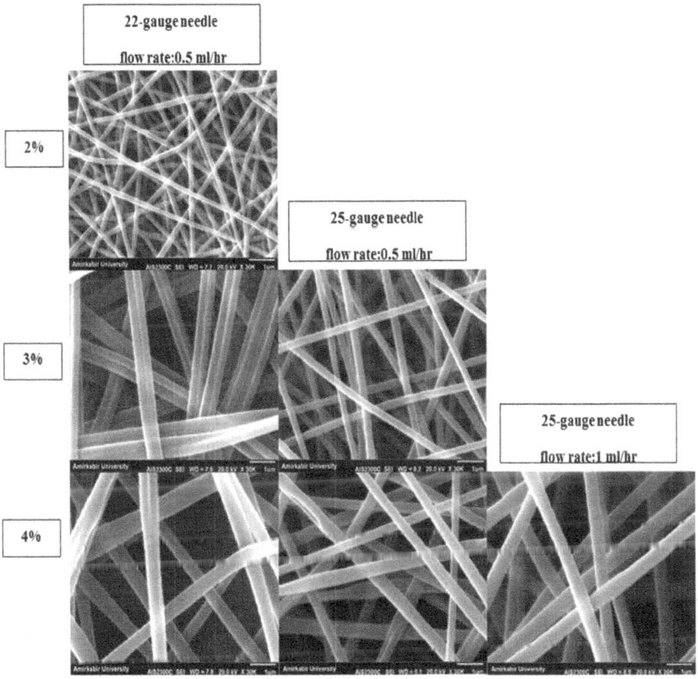

Fig. 1. SEM images of nanofibers with different PMMA concentrations and needle gauges and flow rates

Table 1. Electrospinning conditions and fibers diameter and morphology

PMMA concentration (wt%)	Needle gauge	Flow rate (mL/h)	Average diameter of fiber (nm)	Morphology of fiber
2(a)	22	0.5	166	Cylinder
3(b)	25	0.5	312	Cylinder
4(c)	25	0.5	483	Cylinder
4(d)	25	1	531	Cylinder

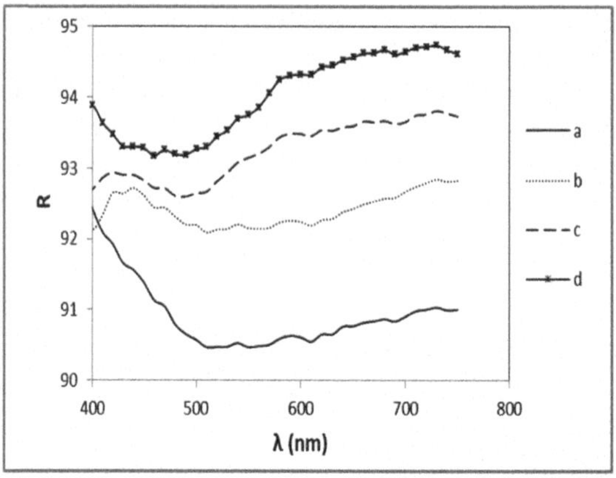

Fig. 2. Spectral reflectance of nanofibers with different diameters

4 Conclusion

In conclusion, PMMA nanofibers with different morphologies and fiber diameters were prepared. For the study of the optical properties, the fibers with similar morphologies and different diameters were selected. It was observed that in the range of 166 nm to 531 nm fiber diameter the reflectance of fibers increased with increasing diameter.

References

1. Yang, Y., Fan, X., Long, Y., Su, K., Zou, D., Li, N., Zhou, J., et al.: J. Mater. Chem. **19**, 7290–7295 (2009)
2. Chang, Ch., Huang, Ch., Chang, Y., Kuo, Ch.: Opt. Express **18**, 174–184 (2010)
3. Piperno, S., Lozzi, L., Rastelli, R., Passacantando, M., Santucci, S.: Appl. Surf. Sci. **252**, 5583–5586 (2006)
4. Li, L., Jiang, Zh., Li, M., Li, R., Fang, T.: RSC Adv. **4**, 52973–52985 (2014)

Effect of Compatibilizers on Polyamide 6 and Styrene-Butadiene Rubber Blend: Graphene Oxide and Glycidyl Methacrylate

Ramin Mirzaee and Ahmad Aref Azar(✉)

Department of Polymer Engineering and Color Technology, Amirkabir University of Technology, Tehran, Iran
arefazar@aut.ac.ir

Abstract. In this study, two types of compatibilizer were used for improving the mechanical and morphological properties of polyamide 6 (PA6)/Styrene-butadiene rubber (SBR). Glycidyl methacrylate (GMA) was grafted onto SBR chain through a free radical polymerization for improving its miscibility with PA6 because epoxy groups of the GMA can be cured by the amine groups of PA6 chains, so that these two polymers can be strongly connected by covalent bonds. As GMA has less activity toward SBR, styrene was used as a co-monomer according to the fact that styryl macroradicals have better activity toward GMA than SBR chain. 0.25, 0.5, and 1% of graphene oxide (GO) were used as another compatibilizer for reaching higher mechanical properties because of its amphiphilic properties. The toughness of polymer was investigated by tensile test and increased by 83.04% in 0.5% GO. The degree of grafting was measured by titration and Fourier-Transform Infrared Spectroscopy (FTIR).

Keywords: Polyamide6 · Styrene-butadiene rubber · Grafting · Glycidyl methacrylate · Graphene oxide

1 Introduction

PA6 is a tough polymer but its properties decrease at the lower temperatures of ambient so it is blended with rubbers for increasing its toughening property, but this poly amide is incompatible with rubbers However, scientists have tried to solve this problem by using compatibilizers. In this work, with using a compatibilize, SBR was blended with PA6 because of its low price and high strength properties. SBR-graft-GMA compatiblized PA6 and SBR through linking them by covalent bonds. In addition, by adding GO miscibility and modulus of the blend enhanced because of amphiphilic and excessive modulus properties of GO, respectively. GOs can attract polar polymers with the aid of their polar functional groups on GO edges and attract non-polar polymers by their non-functional groups on GO basals [1]. So, SBR-graft-GMA and GO were used simultaneously to obtain maximum compatibilization and toughness properties.

H. Mirzadeh and A. A. Katbab (Eds.): ISPST 2018, *Eco-friendly and Smart Polymer Systems*, pp. 469–472, 2020.
https://doi.org/10.1007/978-3-030-45085-4_114

2 Experimental

2.1 Materials

PA6 F223-D and SBR 1502 were supplied by Akulon and Bandar-e-emam, respectively. GMA and St and dicumyl peroxide (DCP) were supplied by Sigma-Alderich.

2.2 Sample Preparation

SBR was added to an internal mixer at 50 rpm and 100 C for 2 min, subsequently DCP and St were added after 1 min and the reaction was completed after 10 min. The amounts of GMA and St/GMA used were between 1–2 and 0.5–1.5 phr, respectively. These samples were designed by design expert application and the results are shown in Table 1. Then, SBR-g-GMA with different degrees of grafting was added into the PA6/SBR blend in different loadings. Finally, the various amounts GO (0.25, 0.5, and 1%) were incorporated into the sample with the weakest toughness properties for investigating the effect of GO on the mechanical properties of blend.

2.3 Test Methods

Tensile tests were performed at room temperature using dumb-bell shaped 250 ± 20 μm specimens according to ASTM D638.

2.4 Results and Discussion

Grafting samples were prepared and the values of degree of grafting calculated by titration and FTIR are shown in Fig. 1. According to Fig. 1, it is expected that the maximum grafting occurs when the amount of St/GMA is close to unity, because St can enhance the reactivity of GMA toward SBR, and this has an optimum point which can be measured [2]. The amount of gel must be at its lowest level to obtain the best properties. The results are summarized in Fig. 2. Due to the fact that gel acts as a stress concentrator, sample exactly fails at the points the gel is exist. In addition, we found the least, best and moderate degree of grafting for studying the effect of grafting on toughness energy (Fig. 3). Then, we selected the sample with the lowest toughness and incorporated GO at the concentrations of 0.25, 0.5, and 1%. The properties increased with increasing GO concentration up to 0.5%, and then reduced.

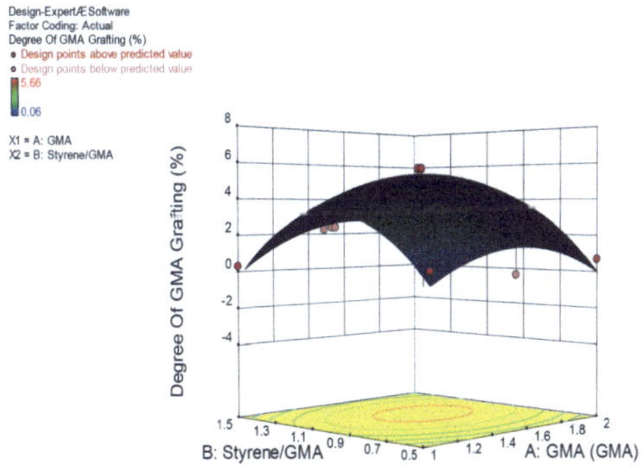

Fig. 1. 3D plot of degree of grafting as a function of St/GMA and GMA content (phr)

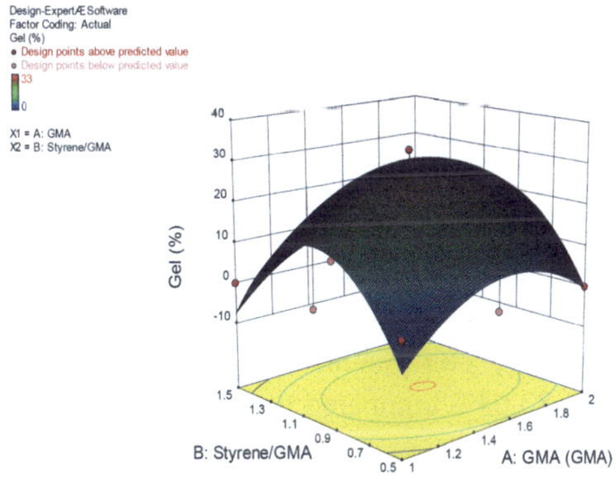

Fig. 2. Gel content as a function of St/GMA and GMA content (phr)

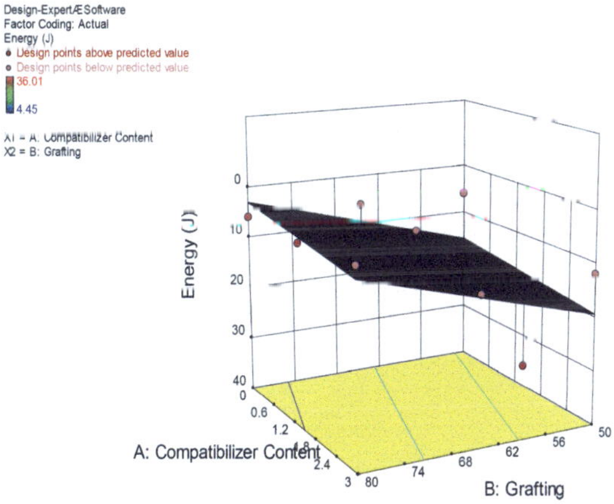

Fig. 3. Toughness as a function of compatibilizer content and degree of grafting

Table 1. Design of experiment 1

Run	A:GMA	B:Styrene/GMA
1	2	1.5
2	2	0.5
3	1.5	1.5
4	1.5	1
5	1	0.5
6	1.5	0.5
7	2	1
8	1	1.5
9	1	1
10	1.5	1
11	1.5	1
12	1.5	1
13	1.5	1

3 Conclusion

With the addition of St, the amount of gel decreased and the degree of grafting improved due to increase of reactivity of GMA with SBR chain. Higher amount of gel led to lower toughness energy. GO was incorporated into the blend and the toughness energy was enhanced by 83.04% compared to the blend sample without GO.

References

1. Cao, Y., Zhang, J., Feng, J., Wu, P.: Compatibilization of immiscible polymer blends using graphene oxide sheets. Acs Nano **7**, 5920–5927 (2011)
2. Hu, G.H.: J. Polym. Sci. A1 **36**(7),1053–1063 (1997)

Investigation of Tensile Strength and Impact Strength of Nanocomposites Based on PP/NBR/HNT

Amir Bakhtiari[1], Faramarz Ashenai Ghasemi[1], Ghasem Naderi[2], and Mohammad Reza Nakhaei[3](✉)

[1] Faculty of Mechanical Engineering, Shahid Rajaee Teacher Training University, Tehran, Iran
[2] Department of Rubber, Iran Polymer and Petrochemical Institute, Tehran, Iran
[3] Faculty of Mechanics and Energy, Shahid Beheshti University, Tehran, Iran
mr_nakhaei@yahoo.com

Abstract. The object of this paper is to study the effect of different levels of halloysite nanotubes (HNTs) loading on tensile strength and impact strength of PP/NBR/HNT nanocomposite in presence of PP-g-MA as compatibilizer. Samples containing 20 wt% of NBR, 1, 3, and 5 wt% of HNTs and 10 wt% of PP-g-MA were produced using counter-rotating twin-screw extruder. HNTs enhanced both of the mechanical properties. At all levels of loading, tensile strength plot showed an ascending trend, however in case of impact strength, the optimum level of HNTs was achieved at 3 wt%. Scanning electron microscopy was utilized to figure out the effect of HNTs loading on the way the elastomer phase dispersed into the continuous phase or the matrix.

Keywords: Nanocomposites · PP/NBR · Halloysite nanotubes · Tensile strength · Impact strength

1 Introduction

As one of the most widely used thermoplastics, polypropylene has received considerable attention for several decades and it is due to good processability and recyclability, light weight and low price of this engineering-commodity material. However limited impact strength of this polymer has been an obstacle for its utilization in industries which require balanced mechanical properties. An efficient way to overcome this drawback is incorporation of a rubbery phase, but at the expense of its mechanical properties like tensile strength and stiffness. Maintaining the balance needed for engineering application could be achieved by reinforcing these blends with fillers like talc, carbon fiber, glass fibers and nanofillers like CNTs and organoclays. Halloysite nanotubes have received considerable attention, recently. They have hollow tubular structure and they could be alternatives to multi-walled carbon nanotubes (MWCNTs) because of their much lower cost in comparison with them [1]. PP/NBR blends have oil resistant property of NBR and good processing and other aforementioned properties of

H. Mirzadeh and A. A. Katbab (Eds.): ISPST 2018, *Eco-friendly and Smart Polymer Systems*, pp. 473–476, 2020.
https://doi.org/10.1007/978-3-030-45085-4_115

PP. However due to their dissimilar polarity and different surface energy, they are incompatible [2]. In order to promote the interaction between these polymers, compatibilizers like PP-g-MA and PP-g-PH have been utilized [3]. George et al. [3] reported that the dipolar interaction between PP-g-MA and NBR causes an increase in interfacial adhesion between PP and NBR phases. Kubade et al. [4] concluded that using PP-g-MA improves the interaction of PP/ABS and also results better dispersion of HNTs. This study is an attempt to figure out the effect of different levels of HNTs loading on tensile strength and impact strength of PP/NBR/HNTs in presence of PP-g-MA.

2 Experimental/Theoretical

Polypropylene (PP-Z30S) was supplied by Marun Petrochemical Company (Iran), having melt flow index of 25 g/10 min at 230 °C and density of 0.9 g/cm^3. Nitrile rubber (KNB-35L) containing 34 wt% of acrylonitrile was purchased from Kumho (Korea). HNTs were supplied by Imerys Tableware, Asia Limited, New Zealand, having density of 2.55 g/cm^3 and brightness of 98.9%. PP-g-MA was supplied by Karangin Company (Iran) having 1 wt% of maleic anhydride content, density of 0.9 g/cm^3 and MFI of 10 g/10 min. Melt blending of materials was carried out using counter-rotating twin-screw extruder Collin with temperature profile of 165–190 °C and screw speed of 60 rpm. Injection molding of pellets was performed by Aslanian injection molding machine with temperature of 190°C to 210°C. Tensile and impact tests have been done according to ASTM D638, with crosshead speed of 50 mm/min, and ASTM D256 respectively by Zwick/Roell–Z100 and Zwick ML2953. Cryogenic fracture surfaces of the samples were etched by acetone to extract the NBR phase, then gold coated and their morphology was studied by VEGATESCAN scanning electron microscope. The number average diameter of rubber particles was determined by the following equation where N_i and D_i stand for number of particles and their diameters, respectively.

$$D_n = \frac{\sum N_i D_i}{\sum N_i} \tag{1}$$

3 Results and Discussion

Figure 1 shows the effect of HNTs on the tensile strength of nanocomposite, as can be seen, enhancing effect is continuous. Enhancing effect is attributed to the load transferring from matrix to HNTs and this phenomenon is also affected by better adhesion between HNTs and matrix in presence of PP-g-MA. Figure 3 shows that by increasing nanofiller's content, rubber particle size decreases from 2.41 μm in sample containing

1 wt% of HNTs to 2.02 μm in sample containing 3 wt%. This could be due to effect of HNTs on the viscosity ratio of the polymer phases. HNTs loading increases the melt viscosity of the matrix, and consequently decreases the difference between viscosities of the phases [5]. As can be seen in Fig. 2, HNTs reinforce impact strength of the nanocomposites till 3 wt% of loading, after that, it shows a slight decrease that might be a result of the stress concentration points which have been made by HNTs. It is believed that HNTs could hinder the propagation of crack due to their intrinsic toughening properties [6].

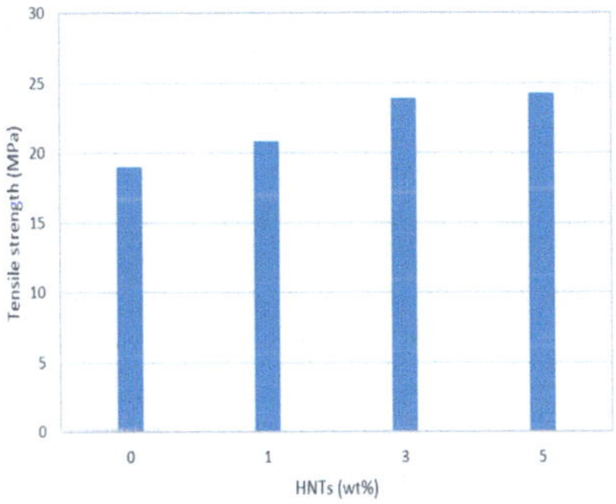

Fig. 1. Effect of HNTs on tensile strength

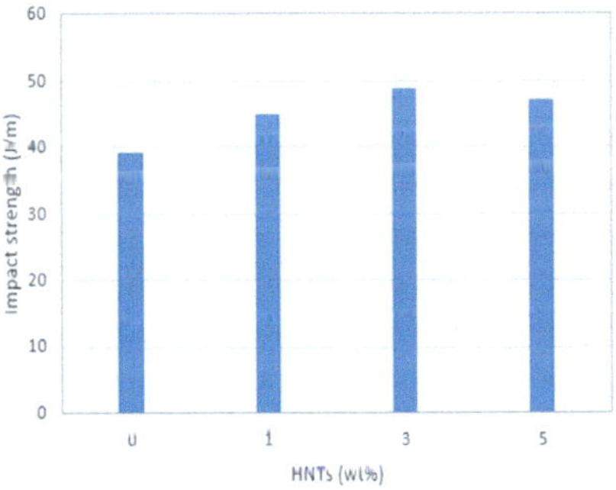

Fig. 2. Effect of HNTs on impact strength

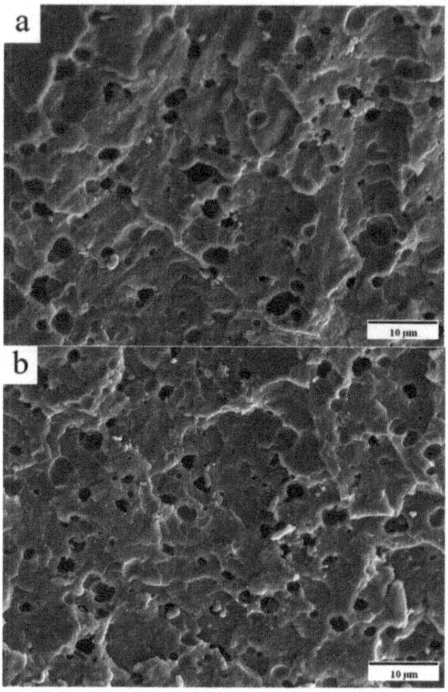

Fig. 3. SEM images of cryogenically fractured specimens: (a) 1 wt% of HNTs and (b) 3 wt% of HNTs

4 Conclusion

In this study, reinforcing effect of HNTs on tensile strength and impact strength of PP/NBR/HNT nanocomposite was analyzed. In case of tensile strength, at 5 wt% of nanofiller content (in presence of 10 wt% of PP-g-MA), an increase of nearly 29% in comparison with PP/NBR blend (80/20 wt%) was seen. However for impact strength, ascending trend was continued till 3 wt% of loading and after that due to more stress concentration points, a slight decrease was observed.

References

1. Liu, M., Jia, Z., Jia, D., Zhou, C.: Prog. Polym. Sci. **39**(8), 1498–1525 (2014)
2. Soares, B.G., Almeida, S.M., Deepa Urs, M.V., Kumaraswamy, G.N., Ranganthaiah, C., Siddaramaiah, P., Mauler, R.J.: Appl. Polym. Sci. **102**, 4672–4681 (2006)
3. George, S., Joseph, R., Thomas, S.: Polymer **36**, 4405–4416 (1995)
4. Kubade, P., Tambe, P.: Compos. Interfaces **23**(5), 433–451 (2016)
5. Paran, S.M.R., Naderi, G., Ghoreishy, M.H.R.: Polym. Compos. **38**, 451–461 (2017)
6. Deng, S., Zhang, J., Ye, L., Wu, J.: Polymer **49**, 5119–5127 (2008)

Synthesis and Characterization of Fe_3O_4 Magnetite Nanoparticles Coated by Polyvinylpyrrolidone

Alireza Heiran[1], Shadi Hassanajili[1(✉)], and Mehdi Escrochi[1,2]

[1] Department of Chemical Engineering, School of Chemical and Petroleum Engineering, Shiraz University, Shiraz, Iran
ajili@shirazu.ac.ir
[2] IOR/EOR Research Institute, Shiraz University, Shiraz, Iran

Abstract. The purpose of this study was to synthesize polyvinylpyrrolidone (PVP) coated with iron oxide nanoparticles (Fe_3O_4@PVP NPs) and to evaluate the effect of PVP content on its properties. The Fe_3O_4@PVP nanoparticles were synthesized by co-precipitation method and characterized by field-emission scanning electron microscopy (FE-SEM), Fourier-transform infrared spectra (FT-IR) and vibrating sample magnetomcter (VSM). Results indicated successful synthesis of coated nanospheres with supermagnetic properties. The core size of the nanoparticles was about 17 nm and the overall size was around 40 nm. The magnetization of NPs was also changed with PVP content and the highest saturation magnetization was observed for Fe_3O_4@PVP 0.32.

Keywords: Nanoparticle · Polymer · Fe_3o_4 · PVP · Coating

1 Introduction

Nanotechnology is being considered for use in many fields such as information technology, medicine and food industries, heat transfer enhancement, energy, and environmental science targeted to provide cleaner and more efficient energy supplies and uses [1, 2]. Nanotechnology is characterized by the use of very small particles (<100 nm), called nanoparticles (NPs).

Among nanoparticles, magnetic NPs (e.g. Fe_3O_4 magnetic nanoparticles) have attracted much interest due to their several fascinating characteristics, such as superparamagnetic behavior and the unique advantage of easy separation under external magnetic fields, biocompatibility, biodegradation and potential non-toxicity [3]. To this, numerous studies have been done in order to maximize the potential usage of magnetic NPs in various fields of applications.

W. Jiang et al. have been used Fe_3O_4 NPs as a promising choice for arsenic sequestration because it could effectively capture arsenic species from water [4]. H. Soleimani et al. synthesized ferromagnetic NPs and used a dispersion of 0.01 wt% of the prepared magnetic materials in distilled water for core flooding test to validate its feasibility in enhanced oil recovery (EOR). Their results showed an almost 20% increment in the recovery of oil with the application of electromagnetic waves [4]. In general,

H. Mirzadeh and A. A. Katbab (Eds.): ISPST 2018, *Eco-friendly and Smart Polymer Systems*, pp. 477–480, 2020.
https://doi.org/10.1007/978-3-030-45085-4_116

the instability of Fe_3O_4 NPs in different base fluids does not allow their use without surface modification. Two types of stabilizing agents are commonly used on the surface of nanoparticles. Electrostatic agents and steric agents. Steric agents stabilize the nanoparticles by physical hindrance. Polymers with long hydrophilic chains, such as polyethylene glycol (PEG) and polyvinyl alcohol (PVA), are commonly used to avoid aggregation at physiological conditions [6]. In this study, we explained chemical synthesis and characterization of polyvinylpyrrolidone coated Fe_3O_4 (Fe_3O_4@PVP) nanoparticles. The addition of PVP during synthesis changes the particle morphology from compact to porous with chains of small iron oxide nanocrystals at the surface of the particles due to the steric repulsion of PVP ligands. The addition of PVP also reduces average particle size by preventing agglomeration. In addition, the presence of PVP restricts crystal growth of iron oxide and limits oxidation [7]. In this work three different Fe_3O_4@PVP nanoparticles with different ratios of PVP coating layer to magnetic core were synthesized and characterized by field-emission scanning electron microscopy (FE-SEM) and Fourier transform infrared spectroscopy (FT-IR) and their magnetic properties were measured using vibration sample magnetometry (VSM).

2 Experimental

2.1 Preparation of Fe_3O_4@PVP Nanoparticles

At First, 0.5 g of $FeCl_2$ and 1.35 g of $FeCl_3$ and three different contents of PVP (0.32, 0.745, and 1.74 g) were dissolved in 600 mL of nitrogen degassed distilled water in a three-necked flask. Then ammonia solution was added dropwise to the solution under vigorous stirring at 70 °C. The obtained Fe_3O_4@PVP particles were separated with a magnet and washed with ethanol. Finally, the precipitant was dried at 65 °C for 6 h. The schematic of nanoparticles and PVP Structure is shown in Fig. 1.

3 Results and Discussion

3.1 Characterization of Synthesized Nanoparticles

The FTIR spectra of Fe_3O_4@PVP (0.745 g) are shown in Fig. 2. In this figure, the transmittance peaks of Fe_3O_4 NPs are at 591 cm^{-1} and 3402 cm^{-1}. The peak at 591 cm^{-1} shows the Fe-O characteristic vibration. The peak at 3402 cm^{-1} depicts -OH groups on the surface of Fe_3O_4 NPs. The presence of CO band at 1660 cm^{-1} which is shifted from 1672 cm^{-1} for pure PVP to this wavenumber (results not shown) indicated the successful PVP coating and interaction between CO groups of PVP and free OH on the surface of Fe_3O_4 nanoparticles. Figure 3 shows the FE-SEM images of Fe_3O_4 nanoparticles coated by PVP with three different contents. Morphological observations through FE-SEM proved Fe_3O_4@PVP nanoparticles to be spherical and relatively mono-dispersed in the range of 30–40 nm.

The magnetic properties were measured at room temperature using a vibration sample magnetometer (VSM) are shown in Fig. 4. It can be seen that all synthesized NPs showed supermagnetic behavior. Increasing the amount of PVP from 0.32 g to 0.745 g could not change the magnetic power of synthetized NPs, but further increase of PVP reduced magnetic saturation value of Fe_3O_4@PVP NPs from about 20 emu/g to 12 emu/g.

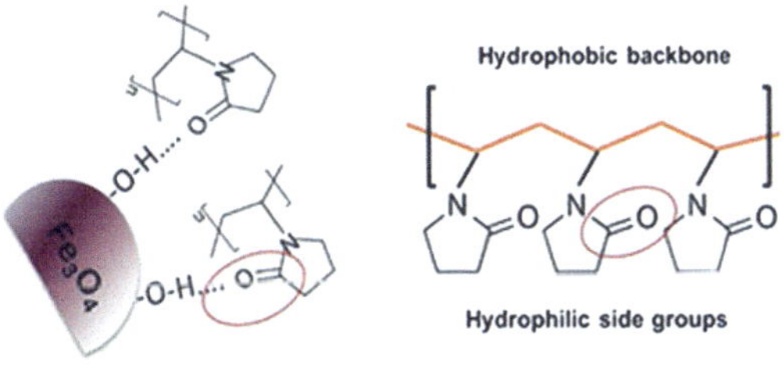

Fig. 1. Schematic of Fe_3O_4@PVP nanospheres and PVP chemical formula

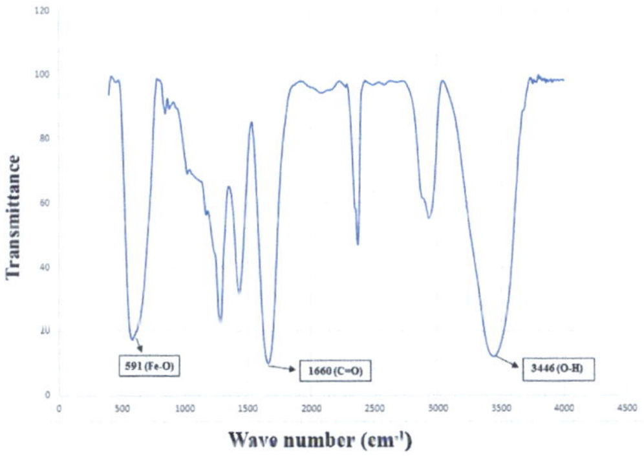

Fig. 2. FTIR spectra of Fe_3O_4@PVP

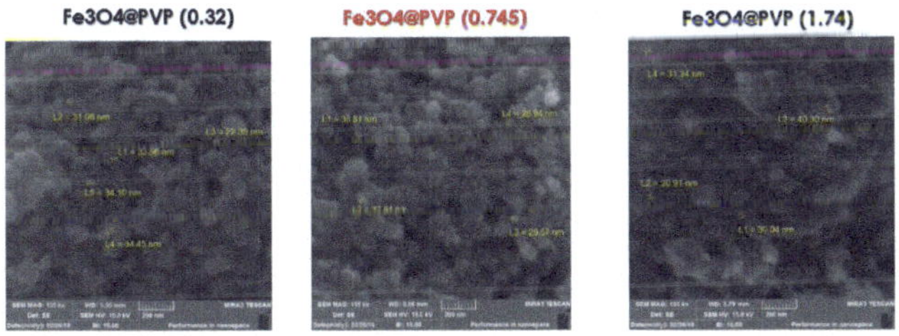

Fig. 3. FE-SEM of three kinds of Fe_3O_4@PVP nanoparticles

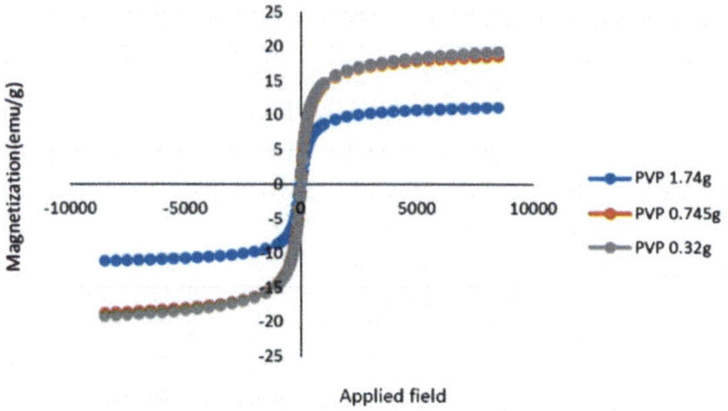

Fig. 4. Magnetization curves of Fe_3O_4@PVP

4 Conclusion

In the present work, Fe_3O_4@PVP nanospheres with three different PVP contents were synthesized and analyzed by FTIR spectra, FE-SEM and VSM. Results indicated successful synthesis of coated NPs. Results showed that PVP molecules prevent the aggregation of the Fe_3O_4 nanoparticles and restrict their size by utilizing the interaction between the carbonyl groups and free OH groups on the surface of Fe_3O_4 NPs.

References

1. Wen, D., Lin, G., Vafaei, S., Zhang, K.: Particuology **7**, 141–150 (2009)
2. Tang, S.C.N., Lo, M.C.I.: Water Res. **47**, 2613–2632 (2013)
3. Huang, Z., Tang, F.: J. Colloid. Interf. Sci. **275**, 142–147 (2004)
4. Soleimani, H., Yahya, N., Ahmad Lattif, N.R., et al.: J. Nano. Res. **26**, 111–116 (2014)
5. Jiang, W., Chen, X., Niu, Y., Pan, B.: J. Hazard Mater. **243**, 319–325 (2012)
6. Maurizi, L., Papa, A.L., Dumont, L., Bouyer, F., Walker, P., Vandroux, D., Millott, N.: J. Biomed. Nanotechnol. **11**, 126–136 (2015)
7. Seo, K., Sinha, K., Novitskaya, E., Graeve, O.A.: Mater. Lett. **215**, 203–206 (2018)

Cathodic Disbondment of Unpigmented Epoxy Coatings: Effect of Thickness, pH, and Curing Conditions

Shirin Ardeshirian Sharifabadi, Bahar Hassanpour, and Mohammadreza Mohammadzadeh Attar(✉)

Faculty of Polymer Engineering and Color Technology, Amirkabir University of Technology, Tehran, Iran
attar@aut.ac.ir

Abstract. Due to increasing usage of metals, it is very important to protect them against corrosion and increase the service life of the components. This paper examines the cathodic disbondment of unpigmented epoxy resin. Disbondment of coating occurs when coatings in a cathodic protection system interact either chemically or physically, ultimately causing corrosion beneath the coating. In this paper, the cathodic disbondment of epoxy resin applied on steel sheets, which starts from around a fixed-diameter cavity, is studied through a sacrificial magnesium anode and an electrolyte with 3% salt in a fixed amount of time. Thickness, pH, and curing condition are considered as variable parameters. During the experiment, it was observed that the cathodic disbonding rate of this resin will be increased with a decrease in the pH of the electrolyte, thickness reduction, and also incomplete curing of epoxy resin. Due to importance of corrosion control in the industry, especially pipelines, in this paper, we examined the factors affecting cathodic disbondment.

Keywords: Unpigmented epoxy resin · Cathodic disbondment · Cathodic protection

1 Introduction

Cathodic protection (CP) is an old method of corrosion prevention, but it is still used in a wide array of environments, from steel pipelines to home water heaters. This method involves making that metal the cathode of an electrochemical cell, which often means connecting the metal to be protected to a sacrificial anode, which corrodes instead. The metallic surfaces exposed to an electrolyte have a multitude of microscopic anodic and cathodic sites. Where anodes are more electronegative than cathodes, a potential difference is created between them, allowing corrosion to occur [1, 2]. In CP systems, current passing through the metal can be free hydroxyl ions, causing coating disbondment. This current passes only through metal coating faults, and the amount of the current flow depends on the size of the coating fault. If the size of the fault increases, the current forces the coating away from the metal, resulting in a voltage drop at the interface, causing cathodic disbondment. If there is no coating fault, then cathodic disbondment does not occur. Various factors could affect cathodic disbonding of a

H. Mirzadeh and A. A. Katbab (Eds.): ISPST 2018, *Eco-friendly and Smart Polymer Systems*, pp. 481–484, 2020.
https://doi.org/10.1007/978-3-030-45085-4_117

coating, e.g., the polarization potential, temperature, cation mobility, and concentration, dissolved oxygen concentration, type of coating binder, type and concentration of pigments and corrosion inhibitors, formulation of the product, surface preparation, cure percentage, and coating film thickness [3–9].

2 Experimental

In this study, a mixture of epoxy resin and its hardener with a ratio of 3:1 was applied to steel sheets by an applicator. After curing the applied film at 100 °C, a hole with 6 mm diameter was applied on center of a 5 cm^2 window on the film in each sheet. The area around the window was covered with beeswax/colophony to prevent the possibility of leakage of electrolyte to other areas except for the window. Also a wire was applied to each sheet in order to set the connection between the sheet and the sacrificial magnesium anode. The electrolyte was prepared with solution of 3% NaCl in water. HCl was used to control the decreasing pH of the solution.

3 Results and Discussion

First, the thickness of the applied films was considered as a variable. The fully cured epoxy films were immersed in electrolyte with pH 7 for 1 h at 25 °C. Figure 1 shows the changes of the amount of disbondment with increasing the thickness of the film.

Another parameter considered as variable was pH of the electrolyte. By adding drops of HCl into the electrolyte, we decreased the pH. The fully cured films with the same applying thickness were separated to three categories. The first, second and third category, respectively, were immersed in electrolytes with pH 7, 5, and 3. Figure 2 shows the changes of disbondment with increasing the pH of the electrolyte.

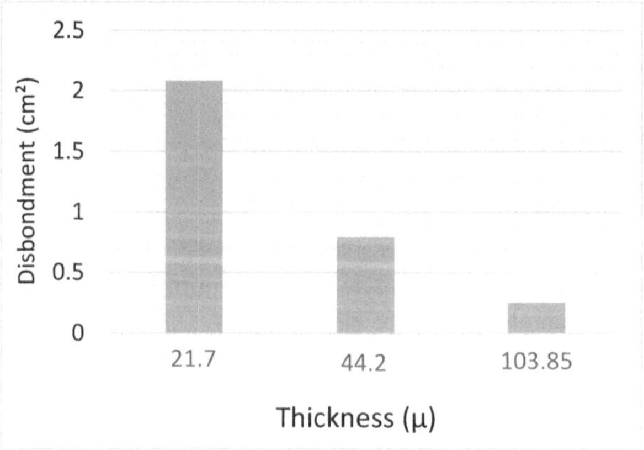

Fig. 1. Changes of disbondment (cm^2) by increasing the thickness (µ) of film

At last, the rate of curing was considered as a variable. In this case, epoxy resin was applied to the sheets with the same applying thickness and then they were separated into three categories. The first, second and third category, respectively, were cured for 1 h, 30 and 5 min. Figure 3 shows the changes of disbondment for incomplete cured film, semi-cured film and fully cured film.

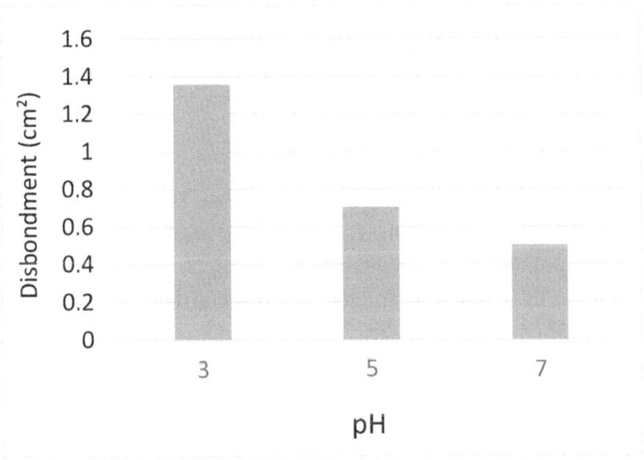

Fig. 2. Changes of disbondment (cm^2) by increasing the pH of electrolyte

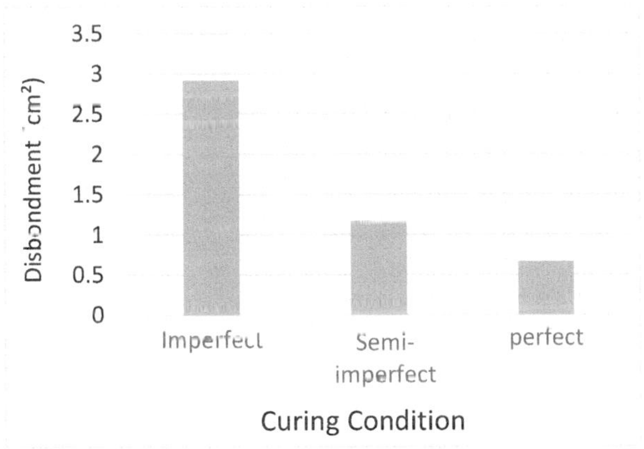

Fig. 3. Changes of disbondment (cm^2) for incomplete cured film, semi-cured film and fully cured film by increasing time of curing (min)

4 Conclusion

The purpose of this work was to determine the cathodic disbonding rate of unpigmented epoxy resin. In this study, thickness, pH, and curing rate were considered as variable parameters. We observed that the cathodic disbonding rate of epoxy resin was increased with a decrease in the pH of the electrolyte, thickness reduction, and also incomplete curing of the epoxy resin. These were resulted from an increase in pH at interface, where disbonding mechanisms could have taken place.

References

1. Glover, C.F., Richards, C.A.J., Williams, G., McMurray, H.N.: Corros. Sci. **136**, 304 (2018)
2. Izadi, M., Shahrabi, T., Ramezanzadeh, B.: Appl. Surf. Sci. **440**, 491 (2018)
3. Sorensen, P.A., Dam-Johansen, K., Weinell, C.E., Kiil, S.: Prog. Org. Coat. **68**, 283 (2010)
4. Hamade, R.F., Seif, C.Y., Dillard, D.A.: Int. J. Adhes. Adhes. **27**, 108 (2007)
5. Song, F.M., Sridhar, N.: Corros. Sci. **50**, 70 (2008)
6. Chen, X., Li, X.G., Du, C.W., Cheng, Y.F.: Corros. Sci. **51**, 2242 (2009)
7. Reynolds, G.J., Barrett, Z.S., McMurray, H.N., Williams, G.: Corros. Sci. **70**, 82 (2013)
8. Ramezanzadeh, B., Attar, M.M.: Corrosion **69**, 793 (2013)
9. Harun, M.K., Marsh, J., Lyon, S.B.: Prog. Org. Coat. **54**, 317 (2005)

Effect of Ultrasound Stimulation on the Chondrogenic Activity of MSCs on a Hybrid Hydrogel Scaffold

Ziba Naghizadeh[1], Akbar Karkhaneh[1(✉)], and Arash Khojasteh[2(✉)]

[1] Biomedical Engineering Faculty,
Amirkabir University of Technology (Tehran Polytechnic), Tehran, Iran
a.karkhaneh@aut.ac.ir

[2] Department of Tissue Engineering and Applied Cell Sciences,
School of Advanced Technologies in Medicine,
Shahid Beheshti University of Medical Sciences, Tehran, Iran

Abstract. The objective of this study was to determine the effects of low-intensity pulsed ultrasound (LIPUS) on chondrogenic activity and gene expression of mesenchymal stem cells (MSCs) seeded on alginate oxide (AD) - carboxymethyl chitosan (CMC) – polyethylene glycol (PEG) hydrogel for cartilage tissue engineering. DAPI staining for evaluating cell viability of groups was done. DMMB and real-time PCR analysis for assaying cartilage-specific extracellular matrix (ECM) gene markers were carried out. According to the results, LIPUS at a resonant frequency of 40 kHz for 7 days with 5 min has the most effect on GAG secretion and SOX9 expression compared other times.

Keywords: Injectable · Hydrogel · In situ crosslinked · LIPUS · Cartilage tissue engineering

1 Introduction

Cartilage tissue has a very poor ability to repair. At present, no clinical method is capable of fully restoring cartilage function in the long-term; therefore, tissue engineering is believed to be promising for cartilage repair. Low-intensity pulsed ultrasound (LIPUS) is used for bone fractures healing and is used as an alternative treatment for cartilage repair, hence it is documented as a clinical application [1].

2 Experimental

Alginate oxide (AD)-carboxymethyl chitosan (CMC) - polyethylene glycol (PEG) hydrogel was synthesis according to our previous work [2, 3]. The MSCs were cultured on hydrogels in five groups for 7 days, and then simulated with LIPUS at a resonant frequency of 40 kHz for 7 days according to the study groups: group 1 (5 min), group 2 (10 min), group 3 (15 min), group 4 (20 min), group 5 (25 min), and group 6 (without ultrasonic). Cells survived on the hydrogel done after 3 days by staining the acridine orange (AO, 30 μg/mL) and propidium iodide (PI, 50 μg/mL). Proteoglycan content of

H. Mirzadeh and A. A. Katbab (Eds.): ISPST 2018, *Eco-friendly and Smart Polymer Systems*, pp. 485–487, 2020.
https://doi.org/10.1007/978-3-030-45085-4_118

the MSCs cultured on the surface of the hydrogel was obtained by calculating the number of GAGs using DMMB assay at 530 nm. After isolation of RNA and cDNA synthesis, all groups were analyzed to express SOX9, COL2 and COL1 using real-time PCR method.

3 Results and Discussion

The distribution of the cells on the hydrogel using AO/PI is shown in Fig. 1. More cellular distribution in group 1 was observed compared to other samples. According to Fig. 2, among groups with ultrasonic application, group 1 had the highest levels of proteoglycan secretion. This group also had the highest expression of SOX9 and COL1 genes. The expression of the SOX9 gene is very important for the induction of cartilage phenotype during skeletal development. Group 1 with ultrasonic applied for 5 min performed the best results in proteoglycans secreting and SOX9 expressing. The increase in ultrasonic application time seems to have had a damaging effect.

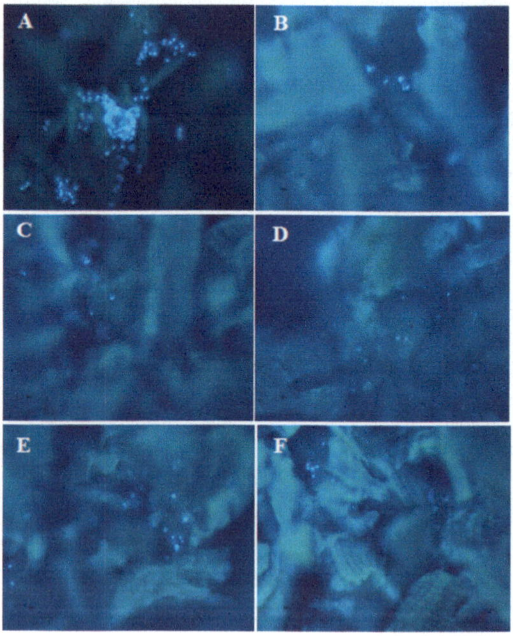

Fig. 1. DAPI staining of MSCs cultured on hydrogels of groups 1, 2, 3, 4, 5, and 6 (control)

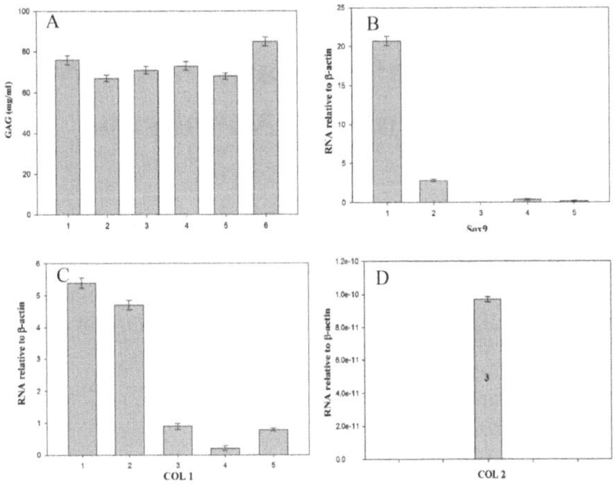

Fig. 2. Total GAG contents and real-time PCR results of mRNA expression of SOX9, COL2, and COL1 (relative to control) for groups 1, 2, 3, 4, 5, and 6 (control)

4 Conclusion

In the current study, the effect of LIPUS at different times on GAG, SOX9, COL2, and COL1 expression during the differentiation of MSCs on AD-CMC- PEG hydrogel was investigated. LIPUS usage at a resonant frequency of 40 kHz for 7 days with 5 min showed the best results compare to other groups.

References

1. Thakurta, S.G.: Long-term culture of human mesenchymal stem cell-seeded constructs under ultrasound stimulation: evaluation of chondrogenesis. Biomed. Phys. Eng. Express **2**(5), 055016 (2016)
2. Naghizadeh, Z., Karkhaneh, A., Khojasteh, A.: Mater. Sci. Eng. C **89**, 256–264 (2018)
3. Naghizadeh, Z., Karkhaneh, A., Khojasteh, A.: J. Biomed. Mater. Res. Part A **106**(7), 1932–1940 (2018)

Effect of Relative Energy Difference (RED) of Coagulation Bath on the Structure of Polysulfone Membrane

Rahim Dehghan, Jalal Barzin(✉), Zahra KordKatooli, and Shahla Mahdavi

Department of Biomaterials, Iran Polymer and Petrochemical Institute (IPPI), P.O. Box 14975-112, Tehran, Iran
j.barzin@ippi.ac.ir

Abstract. Phase inversion is one of the favorite methods for preparation of polysulfone (Psf) membranes. Some main factors such as: solvent and non-solvent exchange rate, phase inversion process, thereby, they can influence membrane structure. This study focused on the relative energy difference (RED) of coagulation bath which extracted from Hansen solubility parameters. Ethanol and water were selected as coagulant and effect of RED in different composition of ethanol/water was investigated. Results showed that at high concentrations of ethanol in coagulation bath, morphology of membranes declined to a sponge like structure and size of macrovoids was decreased due to the reducing of RED. On the other hand, precipitation time and porosity of membranes were increase accordingly.

Keywords: Membrane · Solubility parameter · RED · Morphology · Coagulation bath

1 Introduction

Psf membrane can be readily prepared by phase inversion method [1–3]. Several key factors can influence on the phase inversion process during membrane formation including: solvent and non-solvent systems [4], the composition of polymeric casting solution and coagulation bath and also casting conditions [5]. RED is equal to R_a/R_0, where R_0 is the radius of Hansen sphere and R_a is difference of solubility parameters. If solubility of polymer is considered, the RED should not exceed of 1 and Solubility will increase by decreasing of RED to 0 [6]. The current study focused on the coagulation bath during Psf membrane formation. As shown in Table 1, the RED of water as a famous strong non-solvent for Psf is 4.5 [7]. By choosing of a coagulant with low RED, controlling of solvent and non-solvent exchange rate would be possible. Therefore, ethanol and water at different composition were used as coagulant and the RED of mixed coagulation baths (ethanol/water) were estimated (Table 2). Finally, its effect on the membrane structure, precipitation rate and porosity was investigated.

H. Mirzadeh and A. A. Katbab (Eds.): ISPST 2018, *Eco-friendly and Smart Polymer Systems*, pp. 488–492, 2020.
https://doi.org/10.1007/978-3-030-45085-4_119

Table 1. Hansen solubility parameters of all component in membrane formation [7]

Material	δ_d ($MPa^{1/2}$)	δ_p ($MPa^{1/2}$)	δ_H ($MPa^{1/2}$)	RED
PSF	19.7	8.3	8.3	–
DMAC	16.8	11.5	10.2	0.9
Ethanol	15.8	8.8	19.4	1.7
Water	15.5	16	42.3	4.5

Table 2. RED of different mixed coagulation bath

Solution composition	Water/ethanol (%)	RED	Porosity (%)
Psf 15%	100/0	4.5	30.48
	90/10	3.76	32.65
	70/30	2.87	37.13
	50/50	2.36	38.52
	30/70	2.03	45.34
	10/90	1.79	49.87

The solubility parameter of a binary mixture is calculated based on the following equation [8]:

$$\delta_{12} = \frac{X_1 V_1 \delta_{i1} + X_2 V_2 \delta_{i2}}{X_1 V_1 + X_2 V_2} \tag{1}$$

Where, X is the molar fraction, V represents molar volume that its value for water is 18 mL/mole and for ethanol is 58 mL/mole, and δ is solubility parameters.

2 Experimental

2.1 Membrane Preparation

Membrane prepared by phase inversion technique. Polysulfone with 15 wt% in DMAC dissolved at room temperature for 24 h and degassed overnight. Then it cast on a smooth glass plate as support and then immediately immersed in a mixed coagulation bath containing water/ethanol as coagulant with ratio of 100/0, 90/10, 70/30, 50/50, 30/70, 10/90, and 0/100.

2.2 Morphology Observation

After sample preparation in liquid nitrogen and gold sputtering, morphology of Psf membrane in different mixed coagulation bath was investigated by scanning electron microscope (SEM) at 5.0 Kx magnification.

2.3 Porosity

Porosity of membrane calculated after membrane sample weighing into dry and wet states and finally calculated according to the following equation:

$$\varepsilon(\%) = \frac{M_w - M_d}{\rho A l} \quad (2)$$

M_w is the weight of membrane at wet state, M_d is the weight of membrane at dry state, ρ is water density, A, represent membrane effective area and l, is membrane thickness.

3 Results and Discussion

Kinetic of phase inversion is strongly influenced membrane formation, and also solubility parameters can control the kinetic of phase inversion process. Hansen solubility parameters are useful to predicting solubility behavior of polymers. RED is an interesting factor of Hansen solubility parameters which can assessed the polymer-solvent interaction. In current study, RED was calculated for different coagulation baths based on the Eq. (1). By increasing ethanol concentration in the coagulation bath, RED was decreased to lower level. SEM study showed that by increasing of ethanol, morphology of PSf membrane declined to a sponge like structure due to reducing of RED. At high concentrations of ethanol (70% and 90%) in coagulation bath, a half spongy structure achieved for membranes. while at low concentrations of ethanol, finger like morphology was developed (Fig. 1).

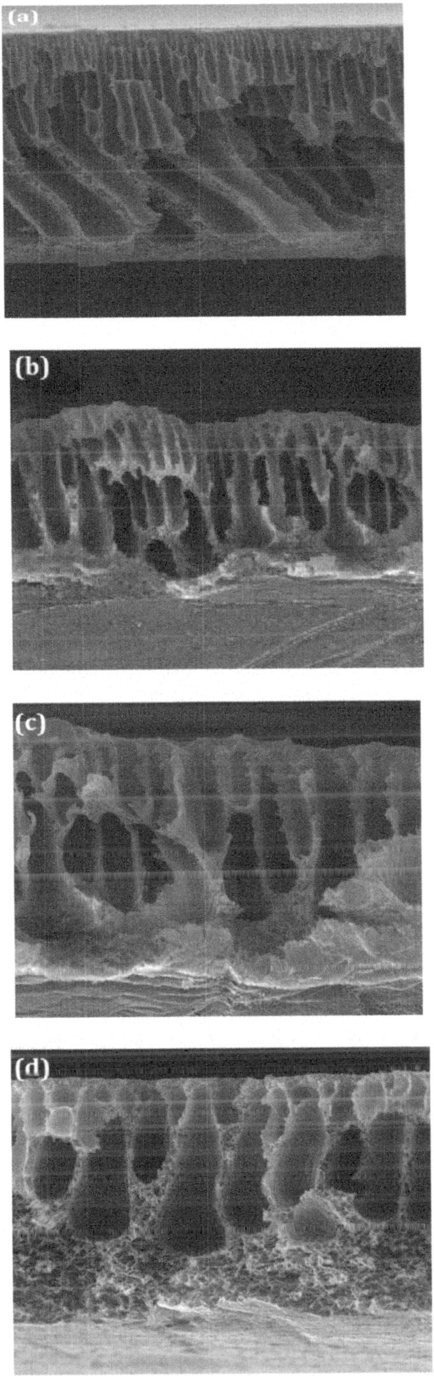

Fig. 1. Effect of mixed coagulation bath on the membrane cross section: (a) 0% ethanol, (b) 30% ethanol, (c) 50% ethanol, (d) 70% ethanol, and (e) 90% ethanol

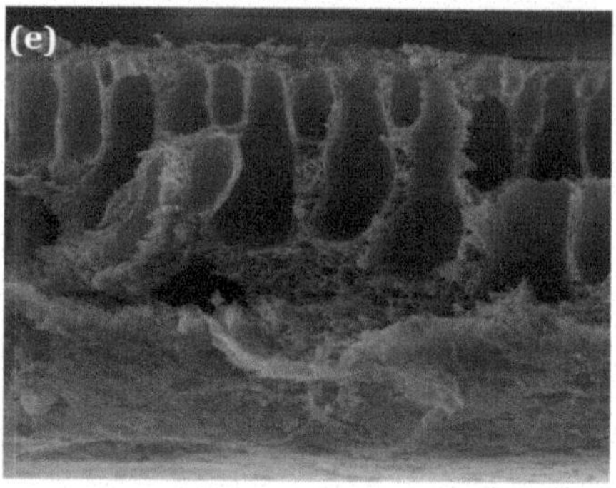

Fig. 1. (*continued*)

References

1. Ishihara, K., Fukumoto, K., Iwasaki, Y., Nakabayashi, N.: Modification of polysulfone with phospholipid polymer for improvement of the blood compatibility. Part 1. surface characterization. Biomaterials **20**, 1545–1551 (1999)
2. Zhao, C.S., Xue, J.M., Ran, F., Sun, S.D.: Prog. Mater Sci. **58**, 76–150 (2012)
3. Yue, W.W., Li, H.J., Xiang, T., Qin, H., Sun, S.D., Zhao, C.S.: Grafting of zwitterion from polysulfone membrane via surface-initiated ATRP with enhanced antifouling property and biocompatibility. J. Membr. Sci. **446**, 79–91 (2013)
4. Barzin, J., Sadatnia, B.: Correlation between macrovoid formation and the ternary phase diagram for polyethersulfone membranes prepared from two nearly similar solvents. J. Membr. Sci. **325**, 92–97 (2008)
5. Deshmukh, S.P., Li, K.: Effect of ethanol composition in water coagulation bath on morphology of PVDF hollow fibre membranes. J. Membr. Sci. **150**, 75–85 (1998)
6. Hansen, C.M.: Hansen Solubility Parameters: A User's Handbook, 2nd edn. CRC, New York (2002)
7. Guillen, G.R., Pan, Y., Li, M., Hoek, E.M.: Preparation and characterization of membranes formed by nonsolvent induced phase separation: a review. Ind. Eng. Chem. Res. **50**, 3798–3817 (2011)
8. Xu, Z.L., Qusay, F.A.: Polyethersulfone (PES) hollow fiber ultrafiltration membranes prepared by PES/non-solvent/NMP solution. J. Membr. Sci. **233**, 101–111 (2004)

Improvement of Delamination Resistance in Glass/Phenolic Composite Laminates with Nano-interlayers

Hossein Ipakchi(✉), Masoud Esfandeh, Amir Masoud Rezadoust, and Mohammad Rezaei

Iran Polymer and Petrochemical Institute, P.O. Box 14975-112, Tehran, Iran
h.ipakchi@ippi.ac.ir

Abstract. Generally, phenolic multi-layered composites are not resistant to delamination and that is because of the brittle matrix in their composition. One of the best ways for toughening the multi-layered composites is the use of interleaved thermoplastic flexible polymers. In this research, PVB (Polyvinyl Butyral) electro-spun nanofiber (average diameter = 400 nm) has been used as interleaving in various thicknesses. The results showed that a 30 μm thickness is the optimal value for these glass fiber composites.

Keywords: Phenolic · Glass fiber · Polyvinyl butyral · Delamination · Electospinning

1 Introduction

Phenolic resin is the earliest type of industrial and thermoset resins which has significant usage in different applications. Because of desired properties, the resin has a widespread use in manufacturing of industrial insulators, thermal shields and ablative composites [1]. In addition, disadvantages like poor impact resistance, high shrinkage and lack of proper cohesion between its layers are limiting factors in its applicability. Also, low delamination resistance is one of the main problems of phenolic resins. Layers debonding is an important mechanism of degradation which causes breakdown between matrix and reinforcement. The origin of layers debonding is lack of reinforcement in the thickness direction. Different methods have been proposed by researchers to improve delamination in composite structures. Saghafi et al. had a study on epoxy-glass composites interleaved with nylon 6,6 electrospun nanofibers and the results illustrated the mode I & II fractures has risen up to 62 and 109%, respectively [2]. Also Hamer et al. have investigated the influence of interlaminar electrospun nylon 6,6 nanofiber in epoxy-carbon multi-layers in optimal condition and the test outcomes showed 332% increase in Mode I fracture [3]. Furthermore, Palazzetti had a research on the interleaving nylon 6,6 nanofibers presence in epoxy-carbon multilayers and the effects on geometrical features such as layers thickness, nanofiber's orientation and its thickness. They reported interlaminar thickness has no tremendous effect on Mode II fracture. But at Mode I fracture enhancement of thickness up to an optimum value

H. Mirzadeh and A. A. Katbab (Eds.): ISPST 2018, *Eco-friendly and Smart Polymer Systems*, pp. 493–496, 2020.
https://doi.org/10.1007/978-3-030-45085-4_120

improved mechanical properties. Also, oriented nanofiber has shown a better performance in Mode I fracture. At Mode I fracture, random orientation has shown better properties and ultimately lowering thickness of nanofiber showed better results in both Mode I and Mode II [4].

2 Experimental

2.1 Electrospinning Process

To product electrospun nano-webs, a semi-automatic electrospinning machine (manufactured by Asia Nano-structure, Iran, Model RN/K) was utilized. Also, to prepare polymeric solution, Polyvinyl Butyral (PVB, B30H, M_w = 30,000 g/mol, T_g = 68 °C, Mowital, Kurary specialties for Europe) was dissolved in Ethanol (98%, Merck). Then the solution was stirred by a magnetic stirrer (speed = 200 rpm) for 12 h at ambient temperature to attain better polymer-solvent interaction. Electrospinning factors are listed below: voltage = 25 kV, collector distance = 15 cm, Concentration = 12%, feed rate = 5 mm/min, collector speed = 500 rpm).

The PVB nano-webs were electrospun in the thicknesses of 30, 90, and 150 μm.

2.2 Preparation of Glass Phenolic Pre-pregs

To prepare glass/phenolic pre-pregs, hand lay-up method was applied. For this aim, phenolic resin (resole type, IL800, solid contents = 80 ± 3%) was taken from Resitan Co. of Iran. Also, glass fiber (E-glass, plain woven fabric, density = 200 g/$m^{2,}$ Sichuan Co., China) was utilized. Weight percent of both of resin and fiber was considered 50%. To B-stage pre-pegs, they were held in an oven at various temperatures and times. Then flow test in accordance with ASTM D-5331 was done on the pre-pregs in the optimal condition (Temperature = 90 °C, time = 90 min) and flow-ability was determined equal to16 ± 1%.

2.3 Multi-layers Composites Preparation

To manufacture glass/phenolic multi-layer specimens, 14 glass/phenolic plies were stack on each other. Then a 14 μm thickness Teflon (PTFE) sheet was set between mid-layers (7th and 8th). After manipulation by vacuum bag process the specimens were cured by means an autoclave. To assess mechanical properties, obtained composites were tested according to three bending flexural test based on ASTM D790. To prepare notched three bending flexural test specimens in accordance with ASTM D7905 a plastic mid-layer was set between pre-preg plies before loading in presence of vacuum.

3 Results and Discussion

The effect of thicknesses of PVB electrospun interlaminates has been shown by means of tension- strain curve (Fig. 1). Regarding the curve, it is notable that increment of interlaminates thickness leads to decline of G_{IIc} and ultimate strength in most of specimens (Table 1). The main cause of G_{IIc} decrement is lack of enough inter-laminar resin in the impregnate PVB nanofibers (Fig. 2). The modulus values of PVB and resole were measured 18.88 and 358 MPa, respectively. Hence, the stiffness of multi-layered composite was limited by resole's strength. PVB is flexible and its strain at breaking point is reported 110% and for resole it is only about 1–2%. When PVB nanofiber enters into tough inter-laminar region owing to its poor crack resistance, nanofiber's dimension changes rapidly and goes into plastic region and absorbs more energy from crack to develop. In Fig. 1, it is obvious that the tension-strain curves of nano-modified fiber specimens at optimal thickness (30 μm) contain wider maximum tension peaks rather than non-modified specimens. It reveals the existence of mechanisms that devour cracks and absorb energy to hinder downfall of force.

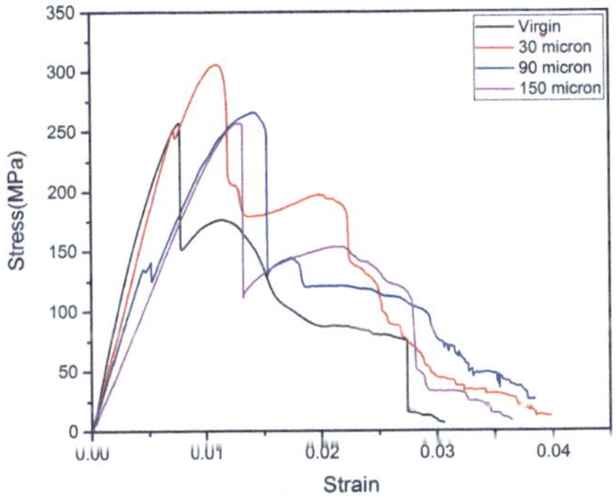

Fig. 1. Tension strain curves of PVB nano-webs modified specimens in different thicknesses

Table 1. Results of end notched flexural (ENF) test of PVB electrospun interleaved specimens

Specimen	Modulus (GPa)	σ_{max} (MPa)	G_{IIc} (J/cm^2)
Virgin	25	285.65	1142.3
30 μm	24	305.63	1855.92
90 μm	21.5	265.6	1544.6
150 μm	23.3	258	1428.2

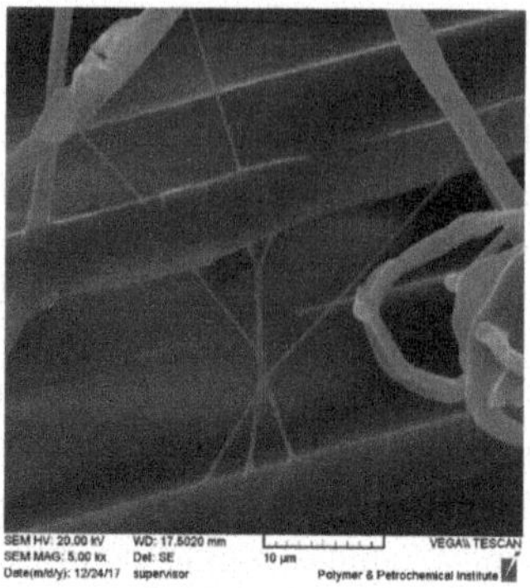

Fig. 2. SEM image of fracture surface of PVB nano-webs modified composite (thickness = 90 μm)

4 Conclusion

Increasing nano-webs thickness leads to decrease the final strength of multi-layered composite. But its strain at breaking point increases due to decreasing modulus resulted from increasing nano-layer's thickness. Also, addition of nanofiber as interlaminate in multi-layered composites caused enhancement of G_{IIc} in most of specimens compared to their reference counterparts. Finally, a 30 μm thickness was chosen as optimal value.

References

1. Pilatom, L.: Phenolic Resins: A Century of Progress. Springer, Berlin Heidelberg (2010)
2. Saghafi, H., Palazzetti, R., Zucchelli, A., Minak, G.: Influence of electrospun nanofibers on the interlaminar properties of unidirectional epoxy resin/glass fiber composite laminates. J. Reinf. Plast. Compos. **34**(11), 907–914 (2015)
3. Hamer, S., Leibovich, H., Green, A., Avrahami, R.: Mode I and Mode II fracture energy of MWCNT reinforced nanofibrilmats interleaved carbon/epoxy laminates. Compos. Sci. Technol. **90**(5), 48–56 (2014)
4. Palazzetti, R., Yan, X., Zucchelli, A.: Polym. Polym. Compos. **16**(2), 137–150, 2014
5. Hajian, M., Reisi, M.R., Koohmareh, G.A., Jam, ARZ: Preparation and characterization of Polyvinylbutyral/Graphene Nanocomposite. J. Polym. Res. **19**(10) (2012)
6. Kaynak, C., Cagatay, O.: Rubber toughening of phenolic resin by using nitrile rubber and amino silane. Polym Test **25**(3), 296–305 (2006)
7. Kheirkhah Barzoki, P., Rezadoust, A.M., Latifi, M., Saghafi, H.: The experimental and numerical study on the effect of PVB nanofiber mat thickness on interlaminar fracture toughness of glass/phenolic composites. Eng. Fract. Mech. **194**, 145–153 (2018)

Biological and Antibacterial Properties of Catechin-Incorporated PLA/Gelatin Hybrid Microfibers

Marziyeh Ranjbar Mohammadi(✉), Mohammad Nouri-Taba, and Ehsan Yousefi

Textile Department, University of Bonab, Bonab, Iran
m.ranjbar@bonabu.ac.ir

Abstract. The main purpose of this study was fabrication of catechin (green tea extract) loaded PLA/gelatin microfibers and investigation about antibacterial and cell culture behavior of produced fibers. For this, catechin was loaded in gelatin or PLA or both polymeric solutions and electrospinning with two different nozzles was used for scaffold production. SEM analysis showed smooth fibers with no cracks for catechin incorporated fibers. Loading catechin increased hydrophilic nature of mat. Antibacterial analysis against gram positive and gram negative bacteria confirmed that fibers with catechin can inhibit bacteria growth.

Keywords: Catchtin · Poly(lactic acid) · Gelatin · Electrospining · Antibacterial

1 Introduction

Wounds with high amount of skin loss need immediate envelopment with a dressing primarily to support the wound. Protection can be accomplished by covering the wound with wound dressings [1]. Infectious diseases caused by bacteria are still one of the biggest health problems around the world that endanger millions every year [2]. In the United States alone, about a quarter of patients die from the need for organ transplants, while in the United States, the annual cost of treating hospital infections is expected to reach $ 5 billion [3]. In 2001, the global wound healing market in the United States was estimated at $ 13156 million with an annual growth rate of 15% [3]. Since ancient times, for effective treatment of ulcers, several materials have been used to cover the wound and also prevent infection. In this regard, honey paste, vegetable fibers and animal fats were used as wound covering materials [4]. According to experts, skin lesions created in the body of people affect their psychological and social aspects of their life. Recently, it is very effective to create antibacterial wound dressing for treating ulcers. The medical applications of protein containing fibers have been widely discussed over the past decades. However, loading natural and antibacterial materials in fiberous structure has been used frequently. Green tea extracts contain catechins along with caffeine, vitamins, amino acids and theanine. Catechin is a nontoxic, polyphenolic antioxidant plant metabolite that belongs to the flavonoid family, and that has been shown to possess antibacterial, antioxidant, and other physiological functions [5]. PLA

H. Mirzadeh and A. A. Katbab (Eds.): ISPST 2018, *Eco-friendly and Smart Polymer Systems*, pp. 497–501, 2020.
https://doi.org/10.1007/978-3-030-45085-4_121

is a biodegradable and bioabsorbable, renewably derived thermoplastic polyester extensively studied over the last several decades [6]. On the other hand, Gelatin is a natural biopolymer derived from collagen by controlled hydrolysis. Because of its many advantages, such as its biological origin, low price, biodegradability, biocompatibility and commercial availability, gelatin has been widely used in the pharmaceutical and medical fields [7]. Electrospinning has been investigated as an effective and easiest method for the production of nano and micro scale fibers. Nowadays, composite electrospun fibers are used in wound dressing's composition because of their major advantages [8]. In this paper, for the first time, production of hybrid fibrous scaffolds from gelatin, PLA and catechin was investigated. It seems that, this composition has proper biological and antibacterial properties and would show high capability in healing process of skin wounds.

2 Experimental

2.1 Materials and Methods

At first, gelatin was dissolved in acetic acid for preparation of solution with concentration of 18 (w/v%). Poly (actic acid) was dissolved in dichloromethane to get solution with concentration of 1 (w/v%). After stirring this solution for 12 h, the homogenous compositions were produced at room temperature. For fabrication of hybrid fibers, catechin was loaded in gelatin or PLA or both of the solutions with concentration of 5 wt% (according to dry weight of dry polymer used for fiber production). Distance between both nozzles and collector was about 14 cm. Applied voltage and feed rate for gelatin fiber production were 17 kV and 0.7 mL/h. Poly lactic acid fibers were fabricated with the applied voltage of 15 kV and feed rate of 1 mL/h. 14 cm and 15 kV, respectively.

2.2 Cell Attachment Study

In order to examine the cell adhesion on the fabricated mats, 2×10^4 cells/ml were seeded on each disinfected mat and incubated for 3 days. Afterwards, the culture medium was removed, samples were fixed with 4 (v/v%) glutaraldehyde solution in PBS at 4 °C for 30 min, and finally dehydrated in a series of ethanol/distilled water solution (10% ethanol increments; each step 10 min). The morphology of the attached cells on dried electrospun scaffolds was observed by SEM. For immunofluorescence staining Cells were fixed with 3.7% paraformaldehyde (PFA) for 10 min, washed in PBS (phosphate buffered saline), permeabilized for 30 s with cold Acetone/Ethanol at −20 °C and then blocked with 1% BSA (bovine serum albumin) for 30 min. The cells were washed with PBS and subsequently incubated with primary F-actin antibodies (Sangon) (1:200 v:v dilutions in 1% BSA) for 1 h at 37 °C, and followed by anti-Rabbit-cy3 secondary (Sangon) antibody (1:300 v:v dilution in 1% BSA solution) for 1 h at 37 °C, DAPI (1:1000 dilute) to stain for F-actin and nuclei. Cells were imaged with an inverted fluorescence microscope (Olympus CKX41) which was equipped with blue, green & UV filters.

With increasing catechin, the antibacterial property increases.

2.3 Antibacterial Test

The antimicrobial activity of the samples against gram-negative bacteria (*E. coli*) and gram positive (*S. aureus*) is determined by the diffusion process antibiogram (Fig. 1).

3 Results and Discussion

3.1 Morphology of Fibers

The morphology of different compositions of fibers were shown in Fig. 2. Our results showed that, all of the compositions have smooth and uniform morphology and the size of the fibers is micro.

3.2 Antibacterial Analysis

To do this, a sterile swab dipped into a microbial suspension is flushed (pressing the swab to the side of the tube) and the culture media are cultivated in the form of a lawn. The specimens are placed in an incubator for 1 day at 37 °C for 1 day. It should be noted that gentamicin disks are used as antibiotics. Scaffolds form an antibacterial property, a far-reaching, non-growth area. Also, for culture counts testing (CFU), a broth medium is used to remove microbial suspensions from 3 to 4 pure colonies from 24 h culture of aureus and *E. coli* with loop, and then added to the broth solution and after The vortexing will be placed inside the incubator for 24 h. Then add the solution's opacity to 0.8. The bacterial broth solution is exposed to the prepared specimens for 24 h. Then, 10 μL of the solution is removed and agar is grown for 24 h to grow bacterial colonies inside the incubator. After 24 h, the colonies comprised by the colonometer are counted, shown in Fig. 3.

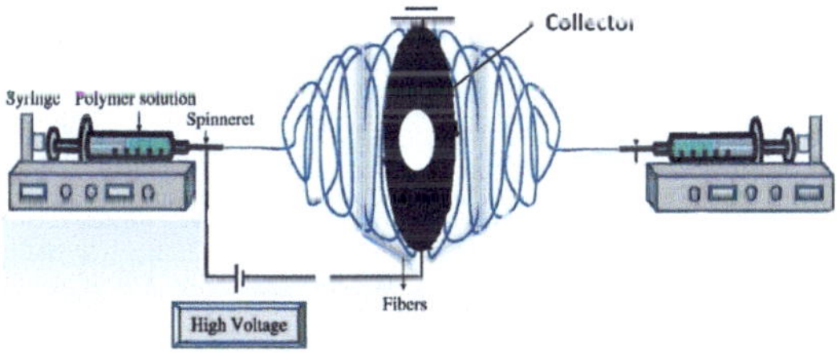

Fig. 1. An overview of the electro-spinning device and how it works

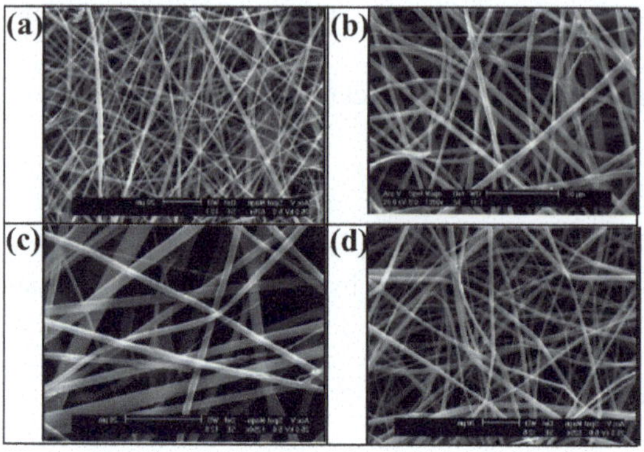

Fig. 2. The morphology of prepared fibers. (a) PLA/gelatin, (b) (pla+catchin) and gelatin, (c) pla and (gelatin+catchin), and (d) (pla+catchin) and (gelatin+catchin)

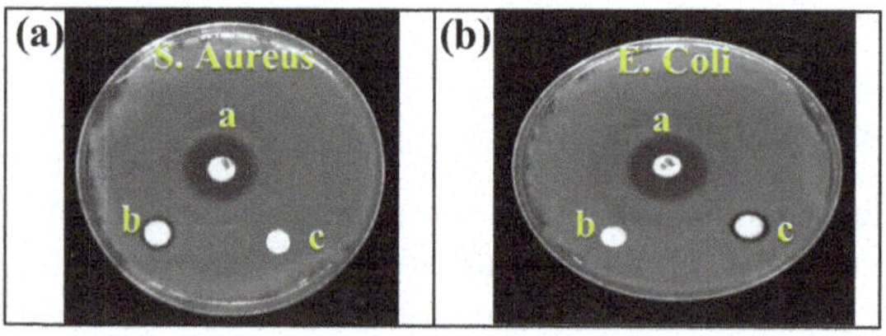

Fig. 3. Antibacterial test of samples: (a) control, (b) PLA-catechin/gelatin-catechin, and (c) PLA/gelatin

4 Conclusion

Catechin was found to contain anti-bacterial and antioxidant drugs in electrospun hybrid structures containing PLA and gelatin. According to the SEM analysis, it is observed that with the increase of catechin, the average amount of diameter contents has also increased.

References

1. Gu, S.-Y., et al.: Electrospinning of gelatin and gelatin/poly (l-lactide) blend and its characteristics for wound dressing. Mater. Sci. Eng. C **29**(6), 1822–1828 (2009)
2. Sun, H., et al.: Graphene quantum dots-band-aids used for wound disinfection. ACS Nano **8**(6), 6202–6210 (2014)
3. De Faria, A.F., et al.: Antimicrobial electrospun biopolymer nanofiber mats functionalized with graphene oxide–silver nanocomposites. ACS Appl. Mater. Inter. **7**(23), 12751–12759 (2015)
4. Zahedi, P., et al.: A review on wound dressings with an emphasis on electrospun nanofibrous polymeric bandages. Polym. Adv. Technol. **21**(2), 77–95 (2010)

5. Hara-Kudo, Y., et al.: Antibacterial action on pathogenic bacterial spore by green tea catechins. J. Sci. Food Agric. **85**(14), 2354–2361 (2005)
6. Rasal, R.M., Hirt, D.E.: Micropatterning of covalently attached biotin on poly (lactic acid) film surfaces. Macromol. Biosci. **9**(10), 989–996 (2009)
7. Lee, C.H., et al.: Nanofiber alignment and direction of mechanical strain affect the ECM production of human ACL fibroblast. Biomaterials **26**(11), 1261–1270 (2005)
8. Tohidi, S., Ghaee, A., Barzin, J.: Preparation and characterization of poly (lactic-co-glycolic acid)/chitosan electrospun membrane containing amoxicillin-loaded halloysite nanoclay. Polym. Adv. Technol. **27**(8), 1020–1028 (2016)

Janus Fiber Fabrication Using Electrospinning Process

Khadijeh Keshavarz Bahaghighat
and Mohammad Hossein Navid Famili(✉)

Chemical Engineering Faculty, Polymer Engineering Department,
Tarbiat Modares University, Tehran, Iran
nfamily@modares.ac.ir

Abstract. Arguably, one of the most versatile approaches toward anisotropic particles, cylinders, fibers, and fiber scaffolds is based on electrohydrodynamic (EHD) co-jetting. Electrospraying and electrospinning processes have been applied to polymer solutions to establish straightforward methods for the preparation of nano-/microparticles using high electrical voltages as driving forces for the transport of polymer solutions [1]. In this study, we have obtained the electrohydrodynamic method of Janus fiber. The precursor solutions of these fibers include polyvinylpyrrolidone (PVP) in blue and yellow, with a flow rate of 0.3 mL/h at 8 kV.

Keywords: Janus · Electrospinning · Fiber · Polymer

1 Introduction

Recently, electrospun fibers have attracted great attention due to their potential applications in tissue engineering scaffolds, filtration membranes, composite fibers, transparent conductive nanofibers, phosphorous, biocatalysts, etc. [2]. Driven by the widespread interest in nanofiber applications, considerable amount of research has aimed to gain better understanding of the electrospinning process, which is a potential process for the preparation of polymer fibers with diameters ranging from micro to nanometer scale, depending on the spinning conditions [3]. These techniques typically involve the transport of polymer solutions through a metal capillary. Upon application of a critical voltage, the droplet at the end of a metal capillary is distorted into a Taylor cone, and a fluid jet is emitted from the cone apex [4].

2 Experimental

Two polymer solutions were prepared separately in ethanol: 1) PVP 30% in ethanol and blue dyes 2) PVP 30% in ethanol and green dyes. The polymer solutions were loaded separately into 5 mL syringes. The syringe was attached to a needle with two 16G needles. A high voltage variable voltage supply that could provide DC voltage up to 30 kV was used. One end was attached to the needle and another to the collector, which was attached to a plate of stainless steel that was placed on an aluminum sheet. In this study, PVP with a molecular weight of 30 k was used. Then the electrospinning

H. Mirzadeh and A. A. Katbab (Eds.): ISPST 2018, *Eco-friendly and Smart Polymer Systems*, pp. 502–504, 2020.
https://doi.org/10.1007/978-3-030-45085-4_122

process was performed at a flow rate of 0.3 mL/h, a voltage of 8 kV, and a gap of 16 cm between the needle and the collector. The fibers were then collected from the collector and transferred to the container.

3 Results and Discussion

Infusion of the microchannel device with two 30 wt% PVP solutions, each loaded with a different colored dye, resulted in the formation of micron to submicron-sized fibers upon application of an 8.5 kV electric potential. By capturing the needle output during the electrospinning process, it was observed that the solutions were placed together and not mixed (Fig. 1). The electrospinning process is performed well by the formation of a Taylor cone in the needle output. In addition, the fibers collected on the collector cause a two-dimensional fiber together, on a macro scale, to give us a green sheet. The formation of fiber on the collector was also confirmed by optical microscopy (Fig. 2 and Table 1).

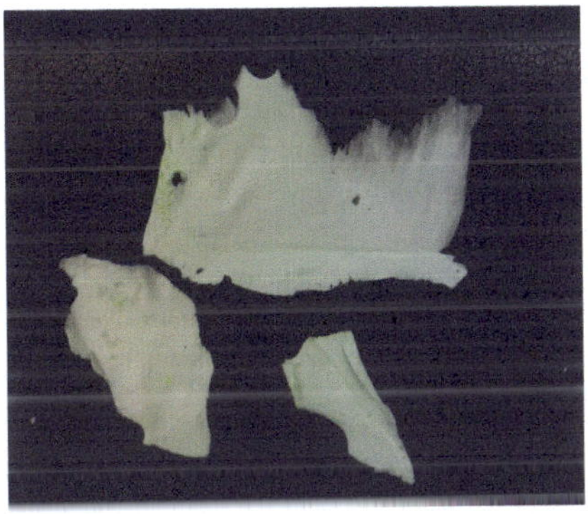

Fig. 1. The fibers obtained from the electrospinning process

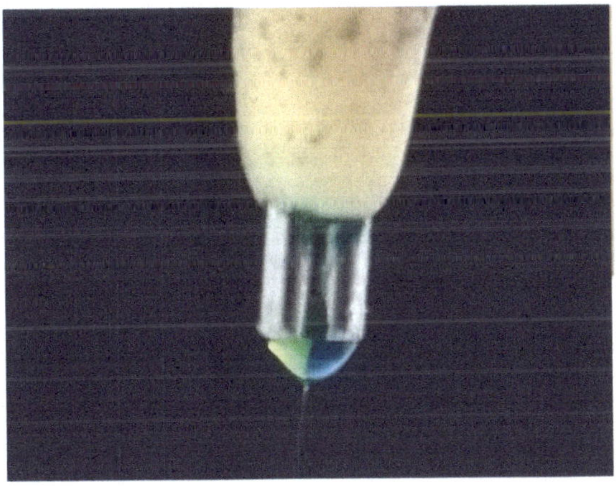

Fig. 2. Photograph of a typical Taylor cone and jet ejection during EHD co-jetting. The interface between the two aligned polymer solutions is clearly observed

Table 1. Processing condition used in electrospinning method

	PVP concentration (w/v%)	Feed rate (mL/h)	Gauge
1	30	0.3	16
2	30	0.3	16

4 Conclusion

Using two 30% PVP solutions in two different colors and a binary needle, PVP fibers were obtained through EHD process involving two different parts. This was confirmed by a picture taken from the needle during the electrospinning process.

References

1. Rahmani, S: Multifunctional Drug Carriers with Programmable Properties, Ph.D. thesis, University of Michigan, pp. 6–16 (2015)
2. Miftahul Munir, M., Suryamas, A.B., Iskandar, F., Okuyama, K.: Scaling law on particle-to-fiber formation during electrospinning. Polymer **50**, 4935–4943 (2009)
3. Wang, C., Hsu, C.H., Lin, J.H.: Scaling laws in electrospinning of polystyrene solutions. Macromolecules **39**, 7662 (2006)
4. Jordahl, J.H., Ramcharan, S., Gregory, J.V., Lahann, J.: Needleless electrohydrodynamic cojetting of bicompartmental particles and fibers from an extended fluid interface. Macromol. Rapid Commun. **38**, 1600437 (2017)

Polymer Recycling to Protect Environment

Mechanical Properties of Poly(Lactic Acid)/ Ethylene-Vinyl Acetate Copolymer/Nanoclay Composites as Biodegradable Packaging Films

Mahdieh Zaji and Masoud Frounchi(✉)

Department of Chemical and Petroleum Engineering,
Sharif University of Technology, Tehran, Iran
frounchi@sharif.edu

Abstract. To improve flexibility of poly(lactic acid), PLA, different amounts of ethylene-vinyl acetate copolymer, EVA, with 28% vinyl acetate monomer were added as a toughing agent. The optimum composition of EVA in EVA/PLA blends was determined for preparing the nanocomposites. Nano-sized clay (DK4) was used to improve mechanical and barrier properties of PLA. PLA/EVA blends and PLA/EVA/DK4 nanocomposite films were prepared by solvent casting method. Resultant nanocomposites and blends were characterized by tensile tests. It was found that by adding just 3 wt% of EVA to PLA, elongation at break of PLA increased by approximately 300% while tensile strength decreased only 4%. Then PLA/EVA3 blend was chosen for producing nanocomposite with 1, 3, and 5 pph of nanoclay. Introducing 1 pph nanoclay compensated for the strength decrease due to adding EVA. The elongation at break for the nanocomposites was still higher than that of pure PLA (about 225%). These results show success in toughening PLA.

Keywords: Poly(lactic acid) · Ethylene-vinyl acetate copolymer · Biodegradable · Nanocomposite · Mechanical properties

1 Introduction

Poly(lactic acid), PLA, has been considered for biodegradable packing applications. However, it suffers from low flexibility and gas permeability which are important properties for packaging films. Methods have been proposed like using plasticizers and in-situ polymerization of PLA with other monomers to improve flexibility of PLA. EVA is a flexible copolymer that at high contents of vinyl acetate could interact with PLA. Indeed, according some reports, compatibility and/or miscibility of PLA/EVA blends depend on parameters such as amount of vinyl-acetate in EVA and composition of PLA/EVA blend [1], employing PLA-g-EVA graft copolymer as coupling agent [2, 3]. In this research work we have investigated mechanical, morphological and microstructural properties of PLA/EVA blends in which EVA has 28% VA content. Also nanoclay has been used to adjust mechanical and permeability properties of the blends.

H. Mirzadeh and A. A. Katbab (Eds.): ISPST 2018, *Eco-friendly and Smart Polymer Systems*, pp. 507–510, 2020.
https://doi.org/10.1007/978-3-030-45085-4_123

2 Experimental

2.1 Materials

PLA granules were purchased from a chemical company in China. Ethylene vinyl acetate copolymer with 28% vinyl acetate was purchased from LOTE Company (China). Chloroform used as solvent with more than 99% purity. Nanosized organoclay with the trade name of Nanolin DK4 (alkyl ammonium modified montmorillonite) with size of 1–100 nm and aspect ratio of 100–1000 supplied from Zhejiang clay chemicals, China.

2.2 Sample Preparation

PLA and EVA were dissolved in chloroform as solvent to prepare blend solutions in different proportions of two polymers. Then films of PLA/EVA blends were prepared by solution casting method. The solutions were poured into Petri dishes and underwent a drying process. For preparing nanocomposites, DK4 first suspended in chloroform and then added to PLA/EVA solutions and mixed in a magnetic stirrer for 1 h and then for 1 min in an ultrasonic bath. Nanocomposite films were prepared similarly by solution casting. Finally, the blend and nanocomposite films were put in a vacuum oven for 1 h.

2.3 Characterization

Mechanical properties of films were characterized by HIWA200 tensile testing machine. Films were cut into 8×1.5 cm with approx. 0.12 mm thickness. SEM and XRD tests were used to explore the morphology and state of dispersion of nanoclay particles.

3 Results and Discussion

Tensile test was performed on pure specimens, blends, and nanocomposites. Figure 1 shows results for the blends. Tensile strength decreased by increasing EVA amount, which was expected due to the flexibility of EVA. However, all three PLA blends (3, 5, and 7% EVA) had good elongation and tensile strength at break. On the other hand, tensile strength and elongation at break in PLA/EVA5 and PLA/EVA3 were higher than PLA/EVA7 indicating the limited miscibility of EVA in PLA. It was unexpected result that the ultimate elongation of PLA/EVA blends were greater than pure EVA. Combination of higher elongation of PLA/EVA blends with ultimate strength comparable to that of pure PLA produced blends with extraordinary toughness and flexibility. This demonstrates success in PLA toughening with EVA, without sacrificing the tensile strength of PLA. These results imply strong interactions and good compatibility between PLA and EVA at low EVA compositions. Indeed, in PLA/EVA3 blend highest elongation was obtained. This results from proper distribution of the dispersed EVA phase in PLA matrix in this composition. Finally, due to high tensile strength and elongation of PLA/EVA3 blend, it was selected to make nanocomposites. As shown in Fig. 2, by increasing DK4, tensile strength increased and elongation decreased compared to PLA/EVA3 blend. In PLA/EVA3/C1 as the amount of organomodified DK4 was low the nanoparticles were well-dispersed in matrix and showed exfoliated structure as observed by SEM images and XRD spectra. In this sample elongation at break was not significantly reduced by adding DK4 (14.84% compared to the PLA/EVA3 blend). On the other hand, in higher amount of DK4 (3 and 5 pph) there was an intercalated structure. In these amounts of nanoclay there was a significant decrease in elongation at break (65 and 98% respectively compared to the PLA/EVA3 blend). Increase in tensile strength of nanocomposites compared to the corresponding blends indicates a strong interaction between polymers and nanoparticles that resulted in formation of exfoliated and intercalated structures in the nanocomposites. The SEM images showed uniform dispersion of nanoparticles with low agglomeration of nanoclay particles. The results show that the presence of nanoparticles offset the reduction in tensile strength of PLA by adding EVA while the elongation in 1 pph of DK4 remains at a desirable level.

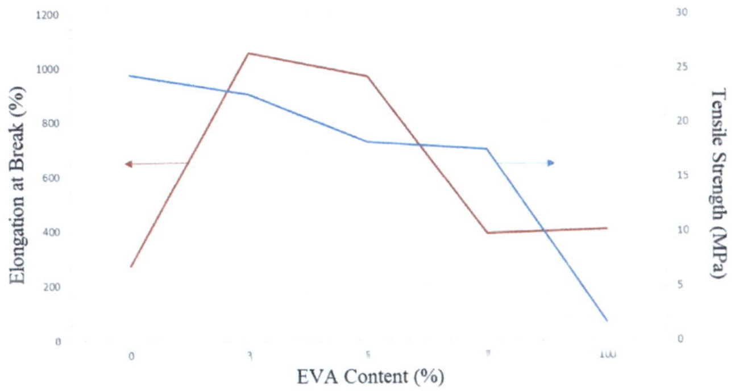

Fig. 1. Tensile strength and elongation at break of blends

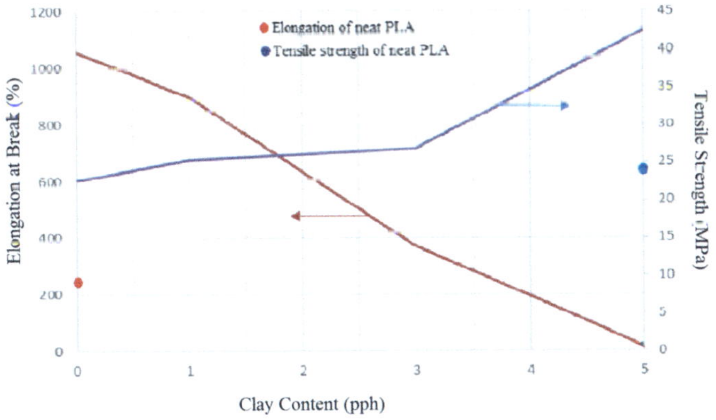

Fig. 2. Tensile strength and elongation at break of nano-composites

4 Conclusion

By adding just 3 wt% of EVA to PLA, elongation at break increased approximately 300% while tensile strength decreased only 4%. PLA/EVA3 was selected for making nanocomposites with 1, 3, and 5 pph of nanoclay. Introducing 1pph nanoclay compensated the decrease in PLA tensile strength upon adding EVA. On the other hand, elongation at break was remarkably 225% higher than that for the pure PLA.

References

1. Ma, P., Bogaerds, D.G.H., Goossens, J.G.P., Spoelstra, A.B., Zhang, Y., Lemstra, P.J.: Toughening of poly(lactic acid) by ethylene-co-vinyl acetate copolymer with different vinyl acetate contents. Eur. Polym. J. **48**(1), 146–154 (2012)
2. Moura, I., Botelho, G., Machado, A.V.: Characterization of EVA/PLA blends when exposed to different environments. J. Polym. Environ. **22**(1), 148–157 (2014)
3. Moura, I., Nogueira, R., Bounor, V., Machado, A.V.: Synthesis of EVA-g-PLA copolymers using transesterification reactions. Mater. Chem. Phys. **134**(1), 103–110 (2012)

Catalytic Degradation of Linear Low-Density Polyethylene Over USY Catalyst: Effect of Catalyst to Polymer Ratio

Mohammad Ghashghaee[1(✉)], Mohammad Reza Omidkhah[2], and Samira Shirvani[3]

[1] Faculty of Petrochemicals, Iran Polymer and Petrochemical Institute, P.O. Box 14975-112, Tehran, Iran
m.ghashghaee@ippi.ac.ir

[2] Chemistry and Chemical Engineering Research Center of Iran, P.O. Box 14335-186, Tehran, Iran

[3] Department of Chemical and Biological Engineering, University of British Columbia, 2360 East Mall, Vancouver, BC V6T 1Z3, Canada

Abstract. Catalytic degradation of linear low-density polyethylene (LLDPE) has been investigated using non-isothermal thermogravimetric analyses of the polyolefin samples mixed with different amounts of ultra-stable Y (USY) zeolite catalyst to assess the effect of catalyst to polymer ratio (0, 0.2, 0.6, and 1) on several events of catalytic decomposition, reflected as the onset, maximum rate and completion temperatures. The obtained results revealed that upon increasing of catalyst to polymer ratio from 0 to 1, the onset temperature decreased by more than 150 K while the decrease for the maximum-rate and completion temperatures were approximately 80 and 70 K, respectively. The results presented useful data for the reactor design of the plastic waste degradation process.

Keywords: Plastic waste · Catalytic degradation · Polyolefin · TGA · USY

1 Introduction

During the last decades, there has been a growing interest in consumption of plastic materials due to their astonishing features such as versatility, low cost, portability, thermal and mechanical strength, etc. Unfortunately, the amounts of plastic wastes, which bring severe environmental problems have also been increased accordingly [1]. Despite the serious disposal concerns raised for these polymeric materials, they also indicate promising potential in producing valuable chemicals and liquid fuels through thermal/catalytic decomposition pathways. Catalytic degradation of plastic wastes is considered as a reliable and selective approach to convert these wastes to value-added chemicals [2]. By far, zeolites of different structures, e.g., MFI, FAU, BEA, etc. have been applied to this pyrolysis scheme.

To get more insights into the degradation steps, thermogravimetric analysis is often employed, which provides beneficial results such as the starting temperature (the onset temperature) for the plastic waste decomposition as well as the maximum-rate and

H. Mirzadeh and A. A. Katbab (Eds.): ISPST 2018, *Eco-friendly and Smart Polymer Systems*, pp. 511–514, 2020.
https://doi.org/10.1007/978-3-030-45085-4_124

completion temperatures in addition to the kinetic parameters such as activation energy and pre-exponential factor [3].

In this study, we aimed to investigate the catalytic degradation of LLDPE, a widely employed polymer in various industries, with the USY zeolite as an acidic catalyst through the non-isothermal thermogravimetric experiments. More precisely, several catalyst-to-feed (Cat/Feed or Cat/Waste) ratios were used to explore the effects of the amount of catalyst applied in the LLDPE degradation.

2 Experimental

LLDPE was provided from Amirkabir Petrochemical Complex with the specifications listed in Table 1. A USY zeolite sample was also used as the catalyst with the main characteristics mentioned in Table 1. More details of the USY characterization can be found elsewhere [4].

Table 1. Characteristics of LLDPE polymer and USY catalyst

Polymer waste (LLDPE) characteristics	Value
Density (kg/m^3)	920
Tensile at yield MD/TD (MPa)	10/11
MFR (190 °C/2.16 kg) (g/10 min)	0.5
Catalyst (USY) characteristics	Value
BET surface area (m^2/g)	637.6
Total pore volume (cm^3/g)	0.3112
Total acidity (mmol NH_3 per gram of USY)	0.68

In a typical procedure, the LLDPE granules were crushed into fine powders. Then, the powder was mixed with a specific amount of USY catalyst to four Cat/Feed ratios of 0, 0.2, 0.6, and 1. The samples were further analyzed using the non-isothermal degradation experiments on a Netzsch–TGA 209F1 apparatus. The experiments involved heating of the samples from room temperature to nearly 1273 K with a heating rate of 10 °C/min in nitrogen atmosphere.

3 Results and Discussion

Figure 1 depicts the TG/DTG thermographs of the four samples with different Cat/Feed ratios at the heating rate of 10 °C/min. The results of the onset (T_{onset}), maximum-rate (T_{max}) and completion (T_{end}) temperatures can be observed in this figure. As is evident, by increasing the amount of the catalyst, the maximum-rate temperature of the degradation shifted to lower values (see Fig. 1b). This transfer was more severe in the case of 0.2 and 0.6 and was maximum at the Cat/Feed ratio of 1. The corresponding value for the Cat/Feed ratio of 1 was very close to that of 0.6, however. Whereas the decomposition process started at 673 K, reached its maximum at 763 K, and completed at about 773 K for the original LLDPE waste, it was almost commenced at 520 K, reached the highest rate at about 683 K, and completed at 703 K while for the Cat/Feed ratio of 1. Figure 2 shows the onset, maximum-rate and completion temperatures for all samples. For samples with higher amounts of catalyst, more weight losses were observed below 473 K, which could be attributed to the physically bonded water molecules or weak hydroxyl groups on the surface of the USY catalyst (see Fig. 1c).

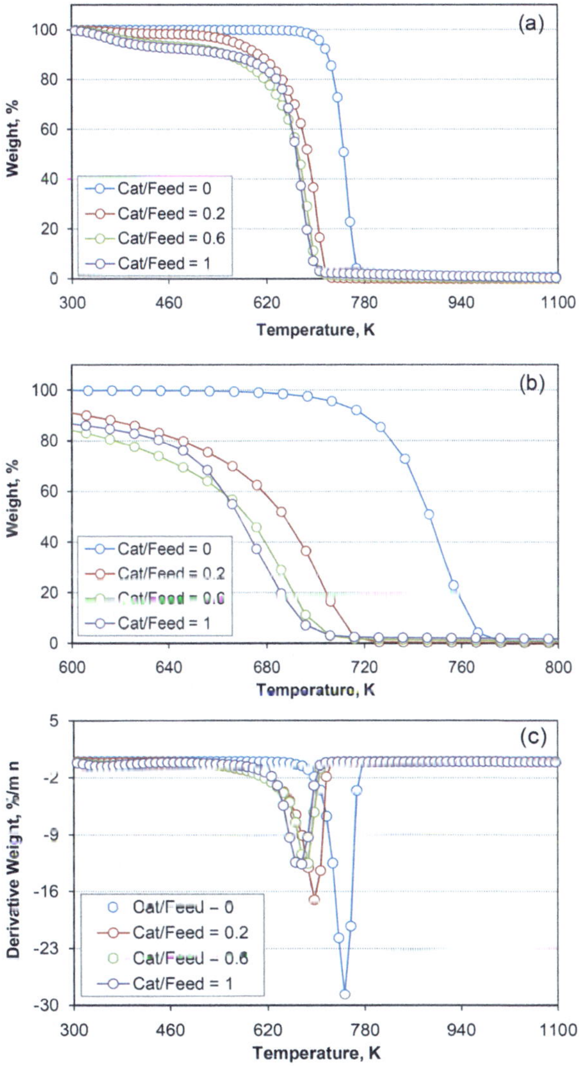

Fig. 1. TG/DTG thermographs for the polymer degradation: (a) TGA overall view, (b) TGA enlarged view, and (c) DTG

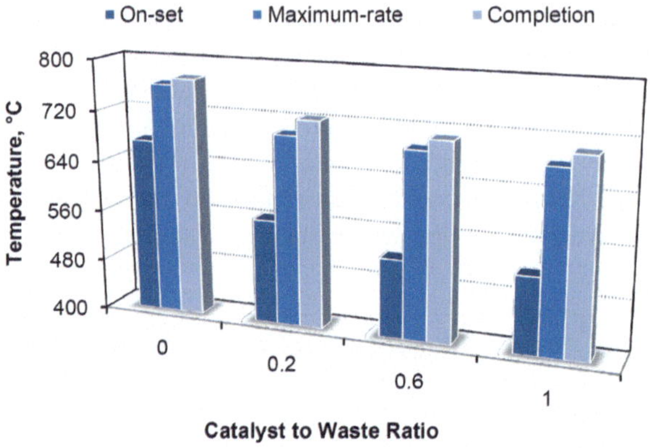

Fig. 2. Onset, maximum-rate, and completion temperatures in terms of Cat/Waste ratio

4 Conclusion

Four different samples were applied to the non-isothermal thermal and catalytic degradation of LLDPE by the USY catalyst. The influence of different Cat/Feed ratios in decreasing the decomposition temperatures at different stages was discussed. The results will be helpful in the effective design of the associated process.

Acknowledgment. The assistance received from Mr. Farid Hariri is gratefully acknowledged.

References

1. Sharuddin, S.D.A., Abnisa, F., Daud, W.M.A.W., Aroua, M.K.: A review on pyrolysis of plastic wastes. Energy Convers. Manag. **115**, 308–326 (2016)
2. Almeida, D., Marques, M.D.F.: Thermal and catalytic pyrolysis of plastic waste. Polímeros **26**, 44–51 (2016)
3. Ghashghaee, M., Shirvani, S.: Two-step thermal cracking of an extra-heavy fuel oil: experimental evaluation, characterization, and kinetics. Ind. Eng. Chem. Res. **57**, 7421–7430 (2018)
4. Shirvani, S., Ghashghaee, M.: Combined effect of nanoporous diluent and steam on catalytic upgrading of fuel oil to olefins and fuels over USY catalyst. Pet. Sci. Technol. **36**, 750–755 (2018)

Effects of Black Content and Type on the Biodegradability of Natural Rubber

Somayyeh Mohammadian[1](✉), Mohammad Barghamadi[2], and Mehri Alimohammadi[1]

[1] Payame Noor University, P.O. Box 19395-3697, Tehran, Iran
s.mohammadian@pnu.ac.ir
[2] Iran Polymer and Petrochemical Institute, P.O. Box 14965-115, Tehran, Iran

Abstract. In this study, biodegradability of compounds based on natural rubber filled with different black content and type, has been investigated. By increasing the amount of black reinforcement, the degradation percent decreased. Furthermore, the blends with the smaller filler were less degraded than that of the larger filler at the same amount of black. The unfilled sample (reference) had the highest final degradation and the highest percentage of extracted sole fraction. While the lowest percentage of sole fraction was seen in the compounds filled with highest amount of the black with the greater surface area. In the same amount of black, the compounds filled with the smaller black, showed the lower sole fraction in comparison with compounds containing bigger black particle size.

Keywords: Natural rubber · Black content · Black type · Bio degradation · Sole fraction

1 Introduction

Rubber products are not easily reusable due to cross-links in their chemical structure. On the other hand, chemicals used during the processing and manufacturing of rubber parts are another source of chemical contamination that may have bad effects on the environmental and human health. So today, it is trying to prevent the accumulation of these materials in nature by finding suitable ways [1]. One of the effective methods for the reclaiming of rubber products is the devulcanization of these materials. During this process, chemical crosslinks are broken and the devulcanized rubber can be reprocessed and cured. During this process, insoluble and non-melting rubber 3D networks form a two-dimensional soluble, processable, soft, recurable and low-modular structure [2]. Natural Rubber (NR), the only natural elastomer in the world, is consumed on a very high scale [3]. So reclaiming this rubber is important to reduce environmental pollution. The most commonly used reinforcing agent or filler in the rubber industry is black. The reinforcing effect of this filler depends on the rubber-filler bonds. As the filler is more active, the strength and number of these interactions are greater between the filler particles and the rubber chains [4]. In this research, the effects of black content and type on the biodegradability of Natural Rubber were studied.

H. Mirzadeh and A. A. Katbab (Eds.): ISPST 2018, *Eco-friendly and Smart Polymer Systems*, pp. 515–519, 2020.
https://doi.org/10.1007/978-3-030-45085-4_125

2 Experimental

Table 1 shows the formulation and naming method of the samples. The compounds were prepared by the two roll mill polymix 200 L from Schwabenthan Company. The mixing operation was performed for 15–20 min at 15 rpm. The specimens were cured before the destruction test. In order to determine the amount of degradation, the weight loss of the samples was calculated after immersing in the activate sludge by sampling at different time until 90 days.

Table 1. Formulation of prepared compounds

Ingredients	Content (phr*)							Supplier
SMR[1]-20	100	100	100	100	100	100	100	Malaysian rubber
CB[2]-N330	0	10	20	30	0	0	0	Pars
CB[2]-N550	0	0	0	0	10	20	30	
250-Oil	1	1	1	1	1	1	1	Behran
St.Ac[3]	2.5	2.5	2.5	2.5	2.5	2.5	2.5	Unichema
ZnO[4]	5	5	5	5	5	5	5	Rangineh
Sulfur	2	2	2	2	2	2	2	
CBS[5]	2	2	2	2	2	2	2	Bayer

*phr = part per hundred rubber
1: Standard malaysian rubber, 2: Carbon Black,
3: Stearic acid, 4: Zinc oxide, 5: cyclohexyl benzothiazole sulfonamide

3 Results and Discussion

Figure 1 shows the effect of the amount and type of black on the degradation percent of the S compound with time. By increasing the amount of black reinforcement, the degradation percent decreases. Moreover the compound containing smaller blacks (N330) is less damaged than the bigger black (N550). This observation can be analyzed with respect to the black-rubber interactions. By increasing the amount of these interactions and the amount of rubber absorbed on the black reinforcement, the possibility of destruction is reduced due to the inhibitory effect of these bonds. In the compounds with greater amount or smaller black, the greater the amount of these bonds and interactions, therefore, the amount of degradation decreases. Figure 2 shows the effect of different amounts and types of black on the final degradation and after 90 days, respectively. By increasing the amount of black reinforcement, the final degradation is reduced. In addition, compounds filled with smaller black (N330) has less final degradation than the compounds filled bigger black (N550). This trend is due to the black-rubber interactions and rubber adsorbed on black surface. Since the more content or smaller black is used, these bonds and interactions are higher.

The rate of degradation, which is obtained from the slop of the degradation percentage versus time, shown in Fig. 3, decreases significantly as the amount of filler increases. With the addition of black, the black-rubber interactions increase and the rubber degradation becomes harder. As can be seen in figure, bigger black, compared with smaller black, shows a lower percentage of degradation as well due to the less interactions between N550 black with the rubber chains.

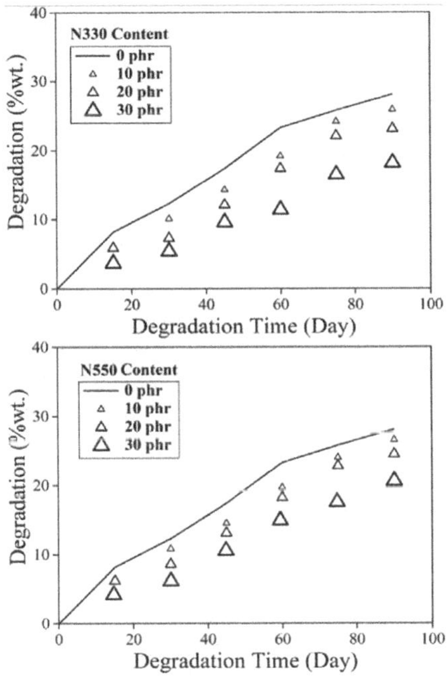

Fig. 1. Degradation percent versus time for compounds

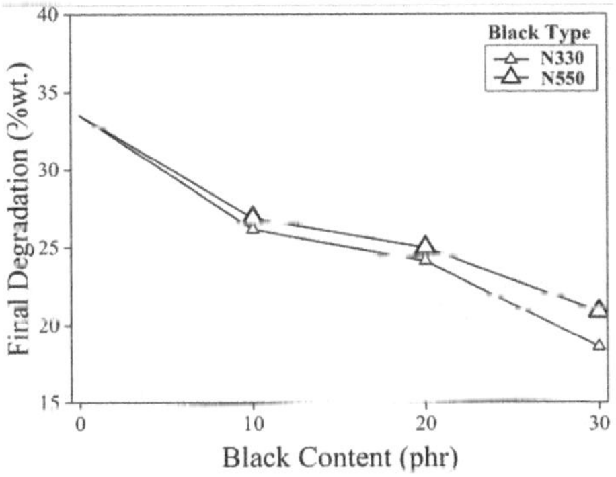

Fig. 2. The effect of black amount and type on the final degradation of the compounds after 90 days

When the degraded compound is placed in the solvent, the part whose crosslinked bands has not been solved in the solvent, which is known as the gel portion, while the part whose crosslinks is broken, it is solved in the solvent called the sole portion. After being placed in the solvent, the sole fraction was obtained by calculating the weight ratio of the sample dissolved in the solvent. Figure 4 shows extracted sole fraction. The sample without black has the highest final degradation and the highest percentage of sole fraction; while the lowest percentage of sole fraction is seen in the highest amount of N330 black. In compound containing black, compounds filled with smaller black, has a lower percentage of sole fraction. As the possibility of degradation alters with the change in the black amount and type, the broken of crosslinks is also subject to the same trend. As shown in Fig. 4, the percentage of sole fraction is declined by increasing the amount of black or reducing the black particle size.

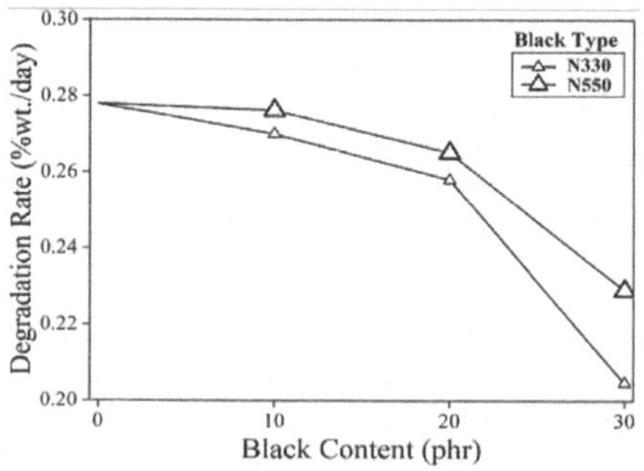

Fig. 3. The effect of black amount and type on the degradation rate of samples

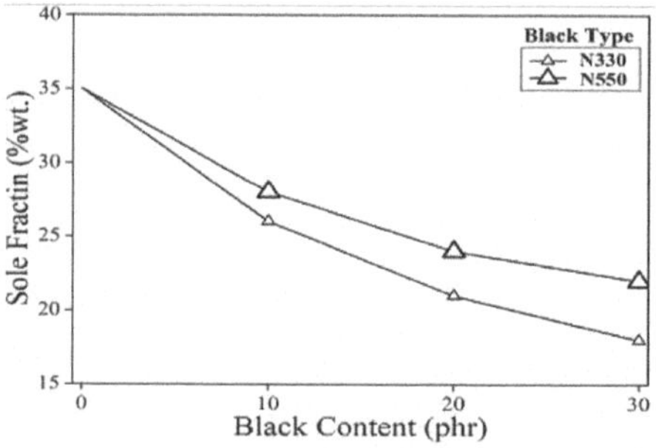

Fig. 4. The effect of black amount and type on the percentage of extracted sole fraction of samples after degradation in 90 days

4 Conclusion

- By increasing the amount of black reinforcement, the degradation percent of NR compounds decreased. Samples filled with smaller black (N330) was less degraded than that of the bigger black (N550), in the same filler content.
- Smaller black containing samples had the higher final degradation, rate of degradation and percentage of sole fraction than the bigger black containing ones. The lowest degradation was seen in the compound with highest amount of N330.

References

1. Yazdani, H., Ghasemi, I., Karrabi, M., Azizi, H., Bakhshandeh, G.R.: Continuous devulcanization of waste tires by using a Co-rotating twin screw extruder: effects of screw configuration, temperature profile, and devulcanization agent concentration. J. Vinyl Addit. Technol. **19**(1), 65–72 (2013)
2. Karrabi, M., Mohammadian-Gezaz, S.: Iran Rubber Mag. **15**(59), 143–147 (2010)
3. Mohammadian-Gezaz, S., Karrabi, M.: Prog. Rubber Plast. Recycl. Technol. **33**(4), 261 (2017)
4. Karrabi, M., Mohammadian-Gezaz, S.: Study of the cure characteristics and viscoelastic behavior of styrene-butadiene rubber compounds by using a rubber process analyze. J. Vinyl Addit. Technol. **16**(3), 209–216 (2010)

Recycling of Polyethylene Terephthalate (PET) via Glycolysis Method for Synthesis Waterborne Polyurethane

Mehran Dadash Ziyaei, Mehdi Barikani, and Hengameh Honarkar(✉)

Polyurethane and Advanced Polymers Department, Faculty of Science, Iran Polymer and Petrochemical Institute, P.O. Box 14975-112, Tehran, Iran
h.honarkar@ippi.ac.ir

Abstract. Due to the several applications and widespread use of polyethylene terephthalate (PET) in the world and the resulting waste, recycling of this material has a particular importance. In this work, glycolysis of PET waste was carried out using excess ethylene glycol in the presence of zinc acetate as catalyst at 195 °C for 2 h. Weight ratio of catalyst to PET (0.01) and molar ratio of PET to ethylene glycol (1:7.6) have been considered as suitable conditions for depolymerization process. The main products of the glycolytic reaction were bis (2-hydroxyethyl) terephthalate (BHET) and its oligomers or asymmetrically substituted terephthalates. The obtained product was purified then its structure was analyzed by Furrier Transform Infra-Red (FTIR) spectroscopy and the melting point was measured. It is important to point out that this product was used as polyol to produce waterborne polyurethane.

Keywords: Glycolysis · PET · Recycling · BHET · FTIR

1 Introduction

Polyethylene terephthalate (PET) is the most common and consumable fiber in the form of thermoplastic polyester, which was invented in the first decade of 1940. PET is semi-crystalline, with characteristic properties such as high strength, thermal stability, transparency, non-inflammability and safety. Due to the growing demand for textile utility and in food packaging as well as bottle markets for substitute of glass, huge quantity of PET is produced. polymer recycling is essential for a wide quantity of reasons such as conservation of oil decrease of greenhouse gas exhausts, saving of land fill space, protection of one's and benefits associated with reuse [1].

2 Chemical Recycling

Chemical recycling involves the transformation of polymer chain. Chemical recycling processes for PET are divided as follows: a) glycolysis, b) methanolysis, c) hydrolysis, and d) other processes such as aminolysis or ammonolysis [2]. Here, recycling of PET via glycolysis and characterization of final product is considered.

H. Mirzadeh and A. A. Katbab (Eds.): ISPST 2018, *Eco-friendly and Smart Polymer Systems*, pp. 520–523, 2020.
https://doi.org/10.1007/978-3-030-45085-4_126

3 Glycolysis

The glycolysis method may be described as a molecular depolymerization. In this procedure, by using transesterification reaction between PET ester groups and a diol, generally, ethylene glycol (EG) in excess, the monomer under the title of bis (2-hydroxyethyl terephthalate) (BHET) is obtained.

4 Experimental

Glycolysis reaction of PET process without catalyst is so slowly. It takes about 9 h to begin the degradation reaction, and needs 8–9 h to complete. In this work, the optimum values of solvent, catalyst and reaction time parameters have been used to overcome this problem. Since the goal of research is to obtain diol for the synthesis of polyurethane, the complete degradation of PET has not been done because the complete degradation of the PET is caused reduction in the molecular weight of the diol, and consequently falling in the properties of the final synthesized polyurethane.

Glycolysis experiments were carried out in a 500 mL a three necked, flat-bottom glass reactor equipped with a thermometer and a reflux condenser operated at atmospheric pressure. A mechanical stirrer was used to ensure proper mixing and turbulence. Heating mantle provided the device for control temperature for with no external pressure or stirring action. At first, the system heated up to 190–195 °C and the mass began boiling. The EG:PET (w/w) molar ratio was 7.6:1 with 1 (w%) zinc acetate catalyst added at temperature of 196 °C. After 3 h the mixture of reaction cooled to room temperature.

5 Purification of Product

For purification of product hot distilled water was added in excess to the reaction mixture under vigorous agitation since BHET is known to be quite soluble in boiling water. Hot water also dissolved the catalyst employed, and probably to some extent higher oligomers such as dimers and trimers. After adding hot water, the suspension was quickly filtered. The product was separated into solid and aqueous phases using a sintered glass filter. The filtrate contained ethylene glycol, BHET, and small quantities of few water soluble oligomers. This was heated until a clear mixture was obtained. This second filtrate was collected and stored in a refrigerator at 4 °C for 20 h. The obtained material is white crystalline BHET flakes.

6 Results and Discussion

The chemical structure of destruction product (BHET) has been investigated via FTIR analysis and shown in Fig. 1. As can be seen, stretching vibration of O-H, and C-H bonds appeared at 3438 and 2878-2932 cm^{-1}, respectively. The presence of above peaks shows that the depolymerization process has been occurred.

Furthermore, stretching vibration of C=O, C-O (ester), and C-O (alcohol) peaks showed distinct peaks at 1716, 1280, and 1078 cm^{-1}, respectively and the peak at 1505 cm^{-1} is related to aromatic ring of C-H. On the other hand, melting point of the product was observed at 106 °C by melting point meter, which is closely to melting point of BHET. From most of these observations it was concluded that the product from glycolysis is pure BHET. The molecular weight of the product was determined as 304 g/mol (Fig. 2).

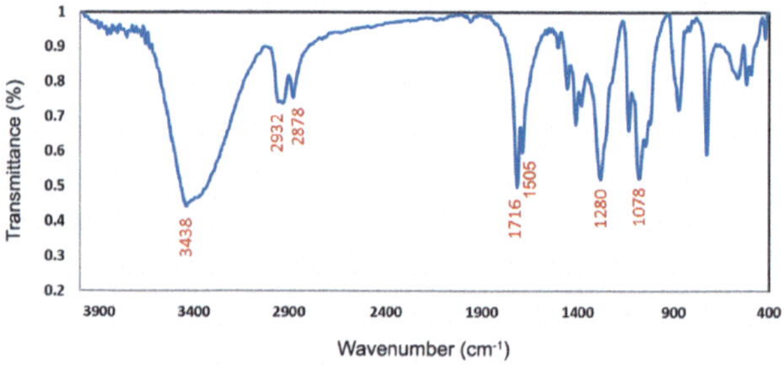

Fig. 1. FTIR spectroscopy of the BHET fraction

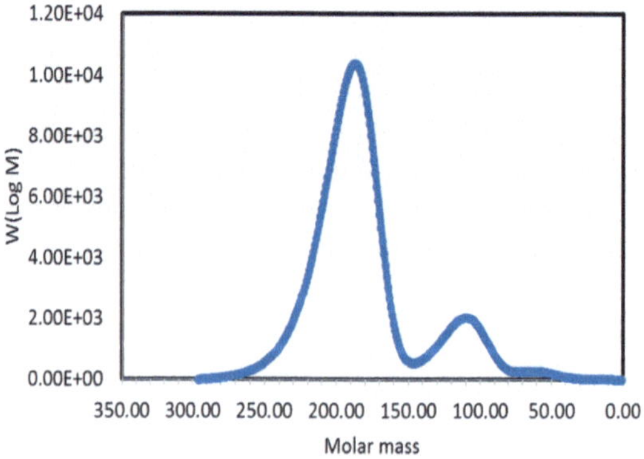

Fig. 2. GPC curve of glycolysis product

7 Conclusion

Glycolysis method due to low cost, and most importantly ability of producing variety of products has attracted more attention. In this study, by employing optimum conditions, degradation and recycling of PET were studied and recycled product was subjected to analysis and identification. Then resultant product was employed as polyol for synthesis of waterborne polyurethane.

References

1. Al-Sabagh, A.M., Yehia, F.Z., Eshaq, G., Rabie, A.M., ElMetwally, A.E.: Egypt J. Petrol **25**, 53–64 (2016)
2. Sinha, V., Patel, M.R., Patel, J.V.: J. Polym. Env. **18**, 8–25 (2010)

Feasibility Study on Producing of Side-by-Side Bicomponent Filament Yarns Produced from Recycled and Fiber Grade PET

Marjan Abbasi[1(✉)] and Richard Kotek[2]

[1] Department of Textile Engineering, Faculty of Engineering, University of Guilan, Rasht, Iran
m.abbasi@guilan.ac.ir
[2] Textile Department, North Carolina State University, Raleigh, NC, USA

Abstract. Side by side bicomponent filament from recycled poly(ethylene) terephthalate from post-consumer bottles and fiber grade PET was extruded. The bicomponent fibers were produced in a melt spinning machine and drawn to improve their mechanical properties. The effects of conditions on the fiber structure and physical properties of resultant fibers were evaluated by the following methods: differential scanning calorimetric (DSC), scanning electron microscopy (SEM), optical microscopy, intrinsic viscosity (IV) and tensile testing. The mechanical properties, such as tensile strength and initial modulus for bicomponent filament showed improvement compared to those for FG-PET. Furthermore, degradation of R-PET molecular structure leads that the viscosities of two polymers are of comparable value.

Keywords: Bicomponent · PET · Recycling · Side by side · Filament

1 Introduction

Currently a wide range of things, such as mineral water, soft drinks, and other foods, are being kept in PET dishes, and many more are being added to the list every day [1]. The accumulation of this non-biodegradable polymer waste has led to serious ecology difficulties [2]. R-PET fiber have lower price by 20% compared to other fibers for the same physical characters. It is clear that cost advantage and being eco- friendly of the fiber is intention for increasing R-PET fiber usage [3]. Bicomponent fibers can be defined as "extruding two polymers from the same spinneret with both polymers contained within the same filament" [4]. These fibers obtain the possibility for applying the properties of two polymers at the same time and, therefore, offer fibers with several new features. The main objective of producing bicomponent fibers is to exploit capabilities not existing in either polymer alone. Previous work has demonstrated the producing filament yarns from used PET bottle and the production was successfully carried out [5]. The aim of this work is to investigate processing behavior of side by side bicomponent filament yarn from recycled PET (R-PET) flake and fiber grade PET (FG-PET) in the melt spinning machine.

H. Mirzadeh and A. A. Katbab (Eds.): ISPST 2018, *Eco-friendly and Smart Polymer Systems*, pp. 524–527, 2020.
https://doi.org/10.1007/978-3-030-45085-4_127

2 Experimental

Virgin fiber grade PET (FG) chips with the IV of 0.6 and density of 1.38 g/cm^3 and recycled PET (R) material came from irregular beverage bottle having the IV of 0.75 and density of 1.383 g/cm^3 were used as the starting materials in this study.

A bicomponent extruder machine with a 288 hole spinneret, 4 denier per filament (dpf), spinning temperature of 285 °C, winding speed of 1600 m/min was employed for production of FG/R filament yarn. In order to evaluate and compare some properties such as shrinkage and birefringence, pure fibers from FG-, R- and BG-PET were also produced under the same conditions.

Intrinsic Viscosity (IV) of PET was measured using 40/60 mixture of 1,1,2,2 tetra chloro ethan and phenol according to ASTM D-4603.

The thermal analyses were performed on a Pyris Series-Diamond differential scanning calorimetry. DSC curves were obtained in the first heating, in the temperature range of 25–270 °C and with a heating rate of 20 °C/min. Crystallinity of samples was also obtained from DSC method was calculated from the following equation:

$$X_{DSC} = \frac{\Delta H_m - \Delta H_c}{\Delta H_{ref}} \times 100 \quad (1)$$

ΔH_m refers to the measured melting enthalpy, ΔH_c is the crystallization enthalpy induced during DSC test and ΔH_{ref} refers to 100% crystalline polymer which in the case of PET equals to 140 ± 20 J/g.

The ultimate strength and elongation of individual fiber sample were measured in a MTS Q-Test according to ASTM D2226.

3 Results and Discussion

BG-PET materials have high melt viscosity, which may affect the spinning uses of the R-PET. The molecular weight of the raw material and spun fibers for the processing effects with intrinsic viscosity (IV) measurement were verified. The results are listed in Table 1. The IV measurements of FG/R confirmed that molecular weights decreased with the recovery process and that molecular weight distribution became narrower, as indicated by MWD values. Degradation of R-PET molecular structure because of thermal, hydrolytic and chemical agents is confirmed by IV analysis [6]. It is needed that the viscosities of two polymers are of comparable value; otherwise, the higher viscosity component will not tendency to rearrange during spinning causing the distortion of the distribution of the components in the cross section of the bicomponent [4].

The results of the thermal analysis in FG- and R-PET are shown in Table 2. The crystalinity of two components is completely different. Table 3 gives the results of the tensile properties of the fibers produced at primary draw ratio.

There is a progressive increase in tenacity and initial modulus of bicomponent filament. Initial modulus is usually highly related to the crystallinity, crystal size and somewhat to the orientation. The mechanical properties, such as tensile strength and initial modulus in bicomponent filament showed improved compared to the FG-PET.

Table 1. Intrinsic viscosity & molecular weight analyses of raw materials and FG/R spun fibers

Properties	R-PET chips	FG-PET chips	FG/R spun fiber
IV, dL/g	0.75	0.6	0.50
M_n^*	33631	24773	19298
M_w	72375	37192	28445
MWD	2.15	1.50	1.47

Table 2. Thermal properties of starting materials

Crystallinity (% XDSC)	Melting enthalpy (J/g)	Melting temperature (°C)	Glass transition temperature (°C)	Starting material
7.58	37.688	253.53	79	Fiber grade PET
24.6	34.37	246.44	76	Recycled PET

Table 3. Tensile properties of FG/R-PET samples with primary draw ratio 2.8

Sample	Tenacity (gf/den)	Modulus (gf/den)	Strain (%)
FG/R	2.3 ± 0.11	57.7 ± 4.68	23.9 ± 2.17

± is the standard deviation

This may be attributed to higher molecular weight of R-PET chips. SEM photography of cross section of bicomponent fiber and optical microscopy was shown in Fig. 1. It shows the cross section of the FG/R by an SEM and optical microscope. There is no clear boundary between the two polymers. Furthermore it shows that there is the proper adhesion between the two components.

Fig. 1. Cross section view of FG/R bicomponent fiber left side, with SEM and right side, with optical microscopy

4 Conclusion

We have successfully developed a spinning process for FG/R-PET bicomponent side by side filament. This yarn exhibits desirable performance compared to FG-PET. R-PET showed improved tensile properties compared to the FG-PET. This may be attributed to higher molecular weight of R-PET chips.

References

1. Gurodatt, K., De, P., Rakshit, A.K., Bardhan, M.K.: J. Appl. Polym. Sci. **90**, 3536–3545 (2003)
2. Shen, L., Worrell, E., Patel, M.K.: Resour. Conserv. Recycl. **55**, 34 (2010)
3. Telli, A., Özdil, N.: J. Eng. Fiber Fabr. **10**, 47–60 (2015)
4. Kikutani, I., Radhakrishnan, J., Arikawa, S., Takaku, A., Okui, N., Jin, N., Niwa, F., et al.: J. Appl. Polym. Sci. **62**, 1913–1924 (1996)
5. Abbasi, M., Mojtahedi, M.R.M., Kotek, R.: ISPST 2012. Amirkabir University of Technology, Tehran (2012)
6. Elamri, A., Abid, K., Harzallah, O., Lallam, A.: Am. J. Nano Res. Appl. **3**, 11–16 (2015)

Effect of Chemical Composition of Vinyl Acetate/Dibutyl Maleat Copolymers on the Permeation Properties of Blend Membranes Based on Poly(Ethylene Oxide-B-Amide6) Block Copolymer for Carbon Dioxide Separation

Marzieh Shirinia[1], Mahdi Abdollahi[1](✉), and Mohammadreza Omidkhaha[2]

[1] Polymer Reaction Engineering Department, Faculty of Chemical Engineering, Tarbiat Modares University, Tehran, Iran
abdollahim@modares.ac.ir
[2] Chemical Engineering Department, Faculty of Chemical Engineering, Tarbiat Modares University, Tehran, Iran

Abstract. Pebax1657, poly(ethylene oxide-b-amide6) block copolymer and blend membranes containing 30 wt% P(VAc-co-DBM) copolymers with different chemical compositions were prepared by solution casting method. Effect of chemical composition of P(VAc-*co*-DBM) on the structure and permeation properties was studied. Formation of hydrogen bonding between NH groups of the PA6 chains and carbonyl groups of VAc-based copolymer chains was confirmed by FT-IR analysis. The permeation results showed that permeability and selectivity increased for all blend membranes simultaneously. Selectivity of blend membranes was increased significantly by increasing the DBM content of VAc-based polymer. Increase in the selectivity can be attributed to the increased solubility of CO_2 gas without any significant solubility change in the CH_4 solubility.

Keywords: Pebax1657-based blend membrane · Vinyl acetate-co-dibutyl maleat copolymer · Permeation of carbon dioxide and methane

1 Introduction

Among the materials with high CO_2 affinity, Pebax1657 is a thermoplastic elastomer suitable for the CO_2/CH_4 separation. It is a multiblock copolymer containing linear chains of rigid polyamide segments interspaced with flexible polyether segments. The hard polyamide (PA6) blocks supply mechanical strength and the presence of the polar ethylene oxide (EO) group, increases the affinity for CO_2, allowing a good CO_2/non polar (H_2, N_2, and CH_4) species separation and permselectivity [1]. Poly(vinyl acetate-*co*-dibutyl maleat), due to the existence of polar acetate groups in its structure is considered as the most CO_2-philic oxygenated hydrocarbon-based polymer to date [2].

H. Mirzadeh and A. A. Katbab (Eds.): ISPST 2018, *Eco-friendly and Smart Polymer Systems*, pp. 528–532, 2020.
https://doi.org/10.1007/978-3-030-45085-4_128

In the present paper, Pebax1657 and blend membranes containing 30 wt% P(VAc-co-DBM) copolymers with different chemical compositions were prepared by solution casting method and effect of P(VAc-*co*-DBM) chemical composition on their structure and permeation properties was studied.

2 Experimental Materials

Pebax1657 (Fig. 1) was received from Arkema, P(VAc-co-DBM) with different mole fraction of vinyl acetate (F_{VAc}) 87, 72, and 50, (P(VAc-co-DBM) with $F_{VAc} = 0.87$ (VD87)), and ethanol was bought from Merck. CH_4 and CO_2 gases with the purity above 99.99% were provided by Technical Gas Services Inc, Iran.

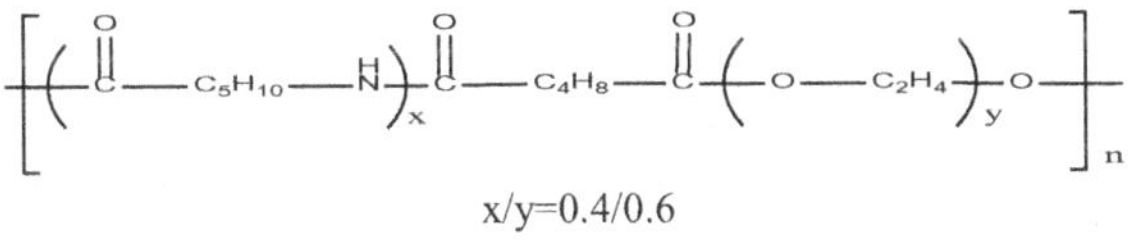

Fig. 1. Shematic of the chemical structures of pebax1657 [2]

3 Preparation of Blend Membrane

3 wt% Pebax solution in 30/70 (wt/wt) mixture of water and ethanol, respectively, was prepared at 80 °C for 3 h. Then given amount of P(VAc-co-DBM) was added to the solution and stirred for additional 2 h. The resultant solution was poured in a petri dish and dried in a vacuum oven at 80 °C for 24 h.

4 Gas Permeation Measurement

The permeation coefficients (P) of the membranes were measured by means of the constant volume method as follows:

$$P = S \times D = \frac{1}{76}\frac{V}{A}\frac{273.15}{T}\frac{l}{P}\frac{dP}{dt} \quad (1)$$

in which D (cm^2/s) and S ($cm^3/cm^3.bar$) indicate gas diffusion and solubility coefficients, respectively, l (cm) is the membrane thickness, V (cm^3) indicates the chamber volume, T (K) is the temperature, A (cm^2) indicate an area of the membrane which gas permeate from it, P (cm.Hg) is the pressure of feeding gas (3 bar in the present work) and dP/dt (cm.Hg/s) indicates slope of the gas permeated from membrane versus time under steady state condition which was obtained from experimental data of down-stream pressure versus time curves for both CO_2 and CH_4 gases.

Then, diffusion coefficient (D) can also be calculated with time-lag method using Eq. (2):

$$D = \frac{l^2}{6\theta} \tag{2}$$

In which θ(s) indicates time lag of permeate, i.e. intercept on the time axis in the experimental downstream pressure versus time curves. From P and D data at hand, solubility coefficient (S) can now be estimated using Eq. (1).

5 Results and Discussion

Intermolecular interactions between the ingredients in the blend membranes were studied using FT-IR spectroscopy (Fig. 2). Characteristic absorption bands of Pebax1657 appear at 3298 (strong peak, stretching vibration of the amide NH group), 1636 (strong peak, stretching vibration of the amide carbonyl group), 1728 (weak peak, stretching vibration of the carbonyl groups of ester linkage between the PA6 and PEO blocks) and 1095 cm^{-1} (strong peak, stretching vibration of the ether C-O group). Comparing FT-IR spectra of pure Pebax and blend membranes showed that wave number of characteristic peaks of the Pebax changes by adding P(VAc-*co*-DBM) copolymer, indicating that there are significant intermolecular interactions between the copolymers and Pebax. In other words, formation of hydrogen bonding between NH groups of PA6 chains and carbonyl groups of copolymer chains were confirmed, resulting in the compatibility between the dispersed and matrix phases in the blend membranes.

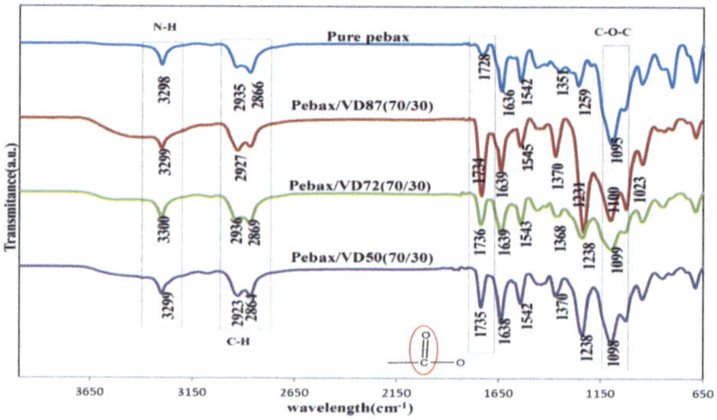

Fig. 2. FTIR spectra of Pebax1657 and blend membranes

Results of permeability as well as solubility and diffusion coefficients and ideal selectivity for Pebax and blend membranes are given in Table 1. Considering the results, permeability of the CO_2 increased by adding VAc-based copolymer. In Pebax/VD87 and Pebax/VD72 membranes, the permeability enhancement can be attributed to the simultaneous increase in the solubility and permeability coefficient, and due to the fact that CH_4 permeability does not increase significantly, the selectivity has also improved. There is a significant decrease in diffusion coefficients of the Pebax/VD50, which reduces the CH_4 permeation and hence selectivity reaches its maximum. According to the results, increasing the mole fraction DBM in the copolymer results in the stronger polymer-CO_2 interactions and higher solubility of the CO_2 without any significant change in CH_4 solubility.

Table 1. CO_2 and CH_4 permeabilities and CO_2/CH_4 selectivity values obtained for pure Pebax and various blend membranes (containing 30% wt additive) at temperature of 30 °C and at pressure of 3 bar

Pebax/copolymer (70/30)	P_{CO_2} (barrer)	P_{CH_4} (barrer)	α_{CO_2/CH_4}	D_{CO_2} (10^-6 cm^2/s)	S_{CO_2} (cm^3/cm^3.bar)
Pure Pebax	54	3.9	13.8	0.52	0.78
Pebax/P(V/D)-87	101	6.0	16.8	0.61	1.23
Pebax/P(V/D)-72	161	5.8	27.7	0.94	1.29
Pebax/P(V/D)-50	81	2.4	34.0	0.40	1.51

Effect of chemical composition in the blend membranes on the CO_2/CH_4 selectivity is shown in Fig. 3. It is clear from this figure that selectivity increases significantly by increasing the DBM content. Based on the above explanations, increase in the selectivity can mainly be attributed to the increased solubility of CO_2 gas without any significant solubility change in the CH_4 solubility [2].

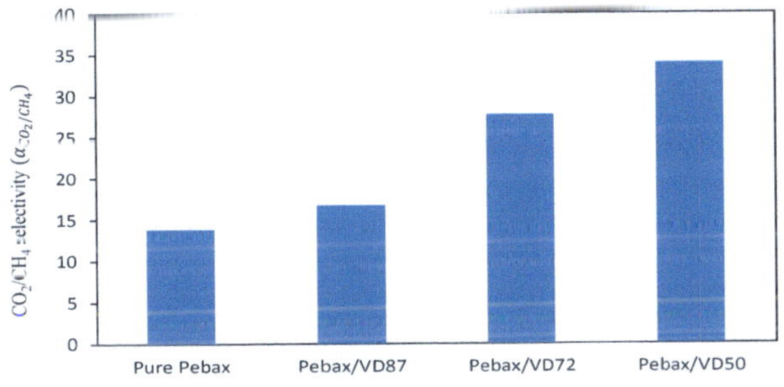

Fig. 3. Effect of Chemical Composition P(VAc-co-DBM) copolymer on the CO_2/CH_4 selectivity of blend membranes from Pebax and copolymer; Pebax/P(VAc co DBM) (70/30) (wt%/wt%)

6 Conclusion

The permeation results showed that permeability and selectivity increased for all blend membranes simultaneously. Selectivity of blend membranes was increased significantly by increasing the DBM content of VAc-based polymer. Increase in the selectivity can be attributed to the increased solubility of CO_2 gas without any significant change in the CH_4 solubility.

References

1. Peyravi, M., Babaluo, A., Ardestani, M.K., Pishghadam, S.R., Hadi, P.: J. Appl. Polym. Sci. **118**(2), 1211–1218 (2010)
2. Abdollahi, M., Khoshbin, M., Biazar, H., Khanbabaei, G.: Polymer **121**, 274–285 (2017)

Synthesis of Pyrolytic Carbon from Polyethylene Terephthalate on Graphite Substrate

Mitra Oliaei(✉) and Mohammad Yousefi

Iran Polymer and Petrochemical Institute, P.O. Box 14975-112, Tehran, Iran
m.oleiaei@ippi.ac.ir

Abstract. Pyrolytic carbon is a type of carbon that has many uses in the aerospace, nuclear, electrical and medical industries due to its unique properties. Usual raw materials for producing it, are mainly hydrocarbon gases and require high temperatures for their thermal decomposition, which consume high cost and energy. In this research, pyrolytic carbon is produced by using polyethylene terephthalate (PET) with the objective of recycling non-degradable plastics and environmental protection. In this study, pyrolytic carbon was synthesized at lower temperatures and without using catalyst. Also, the effect of process factors such as residence time, carrier gas flow and temperature on pyrolytic carbon morphology was investigated. SEM analysis showed pyrolytic carbon structures were made by using PET in the range of 800 °C to 1000 °C. The formation of pyrolytic carbon structures was confirmed by Raman test (presence of D and G peaks) and X-ray diffraction analysis. The optimum residence time was 452 s.

Keywords: Pyrolytic carbon · Polyethylene terephthalate · Waste · Recycling · More valuable materials

1 Introduction

Polymers, including plastics and rubbers are some of the most widely used materials in the present century, which have become an environmental threat due to their high volume and non-degradability [1]. In this study, polyethylene terephthalate as a high-consumption polymer was recycled by thermal decomposition and a higher value added material (pyrolytic carbon), was produced. The pyrolytic carbon structure is between graphite and diamond structure, as a result, its properties are intermediate. The graphite is electricity conductor in the direction of the graphene plates and its insulator in the perpendicular to these plates, while the diamond is very good thermal conductor and electrical insulator. The hardness of the diamond is very high and the hardness of the graphite is very low. According to the process and manufacturing conditions, this combination shows properties in a very different range properties of graphite and diamonds. Pyrolytic carbon unique properties are high thermal resistance and high mechanical strength, high thermal and electrical conductivity (along the layers), high chemical resistance, high melting point, high resistance to fatigue and biocompatibility [2, 3]. To produce this compound, hydrocarbon gases such as methane and other

H. Mirzadeh and A. A. Katbab (Eds.): ISPST 2018, *Eco-friendly and Smart Polymer Systems*, pp. 533–536, 2020.
https://doi.org/10.1007/978-3-030-45085-4_129

hydrocarbons usually decompose at high temperatures (above 1000 °C) [4]. In the present study, polyethylene terephthalate polymers were converted into pyrolytic carbon by thermal decomposition and without catalyst. In this study, in addition to producing more value added products, reducing costs and energy consumption, reacting to lower temperatures, and introducing a new source for pyrolytic carbon, helps to reduce environmental problems.

2 Experimental/Theoretical

A hollow quartz tube (L = 70 cm and i.d. = 4 cm) was used as a reactor. Some granules of PET as the raw material were placed inside the reactor. In the next step, a $1 \times 1\ cm^2$ square graphite as a substrate was put in the center of the reactor. Nitrogen as a carrier gas was passed through the reactor. In order to determine the optimum carrier gas flow and temperature, nitrogen gas was passed through the reactor at 25, 50, and 100 mL/min flow rates, and the process run at various temperatures range of 800 °C to 1000 °C. The reaction time was 60 min.

3 Results and Discussion

The SEM images of the product showed that at three flow of gas carrier (25, 50, and 100 mL/min), spherical pyrolytic carbon structures were formed with an average diameter of 1.06 μm (Fig. 1) [5]. Comparison of the combination morphology with the previous article results confirmed the formation of pyrolytic carbon. Also, with decrease in residence time due to the increase in the gas flow, particles size were reduced. The XRD analysis showed the peak of $2\theta = 25.6$ and confirmed formation of pyrolytic carbon (Fig. 2). This peak is the characteristic peak of pyrolytic carbon, which is displaced to smaller angles due to growth of graphene plates gap into graphite. The effect of reaction temperature on the product formation process was investigated in nitrogen gas flow of 25 mL/min and three temperatures of 800, 900, and 1000 °C. Raman analysis was used to investigate the different structures in the product. This test confirmed the formation of a pyrolytic carbon compound (Fig. 3). The presence of two peaks at 1325 cm^{-1} (I_D peak) and 1553 cm^{-1} (I_G peak), which are characteristic peaks of the pyrolytic carbon compound, has proved the formation of pyrolytic carbon at all temperatures. By the intensity ratio of D and G peaks, ($I_D/I_G = 0.87$), the average of crystal length was calculated about 22.40 nm.

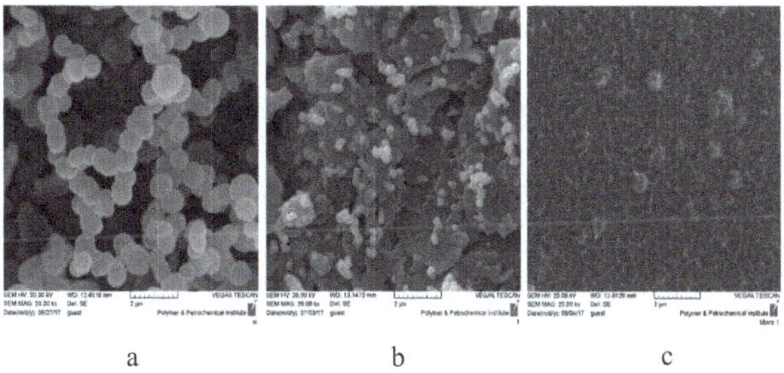

Fig. 1. SEM image of samples at nitrogen flow of: (a) 25 mL/min, (b) 50 mL/min °C, and (C) 100 mL/min °C

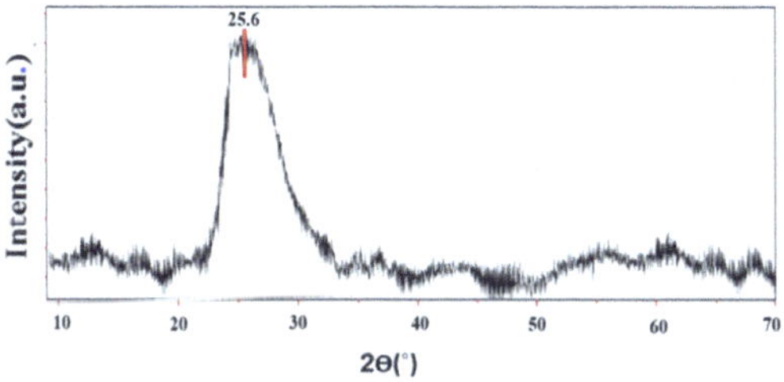

Fig. 2. Sample XRD test at 25 mL/min

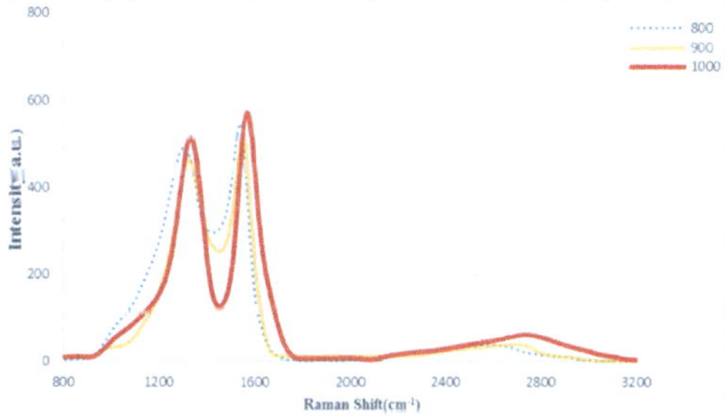

Fig. 3. Sample Raman test at 800, 900, and 1000 °C

4 Conclusion

In this process, pyrolytic carbon was produced without the use of catalyst. Reducing costs and energy consumption and reacting at lower temperatures are profits of this process. Another feature of this method is the use of non-degradable polymer materials in nature to produce useful materials. The results showed that under the experimental conditions, this compound was formed at temperature range of 800 °C to 1000 °C and gas flow rates of 25, 50, and 100 mL/min. According to the obtained SEM images, with increase of carrier gas flow, the size of the particles decreases. Gas flow of 25 mL/min is optimal quantity for product preparation.

References

1. Raman Sharma, K., Sharma, M.: Int. J. Waste Resour. **04**(01), 1–6 (2014)
2. López-Honorato, E., Meadows, P.J., Xiao, P., Marsh, G., Abram, T.J.: Nucl. Eng. Des. **238**(11), 3121–3128 (2008)
3. Andrieu, M.: Carbon. J. **40**, 7–24 (2002)
4. Bourrat, X.: World Conference on Carbon, Biarritz, France, vol. 11 (2009)
5. Sheikholeslami, Z.S., Yousefi, M., Imani, M., Joupari, M.D.: J. Mater. Sci. **53**(2), 959–976 (2018)

Polymers for Smart and Sustainable Future

Alkanolamine-Based Adduct as Latent Accelerator for Epoxy/Dicyandiamide: Statistical Approach to the Relationship Between Adduct Ingredients and Cure Characteristic of One-Pot System

Mohammadnabi Hesabi, Ali Salimi(✉), and Mohammad Hosain Beheshty

Faculty of Polymer Processing, Iran Polymer and Petrochemical Institute, P.O. Box 14975/112, Tehran, Iran
a.salimi@ippi.ac.ir

Abstract. Through the addition reaction of epoxy, phthalic anhydride and different types of alkanolamine, we developed adducts as latent accelerators for one-pot epoxy/dicyandiamide system. The analysis of variance (ANOVA) technique was employed to evaluate the relationship between the adduct ingredients, namely alkanolamine type, alkanolamine and phthalic anhydride content, and cure peak temperature (T_p) of one-pot system. Results showed that the T_p was affected by all three factors while the degree of influence of alkanolamine type on accelerating the curing reaction was higher than that of the others. Moreover, increasing the alkanolamine amount promoted while higher amount of phthalic anhydride demoted the accelerating ability of accelerator.

Keywords: Latent accelerator · Alkanolamine · One-pot epoxy system · DSC · ANOVA

1 Introduction

Some application areas of epoxy resin like prepreg manufacturing necessitate using one-pot epoxy system. Latent curing agents are commonly utilized in one-pot system. Most latent curing agents can be classified into encapsulated, complex, ionic and solid curing agents [1, 2]. Generally in all one-pot epoxy systems achieving long storage time, low curing temperature and high mechanical strength simultaneously is a real challenge. Solid curing agents like Dicyandiamide (Dicy) are thermally latent as a result of their solid states at room temperature. Dicy melts and dissolves into the epoxy resin near to its melting points, around 190–200 °C, to initiate polymerization. Using tertiary amine accelerators to improve the disadvantage of the low reactivity and reduce the high curing temperature of epoxy resin cured by Dicy is a classic way that destroys its long storage life although promoting the reactivity [3]. So Herein we have demonstrated alkanolamine-based adducts solid latent accelerators for epoxy/dicy system to develop a one-pot system with long shelf life at ambient temperature and low

H. Mirzadeh and A. A. Katbab (Eds.): ISPST 2018, *Eco-friendly and Smart Polymer Systems*, pp. 539–542, 2020.
https://doi.org/10.1007/978-3-030-45085-4_130

curing temperature simultaneously. In this work, the effect of adducts ingredients changes on their accelerating performance in one-pot system has been investigated statistically.

2 Experimental

Epoxy resin of diglycidyl ether of bisphenol-A (Epikote 828) with an epoxy molar mass of 184–190 g/eq was supplied by Momentive Co. (Columbus, OH, USA). The curing agent dicyandiamide (Dicy) with the melting point of 209–212 °C and average particle size of 10 μm was from Alz Chem Co. (Germany). 2-dimethylaminoethanol (A1), (Dimethyl- aminomethyl) phenol (A2), 2, 4, 6-tris (dimethylaminomethyl)-phenol (A3), and phthalic anhydride with molecular weight of 89, 151, 265, and 148 g/mol, respectively were purchased from Sigma-Aldrich (Germany). To produce addition compounds each alkanolamine was mixed gently with epoxy and phthalic anhydride at room temperature. Then, the mixtures were heated at 120 °C for 1 h. Adducts were cooled at room temperature, crashed and grinded to powders with average particle size of 100 μm. Different ratios of alkanolamine and phthalic anhydride were used (Table 1). One-pot epoxy resin samples were prepared via mixing of the resin (100 phr), DICY (8 phr), and amine-based adduct (2.5 phr). Curing properties were investigated in a differential scanning calorimeter, DSC (Mettler Toledo DSC1, Switzerland) at the heating rate of 10 °C/min in the temperature range of 35–300 °C.

Table 1. Effect of adducts' ingredients changes on the T_p of one-pot systems

Alkanolamine type	Phthalic anhydride amount ($E_H/E_{epoxide}$)			
	1.2		2	
	Alkanolamine amount ($E_H/E_{epoxide}$)		Alkanolamine amount ($E_H/E_{epoxide}$)	
	0.8	1.5	0.8	1.5
	T_p			
A1	167.0 ± 0.1	150.3 ± 0.3	169.9 ± 0.3	168.5 ± 3.0
A2	170.0 ± 0.2	162.4 ± 0.9	173.8 ± 0.2	164.6 ± 0.5
A3	154.1 ± 0.3	139.8 ± 1.4	167.2 ± 2.5	145.9 ± 2.4

3 Results and Discussion

Statistical methods should be used to analyze the data so that results and conclusions are objective rather than judgmental in nature. For this purpose, an experimental design base on general full factorial method with three treatment factors of alkanolamine type (at three levels), alkanolamine amount (at two levels) and phthalic anhydride amount (at two levels), with two replications, was employed (Table 1) and the response (T_p) was interpreted by means of the analysis of variance (ANOVA) technique. Before any inferences are drawn from ANOVA tables, the main underlying assumptions, namely normality and independence, through the standard ANOVA process should be checked. Figure 1 shows the normal probability plots and the plot of variation of residuals versus observation order for T_p. The distributions of residual errors along the normal demonstrated that the normality assumption is closely satisfied. Furthermore, the random distribution of residuals against the observation order indicated that there was no reason to suspect any violation of the independence assumption [4]. Table 2 summarizes the results of ANOVA for T_p response of the epoxy/Dicy/latent accelerator systems. The analysis was performed using the confidence interval of 99%. Since the *F values* of alkanolamine type, alkanolamine amount and phthalic amount in Table 2 are much larger than the critical significance test values, $F_{0.01,1,12} = 9.33$ and $F_{0.01,2,12} = 6.93$, the null hypothesis is rejected. It means that these three factors significantly affect T_p and the effectiveness of adducts in accelerating the curing reaction. It is obvious from Table 2 that the degree of influence of alkanolamine type on accelerating the curing reaction is higher than the other factors. Furthermore, in Table 2, the F-value for the interactions of all factors except of alkanolamine-phthalic amount are larger than the critical significance test values, indicating some interactions between these factors. According to the results of Table 1, the accelerating efficacy of adducts showed the following trend: A3-based > A1-based > A2-based. Moreover the results of Table 1 show that increasing the alkanolamine amount promotes while higher amount of phthalic anhydride demotes the accelerating ability. More quantity of alkanolamine increases the amine concentration in the system and heightens the probability of the nucleophilic attack which can result in more reactivity. But increasing the phthalic anhydride, especially higher than the equivalent amount, can

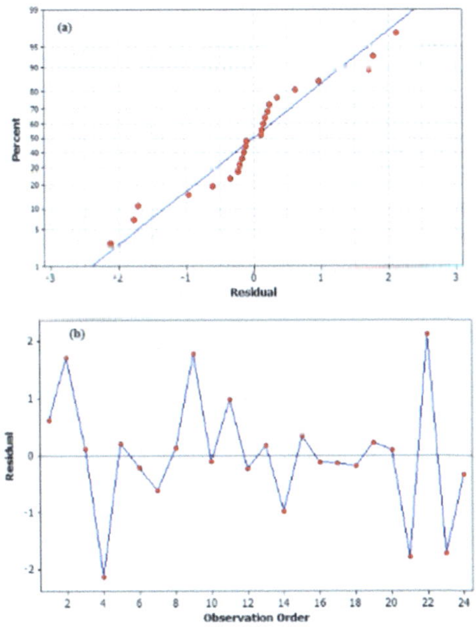

Fig. 1. (a) Normal probability plot of the residuals and (b) variation of residuals T_p

lead to some unreacted anhydride groups. The unreacted phthalic anhydride may form hydrogen bonds with nitrogen atoms and hydroxyl groups of latent accelerators which may stabilize the active groups. This may improve the hindrance effect of active sites and hence reducing their accelerating abilities.

Table 2. Analysis of Variance (ANOVA) for T_p response

Source of variation	Sum of squares	DoF	Mean square	*F* value	*P* value
Alkanolamine type (AT)	1114.27	2	557.13	272.06	0.000
Alkanolamine amount (AA)	829.79	1	829.79	405.20	0.000
Phthalic amount (PA)	358.52	1	358.52	175.07	0.000
AT*AA interaction	109.94	2	54.97	26.84	0.000
AT*PA interaction	67.50	2	33.75	16.48	0.000
AA*PA interaction	7.48	1	7.48	3.65	0.008
AT*AA*PA interaction	134.86	2	67.43	32.93	0.000
Error	24.57	12	2.05	–	–
Total	2646.93	23	–	–	–

4 Conclusion

In conclusion, we have designed an alkanolamine based latent accelerator for epoxy/dicy one-pot system and investigated the effect of its ingredients changes on T_p of one-pot system. Results showed that all three factors, namely alkanolamine type, alkanolamine and phthalic anhydride content, affected T_p which the first factor had the greater impact.

References

1. Liu, X.D., Sudo, A., Endo, T.: J. Polym. Sci. Part A: Polym. Chem. 250–256 (2011)
2. Rahmathullah, M.A.M., Jeyarajasingam, A., Merritt, B., Van Landingham, M., McKnight, S. H., Palmese, G.R.: Macromolecules 3219–3221 (2009)
3. Feng, L., Wang, Y., Wang, Y., Liu, H., Zhao, J.: J. Appl. Polym. Sci. 1895–1900 (2012)
4. Montogomery, D.C.: Design and Analysis of Experiments, 5th edn. Wiley, New York (2001)

Fixation of Carbon Dioxide Using Epoxy Resin: A Thermodynamics and Molecular Simulation Study

Mahtab Shabani, Seyed Tahmoures Keshavarz, Farhid Farahmandghavi, and Mohammad Imani(✉)

Iran Polymer and Petrochemical Institute, P.O. Box 14975/112, Tehran, Iran
m.imani@ippi.ac.ir

Abstract. Carbon dioxide is a non-polar chemical compound which is seriously suspected for global warming. Fixation of CO_2 using epoxy resin results in cyclic carbonate compounds. On the other hand, cyclic carbonate compounds react with diamines to provide non-isocyanate polyurethanes in response to the environmental concerns about polyurethanes made of isocyanates. In this study, CO_2 fixation using epoxy resin to form cyclic carbonate and its subsequent reaction with diethylene triamine (DETA) to form non-isocyanate polyurethane is simulated molecularly. Calculated enthalpies and Gibbs free energies of the reactions revealed that CO_2 fixation is thermodynamically possible using epoxy resin and product became more viscos, but the reaction with DETA may result in incomplete curing.

Keywords: CO_2 fixation · Molecular simulation · Epoxy resin · Non-isocyanate polyurethane · Thermodynamics

1 Introduction

Carbon dioxide is a non-polar, chemically reactive molecule with a high potential to react with epoxide to afford the five-membered cyclic carbonates. The reaction not only fixes CO_2 as an important environmental pollutant but also results in cyclic carbonates. Cyclic carbonates can serve as valuable intermediates for synthesis of polycarbonates, electrolytes for polymer lithium ion batteries, non-isocyanate poly-urethanes or green solvents [1]. Cyclic polycarbonates can be utilized to fabricate non-isocyanate polyurethanes after reaction with diamines [2].

Although CO_2 fixation reaction condition and mechanism is widely studied but it worth to simulate the thermodynamics and kinetics of this reaction. Here, CO_2 fixation reaction with an epoxy resin and consequent reaction with a diamine to produce non-isocyanate polyurethane were simulated molecularly and the reactions thermodynamics were studied.

H. Mirzadeh and A. A. Katbab (Eds.): ISPST 2018, *Eco-friendly and Smart Polymer Systems*, pp. 543–545, 2020.
https://doi.org/10.1007/978-3-030-45085-4_131

1.1 Theoretical

Full geometry optimization of chemical structure for all of the reactants and product molecules (Fig. 1) were carried out using the Gaussian 09W [3] software at HF/3–21 g level. Frequency calculation was done at the same level. Enthalpy (H) and Gibbs free energy (G) of all of the structures were extracted from output file. ΔH and ΔG of the reactions were calculated using Hess equation (Table 1).

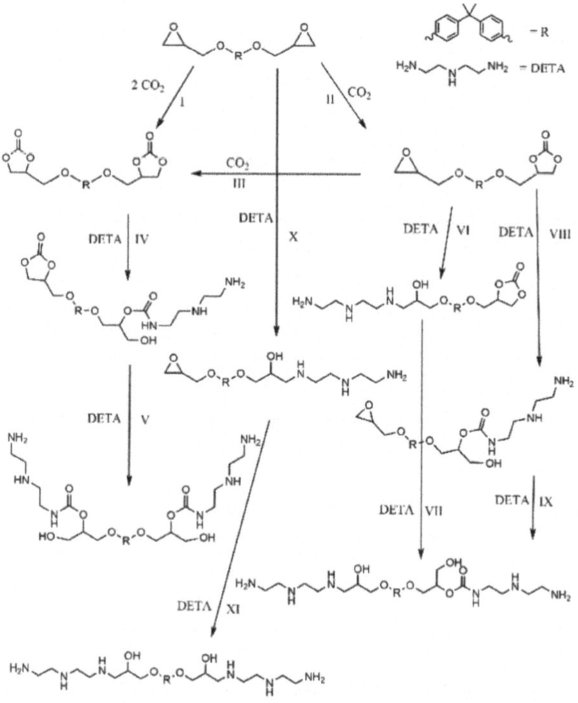

Fig. 1. CO_2 fixation reaction pathway before and after curing

Table 1. ΔH and ΔG calculated for all reactions

Reaction #	ΔH (kJ/mol)	ΔG (kJ/mol)
I	−333.96	−230.70
II	−167.16	−115.54
III	−166.80	−115.16
IV	−85.16	−33.44
V	−39.72	16.50
VI	−161.12	−104.47
VII	−123.93	−55.43
VIII	−108.97	−53.33
IX	−176.08	−106.56
X	−184.12	−128.44
XI	−177.33	−124.98

2 Results and Discussion

Epoxy conversion to cyclic carbonates (I, II, and III) is thermodynamically possible according to Table 1 in contrast to a decrement in entropy due to the conversion of 2 or 3 molecules to a single molecule. This may be due to the ring pressure in three-membered epoxy ring turning it to a very reactive species with a negative ΔG for turning into a cyclic carbonate.

The same analysis for the epoxy and cyclic carbonate reaction with diethylene triamine (DETA) supports more probability for epoxy ring reaction which can be attributed to the epoxy ring internal pressure in contrast to 5-membered cyclic carbonate and ending to more thermodynamically stable products. However, conversion of all cyclic carbonates to urethane linkage is not thermodynamically favorable leaving to leave DETA partially unreacted.

Viscosity is governed by the strength of intermolecular forces and especially by the molecular geometry of liquids. Liquids whose molecules are polar or can form hydrogen bonds are usually more viscous than nonpolar substances [4]. CO_2 (0.0000 Debye) is a non-polar molecule due to its structural symmetry, Reaction of CO_2 with epoxy ring (0.5685 Debye) ends to asymmetric di- and mono- cyclic carbonates (3.1367 and 4.8150 Debye, respectively). It can be expected for cyclic carbonates to be more viscose than the starting materials due to its higher polarity and strong interlayer interactions. Incomplete curing can be inferred when viscosity increases and ΔG is low or positive.

3 Conclusion

CO_2 fixation by epoxy ring is thermodynamically possible and produces more stable products than epoxy resin. Cyclic carbonates are more polar and it is expected to increase viscosity.

On the other hand, study of curing reveals that epoxy group tends to react with DETA more than cyclic carbonate and curing of fully or partially carbonated resin may be incomplete.

References

1. Sun, J., Zhang, S., Cheng, W., Ren, J.: Tetrahedron Lett. **49**, 3588–3591 (2008)
2. Guan, J., Song, Y., Lin, Y., Yin, X., Zuo, M., Zhao, Y., Tao, X., Zheng, Q.: Ind. Eng. Chem. Res. **50**, 6517–6527 (2011)
3. Frisch, M.J., Trucks, G.W., Schlegel, H.B., Scuseria, G.E., Robb, M.A., Cheeseman, J.R., Scalmani, G., et al.: Gaussian 09 Software. Gaussian Inc, Wallingford (2009)
4. John, M.W., Vitz, E., Shorb, J.: ChemPRIME: An Online Wiki Textbook with Exemplars, vol. 242. American Chemical Society, Washington, DC (2011)

Preparation of Polyether Sulfone Ultrafiltration Membrane and Investigation of Factors Affecting Its Performance for Hospital Wastewater Treatment

Seyed Hasan Maghoul(✉) and Ali Niatouri

Isfahan University, Isfahan, Iran
seyedhassanmaghoul@gmail.com

Abstract. Hospital wastewater may contain high levels of pathogen microorganisms, drug and laboratory residues, metabolites from medicinal substances, radioactive elements, or other hazardous chemicals. Such pollutants has made hospital wastewater one of the most dangerous materials for the health of humans and the environment. In this research, the construction and evaluation of the function of poly-sulfone ultrafiltration membrane in the treatment of hospital wastewater were discussed and evaluated. In the first part of this study, the poly-sulfone ultrafiltration membrane was constructed by phase inversion method, using water-soluble and N-methyl-pyrrolidone (NMP) solvent. In the second part, an ultrafiltration system was designed and developed, then the membrane was evaluated by this system. The flux passing through the membrane was measured at pressures of 3 and 5 bar with ultrafiltration system. The effect of concentration, thickness, coagulation bath temperature and opacity on flux was discussed in this research after the measurement.

Keywords: Membrane · Ultrafiltration · Polyether sulfone · Purification of waste water · Flux parameter

1 Introduction

Hospital sewages is important because of the presence of pathogenic microorganisms, radioactive substances and pharmaceutical waste. Their disposal into a non-purified environment can pose serious risks to public health. Due to the increasing standards of sewage treatment and reuse of wastewater, a high efficiency purification process is required. Since the hospitals of the country are often built in urban centers and do not have the space to create new refineries or upgrade inefficient systems, the use of new systems is felt more and more. Therefore, it is possible to select a suitable method for separation by considering different factors such as access to equipment, construction costs, energy and also the separation targets in the process. In this regard, membrane processes are of great importance with the benefits described below.

Some important features of membrane processes include: reducing energy consumption due to lack of phase change, low volume and no need for large space, variation in shape and size, low pressure drop and high mass transfer, high separation

H. Mirzadeh and A. A. Katbab (Eds.): ISPST 2018, *Eco-friendly and Smart Polymer Systems*, pp. 546–549, 2020.
https://doi.org/10.1007/978-3-030-45085-4_132

efficiency for dilute solutions, simple design, and ecofriendly. All these benefits make this method more distinct from other isolated methods. In this study, a polyether sulfone membrane was constructed and evaluated by phase inversion method in ultrafiltration scale for purification of hospital wastewater.

2 Experimental/Theoretical

The first step of preparing layer membrane is the preparation of polymer solution. In order to prepare a polymer solution, the polymers are first placed in the oven for 24 h in order to remove any probable adsorbed water particles onto the surface of the grains. Then, the polymeric solution components made up of polyether sulfone dissolved in NMP solvent were incubated at 50 °C for 24 h in a magnetic stirrer to obtain a clear, non-bubble solution. Using the phase inversion method, prepared solutions for disinfection were placed in a vacuum oven for 3 h. After preparing a homogeneous, clear, and non-bubble polymer, uniform and thin layers of polymeric films with a thickness of 200 μm on flat glass plates were prepared by hand-made steel strainer. After shaping the polymer solution, the glass surface was made in a thin uniform film. Then, they were transferred to a coagulation bath containing an anti-solvent of water. After the membrane was made in the bath and separated from the glass surface, the membrane washed with distilled water and stored in a distilled water container for 24 h in order to completely remove the solvent from its structure. The membrane is then inserted into the preparation stage for stabilization of the structure, using two baths of anti-organic solvent, methanol and normal hexane, each one of them was used for 15 min. At the last stage, the membrane is dried by a vacuum oven at 30 °C.

3 Results and Discussion

Flux parameter which is used to evaluate the membranes shows the output amount of the membrane per area and time unit. The rejection parameter is used to indicate the membrane's efficiency in preventing the passage of materials in the feed. To measure the flux of pure water passing through the membrane, it was placed inside a module of 5 cm in diameter and a pressure difference by a vacuum pump was applied. By measuring the mass of water passing through the membrane at specified time intervals and converting mass to volume of water, the PWP value was calculated. The applied pressure was 3 bar at 25 °C. As shown in the diagrams, with increasing time, the flux has decreased. One reason can be clogging of the cavities by suspended particles in the wastewater. This decrease in flux reflects the fact that the membrane made both the task of particle separation correctly and the selectivity. Also, membrane has isolated the particles in the wastewater from pure water. Moreover, transition flux indicates the amount of porosity under the layers and the cross-section of the cavities. With continued operation and increased shelf-life of the membrane under pressure, gradually the volume of its inflow decreased, and thus the existing openings and cavities in the membrane structure also became smaller. Finally, the flux has been reduced in over time (Fig. 1).

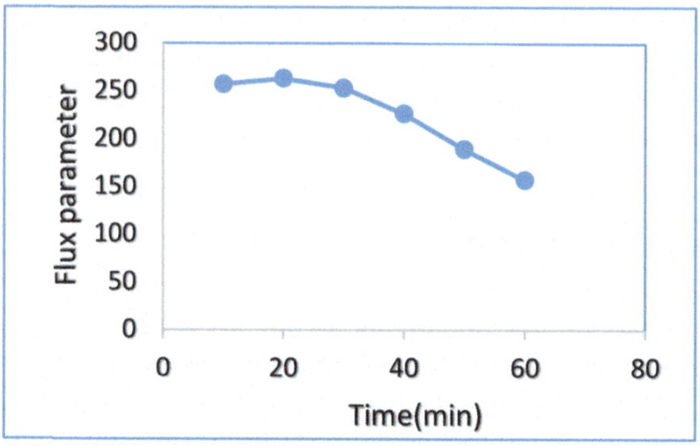

Fig. 1. Flux passing through membrane

In this study, the turbidity of sewage was measured before and after passing through the membrane, the results are shown in Fig. 2. According to the chart, which indicates the performance of the membrane, the turbidity has decreased significantly, which reveals the selectivity of the corresponding membrane that has the ability to remove bacteria and viruses and larger particles in comparison to the size of the water molecules.

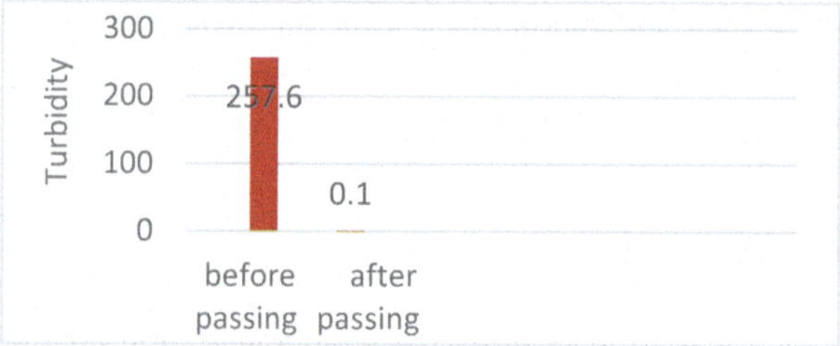

Fig. 2. Turbidity results of hospital sewage before and after passing through membrane

4 Conclusion

Membrane separation processes, with various advantages, can act as an alternative to filtration, chemical filtration systems, and many more. Structural and geometric properties of membranes have led to their flexibility in various applications such as reverse osmosis processes, ultrafiltration, microfiltration, nanofiltration, and so on. Replacement of membrane processes leads to improved quality of products and increased process efficiency and allows the expansion of these processes under different process conditions, including temperature, pressure and pH. These features have made the membrane separation process a successful technology in various industries. Flux indicates the amount of porosity and level of porosity under the layers and in the cross-section of the cavities. With the continuity of operation and the increase of the survival time of the membrane under pressure, gradually the volume of its inflow decreased, and therefore the existing openings and shafts in the membrane structure were also smaller.

References

1. Srikanth, G.: Water Cond. Purif. Mag. **50**, 1–4 (2008)
2. Sharma, R.: Membrane Filtration, pp. 1–95 (2008)
3. Bake, R.W.: Membrane Technology and Applications, vol. 588. Wiley, England (2004)
4. Naylor, T.: Membr. Sci. **60**, 1–2 (2008)

Study on the Viscoelastic Behavior of Nitrile Rubber/Polyvinyl Chloride Compounds Modified with Different Levels of Nanolayered Silicate

Mohammad Barghamadi[1], Mohammad Karrabi[1(✉)], Mir Hamid Reza Ghoreishy[1], and Somayyeh Mohammadian[2]

[1] Polymer and Petrochemical Institute, P.O. Box 14965-115, Tehran, Iran
m.karabi@ippi.ac.ir
[2] Payame Noor University, P.O. Box 19395-3697, Tehran, Iran

Abstract. In this study, the effects of nanolayered Silicate and acrylonitrile content on the viscoelastic behavior of NBR/PVC based nanocomposites were investigated. By increasing the nanoparticle and acrylonitrile content in the samples, the maximum elastic (S') and viscous (S'') vulcanization torques were increased. In the samples containing higher acrylonitrile groups, the peak of S'' shifted to the shorter times, while by adding the nanoparticles to the compound, this peak shifted to the longer times. Drop in S'' (DS'') declined by raising nanoparticle content. Samples with higher acrylonitrile content, showed the lower DS''. Elastic torque at S''_{max} (S' @ $t_{S''max}$), increased by increasing nanoparticle content. In the stress relaxation test, the ratio of the long time drops in modulus (DG_l) to the short term drop in modulus (DG_s) was corresponded with declining in the DS'' of cure test.

Keywords: Nanolayered silicate · Acrylonitrile content · Elastic torque · Viscous torque · Stress relaxation

1 Introduction

Rubbers are polymers with a wide range of applications in the industry, but these materials in the gum state are usually not strong and don't returns to primary condition under apply a high deformation. Thus the rubbers have been vulcanized to form the crosslinks between their chains and increase the elasticity. The vulcanization is important reaction to obtain the optimum properties of rubber based materials [1, 2]. Polymeric nanocomposites are the excellent interest, fastly growing new class of materials instead of the conventional loaded polymers or the blends. The price of polymer nanocomposites are currently affected by the improvement of mechanical properties, additionally fee-delivered houses without sacrificing the Intrinsic processability and physico-mechanical properties of the polymer [3]. Various attempts have been reported about the use of nanolayered silicate nanocomposites especially on the mechanical properties and rheology [4, 5]. In this research, the effect of nanolayered

H. Mirzadeh and A. A. Katbab (Eds.): ISPST 2018, *Eco-friendly and Smart Polymer Systems*, pp. 550–554, 2020.
https://doi.org/10.1007/978-3-030-45085-4_133

silicate on curing and viscoelastic behavior of nanocomposites based nitrile rubber (NBR)/polyvinyl chloride (PVC), were studied, naming a viscoelastic viewpoint.

2 Experimental

70 phr of two types of nitrile rubber with 33% and 45% acrylonitrile groups from Polimeri Europa CO., Italy, 30 phr of polyvinyl chloride supplied by Bandar Imam, Iran, 10 phr of Dioctyl phthalate (DOP) from LG CO., South Korea, 0–1.5 phr of nanolayered Silicate from Sigma-Aldrich CO., Canada, 1 phr cyclohexyl benzothiazole sulfonamide (CBS) from Vulkacit CO., China, 0.25 phr tetramethyl thiuram disulfide (TMTD) from Bayer CO., Germany, 1 phr sulfur from Rangineh Pars CO., Iran, 3.5 phr zinc oxide (ZnO) from Rangineh Pars CO., Iran and 1.4 phr of stearic acid from Unichema CO., Malaysia were used.

NBR was mixed with nanolayered Silicate on two roll mill. Then, NBR/nanolayered Silicate and PVC were added to an internal mixer with the other ingredients at 150 °C with a rotary speed of 50 rpm. Finally, the vulcanization system was added after cooling down to the temperature of 40 °C.

A rubber processing analyzer (RPA 200) from Alpha Technologies Co. (United Kingdom) was used to evaluate the vulcanization (Temperature = 150 °C, strain = 6.98% and frequency = 1.68 Hz) and relaxation behavior (Temperature = 50 °C, strain = 69.75% and torsion mode).

3 Results and Discussion

Elastic and viscous vulcanization torques of samples are given in Fig. 1. As can be seen, with increasing the amount of nanolayered silicate in both samples containing different percentage acrylonitrile, the elastic torque increases at the end of the vulcanization process because of the high specific surface area and platelet morphology of this reinforcement. Furthermore, by presence of nanolayered silicate in the matrix, the higher vulcanization rate and the lower vulcanization time compared with the unfilled samples were achieved. The peak of viscous vulcanization torque is related to the mobility of polysulfide pendant groups. The maximum mobility of pendant groups occurred in the beginning of the crosslinking known as S''_{max}. The difference between S''_{max} and the S'' at the end of test can be referred as DS'', that is attributed to the viscoelastic behavior of compound. Another important parameter is curing state at S''_{max} related to the formation and consumption of aforsaid pendant groups. According to the Fig. 2a, the more polarity of samples containing a higher percentage of acrylonitrile and the tendency of these groups to react with functional groups on the surface of nanolayered silicate, result in a steeper slope in the elastic and viscous torque curves. In Fig. 2b, the percentage drop of the viscous torque and the percentage of vulcanization in the maximum viscous torque are given. By increasing the amount of nanolayered silicate, as seen, the amount of drop has decreased.

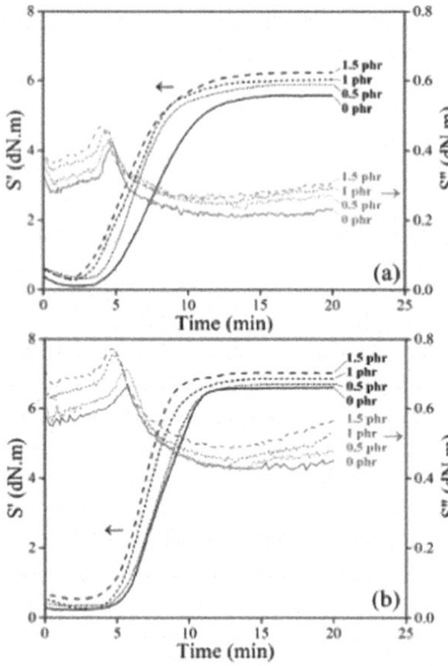

Fig. 1. Elastic and viscous vulcanization torques: (a) NBR with 33% ACN and (b) NBR with 45% ACN

In Fig. 2b, the vulcanization percentage of the samples is plotted against the amount of nanolayered silicate. The time of maximum viscous torque, with an increasing in the amount of nanolayered silicate, is shifted to lower times, or the scorch time is decreased and the sample begins to vulcanize at lower times. That is due to the participation of functional groups on the surface of nanolayered silicate in form of crosslinks and accelerates the curing. The rate of curing is higher in samples containing more acrylonitrile group content, due to the greater polarity of this rubber. So at a constant time, the curing state at S''_{max} (%cure @ $t_{S''max}$) increased.

Figure 3a, represents a schematic of relaxation modulus of the samples. In general, loss modulus is composed of two short-term and long-term regions. The loss modulus in the short-term and long-term regions are represented by the DG_s and DG_l, respectively. These two losses indicate the viscoelastic behavior. In Fig. 3b, the DG_l/DG_s parameter for the vulcanized samples is plotted versus content of nanolayered silicate. As seen, by increasing the nanoamount and decreasing the acrylonitrile groups in the samples, this parameter increased.

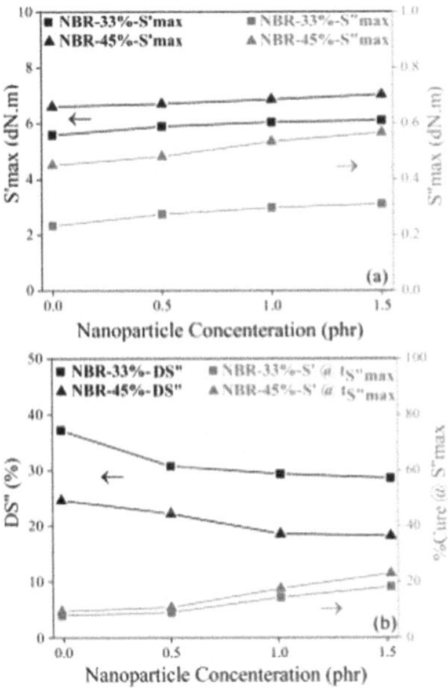

Fig. 2. (a) S'_{max} and S''_{max} and (b) DS'' and %cure @ $t_{S''max}$

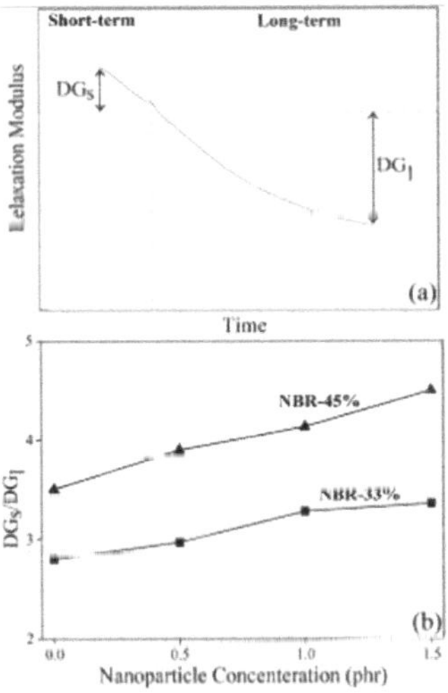

Fig. 3. (a) Schematic of relaxation modulus versus time and (b) DG_l/DG_s versus nanolayered silicate loadings

4 Conclusion

Adding nanolayered Silicate, resulted in an increase in S'_{max}, S''_{max}, curing state @ $t_{S''max}$ and decrease in DS″. Samples containing higher acrylonitrile content. By increasing the nanofiller content, DG_l/DG_s increased. Samples containing higher acrylonitrile groups had higher DG_l/DG_s parameter.

References

1. Mark, J.E., Erman, B., Roland, M.: Academic (2013)
2. Karrabi, M., Mohammadian-Gezaz, S.: J. Vinyl Add. Technol. **16**(3), 209–216 (2010)
3. Thomas, S., Stephen, R.: Wiley, Chichester (2010)
4. Mohammadian-Gezaz, S., Karrabi, M.: Prog. Rubber Plast. Recycl. Technol. **33**(4), 261 (2017)
5. Karrabi, M., Mohammadian-Gezaz, S.: Iran. Polym. J. **20**, 15–27 (2011)

Temperature Dependency of the Payne Effect in NBR/PVC Blends

Mohammad Barghamadi[1], Mohammad Karrabi[1(✉)], Mir Hamid Reza Ghoreishy[1], and Somayyeh Mohammadian[2]

[1] Polymer and Petrochemical Institute, P.O. Box 14965-115, Tehran, Iran
m.karabi@ippi.ac.ir
[2] Payame Noor University, P.O. Box 19395-3697, Tehran, Iran

Abstract. In this study, the effects of temperature and also the presence of PVC on the Payne effect of NBR based compounds were investigated. Increasing the temperature, served to decrease the drop of the shear storage modulus ($\Delta G'$) in the strain sweep test. The $\Delta G'$ for the sample containing PVC show higher value than that for the rubber compound. The slop of modulus drop after Payne effect (A_d) declined by raising the temperature, especially in blend sample. Critical strain (S_c) increased by increasing temperature. In other words, less dependency of modulus on the strain is decreased at higher temperatures. Blend samples due to the presence of PVC phase are more sensitive to the shear than rubber compound and non-linearization occurs in lower strains.

Keywords: NBR · Blend · Payne effect · Shear modulus · Critical strain

1 Introduction

Blending is a direct and inexpensive way to combine the useful properties of polymers and minimize their weaknesses. PVC is one of the most widely used polymers employed for blending with NBR [1]. NBR/PVC blend exhibits good oil fuels and solvents resistance. In addition, it has excellent abrasion resistance, heat resistance and dimensional stability [2]. One of the most prominent behaviors that distinguish polymers from other materials, such as metals and ceramics, is the rheological properties and exclusively of the viscoelastic behavior of these materials. In general, viscoelastic behaviors in polymers include linear and nonlinear. The most important problem of predicting the properties of polymers is the occurrence of nonlinear viscoelastic behaviors [3]. One of the non-linear behaviors is the drop of the storage modulus (G') by increasing the strain amplitude at constant temperature and frequency, which is called the Payne effect. Over the years, it's well-known that rubbers are very sensitive to temperature variations [4]. The study of the combination of two variables i.e. temperature and strain amplitudes on the behavior of rubber based materials, especially in industrial applications, is important. The most commonly used device for analyzing the viscoelastic behavior of rubber-based compounds is the rubber process analyzer (RPA) [5].

H. Mirzadeh and A. A. Katbab (Eds.): ISPST 2018, *Eco-friendly and Smart Polymer Systems*, pp. 555–559, 2020.
https://doi.org/10.1007/978-3-030-45085-4_134

2 Experimental

NBR with 45% acrylonitrile groups from Polimeri Europa CO., Italy, PVC supplied by Bandar Imam, Iran, Dioctyl phthalate (DOP) from LG CO., South Korea, cyclohexyl benzothiazole sulfonamide (CBS) from Vulkacit CO., China, tetramethyl thiuram disulfide (TMTD) from Bayer CO., Germany, sulfur and zinc oxide (ZnO) from Rangineh Pars CO., Iran and stearic acid (SA) from Unichema CO., Malaysia were used that the values of each are given in Table 1. Samples were prepared in an internal mixer at 150 °C with a rotary speed of 50 rpm. RPA from Alpha Technologies Co. (UK) was used to evaluate the strain sweep behavior at four different temperatures of 40, 60, 80, and 100 °C.

Table 1. Formulation of rubber compound and blend

Ingredients	Content (phr)	
	N	NP
NBR	100	70
PVC	–	30
DOP	8	10
ZnO	5	3.5
SA	2	1.4
S	1.4	1
CBS	1.4	1
TMTD	0.3	0.25

3 Results and Discussion

The schematic diagram of the storage modulus-strain for rubber based materials is shown in Fig. 1. At low strains, linear viscoelastic behaviors are observed, while at high strains, the changes are related to the non-linear viscoelastic ones. Elastic responses are related to reversible stretching of the entanglements or the interactions between polar groups in the linear region. On the other hand, viscous responses are related to disentanglements, slippage of the polymeric chains, or even chain scission in the nonlinear region. The transition strain which is named the critical strain (S_c) has occurred at the changing of linear to the non-linear region. In the strain sweep test, other important parameters are the $\Delta G'$ (a drop of the shear storage modulus) and the slop of modulus drop after Payne effect (A_d).

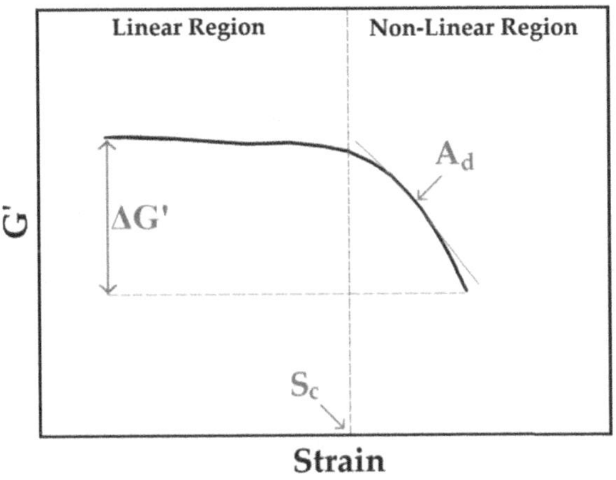

Fig. 1. Schematic representation of Payne effect and strain sweep parameters

In Fig. 2, ΔG′ is plotted versus temperature for rubber (N) and blend (NP) samples. ΔG′ of the NP is much higher than the N sample in any temperature. For the NP blend in comparison with N compound, in addition to chain entanglements, there is an interaction between the acrylonitrile groups in the NBR and the chlorine groups in the PVC, and thus, with increasing the strain, two phenomena i.e. disentanglements and weakening of interactions have occurred and the drop of the modulus is more pronounced. Moreover, the ΔG′ for NP sample at higher temperatures is reduced. In these temperatures, because of the chain movement, most of the viscous responses have taken place and only in large strains, the slight ΔG′ has occurred.

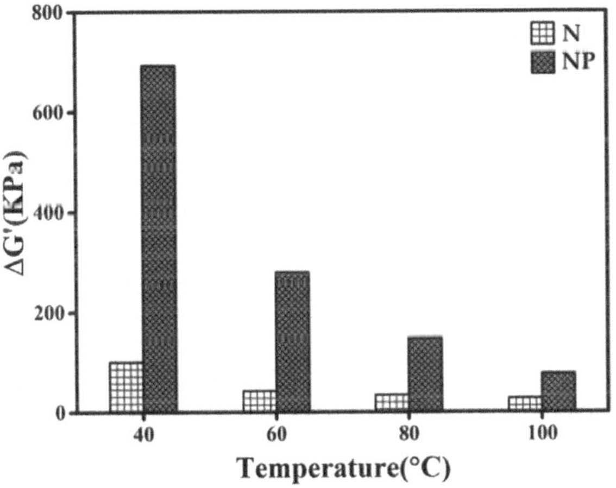

Fig. 2. Drop of the shear storage modulus of samples as a function of temperature

Figure 3, demonstrates the A_d as a function of temperature for N and NP samples. As can be seen, in all the samples, with raising temperature, the A_d has declined, which is more noticeable in the NP sample, because of the lower molecular weight of the PVC phase, and exhibits more dependence on temperature.

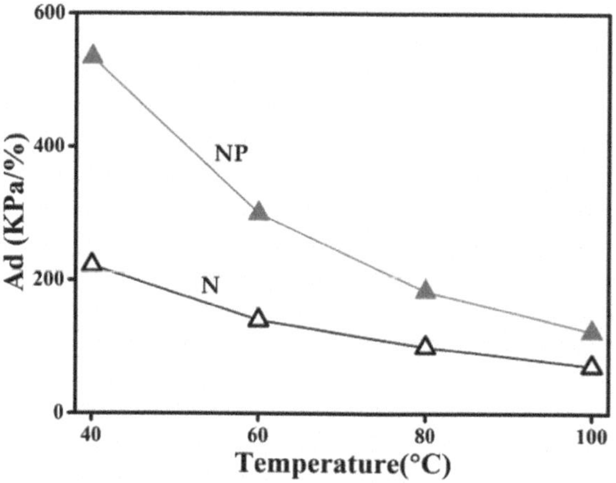

Fig. 3. Temperature dependency of the A_d for N and NP samples

The S_c of the samples for different temperatures is shown in Table 2. In all samples, the S_c increases with increasing temperature. At lower temperatures, low shear strains have the ability to make changes in the structure of the samples and alter the Payne effect, while at higher temperatures, the samples are softened and a higher strain value is required for an effective change in the structure. This result illustrates the less dependence of modulus on the strain at higher temperatures. NP sample is more sensitive to shear than rubber compound, due to the presence of PVC.

Table 2. S_c data of samples derived from G′-Strain diagram

Temperature (°C)	S_c (%)	
	N	NP
40	50.92	4.84
60	53.42	12.86
80	57.44	22.25
100	58.62	23.78

4 Conclusion

By raising the temperature, the ΔG′ of the samples decreased. NP sample exhibited more drop shear modulus than N ones. Sample containing PVC had higher A_d in comparison with N sample. Moreover, A_d showed a decreasing trend with increasing temperature. In the case of NP sample, the Payne effect occurred at lower strains.

References

1. Markovic, G., Visakh, P.M.: Springer **4**, 89–105 (2016)
2. Martinez, L., Alvarez, L., Huttel, Y., Mendez, J., Roman, E., Vanhulsel, A.: Vacuum **81**(11), 1489–1492 (2007)
3. Rendek, M., Lion, A.: ZAMM-J. Appl. Math. Mech. **90**(5), 436–458 (2010)
4. Karrabi, M., Mohammadian-Gezaz, S.: J. Vinyl Add. Technol. **16**(3), 209–216 (2010)
5. Karrabi, M., Mohammadian-Gezaz, S.: Iran. Polym. J. **20**(1), 15–27 (2011)

Numerical Evaluation of Effective Properties of PZT-5A/Epoxy Smart Piezoelectric Composite

Sobhan Sharafkhani and Mehrdad Kokabi(✉)

Department of Polymer Engineering, Faculty of Chemical Engineering, Tarbiat Modares University, P.O. Box 14115-114, Tehran, Iran
mehrir@modares.ac.ir

Abstract. Here, finite element method (FEM) has been employed to determine the effective properties of the active fiber composite (AFC) consists of a circular cross-section piezoelectric fiber (PZT-5A) in a non-piezoelectric polymeric matrix (epoxy) as a representative volume element (RVE). Six simulation runs with adjusted boundary conditions and appropriate mechanical-electrical loading were executed. Boundary conditions firmly control the simulation process. Homogenization approach was used to obtain the average value of the coefficients in the whole of RVE. Finally, eleven electromechanical coefficients including; six stiffnesses (C_{11}^{eff}, C_{12}^{eff}, C_{13}^{eff}, C_{33}^{eff}, C_{44}^{eff}, and C_{66}^{eff}), three piezoelectric (e_{13}^{eff}, e_{33}^{eff}, and e_{66}^{eff}) and two dielectric constants (ε_{11}^{eff}, ε_{33}^{eff}) were evaluated. The adequate accuracy of the analysis was approved by comparing the calculated coefficients with the reported analytical and numerical data.

Keywords: Active fiber composites · Piezoelectric · Finite element method · Simulation · PZT-5A/Epoxy

1 Introduction

Smart piezoelectric composites which possess the electromechanical energy conversion ability have attracted widespread interest for sensor and actuator applications [1]. Among them, AFCs consist of piezoelectric ceramic fibers in a soft polymeric matrix are extensively used in smart systems [2]. FEM is an accurate and effective numerical approach to describe and optimize the electromechanical behavior of smart composites. Recently, Solyaev et al. [3], calculated the effective properties of PZT-7A/epoxy composite with spherical inclusions via FEM. In this work, the effective properties of a RVE, including PZT-5A fiber in epoxy matrix has been analyzed based on FEM using C3D20E quadratic piezoelectric element in ABAQUS™. Boundary conditions and electrical-mechanical loading for each run were precisely applied.

H. Mirzadeh and A. A. Katbab (Eds.): ISPST 2018, *Eco-friendly and Smart Polymer Systems*, pp. 560–564, 2020.
https://doi.org/10.1007/978-3-030-45085-4_135

2 Theoretical

Equation (1) shows the constitutive equation of linear piezoelectric materials in the matrix form.

$$\begin{Bmatrix} \{T\} \\ \{D\} \end{Bmatrix} = \begin{bmatrix} [c]^E & -[e] \\ [e]^t & [\varepsilon]^s \end{bmatrix} \begin{Bmatrix} \{S\} \\ \{E\} \end{Bmatrix} \tag{1}$$

Where, T (N/m^2) and S denote the second-rank stress and strain tensor, respectively. E (V/m) is the electric potential field and D (C/m^2) indicates the electrical displacement field. C (GPa) represents the fourth-order elasticity tensor at a constant electric field, e denotes a third-order piezoelectric coupling tensor and ε is the second-order dielectric tensor under a constant strain field. The superscript t implies the transpose matrix. Figure 1 represents the RVE of epoxy matrix containing unidirectional PZT-5A fiber oriented in Z (3) axis and details of faces. Based on the homogenization method, the electrical and mechanical properties of RVE (bar values) will be obtained from average properties of the composite as follows:

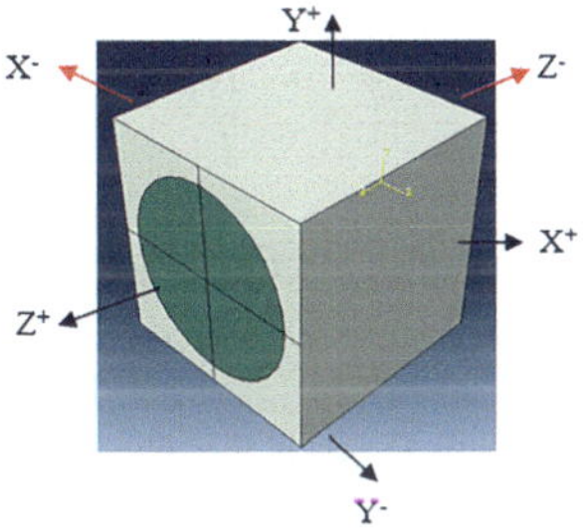

Fig. 1. PZT-5A/Epoxy cubic RVE and corresponding faces

$$\begin{aligned} \bar{T}_{ij} &= \frac{1}{V}\sum_{k=1}^{K} T_{ij}^{k} V^{k} & \bar{S}_{ij} &= \frac{1}{V}\sum_{k=1}^{K} S_{ij}^{k} V^{k} \\ \bar{D}_{i} &= \frac{1}{V}\sum_{k=1}^{K} D_{i}^{k} V^{k} & \bar{E}_{i} &= \frac{1}{V}\sum_{k=1}^{K} E_{i}^{k} V^{k} \end{aligned} \tag{2}$$

Where, K refers to the total number of the unit cell elements. V^k is the volume of the kth element and V is the volume of the unit cell. The volume fraction of circular cross-section fiber was considered 55.5%. Six simulation runs are needed to determine the following eleven electromechanical coefficients. C_{13}^{eff} and C_{33}^{eff} were calculated by setting zero normal displacements to the X$^+$, X$^-$ Y$^+$, and Z$^-$ surfaces (Fig. 1). Also, positive normal displacement applied to the Z$^+$ surface and zero electric potential prescribed on all surfaces ($\bar{E}_1 = \bar{E}_2 = \bar{E}_3 = 0$). So:

$$C_{13}^{eff} = \frac{\bar{T}_{11}}{\bar{S}_{33}} \quad C_{33}^{eff} = \frac{\bar{T}_{33}}{\bar{S}_{33}} \tag{3}$$

If the normal displacement of X^-, Y^+, Y^-, Z^+, and Z^- surfaces was fixed and the X^+ surface exposed to the normal displacement, C_{11}^{eff} and C_{12}^{eff} could be given:

$$C_{11}^{eff} = \frac{\bar{T}_{11}}{\bar{S}_{11}} \quad C_{12}^{eff} = \frac{\bar{T}_{22}}{\bar{S}_{11}} \tag{4}$$

For providing e_{13}^{eff}, e_{33}^{eff}, and ε_{33}^{eff} normal displacement must be equal to zero for all RVE surfaces. Positive voltage enforced to the Z^+ surface. However, zero electrical potential was considered at other surfaces (Eq. (5)).

$$e_{13}^{eff} = \frac{\bar{T}_{13}}{\bar{E}_3} \quad e_{33}^{eff} = -\frac{\bar{T}_{33}}{\bar{E}_3} \quad \varepsilon_{33}^{eff} = \frac{\bar{D}_3}{\bar{E}_3} \tag{5}$$

ε_{11}^{eff} can be assessed similar to the previous section. However, electrical potential was applied to the X^+ surface (Eq. (6)):

$$\varepsilon_{11}^{eff} = -\frac{\bar{D}_1}{\bar{E}_1} \tag{6}$$

C_{66}^{eff} is another unknown electromechanical coefficient to be calculated (Eq. (7)). In this case, parallelism conditions must be established [2]. Z-displacement of all nodes considered to be zero. Zero electrical potential prescribed on all surfaces. Furthermore, equal and opposite shear forces have been imposed on Y^+ and Y^- surfaces in x-direction and on X^+ and X^- surfaces in y-direction.

$$C_{66}^{eff} = \frac{\bar{T}_{12}}{\bar{S}_{12}} \tag{7}$$

Finally, for determination of e_{15}^{eff} and C_{44}^{eff} (Eq. (8)), x-displacement of all nodes considered to be zero. Shear forces have been applied on Y^+ and Y^- surfaces at z-direction and on Z^+ and Z^- surfaces at y-direction.

$$e_{15}^{eff} = \frac{(-\bar{E}_2.\varepsilon_{11} + \bar{D}_2)}{\bar{S}_{23}} \quad C_{44}^{eff} = \frac{(\bar{T}_{23} + \bar{E}_2.e_{15}^{eff})}{\bar{S}_{23}} \tag{8}$$

3 Results and Discussion

Figure 2 depicts the FEA results (run 1), including T_{11}, T_{33}, and S_{33} that are required in Eq. (3). Similarly, all eleven effective parameters were determined for the RVE via six different runs. Our results are seen in the column 2 of Table 1. The column 3 indicates the analytical solution of the observed RVE presented by Berger et al. [1]. The difference between their analytical data and our results are shown in the column 4. The columns 5 and 6 imply the simulation output generated by Medeiros et al. [2] and the difference between their work and our results, respectively. The outputs are in the order of nano and giga which indicates the strong dependency of the model to the applied boundary conditions. As is seen in column 4, the main difference between Berger analytical data and our results are devoted to C_{11}, C_{12}, C_{66}, and e_{15}. However, other coefficient errors are less than 1.5%. Medeiros et al. [2] results also confirm this difference pattern. The error between our work and both reported analytical data are in the acceptable range which validate the accuracy of calculations.

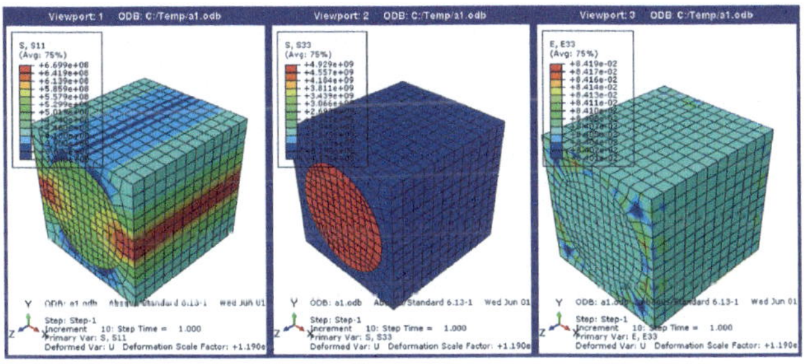

Fig. 2. FEA results of run 1, including T_{11}, T_{33}, and S_{33}

Table 1. Effective properties of PZT-5A/epoxy RVE

Effective properties		Our results	Berger et al. [1]	Error (%)	Medeiros et al. [2]	Error (%)
C_{11}	(GPa)	10.765	9.500	13.30	10.856	0.83
C_{12}		4.639	5.600	17.10	4.666	0.58
C_{13}		6.059	6.000	0.98	6.043	0.26
C_{33}		35.110	35.000	0.30	35.130	0.06
C_{44}		2.194	2.200	0.27	1.969	11.42
C_{66}		1.526	2.000	23.70	1.536	0.60
e_{13}	(C/m^2)	−0.260	−0.260	0.00	−0.258	0.80
e_{15}		0.0248	0.0200	24.00	0.0195	27.00
e_{33}		10.851	11.000	1.30	10.864	0.12
ε_{11}	(nF/m)	0.284	0.280	1.40	0.287	1.04
ε_{33}		4.266	4.200	1.50	4.270	0.10

4 Conclusion

Numerical approach has been applied to predict the effective properties of PVDF-5A/epoxy orthotropic and transversely isotropic smart piezoelectric composite. Boundary conditions which control the simulation outputs were expressed in details for each run. Simulation results were in a considerable agreement with the reported analytical and numerical data.

References

1. Berger, H., Kari, S., Gabbert, U., et al.: Int. J. Solids Struct. **42**, 5692–5714 (2005)
2. De Medeiros, R., Rodríguez-Ramos, R., Guinovart-Díaz, R., et al.: J. Intell. Mater. Syst. Struct. **26**, 101–118 (2015)
3. Solyaev, Y., Lurie, S.: Compos. Struct. (2018, in Press)

Prediction of a Typical Amorphous Shape Memory Polymer Behavior Considering Both Glassy and Rubbery Phases

Ghazaleh Alamdarnejad and Mehrdad Kokabi(✉)

Department of Polymer Engineering, Faculty of Chemical Engineering, Tarbiat Modares University, P.O. Box 14115-114, Tehran, Iran
mehrir@modares.ac.ir

Abstract. In this study, mathematical prediction and validation of shape memory thermomechanical behavior of a typical amorphous polymer was formulated considering both rubbery and glassy phases. The constitutive model with different hyperelastic functions for rubbery and glassy phases was developed using continuum mechanics approach. A typical uniaxial cycle of deformation was considered. The theoretical results exhibited a good agreement with experimental data.

Keywords: Shape memory polymer · Amorphous polymer · Modeling · Thermomechanical behavior

1 Introduction

Shape memory polymers (SMPs), are polymers that return to their original shape even after undergoing large deformation, on the activation of heating above their recovery temperature [1]. In order to model the thermomechanical behavior of a typical amorphous SMPs, its behavior should characterize both beyond and beneath the glass transition temperature (T_g). Beyond T_g, the amorphous polymer behaves like a rubber, while it is in a glassy state beneath T_g. Few models have developed for amorphous SMPs, among them, 1-D constitutive modeling accomplished by Liu et al. [2], appears to be the first effort that could capture the overall thermomechanical behavior at small deformations. Khanolkar et al. [3] also developed a model for amorphous SMPs using the theory of natural configurations. The model took into account the effect of thermal expansion of polymers on the stress-strain response. They assumed both rubbery and glassy phases as a neoHookean solid. In this work, the constitutive modeling of amorphous SMP performed for rubbery and glassy phases based on a uniaxial experiment in tension mode. Different hyperelastic functions were considered for each phase due to different behavior of polymer chains beyond and beneath the T_g.

H. Mirzadeh and A. A. Katbab (Eds.): ISPST 2018, *Eco-friendly and Smart Polymer Systems*, pp. 565–568, 2020.
https://doi.org/10.1007/978-3-030-45085-4_136

2 Theoretical

The modeling of amorphous SMPs could delineate into four parts namely; the rubbery phase (at a temperature above T_g), the glassification process (on cooling to below T_g), the glassy phase (at a temperature below T_g) and backing to the rubbery phase (on heating back to above T_g). Each of these parts should address separately.

Rubbers usually model as incompressible hyperelastic materials. A variety of constitutive equations are available in the literature to model polymers with rubber-like elasticity, such as the neoHookean model, the Mooney-Rivlin model, etc. Here, two hyperelastic models (neoHookean and Yeoh models) compared for rubbery phase, and the neoHookean model considered for the glassy one. For such a case, the equation of total stress for an amorphous SMPs is:

$$\mathbf{T} = -p\mathbf{I} + (1-\alpha)2\rho\mathbf{F}_r\frac{\partial\psi_r}{\partial\mathbf{B}_r}\mathbf{F}_r^T + \int_{t_1}^{t_2} 2\rho\mathbf{F}_g\frac{\partial\psi_g}{\partial\mathbf{B}_g}\mathbf{F}_g^T\frac{d\alpha}{d\tau}d\tau \tag{1}$$

where, p is the Lagrange multiplier due to the constraint of incompressibility, α is the amount of glassy phase, $\mathbf{F}_r$ and $\mathbf{F}_g$ are the deformation gradient in rubbery and glassy phases, ψ_r and ψ_g are the Helmholtz potential or the stored energy of rubbery and glassy phases and $\mathbf{B}_r$ and $\mathbf{B}_g$ are the left Cauchy-Green tensor for rubbery and glassy phases, respectively. The first part of the equation represents the rubbery phase and the integration part indicates the transition of the rubbery phase to the glassy one, passing through the T_g. During the loading process, the temperature of the sample is above the T_g and the SMP is totally in its rubbery state, hence, at a uniaxial extension, the stress in the direction of extension can represent by Yeoh model:

$$\sigma_{11} = 2\left[C_{10} + 2C_{20}(I_{B_a}-3) + 3C_{30}(I_{B_a}-3)^2\right]\left(\Lambda(t) - \frac{1}{\Lambda(t)^2}\right) \tag{2}$$

Where, σ_{11} is the nominal stress, Λ is the stretch ratio, C_{10}, C_{20} and C_{30} are constants in Yeoh model. This addresses as the first stage of shape memory cycle. During the phase transition, the polymer is first in the rubbery phase and gradually converts to the glassy phase when the temperature passing the T_g. So, the glassy phase due to rigidity considers as a neoHookean solid. Thus, the total stress in the second stage of shape memory cycle is as follow:

$$\begin{aligned}\sigma_{11}(t) &= 2(1-\alpha(t))\left[C_{10} + 2C_{20}(I_c-3) + 2C_{30}(I_c-3)^2\right]\left((\Lambda_r(t))^2 - \frac{1}{\Lambda_r(t)}\right) \\ &+ \mu_g\int_0^t\left(\frac{\Lambda_g^2(\tau)}{b^2(t)} - \frac{1}{b^2(t)\Lambda_g(\tau)}\right)\frac{d\alpha}{d\tau}d\tau\end{aligned} \tag{3}$$

where, b is the thermal strain and μ_g is the shear modulus of the glassy phase. In the third stage of shape memory cycle after unloading, the glassification process completes and the total stress in the polymer is:

$$\sigma_{11}(t) = 2(1-\alpha_0)\left[C_{10} + 2C_{20}(I_c - 3) + 2C_{30}(I_c - 3)^2\right]\left((\Lambda(t))^2 - \frac{1}{\Lambda(t)}\right) + \frac{1}{b^2}G_g\int_{t_s}^{t_f}\left(\Lambda_g^2(\tau) - \frac{1}{\Lambda_g(\tau)}\right)\frac{d\alpha}{d\tau}d\tau \tag{4}$$

After unloading, the polymer is in its fixed temporary shape. Recovering the original shape accomplishes by heating the polymer to above the T_g. Indeed, unlike the second stage, the SMP is initially in its glassy phase, gradually converts to the rubbery phase by increasing the temperature beyond the T_g. During the transition process from glassy to rubbery phase, the polymer is in the stress-free state as follow:

$$\sigma_{11} = 0 = 2(1-\alpha(t))\left[C_{10} + 2C_{20}(I_C - 3) + 3C_{30}(I_C - 3)^2\right]\left((\Lambda(t))^2 - \frac{1}{\Lambda(t)}\right) + \mu_g\int_{t_s}^{t^*}\left(\frac{\Lambda_g^2(t)}{b^2(\mathrm{t})} - \frac{1}{b^2(\mathrm{t})\Lambda_g(t)}\right)\frac{d\alpha}{d\tau}d\tau \tag{5}$$

3 Results and Discussion

Figure 1 shows the 3-D stress-strain-temperature diagram illustrating the thermomechanical behavior of a typical amorphous SMP in a shape memory full cycle. As observed, on loading beyond T_g (stage 1) the stress and strain go up. By cooling to beneath T_g at a fixed strain (stage 2), the stress goes up as the glassy phase is born in a stressed state. On unloading (stage 3), the stress decreases to zero while only a small amount of strain is recovered thus helping in retaining the temporary shape. On heating the sample above its T_g (stage 4), the SMP recovers its original shape. Figure 2 illustrates the comparison between the experimental data originated by Liu et al. [2], and the Khanolkar et al. [3] model, also the developed model at small strain ($\sim 10\%$) in the first stage (Fig. 2(a)) and fourth stage (Fig. 2(b)) of shape memory full cycle. It is obvious that the proposed model can reasonably predict the shape memory behavior of a typical amorphous SMP.

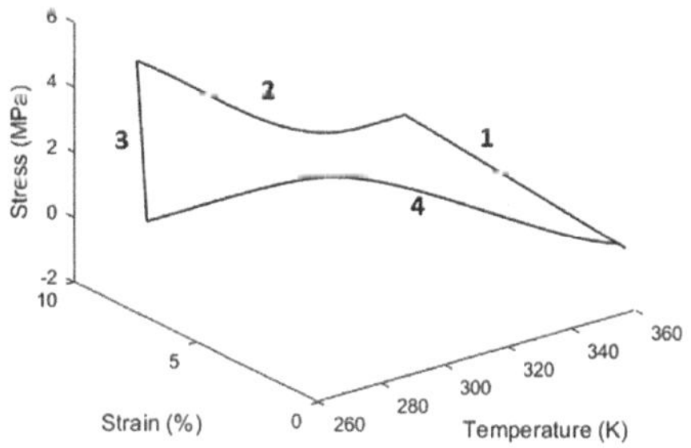

Fig. 1. 3-D Shape memory behavior of amorphous SMP obtained by proposed model

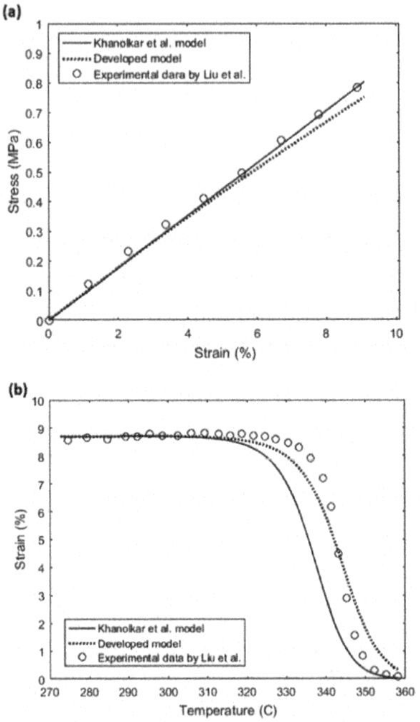

Fig. 2. Comparison of experimental data with two models in the first stage (a) and forth stage (b) of shape memory full cycle

4 Conclusion

In this work, we formulated constitutive equations to predict the thermomechanical behavior of a typical amorphous SMP. The rubbery and glassy phases had different behavior, thus various hyperelastic models used to predict the behavior of SMP in the transition stages (second and fourth stages). The proposed model exhibited good agreement with experimental data.

References

1. Jiang, Z., Xio, Y., Kang, Y., et al.: Appl. Mater. Interfaces **9**, 20276–20293 (2017)
2. Liu, Y., Gall, K., Dunn, M.L., et al.: Int. J. Plast. **22**, 279–313 (2006)
3. Khanolkar, M.: Constitutive modeling of glassy shape memory polymers. Ph.D. thesis, New Jersey Institute of Technology (2010)

Preparation of PVDF Nanowire by Infiltration Method into Lumina Nanopores

Hosna Soleymani[1(✉)], Mohamad Mehdi Abolhasani[1], Mohamad Normohamadi[2], and Mohsen Ashjari[1]

[1] Department of Chemical Engineering, University of Kashan, Kashan, Iran
hosna_soly85@yahoo.com
[2] Faculty of Physics, University of Kashan, Kashan, Iran

Abstract. Poly(vinylidene fluoride) (PVDF) homopolymer nanowire array with the desired ferroelectric β-phase is fabricated in an anodic alumina porous (AAP) template. The results indicate that the hydrogen bonds between the AAP surface hydroxyl groups and PVDF molecules promote the formation of the β-phase. The mechanism of polymer infiltration into the pores is also very important in order to obtain the desired one-dimensional nanostructures and to control their final morphology. In this study, infiltration method was explored with the aim to obtain different 1DPNs. It was shown that, with these infiltration methods, it is possible to obtain nanowire composed of poly(vinylidene fluoride).

Keywords: Poly(vinylidene fluoride) · Anodic alumina porous (AAP) · Nanowire · Infiltration · Morphology

1 Introduction

Scaling the size of materials down to the nanometer dimension is present innovative opportunities not only to study their behavior at the nanoscale but also to discover new useful properties, which can widely extend the range of conceivable (practical) applications. One of the common strategies to generate one-dimensional (1-D) nanostructures is a use of a template filled with the desired material [1, 2]. In particular, to form polymeric nanowires, the template-wetting attitude from a melt polymer or solution is commonly used [3]. In such a form, the polymer outspread on a porous templating substrate to form a thin surface film, thus covering the pore walls in the initial stages of wetting. Undeniably, the polymer preferentially nucleates and grows on the surface of the pore walls, resulting in tubes at short (not long) infiltration times. Complete filling of the pores takes place later, and is normally hindered by thermal quenching in the case of melts or by solvent evaporation in the case of solutions; thus, controlled temperature and pressures or a higher amount of feeding material to fill the pores are required [3, 4]. The final structure, either a nanotube or a nanowire, shows the same size broadcast (distribution) and shape of the templating matrix. Thus, by appropriately selecting a monodispersed size distribution, as well as aligned and long pores of the templating matrix one can obtain ordered and monodispersed arrays of nanowires. Several authors have already studied the wetting template in anodic alumina porous (AAP) membrane in order to produce arrays of 1-D ceramic and metallic

H. Mirzadeh and A. A. Katbab (Eds.): ISPST 2018, *Eco-friendly and Smart Polymer Systems*, pp. 569–572, 2020.
https://doi.org/10.1007/978-3-030-45085-4_137

nanowires, showing ferroelectric, dielectric, or conductive properties. Porous alumina exhibits a high surface energy, completely wettable by liquids, i.e., polymer melts or solutions [2].

Poly(vinylidene fluoride), (PVDF), is one of the most attractive polymers owing to its remarkable pyro-, piezo- and ferroelectroactive properties [5, 6]. These properties stem from its unique polymorphism, which gives rise to its extraordinary mechanical properties, high thermal and chemical stability. Among the four significant crystalline phases α, β, δ, and δ [7] the electroactive β-phase is most utilized for sensing, actuation [8] and nanogenerator applications [9]. During phase inversion, the PVDF films are formed by quickly quenching the cast films into a nonsolvent bath to induce liquid–solid and liquid–liquid phase separation events [10].

2 Experimental

Tailored polymer nanofibers can be prepared by choosing the suitable AAP template, and the appropriate infiltration method and conditions. Depending on the process involved, polymer infiltration methods can be classified in three groups: methods based on wetting phenomena, methods based on vacuum, and methods based on rotation. A solution containing 30 wt% poly(vinylidene fluoride) (PVDF, *M*w) 288,000 g/mol (Sigma-Aldrich) was prepared. For solution wetting, the AAP membrane was placed on the top of a drop of the solution wetting at 60 °C for 20 h. The alumina membranes and PVDf nanowaiers were immersed immediately in a deionized water bath held at various quenching temperatures (0 °C), for 30 min to remove DMF. To characterize the morphology of the polymer nanoarray the template was then removed.

3 Results and Discussion

To characterize the morphology of the polymer nanoarray the template was then removed. Parallel aligned PVDF nanofibers were revealed by scanning electron microscopy (SEM), as shown in Fig. 1. Some nanowires appear to have been pulled out of the pores during the cleaving process.

Scanning XRD with a 1 μm diameter beam was accomplished along the cross section of the sample with diffraction patterns collected at regular intervals between the residual PVDF film (bulk) and the nanowire array. The sample's nanowire axis was carefully aligned perpendicular to the X-ray beam. Figure 2 shows the diffraction pattern of the residual PVDF film and PVDF nanowire inside porous alumina. The peak at $2\theta = 19.7°$ corresponding to diffraction in (200) and (110) planes represents the polar β-phase, and is dominant in the nanowire XRD spectrum. The XRD pattern of the film shows a mixture of α-phases in comparison. The right panel is a focus graph of the XRD patterns between 35° and 45°.

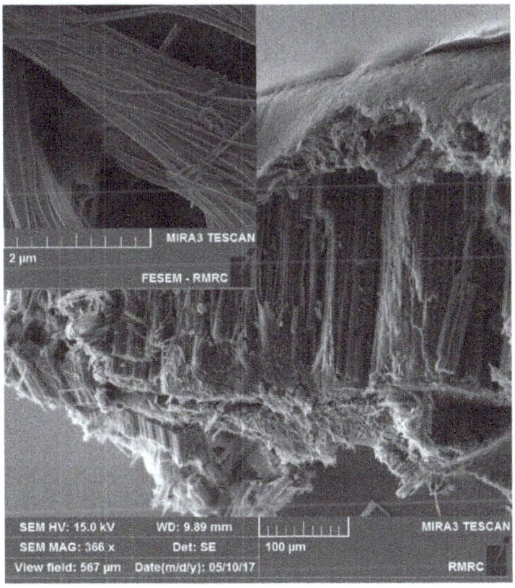

Fig. 1. Template-grown P(VDF) nanowires. Aligned nanowires in AAP template viewed within a cleaved cross-section

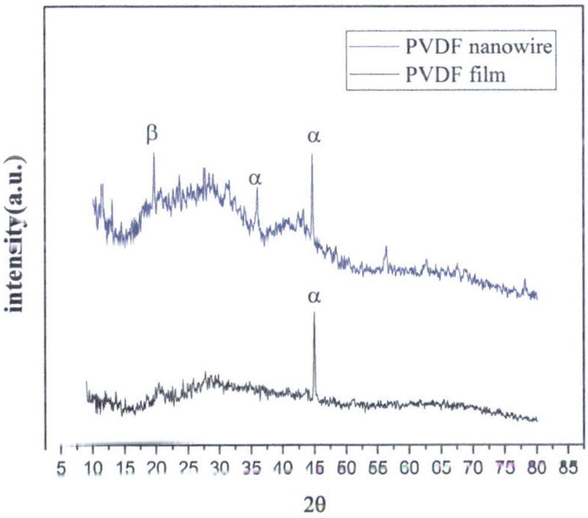

Fig. 2. Characterization of structure and morphology of template-grown nanowires. XRD patterns from PVDF nanowires and PVDF film of the same composition

4 Conclusion

In conclusion, the present work shows that solution template wetting is a versatile method to produce arrays of isolated β-type nanowires connected by a supporting film. The AAP walls template polymer crystallization orienting the lamellar crystals flat-on growing along the nanowires long axis. The method is based upon a phase transition induced by the confinement of PVDF into nanoporous and by its interaction with the walls of the AAP membrane. The advantage of this method using porous anodic aluminum oxide as a template is the high versatility of the method, allowing the preparation of both arrays of ferroelectric nanorods supported by a paraelectric polymer film for applications mainly in photonics and electronics and high aspect ratio free 1D nanostructures, when removing the residual film, for applications in sensors based on piezoelectric properties.

References

1. Xia, Y., Yang, P., Sun, Y., Wu, Y., Mayers, B., Gates, B., Yin, Y., et al.: Adv. Mater. **15**, 353–389 (2003)
2. Steinhart, M., Wendorff, J.H., Wehrspohn, R.B.: Chem. Phys. Chem. **4**, 1171–1176 (2003)
3. Steinhart, M., Wendorff, J.H., Greiner, A., Wehrspohn, R.B., Nielsch, K., Schilling, J., Choi, J., et al.: Science **296**, 1997 (2002)
4. Cao, G., Liu, D.: Adv. Colloid Interface Sci. **136**, 45–64 (2008)
5. Vinogradov, A., Holloway, F.: Ferroelectrics **226**,169–181 (1999)
6. Gomes, J., Nunes, J.S., Sencadas, V., Lanceros-Méndez, S.: Smart Mater. Struct. **19**(6) (2010)
7. Soin, N., Shah, T.H., Anand, S.C., Geng, J., Pornwannachai, P.W., Mandal, D., Reid, S., et al.: Energy Environ. Sci. **7**, 1670–1679 (2014)
8. Sharma, T., Aroom, K., Naik, S., Gill, B., Zhang, J.X.: Ann. Biomed. Eng. **41**, 744–751 (2013)
9. Dongjin, K., et al.: RSC Adv. **5**, 10662–10666 (2015)
10. Soin, N., Boyer, D., Prashanthi, K., Sharma, S., Narasimulu, A.A., Luo, J., Shah, T.H., et al.: Chem. Commun. **51**, 8257–8260 (2015)

Comparison Study of Carbon Nanotube and Carbon Black Filled Polyamide6/Polyethylene as a Tissue Equivalent Material

Farah Khoylou(✉), Azam Akhavan, and Faranak Naimian

Radiation Applications Research School, Nuclear Science and Technology Research Institute, 11365-3486, Tehran, Iran
fkhoylou@aeoi.org.ir

Abstract. Tissue equivalent polymeric materials should be electrically conductive like the real tissue. Composites based on polyamide 6/polyethylene filled with two different conductive additives, carbon nanotubes (CNT) and carbon black (CB), were prepared. Electrical conductivity measurements revealed that the PA/PE/CNT composite has been reached to percolation threshold with 3% CNT which is much lower than 16% CB in PA/PE/CB composite. After percolation, the CNT filled composites showed lower electrical conductivity than CB filled ones because of the low tunneling effect in narrow cylindrical CNTs in comparison with spherical CB. However, electrical conductivity of both composites is in the range of semi conductive materials which are suitable for using as a tissue equivalent material. Lower amount and high strength of CNT offered better mechanical properties for the PA/PE/CNT composite.

Keywords: Polyamide6/polyethylene · Carbon nanotube · Carbon black · Electrical conductivity · Tissue equivalent

1 Introduction

An accurate evaluation of the effects of radiation on living tissue can be obtained by microdosimetry in tissue equivalent materials. These materials should have elemental composition and electrical conductivity similar to human tissue. From the standpoint of elemental composition, polymers are a good substitute for these materials but they are non-conductive [1]. By adding a carbon based conductive filler such as carbon black (CB), carbon nanotube (CNT), graphite or metal powder, polymers can be made electrically conductive.

Among these fillers, CB has been the most widely used. However, a high loading is usually needed to reach the requirement electrical conductivity, which in turn is detrimental to the mechanical properties of composites [2].

CNTs have attracted much attention as a conducting filler for conducting composites on account of their prominent advantages [3, 4]. The high aspect ratio of CNTs

H. Mirzadeh and A. A. Katbab (Eds.): ISPST 2018, *Eco-friendly and Smart Polymer Systems*, pp. 573–576, 2020.
https://doi.org/10.1007/978-3-030-45085-4_138

make it possible to be in contact with each other inside the polymer and provides electrical conductivity at low CNT loading.

The PA6/PE blend with hydrogen, carbon, nitrogen and oxygen elements has chosen as a tissue equivalent material [5]. In this work, for creating electrical conductivity in PA/PE blend, the effect of CB and CNT conductive fillers were compared. Also, mechanical properties of conductive composites were studied.

2 Experimental

Polyamide 6 and high density polyethylene were purchased from the Kolon Company (Korea) and Bandar Emam petrochemical Co. (Iran), respectively. The carbon nanotube with outer diameter of 10–20 nm, length of 10–30 μm and specific surface area of 200 m^2/g was obtained from Tarbiat Modarres University. The CB master batch 940 was prepared from Ariya master batch Co., Iran.

Polyamide 6 granules were dried in the vacuum oven at 80 °C for 16 h. The PA6/PE/CNT and PA6/PE/CB composites with PA/PE ratio of 1:1.5, 1–4 wt% carbon nanotube and 10–18 wt% carbon black were prepared by melt compounding in a brabender internal mixer at 240 °C with the rotational speed of 70 rpm. Sheet samples with 2 mm thickness were prepared with Dr. Collin hot press at 240 °C to obtain specimens for testing.

The electrical conductivity of the samples was measured by four point method by ITECH power supply and 5491A multimeter. For composites with electrical conductivity lower than 10^{-6} S/cm, Keithley electrometer were used.

Mechanical properties of the composites were measured using a tensile testing machine from Hiwa Co., Iran. The test speed was 50 mm/min. The tensile test was done at room temperature.

3 Results and Discussion

The electrical conductivity of the composites as a function of CNT and CB loading is presented in Fig. 1. The PA/PE/CNT composite has been reached to percolation threshold with 3% CNT which is much lower than 16% CB in PA/PE/CB composite. This is due to the high aspect ratio of CNTs which cause to easier formation of the conductive network that facilitate the electrical conduction through the composite.

Also, in Fig. 1 it is seen that after the percolation threshold, the electrical conductivity of CB filled composite reach to 6.25×10^{-4} S/cm which is more than 3×10^{-6} S/cm obtained for CNT filled ones. This could be due to the effect of spherical geometry of CB on electron transfers between adjacent particles in comparison with narrow cylindrical carbon nanotubes. This behavior which is called tunneling effect is presented in Fig. 2 for CB and CNT particles. In addition to the geometric shape, the high content of CB in PA/PE/CB composite makes the conductivity more affected by tunneling mechanism.

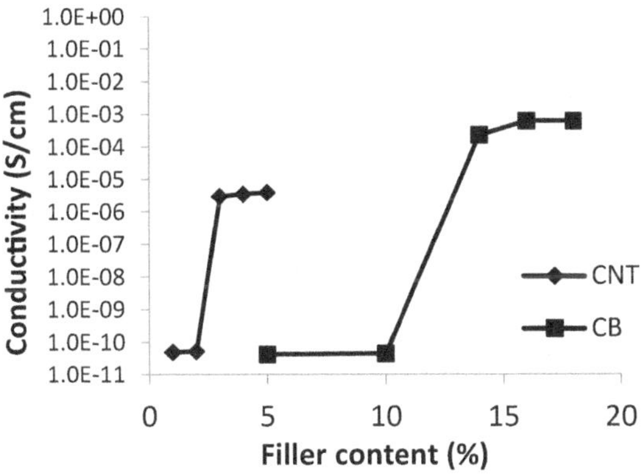

Fig. 1. Electrical conductivity of CB and CNT filled composites

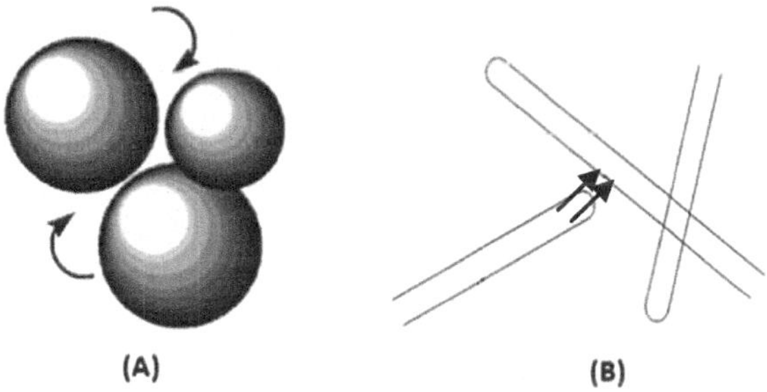

Fig. 2. Tunneling effect in carbon black: (A) and carbon nanotube (B)

Although after the percolation threshold, the CNT filled composites have lower electrical conductivity than CB filled ones, but electrical conductivity of both composites is in the range of semi conductive materials which are suitable for using as a tissue equivalent material.

Mechanical properties of 3% CNT and 16% CB filled conductive composites are shown in Table 1. The results indicate that the lower loading level and high strength of CNT has led to a higher tensile strength in PA/PE/CNT composite. Also, the modulus of the composite with 3% CNT is similar to that of the composite with 16% CB.

Table 1. Mechanical properties of CB and CNT filled conductive composites

Composite	Tensile strength (MPa)	Modulus (MPa)
%16 CB	14.5 ± 1.2	897 ± 47
%3 CNT	20.7 ± 0.3	877 ± 34

4 Conclusion

Comparison of the role of carbon black and carbon nanotube on electrical conductivity of PA6/PE revealed that the large aspect ratio of CNTs exhibited very low percolation threshold for CNT filled composite in comparison with CB filled one. Although after the percolation threshold, the CNT filled composite has lower electrical conductivity but it is in the range of semi-conductive materials. Lower loading level and high strength of CNT offered better mechanical properties for the conductive PA/PE/CNT composite.

References

1. Braby, L.A., Johnson, G.W., Barthe, J.: Radiat. Prot. Dosimetry **61**, 351–379 (1995)
2. Zhang, W., Dehghani-Sanj, A.A.: J. Mater. Sci. **42**, 408–3418 (2007)
3. Tanasa, F., Zanoaga, M., Mamunya, Y.: International Conference Science Papers (2015)
4. Mamunya, Y., Levchenko, V., Boiteux, G., Seytre, G., Zanoaga, M., Tanasa, F., Lebedev, E.: Polym. Compos. (2015). https://doi.org/10.1002/pc.23434
5. International Commission on Radiation Units and Measurements. Tissue substitutes in radiation dosimetry and measurement. ICRU report 44 (1964)

Producing Porous Polyacrylonitrile Fibers Using Wet-Spinning Method for Making Carbon Fibers

Ehsan Samimi[1](✉), Mohammad Mehdi Abolhasani[1], and Shahram Arbab[2]

[1] Department of Chemical Engineering, University of Kashan, Kashan, Iran
ehsansamimi8@gmail.com
[2] Department of Textile Engineering, Amirkabir University of Technology, Tehran, Iran

Abstract. Investigation of the increment in the porosity and diameter of polyacrylonitrile (PAN) fibers during the wet-spinning process has great importance in producing carbon fibers and textile industries. In this study, the influence of several factors such as stretching and flow rate and also the presence of non-solvent on the porosity and diameter of the fibers have been investigated. PAN used in certain quantity as polymer, with DMF (dimethylformamide) as solvent and distilled water as non-solvent were combined under certain temperature to make dope wet-spinning process. At the final stage, the PAN fibers were ejected from the spinning machine under the various flow rates and porosity was generated in the fibers and then the fibers related to a sample were stretched by pulling device.

Keywords: Polyacrylonitrile (PAN) · Porosity · Diameter · Wet-spinning · Morphology

1 Introduction

Carbon fibers have enjoyed increasing use in a broad variety of high-tech applications, including material and devices for the energy, biomedical, electronics, and structural sectors [1]. The need for manufacturing industries to produce high-tech mechanical synthetic fibers, such as high module, high tensile, low density and low weight, has led to the production of carbon fibers. The properties of carbon fibers are strongly affected by the structure and physical properties of the precursor fibers. PAN is the dominant precursor for production carbon fibers; each has its unique advantages, such as high tensile and the production method is more convenient and cheaper as well as better properties of the produced fiber than other precursors [2]. The PAN fibers used in carbon fiber industry are mainly produced by wet-spinning process [3]. Wet-spinning is an interesting example of immersion precipitation and includes phase-separation and various transport phenomena. As soon as the viscose polymeric solution is extruded into the coagulation bath, phase-separation begins due to solvent/non-solvent exchange and polymer chains precipitate as porous fibrils [4].

H. Mirzadeh and A. A. Katbab (Eds.): ISPST 2018, *Eco-friendly and Smart Polymer Systems*, pp. 577–580, 2020.
https://doi.org/10.1007/978-3-030-45085-4_139

2 Experimental

Materials used in this research: Poly(acrylonitrile-ran-methylacrylate) (94 wt% of acrylonitrile monomer) was provided by Polyacryl Co. (Isfahan, Iran). Dimethylformamide (DMF, Merck Ltd.). Polymer values are described, $Mn = 36{,}000$ g/mol, $Mv = 70{,}000$ g/mol, $Mw = 116{,}000$ g/mol, and PDI = 3.2 were obtained. The obtained values agreed well with the values provided by the supplier. The process of making spinning solution (dope) and the operating conditions of the spinning fibers is stated here, at first, polymer powder swelled in DMF to obtain a 10% homogeneous polymer solution and in one sample 4 vol.% non-solvent was added. The prepared dope was then extruded by a metering pump through a single-hole spinneret into a coagulation bath containing 100 wt% water at room temperature. The fibers were prepared at two frequencies of 4.5 Hz, 10 Hz related to the metering pump to generate a flow rates of 0.522 cc/min and 1.335 cc/min, respectively. The numbers listed were calculated using the RPM formula of the wet-spinning device. Spinning conditions are reported in Table 1. At the last step, one sample of the fibers were stretched at room temperature by pulling device (sample b in Table 1). The morphology of all samples was observed by Field emission electron microscopy (FE-SEM). Fibers were freeze-fractured in liquid nitrogen. Fractured surface was then coated with a thin layer of gold and scanned with an electron microscope. Overall porosity of fibers was assessed by measuring the weight and diameter from randomly selected regions of fibers. Fiber diameter was determined by FE-SEM. The overall fiber porosity was calculated using the following equation:

$$Overall\,porosity\ (\%) = \frac{(\pi R^2 L - m\rho_p^{-1}) \times 100}{\pi R^2 L}$$

Where, R, L, m, and ρ_p are radius, length, mass, and density of the sample, respectively.

Table 1. Spinning conditions

Sample	Flow rate (cc/min)	Non-solvent (vol%)	Stretching
a	0.522	0	–
b	0.522	0	×
c	0.522	4	–
d	1.335	0	–

3 Results and Discussion

The morphology of fibers with different spinning conditions are shown in Fig. 1 with various magnifications. All fibers had almost a circular cross-section and all fibers possessed a dense center and finger-like channels in radial direction. Three factors including stretching after spinning, non-solvent, and flow rate have affected the diameter and porosity of the fibers. We first investigate the influence of these factors on the fiber diameter. Stretching of the fibers by the pulling device was found to be the most important factor in fiber diameter variations, which significantly reduces the diameter of the fiber by about 83% compared to the prototype, (Fig. 1(a, b)). Also this factor has affected the porosity of the fiber, it can be seen in Fig. 2 that there has been a 4.58% reduction in the porosity of the fibers by stretching. Considering the addition of non-solvent as the second factor, it was concluded that by adding distilled water, the diameter of the fibers has increased about 52% compared to the prototype (Fig. 1(a, c)). Furthermore, it was understood that the addition of non-solvent has caused the porosity of the fibers to decrease about 0.25% (Fig. 2). As for the third factor, through increasing the flow rate, the shear rate rises, the shear tensions are increased, so the dope will be more swollen after leaving the spinneret and it can be seen that the fiber diameter increases about 8% compared to the prototype (Fig. 1(a, d)). Also, Fig. 2 illustrates that the porosity of the fibers decreased about 1.04% through the flow rate increase.

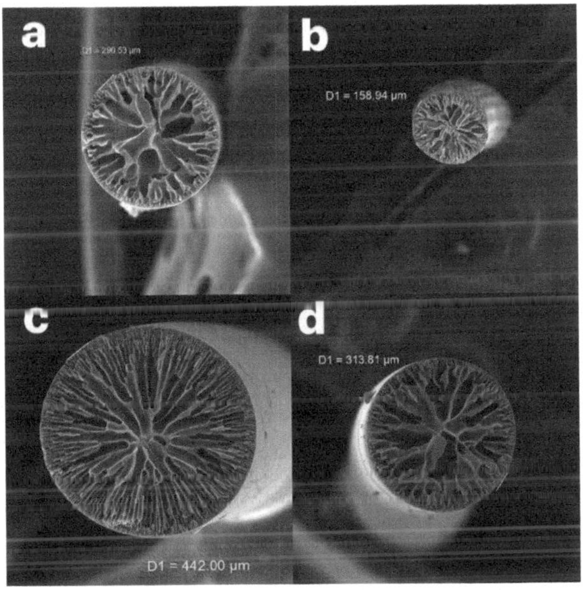

Fig. 1. FESEM micrographs of fibers' cross-section; all of samples are shown with magnification (350×). Sample codes are according to Table 1

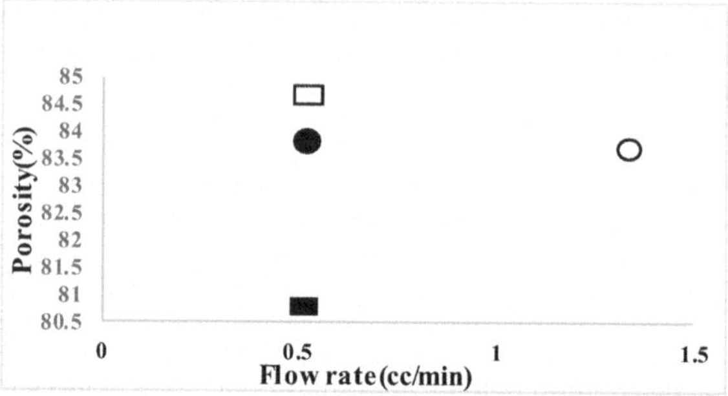

Fig. 2. Fibers' porosity percent versus frequency: open rectangle a, filled rectangle b, filled circle c, open circle d. Sample codes are according to Table 1

4 Conclusion

The overall variations in fiber diameter and porosity were controlled by three factors named as stretching, non-solvent, flow rate. The most important factor among the mentioned factors was found to be the stretching after spinning and the addition of non-solvent and increasing the flow rate in the following, which caused 83% decrease, 52% increase, 8% increase in the fiber diameter, respectively. Regarding the porosity of the fiber, the mentioned factors caused the porosity reduction, so that by increasing the stretch, adding non-solvent, the increase in flow rate, a decrease in the porosity was obtained by 4.58%, 0.25%, 1.08%, respectively. In case of porosity, stretching was known as the most important factor which affects the fiber diameter.

References

1. Papkov, D., Goponenko, A., Compton, O.C., Moravsky, A., Li, X.-Z., Nguyen, S.T., Dzenis, Y.A.: Adv. Funct. Mater. **23**, 5763–5770 (2013)
2. Xiang, C., Behabtu, N., Liu, Y., Gi Chae, H., Young, C.C., Genorio, B., Zhang, C., et al.: ACS Nano **7**(2), 1628–1637 (2013)
3. Frank, E., Ingildeev, D., Buchmeiser, M.R.: Structure and Properties of High-Performance Fibers, pp. 7–30. Woodhead Publishing, Cambridge (2017)
4. Arbab, S., Noorpanah, P., Mohammadi, N., Zeinolebadi, A.: Polym. Bull. **66**, 1267–1280 (2011)

Shape Memory Behaviour of Waterborne Polyurethane-Based Aerogel

Mohsen Teymouri, Reza Akbari, and Mehrdad Kokabi(✉)

Department of Polymer Engineering, Faculty of Chemical Engineering, Tarbiat Modares University, P.O. Box 14115-114, Tehran, Iran
mehrir@modares.ac.ir

Abstract. In this work, shape memory behaviour of waterborne polyurethane-based aerogel was investigated using dynamic mechanical analysis technique. The porosity and pore-size distribution of polyurethane aerogel also determined by a mercury porosimetry test. The results showed the prepared aerogel had 81% porosity with average pore diameter around 10 nm, also a shape fixing ratio and shape recovery ratio of 98% and 48%, respectively.

Keywords: Shape memory effect · Waterborne polyurethane · Aerogel · Thermomechanical behaviour

1 Introduction

Aerogels are an interesting class of materials, which form by the extraction of the solvent from a hydrogel while conserving the porous structure via shape stability. This gives a range of attractive properties such as; low density, high surface area, low thermal conductivity, and low dielectric constant [1]. The drawback of mineral aerogels is fragility and brittleness. Recently, flexible aerogels made from polymers such as; polyurethane [2, 3], polystyrene [4] and polyethylene [5]. These aerogels showed desirable toughness and deformability [1]. Such features indicate that it is possible to access shape-memory aerogels that are capable of "memorize" a permanent shape and could fixate a temporary shape under suitable conditions. Subsequently, a trigger (e.g., heat or light) will lead to a transformation of the temporary shape to the memorized permanent shape [6]. Michal et al. [1] investigated the thermoresponsive shape-memory polymeric aerogels that produced from thiol–ene networks. The resulting aerogel had porosity ca. 72% with a shape fixing and shape recovery of ca. 99% and 98%, respectively.

In this work, shape memory behaviour of waterborne polyurethane-based aerogel was investigated in comparison with its gel counterpart.

H. Mirzadeh and A. A. Katbab (Eds.): ISPST 2018, *Eco-friendly and Smart Polymer Systems*, pp. 581–584, 2020.
https://doi.org/10.1007/978-3-030-45085-4_140

2 Experimental

2.1 Materials and Methods

Poly(ethylene glycol), (PEG), with a molecular weight of 4000 g/mol, dimethylol propionic acid (DMPA), triethylamine (TEA), ethylenediamine (EDA) and Dibutyltin dilaurate (DBTDL) (as a catalyst) without further purification were obtained from Merck, Germany. 4,4′-Methylenebis (cyclohexyl isocyanate), (H12MDI), (the mixture of isomers, 90%) was purchased from Sigma Aldrich, USA. 1-Methyl-2-pyrrolidone (NMP) was obtained from Daejung, China.

Waterborne polyurethane was synthesized by the polyaddition reaction. PEG and DMPA were first dried and degassed separately at 95 °C under a static vacuum overnight before use. The PEG was placed in a four-necked round-bottom flask equipped with a thermometer, a stirrer, an inlet of dry nitrogen, a condenser, and an oil bath. Then DMPA/NMP was added to the flask, and the mixture stirred moderately at 85 °C. H12MDI and DBTDL (as a catalyst) summed to the flask under moderate stirring for 3 h. Then TEA was affixed to the reaction mixture for neutralizing the carboxyl group of the NCO-terminated polyurethane prepolymer. The neutralized prepolymer was chain-extended by dropping in EDA at 85 °C for 1.5 h. To prepare a polyurethane aerogel, the polymer was placed in distilled water for 48 h to have a polyurethane hydrogel. Finally, the polyurethane hydrogel was placed in freeze-dryer for 24 h to obtain the polyurethane aerogel.

The bulk density of polyurethane (ρ_s) and polyurethane aerogel (ρ_a) determined by measuring the length and width of quadric samples. The porosity (P) was then calculated by:

$$P = \frac{\rho_s - \rho_a}{\rho_s} \times 100 \tag{1}$$

Nanoscale porosity studied by mercury porosimetry test. The measurements performed on a Thermofinnigan Pascal 440 porosimeter, Italy.

DSC carried out on a DSC 200F3 Netzsch, Germany. In order to eliminate any unknown thermal history of the samples, heating and cooling repeated from −50 °C to 200 °C. The process was carried out at a heating and cooling rate of 5 °C.min^{-1}.

The shape memory properties of polyurethane and polyurethane aerogel measured by DMA 242C Netzsch, Germany.

3 Results and Discussion

According to the Eq. (1), the polyurethane aerogel has a porosity of ca. 81%. Pore size distribution (PSD) of polyurethane aerogel was obtained from mercury porosimetry, Fig. 1. It shows that the average pore diameters of aerogel is around 10 nm.

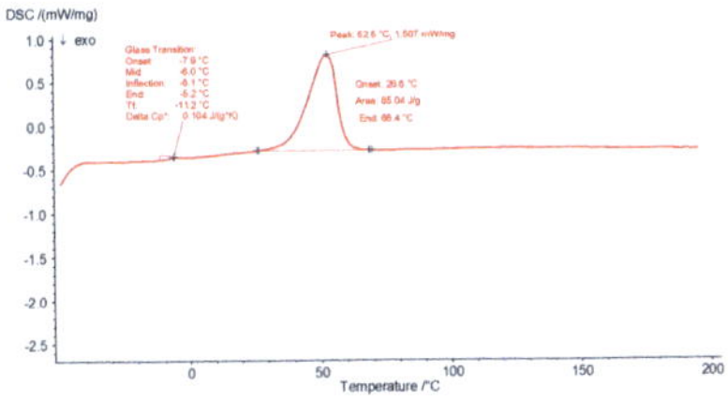

Fig. 1. DSC curve of polyurethane

To determine the transition temperature of the shape memory polyurethane, DSC test was performed (Fig. 2). Polyurethane shows a melting temperature (T_m) at ~53 °C and glass transition (T_g) at ~−6 °C. According to the above data, the switch temperature was selected at 53 °C.

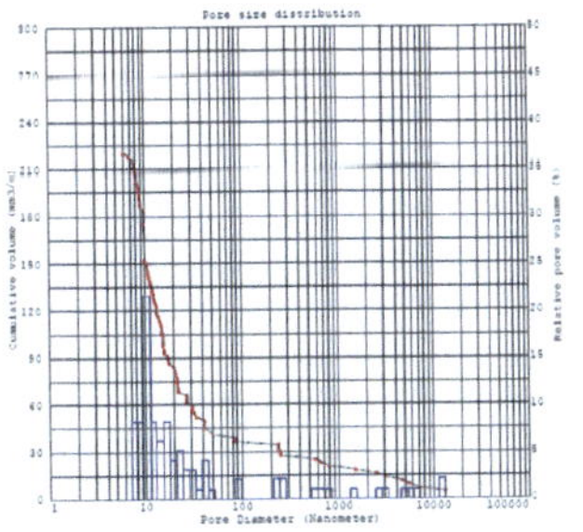

Fig. 2. Pore-size distribution determined by mercury porosimetry for polyurethane aerogels

To investigate the shape memory behaviour of aerogel systems, a four steps procedure performed at a constant heating/cooling rate of 5 °C.min^{-1}. First, the sample was heated to 56 °C (above the switch temperature) and stretched to a certain strain (ε) under a constant force (Step 1). Then, the deformed sample was cooled to 10 °C while the force was kept fixed (Step 2). The force then removed and the temporary strain (ε_{temp}) was measured (Step 3). Finally, the sample was reheated to 56 °C and kept at 56 °C for 20 min, then the recovery strain (ε_{rec}) was recorded (Step 4). The shape

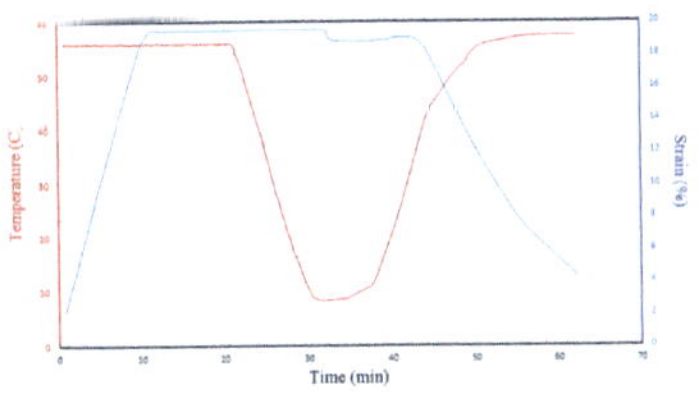

Fig. 3. The thermal-mechanical curve of polyurethane

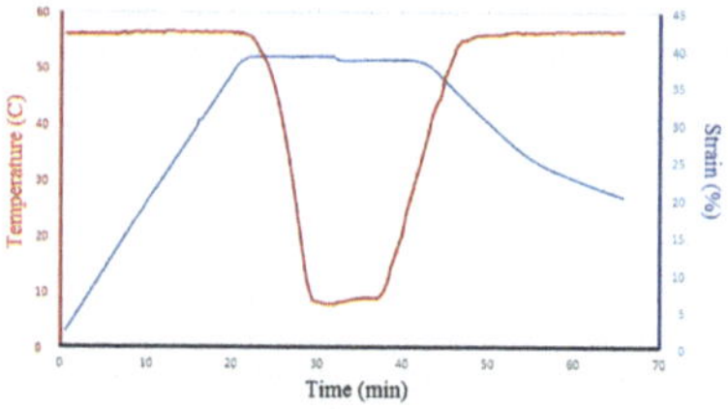

Fig. 4. The thermal-mechanical curve of polyurethane aerogel

fixing ratio (R_f) and shape recovery ratio (R_r) of polyurethane aerogel and its gel counterpart calculated from Figs. 3 and 4 according to the following equations:

$$R_f = \frac{\varepsilon_{temp}}{\varepsilon} \tag{2}$$

$$R_r = \frac{\varepsilon - \varepsilon_{rec}}{\varepsilon} \tag{3}$$

Strain and temperature curves versus time in tension mode depicted. The values of shape fixing ratio and shape recovery ratio of polyurethane aerogel were 98% and 48%, while these values for polyurethane gel were 96% and 78%, respectively. The results indicated the shape fixing ratio of polyurethane aerogel was more than polyurethane gel, but its shape recovery ratio was less than polyurethane gel. This behaviour attributed to the reduction of the modulus of aerogel due to the presence of more than 80% porosity in it.

4 Conclusion

The shape memory behaviour of waterborne polyurethane-based aerogel was studied. The obtained aerogel had a porosity of ca. 81% with average pore diameters of around 10 nm. According to DSC results, the switch temperature of shape memory polyurethane was at 53 °C. According to DMA results, the shape fixing ratio and shape recovery ratio of polyurethane aerogel were 98% and 48%, which were less than polyurethane gel due to the presence of 81% porosity.

References

1. Michal, B.T., Brenn, W.A., Nguyen, B.N., et al.: Chem. Mater. **28**, 2341–2347 (2016)
2. Rigacci, A., Marechal, J.C., Repoux, M., et al.: Cryst. Solids **350**, 372–378 (2004)
3. Diascorn, N., Calas, S., Sallée, H., et al.: J. Supercrit. Fluids **106**, 76–84 (2015)
4. D'Aniello, C., Daniel, C., Guerra, G.: Macromolecules **48**, 1187–1193 (2015)
5. Daniel, C., Longo, S., Guerra, G.: Polyolefins J. **2**, 49–55 (2015)
6. Hager, M.D., Bode, S., Weber, C., et al.: Prog. Polym. Sci. **49–50**, 3–33 (2015)

Surface Modification of Cellulose Nanocrystal Through Dendrimer with Amine End Groups

Ali Mianehro[1(✉)], Somaye Akbari[1], Atefeh Solouk[2], and Seid Ali Ahmadi[3]

[1] Department of Textile, Amirkabir University of Technology, Tehran, Iran
ali.mianehro@aut.ac.ir

[2] Department of Biomedical Engineering, Amirkabir University of Technology, Tehran, Iran

[3] Iran University of Medical Sciences, Tehran, Iran

Abstract. Cellulose nanocrystal with the presence of a large number of chemical functionalities within its structure, provides a unique platform for significant surface modification through various chemistries. Using dendrimer for this purpose is a unique choice that not only does shield hydroxyl groups and change surface charge but also introduce several functional groups on its surface. In this work surface modification of cellulose nanocrystal through dendrimer with amine end groups by physical adsorption technique was conducted to decrease its negative charge while increasing surface functional groups and its antibacterial properties. The result was analyzed by Fourier-transform infrared spectroscopy (FTIR) and X-ray photoelectron spectroscopy (XPS) and Zetasizer instrument. It was found that due to strong hydrogen bonds the result of physical adsorption reaction is comparable with chemical grafting and is an efficient route for surface modification of cellulose nanocrystal.

Keywords: Cellulose nanocrystal · Surface modification · Dendrimer · Polyamidoamine · Physical adsorption

1 Introduction

Cellulose constitutes the most abundant renewable polymer resource produced in the biosphere with an annual production estimated to be over 7.5×10^{10} tons. This fascinating and almost inexhaustible sustainable polymer possesses remarkable chemical and physical attributes allowing its use in different forms for a wide range of materials and products [1]. What has not been known until relatively recently is that when cellulose fibers are subjected to acid hydrolysis, the fibers yield defect-free, rod-like crystalline residues. Cellulose nanocrystals (CNCs) have garnered in the materials community a tremendous level of attention that does not appear to be relenting [2]. These biopolymeric assemblies warrant such attention because of their inherent renewability and sustainability in addition to their abundance and their unsurpassed quintessential physical and chemical properties such as high surface area (~ 250 m^2/g), high tensile strength (7500 MPa), high stiffness (Young's modulus up to 140 GPa), and an abundance of surface hydroxyl groups. These surface hydroxyl groups offer a

H. Mirzadeh and A. A. Katbab (Eds.): ISPST 2018, *Eco-friendly and Smart Polymer Systems*, pp. 585–588, 2020.
https://doi.org/10.1007/978-3-030-45085-4_141

facile platform for chemical modification [3]. Treatment with cations can be performed to render cellulose nanoparticles cationic by introducing positive charges on their surface [4]. Using dendrimer for this purpose is a unique choice that not only does shield hydroxyl groups and change surface charge but also does introduce several functional groups on its surface. In this work surface modification of CNCs through Polyamidoamine (PAMAM) by physical adsorption technique was conducted and the zeta-potential of CNC before and after modification was determined using the Zetasizer instrument also Modification of the surface was confirmed by FTIR and XPS spectroscopy.

2 Experimental

0.025 g of CNC was stirred continuously with 7.365 g of dendrimer aqueous solution 2 w/v% for 4 days at room temperature then the mixture was dialyzed for 3 days to remove unreacted and free dendrimers from the solution. The CNC suspension was dispersed by ultrasonic treatment and finally, the powder was obtained by freeze-drying.

Surface cationization of CNC was conducted involving the surface coverage by cationic polymer chains from “physical adsorption”. The physical adsorption of dendrimer on the surface of CNC was achieved by the electrostatic interactions between cationic polymer chains and negative sulfate groups.

3 Results and Discussion

The average zeta-potential before modification was −40 but after modification it was +3.7 that shows hydroxyl groups are shielded and surface charge completely changed. FTIR spectroscopy analysis was used to investigate the surface modification on cellulose nanocrystals which is shown in Fig. 1, CNCs are characterized by the typical bands of the cellulose skeleton in the region 1000–1200 cm^{-1} assigned to the carbohydrate ring. Bands at 1160, 1110, and 1060–1030 cm^{-1} correspond to C–O–C antisymmetric stretching, ring asymmetric stretching, and C–O stretching, respectively. The large band at 1635 cm^{-1} is related to the OH bending of water tightly absorbed by the cellulose. Other peaks at 1460, 2869, and 2927 cm^{-1} relative to CH_2 groups of the alkyl moiety are visible in the spectra of modified CNC. The emergence of new peak at 1560 cm^{-1} for modified CNC provided clear evidence of presence of amide groups of the core of PAMAM on the surface of CNCs. Also, XPS spectroscopy was conducted. The deconvolution of N 1 s level shown in Fig. 2, revealed the presence of two components. The peak ~400 eV was attributed to a nitrogen atom linked to a carbon atom by a single bond and the peak ~398 was assigned to amine groups. the deconvolution of C 1 s level resulted peak ~283 that is for C–C and/or C–H, peak 285 is attributed to C–N and peak 286 is assigned to C = O and O–C–O bonds. thus XPS results approve presence of dendrimer on the surface of CNCs. These results show remarkable efficiency and sustainability of physical adsorption technique and its potential application as host for guest particles.

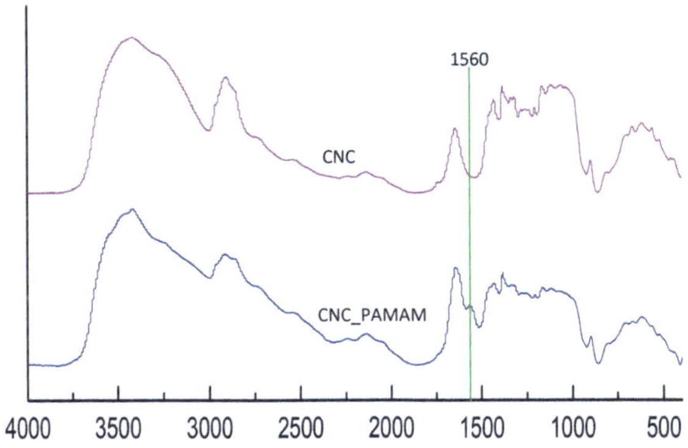

Fig. 1. FTIR spectra of pristine CNC and modified CNC

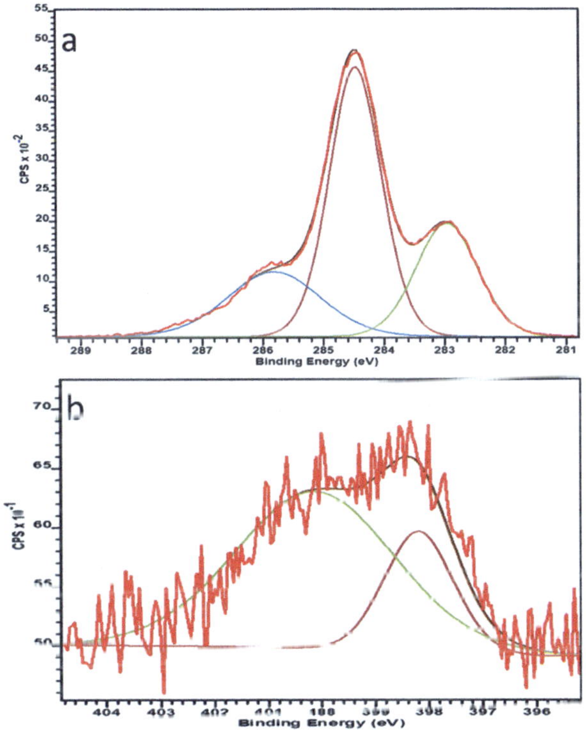

Fig. 2. (a) C 1 s and (b) N 1 s XPS spectrum of CNC-PAMAM

4 Conclusion

In this study, we have prepared modified CNCs by PAMAM through physical adsorption technique. This method was proved that is remarkably efficient and gives new traits to CNCs such as improvement of its host guest properties, changing its surface charge and reducing its repulsion with anionic particles and introducing several amine groups on the surface of CNCs. This product practically can be used for drug delivery, wastewater treatment, reinforcer nanoparticle, etc.

References

1. Youssef, H.: Chem. Soc. Rev. **43**(5), 1519–1542 (2014)
2. Youssef, H., Lucia, L.A., Rojas, O.J.: Chem. Rev. **110**(6), 3479–3500 (2010)
3. Nathan, G., Mohammed, N., Tang, J., Chiu Tam, K.: Curr. Opin. Colloid Interface Sci. **29**, 32–45 (2017)
4. Goraltchouk, A., Freier, T., Shoichet, M.S.: Biomaterials **26**, 7555–7563 (2005)

Effect of Surface Treatment on Polyethylene Separator's Properties of Lithium-Ion Batteries

Mohammad Hadi Moghim[1,2(✉)], Ashkan Nahvi Bayani[1,2], Abdol Majid Ghasemi[1], and Rahim Eqra[1]

[1] Department of Energy Storage, Institute of Mechanics, Shiraz, Iran
moghim_hadi@shirazu.ac.ir

[2] Department of Materials Science and Engineering, Engineering School, Shiraz University, Shiraz, Iran

Abstract. In the current research, polyethylene and alumina coated polyethylene (AC-PE) Li-ion battery separators were studied and the effect of surface treatment on the properties of polyethylene separator was investigated. Tensile test, contact angle measurement and thermal shrinkage investigation were done to clarify the role of ceramic coating on the mechanical and thermal properties, and wettability of PE separator surface. Results showed that the contact angle of modified polyethylene separator was obtained much lower (69°) with respect to unmodified one (118°), which showed better wettability of PE surface with the incorporation of Al_2O_3. Tensile strength of PE and AC-PE was also obtained 156.2 and 142.4 MPa, respectively, and alumina coated polyethylene showed much brittle mechanical behavior. Finally, Al_2O_3 coated separator had obvious thermal shrinkage resistance with respect to PE separator, as shrinkage changed from 60.3% to 12.4% of separators original size.

Keywords: Lithium-ion · Battery · Separator · Surface treatment

1 Introduction

Lithium-ion batteries have received considerable attention as the main power sources on storage energy market, from portable electronic devices to electric vehicles [1, 2].

Separators as a main component, play a key role in Li-ion batteries. Their function is to keep the positive and negative electrodes apart to prevent electrical short circuits and at the same time allow rapid transport of ionic charge carriers that are needed to complete the circuit during the passage of current in an electrochemical cell [3].

Separator structure and properties play an important role in determining the battery performance, including cycle life, safety, energy density, and power density, through influencing the cell kinetics [1].

The microporous polyolefins (PP, PE) are widely used in lithium-ion batteries as membrane separator. To overcome the drawbacks of conventional polyolefin separators and also improve their thermal stability (due to safety problems), variety of approaches have been studied [4, 5]. These methods include the incorporation of inorganic

H. Mirzadeh and A. A. Katbab (Eds.): ISPST 2018, *Eco-friendly and Smart Polymer Systems*, pp. 589–592, 2020.
https://doi.org/10.1007/978-3-030-45085-4_142

materials into organic membranes to obtain composite separators and use of highly thermal stable organic or inorganic membranes. Coating commercial PP and PE membranes with a thin layer of inorganic particles has been considered as one of most effective and inexpensive ways [4].

In the current research, different properties of two polyethylene (PE) and alumina coated polyethylene (AC-PE) separators are studied.

2 Experimental

Separators were prepared with the specifications given in Table 1.

Table 1. Separators specifications

Materials	Thickness (μm)	Porosity (%)
PE	20	48
AC-PE	24	48

Contact angle measurement, tensile test and thermal shrinkage properties were investigated to characterize different properties of separators.

3 Results and Discussion

The hydrophilic characteristic of the PE separators was investigated by a sessile drop method in which measurement of contact angles is done when solid and liquid interact. Figure 1 shows the water contact angle images of PE and AC-PE membranes, which are 118° and 69°, respectively. The contact angle of modified polyethylene separator was much lower, which showed better wettability of PE surface with the incorporation of Al_2O_3.

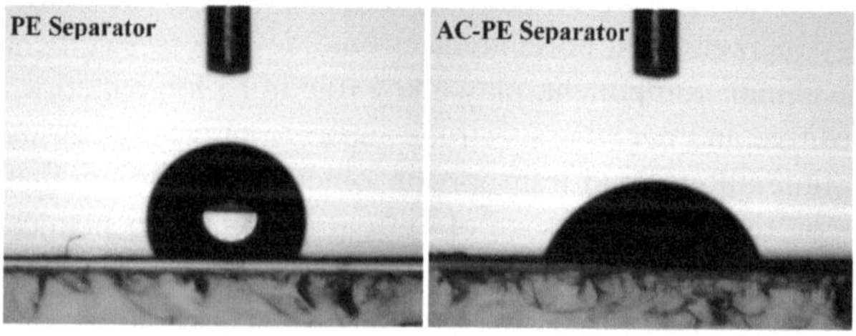

Fig. 1. Results of contact angle measurement

Tensile tests were carried out to analyze the effect of surface treatment on mechanical properties of PE membrane along machine direction, and the results are presented in Fig. 2. As shown, alumina coated polyethylene showed brittle mechanical behavior with respect to PE. Tensile strength was also obtained 156.2 and 142.4 MPa respectively for PE and AC-PE, respectively.

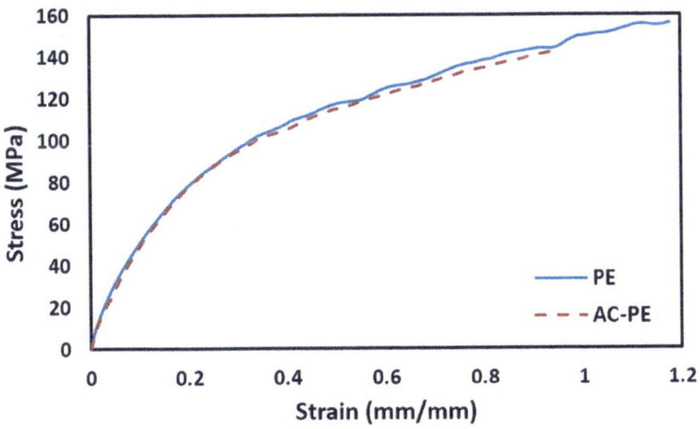

Fig. 2. Tensile stress-strain curve of PE and AC-PE separators along machine direction

Thermal shrinkage properties of polyethylene and alumina coated polyethylene are shown in Fig. 3. As temperature was increased from 110 °C to 130 °C, separators shrunk sharply. PE separator shrunk to 60.3% of its original size at 130 °C. Moreover, Al_2O_3 coated separator had obvious thermal shrinkage resistance with respect to PE separator. The Al_2O_3 composite separator only shrunk to about 12.4% of its original size at 130 °C.

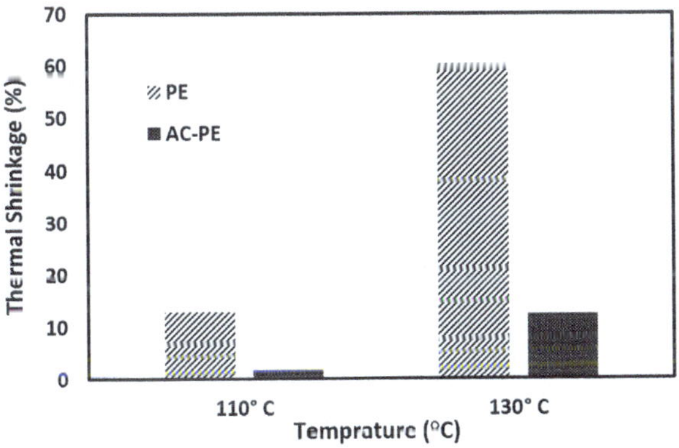

Fig. 3. Thermal shrinkage of PE and AC-PE separators at different temperatures

4 Conclusion

Tensile test, contact angle measurement and thermal shrinkage investigation were done to investigate the mechanical properties and wettability of polyethylene and alumina coated polyethylene Li-ion separators. AC-PE separator demonstrated better wettability with respect to PE surface as the contact angle improved sharply from 118° to 69°. Tensile strength of PE and AC-PE was also obtained 156.2 and 142.4 MPa, respectively. It showed that percent elongation was decreased with the presence of inorganic coating. Finally, Al_2O_3 coated separator had obvious thermal shrinkage resistance with respect to PE separator, as shrinkage changed from 60.3% to 12.4% of separators original size.

References

1. Lee, H., Yanilmaz, M., Toprakci, O., Fu, K., Zhang, X.: Energy Environ. Sci. https://doi.org/10.1039/c4ee01432d
2. Yang, M., Hou, J.: Membranes **2**, 367–383 (2012)
3. Arora, P., Zhang, Z.: Chem. Rev. **104**, 4419–4462 (2004)
4. Yu, L., Jin, Y., Lin, Y.S.: RSC Adv. **6**, 40002–40009 (2016)
5. Kim, J.Y., Lim, D.Y.: Energies **3**, 866–885 (2010)

Copolymerization of Styrene and 1-Hexene via Ziegler-Natta and Atom Transfer Radical Polymerization Methods

Mohammad Reza Rostami Darounkola(✉), Naeimeh Bahri-Laleh, Mehdi Nekoomanesh-Haghighi, and Sanaz Rahmatiyan

Polymerization Engineering Department, Iran Polymer and Petrochemical Institute (IPPI), P.O. Box 14965/115, Tehran, Iran
m.rostami@ippi.ac.ir

Abstract. Polymerization of 1-hexene and its copolymerization with styrene were done via Ziegler-Natta catalyst in bulk medium at 60 °C. The results showed that 1-hexene was well polymerized with this type of catalyst, but increasing the molar ratio of styrene in the reaction feed led to decrease the catalyst activity and, consequently, a reduction of the overall conversion. Compared to this catalytic reaction, the atom transfer radical copolymerization of styrene and 1-hexene was carried out with CCl_4 initiator and CuCl/N, N, N′, N′, N″-pentamethyldiethylenetriamine (PMDETA) catalyst system in bulk at 80 °C. The results of conversion showed that styrene has a high activity in the atom transfer radical polymerization. By increasing the molar ratio of styrene monomer in the feed, the overall conversion of the reaction increased. Thus, increasing the concentration of 1-hexene can decrease the activity of macro-radicals. The structure of various synthesized copolymers was characterized by ^{1}H NMR method. Thermal properties of the synthesized poly 1-hexene and its copolymers with styrene were determined by DSC method. The results showed that the synthesized polymers via Ziegler-Natta have glass transition temperature in the range of −40 °C to −45 °C.

Keywords: Ziegler · Natta · Atom transfer radical polymerization · Copolymer · 1-hexene

1 Introduction

Recently designing and synthesis of novel polymers with various structures Has a direct effect on their final properties. Among the various synthesized polymers, the synthesis of Styrene-based copolymers was considered [1, 2]. Copolymerization reaction between styrene and heavy alpha olefin monomers were done by various catalysts. Alpha olefins were polymerized by coordination catalysts but styrene can easily polymerized by the atom transfer catalysts [3, 4]. In this research, various styrene/1-hexen copolymers were synthesized by Ziegler-Natta and atom transfer radical polymerization. The molecular structure of the synthesized copolymers was investigated using ^{1}H NMR method. Furthermore, the thermal properties of the synthesized copolymers were also determined by DSC method.

H. Mirzadeh and A. A. Katbab (Eds.): ISPST 2018, *Eco-friendly and Smart Polymer Systems*, pp. 593–596, 2020.
https://doi.org/10.1007/978-3-030-45085-4_143

2 Experimental/Theoretical

In this paper, styrene (St) and 1-hexene monomers (all industrial grade; ≥98%) were used after purification. PMDETA ligand, CCl_4 initiator, and CuCl catalyst (All perches from Merck) were used as received.

2.1 Ziegler-Natta Polymerization of Poly1-Hexene and Its Copolymers

Coordination polymerization of 1-hexene and styrene was carried out with a required amount of triethyl aluminum and then $TiCl_4/Mg(OEt)_2$ in dried toluene solvent at 60 °C. The reaction mixture after 90 min was precipitated and purified. The sample was then studied by ^1H NMR spectroscopy and DSC techniques.

2.2 Atom Transfer Radical Polymerization

Atom transfer radical copolymerization of styrene and 1-hexene was carried out with CCl_4 initiator and CuCl/N, N, N′, N′, N″-pentamethyldiethylenetriamin (PMDETA) catalyst system in bulk at 80 °C according to a procedure in Ref [5].

3 Results and Discussion

3.1 Ziegler-Natta Polymerization Reactions of Styrene/1-Hexene

Copolymerization reactions of 1-hexene/styrene at various molar values of the monomers were done via Ziegler-Natta catalyst at 60 °C. As shown in Table 1, the values of poly styrene and poly 1-hexene conversion are calculated to be 1 and 40%, respectively. Furthermore, by increasing the mole fraction of 1-hexene in the feed, the overall conversion of the reaction increases. The copolymer composition of various synthesized 1-hexene/styrene copolymers are determined and summarized in Table 1. The mole fraction of styrene in the synthesized copolymers is in the range of 1.7–5.4% until 50 mol percent of styrene monomer in the feed. Furthermore, increase in mole percent of styrene in the feed to 80 leads to significant increase in mole percent of this monomer in copolymer.

Table 1. Reaction conditions of 1-hexene and styrene copolymerization reactions with Ziegler-Natta catalyst at 60 °C

Reaction	St in feed (mol%)	1-He in feed (mol%)	Conver (%)	1-He in copolymer (mol%)	St in copolymer (mol%)
1	0	100	40	100	0
2	10	90	27	98.3	1.7
3	19.4	80.6	19.6	97	3
4	50	50	10.7	94.6	5.4
5	80.6	19.4	5.2	28	72
6	100	0	1	0	100

Time of the reaction is 90 min and co catalyst to catalyst ratio is equal to 200

Thermal behavior of the synthesized 1-hexene/styrene homo and copolymer are studied by DSC method. As shown in Fig. 1, the glass transition temperatures of the synthesized poly styrene and poly 1-hexene are 75 and −44.6 °C, respectively. The glass transition temperature of the copolymers with 1.7–5.4 mol percent of styrene is low and in the range of −42.4 °C to −40.3 °C. In return, the glass transition temperature of the synthesized copolymer with 28 mol percent of styrene is obtained to be 61.6 °C.

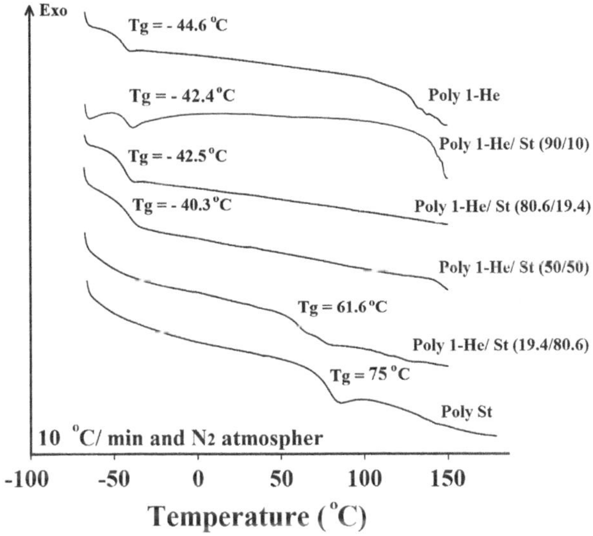

Fig. 1. DSC thermograms of the synthesized polymers and copolymers of styrene and 1-hexane via Ziegler-Natta catalyst

The atom transfer radical copolymerization of styrene and 1-hexene are done with mole fraction of 0.2–0.8 of styrene in the feed. In all reactions, the mole ratios of monomers, CuCl catalyst, carbon tetrachloride initiator and N, N, N′, N″, N″-Pentamethyldiethylenetriamine (PMDETA) ligand were constant and equal to 100/1/1/2 and reactions carried out at 80 °C. As shown in Table 2, all the reactions were done at 1–2 h and overall conversion increased with styrene mole percent in the feed.

Table 2. Styrene/1-hexene copolymerization reaction conditions via ATRP at 80 °C

Reaction	St in feed	1-He in feed	Time (min)	Conversion (%)
1	0.2	0.8	120	6.85
2	0.4	0.6	120	10.92
3	0.6	0.4	120	–
4	0.8	0.2	60	9.8

Molar ratios of monomers, CuCl catalyst, carbon tetrachloride initiator and N, N, N′, N″, N″-pentamethyldiethylenetriamine ligand are equal to 100/1/1/2 and are constant in the reactions.

4 Conclusion

The catalyst type has direct effect on overall conversion of styrene/1-hexene copolymerization reactions. In polymerization reaction with Ziegler-Natta catalyst the overall conversion of the reaction increases with mole fraction of 1-hexene in the feed, indicating the higher activity of catalyst active sites for polymerization of this monomer. The composition of styrene in copolymer was 5.4 in 50/50 mol percent of St/1-H in the feed. In contrast, in ATRP reactions, the overall conversion increased with styrene mole percent in the feed which confirmed the lower activity of 1-hexene in ATRP.

References

1. Gonza, J.A., Leibler, L.: Macromolecules **41**, 9823–9830 (2008)
2. Hanifpour, A., Bahri-Laleh, N., Nekoomanesh-Haghighi, M., Mirmohammadi, S.A.: J. Appl. Polym. Sci. **133**, 43882–43889 (2016)
3. Rahmani, S., Abbasian, M., Moghadam, P.N., Entezami, A.A.: J. Appl. Polym. Sci. **104**, 4008–4014 (2007)
4. Franco, M., Rabagliati, R.A., Cancino, A.M., Sebastián, M.G.: Europ. Polym. J. **41**, 1013–1019 (2005)
5. Semsarzadeh, M.A., Rostami-Daronkola, M.R.: Iran. Polym. J. **15**, 829–839 (2006)

A Multi-objective Optimization Strategy for Controlling the Structural Properties of Lightweight Mutiphasic PE/EVA Foams

Sirwan Ghavami, Mohammad-Hasan Khademi, Farkhondeh Hemmati(✉), Ali Fazeli, and Jamshid Mohammadi-Roshandeh(✉)

Caspian Faculty of Engineering, College of Engineering, University of Tehran, Tehran, Iran
{f.hemmati,roshandeh}@ut.ac.ir

Abstract. In this work, a mathematical approach based on the combination of response surface methodology and desirability function was applied to optimize the foaming extrusion process of polyethylene/ethylene vinyl acetate copolymer blends in order to control the structural properties of them, which is of great importance for their applications in sectors like energy management. Continuous foaming extrusion of these blends has complexity due to the mutual effects of material and processing parameters as well as the phase separation phenomena. The findings show that the employed mathematical strategy can be a useful tool for muti-objective optimization of these parameters.

Keywords: Cellular foams · Lightweight · Extrusion · Optimization · Desirability function

1 Introduction

In the present world, polymeric foams have very extensive applications in many sectors such as construction, automotive, aeronautics, packaging, biotechnology and energy management. Due to the rising public awareness of the utmost importance of sustainability in the future, cellular materials having enhanced properties have been developed in order to diminish the consumption of energy and raw materials of synthetic polymers [1]. Understanding the correlation of influential factors in foaming processes with cellular morphology and final properties is a matter of the greatest significance to obtain polymeric foams with improved performances [2]. Therefore, the objective of this work is to study the interrelationship of some major parameters in continuous extrusion foaming process with microstructure in mutiphasic PE/EVA blend foams by using a hybrid mathematical model based on the combination of response surface methodology (RSM) and desirability function (DF) capabilities.

H. Mirzadeh and A. A. Katbab (Eds.): ISPST 2018, *Eco-friendly and Smart Polymer Systems*, pp. 597–601, 2020.
https://doi.org/10.1007/978-3-030-45085-4_144

2 Experimental and Theoretical

PE (LD-LH0075 grade, Bandar Imam Petrochemical Co.), EVA (Seetec VS430, Lotte Chemical Co.) and azodicarbonamide (ADA) (Cellcom-7000DB, Kumyang Co.) as chemical blowing agent were used to prepare PE/EVA foams. Extrusion foaming process was performed using a microlab twin-screw extruder (Rondol, UK) at screw speed of 150 rpm. Three different influential parameters in foaming process were considered including extruder temperature profile (3 levels), PE content (3 levels) and ADA loading (2 levels). The applied temperature profiles were as follows: 140 °C to 170 °C (155 °C), 155 °C to 175 °C (170 °C) and 170 to 190 (185 °C). The void fraction of foams was determined by measuring the density of samples according to ASTM D792 and ASTM D1622. The morphology of cells was quantitatively analyzed by using micrographs obtained by AIS-2100 SERON scanning electron microscopy (SEM). The combination of RSM and DF capabilities provides a particularly useful tool for analysis, prediction and optimization of systems with great complexities like multiphasic polymeric foams during bubble nucleation, growth and stability in continuous extrusion process [3]. Phase separation of PE/EVA blends after exiting the die is an additional complexity in controlling the foam morphology. Thus, we applied Miscellaneous designs of RSM for analysis and prediction of the foam structural properties along with DF approach for multi-objective optimization of them.

3 Results and Discussion

The SEM micrographs of the prepared PE/EVA foams at fixed level of ADA at 4 and 8 phr are shown in Figs. 1 and 2, respectively. The comparison of bubble morphologies in different images demonstrates the profound impact of PE and ADA content as well as extruder temperature profile on the foam microstructure. The chemical foaming process in twin-screw extruder for PE/EVA blends with upper-critical solution temperature (UCST) phase behavior is seriously complex to control the foam microstructure, especially considering the phase separation and crystallization phenomena after exiting the die. RSM analysis is found as an acceptable tool for prediction and optimization of complex processes having different influential parameters with strong interactions [3]. In RSM, the quadratic model has the best fit among others for prediction the studied PE/EVA foam characteristics including density, void fraction, number-average diameter of bubbles (D_n), volume-average diameter of bubbles (D_v), cell size distribution (D_v/D_n) and cell density. The optimum foam morphology is attained at 25 wt% of EVA and 8 phr of ADA, foamed at temperature profile of 170 °C. For producing lightweight PE/EVA foams with improved properties, it is of great importance to control all physical and morphological characteristics of foams. Figure 3 shows the individual optimization of void fraction along with multi-objective optimization of (void fraction, cell density), (void fraction, cell density, Dn), (void fraction, cell density, cell size distribution) and (void fraction, cell density, Dn, cell size distribution) given by RSM analysis and desirability function. When multiple responses are optimized simultaneously, the operating window (regions with yellow color) can be found wherein the desired targets are concurrently met. As can be seen in Fig. 3, the operating window has larger area for PE/EVA foams containing 8 phr of ADA. The overlay plots illustrated in Fig. 3 (e) prove that the lightest foams with small cells, narrow cell size distribution and high cell density can be prepared for PE/EVA blends at larger PE and ADA content and moderate temperature profiles about 170 °C.

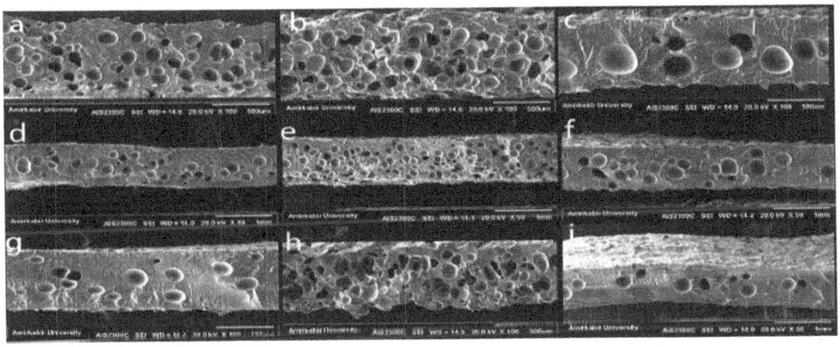

Fig. 1. SEM micrographs of PE/EVA foams at ADA fixed content of 4 phr; Each row belongs to the foams with fixed level of PE: (1st) 25 wt%, (2nd) 50 wt%, (3rd) 75 wt%; Each column belongs to the foams prepared at an unchanged temperature profile: (1st) 155 °C, (2nd) 170 °C, (3rd) 185 °C.

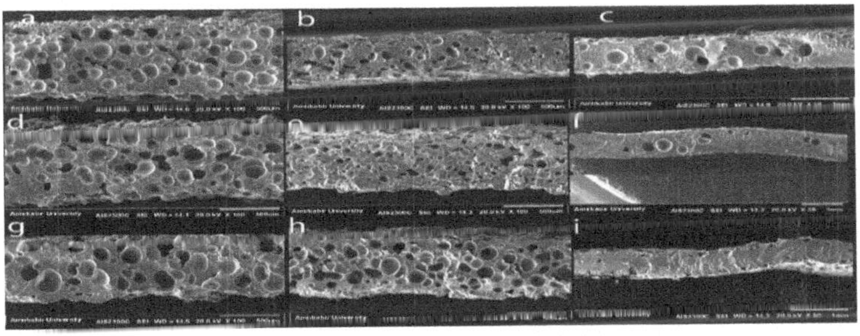

Fig. 2. SEM micrographs of similar PE/EVA foams to Fig. 1 with an exception of ADA content, which is 8 phr for these images

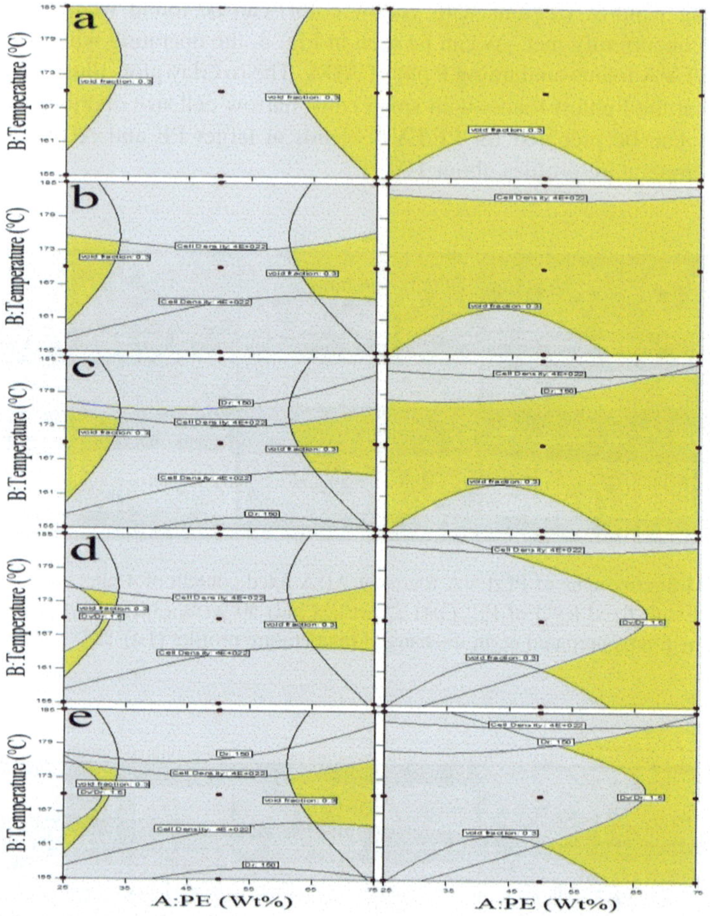

Fig. 3. Overlaid plots of RSM showing the optimized regions at constant ADA level (left column: ADA = 4 phr, right column: ADA = 8 phr) for controlling: (a) void fraction, (b) void fraction and cell density, (c) void fraction, cell density and D_n, (d) void fraction, cell density and D_v/D_r, and (e) void fraction, cell density, D_n and D_v/D_n in the range of void fraction > 30%, cell density > 4×10^{22} cells.m^{-3}, 60 μm < D_n < 150 μm and 1 < D_v/D_n < 1.5

4 Conclusion

In this study, lightweight foams with void fractions about 50% based on PE/EVA blends having UCST phase behavior were prepared using continuous extrusion process. The material and processing parameters of the foaming process were simultaneously optimized by applying RSM analysis and desirability function.

References

1. Patankar, S.N., Kranov, Y.N.: Mater. Sci. Eng. Part A **527**, 1361–1366 (2010)
2. Lee, S.T.: Introduction: Polymeric Foams, Mechanisms, and Materials in Polymeric Foams. CRC Press, Boca Raton (2004)
3. Ataeefard, M., Shadman, A., Saeb, M.R., Mohammadi, Y.: Appl. Phys. Part A **122**, 726 (2016)

Fabrication of Photonic Crystal Heterostructures Free of Interface Imperfection Based on Plasma Assisted Versatile Deposition

Zahra Sadat Azizi[1(✉)], Mohammad Mehdi Tehranchi[2], and Saeed Pourmahdian[1,3]

[1] Laser and Plasma Research Institute, G. C., Tehran, Iran
Zahra.s.azizi@gmail.com
[2] Department of Physics, Shahid Beheshti University, Tehran, Iran
[3] Polymer Engineering Department, Amirkabir University of Technology, Tehran, Iran

Abstract. In this study, we proposed a facile one-step thermal self-assemble method for depositing a photonic crystal heterostructure (PCH) on a flexible polymer substrate which can filter a determined wavelength band from a broad spectrum illumination and provide transmission of an adjustable wavelength. Method includes improving surface hydrophilicity property of the polymer substrates using the application of cold plasma treatment and using one-step thermal treatment for deposition of PCH. The plasma treatment successfully increased the hydrophilicity of the surface of the heat-sensitive polymer substrates. The device resulting from the method has defect-free junction structure that contains at least two photonic crystals with different photonic crystal lattice constants. The PCHs led to a broadband reflection as well as creating low-loss waveguides, high-performance filters, frequency selectors, and chemo/biosensors.

Keywords: Colloidal deposition · Flexible substrate · Interface imperfection · Photonic crystal · Plasma treatment

1 Introduction

Over the past several decades, a number of methods have been used to fabricate photonic crystals (PCs) composed of monodisperse spheres [1, 2]. However, these PCs are in limited use because of the only one transmission peak for the PCs. In order to resolve this problem, complex PCs such as formed PC heterostructures and embedded artificial defects are designed for modulation of light [3]. Currently, most of these crystal films have been prepared by an evaporation process, which results in increased manufacturing time [4]. The crystals may be rapidly prepared, but it is difficult to manipulating thick crystal with high reflectance and there is not any way to adjust lattice constant to have various reflection colors using one size particles.

H. Mirzadeh and A. A. Katbab (Eds.): ISPST 2018, *Eco-friendly and Smart Polymer Systems*, pp. 602–605, 2020.
https://doi.org/10.1007/978-3-030-45085-4_145

To address the above-described problems, we proposed plasma assisted self-assembly method in short duration and low thermal treatment. This one step method fabricated PCHs with broad band width spectral reflection resulting from band gaps overlapping of upper and lower photonic crystal layers. The effect of fabrication route on the spectral reflection of the samples was ascertained through the optical spectroscopy measurements.

2 Experimental

Anionic PS spheres of 260 nm ± 10 nm diameter were prepared by appreciate percent of poly(vinylpyrrolidone), styrene monomer, and potassium persulfate initiator. Afterward, sample was gently stirred (20 min and 300 rpm) at room temperature to deionized water in a three-neck round bottom flask in a water bath. The reaction mixture was deoxygenated by bubbling nitrogen gas at room temperature for 30 min. The temperature of mixture was gradually increased to 75 °C, and kept at this temperature (75 ± 5 °C) for 8 h. At the first, for improving the surface hydrophilicity property of the glass or polymer substrates, a super-arc (Plasmatech-17A) was conducted. Drying the colloidal suspension on cover glass substrate, PC film was made in two ways. The ordinary photonic crystal was dried according to slow sedimentation self-assembly process in an oven and a special rapid thermal procedure employed to produce the PCH. To ensure the sample topography and structure, scanning electron microscopy (SEM) was conducted. Moreover, the optical properties of the samples were measured by UV-visible reflection spectroscopy.

3 Results and Discussion

In order to investigate the difference between an ordinary photonic crystal and the photonic crystal heterostructure, the plasma treatment was employed to prepare a hydrophilic flexible substrate for deposition of both ordinary photonic crystal and the photonic crystal heterostructure. Figure 1 illustrates UV-vis reflection spectrum of the ordinary oven dried PC (filled triangle dots) and PCH (opened rectangular dots) both comprising polystyrene spheres of 260 nm in diameter. The fast thermal approach leads to 3 × increase in photonic band gap width of the PCH compared to ordinary photonic crystal enhanced. The transmittance spectrum of the PCH (thick line) confirms the broader photonic bang gap, too. In addition, the formation of PCH is indicated by appearance of a distinct sharp dips or a wide photonic band gap with a defect (tunneling) mod in the UV-vis reflection spectrum. This appearance can be confirmed with a cross section SEM image of the sample.

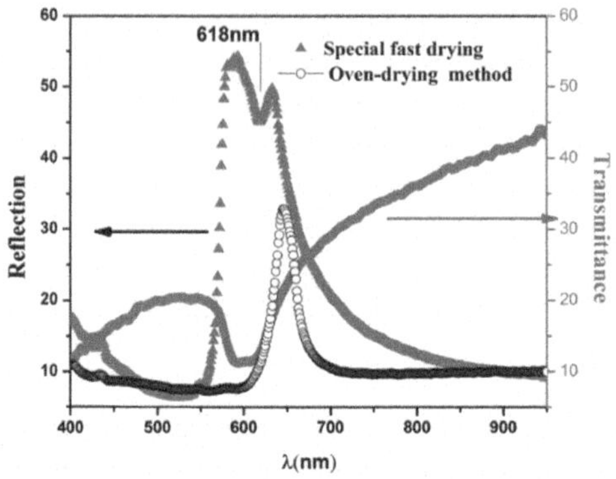

Fig. 1. Corresponding reflectance spectra of the ordinary oven dried photonic crystal (filled triangle dots) and photonic crystal heterostructures (opened rectangular dots) both comprising polystyrene spheres of 260 nm in diameter, and in the right side of panel, the transmittance spectra of photonic crystal heterostructures (thick line)

Scanning electron microscope images illustrating a lateral view (Fig. 2) of PCH comprises polystyrene spheres of 260 nm in diameter. The image shows a highly ordered photonic crystal and possesses a close-packed face-centered-cubic structure with the (1 1 1) planes oriented parallel to the substrate. From the cross-sectional views, it can be seen that the photonic heterostructures composed two stacks of photonic

Fig. 2. Scanning electron microscope image: illustrating a lateral view of photonic crystal heterostructure comprising polystyrene spheres of 260 nm in diameter

crystals with two distinct lattice constants. It comfier the existence of two well-ordered stacks of photonic crystals without any imperfection in their interface of the heterostructure. Further experimental analyses shows the difference of the lattice constants between upper layers and down layers of PCH is around 18 nm. According to theoretical calculations, this difference is agreed with the corresponding photonic band gap shift.

4 Conclusion

The result of a facile one-step thermal method in depositing PCH film stacks on flexible substrates has been reported. The spectral reflectometry and the SEM image have shown well-arranged PCH. The plasma treatment successfully increased the hydrophilicity of the flexible substrate surface. The resulting defect-free junction sample contains at least two photonic crystals with different photonic crystal lattice constants. The PCH led to a broadband reflection, high-performance filters, frequency selectors, and chemo/biosensors.

References

1. Jiang, Q., Li, C., Shi, S., Zhao, D., Xiong, L., Wei, H., Yi, L.: J. Non-Cryst. Solids **358**, 1611–1616 (2012)
2. Xue, F., Meng, Z., Qi, F., Xue, M., Qiu, L.: Colloid Polym. Sci. **294**, 479–482 (2016)
3. Kedia, S., Srinivas Reddy, M., Vijaya, R.: Opt. Photonics J. **2**, 242–248 (2012)
4. Cai, Z., Jun Liu, Y., Teng, J., Lu, X.: Soc. J. Appl. Mater. Interf. **4**, 5562–5569 (2012)

Shape Memory Behavior of Polyacrylamide Hydrogel

Sonia Norouzi Esfahany, Ghazaleh Alamdarnejad, and Mehrdad Kokabi(✉)

Department of Polymer Engineering, Faculty of Chemical Engineering, Tarbiat Modares University, P.O. Box 14115-114, Tehran, Iran
mehrir@modares.ac.ir

Abstract. In this work, polyacrylamide hydrogel prepared by free radical polymerization and its shape memory behavior are investigated using dynamic mechanical analysis technique. The switch (transition) temperature of the amorphous hydrogel containing 85% water was obtained around ~−7 °C. The results indicated that the shape fixing ratio and shape recovery ratio of the system were 99% and 82%, respectively.

Keywords: Hydrogel · Shape memory · Thermoresponsive · Polyacrylamide

1 Introduction

Polymeric hydrogels as three-dimensional networks have capability to absorb large amounts of water. Hydrophilicity and cross-linking density of hydrogel could control the amount of absorbed water. Polymeric hydrogels are generally amorphous in nature [1]. Polyacrylamide (PAAm) hydrogel is an amorphous hydrogel, which shows a reversible transition at its glass transition temperature (T_g). It has the ability to fix the programming temporary shape beneath the T_g and to recover its original shape beyond the T_g. PAAm hydrogel exhibits shape memory behavior by undergoing an ordered-disordered transition using heat as external stimuli [2].

In this work, polyacrylamide hydrogel was prepared by free radical polymerization of acrylamide (AAm) as a monomer with *N, N'*-methylene bisacrylamide (MBAAm) as a crosslinking agent, ammonium persulfate (APS) as an initiator and tetra ethylene methylene diamine (TEMED) as an accelerator. Then the shape memory behavior of the prepared hydrogel was investigated.

2 Experimental

2.1 Materials

AAm, MBAAm, APS, and TEMED without further purification purchased from Merck, Germany.

H. Mirzadeh and A. A. Katbab (Eds.): ISPST 2018, *Eco-friendly and Smart Polymer Systems*, pp. 606–608, 2020.
https://doi.org/10.1007/978-3-030-45085-4_146

2.2 Hydrogel Synthesis

PAAm hydrogel containing 85% water was synthesized by the simultaneous radical polymerization and crosslinking of AAm. The polymer mixture contained 15 w/v% of monomer (AAm) in distilled water together with the crosslinker (MBAAm) and redox initiator (APS and TEMED). The concentration of MBAAm was fixed at 0.5 wt%. Hydrogel sample was prepared using 0.1 mL of initiator solution (0.05 g.mL^{-1} water) and 0.125 mL of accelerator solution (0.01 g.mL^{-1} water). Each 10 mL solution of reactants was prepared in distilled water. The reaction was carried out at a constant temperature of 40 ± 0.1 °C using a water bath for 24 h.

2.3 Characterization

Dynamic mechanical analysis (DMA) was carried out by DMA 242C Netzsch, Germany, with a heating rate of 5 °C.min^{-1} in the range of −50 °C to 70 °C. The compression mode was used in order to determine the glass transition temperature (T_g). Subsequently, the tension mode was employed for investigating shape memory effect.

3 Results and Discussion

DMA test results in Fig. 1 indicate the transition temperature of the shape memory polyacrylamide hydrogel containing 85% water obtained around ~−7 °C. According to the above data, the switch (transition) temperature of shape memory process was selected −7 °C. To investigate the shape memory behavior of PAAm hydrogel, a four steps procedure was performed at a constant heating/cooling rate of 5 °C.min^{-1}. First, the sample was heated to 30 °C (above the switch temperature) and stretched to a certain strain (ε) under a constant force (Step 1). Then, the deformed sample was cooled to −60 °C while the force kept fixed (Step 2). Then, the force was removed and the temporary strain (ε_{temp}) was measured (Step 3). Finally, the sample was reheated to 30 °C and kept at 30 °C for 1 h, then the recovery strain (ε_{rec}) was recorded (Step 4). The shape fixing ratio (R_f) and shape recovery ratio (R_r) of hydrogel were calculated from Fig. 2 according to the following equations:

$$R_f = \frac{\varepsilon_{temp}}{\varepsilon} \tag{1}$$

$$R_r = \frac{\varepsilon - \varepsilon_{rec}}{\varepsilon} \tag{2}$$

Length displacement and temperature curves versus time were depicted. The values of shape fixing ratio and shape recovery ratio of PAAm hydrogel were 99% and 82%, respectively, which indicated the reasonable shape memory behavior of synthesized PAAm hydrogel.

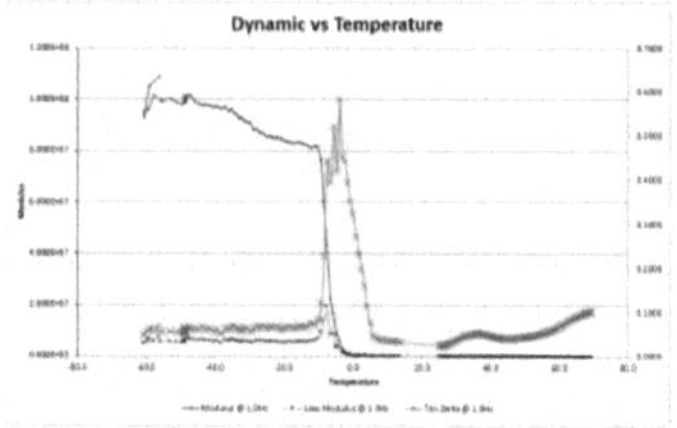

Fig. 1. DMA curve of PAAm hydrogel

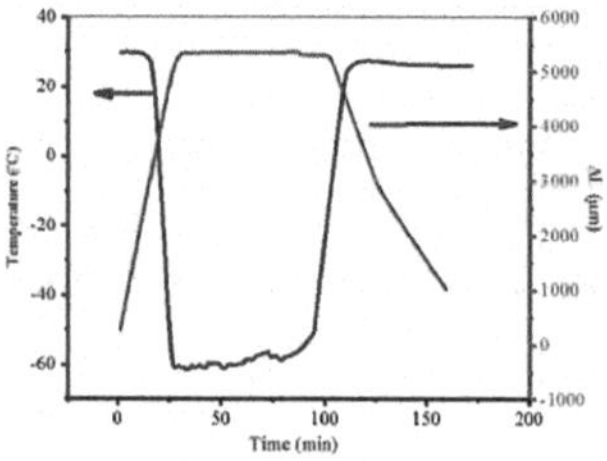

Fig. 2. Thermomechanical behavior of PAAm hydrogel

4 Conclusion

PAAm hydrogel containing 85% water was successfully synthesized by the simultaneous radical polymerization and crosslinking of AAm. The switch temperature of synthesized shape memory hydrogel was around −7 °C. The shape fixing ratio and shape recovery ratio of hydrogel were 99% and 82%, respectively, which indicated the reasonable shape memory behavior of synthesized PAAm hydrogel.

References

1. Lin, X.K., Chen, L., Zhao, Y.P., Dong, Z.Z.: J. Mater. Sci. **45**, 2703–2707 (2010)
2. Kabir, M.H., Ahmed, K., Furukawa, H.: Micr. Eng. **150**, 43–46 (2016)

Determination of Entangled Supramolecular Polymer Networks Dynamics by High-Order Associations of Strong Hydrogen Bonding Groups

Amir Jangizehi, Seyed Reza Ghaffarian(✉), and Mostafa Ahmadi

Department of Polymer Engineering and Color Technology, Amirkabir University of Technology, 15875-4413 Tehran, Iran
sr_ghaffarian@aut.ac.ir

Abstract. Dynamics of entangled polymer chains in the melt state are deliberately excluded in most of the studies on supramolecular polymer networks by utilizing non-entangled precursor chains. Relaxation of the system mainly depends on the dissociation of the associative groups in latter case and non-entangled chains deliver nothing to resist afterwards. Conversely, in an entangled system relaxation of polymer chains and dissociation of associative groups can occurred parallel. Supramolecular networks based on an entangled precursor polymer with different levels of strong associating ureidopyrimidinone, UPy, groups are synthesized to screen the corresponding effects on the dynamics of the system. Binary-associated UPy groups phase separate into collective assemblies by stacking and form high-order, needle-like domains at higher UPy contents. Relaxation of polymer chains is significantly hindered due to the trapping of polymer segments between UPy stacks. Above a certain threshold of UPy content (~4 mol%), the plateau level as well as final relaxation time of networks show a significant jump, which is attributed to the onset of high-order association of UPy groups.

Keywords: Supramolecular networks · High order associations · Chain dynamics · Chain entanglements · Ureidopyrimidinone groups

1 Introduction

Supramolecular polymer networks are macromolecular structures in which a fraction of covalent bonds are replaced with transient associations, such as ionic attractions, metal–ligand coordinations, hydrogen bondings or π–π interactions [1, 2]. Such transient bonds are in the dynamic equilibrium between associated and dissociated states. Hydrogen bonding moieties are widely used for designing supramolecular polymers, since they demonstrate very broad range of strength, which can be controlled by the number and the configuration of the interactions. One of these promising moieties is Ureidopyrimidinone, UPy, group which self-assembles with high strength, quadruple hydrogen bonds [3]. To rationally design such materials for a specific application, like self-healing [4]. It is essential to understand their dynamic properties. The dynamic

H. Mirzadeh and A. A. Katbab (Eds.): ISPST 2018, *Eco-friendly and Smart Polymer Systems*, pp. 609–612, 2020.
https://doi.org/10.1007/978-3-030-45085-4_147

properties of an entangled supramolecular polymer networks in the melt state is determined by the interplay of chain entanglement and life-time of binary/collective assembly of associative groups. However, the contribution of the second parameter in a well-entangled system and in the melt state has not been well studied. We have therefore designed a well-entangled supramolecular system in which the number of associative groups is systematically varied. Supramolecular polymer networks based on poly(n-butyl acrylate-ran-hydroxyl ethyl acrylate), P(BA-ran-HEA) bearing different contents of UPy groups, and P(BA-ran-HEA-UPy) are synthesized and the influence of strong binary and collective associations on thermal and rheological properties are studied.

2 Experimental

Synthesis of PBA homopolymer and P(BA-ran-HEA) copolymers have been reported elsewhere, recently [5] and summarized in Scheme 1. Thermal, mechanical and dynamic properties of the networks were studied by differential scanning calorimetry (DSC), dynamic mechanical thermal analysis (DMTA), and oscillatory shear rheology experiments.

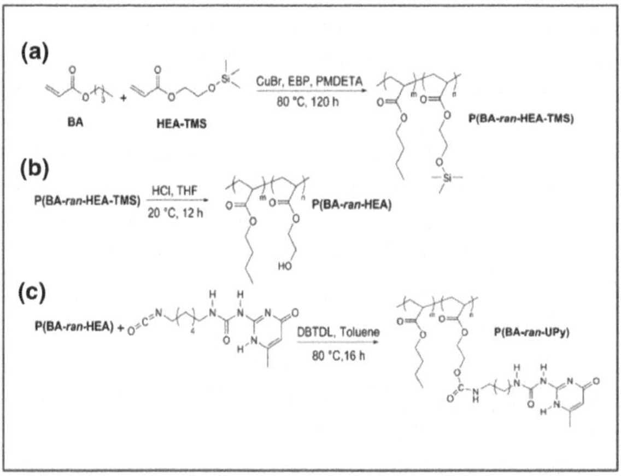

Scheme 1. Preparation of P(BA-ran-HEA-UPy) supramolecular network. (A) P(BA-ran-HEA-TMS) is synthesized by ATRP, (B) and then the trimethylsiloxy protection is removed by concentrated HCl to obtain P(BA-ran-HEA), and (C) supramolecular polymer networks prepared by functionalization of random copolymers with UPy groups

3 Results and Discussion

P(BA-ran-HEA-UPy) supramolecular networks form sea-island morphology, in which a fraction of binary-associated UPy groups phase separate from the polymer matrix. The proportion of these different UPy structures can be tuned by the thermal treatment of the sample. Evolution of the storage, E', and loss, E", moduli of the supramolecular networks upon heating is shown in Fig. 1(a). Above glass transition temperature, two major relaxations can be observed, Endo1 and Endo2. The temperatures at which these relaxations occur are correlated to the endothermic processes observed in DSC. There is a maximum in the Log tanδ of the networks containing 2 and 3 mol% of UPy groups around −75 °C (Fig. 1(b)), which shows that these networks have a relaxation mechanism at temperature lower than glass transition temperature of polymer backbone. This maximum could be related to the local motion of the backbone.

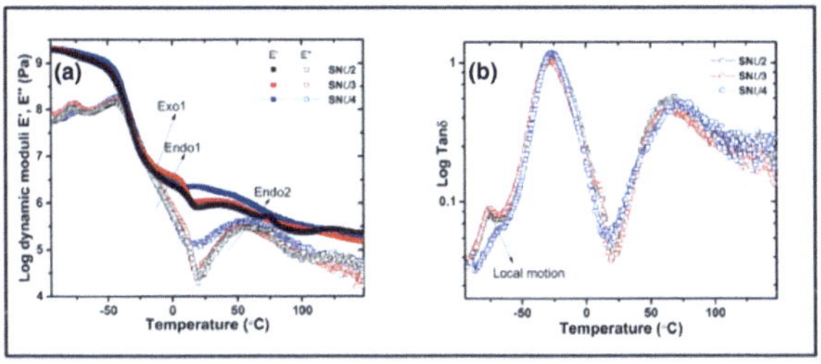

Fig. 1. Temperature sweep analysis of dynamic mechanical properties of P(BA-ran-HEA-UPy) supramolecular networks containing different UPy mol%: 2 (SNU2), 3 (SNU3), and 4 (SNU4). (A) Evolution of storage and loss moduli and (B) the damping factor curves, tanδ, vs. temperature

The terminal relaxation of the chains is considerably hindered by inclusion of supramolecular moieties. At 35 °C (Fig. 2(a)) while PBA precursor is flowing with phenomenological scaling laws of G′ ~ (frequency)^2 and G″ ~ (frequency), these exponents shift to *0.5* in all P(BA-*ran*-HEA-UPy) networks. Consequently, only a small fraction of stress is relaxed after 100 s. It is clear from the viscoelastic data that the supramolecular networks have significantly higher plateau modulus compared to the PBA precursor, which confirm the network like structure in all samples. Figure 2(b) (T = 75 °C) illustrates once again that a low frequency second plateau is observed in G′-frequency curve, which levels up as the sticker content increases in the system. The transition region between the two plateaus shows a slope of 0.5. By comparing the shape of viscoelastic curves at 35 and 75 °C, it can be stated that the networks demonstrate strong thermo-rheological complexity, as anticipated.

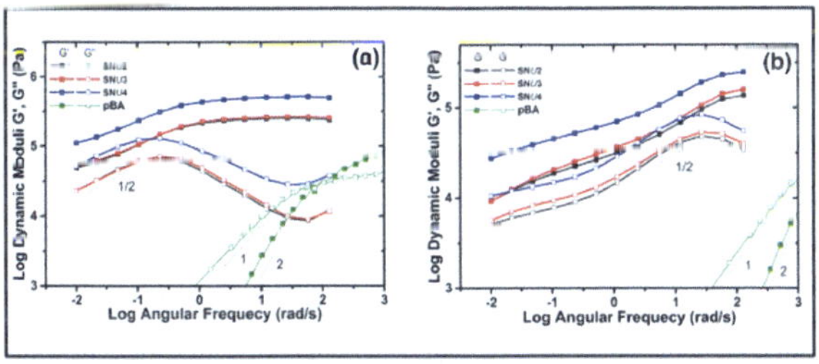

Fig. 2. Storage, G′, and loss, G″, moduli of P(BA-ran-HEA-UPy) supramolecular networks containing 2 (SNU2), 3 (SNU3), and 4 mol% (SNU4) of UPy groups against angular frequency at (A) 35°C and (B) 75°C. The data of precursor chains, PAB, are shown for comparison. Dashed lines indicate various slopes

4 Conclusion

Entangled supramolecular polymer networks containing strong associative UPy groups show higher order association process. P(BA-ran-HEA) with relatively high molar masses and narrow polydispersities were synthesized via ATRP and subsequently functionalized by UPy at three different contents of 2, 3, and 4 mol%, to create supramolecular networks via strong multiple hydrogen bonds. Binary-association of UPy groups phase separate into collective assemblies and form structures with different morphologies and sizes, which is revealed by optical microscopy and DSC results. Above a certain threshold of UPy content ($\sim$4 mol%), this phase separation leads to the creation of high-order domains with needle-like shape, which are more stable than other structures. Samples with 2 and 3 mol% of UPy groups had similar thermal and rheological behavior which was different from the one with 4 mol% due to the lack of high-order association of UPy groups. The storage moduli for all networks show two plateaus in LVE regime with transition slope of 0.5.

References

1. Yang, L., Tan, X., Wang, Z., Zhang, X.: Chem. Rev. **115**, 7196–7239 (2015)
2. Wei, P., Yan, X., Huang, F.: Chem. Soc. Rev. **44**, 815–832 (2015)
3. Folmer, B.J., Sijbesma, R., Versteegen, R., Van der Rijt, J., Meijer, E.: Adv. Mater. **12**(4), 874–878 (2000)
4. Yang, Y., Urban, M.W.: Chem. Soc. Rev. **42**, 7446–7467 (2013)
5. Jangizehi, A., Ghaffarian, S.R., Ahmadi, M.: Polym. Adv. Technol. **29**, 726–735 (2017)

Investigating the Physico-Mechanical Properties of Elastomers Based on EPDM/Kevlar/Silica: Effect of Cork Filler and Curing Condition

Tohid Farajpour, Mohamad Javad Babaei-Velin(✉), Hadi Rezaei-Vahidian, Ghodratoallah Mohammadzamani, and Saeed Moradi

Space Transportation Research Institute, Iranian Space Research Center, Tehran, Iran
mj.babaei86@gmail.com

Abstract. Fiber filled ethylene propylene diene (EPDM) elastomers are widely used as internal insulator in aerospace industries. Kevlar fibers, silica and flame retardants are used in most heat insulators. In this study, in order to prepare lightweight EPDM based composites, the effect of adding Cork into EPDM compound was investigated. The results showed that the addition of Cork into EPDM compounds reduced the density and elasticity of cured elastomer. On the other hand, Cork content and curing condition (time and temperature) affected the tensile strength and elongation at break. Therefore, through investigating the effect of different proportions of the Cork in the compound, it became clear that the compound filled with 15 phr Cork had optimum physical and mechanical properties. Tensile and density test results showed that the density, tensile strength and elongation of mentioned sample were 1.1951 g/cm^3, 6.3 MPa and 272%, respectively.

Keywords: EPDM · Cork · Mechanical properties

1 Introduction

Ethylene-propylene diene terpolymer (EPDM) rubber is a terpolymer of ethylene, propylene and a non-conjugated diene. It has thermal conductivity and glass transition temperature of about 0.25 Wm^{-1} K^{-1} and −50 °C, respectively, which can introduce it as a suitable heat insulator binder for aerospace application. In addition, the elastomeric materials used as insulator in aerospace industries should have been Low density (light weight) and suitable mechanical properties. In addition, Cork is a lightweight and elastic cellular material that composed of small cells about 42 × 10^6 per cubic centimeter. This cellular structure has a strong effect on mechanical and physical properties of Cork filled material [1–3].

H. Mirzadeh and A. A. Katbab (Eds.): ISPST 2018, *Eco-friendly and Smart Polymer Systems*, pp. 613–616, 2020.
https://doi.org/10.1007/978-3-030-45085-4_148

2 Experimental

Kevlar pulp and EPDM rubber (NORDEL4640) supplied from DuPont Co., Ultrasil VN3 supplied from Degussa Co., Cork powder with particle size of 0.2–0.5 mm purchased from Advance Cork International company (India), zinc stearate (Cognis Co., Malaysia) and MBTS (Volcacit DM) were used in this study. The basic formulation used in the samples A-I contained of a mixture of 100 phr EPDM, 20 phr Kevlar pulp, 1.5 phr sulfur as a curing agent, 2.2 phr MBTS as an accelerator, and 7 phr zinc stearat as an activator. These components were used in all formulations and effect of Cork and silica content were investigated for samples listed in Table 1.

Table 1. EPDM composite compounds and mechanical/density test results (SC: sample code, d: density, σ: tensile strength, and ε: elongation at break)

SC	Cork (phr)	Silica (phr)	d (g/cm^3)	σ (MPa)	ε (%)
A	0	25	1.2147	7.6	453
B	5	25	1.2075	6.7	396
C	20	25	1.2080	6.8	218
D	0	20	1.2023	6.8	470
E	5	20	1.2035	6.5	364
F	10	20	1.1977	6.5	302
G	15	20	1.1951	6.3	272
H	20	20	1.1978	7.1	251
I	25	20	1.2017	7	199

Compound of EPDM rubber was prepared with a calendar. Firstly, crude EPDM entirely was soften in the calendar heated with hot water. After that silica and Cork were added gradually in several stages. In the subsequent stages curing agent was added and mixed homogeneously. Rubber sheets were produced from final compound in 0.15 * 0.15 * 0.003 m, under press pressure of 15 MPa, cured at 170 °C for 15 min. Tensile strength, elongation at break and modulus were measured according to ASTM D412 by using an Instron machine (model: Inspekt 50) at the tensile speed of 500 mm/min. Density test was carried out according the Archimedes method (ASTM D-792).

3 Results and Discussion

Ultrasil and Cork were used in the formulations as components for promoting mechanical properties and reducing structure weight, respectively. Cork is not a reinforcement filler and it is used only to reduce density and thermal conductivity. By adding the Cork to the mixture, the amount of filler in the mixture increases and the flexibility of the prepared composites reduces. As a result, with increasing amount of Cork, elongation and density were sharply decreased. With increasing Cork content, the density of all compounds was reduced. Cork cells are hollow, containing a gas similar to air in their interior which plays an important role in lowering the density of Cork (Fig. 1). Of course, in higher levels of Cork, we see a rise in density, probably due to the structural alteration and fracture of a part of the Cork honeycomb structure under pressure during the mixing process. At high temperature and pressure, the gas in interior emitted and density of the Cork powder increased, which ultimately leaded to increase in density of the cured samples. The density of Cork under high curing pressure and temperature can be increased from 0.5 to 0.98 g/cm^3 [3]. The results showed that the density of sample G decreased about 1.6% compared to the A Besides, tensile strength and elongation at break were in acceptable range to use as aerospace elastomeric insulator.

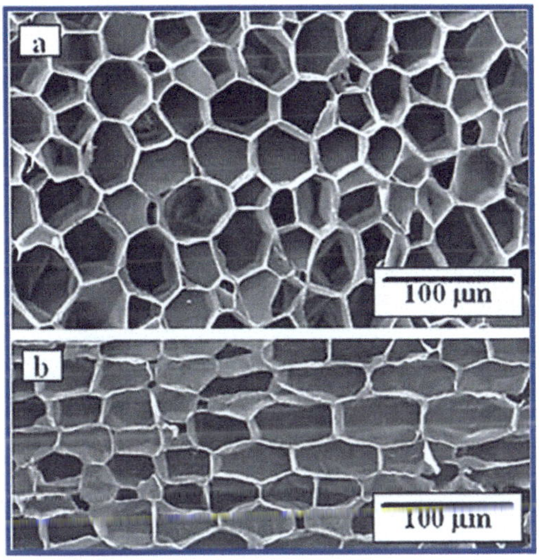

a radial section; *b* tangential section

Fig. 1. Cellular structure of Cork [4]

In order to determine the appropriate curing conditions, the elongation of sample G was investigated at different curing times and temperatures (Fig. 2). The results showed that optimum condition of curing was 170 °C for 15 min. As mentioned before, the curing condition of the sample was suitable to obtain good mechanical properties and flexibility.

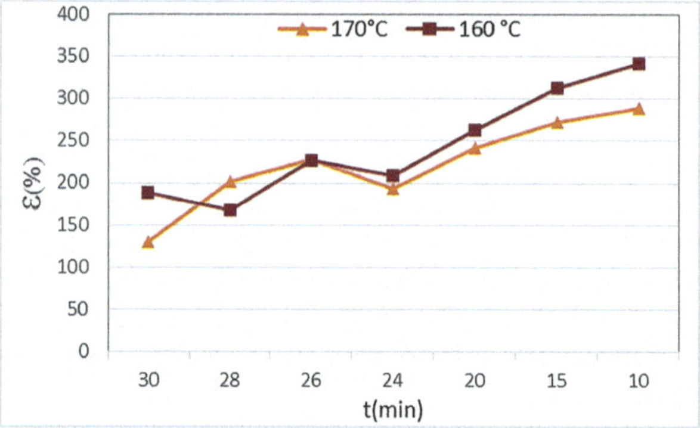

Fig. 2. Effect of curing time and curing temperature on the elongation of sample G

4 Conclusion

Cork powder acts as the most important lightweight filler into the rubber insulation. Comparison between different formulations showed that EPDM composites containing 20 phr silica and 15 phr Cork had suitable physico-mechanical properties in order to use as aerospace insulator. Cork loading could significantly decreases the density under optimum mixing condition.

References

1. Gil, L.: Materials **8**, 625–637 (2015)
2. Oliveira, F.R., Silva, E.A.A., do Carmo, S.N., Steffens, F., de Valadares Souto, A.P.G.: Adv. Mater. Sci. Eng. (2014). https://doi.org/10.1155/2014/685829
3. Gul, J., Saleemi, R., Mirza, S., Feroze, N., Mansha, M.: Plast. Rub. Compos. **39**, 28–32 (2010)
4. Silva, S.P., Sabino, M.A., Fernandes, E.M., Correlo, V.M., Boesel, L.F., Reis, R.L.: Inter. Mater. Rev. **50**(6), 345–365 (2005)

Effect of Peroxides Mixtures on the Molecular Structure of Ethylene-1-Octene Copolymer in Reactive Blending

Mahdi Mahmoodabadi, Shervin Ahmadi(✉), and Hossein Ali Khonakdar

Iran Polymer and Petrochemical Institute, P.O. Box 14965-115, Tehran, Iran
sh.ahmadi@ippi.ac.ir

Abstract. Various methods have been examined to achieve high performance polymers, among them, chain branching method has shown excellent potential for production of polymers suitable for new applications. This work aims at investigating the effects of different mixtures of peroxides on the molecular structure of ethylene-1-octene copolymer during reactive blending. The torque of mixing has been applied as an indicative of a change in the molecular structure. Moreover, the gel content test is used to investigate the partial crosslinking of the samples. In this research, the response surface method (RSM) according to a box Behnken statistical design was applied to investigate the effect of three peroxides mixtures with different half-life times and radical stability on the molecular structure. The method suggests the optimized formulation for maximum branching with little gel content.

Keywords: Ethylene-1-octene · Peroxide modification · Molecular structure · Reactive blending · Response surface method

1 Introduction

Ethylene/α-olefin copolymer with high comonomer content is one of the most widely-used thermoplastic elastomers (TPEs). Compared with other TPE types, polyolefin-based thermoplastic elastomers have better chemical resistance, lower density, excellent weatherability and lower cost for elasticity formation. Ethylene/α-olefin random copolymer elastomer (POE) is a commercially available type of such materials with good compatibility with commercial polyolefins such as polypropylene [1].

The modification of polymer molecular structure after polymerization is a common method to enhance the processability and/or properties of polymers and adapt them to a specific application. Peroxide modification of polymers is one of the most inexpensive available methods to control the molecular structure of saturated as well as unsaturated polymer chains. Chain scission, long and short branching, chain extension and crosslinking are the probable reactions in the presence of peroxides. The relative share of each of these structures varies with process parameters, such as time, temperature, mixing rate, the presence and percentage of comonomer and peroxide type and content.

H. Mirzadeh and A. A. Katbab (Eds.): ISPST 2018, *Eco-friendly and Smart Polymer Systems*, pp. 617–620, 2020.
https://doi.org/10.1007/978-3-030-45085-4_149

Su et al. [2] studied the effect of 1,6-hexanediol diacrylate (HDDA) comonomer and various types of peroxides on the branching efficiency of polypropylene during reactive extrusion. They showed that peroxides with lower decomposition temperatures and more stable radicals after decomposition promoted the branching reaction, leading to the modified polypropylenes with higher branching level. Golriz et al. [3] investigated the effect of process conditions such as mixing time, temperature and peroxide concentration on the formation of LCB on linear low-density polyethylene (LLDPE). By the application of response surface method, they found that the decreasing order of significance of the selected process variables were concentration, temperature and mixing time, ectively.

The aim of this research is to investigate the effect of different mixtures of peroxides on the structure of ethylene-1-octene copolymer.

2 Experimental

Methyl ethyl ketone peroxide (MEKP), 1,1-bis(tert.butyl peroxy)3,3,5-trimethyl cyclohexane (TMCH) and 2,5-dimethyl-2,5-di(tert.butylperoxy)hexane (DHBP) were used as three different peroxides with different half-life times.

A laboratory internal mixer (Brabender W50) was used at 150 °C and 100 rpm for 10 min to mix ethylene-1-octene copolymer and peroxides. The peroxides were injected into the chamber using syringe 3 min after the addition of polymer. All of the torque plots against time showed a plateau before 10 min.

The samples were examined for gel content according to ASTM D2765. Based on this standard the extraction in boiling xylene was the basis of all the analysis.

3 Results and Discussion

Samples were mixed in the internal mixer according to the formulations shown in Table 1, designed by the response surface method according to a box Behnken statistical design.

The increase in the torque of the mixing after the addition of peroxides has been attributed to the increase of polymer viscosity as a result of the reactions including long and short chain branching, chain extension and crosslinking. Therefore, the maximum torque of the mixing has been assigned as an indication of branching or crosslinking reactions and chosen as the response parameter in box Behnken model to be maximized. Since branching is the desired reaction, the gel content is applied as another response parameter to be minimized.

As shown in Fig. 1, the torque of the mixing is increased by TMCH, decreased by MEKP and first increased and then decreased by DHBP. According to model optimization, it is predicted that a formulation containing 2 wt% TMCH and 0.5 wt% DHBP results in the maximum torque. The experiment results showed good accordance with the prediction.

Table 1. Samples formulations, mixing torques and gel contents

Peroxide content (wt%)			Torque (N.m)	Gel Content (%)
MEKP	TMCH	DHBP		
0	1	0	21.2	1.41
2	1	2	31.8	48.95
1	1	1	21.2	18.64
1	1	1	22.3	35.18
2	0	1	17.4	0.58
0	2	1	46.7	34.88
1	2	0	24.8	0.99
0	1	2	26.6	54.47
2	2	1	24.2	19.81
1	0	2	30.4	48.07
2	1	0	11.1	1.12
1	0	0	10.5	0.5
0	0	1	29.9	39.32
1	2	2	24.2	52.4
1	1	1	22.8	28.44

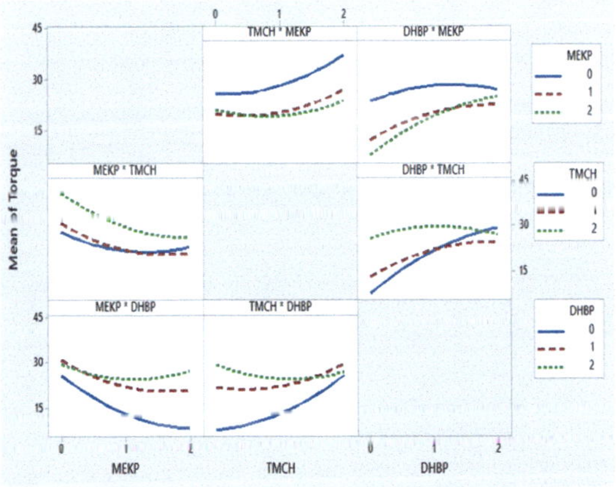

Fig. 1. Interaction plots for torque against hybrids of MEKP, TMCH, and DHBP

To eliminate the effect of crosslinking from torque increase, the box Behnken model was applied for two response parameters to maximize torque and minimize gel content simultaneously. The optimized formulation consists of 2 wt% MEKP and 0.65 wt% DHBP resulting in the torque to be 17.5 N.m with no gel content. It can be attributed to the formation of short and long branches on ethylene-1-octene copolymer without crosslinking. The tendency of MEKP to form branches on the polymer can be attributed to its low half-life time. While electron-donor methyl groups of DHBP may be the reason of high stability of the produced radicals and so higher branch contents. These results are in accordance with the data reported by Su et al. [2].

4 Conclusion

Branched and crosslinked ethylene-1-octene copolymers were produced using MEKP, TMCH and DHBP peroxides from 0 wt% to 2 wt% during reactive blending. The box Behnken statistical design was applied to optimize the formulation in order to obtain branched ethylene-1-octene copolymer with minimum gel content. The formulation containing 2 wt% MEKP and 0.65 wt% DHBP showed the desired results.

References

1. Liu, W., et al.: Polym. (U.K.) **72**, 118–124 (2015)
2. Su, F.-H., Huang, H.-X.: Polym. Eng. Sci. **50**(2), 342–351 (2010)
3. Golriz, M., et al.: Macromol. Mater. Eng. **299**(2), 154–164 (2014)

Effect of Nanographene on the Curing Behavior and Rheological Properties of NBR/Phenolic Compounds

Sara Shafiee, Mohammad Karrabi(✉), and Mir Hamid Reza Ghoreishy

Iran Polymer and Petrochemical Institute, P.O. Box 14965-115, Tehran, Iran
m.karabi@ippi.ac.ir

Abstract. In the recent years, many researchers have attended to reinforce rubbers by nanographene. The aim of this work is to study the effect of nanographene on the curing behavior and rheological properties of NBR/phenolic blends. Nanocomposites containing NBR, novolac resin and nanographene as reinforcing agent (0.5, 1, and 1.5 phr) were prepared on a two-roll mill. Curing behavior and rheological properties are investigated by Rubber Processing Analysis (RPA) and strong role of nanographene on these properties was demonstrated. So, the results showed that nanographene increased the scorch time of the samples in comparison with blends without nanographene. Moreover, the difference between the highest and the lowest torque (M_H-M_L) increased with addition of nanographene. Interestingly, it was observed that by increasing nanographene content, viscosity reduced.

Keywords: Nitrile/phenolic blends · Grapheme · Curing torque · Crosslink density · Rheological properties

1 Introduction

The use of low amounts of nanoparticles to reinforce the properties of polymers is very common. Moreover, nitrile rubber is one of the common elastomers which has a wide range of applications because of its mechanical and rheological properties. The blend of nitrile rubber and inexpensive thermal resistant phenolic resins has wide applications in automobile, aerospace, sealing and cables. In this research, the rheological properties and curing behavior of the nitrile rubber/phenolic resin blends containing nanographene are investigated.

Mirabedini et al. [1] studied the viscoelastic behavior of NBR/Phenolic compounds and showed that the difference between the highest and lowest torque of curing (M_H-M_L) increases after addition of phenolic resin, which is indicative of the increase in crosslink density and the formation of IPNs.

Wu et al. [2] studied vulcanization kinetics of natural rubber (NR)/graphene compounds and showed that the induction period of the vulcanization process is drastically suppressed at low graphene loading and then shows a gentle decrease, meanwhile the vulcanization rate is enhanced at first and subsequently shows a

H. Mirzadeh and A. A. Katbab (Eds.): ISPST 2018, *Eco-friendly and Smart Polymer Systems*, pp. 621–624, 2020.
https://doi.org/10.1007/978-3-030-45085-4_150

depression. The combination effect of these two aspects leads to a prominent reduction in optimum cure time at first and then a slight increase. The aim of this work is to investigate the curing behavior and rheological properties of NBR/Phenolic compounds in the presence of nanographene.

2 Experimental

The compounds were prepared on a two-roll mill (Polymix 200 L, Schwabenthan Co., Germany) at a speed of 15 rpm and a friction ratio of 1.6 for 20 min. It should be noted that when adding phenolic resin to the mixture, the roller temperature was set at 80 °C. Finally, the compounds were volcanized at 160 °C with pressure of 150 kg/cm^2. Sample information is given in Table 1.

Table 1. Compounds formulations

Ingredients	Type	Parts per hundred rubber (phr)			
		N_3P	$N_3PG_{0.5,1,1.5}$		
NBR	KNB 35L	70	70		
Resin	Novolac	30	30		
Reinforcing agent	Graphene	0	0.5	1	1.5
Softner	250 oil	10	10		
Stabilizer	4010 NA	10	10		
Activator	Zinc oxide	4	4		
	Stearic acid	2	2		
Curing system	CBS	0.8	0.8		
	MBT	0.7	0.7		
	Sulfur	1.5	1.5		

A rubber processing analyzer (RPA 2000) from Alpha Technologies Co., UK was used to evaluate the rheological behavior. Time sweep test was performed under 7% strain, 50 °C and frequency up to 30 Hz. Curing test was done at 160 °C under the strain of 7% and frequency of 2 Hz.

3 Results and Discussion

Figure 1 shows the complex viscosity versus time. The viscosity at the beginning of the graph is shown by η_i and at the end of the graph by η_f. The difference between the mentioned parameters is considered as a viscosity drop and is displayed with Dη*. Also, the inflection point of the graph where the slope change occurs suddenly is shown with t_{in}.

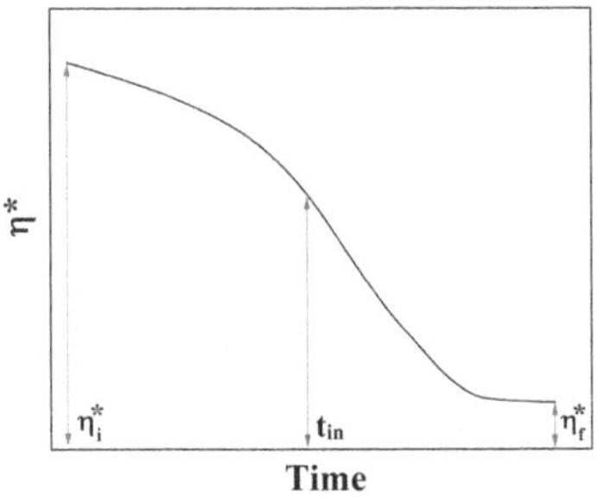

Fig. 1. Shematic of complex viscosity versus time

The change in these parameters with filler content is shown in the Fig. 2. By incorporating nanoparticles, the viscosity drop has increased, which is more tangible in larger amounts of the nanoparticle. This is because of the presence and increase of nanoparticles by increasing the viscosity of the compound. Also, by adding nanoparticles to the polymer matrix, especially for higher values of the nanoparticles, it is transferred to longer periods. This translates the viscosity drop across specific strains to longer times. As the nanoparticles increase, under certain strains due to the physical filler-filler breakdown, the amount of viscosity decreases, especially for compound containing 1.5 phr of nanographene [3].

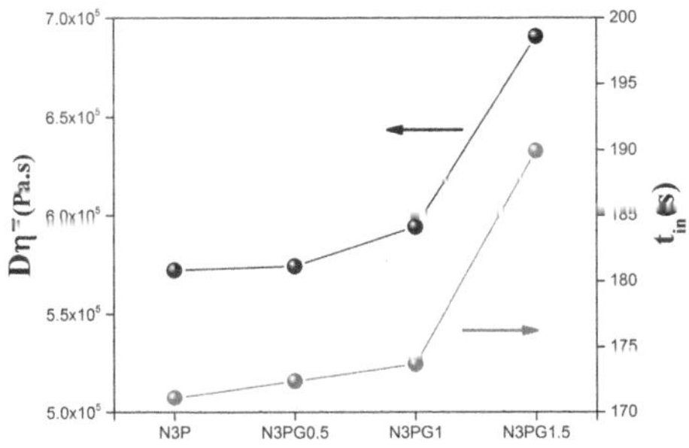

Fig. 2. Changes of viscosity drop and t_{in}

Figure 3 illustrates the curing torque in terms of time. As can be seen, with increasing nanoparticles up to 0.5 phr, the torque has increased and then decreased, which indicates the nanoparticles agglomerated for 1 and 1.5 phr, and it prevents the formation of crosslinks in higher quantities. Adding nanographene especially in 1 and 1.5 phr increases the scorch time. It can be attributed to the plate morphology of nanographene preventing crosslinking. By increasing nanoparticle amount, scorch time increases because nanoparticle morphology prevents the formation of crosslinking points.

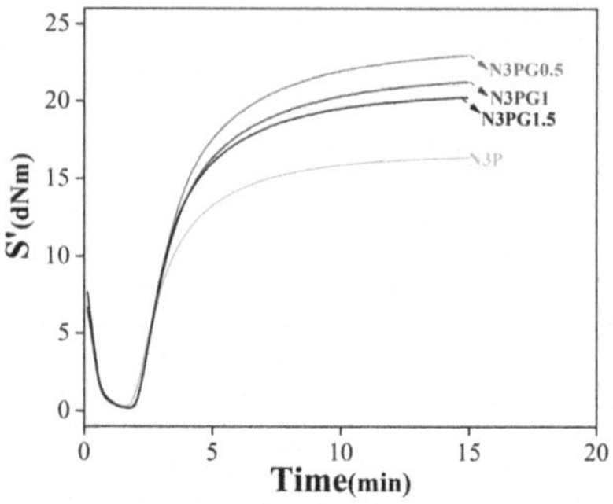

Fig. 3. Curing plots

Regarding to Table 2, by adding nanographene, the curing rate increases because the heat transfer from the RPA to the nanocomposites is improved and the formation of crosslinks occurs more rapidly. The amount of MH–ML first increases significantly after adding nanographene and then becomes constant, which can be attributed to the agglomeration of nanographene in amounts higher than 1 phr [4].

Table 2. Curing characteristics of compounds

Sample	t_{90}	t_{10}	M_H-M_L	K
N_3P	6.64	0.6	16.109	2.18
$N_3PG_{0.5}$	7.59	0.52	22.842	2.60
N_3PG_1	7.75	2.27	21.050	3.11
$N_3PG_{1.5}$	7.03	2.27	20.047	3.40

4 Conclusion

The relationship between curing behavior and rheological properties of NBR/phenolic blend in the presence of nanographene was investigated. It was shown that the viscosity decreased drastically in high loadings of nanographene. The scorch time of the blends and curing rate increased by adding nanographene. The difference between the highest and lowest torque in the curing plot first increased by an increase in nanographene content and then fixed in a constant value.

References

1. Mirabedini, A.S., Karrabi, M., Ghasemi, I.: Iran Polym. J. (Eng. Edit.) **22**(1), 25–32 (2013)
2. Wu, J., et al.: Polym. (Guildf) **54**(13), 3314–3323 (2013)
3. Tang, M., Xing, W., Wu, J., Huang, G., Li, H., Wu, S.: Chin. J. Polym. Sci. **32**(5), 658–666 (2014)
4. Saatchi, M.M., Shojaei, A.: Polym. Int. **61**(4), 664–672 (2012)

Encapsulation of Epoxy Resin Within Nanocapsules via Facile and Cost Efficient Electrospray Method

Alireza Safdari(✉), Saied Nouri Khorasani, and Rasoul Esmaeely Neisiany

Department of Chemical Engineering, Isfahan University of Technology, 84156-83111 Isfahan, Iran
a.safdari@ce.iut.ac.ir

Abstract. Corrosion of metals is an important issue in various industries, causing heavy maintenance costs and other problems which have always attracted researchers' attention to minimization of these expenses. Polymer coatings are used for corrosion prevention; however, these coatings face practical difficulties and lose their functionality after any damages. Therefore, self-healing coatings have been proposed. In this research, a self-healing agent was encapsulated in core-shell nanocapsules via electrospray method. For this purpose, low viscosity epoxy resin was stored in styrene acrylonitrile nanocapsules using emulsion electrospray. The morphology of the prepared capsules was studied by scanning electron microscopy (SEM), confirming the fabrication of capsules with the average size of 870 ± 320 nm. Furthermore, Fourier-transform infrared spectroscopy (FTIR) results confirm the successful encapsulation of epoxy within the nanocapsules without any chemical reaction between core and shell.

Keywords. Self-healing · Encapsulation · Electrospray · Epoxy resin · Styrene acrylonitrile

1 Introduction

Self-healing polymeric materials are one of the advancements of automatic damage repair in this decade and many researches have highlighted their application [1]. Self-healing polymers feature a capability of controlled release of active agents [2]. Active agents are stored in a capsule or containment shell and placed in polymer matrix. The benefit of these capsules is controlled release of active agents [3]. When the coating is damaged or cracked, these agents are released at the damage point and the surface is automatically healed [4].

According to the previous researches the best approach towards self-healing materials is the releasing healing agent which could be stored in microcapsules, nanocapsules, micro channels, or Nano channels [5]. One of the major issues in this issue is the encapsulation process which is significantly expensive and time consuming. Electrospray is one of the methods used for core-shell nanocapsule preparation. However, no dedicated research has been conducted using this method for development

H. Mirzadeh and A. A. Katbab (Eds.): ISPST 2018, *Eco-friendly and Smart Polymer Systems*, pp. 625–628, 2020.
https://doi.org/10.1007/978-3-030-45085-4_151

of self-healing coating. The following research is concentrated on the encapsulation of epoxy resins (as healing agent) via electrospray technique for self-healing purposes.

2 Experimental

Styrene acrylonitrile (SAN) was purchased from Sigma-Aldrich and DMF (N,N-dimethylformamide 99.8%) was provided by DaeJung. EPON 828 as a commercial epoxy resin, viscosity 11–15 Pa · s was purchased from Hexion and mixed with diglycidyl ether of 1.6-hexanediol (ED 180) as the reactive diluent, viscosity 0.015–0.025 Pa.s, which was provided from Inchem Ltd. SAN was firstly solved in a DMF solution containing 4 wt% of SAN acting as the shell. Afterward, the diluted epoxy resin was added to the solution and magnetically stirred for more 2 h.

Finally the solution was sonicated for more 10 min. For electrospray, the solution was fed to 1 mL synergies. The optimum electrospray parameters conditions were determined by trial and error and obtained as 0.3 $mL.h^{-1}$ speed, and 15 cm distance of the collector from the needle tip. The applied voltage was adjusted at 24 kV. The morphology of the prepared capsules was investigated using a scanning electron microscopy (SEM, Philips, XL 30). In addition, FTIR analysis was conducted on nanocapsules to ensure the absence of any reaction between the core and the shell.

3 Results and Discussion

Figure 1 shows the SEM images of the prepared capsules while Fig. 2 displays the capsule diameters distribution. The images were analyzed using Image J. Averaged over 50 measurements; the average capsule diameter was measured to be 870 nm. The standard deviation of capsule diameters was further determined to be 320 nm. Therefore, it can be said that the employed procedure and materials provided the conditions for preparation of capsules in the range of nano scale.

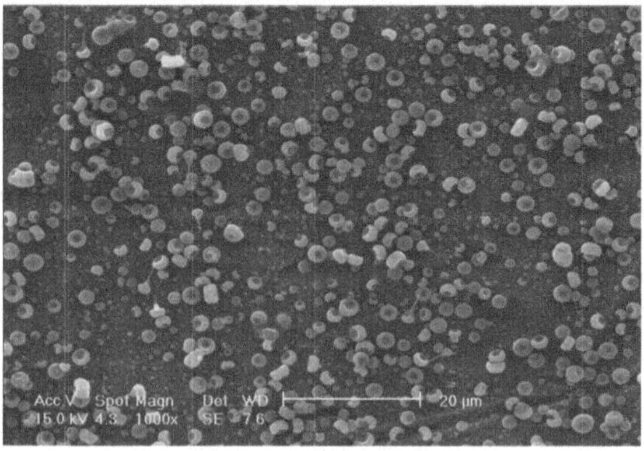

Fig. 1. SEM micrograph of the prepared SAN nanocapsules containing epoxy resin

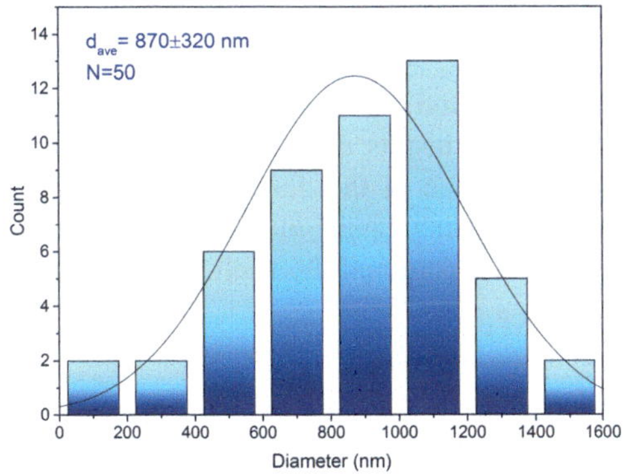

Fig. 2. Diameter distribution of the prepared SAN nanocapsules containing epoxy resin

The chemical structure of the epoxy resin agent containing nanocapsules was investigated by FTIR. Figure 3 shows the FTIR spectra of neat SAN nanocapsules, neat epoxy resin, and crushed core-shell nanocapsules with epoxy core. For neat SAN nanocapsules, an absorbance band is observed at 2245 cm^{-1} attributed to the acrylonitrile nitrile ($-C \equiv N$) stretching vibration, another absorbance band at 1600 cm^{-1} assigned to the styrene ring stretching vibration. For the neat epoxy resin, absorbance peaks at 1250 and 910 cm^{-1} corresponded to the C–O stretching vibration and deformation band of the oxirane group respectively. The crushed core-shell nanocapsules FTIR spectrum confirmed the absorbance peaks of both SAN and epoxy resins, and tellingly does not contain any new absorbance peaks, leading to the conclusion that encapsulation of epoxy resin in SAN nanocapsules was merely a physical phenomenon without any undesirable chemical reaction.

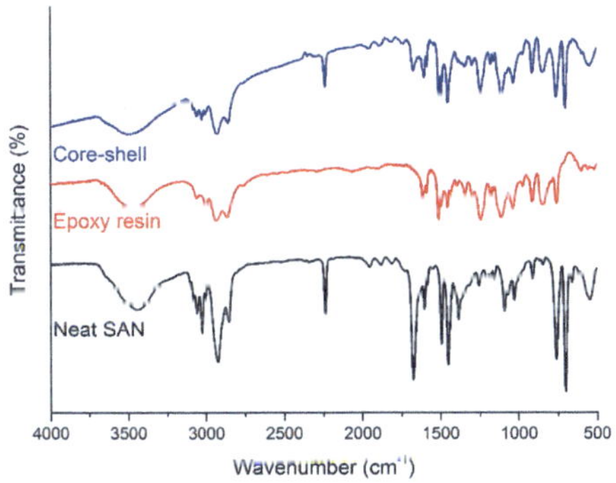

Fig. 3. FTIR spectra of the neat SAN nanocapsules, neat epoxy resin, and core-shell nanocapsules

4 Conclusion

Electrospray is a suitable method for encapsulation of the healing agent inside nanometer-sized capsules. SEM results demonstrated the spherical shape of the prepared capsules and the nano range of the diameter. FTIR results also confirmed the presence of both shell and healing agent in the capsules and also proved that no reaction took place between the core and the shell.

References

1. Neisiany, R.E., Khorasani, S.N., Yoong Lee, J.K., Ramakrishna, S.: RSC Adv. **6**, 70056 (2016)
2. Esmaeely Neisiany, R., Lee, J.K.Y., Nouri Khorasani, S., Bagheri, R., Ramakrishna, S.: J. Ind. Eng. Chem. **59**, 456 (2018)
3. Safaei, F., Khorasani, S.N., Rahnama, H., Neisiany, R.E., Koochaki, M.S.: Prog. Org. Coating **114**, 40 (2018)
4. Ataei, S., Khorasani, S.N., Torkaman, R., Neisiany, R.E., Koochaki, M.S.: Prog. Org. Coating **120**, 160 (2018)
5. Neisiany, R.E., Lee, J.K.Y., Khorasani, S.N., Ramakrishna, S.: Polym. Test **62**, 79 (2017)

Solvathochromism Effect of a New Synthesized Conjugated Organic Compound in Different Environments and Types of Polymers

Fateme Miri[1], Saeideh Gorji Kandi[1(✉)], and Farhad Panahi[2]

[1] Faculty of Polymer Engineering and Color Technology, Amirkabir University of Technology, Tehran, Iran
s.gorji@aut.ac.ir
[2] Mahshahr Campus, Amirkabir University of Technology, Mahshahr, Iran

Abstract. Fluorescence is widely used in optical sensors and cellular imaging to detect cancer types and security issues. The effect of the environment, including the effect of the solvent, the protonation, the polarity of the conjugate fluorescence organic compounds, and the addition of the donor-acceptor substitutions to the derivatives of distyrylbenzene and the polymers combined with conjugated organic compounds, change the behavior of the optical behavior. Distyrylbenzene derivatives have fluorescence properties. Adding the various substitutions of the electron-donor and the electron withdrawing to these derivatives changes the emission spectrum. These compounds have a different color in different polarity solvents and polymers, which can be seen solvathochromism effect. By increasing polarity, the emission spectrum will have a shift toward red colors.

Keywords: Distrylbenzene derivatives · Polarity · Solvathochromism effect · Red shift · Luminescent film

1 Introduction

Conjugated polymers are organic semiconductors, which semiconductivity is due to molecular orbital π-electron delocalized along polymeric chains. Their main advantage over non-polymer organic semiconductors is that their reaction with the polymer creates a strong bond. Fluorescent polymer specimens are found in numerous applications such as multi-color diffusion spectra, barcodes, and photonic crystals, and as a standard in fluorescent-based systems such as cellular storage, cell identification and cell imaging [1].

Most organic fluorescent molecules have high-resonance conjugate systems. For example, aromatic hydrocarbons exhibit intense fluorescence of several circles [2, 3].

McGrier et al. have designed dyes in which the geometric overlap of HOMO and LUMO is minimized. These dyes, cruciforms (XF), consist of two independent but centrally connected molecular axes, which carry electron-donating and electron-withdrawing substituents, respectively. It was done by attaching the essential parts of the groups of auxochromic, which are electron accepting or donating substitutes, and

H. Mirzadeh and A. A. Katbab (Eds.): ISPST 2018, *Eco-friendly and Smart Polymer Systems*, pp. 629–632, 2020.
https://doi.org/10.1007/978-3-030-45085-4_152

regulates the spectra of absorption and emission. The auxochrome groups are directed toward the red shift of the emission spectrum. The deprotonation of symmetric hydroxy fluorophores allows the redshift [4].

Today, there is a great interest in the use of distyrylbenzene (DSB) derivatives in organic devices, such as organic light emitting diodes (OLEDs), due to the presence of several important factors in these compounds. One of these important factors that can be mentioned is that DSB derivatives have high fluorescence [5]. The complex of stilbene and distyrylbenzene fluorophores have potentially significant light intensity coupling systems due to the potential of optoelectronics and it has sensors, which include poly(phenylvinylamine) (PPVs) in organic light emitting diodes.

Distyrylbenzene (DSB) derivatives exhibit a blue emission spectrum in toluene solvent (maximum wavelength = 457 nm). When the solvatochromic shift in the spectra of absorption of these derivatives is observed, the diffusion spectrum depends on the solvent type [2].

The purpose of this paper is to investigate the optical behavior of a conjugated organic compound based on stilbene in environments with different polarity.

2 Experimental

We used cruciform of a new synthesized conjugated organic compound as shown in Fig. 1. The molecules of this compound are a donor-π-acceptor system (SO2Me as an acceptor group) and $(CH_3CH_2)N(CH_2CH_2OH)$ as a donor group in the opposite ends of the chromophore [5].

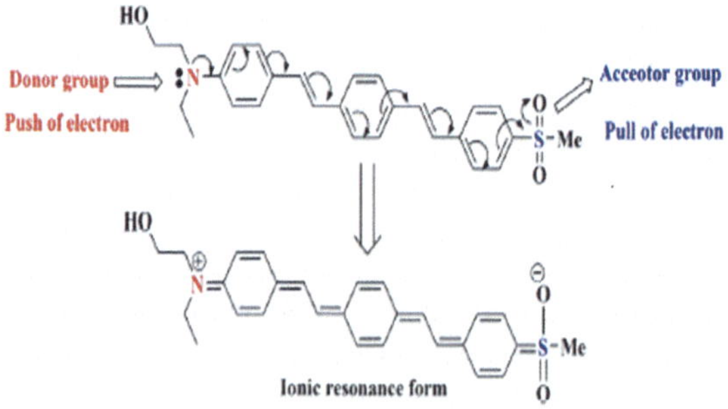

Fig. 1. Push-pull effect of structure [5]

Six different solvents containing toluene (T), dichloromethane (DCM), ethylacetate (EA), tetrahydrofuran (THF), methanol and N,N-dimethylformamide (DMF) purchased from Merck company, Germany, were used to prepare 10^{-5} molar solutions. Then, we placed the solutions under the UV light (λ = 360 nm) and daylight simulator, and took photos from samples by a Nikon camera with high-speed shutter and high ISO under both conditions as shown in Fig. 2.

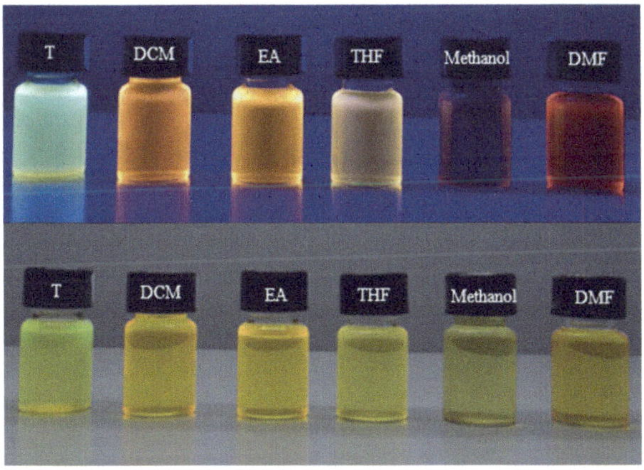

Fig. 2. Photos of solutions with different polarities (polarity increases from left to right), (top) under UV lamp (λ = 360 nm) (bottom) the daylight simulator, by Nikon's camera

Poly(methyl methacrylate) and polyvinyl alcohol were added to these solutions. 20 wt% of the solutions were drop casted to make free films, which then observed under UV lamp.

3 Results and Discussion

As shown in Fig. 2, all specimens in the different solvents under the daylight simulator have different yellow color, but under the UV light, a sensible color change is observed by changing the polarity of the solution. As the polarity increases, a color change is observed from green to red. Solvatochromism is the ability of a chemical substance to change color due to a change in solvent polarity. In these compounds, there is a positive solvatochromism due to the bathochromic shift (red).

The push-pull effect must be expected in this molecule. Also, the HOMO-LUMO energy gap is significantly reduced in this system.

By investigation free films under UV light, red-shift could be observed by changing the polarity of the polymer.

4 Conclusion

Conjugated organic compounds in both daylight and UV are sensitive, but the property of luminescence can be observed under UV condition. The increase in the polarity of solvent and polymers leads to bathochromic shifts and lower energy levels.

References

1. Sonawane, S., Asha, S.: ACS Appl. Mater. Interf. **5**, 1205–1214 (2013)
2. Zucchero, A., Tolosa, J., Tolbert, L., Bunz, U.: Chem. Eur. J. **15**, 13075–13081 (2009)
3. Freudenberg, J., Kumpf, J., Schafer, V., Sauter, E., Worner, S., Brodner, K., Dreuw, A., et al.: J. Org. Chem. **78**, 4949–4959 (2013)
4. McGrier, P.L., Solntsev, K., Miao, Sh., Tolbert, L., Miranda, O., Rotello, V., Bunz, U.: Chem. Eur. J. **14**, 4503–4510 (2008)
5. Sharbati, M.T., Panahi, F., Nekoei, A.R., Emami, F., Nikam, Kh.: J. Photon. Energy **4**, (043599)1–(043599)11 (2014)

MWCNT/PDMS Mixed Matrix Membrane for Oxygen Enriched Air Production

Mohammad Fazel Hasanzade[1], Fereidoon Mohammadi[1(✉)], Seyed Ali Ghadimi[1], and Maryam Jalaei[2]

[1] Petrochemical Department, Iran Polymer and Petrochemical Institute, P.O. Box 14975-112, Tehran, Iran
f.mohammadi@ippi.ac.ir
[2] Processing Department, Iran Polymer and Petrochemical Institute, P.O. Box 14975-112, Tehran, Iran

Abstract. Mixed matrix membranes (MMMs) composed of MWCNT and a rubbery polymer, polydimethylesiloxane, were prepared by priming method. Investigation by FESEM showed that agglomeration of carbon nanoparticles increased with its weight fraction exceeding 0.7 wt%. The O_2/N_2 gas separation performance by using the PDMS/MWCNT membranes has also been carried out. In consequence, the optimized membrane sample contains 0.6 wt% MWCNT with 3.13 as selectivity.

Keywords: Gas separation · Mixed matrix membrane · Oxygen · Nitrogen · Polydimethylsiloxane · MWCNT

1 Introduction

Commercial gas separations include O_2/N_2, CO_2/CH_4, H_2/N_2, He/air, H_2O/air and H_2/CH_4. The separation of oxygen from the air is widely used in the industry and environmental applications, in order to use in medical devices, steel, and chemical manufacturing as well as in the combustion enhancement of natural gas and coal gasification [1]. One of the common methods of oxygen separation is application of polymeric membranes owing to their efficiency and economical reasons. One of the common membranes used for O_2/N_2 separation is PDMS due to its biocompatibility, flexibility, low cost, and ease of fabrication [2, 3]. Carbon nano tube was also used because of its prevention of crossing the gas across the membrane to extend the gas path. In this study, O_2/N_2 selectivity and morphology of the PDMS/MWCNT mixed matrix membranes (MMMs) were evaluated.

2 Experimental

Polydimethylsiloxane (Sylgard 184) was purchased from Dow corning. Carbon nanotubes (nc7000) was obtained from Nanocyl and tetrahydrofuran was purchased from Merck.

H. Mirzadeh and A. A. Katbab (Eds.): ISPST 2018, *Eco-friendly and Smart Polymer Systems*, pp. 633–636, 2020.
https://doi.org/10.1007/978-3-030-45085-4_153

MWCNTs were sonicated in 8 mL THF for 10 min using a horn sonicator (Sonoplus HD 7300, horn diameter 3 mm, Bandelin electronic GmbH, Berlin, Germany). Then 0.1% of known amount of PDMS (0.1% to 1%) was prepared mixing with MWCNT/THF solution due to wetting MWCNT by polymer to prevent reagglomeration of the MWCNT. Briefly, remaining of the PDMS was dissolved in 8 mL THF for 3 h while stirring to ensure complete dissolution of the polymer. Finally, The MWCNT/THF solution was added to the polymer. The prepared solution was put on the stirrer for 2 h and promoted by sonication using a 70 W probe sonicator for 5 min. Then, curing agent in mixing ratio of 10:1 of polymer was added to the ultimate solution.

3 Results and Discussion

Morphology and gas separation performance of the PDMS/MWCNT MMMs were determined using cross-section FESEM and gas permeability analyzer device designed for this study.

To analyze the selectivity of the samples, each prepared membrane sample was placed in the cell of the device and the values for permeation of O_2 and N_2 were recorded by single gas method separately. According to Fig. 1, in terms of selectivity, membranes with 0.6 and 1.0 wt% MWCNT have the best and the worst performance, respectively. The reason for this behavior is likely due to a better dispersion of the carbon nano particles in the PDMS matrix for 0.6 wt% sample. As previously stated presence and dispersion of MWCNT lead to extend the gas path and increased selectivity. In order to investigate morphology of the prepared membranes, their cross-section FESEM pictures were taken.

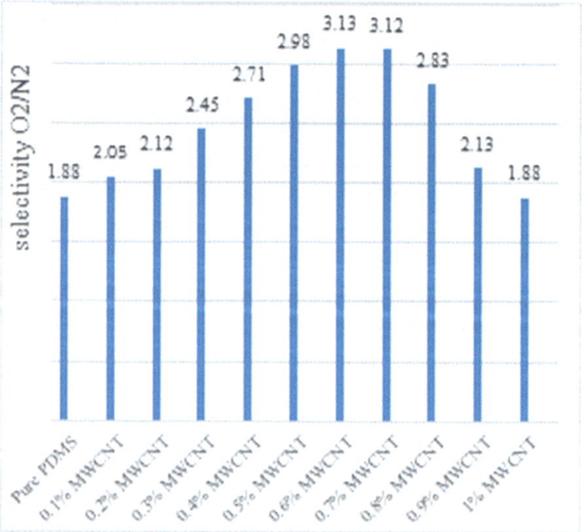

Fig. 1. Selectivity of samples prepared with different weight fractions of MWCNT in PDMS

As seen in Fig. 2a and 2b, MMMs containing 0.6 wt% MWCNT as optimal sample has a good dispersion which is likely due to the priming method which prevents reagglomeration. By increasing the weigh fraction of MWCNT up to 1.0 wt%, agglomeration of carbon nanotubes is observed in Fig. 2c and 2d. Agglomeration of Nano particles can be prevented by using priming method as previously reported [4]. By using this technique no agglomeration was observed up to 0.7 wt% carbon Nano particles in this study. However, after that, some signs of agglomerations were detected.

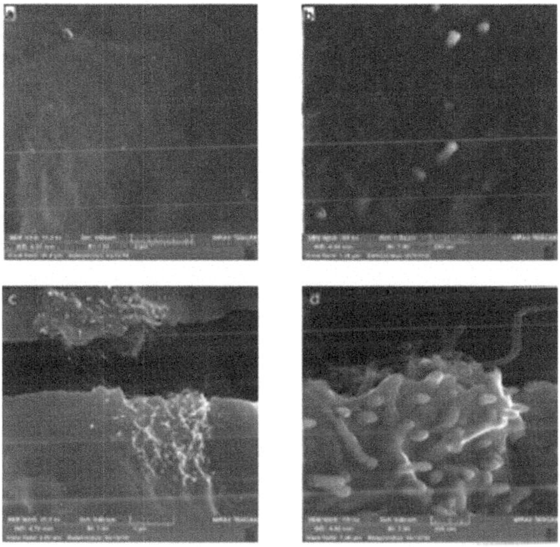

Fig. 2. FESEM images of MWCNT/PDMS MMMs: (a), (b) 0.6 wt% MWCNT, (c), (d) 1 wt% MWCNT

4 Conclusion

The addition of MWCNT to PDMS matrix showed different effects on membrane properties, depending on the Nano CNTs content. Low loading of nano-carbon tubes caused a decrease in free volume fraction in PDMS with increased selectivity. For MWCNT contents more than 0.7 wt%, the increase in membrane average size of the free volume cavities caused an increase in permeability with a decrease in selectivity.

References

1. Burdyny, T., Struchtrup, H.: Energy **35**(5), 1884–1897 (2010). Elsevier Ltd.
2. Zhan, X., Li, J., Huang, J., Chen, C.: Chin. J. Polym. Sci. **27**(4), 533 (2009)

3. De Menezes Atayde, C., Doi, I.: Phys. Status Solidi. Curr. Top Solid State Phys. **7**(2), 189–192 (2010)
4. Aroon, M.A., Ismail, A.F., Matsuura, T., Montazer-Rahmati, M.M.: Sep. Purif. Technol. **75**(3), 229–242 (2010). Elsevier BV

Modeling of Shape Memory Behavior of Semi-crystalline Polymer Considering Both Amorphous and Crystalline Phases

Reza Akbari and Mehrdad Kokabi(✉)

Department of Polymer Engineering, Faculty of Chemical Engineering, Tarbiat Modares University, P.O. Box 14115-114, Tehran, Iran
mehrir@modares.ac.ir

Abstract. Shape memory polymers easily form different temporary shapes and recover their permanent shape even after undergoing large deformation. In this work, shape memory behavior of a typical semi-crystalline polymer is modeled and the constitutive equations for the original amorphous and the semi-crystalline phases are derived. The amorphous and crystalline phases were characterized by an Ogden and a neo-Hookean response, respectively. The model was used to simulate a typical uniaxial cycle of deformation. The results demonstrated very good agreement between simulation and experimental data.

Keywords: Shape memory polymer · Crystalline polymer · Modeling · Thermomechanical behavior

1 Introduction

Shape memory materials represent an important class of smart materials with the ability to return from a temporary shape to the permanent one. Such a property, known as shape memory recovery, is generally induced by an external stimulus such as heat, electricity or magnetism and allows the shape memory materials, use in a wide range of applications [1]. The mechanism responsible for the shape memory of the polymer is the fact that the polymer has a semi-crystalline state. In this case, melting temperature, Tm, is the recovery temperature, T_r, and such polymers are called semi-crystalline shape memory polymers [2]. In their later work, Barot et al. [2] extended the model to fit a full thermodynamic framework and derived constitutive equations that were invariant in their description of mechanical stress and strain. Their model is fully three-dimensional and fulfills general physical requirements such as frame invariance and the second law of thermodynamics [3]. They assumed both amorphous and crystalline phases as a neoHookean solid. In some cases, the model is not in a good agreement with experimental data. In this work, different hyperelastic functions are considered for amorphous and crystalline phases due to different behavior of polymer chains in each phase. The model was used to simulate a typical uniaxial cycle of deformation.

H. Mirzadeh and A. A. Katbab (Eds.): ISPST 2018, *Eco-friendly and Smart Polymer Systems*, pp. 637–640, 2020.
https://doi.org/10.1007/978-3-030-45085-4_154

2 Theoretical

The modeling of semi-crystalline shape memory polymers could delineate into four parts; namely the rubbery phase (at a temperature beyond T_r), the semi-crystalline phase (at a temperature beneath T_r), the crystallization process (on cooling to below T_r) and the melting process (on heating back above T_r). Each of these parts should address separately.

Above the recovery temperature, the semi-crystalline shape memory polymer behaves like an isotropic rubber and models as an incompressible isotropic hyperelastic material. Crystallization is a time-dependent process and takes place in a gradual manner. To model the shape memory behavior of semi-crystalline polymer, beyond the recovery temperature, we have to make constitutive assumptions for this mixed region. The newly formed crystalline phase is assumed to be an elastic solid. For such a case, the equation of total stress for a semi-crystalline shape memory polymer is:

$$\boldsymbol{T} = -pI + (1-\alpha)2\rho \boldsymbol{F}_a \frac{\partial \psi_a}{\partial \boldsymbol{C}_a} \boldsymbol{F}_a^T + \int_{t_1}^{t_2} 2\rho \boldsymbol{F}_c \frac{\partial \psi_c}{\partial \boldsymbol{C}_c} \boldsymbol{F}_c^T \frac{d\alpha}{d\tau} d\tau \tag{1}$$

Where, p is the Lagrange multiplier due to the constraint of incompressibility, α is the amount of crystalline phase, $\mathbf{F}_a$ and $\mathbf{F}_c$ are the deformation gradient in amorphous and crystalline phases, ψ_a and ψ_c are the Helmholtz potential of amorphous and crystalline phases and $\mathbf{C}_a$ and $\mathbf{C}_c$ are the right Cauchy-Green tensor for amorphous and crystalline phases, respectively. The first part of the equation represents the amorphous phase and the integration part indicates the transition of the amorphous phase to the crystalline one, by passing through the T_m. The amorphous and crystalline phases could characterize by an Ogden and a neo-Hookean response, respectively.

In a uniaxial extension, the shape memory polymer stretches beyond the transition temperature (step 1) and cools down while keeping the length fixed (step 2). During the cooling process, crystallization takes place. By completing the crystallization process, the polymer unloads (step 3), then reheats. The crystals melt and the crystalline shape memory polymer reverts back to the original amorphous phase (step 4). According to the Eq. (1), in a uniaxial extension, the stress for each step (1 to 4) in a cycle is as follow (Eqs. (2) to (5)):

$$N_{11}(t) = \frac{2\mu_1}{\alpha_1}\left(\Lambda(t)^{\alpha_1 - 1} - \left(\frac{1}{\sqrt{\Lambda(t)}}\right)^{\alpha_1 + 2}\right) \tag{2}$$

$$N_{11}(t) = 2(1-\alpha)\frac{\mu_1}{\alpha_1}\left(\Lambda(t)^{\alpha_1 - 1} - \left(\frac{1}{\sqrt{\Lambda(t)}}\right)^{\alpha_1 + 2}\right) \tag{3}$$

$$N_{11}(t) = 2(1-\alpha_0)\frac{\mu_1}{\alpha_1}\left(\Lambda(t)^{\alpha_1-1} - \left(\frac{1}{\sqrt{\Lambda(t)}}\right)^{\alpha_1+2}\right) + \mu_c\Lambda(t)L_1 - \mu_c\left(\frac{1}{(\Lambda(t))^2}\right)L_2 + \mu_{c1}(\Lambda(t))^3L_3 - \mu_{c1}\Lambda(t)L_1 \tag{4}$$

$$N_{11}(t) = 0 = 2(1-\alpha(t))\frac{\mu_1}{\alpha_1}\left(\Lambda(t)^{\alpha_1-1} - \left(\frac{1}{\sqrt{\Lambda(t)}}\right)^{\alpha_1+2}\right) + \mu_c\Lambda(t)L_1 - \mu_c\left(\frac{1}{(\Lambda(t))^2}\right)L_2 + \mu_{c1}(\Lambda(t))^3L_3 - \mu_{c1}\Lambda(t)L_1 \tag{5}$$

Where, N_{11} is the nominal stress, Λ is the stretch ratio, α_1 and μ_1 are constants for Ogden model. μ_c and μ_{c1} are the shear modulus of crystalline phase. L_1; L_2 and L_3 are integrals that are defined through:

$$\begin{aligned} L_1 &= \int_{t_s}^{t^*} \frac{1}{(\Lambda(\tau))^2}\frac{d\alpha}{d\tau}d\tau \\ L_2 &= \int_{t_s}^{t^*} \Lambda(\tau)\frac{d\alpha}{d\tau}d\tau \\ L_3 &= \int_{t_s}^{t^*} \frac{1}{(\Lambda(\tau))^4}\frac{d\alpha}{d\tau}d\tau \end{aligned} \tag{6}$$

3 Results and Discussion

Figure 1 shows the stress versus strain for a uniaxial extension, illustrating the thermomechanical behavior of a semi-crystalline shape memory polymer in a full cycle. In this Fig. the experimental data (Barikani et al. [4]) are compared with two models (Barot et al. [2] and our proposed model).

As observed, on loading beyond T_m (step 1) the stress and strain go up. In this step, the experimental data predicted with our model show better result than those obtained by Barot et al. model. In step 2, by cooling to beneath T_m at a fixed strain, the stress goes down as the crystalline phase is born in a stressed state. On unloading (step 3), the stress decreases to zero. On reheating beyond T_m (step 4), the shape memory polymer recovers its original shape. The experimental data are in good agreement with Barot et al. and our models. In steps 2 to 4, the crystalline phase is dominant in the polymer, so, in both models, the crystalline phase has to consider by neo-Hookean model.

Therefore, our proposed model can reasonably predict the shape memory behavior of a typical semi-crystalline polymer.

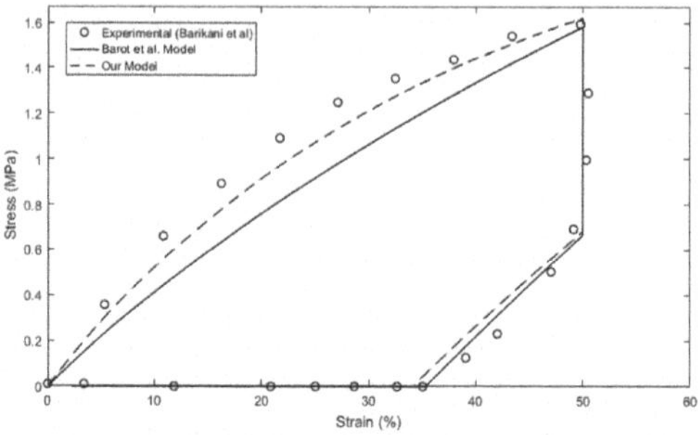

Fig. 1. Plot of nominal stress versus strain for a uniaxial extension. Crystallization process takes place at constant strain. O indicates experimental data (Barikani et al.). Solid line demonstrates Barot et al. model and dash line shows our proposed model

4 Conclusion

In this work, we formulated constitutive equations to predict the thermomechanical behavior of a typical semi-crystalline shape memory polymer. The amorphous and crystalline phases had different behaviors, thus the amorphous phase and crystalline phase were characterized by an Ogden and a neo-Hookean response, respectively. The proposed model exhibited good agreement with experimental data.

References

1. Scalet, G., Auricchio, F., Bonetti, E., et al.: Int. J. Plast. **67**, 127–147 (2015)
2. Barot, G., Rao, I.J., Rajagopal, K.R.: Inter. J. Eng. Sci. **46**, 325–351 (2008)
3. Hu, J., Zhu, Y., Huang, H., et al.: Prog. Polym. Sci. **37**, 1720–1763 (2012)
4. Momtaz, M., Razavi-Nouri, M., Barikani, M.: J. Mater. Sci. **49**, 7575–7584 (2014)

Synthesis and Characterization of Linseed Oil Microcapsules with Poly(Melamine-Urea-Formaldehyde) Shell

Ashkan Karami(✉) and Atiyeh Adelinia

Department of Polymer Engineering and Color Technology,
Amirkabir University of Technology, Tehran, Iran
ashkan.k@aut.ac.ir

Abstract. In this project, it has been attempted to encapsulate linseed oil by in-situ polymerization as the core of the poly (melamine-urea-formaldehyde) which is capable to be used in smart self-healing coatings. Sodium lauryl sulfate had the role of anionic emulsifier and polyvinylpyrrolidone with molecular mass of 40000 g/mol was used as stabilizer with electrostatic and steric hindrance functions. In this work, various factors and conditions were studied on the morphology of the obtained microcapsules by optical microscopy. These factors and conditions include change in stabilizer amount, reaction time, process temperature, mechanical mixer blade type, emulsifier and stabilizer type.

Keywords: Microcapsule · Self healing · Smart coating · Linseed oil · Poly(melamine-urea-formaldehyde)

1 Introduction

Utilizing organic coatings is a common economic way in order to protect substrate. However, these coatings may be damaged due to mechanical or chemical reasons which results in micron-sized cracks [1]. Formation of these micro cracks may occur in brittle polymers and structural composites. Micro cracks in composites may cause fiber/matrix debonding which is hard to recognize [2]. Micro cracks formation in coatings leads to drastic drop in mechanical properties. This problem can be modified by self healing property. There are two main ways to obtain these smart coatings with self healing property. intrinsic self-healing by coating's healing ability and extrinsic self-healing by embedded healing agent in coating. Extrinsic self-healing includes microcapsule usage and vascular self-healing. In vascular method healing agent exposed on cracks by micro channels. Among these methods, embedding encapsulated healing agents in coating is more applicable [1, 3]. In this project, linseed oil microcapsules with poly(melamine-urea-formaldehyde) shell were successfully prepared and the effect of different process parameters and materials on microcapsules properties was observed by optical microscopy.

H. Mirzadeh and A. A. Katbab (Eds.): ISPST 2018, *Eco-friendly and Smart Polymer Systems*, pp. 641–643, 2020.
https://doi.org/10.1007/978-3-030-45085-4_155

2 Experimental

Materials used in this project consist of melamine, urea and formaldehyde solution in order to form the shell. Linseed oil was consumed as core. Polyvinyl pyrrolidone (PVP K-30) and sodium lauryl sulfate (SLS) were used as stabilizer and emulsifier, respectively. For comparing stabilizer and emulsifier type, polyvinyl alcohol (PVA Mn = 50000 g/mol) and sodium dodecyl sulfonate (SDS) were also used.

Two different types of mechanical mixer blades (half-moon and three-blade) were used to observe blade type effect on microcapsule synthesis.

Also, the effect of stabilizer amount, reaction time and process temperature were investigated. Table 1 shows the synthesized samples. Sample no. 1 which is shown in Fig. 1a, was synthesized based on the work of Khalaj et al. [4]. This sample is the reference one and other samples are compared with it.

Table 1. Synthesized samples

No	Emulsifier solution	Stabilizer solution	Blade type	Temp (°C)	Time (h)
1	0.5 wt% SLS	6.66 wt% PVP	Three blade	80	5
2	0.5 wt% SLS	10 wt% PVP	Three blade	80	5
3	0.5 wt% SLS	6.66 wt% PVP	Half moon	80	5
4	0.5 wt% SLS	6.66 wt% PVP	Three blade	80	4
5	0.5 wt% SLS	6.66 wt% PVP	Three blade	80	6
6	0.5 wt% SLS	6.66 wt% PVP	Three blade	60	5
7	0.5 wt% SLS	6.66 wt% PVP	Three blade	100	5
8	0.5 wt% SDS	6.66 wt% PVA	Three blade	80	5

3 Results and Discussion

After visual observation of oil in water emulsion stability via 6.66, 10, 13.33, and 16.66 wt% of PVP, 10 wt% of it was selected as optimum amount of stabilizer. Hence, synthesized microcapsules by 6.66 and 10 wt% were compared. As it is shown in Fig. 1b, smaller spheres were obtained by optimum amount of PVP which can be related to more stability of oil drops in water during polymerization and shell formation. In addition, mechanical mixer blade plays a significant role in microcapsule formation. Undesired product by using half-moon blade can be explained by its low mass transfer and low shear rate.

Lower reaction time (less than 5 h) led to rough surface which indicated insufficient time for shell formation. Although, more reaction times resulted in network formation of microcapsules.

Furthermore, optical microscopy images of synthesized microcapsules in lower temperature (less than 80 °C) showed rupture in shell which can be correlated to low mechanical strength as a result of inappropriate polymerization. However, no

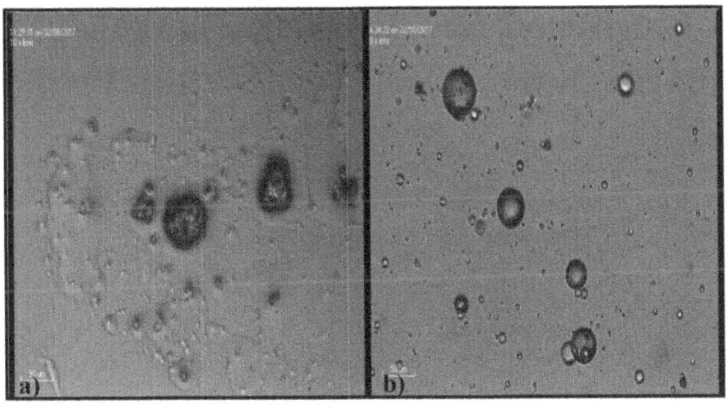

Fig. 1. Optical microscopy images of synthesized microcapsules (a) reference sample and (b) second sample

microcapsule was obtained in higher temperature that can be related to linseed oil's smoke point. At last, SDS and PVA did not perform like SLS and PVP which led to undesired product.

4 Conclusion

Linseed oil was successfully encapsulated by poly(melamine-urea-formaldehyde) utilizing in-situ polymerization. These microcapsules can be used in smart self-healing coatings. The effect of process parameters and materials on microcapsules properties was also characterized.

Results of optical microscopy images indicated that 0.5 wt% SLS and 10 wt% PVP were optimum amounts and types of emulsifier and stabilizer, respectively. Also, polymerization at 80 °C for 5 h by three-blade mechanical mixer was optimum condition in this study.

References

1. Thanawala, K., Khanna, A.S., Raman, R.K.S.: NACE Int. **7310**, 1–9 (2016)
2. Kessler, M.R., Sottos, N.R., White, S.R.: Compos. Part A Appl. Sci. Manuf. **8**, 743–753 (2003)
3. Song, Y.-K., Chung, C.-M.: Polym. Chem. **18**, 40–49 (2013)
4. Khalaj Asadi, A., Ebrahimi, M., Mohseni, M.: Pigment Resin. Technol. **4**, 1–76 (2016)

Bio-Based Self-healing Thermoset Polyurethane Based on Pseudo-Tannin

Adel Soleimani, Mobina Khakbaz, Negar Tavakoli, and Gity Mirmohamad Sadeghi(✉)

Department of Polymer Engineering and Color Technology, Amirkabir University of Technology, P.O. Box 15875-4413, Tehran, Iran
gsadeghi@aut.ac.ir

Abstract. In this research, we have produced a thermoset polyurethane that has self-healing properties above 110 °C. The reaction of hydroxyls bonded to the benzene ring, with isocyanate, is capable of being isolated from isocyanate at high temperature and rejoin at low temperatures. We used bio-based pseudo-tannin material that comes from walnut trees and gives us the ability to self-healing. This mechanism led us to reach about 70% of the initial strength and 80% of the initial elongation at break after healing.

Keywords: Self-healing · Polyurethane · Thermoset · Pseudo-tannin · Bio-based polyurethane

1 Introduction

So far, several methods for the synthesis of self-healing polymers have been used, including extrinsic and intrinsic methods, but the first method has been less considered because the number of healing times can be initiated and it is dependent on the number of capsules placed in the material and fails to repair macroscopic damage. Due to these weaknesses, intrinsic self-healing, such as cross-linking, was further taken into account, noncovalent bonding were much weak for applications. For this reason, reversible covalent bonds have been receiving increasing attention.

All of these reactions need special groups, rigorous operation condition or restricted to a small damage zone and therefore is limited in its range of applications. In this study, we synthesis thermoset polyurethane based on polyester diol and pseudo-tannin with better mechanical properties and need lower temperature for self-healing from ether-based polyurethane [1–4].

2 Experimental

We synthesized as follows: polyester diol and TDI (toluene diisocianate) and propyl 3, 4, 5 trihydroxybenzoate were reacted with each other in ratio 1/3/2 equivalent weight after that the temperature of the mixture was raised to 100 °C, the bubbles were eliminated under vacuum conditions (0.030 MPa), the mixture was placed at 120 °C for 10 h, and cooled for 2 h.

H. Mirzadeh and A. A. Katbab (Eds.): ISPST 2018, *Eco-friendly and Smart Polymer Systems*, pp. 644–647, 2020.
https://doi.org/10.1007/978-3-030-45085-4_156

We used FTIR (Fourier transform infrared spectroscopy analysis) and ATR-FTIR for characterization of prepolymer and polyurethane.

Also, dynamic mechanical properties of the samples were investigated by DMA (dynamic mechanical analysis). The tensile properties of the original and self-healed samples were tested to measure the healing ability of the sample. And the dissolution experiments verified that the urethane bonds were decomposed (Fig. 1).

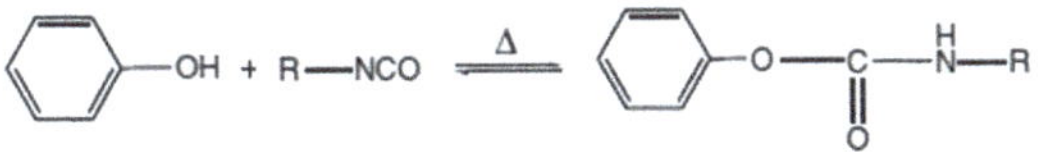

Fig. 1. Reversible reaction

3 Results and Discussion

The Fourier transformed infrared spectra of TDI, polyester diol, NCO terminated polyurethane prepolymer, and polyurethane cross-linked with tannin are shown in Fig. 2. For tannin, the characteristic peak at 3500 cm^{-1} is the transmittance of the phenolic hydroxyls. And for prepolymer the peak at 2270 cm^{-1} points out the iso-cyanate vibrations. The FTIR analysis showed that in PU sample the disappearance of the NCO peak at 2270 cm^{-1} and the appearance of N-H peak at 3340 cm^{-1} imply the synthesis of polyurethane.

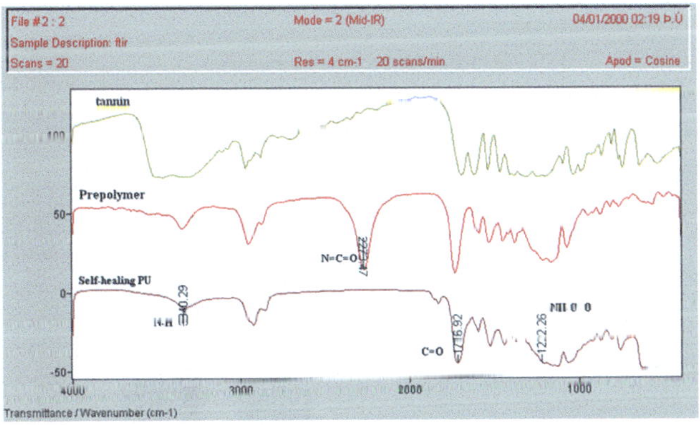

Fig. 2. Infrared spectroscopy (transmittance) of the prepolymer, pseudo-tannin and polyurethane

To verify the decomposition of urethane bonds in polyurethane, the dissolution experiments were employed. Swelling ratio of the sample was significantly increased with temperature which demonstrated decomposition of the urethane bonds (Fig. 5).

DMA test verified that the decomposition of the urethane bonds occurs at 110 °C, and its glass transition temperature (T_g) is 16.9 °C. The urethane bonds disconnected at approximately 110 °C and storage modulus decreased (Fig. 3).

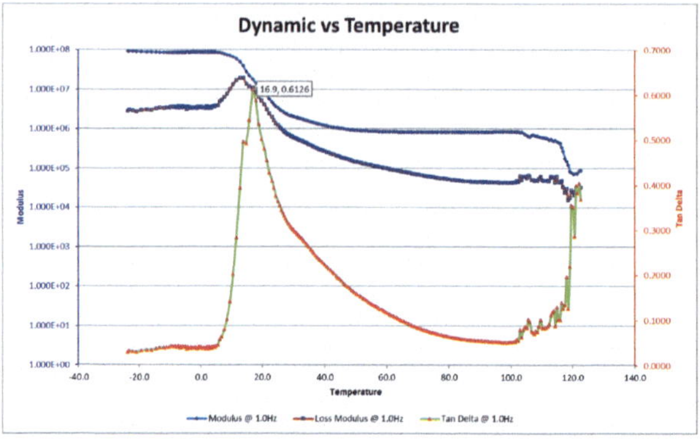

Fig. 3. Storage/loss modulus and tanδ profiles of the PU sample measured using DMA

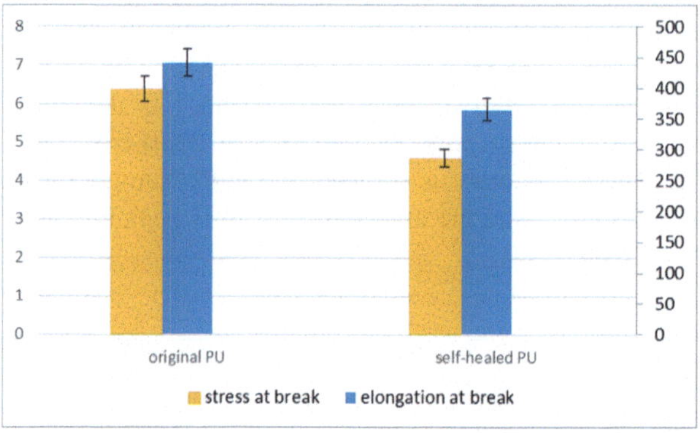

Fig. 4. Tensile strength and elongation at break of the original and self-healed samples

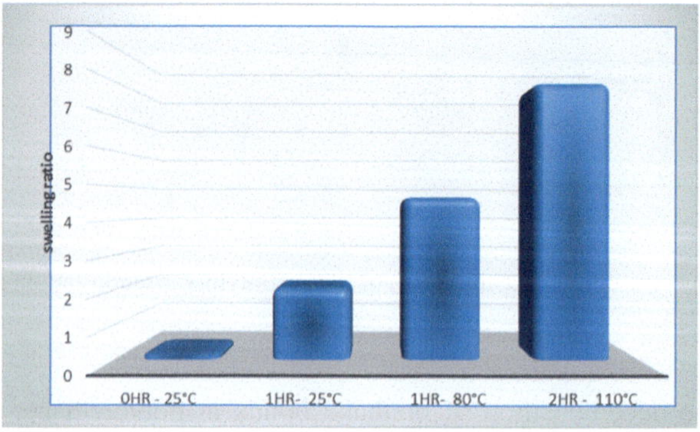

Fig. 5. Swelling of the polyurethane sample in DMF

The tensile properties (including tensile strength and elongation at break) of the original and self-healed samples were tested. The sample was cut with scissors and the surfaces were placed in contact and heated to 110 °C for 10 h in oven for healing (Fig. 4).

4 Conclusion

We produced thermoset polyurethane that could repair itself at 110 °C, maintaining a value of 70% of its initial strength and 80% of elongation at break.

References

1. Zhu, D.Y., Rong, M.Z., Zhang, M.Q.: Prog. Polym. Sci. **49**, 175–220 (2015)
2. Zhong, N., Post, W.: Compos. Part A Appl. Sci. Manuf. **69**, 226–239 (2015)
3. Lv, Z., Chen, H.: Comput. Mater. Sci. **68**, 81–89 (2013)
4. Cao, S., et al.: Polym. J. **49**(11), 775 (2017)

Effect of Chain Extender on the Properties of Polysulfone-Based Membrane as a Candidate for Fuel Cell Proton Exchange Membrane

Maryam Mohammadi, Maryam Oroujzadeh, and Shahram Mehdipour-Ataei(✉)

Iran Polymer and Petrochemical Institute, P.O. Box 14975-112, Tehran, Iran
s.mehdipour@ippi.ac.ir

Abstract. Proton exchange membrane is a key component in fuel cell system with major impact on the performance. So tuning polymer structure to obtain the best properties and performance has been attracted many attentions. In this research two types of sulfonated polysulfone (sPSU) based on bisphenol AF, bisfluorophenyl sulfone (BFPS), and sulfonated bisfluorophenyl sulfone (sBFPS) in the presence or absence of decafluorobiphenyl (DFBP) as chain extender with 30 wt% degree of sulfonation (DS) were synthesized. FTIR and ^{1}H NMR analyses, DSC, TGA, and tensile tests confirmed successful synthesis and acceptable properties of sulfonated monomer and polymers. Suitable membranes were obtained from the prepared polymers. Water uptake and IEC measurements, MEA preparation, and performance of membranes were evaluated. Besides these, presence or absence of DFBP in two synthesized structures was compared.

Keywords: Polysulfone · Synthesis · Chain extender · Fuel cell

1 Introduction

PEMFCs have recently received significant attention as a green alternative power source to replace conventional combustion engines that cause problems in the form of air pollution to global environmental damage. Among many parts of a PEMFC system, the polymer electrolyte membrane plays an important role in the transportation of protons determining fuel cell performance. Many researchers have tried to replace the traditional perfluorosulfonic acid (PFSA) membrane, such as Nafion®, Flemion®, and Aquivion® with hydrocarbon membranes because of the high cost, high gas leakage, and low thermal stability of the former. Among the reported polymers, sulfonated poly (arylene ether)sulfone (SPAES) is one of the most studied materials due to their simple synthetic routes and versatile polymer structures. Several studies have shown a large difference in physical and chemical properties of SPAES polymers with a slight alternation in monomer structure [1] and some researches on effect of monomer type in polymer structure, and fluorinated monomer on the properties of synthesized polymers have been done [2–4]. In this study, two different types of polymer based on sPSU comprising BFPS, sBFPS and bisphenol AF, with and without DFBP as a highly

H. Mirzadeh and A. A. Katbab (Eds.): ISPST 2018, *Eco-friendly and Smart Polymer Systems*, pp. 648–651, 2020.
https://doi.org/10.1007/978-3-030-45085-4_157

fluorinated monomer, were synthesized. Subsequently, effect of DFBP on some properties of polymers for application in PEMFC was studied.

2 Experimental

2.1 Chemicals

BFPS, sBFPS (sulfonated BFPS) and bisphenol AF as monomers, DFBP as chain extender, K_2CO_3 as weak base in polycondensation reaction, anhydrous dimethyl acetamide (DMAc) and toluene as solvent and azeotroping agent respectively, and oleum 30% as sulfonating agent were supplied by Merck company. All other materials and solvents were analytical grade.

2.2 Analysis and Tests

FTIR and ^{1}H NMR spectrometry, TGA (heating from room temperature to 700 °C with heating rate of 10 °C/min), DSC (heat/cool/heat pattern initiated from room temperature with a rate of 5 °C/min, N_2), tensile test (according to ASTM D-882, extension rate: 5 mm/min), IEC (calculated using Eq. (1)) and water uptake measurements according to standard procedures were performed. To acidify membranes, H_2SO_4 solution (4 M) for 48 h was used, and then washed with DI water several times.

$$\mathrm{IEC} = \frac{V_{\mathrm{NaOH}} \times M_{\mathrm{NaOH}}}{\text{weight of dry sample}} \tag{1}$$

3 Results and Discussion

3.1 Synthesis and Characterization of Sulfonated Monomer and Random Copolymers

Synthesis of sulfonated BFPS (sBFPS) was done via electrophilic reaction in the presence of oleum. ^{1}H NMR and FTIR spectra confirmed successful synthesis and complete purification of monomer [5].

To synthesize two structurally different polymers two related methods based on polycondensation reaction were used. The first: typical procedure for sPSU based on reaction of BFPS, sBFPS, and Bisphenol AF according to Scheme 1. The second: sPSU based on reaction of BFPS, sBFPS, bisphenol AF, and DFBP (highly fluorinated monomer) in order to studying DFBP presence as chain extender on properties, according to synthesis route demonstrated in Scheme 2. In both routes, sBFPS contents were tuned to obtain polymers with 35 wt% degree of sulfonation.

FTIR spectra of samples showed sPSU related bands and confirmed formation of sPSU [5]. As demonstrated in synthetic routes, effect of reactive DFBP on decreasing time and temperature of polymerization reaction was obvious.

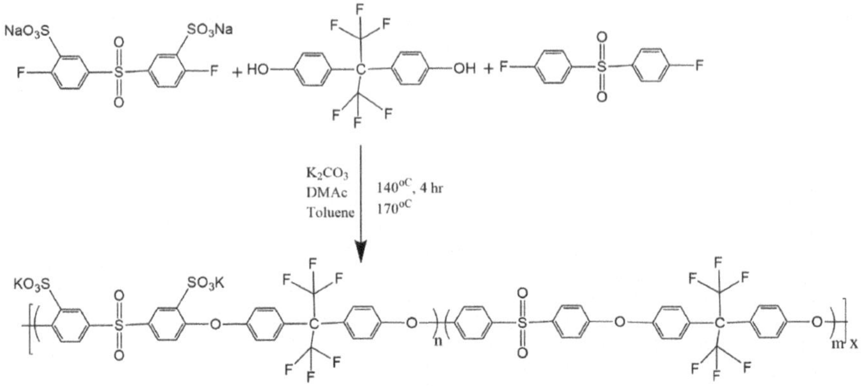

Scheme 1. Synthesis route for typical sPSU

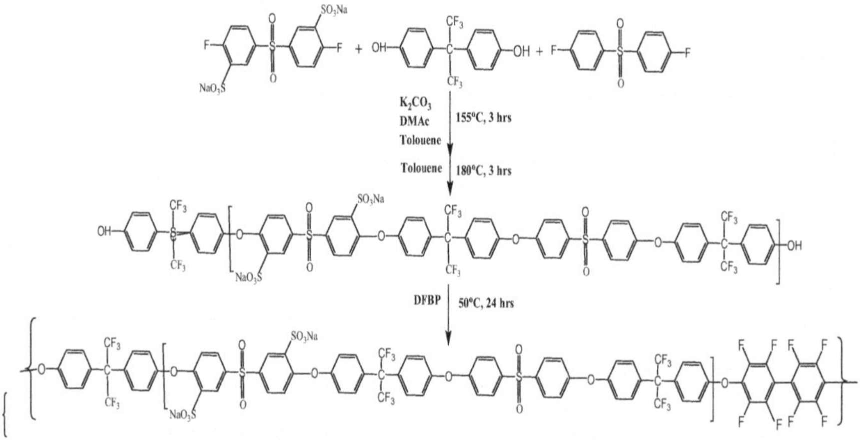

Scheme 2. Synthesis route for chain-extended sPSU

3.2 Membrane Preparation

Membranes were prepared using solution casting of 15 w/v% of polymers solution in DMAc. Prepared membranes as seen in Fig. 1 were transparent and flexible with a thickness of 40-50 μm. As shown in Fig. 1 and also in mechanical properties results (Table 1), chain-extended sPSU had better film formation ability due to better mechanical properties comprising from DFBP presence.

For assessment of membranes the water absorption, weight change of polymers after 24 h at 25 °C, was calculated (Table 1). For long-term stability in aqueous media, weight change of samples for one month was followed and no more change observed. To measure IEC, certain amounts of acidified membranes were placed into a NaCl (2 M) solution for 24 h at 25 °C, in the following titrated by NaOH (4 mM) solution and finally according to Eq. (1), IEC values were calculated (Table 1). Both type of polymers exhibited acceptable properties. By comparing two polymers it can be said that with regard to the same DS, the values of IECs were approximately equal (the small difference was related to WU different values related to hydrophobic F atoms). T_g of the chain-extended sPSU was higher than that of typical sPSU as well as mechanical properties, comprising from F atoms presence in the chain-extended sPSU.

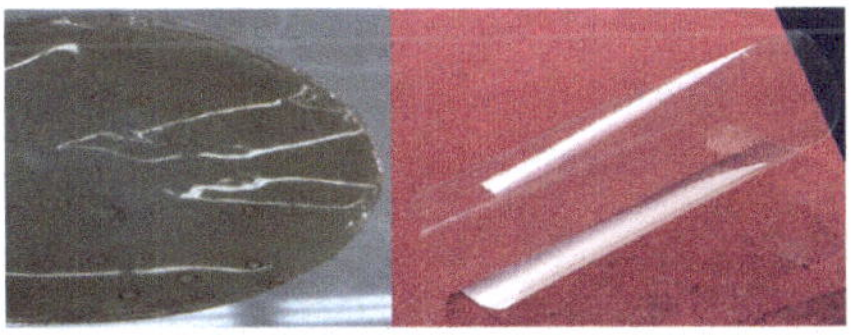

Fig. 1. Left (typical and right) chain-extended sPSU membrane

Table 1. Measured properties of polymers

Polymer	DS (%)	IEC (meq/g)	WU (%)	T_g (°C)	Tensile strength (MPa)	Elongation (%)
Chain-extended sPSU	35	0.9	9.68	272.6	54.5	5.98
Typical sPSU	35	1.05	14.5	231	21.83	4.67

4 Conclusion

Two types of sPSU with and without chain-extender were successfully synthesized and characterized. Effect of DFBF on polymerization reaction procedure and properties of polymers was examined. In brief, it was concluded that the prepared membranes from DFBP containing polymer was better choice for further study as a candidate for PEM in fuel cell application.

References

1. Lee, K.H., Cho, D.H., Kim, Y.M., Moon, S.J., Kim, J.F., Lee, Y.M.: J. Membr. Sci. **535**, 35–44 (2017)
2. Gindt, B.P., Tang, Z., Watkins, D.L., Abebe, D.G., Seo, S., Tuli, S., Fujiwara, T.: J. Membr. Sci. **532**, 58–67 (2017)
3. Ghassemi, H., McGrath, J.E., Zawodzinski Jr., T.A.: Polymer **47**(11), 4132–4139 (2006)
4. Mack, F., Gogel, V., Jörissen, L., Kerres, J.: J. Power Sour. **255**, 223–229 (2014)
5. Oroujzadeh, M., Mehdipour-Ataei, S., Esfandeh, M.: Int. J. Polym. Mater. Polym. Biomater. **64**(6), 279–286 (2015)

A Novel Polysulfone-Based Proton Exchange Membrane: From Polymer Synthesis to Membrane Preparation, MEA Fabrication, and Fuel Cell Performance

Maryam Mohammadi, Maryam Oroujzadeh, and Shahram Mehdipour-Ataei(✉)

Iran Polymer and Petrochemical Institute, P.O. Box 14975-112, Tehran, Iran
s.mehdipour@ippi.ac.ir

Abstract. Polymer electrolyte membrane is one of the key components in fuel cell systems. Finding a competitive alternative to commercial membranes to decrease cost and improve their properties is one the most interesting subjects in fuel cell field research. In this study, a novel category of proton exchange membrane (PEM) was prepared from a thermally stable chain-extended sulfonated polysulfone (sPSU) with 45 wt% degree of sulfonation. After membrane preparation, optimizing ink formulation and membrane electrode assembly (MEA) condition, performance assessment in a fuel cell system was done successfully. Based on the results in the typical test condition (T = 80 °C, RH = 100%) it was revealed that this type of synthetic membrane could be an acceptable alternative for commercially available PEMs.

Keywords: PEM · Polysulfone · Synthesis · MEA · Fuel cell

1 Introduction

A fuel cell is an electrochemical device that directly converts chemical energy to electrical energy. Typical reactants used in a fuel cell are hydrogen or hydrogen rich gas on the anode and oxygen or air on the cathode. Fuel cells are most commonly classified by the type of electrolyte used in the cells: Polymer Electrolyte Membrane Fuel Cell (PEMFC), Alkaline, Phosphoric Acid, Molten Carbonate, and Solid Oxide Fuel Cell. Generally, the choice of electrolyte determines operating condition and also cell performance [1, 2]. Recently proton exchange membranes are under enormous consideration especially in PEMFCs. Most of today's commercial PEMFCs are based on perflourosulfonic acid (PFSA) membranes. PFSA polymers show some disadvantages in fuel cell environment such as limited operating temperature to 80 °C, fuel crossover in methanol fuel cells, and high cost. To overcome the obstacles, alternative hydrocarbon polymers have been investigated [2, 3]. In this study we synthesized chain-extended sPSU-based random copolymer. Subsequently membrane formation, MEA fabrication, and finally single cell performance were examined in order to evaluate its performance as an alternative candidate for commercially available PEMs.

H. Mirzadeh and A. A. Katbab (Eds.): ISPST 2018, *Eco-friendly and Smart Polymer Systems*, pp. 652–655, 2020.
https://doi.org/10.1007/978-3-030-45085-4_158

2 Experimental

2.1 Chemicals

Preparation of monomer and polymer and also all required analyses and tests to evaluate suitability of the synthesized polymer for MEA fabrication and fuel cell test performance were based on our published results [4].

2.2 Membrane Preparation, MEA Fabrication and Cell Performance

Membranes were prepared using solution casting of 20 w/v% of polymers solution in DMAc (Fig. 1). About 0.1 g of Pt/C (20 wt%), 1.332 g of Nafion® solution (5 wt%), 7.568 g of isopropanol, and 1.0 g of DI water were mixed together, and the mixture was sonicated for 2 h to prepare a catalyst slurry [5]. The catalyst slurry was coated onto a gas diffusion layer (GDL) by painting a few times. The prepared GDE (GDL containing catalyst layer (CL)) dried in vacuum oven for 30 min. To fabricate MEA, the prepared GDEs were assembled on both side of membrane by hot-pressing at 150 °C. The active area of the MEA for this process was 2.5 cm × 2.5 cm. The Pt loading of both anode and cathode was 0.4 mg/cm^2 (Fig. 2). The flow rates of hydrogen and oxygen at the both anode and cathode sides were 250 mL/min. The cell test was conducted at 80 °C with 100% RH.

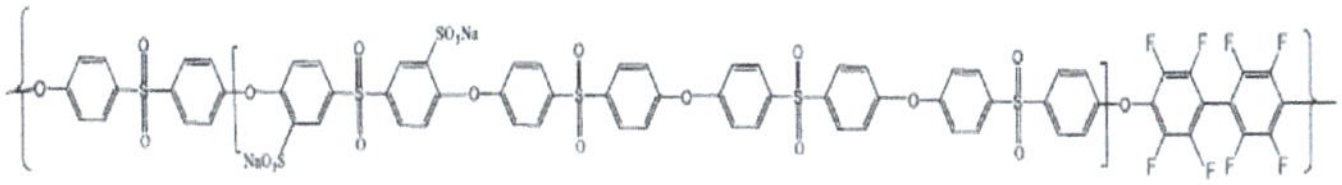

Scheme 1. Structure of sPSU random copolymer

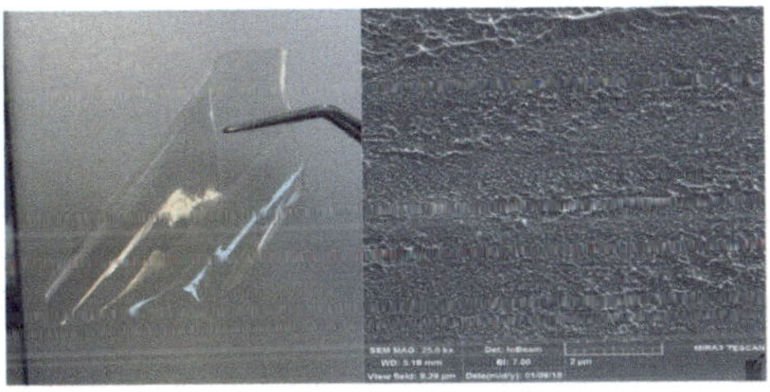

Fig. 1. sPSU solution-cast dense membrane (left) and, its cross-sectional SEM image (right)

3 Results and Discussion

3.1 Polymer Synthesis and Membrane Preparation

Synthesis of sulfonated BFPS (sBFPS) and sPSU-based random copolymer with 45 wt % degree of sulfonation were done according to our explained procedures [4]. Structure of the synthesized copolymer is demonstrated in Scheme 1.

Fig. 2. Fabricated MEA

Prepared membranes were transparent and flexible with a thickness of 40–50 μm. Formation of dense membrane with no defect which is essential for performance evaluation in fuel cell to prevent gas crossover, was confirmed by SEM image (Fig. 1).

3.2 Cell Performance

The polarization curve of chain-extended sPSU (IEC = 1.2) membrane applied to a single cell PEMFC performance test is shown in Fig. 3. OCV was in the acceptable range of about 0.975 V. Maximum power density of 333 mW/cm^2 was attained at current density of about 1000 mA/cm^2. The result of the test was well comparable with polymers based on naphthalene diol isomeries with IEC = 2.2 and Nafion 212 at the same condition [6].

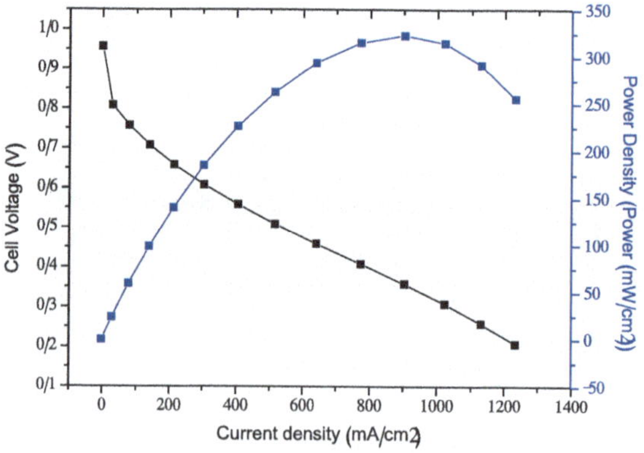

Fig. 3. Polarization curve of fuel cell test

4 Conclusion

In this research, preparation of a dense PEM membrane with no defect from a synthesized chain-extended sPSU-based polymer was done successfully. Ink formulation and MEA preparation condition were optimized for performance evaluation in a fuel cell test station. The obtained maximum power density was comparable with the reported ones. So this type of polymer could be a good candidate as fuel cell PEM.

References

1. Chee, K.P.: Fabrication of Membrane-Electrode-Assembly for Polymer Electrolyte Membrane Fuel Cell, Diss (2007)
2. Hamada, T., Fukasawa, H., Hasegawa, S., Miyashita, A., Maekawa, Y.: Int. J. Hydrogen Ener. **41**(41), 18621–18630 (2016)
3. Oroujzadeh, M., Mehdipour-Ataei, S.: Int. J. Polym. Mater. Polym. Biomater. **65**(7), 330–336 (2016)
4. Mehdipour-Ataei, S., Mohammadi, M., Oroujzadeh, M.: In: 12th International Energy Conference (2018)
5. Nguyen, M.D.T., Yang, S., Kim, D.: J. Power Sour. **328**, 355–363 (2016)
6. Lee, K.H., Cho, D.H., Kim, Y.M., Moon, S.J., Kim, J.F., Lee, Y.M.: J. Membr. Sci. **535**, 35–44 (2017)

Microencapsulation of 2-Ethylhexyl Acrylate with Potential Application in Thread Locking Adhesives

Elahe Adibzade, Seyed Mojtaba Mirabedini(✉), and Morteza Behzadnasab

Iran Polymer and Petrochemical Institute, P.O. Box 14975-112, Tehran, Iran
sm.mirabedini@ippi.ac.ir

Abstract. The main objective of this research was to synthesize polyurethane/polyuria-formaldehyde (PU/UF) microcapsules containing 2-ethylhexyl acrylate (EHA). Bi-layer microcapsules containing EHA monomers were synthesized via interfacial polymerization in an oil/water emulsion. The surface morphology and shape of synthesized microcapsules were evaluated using optical microscopy (OM) and scanning electron microscopy (SEM). FTIR spectroscopy was carried out for characterization of the microcapsule. The results showed that the prepared microcapsules were spherical, with an average diameter about 20-140 and thickness of about 1.5 μm, without any inter-capsule bonding and had a smooth surface. The microcapsules synthesized in this way offer an interesting prospect for applications in thread locking adhesives .

Keywords: Microencapsulation · Smart coating · Thread locking · Polyurethane polyurea formaldehyde

1 Introduction

Microencapsulation is a method for covering active compounds by a thin wall material to protect them from the surrounding environment and unwanted reaction [1]. Recently microencapsulation process has been received intensive attention in the field of two-stage adhesive materials. Application of thread locking adhesives (TLA) is one of the promising methods to improve the mechanical strength of metallic joints, especially during the vibration or movement. TLA not only prevents the metal joints from the loosening, but also by filling the gap between screw and nut (metallic joints), prevents the possible penetration of oxygen and moisture into the gap and delays the occurrence of corrosion. TLA usually contains two types of microcapsules filled with active thermosetting resins and hardener, respectively, which are well-dispersed in a thermoplastic media. In an adhesive system, separating the resin and its hardener by microencapsulation increases the storage time of these two compounds in the same media [2].

Brown et al. [3] in 2012 reported microencapsulation of dicyclopentane using polyurea-formaldehyde (PUF) shell. Barbara et al. [4] in 2013 synthesized bi-layer polyurethane-polyurea formaldehyde (PU/UF) microcapsules containing isophorone

H. Mirzadeh and A. A. Katbab (Eds.): ISPST 2018, *Eco-friendly and Smart Polymer Systems*, pp. 656–659, 2020.
https://doi.org/10.1007/978-3-030-45085-4_159

diisocyanate (IPDI). In all published reports, prepared microcapsules have been used to provide self-healing properties in coatings and composites. In the current study, microencapsulation of 2-ethylhexyl acrylate using polyurethane-polyurea formaldehyde (PU/UF) was evaluated via interfacial polymerization followed by in situ PUF microencapsulation.

2 Experimental

Commercially grade of 2-ethylhexyl acrylate (EHA) was used as a core material. Shell components consisted of urea formaldehyde and prepolymer Desmodur (AK-75, obtained from AEK yung Chemical Co.). Ammonium chloride, resorcinol, and gum Arabic (GA) were obtained from Merck industrial group.

2.1 Preparation of PU/UF microcapsules

Dual-layer microcapsules containing EHA monomers were synthesized via interfacial polymerization oil in water emulsion. PU/PUF microcapsules were synthesized via interfacial polymerization route using Desmodus (AK-75) followed by in situ PUF microencapsulation. 13.5 g GA as a surfactant was dissolved in 120 mL of deionized water under mechanical agitation at ambient temperature for 1 h. In parallel, 2.50 g urea, 0.25 g ammonium chloride, and 0.25 g resorcinol were added to an aqueous solution under mechanical agitation, and the pH was adjusted at 3.50 using a few droplets of diluted HCl. 4 g of Desmodus (AK-75) was added to 20 mL EHA and well-mixed together. The mixture was then dropwise added to the GA solution. After 10 min, a defined amount of aqueous solution of formaldehyde was added to the dispersion, and the temperature was increased to 55 °C. The reaction proceeded under continuous stirring at mixing speed of 1000 rpm for 3 h, the microcapsules were rinsed with deionized water and left to dry for 24 h.

Optical microscopy Zeiss Stemi 305, scanning electron microscopy Vegall Tescan and FTIR techniques Bruker Equinox 55 was used to characterize synthesized microcapsules.

3 Results and Discussion

FTIR spectra of microcapsules and core material are shown in Fig. 1. In the FTIR spectrum of the microcapsule, the absorption peaks at 1647 and 3319 cm^{-1} are attributed to C = C stretching vibration and N-H bending vibration, respectively. The existence of stretching bond of C-N at the region of 1200 cm^{-1} confirms the presence of urea formaldehyde in the wall compound. Absorption peaks at the region of 2923, 1720, and 1190 cm^{-1} which are attributed to the C-H stretch, C = O, and C-O-C respectively, are disappeared in both FTIR spectra of the microcapsule and core material. Figure 2 shows OM and SEM micrographs of synthesized microcapsules. Figure 2a reveals that microcapsules are spherical and uniform. The diameter of the microcapsules was determined about 20–140 μm by analyzing about 20 OM images.

SEM micrographs (Figs. 2b to 2d) of microcapsules show that the resultant microcapsules are spherical, have a smooth surface and without inter-capsule bonding. As shown in Fig. 2d, the microcapsules shell has a non-porous and compact structure, with a shell thickness of about 1.5 μm.

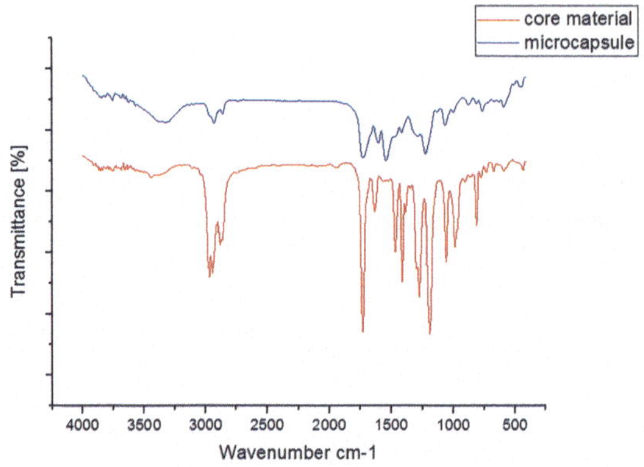

Fig. 1. FTIR spectra of PU-UF microcapsule and core material

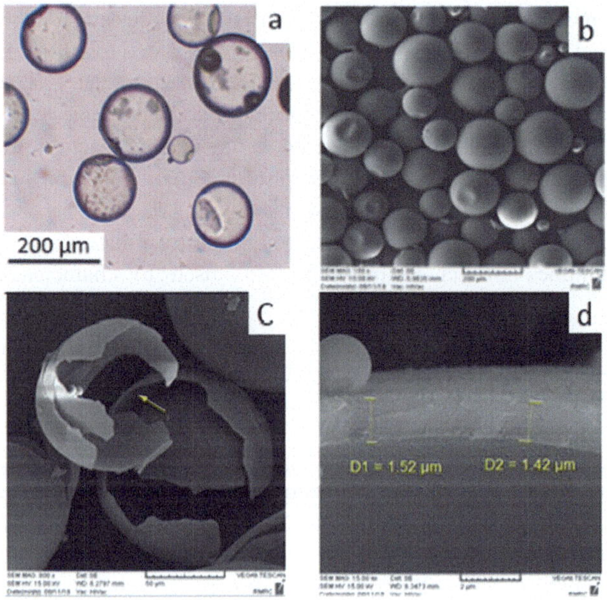

Fig. 2. (a) OM image, (b) SEM micrographs, (c) and (d) ruptured microcapsules prepared at mixing speed of 1000 rpm

4 Conclusion

PU-UF microcapsules containing ETA monomer were successfully synthesized. FTIR spectra of microcapsules and core material confirmed successful microencapsulation of 2-ethylhexyl acrylate. OM image and SEM micrographs showed the microcapsules are spherical and have a smooth surface.

References

1. Dubey, R., Shami, T.C., Bhasker Rao, K.U.: Def. Sci. J. **59**, 82–95 (2009)
2. McFarland, B., Popwell, S., Pojman, J.A.: Macromolecules **39**, 55–63 (2006)
3. Whites, S.R.: Poly(Urea-Formaldehyde) Microencapsulation of Dic Yclopentadien (2003)
4. Di Credico, B., Levi, M., Turri, S.: Eur. Polym. J. **49**, 2467–2476 (2013)

Processing and Reactive Processing of Environmental Polymers

Synthesis, Characterization and Thermal Stability of UV-Curable Hydrogel

Zahra Rahmani[1], Saeed Pourmahdian[1], and Behzad Shirkavand Hadavand[2(✉)]

[1] Department of Polymer, Amirkabir University, Tehran, Iran
[2] Department of Resin and Additives, Institute for Color Science and Technology, Tehran, Iran
shirkavand@icrc.ac.ir

Abstract. In this research, the synthesis, characterization and thermal stability of UV curable hydrogels have been investigated. To this end, urethane acrylate was initially synthesized with isophorone diisocyanate, polyethylene glycol and hydroxyethyl methacrylate. In the next step, the UV curable hydrogel was synthesized with acrylamide, acrylic acid and the urethane acrylate. FTIR spectra proved the structure of urethane acrylate and hydrogel. The study of thermal behavior showed that the synthesized hydrogel is stable to 327 °C.

Keywords: Hydrogel · Urethane acrylate · Thermal stability

1 Introduction

More than fifty years hydrogels have been used in various applications such as biomedical uses, drug delivery systems, contact lenses, tissue scaffold implants and so on [1]. Traditionally, hydrogels were created using chemical cross-linking, however in the last years, creating cross-link via photo polymerization method (UV curing systems) is commonplace [2, 3]. This method, in addition to simplicity, saves energy and results in a shorter time.

Urethane acrylates is one of the most common oligomers indicate some characteristics such as flexibility, chemical and thermal resistance [4]. Long chain of polyol in synthesize of urethane acrylate case to be prepare an elastic oligamer.

In this research, UV curable hydrogel with highly flexible urethane acrylate as crosslinking agent is used. The presence of a high-elastic crosslink agent helps to strengthen the hydrogel when swelling.

2 Experimental

2.1 Materials and Instruments

Acrylamide (Am), acrylic acid (AA), 2-hydroxyethyl methacrylate (HEMA), isophorone diisocyanate (IPDI), benzophenone, tertiary amine triethanolamine (TEA), dibutyltin dilaurate (DBTDL) and acetone were purchased from Merck Co. (Germany).

H. Mirzadeh and A. A. Katbab (Eds.): ISPST 2018, *Eco-friendly and Smart Polymer Systems*, pp. 663–666, 2020.
https://doi.org/10.1007/978-3-030-45085-4_160

Also trimethylolpropane triacrylate (TMPTA) was provided by Aldrich Co. (USA). Polyethylene glycol 1000 was prepared from Daejung of Korea and dried and degassed by vacuum oven at 80 °C for 3 h before use.

2.2 Synthesis of Urethane Acrylate

IPDI (0.2 mol) and polyethylene glycol 1000 (0.1 mol) and 10 mL of acetone were mixed in the 100 mL bottle and 0.2 mL 3 drops of DBTDL as a catalyst were added. This mixture was mixed 50 °C for 2 h. Then 0.2 mol HEMA were added and stirred for about 2 h at 50 °C again. Finally acetone was evaporated from synthesized pre-polymer by vacuum oven.

2.3 Synthesis of Hydrogel

0.126 mol acrylamide and 0124 mol acrylic acid and 10 mL distillated water were mixed. Then 0.015 mol 2-hydroxyethyl methacrylate, 0.2 g trimethylol propane tri-acrylate and 6 wt%. urethane acrylate were mixed. Finally 0.4% by weight ben-zophenone and tertiary amine triethanolamine were added and mixed again.

3 Results and Discussion

3.1 FTIR Analysis of Urethane Acrylate

FTIR spectroscopy of urethane acrylate is shown in Fig. 1. In this figure absorption band for N-H at 2268 and 1551 cm^{-1}, sharp peak around 2917 cm^{-1} due to the C-H stretching vibrations, absorption peak at 1639 cm^{-1} corresponding to the C=C groups of HEMA were observed. The formation of urethane structure was demonstrated from the peaks at 1717 cm^{-1} which are indicative of the C=O group of urethane. The peaks at 3411 cm^{-1} and disappearance of the characteristic peak of isocyanate groups revealed formation of urethane N-H bonds.

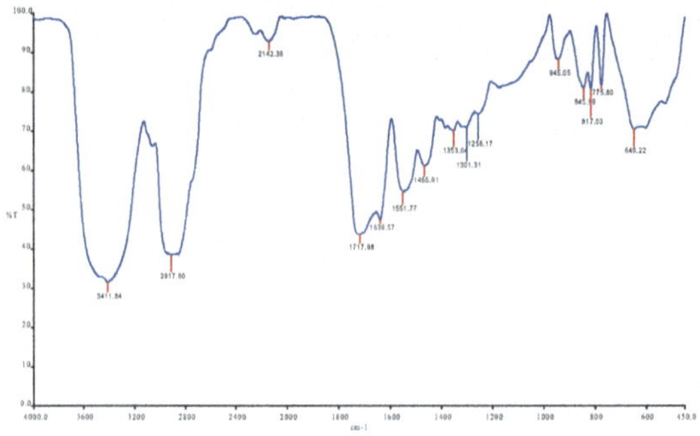

Fig. 1. FTIR spectra of urethane acrylate

3.2 FTIR Analysis of Hydrogel

The FTIR spectrum of synthesized hydrogel is presented in Fig. 2. In this figure, the peak at 3407 cm^{-1} belongs to N-H. The peaks at regions of 1182 and 1614 cm^{-1} are related to C=O band of amine. Also, we can observe the carboxylic group at 2922 cm^{-1} and C-H of CH_2 at 1425 cm^{-1}.

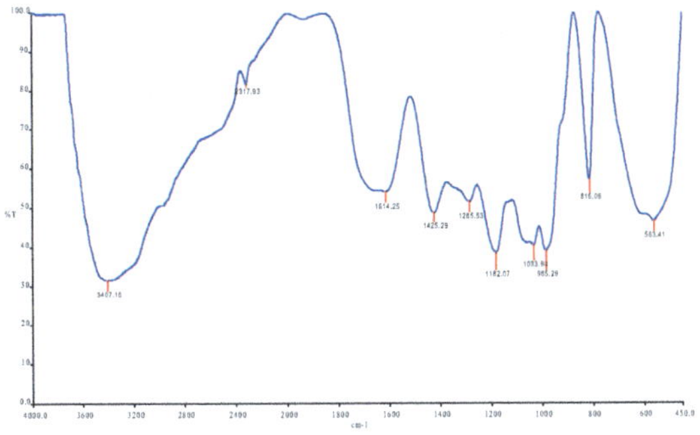

Fig. 2. FTIR spectra of hydrogel

3.3 Thermal Stability of Hydrogel

Figure 3 shows the thermal degradation of the hydrogel. As shown in this figure, slight weight loss occurs at a temperature of 150–100 °C, which is due to the exit of water and none reacted materials. There is another drop in weight at a temperature in the range of 275–200 °C, which can be due to the breakdown of polymetric crosslink. The final and main degradation step starts at 325 °C and continues to 475 °C, which is related to the degradation of the hydrogel structure.

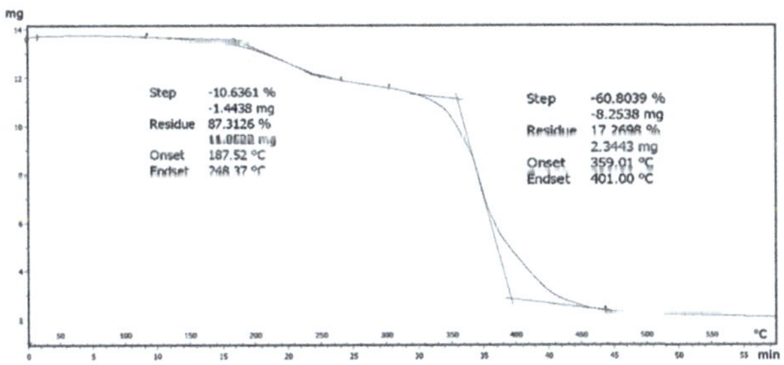

Fig. 3. TGA graph of hydrogel

4 Conclusion

The synthesized hydrogel was cured with photo initiator and ultraviolet irradiation. Investigating the FTIR spectra reflects the correct structure of the hydrogel. The investigation of the thermal behavior of the hydrogel showed its stability up to 375 °C.

References

1. Patil, J.S., Kamalapur, M.V., Marapur, S.C., Kadam, D.V.: Digest. J. Nanomater. Biostruct. **5**, 241–248 (2010)
2. Yao, H., Wang, J., Mi, S.: Polymers **10**, 1–37 (2018)
3. Kure, M., Sfiligoj-Smole, M., Stana-Kleinschek, M.: Mater. Technol. **46**, 87–91 (2012)
4. Madhi, A., Shirkavand Hadavand, B., Amoozadeh, A.: J. Appl. Chem. **12**, 91–97 (2018)

Effect of Polyolefin Addition on the Heat Seal Strength of HDPE Blown Films

Ayat Mehdipour[1(✉)], Hossein Abedini[1], Majid Habibelahi[2], and Morteza Ehsani[1]

[1] Iran Polymer and Petrochemical Institute, P.O. Box 14965-115, Tehran, Iran
ayat.mehdipour@gmail.com
[2] Research and Technology of National Petrochemical Company, P.O. Box 14358, Tehran, Iran

Abstract. Heat seal strength and quality of HDPE blown films with addition of 4 different polyolefins (LLDPE-g-MAH, LDPE-g-MAH, POE-g-MAH, and PP) were investigated. HDPE contained 1 wt% to 5 wt% of each polyolefin was passed though film blowing process followed by heat sealing temperature as low as 140 °C. The sealing time ranged from 1 s to 5 s. Peeling test on films led to seal strength and type of failure considered as seal quality. Type and content of added polyolefin affected both of them. Different thermal properties of polyolefin, as a function of corresponding microstructure, is a key parameter which determined the way in which sealing quality of HDPE was influenced. Besides that, dwell time was found to be a critical parameter specially for sealing quality. All additives improved the heat seal properties, but in their special conditions.

Keywords: Heat seal strength · HDPE · LLDPE-*g*-MAH · LDPE-*g*-MAH · POE-*g*-MAH

1 Introduction

Meka and Stehling have presented a mechanism for heat seal process. In their theory, films are in contact together and heated under squeezing device jaws. Through heating, the crystalline parts of polymer are melted while the exerted pressure expands the contact area of film plies. After sufficient time, diffusion of polymer chains between two surfaces occurs and entanglement takes place. Finally, by cooling, this region undergoes re-crystallization which results in sealing of the two films [1–3].

2 Experimental

EX5 grade of HDPE, maleic anhydride (1.7%)-grafted polyolefins including LDPE, LLDPE, POE and PP of 340 N grade were used. Film production was done in a Brabender blown film with 30 μm thickness. In heat sealing process the produced blown films were sealed by an Aria Heat sealer made in Iran. The films were cut by a Loretzen & Wettr cutter. Temperature used was 140 °C, pressure 2 bar and dwell time were set according to 1, 3, and 5 s. Peeling test was performed by the machine model

H. Mirzadeh and A. A. Katbab (Eds.): ISPST 2018, *Eco-friendly and Smart Polymer Systems*, pp. 667–672, 2020.
https://doi.org/10.1007/978-3-030-45085-4_161

was SANTAM, STM20, with a 200 N load cell. The peel rate was 300 mm/min according to standard ASTM F88–85. Scanning Electronic Microscopy (SEM) was done by the SEM analyze was performed by a Vega XMU-Tescan electronic microscope (Fig. 1).

3 Results and Discussion

3.1 Contact Angle

The results of contact angle test are presented in Table 1 and Fig. 2. Polyolefin-added HDPE decreases the contact angle of produced films. Except PP, other polyolefins are grafted by maleic anhydride. Maleation imparts some polarity to plastics and thereby to a host matrix such as HDPE.

3.2 Peel Test

The results are presented in Tables 2, 3, 4 and Fig. 3. In the neat HDPE sealed films, the seal strength at 140 °C is not good and failure type was 1 or 2, but at 180 °C, the seal is acceptable and failure type is 3 [1, 2, 4]. In the presence of polyolefin-added HDPE, dependency of seal strength extends to other factors like the amount and type of polyolefin. When dwell time is as low as 1 s, at 1 wt% of PP, the seal strength of HDPE film raises, but raising the amount of PP to 5% decreases it sharply to nearly zero. According to the contact angle data, PP is present on the surface of the HDPE film Increasing the dwell time from 1 s to 5 s increases the seal strength of the films containing PP. When LDPE-g-MAH is added to HDPE, seal strength of the film in dwell time of 1 s falls to 1/3 in comparison to HDPE film. Increasing the sealing time from 1 s to 5 s doubles the seal strength of HDPE film, depending on the content of LDPE-g-MAH. Under this condition failure type changes from 1 to 3 because of LDPE lower crystallinity and melting point HDPE [5]. By considering POE-g-MAH instead of LDPE-g-MAH the same behavior is followed by seal strength. At 5 wt% of LLDPE-g-MAH, the seal strength of the films is dramatically reduced to nearly zero in dwell time of 1 s. That reaches over 8 N/0.03 mm by longer dwell time of 5 s. The time of 1 s is not long enough even for diffusion LLDPE-g-MAH molecules.

3.3 SEM Results

For HDPE pure film, longer dwell time leads to more seal strength. This is clear from more rough surfaces in comparison of smoother one attributed to the higher levels of localized deformations for longer dwell time. Actually, at low sealing time, surfaces from the weakest seals contained isolated fractured fibirils, which were very low in diameter. At higher sealing time, 5 s, in Fig. 4(a), the surface fibrils start to being bigger and they are noticeably longer and thicker than those at lower sealing time, 3 s, shown in Fig. 4(b). The same trends and facts were founded for other materials.

Table 1. Contact angle of HDPE films with and without polyolefin

Components	Composition of HDPE/polyolefin (wt%)	Contact angle (°)
HDPE	100/0	93.9
HDPE + PP	99/1	68
	95/5	65.5
HDPE + LLDPE-g-MA	99/1	63.5
	95/5	64.4
HDPE + LDPE-g-MA	99/1	79.7
	95/5	68.2
HDPE + POE-g-MA	99/1	74
	95/5	73.1

Table 2. Seal strength and elongation-at-break for sealed films. For pure HDPE and additive + HDPE films in 1 wt% additive.

Materials	Sealing dwell time (s)			
	1		5	
	Seal strength (N/0.03 mm)	Elongation-at-break (%)	Seal strength (N/0.03 mm)	Elongation-at-break (%)
HDPE	3.9	10.4	6.4	1.3
HDPE + PP	6.5	11.8	11.2	2.9
HDPE + LLDPE-*g*-MAH	4.7	10.0	5.5	27.4
HDPE + LDPE-*g*-MAH	1.2	2.6	8.7	19.5
HDPE + POE-*g*-MAH	1.2	17.5	6.6	1.5

Table 3. Seal strength and elongation-at-break for sealed films. For pure HDPE and additive + HDPE films in 5 wt% additive.

Materials	Sealing dwell time (s)			
	1		5	
	Seal strength (N/0.03 mm)	Elongation-at-break (%)	Seal strength (N/0.03 mm)	Elongation at break (%)
HDPE	3.9	10.4	6.4	1.3
HDPE + PP	0.7	5	2.9	2.1
HDPE + LLDPE-*g*-MAH	0.7	11.1	8.2	2
HDPE + LDPE-*g*-MAH	2.7	10.5	10.9	16.2
HDPE + POE-*g*-MAH	2.3	11.1	9.6	67.2

Table 4. Seal strength and elongation-at-break for sealed films at 140 °C, sealing time 3 s and 3 wt% of additive materials

Materials	Tensile strength (N/0.03 mm)	Elongation-at-break (%)
HDPE	6	1.7
HDPE + PP	10	1.8
HDPE + LLDPE-*g*-MAH	10.9	30
HDPE + LDPE-*g*-MAH	10.8	2.2
HDPE + POE-*g*-MAH	11	16.8

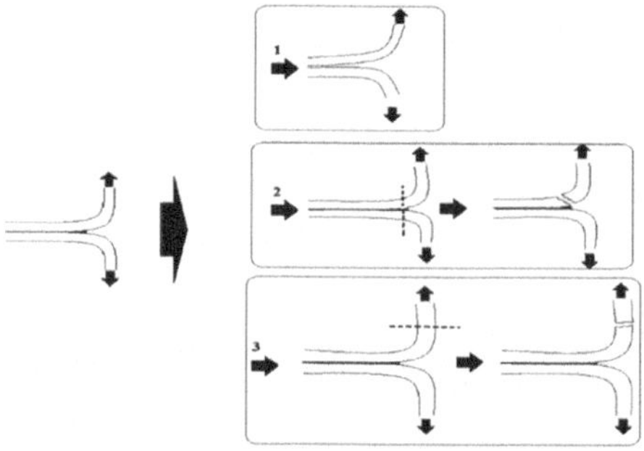

Fig. 1. Failure types in peeling test of heat sealed films: (1) separation of sealed layers, (2) breakage in heat seal, and (3) tearing in films structure

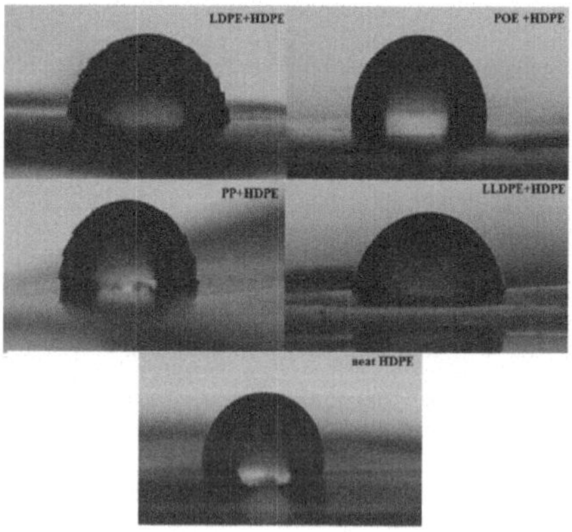

Fig. 2. Images of contact angle test for the produced films

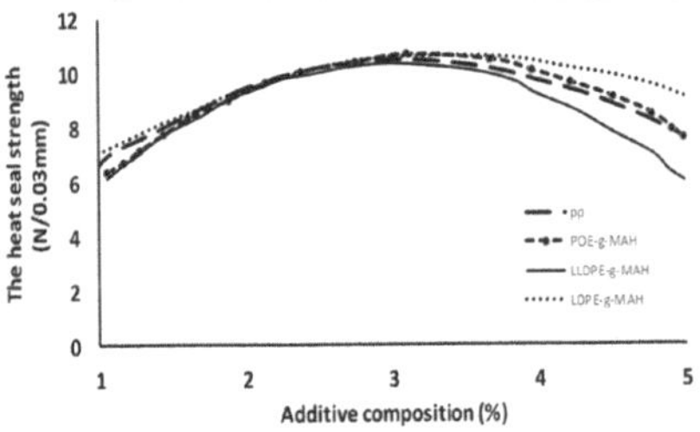

Fig. 3. Heat seal strength versus composition of additive polymers

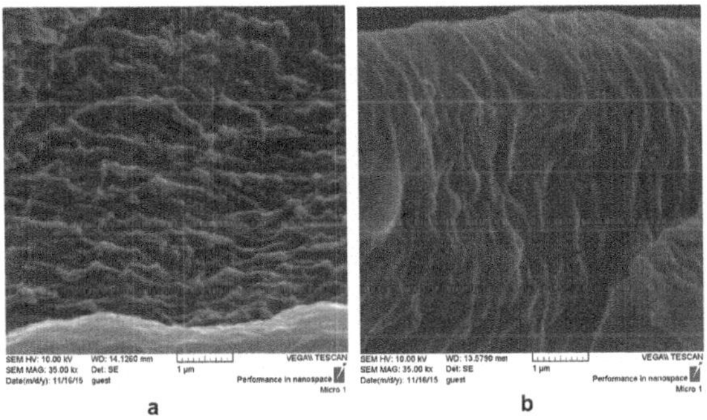

Fig. 4. SEM pictures for HDPE pure films: (a) sealing at 140 °C, 5 s and (b) sealing at 140 °C, 3 s

4 Conclusion

The employed HDPE is a high molecular weight grade that under seal condition of 140 °C could not be heat sealed even at long time of 5 s. In the presence of polyolefin depending on its type and content, the seal strength of HDPE films is influenced even though its amount is small. The results showed that in dwell time of 1 s, the added polyolefin although changed the seal strength of HDPE film but did not affect its type of failure. When the dwell time was set at 5 s, failure type of HDPE improved from 1 to 3 and its strength was raised. In the case of LDPE-g-MA or POE-g-MA with low melting points, the strength as well as the quality of sealed films improved. Nevertheless, for PP as an additive with a high melting point, improvement is made in peeling strength rather than in heat seal quality.

References

1. Mueller, C., Capaccio, G., Hiltner, A., Baer, E.: J. Appl. Polym. Sci. **70**, 2021–2030 (1998)
2. Hashimoto, Y., Yamada, K., Leong, W.Y., Hamada, H.: J. Appl. Polym. Sci. **20**, 273–282 (2011)

3. Meka, P., Stehling, F.C.: J. Appl. Polym. Sci. **51**, 89–103 (1994)
4. Poisson, C., Hervais, V., Lacrampe, M.F., Krawczak, P.: J. Appl. Polym. Sci. **99**, 974 (2006)
5. Tetsuya, T., Hashimoto, Y., Ishiaku, U.S., Mizoguchi, M., Leong, Y.W., Hamada, H.: J. Appl. Polym. Sci. **99**, 513 (2006)

Preparation of Novel Liquid, Solvent-Free, Polyalfaolefin-Based Adhesives

Ahad Hanifpour, Naeimeh Bahri-Laleh(✉), and Mehdi Nekoomanesh-Haghighi

Polymerization Engineering Department, Iran Polymer and Petrochemical Institute (IPPI), P.O. Box 14975/112, Tehran, Iran
n.bahri@ippi.ac.ir

Abstract. An innovative approach for functionalization of polyolefins was developed and newfound polyolefins end-capped with trimethoxysilane (silylated polyolefins) was discovered as a material which can be applicable as an extraordinary polyolefin-based adhesive. For synthesizing mentioned materials, 1-octene as a higher α-olefin was cooligomerized with linear, non-conjugated (1,5-hexadine and 1,7-octadiene) by the metallocene catalyst system of Cp_2HfCl_2/MAO at room temperature. Then, amine-terminated trimethoxysilane (3-aminopropyltrimethoxy silane) in the presence of palladium(II) acetate was reacted with unsaturated bonds containing of synthesized cooligomers by nucleophilic addition. The adhesion properties of resultant adhesives were subjected to different substrates. Obtained adhesives indicated greater tendency to Al substrate. Furthermore, the tensile shear strength of the synthesized adhesives on polar substrates was 2.21% to 2.84% more than that on polar substrates.

Keywords: Polyolefin · Diene · Adhesive · Silane · Adhesion

1 Introduction

Olefin-based polymers have numerous applications due to their being chemically inert, having low density, and low cost. However, polyolefins present non-polar surfaces to which more polar materials may not readily adhere. As such, to apply a layer of paint, adhesive, or the like to a surface comprised of an olefin-based polymer such as polypropylene, a tie layer or other adhesion promoter may be required. In addition, adhesive compositions comprising olefin-based polymers may not readily adhere to polar substrates such as paper, cardboard and the like. To provide effective adhesion between polar and non-polar substances, good bonding strength and/or intimate integrity between the materials must be achieved. A tie layer or adhesion promoter may provide a bridge between a non-polar surface such as isotactic polypropylene, and a more polar surface such as paper [1]. This scenario presents the first attempt to produce α-olefin-based adhesives with higher adhesion properties. To this end, in order to introducing reactive C=C bond on olefin chains, firstly 1-octene was copolymerized (cooligomerized) with 1,5-hexadiene and 1,7-octadiene. In the second step, generated unsaturation bonds were reacted with 3-aminopropyltrimethoxysilane as an adhesive

H. Mirzadeh and A. A. Katbab (Eds.): ISPST 2018, *Eco-friendly and Smart Polymer Systems*, pp. 673–676, 2020.
https://doi.org/10.1007/978-3-030-45085-4_162

promoter in order to modify the C=C groups of cooligomers. Then the adhesion properties of obtained poly1-octene end-capped with 3-aminopropyltrimethoxysilane were evaluated experimentally.

2 Experimental

2.1 Cooligomerization

The cooligomerization was carried out at selected temperature in a 100 mL Schlenk-type reaction flask. After the addition of the monomers, 7 mL of a 1.5 M solution of MAO (10.5×10^{-3} mol) cocatalyst with 1.31×10^{-2} M solution (2.63×10^{-5} mol, stock solution prepared by dissolving 0.01 g of Cp_2HfCl_2 in 2 mL of toluene) of Cp_2HfCl_2 were injected to reactor to initiate the cooligomerization.

2.2 Preparation of the Cooligomer End-Capped with Trimethoxysilane

The palladium(II) acetate complex (1.12 mmol) was suspended in a solution of the cooligomers (10 g) in dry tetrahydrofuran (THF 150 mL) under inert atmosphere. The mixture of reaction was stirred for 30 min at 0 °C, then cooled to the −15 °C. A solution of the appropriate 3-aminopropyltrimethoxysilane (6 mmol) in THF (4 mL) was added dropwise over a period of 10 min. The reaction temperature was maintained for 2 h. The system was then flushed with hydrogen, and the mixture was allowed to warm to room temperature under hydrogen atmosphere.

3 Results and Discussion

Incorporation of 1,5-hexadiene and 1,7-octadiene as 1-butene/cyclopantane and hexene/cycloheptane branches, respectively in the presence of considered catalyst system was confirmed by FT-IR spectroscopy. In these spectra, strong peaks at 2800 and 2900 cm^{-1} are corresponding to the C-H stretching vibration. Furthermore, another weak peak at 3100 cm^{-1} is related to =C-H stretching vibration and one distinct peak at 1650 cm^{-1} (see Fig. 1) corresponded to the stretching vibration of C=C double bond of the comonomer. Bending vibration of CH_2 and CH_3 groups is appeared around 1465 and 1320 cm^{-1}. Tiny and distinct peaks at 960 and 700 cm^{-1} are corresponded to the out of plane vibration of =C-H and C-H bonds [2].

Figure 2 demonstrates the IR spectrum of the obtained 1-octene-based adhesive (oct-adh). FTIR spectrum of the oct-adh sample indicated the introduction of two sharp peaks at 800 and 1150 cm^{-1} which correspond to out of plan vibration of N-H and stretching vibration of C-N bonds, respectively. Furthermore two distinct peaks at 900 and 1050 cm^{-1} are related to stretching vibration of Si-O bond and stretching vibration of C-O bond, respectively.

In order to explore silylated oligomers ability as a polymer of adhesive, the tensile shear bond strength and peeling test were conducted. The results of tensile shear bond strength and adherence of synthesized adhesives are summarized in Table 1.

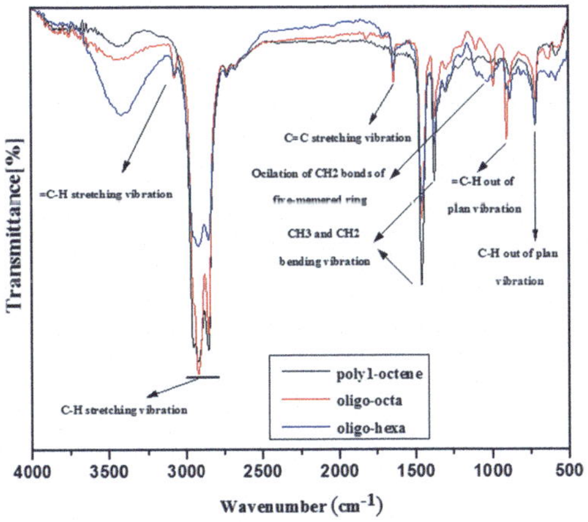

Fig. 1. FT-IR spectra of synthesized 1-octene cooligomers

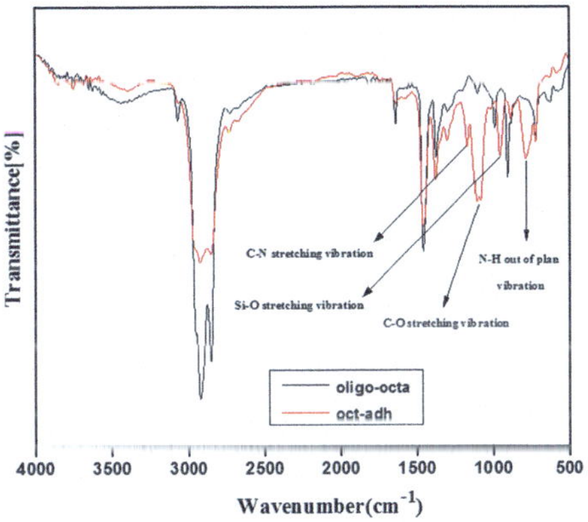

Fig. 2. IR spectrum of synthesized oct-adh sample

By focusing on tensile shear strength of the oct-adh adhesive, it was clearly disclosed that maximum tensile shear strength of the prepared adhesive reached to 0.56, 0.86, 0.41, 1.05, and 1.45 N/mm^2 for PMMA, PC, PS, Nylon, and Al substrates, respectively. About hex-adh adhesive, theses values are 0.47, 0.75, 0.37, 0.91, and 1.15 N/mm^2 for PMMA, PC, PS, Nylon, and Al substrates, respectively.

Table 1. Tensile shear strength and peeling test analysis of the synthesized adhesives

Adhesive	Tensile shear strength (N/mm^2)				
	PMMA	PC	PS	Nylon	Al
hex-adh	0.47	0.75	0.37	0.91	1.15
oct-adh	0.56	0.88	0.41	1.05	1.45

4 Conclusion

Preparation of novel, liquid, solvent-free, polyolefin-based adhesives followed by a two step reaction mechanism. Results showed that, prepared adhesives have high tensile shear bond strength and adherence to Al substrate in comparison to other substrates. Among the employed polymeric substrates, synthesized adhesives disclosed less adherence to PS substrates. Adhesion evaluation of two adhesive samples disclosed that oct-adh adhesive, due to having more C=C unsaturation bond, has high tensile shear bond strength in comparison to hex-adh adhesive sample.

References

1. Charles, R., Sims, L., Brant, P., Jiang, P.: US Patent 10825635 (2004)
2. de Fatima, M., Marques, V., Ramos, D., Rego, J.D.: Eur. Polym. J. **40**, 2583–2589 (2004)

Evaluation of the Mechanical Properties of Poly(Lactic) Acid/ABS Bio-Blend Foams

Miguel Sánchez-Soto[1], Mohammad Reza Kamrani[1,2](✉), Maria Lluisa Maspoch[1], Orlando Santana[1], and Jonathan Cailloux[1]

[1] Centre Català del Plàstic, Universitat Politécnica de Catalunya, Barcelona Tech, Spain
[2] Department of Polymer Science and Technology of Azad University of Shiraz, Shiraz, Iran

Abstract. Foaming is a powerful method that allows lightening of products and reducing the amount of material used. Poly(lactic) acid (PLA) is considered a promising thermoplastic for the replacement of traditional non-biodegradable polymers. In this study, PLA was modified with an epoxide chain extender via reactive extrusion and blended with Acrylonitrile – Butadiene – Styrene (ABS) in the presence of a compatibilizer. The MuCell® process was applied to obtain foams from the previous prepared blends. Results shows that the presence of the epoxide and ABS inhibit the degradation during processing making foaming possible. Under flexural conditions where foamed are designed to work, the mechanical properties of the foamed blends were very similar to the solid counterparts.

Keywords: REX-PLA · ABS · MuCell® · Foaming

1 Introduction

PLA has become an alternative to traditional commodity plastics due to its reasonable price and properties such as high strength and stiffness [1, 2]. Apart of its biodegradability, main reasons for choosing PLA as matrix are the its competitive properties and relatively low cost [5]. However, PLA has some drawbacks such its low melt elasticity and low thermal stability, which in one hand can cause its degradation during processing and on the other limits its wide application. An alternative to diminish the degradation during processing is adding substances capable of inhibiting the PLA degradation paths. Reactive extrusion is a technique that has been used to minimize degradation and increase PLA melt strength [3, 4]. On the other hand, blending is a relatively inexpensive and common route to tailor the properties of the polymers and that can be employed to enhance PLA thermal stability. The present work aims on the manufacturing of PLA/ABS bio-blend foams. Therefore, in this study firstly PLA was modified with a multi-functional epoxide by reactive extrusion obtaining what we called REX-PLA. Then REX-PLA was blended with ABS in the presence of ABS-grafted maleic anhydride (ABS-g-MAH) to enhance the compatibility. Finally, the obtained blends were foamed using nitrogen via the MuCell®

H. Mirzadeh and A. A. Katbab (Eds.): ISPST 2018, *Eco-friendly and Smart Polymer Systems*, pp. 677–681, 2020.
https://doi.org/10.1007/978-3-030-45085-4_163

process studying the final properties of the blends. Through this preliminary research it is demonstrated the possibility of obtaining PLA-based biofoams with enhanced properties in regards to neat PLA ones.

2 Experimental

2.1 Materials

The PLA used in this study was an extrusion grade (Ingeo 4032D) purchased from NatureWorks LLC. It had a D-lactide content of about 2% and a melt flow index of 6.4 ±0.3 g/10 min (210 °C/2.16 kg). SAmfE reactive agent, namely Joncryl-ADR-4300F®, was kindly supplied by BASF (Ludwigshafen, Germany) in flake form. It had a functionality, fn ≈ 12, a glass transition temperature of 56 °C, a MW of 5443 $g.mol^{-1}$ and an epoxy equivalent weight of 433 $g.mol^{-1}$. The ABS used in this study was a Terluran-GP22 (Ineos-Styrolution). The density of this ABS grade was of 1.04 g/cm^3. Finally, a commercial ABS grafted maleic anhydride (ABS–g–MAH) compatibilizer, under the trade name of Bondyram 6000, was purchased from Polyram Group (Israel). The content of grafted maleic anhydride, as previously determined through a FTIR calibration curve, was 0.6 molar.

2.2 Preparation

A twin-screw extruder (Collins Gmbh) was used to mix PLA with SAmfE in a 0.6 wt. % ratio. The obtained product (referred to as REX-PLA) was dry-mixed with ABS and ABS-g-MAH (REX-PLA/ABS/ABS-g-MAH) in the weight ratio of 70/24/6. The extrudates were water-cooled and pelletized prior to be transferred to the injection unit. Finally, dog-bone tensile samples and flexural specimens (nominal dimensions: 127×12.7×3 mm) were foamed by injection moulding (Engel Victory 110) using the MuCell® system and nitrogen as foaming gas. The temperatures from the hoper to the die were set to 180, 190, 200, 210, 220 °C and the mould temperature was maintained at 20 °C. For comparison, neat PLA samples were manufactured using similar processing conditions.

2.3 Morphological and Mechanical Analysis

The structure of produced blends was analyzed by SEM through a JEOL JSM-5610 scanning electron microscope. Samples were prepared using liquid nitrogen to avoid deformation during fracture. Tensile and flexural test were carried out in the previously defined specimens in a Galdabini Sun 2500 universal testing machine at a speed of 1 $mm.min^{-1}$.

3 Results and Discussion

The SEM micrographs of the foams are shown in Fig. 1.

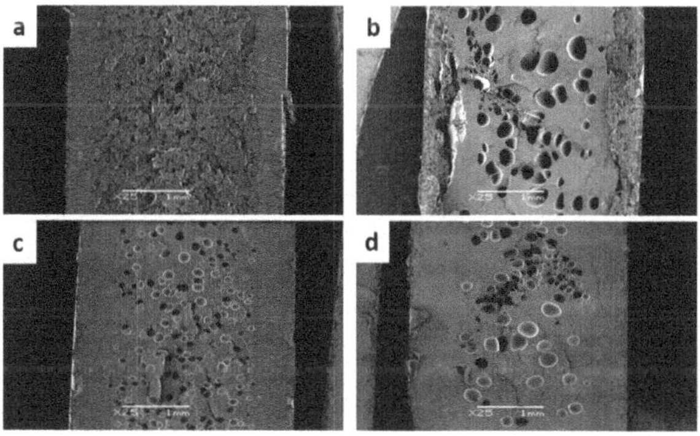

Fig. 1. SEM images of foams: a and b from dog- bone tensile specimens; c and d are from prismatic flexural samples. a, c corresponds to foamed PLA; b, d is REX-PLA/ABS/ ABS-g-MAH6%.

As it can be seen the size of the cells is quite high (in the range of mm) and mainly located in the center of the injected specimens. In the MuCell® process the gas is injected at high pressure having the polymer in the melted state. When the polymer is injected in the mold the pressure drop causes the nucleation and growing of cells. As the gas is forming a unique phase with the polymer, the cells are expected to be evenly distributed along the whole surface. However, this was not the present case. This fact is attributed to a non stable MuCell® injection process due to the small amount of material injected, which was below the limit of the injection machine and leads to an uneven N_2 diffusion. Usually the application of MuCell® requires a considerable amount of material to be processed prior to reaching optimal conditions of temperature and pressure and process stabilization. This requires a large amount of material in contrast to the limited quantity of the blends available. The presence of non-foamed thick skins were also the result of the rapid cooling of mold which inhibits the cell nucleation and growing. The mechanical properties of the blends are shown in Fig. 2 and the numerical values collected in Table 1. As it can be observed the mechanical properties of PLA in foamed samples had a 19.5% of reduction in modulus for the tensile case and 5.2% in flexural in comparison to the solid state with similar behavior for the blends. Logically, the stress concentrator effect of cells and the reduction in the transversal resistant section causes a higher reduction of values in tensile mode in comparison to flexural conditions where the solid skins are able to withstand the applied loads and where foams are designed to work.

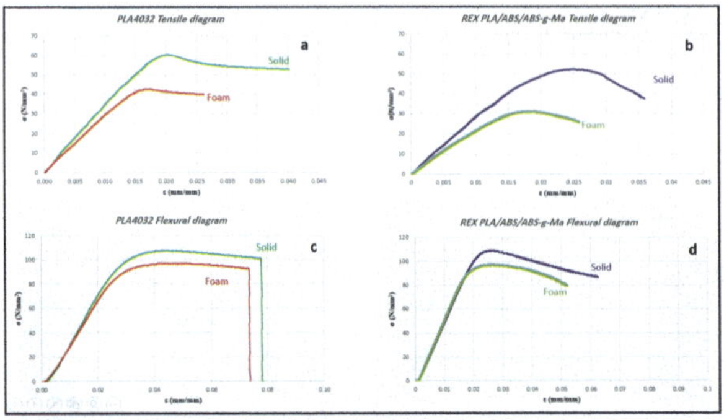

Fig. 2. Tensile and Flexural analysis of foams: a and b are tensile test; c and d are flexural test. a, c is PLA; b, d is REX-PLA/ABS-g-MAH6%.

Table 1. Tensile and flexural values for solid and foamed samples

Test	State	PLA	REX-PLA/ABS-g-MA6%
Tensile modulus (MPa)	Solid	4100 ± 235	3100 ± 193
	Foam	3300 ± 182	2400 ± 110
Flexural modulus (MPa)	Solid	3800 ± 110	4300 ± 160
	Foam	3600 ± 96	3900 ± 123

4 Conclusion

Despite the big size of the cells obtained it is demonstrated the possibility of foaming bio-blends based on PLA. The foamed bioblends exhibit mechanical properties similar to the unfoamed ones in particular in regards to flexural strength where the behavior is very similar to the solid components. The modification of PLA with epoxide and the presence of ABS in the blends stabilize the process and limits the degradation of the blends, facilitating foaming.

Acknowledgments. The authors acknowledge the financial support from the Spanish Government through the Project MAT2016-80045-R (AEI/FEDER, UE).

References

1. Hormnirun, P., et al.: Remarkable stereocontrol in the polymerization of racemic lactide using aluminum initiators supported by tetradentate aminophenoxide ligands. J. Am. Chem. Soc. **126**(9), 2688–2689 (2004)
2. Vink, E.T., et al.: The sustainability of NatureWorks™ polylactide polymers and Ingeo™ polylactide fibers: an update of the future. Macromol. Biosci. **4**(6), 551–564 (2004)

3. Raquez, J.-M., et al.: Biodegradable materials by reactive extrusion: from catalyzed polymerization to functionalization and blend compatibilization. C. R. Chim. **9**(11), 1370–1379 (2006)
4. Raquez, J.M., Narayan, R., Dubois, P.: Recent advances in reactive extrusion processing of biodegradable polymer-based compositions. Macromol. Mater. Eng. **293**(6), 447–470 (2008)
5. Li, Y., Shimizu, H.: Improvement in toughness of poly (l-lactide)(PLLA) through reactive blending with acrylonitrile–butadiene–styrene copolymer (ABS): morphology and properties. Eur. Polym. J. **45**(3), 738–746 (2009)

PET/PLA Blends Crystallization Kinetics

Hasan Azizi Topkanlo[1], Zahed Ahamadi[2(✉)], and Faramarz Afshar Taromi[1]

[1] Department of Polymer Engineering and Color Technology, Amirkabir University of Technology, Tehran, Iran
[2] Department of Chemistry, Amirkabir University of Technology, Tehran, Iran
zahmadi@aut.ac.ir

Abstract. The crystallization behavior and mechanical properties of PET/PLA blends with various amounts of PLA were investigated using wide angle X-ray diffraction (WAXD), differential scanning calorimeter (DSC) and tensile analyses. The crystallization rate and relative crystallinity of the PET/PLA blends were studied by theoretical models of Kissinger, Avrami, Ziabicki and Ozawa. The WAXD analysis showed that the PLA phase was wholly amorphous in all blends after cooling from the melt to ambient temperature. Crystallization behavior assessments on PET/PLA blends suggest that PLA acts as a nucleating agent for PET phase leading to an increase in the initial and peak crystallization temperatures. Kissinger's model showed a rise in activation energy up to 72% for the PET/PLA blends containing 30 wt% PLA. Ziabicki's model gave a minimum value for kinetic parameter in PET/PLA (70/30 w/w) due to the nucleating action of PLA. On the other hand, PLA acted as a retarder for chain segments of PET tending to diffuse through the surface of growing crystals. Therefore, at an optimal composition of PET/PLA, crystallization occurs appropriately. However, an increase in PET content leads to fall in ductility, tensile strength, modulus, elongation-at-break, and fracture toughness of PET/PLA blends.

Keywords: Poly(lactic acid) · Polyethylene terephthalate · Blend · Crystallization · Mechanical properties

1 Introduction

Poly(lactic acid) (PLA) is a biodegradable thermo-plastic aliphatic polyester [1] which is used for manufacturing consumer products [2], biodegradable medical devices [3] and 3D-printed prototypes [4]. In recent years, PLA is developed as one of the highest recognized compostable thermoplastics to replace the traditional synthetic polymers for controlling the environmental pollutions. However, PLA has a low oxygen barrier property with low thermal stability which is not appropriate for industrial operations [5]. Blending is one of the most used and low-cost methods to improve the properties of polymers for development and commercialization of new materials [10]. Several blends of PLA with other polymers have been employed to improve the mechanical properties [1, 2], degradation behavior [3], processability [4] and crystallization [5]. The researchers [6] have reported the effect of several polymers on the film performance of

H. Mirzadeh and A. A. Katbab (Eds.): ISPST 2018, *Eco-friendly and Smart Polymer Systems*, pp. 682–685, 2020.
https://doi.org/10.1007/978-3-030-45085-4_164

PLA binary blends and found that the blending of PLA with polycaprolactone (PCL), polybutylene succinate and ethylene vinyl acetate leads to the improvement of tear resistance. Some researchers [6] have studied the mechanical properties of PLA/PCL blends and found that the PCL has a marked effect on the ductile properties. However, the slight increase in melt temperature of PLA indicates the low miscibility of PLA and PCL. The researchers [6] investigated the effect of natural rubber (NR) on the impact strength of PLA and reported that the highest impact strength could be achieved by NR when masticated at 40 rpm for 15 min. Some researchers [5] prepared the PLA/PCL blends cured with dicumyl peroxide and found that the ultimate tensile strength was significantly improved which the resulted mechanical properties were comparable to those of high impact polystyrene. Polyethylene terephthalate (PET) is one of the most used polymers in packaging industries with low permeability and good mechanical properties [2–4]. Therefore, PET is a good candidate for blending with PLA to achieve a biodegradable material for packaging and manufacturing bottle with appropriate barrier properties. Some researchers [3] have focused on lowering the degradation temperature of PET/PLA blends and found that the PLA concentration has a considerable effect on lowering thermal stability and mechanical properties. The researches [4] have investigated the nonisothermal crystallization of PET/PLA blends and reported that PLA can crystallize in both amorphous and crystalline PLA. However, the degree of PET crystallinity decreases with PLA loading. In another research [5] the miscibility in PET/PLA blend were studied and compared the state of compatibility with PET/chitosan blends. It was found that the miscibility of PLA is higher than that of chitosan and the saturation of PLA into the polymer matrix could be reached 10 wt%.

2 Experimental

2.1 Materials

Poly(lactic acid) (PLA) was purchased from Nature Works LLC, USA, containing 1.5–2% of D isomer (PLA-2002D grade) with onset degradation of 290 °C, molecular weight of 2.71 × 10^5 g/mol and MFI of 12.4 g/10 min. PET bottle grade chips (BG-821) with intrinsic viscosity of 0.82 dL/g and melting point of 248 °C was provided by Shahid Tondguyan Petrochemical Co., Iran.

3 Results and Discussion

Figure 1 reveals the cold crystallization behaviors of PLA, PET, and the blends of PLA/PET containing various PET loadings. As can be seen from this figure, the PLA phase has been crystallized at lower temperatures compared to the PET component during the heating process. However, introduction of PET into PLA leads to a drop in cold crystallization of PLA especially at higher PLA concentrations. Figure 1 indicates that the peak cold crystallization temperature of PLA is dependent on the percent of crystallized PET phase in the PLA/PET blends. So that the PLA phase is crystallized at lower temperatures with higher weight percentage of crystallized PET component.

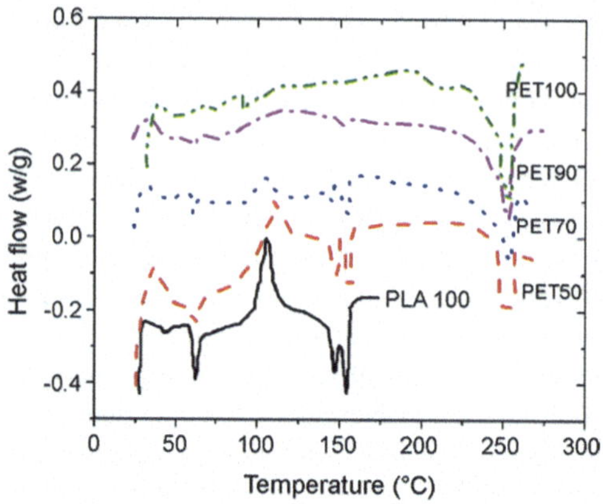

Fig. 1. Cold crystallization behavior of neat PLA, neat PET and PET/PLA blends with different PET weight percentages

3.1 Non–isothermal Crystallization

The effect of PLA on non-isothermal crystallization behavior of PET component in PLA/PET was studied by DSC with various cooling rates. A non-isothermal crystallization behavior of PLA/PET blends at a cooling rate of 10 °C/min is depicted in Fig. 1. Crystallization temperature of PET increased with PLA concentration due to the nucleating effect of PLA phase [26]. Figure 2 shows the non-isothermal crystallization data for PLA/PET blends containing various PET concentrations. As depicted in this figure, the initial and peak crystallization temperatures are decreased with increasing cooling rate from 5 to 15 °C/min.

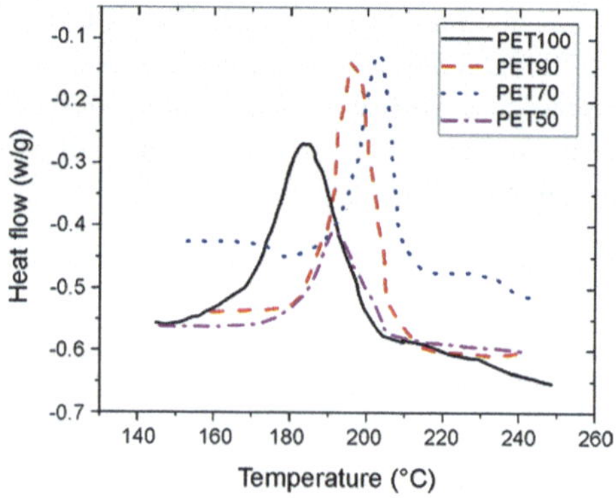

Fig. 2. Non-isothermal crystallization behavior of neat PET and PET/PLA blends having different contents of PET (10 °C/min cooling rate)

4 Conclusion

The applicability of Ozawa's equation for nonisothermal crystallization behavior of PLA/PET has been studied and the results show that the ability of crystallization is increased when temperature is decreased. The PLA/PET blends containing 30 wt% of PLA display the minimum value of Ozawa's index and maximum value of $K(T)$ which indicate a high activation energy of polymer segments to diffuse into the surface of growing crystals. The active parameters of isothermal crystallization measurements of PLA/PET blends have revealed that the required time for crystallization of PET phase decreases with PLA content up to 30 wt%. In this respect, the PLA phase can increase the rate of crystallization of PET phase which has a maximum rate in PLA/PET (30/70 w/w) blends. Mechanical properties of PLA/PET blends containing various PLA loadings indicated that an increase in PET concentration leads to lower ductility and tensile strength of PLA/PET blends. It can be concluded that an increase in PET loadings leads to lower Young's modulus and elongation-at-break of the prepared blends. The results suggest that introducing PET into the blends leads to a brittle behavior in view of fracture toughness.

References

1. Garlotta, D.: J. Polym. Environ. **9**, 63–84 (2001)
2. Datta, R., Henry, M.: J. Chem. Technol. Biotechnol. **81**, 1119–1129 (2006)
3. Tyler, B., Gullotti, D., Mangraviti, A., Utsuki, T., Brem, H.: Adv. Drug Deliv. Rev. **107**, 163–175 (2016)
4. Tymrak, B., Kreiger, M., Pearce, J.M.: Mater. Design **58**, 242–246 (2014)
5. Nampoothiri, K.M., Nair, N.R., John, R.P.: Bioresour. Technol. **101**, 8493–8501 (2010)
6. Fukushima, K., Abbate, C., Tabuani, D., Gennari, M., Camino, G.: Polym. Degrad. Stabil. **94**, 1646–1655 (2009)

Microstructure and Mechanical Properties of Compatibilized Polylactide/Thermoplastic Starch Blends

Seyed Ali Nezamzadeh[1], Zahed Ahamadi[2(✉)], and Faramarz Afshar Taromi[1]

[1] Department of Polymer Engineering, Amirkabir University of Technology, Tehran, Iran
[2] Department of Chemistry, Amirkabir University of Technology, Tehran, Iran
zahmadi@aut.ac.ir

Abstract. Biodegradable polymer blends of polylactic acid/thermoplastic starch (PLA/TPS) were prepared via direct melt blending varying order of mixing of ingredients fed into the extruder. The effect of interface interactions between PLA and TPS in the presence of maleic anhydride (MA) compatibilizer on the microstructure and mechanical properties was then investigated. The prepared PLA/TPS blends were characterized by scanning electron microscopy, differential scanning calorimetry (DSC), tensile, and rheological measurements. Morphology of PLA/TPS shows that the introduction of MA into the polymer matrix increases the presence of TPS at the interface region. The DSC results revealed the reduction of glass transition of PLA with contributions from both TPS and MA. The crystallization temperature was decreased by the addition of MA leading to reduction of overall crystallization of PLA/TPS blend. The mechanical measurements show that increasing MA content up to 2 wt% enhances the modulus of PLA/TPS more than 45% compared to the corresponding blends free of MA compatibilizer.

Keywords: Interface · Maleic anhydride · Mechanical properties · Poly(lactic acid) · Thermoplastic starch

1 Introduction

Polylactic acid (PLA) has been the subject of extensive works because of being biodegradable and compatible with various polymers [1–3]. However, there are a number of limitations from application standpoint that make challenging to use PLA. The improvement of physical and mechanical properties of PLA gives the researchers of diverse fields to more investigate. One of the mostly treated strategies applied in this sense is the blending of PLA with other polymers, especially biodegradable ones [4–6]. Mixture of PLA and thermo-plastic starch, as a biodegradable polymer blend, has extensively been studied in recent years [7–9].

The introduction of starch into the PLA matrix produces a low cost and eco-friendly polymer blend [10]. Ke and Sun [11] investigated the mechanical properties of PLA/starch with various blending ratios and found that the modulus of the PLA/starch

H. Mirzadeh and A. A. Katbab (Eds.): ISPST 2018, *Eco-friendly and Smart Polymer Systems*, pp. 686–689, 2020.
https://doi.org/10.1007/978-3-030-45085-4_165

increases up to 70 wt% of starch concentration. However, the tensile strength took lower values with higher starch loadings. The melting behavior and isothermal crystallization kinetics of PLA/starch has also been addressed by the researchers [12]. The findings show that the introduction of neat starch to the PLA could affect the crystallization rate of PLA, even at low concentrations of starch. On the other hand, hydrophilic nature of starch causes a weak interfacial interaction between this polymer and the hydrophobic polymers like polyethylene (PE) [13], polypropylene (PP) [14], and PLA [15]. One of the most efficient methods to increase the interfacial interactions between two immiscible polymers is the incorporation of a compatibilizer. Zhang and Sun [16] have prepared the PLA/starch blends in the presence of dioctyl maleate (DOM) as a compatibilizer. They found that the addition of low contents of DOM significantly improves the tensile strength of PLA/starch blends. However, the DOM can act as a plasticizer at higher concentrations and may lead to the enhancement of elongation-at-break. The effect of maleic anhydride (MA) on the mechanical properties of PLA/wheat starch in the presence of an initiator such as 2,5-bis(tert-butylperoxy)-2,5-dimethyl hexane was studied elsewhere. The results show an improvement in the PLA/starch interfacial adhesion and mechanical properties. Jariyasakoolroj and Chirachanchai [17] have prepared the silane-modified starch/PLA blends through reactive blending. The prepared blends show reduced glass transition temperature and an increased tensile strength compared to other PLA/starch blends. The aim of this study was to evaluate the effect of MA contents on the microstructure and mechanical properties of PLA blends containing various thermoplastic starch contents. The PLA/TPS blends were prepared through a melt mixing procedure.

2 Experimental

PLA Ingeo 4043D with molecular weight of 1.33×10^5 g/mol, specific gravity of 1.24 g/cm^3, and melting point of about 150 °C was purchased from Natureworks LLC, Cargill, Minneapolis, MN. Natural corn starch was provided by Daesang (South Korea) with molecular weight of 2.14×10^5 g/mol, specific gravity of 1.5 g/cm^3, and glass transition temperature of 223 °C. Maleic anhydride was procured from Sigma-Aldrich (St. Louis, MO) with molecular weight of 98 g/mol, specific gravity of 1.4 g/cm^3, and melting point of 52.6 °C. Ethylene glycol was supplied by Sigma-Aldrich (St. Louis, MO) with molecular weight of 62 g/mol and specific gravity of 1.113 g/cm^3.

3 Results and Discussion

The percentage of grafted MA in the PLA/TPS blends shows a maximum value with the MA content and various TPS loadings. The results show that the amount of grafted MA decreases with the addition of MA compatibilizer from 1 wt% to 2 wt%. However, this parameter shows a maximum value for PLA/TPS blends containing 39 wt% of TPS and 1 wt% of MA compatibilizer. Data shows that the addition of MA at the end of mixing process leads to a decrease in the percent of grafted MA. Figure 1 displays the FTIR spectra of PLA/TPS blends containing various MA and TPS loadings. The effect of MA introduction during the mixing process was monitored by the height of the spectrum of carbonyl group at the region of 1840–1700 cm 21 for the prepared samples. Data also show the measured carbonyl indexes from FTIR spectrum for the prepared blends. Figure 1(a) and the results together confirm that the addition of TPS more than 40 wt% will increase the carbonyl index. Figure 1(b) shows that the introduction of MA into the PLA/TPS blends leads to an increase in the carbonyl index. The blends containing various MA loadings have a maximum value for samples containing 39 wt% of TPS and 1 wt% of MA. Figure 1(c) shows that the addition of MA compatibilizer at the end of mixing process leads to a decrease in carbonyl index. The results suggest that mixing of PLA and MA at the first step of mixing causes a higher carbonyl index due to the more diffusion of MA into the PLA chains.

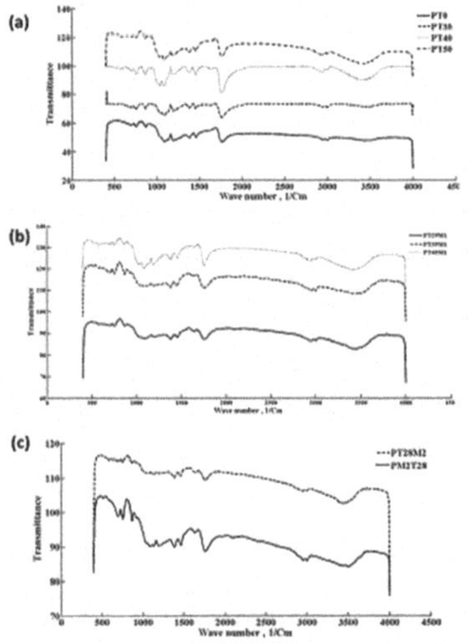

Fig. 1. FTIR spectra of PLA/TPS blends containing various MA and TPS loadings: (a) blends without any MA compatibilizer, (b) blends containing 1 wt% of MA and various TPS loadings, and (c) blends containing 2 wt% of MA and 28 wt% of TPS with different mixing methods

References

1. Lee, S.H., Wang, S.: Compos. Part A **37**, 80 (2006)
2. Lai, S.M., Lan, Y.C.: J. Polym. Res. **20**, 140 (2013)
3. Iwatake, A., Nogi, M., Yano, H.: Compos. Sci. Technol. **68**, 2103 (2008)
4. Yang, S., Madbouly, S.A., Schrader, J.A., Srinivasan, G., Grewell, D., McCabe, K.G., Kesslere, M.R., et al.: Green Chem. **17**, 380 (2015)
5. Pracellaa, M., Haqueb, M.M., Pacib, M., Alvarezc, V.: Carbohydr. Polym. **137**, 515 (2016)
6. Xiong, Z., Yang, Y., Feng, J., Zhang, X., Zhang, C., Tang, Z., Zhu, J.: Carbohydr. Polym. **92**, 810 (2013)
7. Ferri, J.M., Garcia, D.S., Anchez-Nacher, L., Fenollar, O., Balart, R.: Carbohydr. Polym. **147**, 60 (2016)

8. Akrami, M., Ghasemi, I., Azizi, H., Karrabi, M., Seyedabadi, M.: Carbohydr. Polym. **144**, 254 (2016)
9. Khanoonkon, N., Yoksan, R., Ogale, A.A.: Carbohydr. Polym. **137**, 165 (2016)
10. Jantanasakulwong, K., Leksawasdi, N., Seesuriyachan, P., Wongsuriyasak, S., Techapun, C., Ougizawa, T.: Carbohydr. Polym. **153**, 89 (2016)
11. Ke, T., Sun, X.: Cereal Chem. **77**, 761 (2000)
12. Ke, T., Sun, X.J.: Appl. Polym. Sci. **89**, 1203 (2003)
13. Sabetzadeh, M., Bagheri, R., Masoomi, M.: Carbohydr. Polym. **141**, 75 (2016)
14. Martins, A.B., Santana, R.M.C.: Carbohydr. Polym. **135**, 79 (2016)
15. Sanyang, M.L., Sapuan, S.M., Jawaid, M., Ishak, M.R., Sahar, J.: Carbohydr. Polym. **146**, 36 (2016)
16. Zhang, J.F., Sun, X.J.: Appl. Polym. Sci. **94**, 1697 (2004)
17. Zhang, J.F., Sun, X.: Biomacromol **5**, 1446 (2004)

Solvent Effect in 1-Decene Oligomerization Using Silica Supported $AlCl_3$ Catalyst

Sevda Dehghani[1], Mehdi Nekoomanesh-Haghighi[1], Samahe Sadjadi[2], and Naeimeh Bahri-Laleh[1(✉)]

[1] Polymerization Engineering Department, Iran Polymer and Petrochemical Institute, P.O. Box 14975-112, Tehran, Iran
n.bahri@ippi.ac.ir

[2] Gas Conversion Department, Faculty of Petrochemicals, Iran Polymer and Petrochemical Institute, P.O. Box 14975-112, Tehran, Iran

Abstract. In this research, poly1-decene was synthesized by Silica/$AlCl_3$/H_2O catalytic system in the presence as well as absence of toluene as a solvent. The effect of solvent amount on the final product properties was explored. The obtained oligomers were characterized with GPC, TGA, Kinematic Viscosity at 40 and 100 °C (KV_{40}, KV_{100}), Viscosity Index (VI) and ^{1}H NMR methods. The results showed that the molecular weight and kinematic viscosity of 1-decene oligomer decreased as the toluene amount increased. TGA analysis demonstrated that thermal stability of the synthesized oil decreased in the poly1-decene samples synthesized in the presence of toluene. ^{1}H NMR spectroscopy revealed the presence of aromatic ring in the structure of oligomer chains synthesized in the presence of toluene while that for the chains without solvent, the characteristic peaks of aliphatic and alkene type hydrogens were detected.

Keywords: Polyalphaolefins · Oligomerization · Supported $AlCl_3$ catalyst · 1-decene · Solvent

1 Introduction

Polyalphaolefins (PAOs) are hydrocarbon oligomers produced via catalytic oligomerization of linear alpha olefins with carbon numbers of C_6–C_{20} followed by hydrogenation. 1-Decene is the most commonly used starting alphaolefin. With a variety of excellent performance features, including good thermal and oxidative stability, high viscosity index, good low-temperature fluidity, low volatility, low corrosiveness, and high biodegradability, synthetic lubricants found increasingly wide applications under extreme conditions [1]. The representative syntheses for PAOs have been performed via cationic polymerization mechanism by using Lewis acid catalysts such as BF_3 and aluminum compounds. The use of Lewis acid catalysts is very practical for achieving desirable properties. However, their highly corrosive and hazardous nature together with the problem of disposal evoked researchers attempt to find new green processes to circumvent such defects. Stabilizing Lewis acid catalysts on different supports is a very promising way to minimize the costs and environmental impacts of these catalysts [2]. In this study, $AlCl_3$ was fixed on silica support in the presence of CCl_4 as solvent.

H. Mirzadeh and A. A. Katbab (Eds.): ISPST 2018, *Eco-friendly and Smart Polymer Systems*, pp. 690–693, 2020.
https://doi.org/10.1007/978-3-030-45085-4_166

Then, 1-decene oligomerization was conducted with this catalyst and toluene solvent to produce poly1-decene (Pdec) type synthetic oils. Effect of toluene amount on the final PDec properties was studied in detail.

2 Experimental

$AlCl_3$ was fixed on silica support in the presence of CCl_4 as solvent. After the catalyst synthesis, it was employed in 1-decene oligomerization in the presence of different amounts of toluene as solvent at 80 °C for 2 h (Table 1). Finally, the properties of the synthesized oils, including kinematic viscosity at 40 and 100 °C, viscosity index (VI) and molecular weight were compared. The results were shown in Table 1.

Table 1. Molecular weight, kinematic viscosity at 40 and 100 °C and viscosity index (VI) results for the synthesized samples

Sample no.	Tol (mL)	KV_{40} (cSt)	KV_{100} (cSt)	VI	M_n (g/mol)
1	0	86.2	12.9	131	1063
2	5	58.0	8.4	114	643
3	20	20.4	4.4	124	427

Oligomerization conditions: 1-Decene: 50 mL, $AlCl_3$: 2 g, H_2O: 0.5 mL, T: 80 °C, t: 2 h

3 Results and Discussion

Gel permission chromatography (GPC) was used to determine molecular weight of the samples. As can be seen in Table 1, M_n decreases by increasing toluene amount. In the case of the mechanism of the effect of toluene on alpha olefins oligomerization, we can refer to Friedel-Crafts reaction [3]. This reaction leads to the addition of an aromatic ring to the synthesized oil chain, instead of chain growth and increasing molecular weight, which reduces the M_n and viscosity of the final product.

To prove this deduction, Nuclear Magnetism Resonance (^1H NMR) was used to characterize structure of the synthesized oligomers. ^1H NMR spectra of samples 1 and 2 were shown in Fig. 1. The peak at 7.26 ppm is related to the $CDCl_3$ solvent used to solve the samples for ^1H NMR analysis. Distinct peaks of the synthesized PDec (sample 1) in ^1H NMR spectrum were as follows:

^1H NMR: 5.2–5.4 ppm (olefinic **CH**R = **CH**R′), 4.9 5.2 ppm (olefinic **CH**R = CR′R″), 4.5–4.9 ppm (olefinic **CH**$_2$ = CR′R″), 2.0 ppm (allyl, = C-**CH**RR′), 1.96 ppm (allyl, = C-**CH$_2$**), 1.27 ppm (aliphatic **CH**- and **CH**$_2$-), 0.75–0.90 ppm (**CH**$_3$ in the branch and main chain ends).

In the ^1H NMR spectra of samples 2 and 3, according to the presence of toluene solvent in oligomerization, the peak of aromatic C-H related to benzene ring appeared in the range of 6.9–7.2 ppm besides the peaks presented in sample 1 ^1H NMR spectrum.

Thermal gravimetric analysis (TGA) was used to characterize the thermal stability of the sample 1. The graph of TGA analysis for sample 1 was shown in Fig. 2 and the data were collected in Table 2. The results showed that this sample had a high thermal resistance demonstrating good stability against degradation. The temperature of initial degradation temperature (3% weight loss, T_3), temperature of 10% weight loss (T_{10}) and

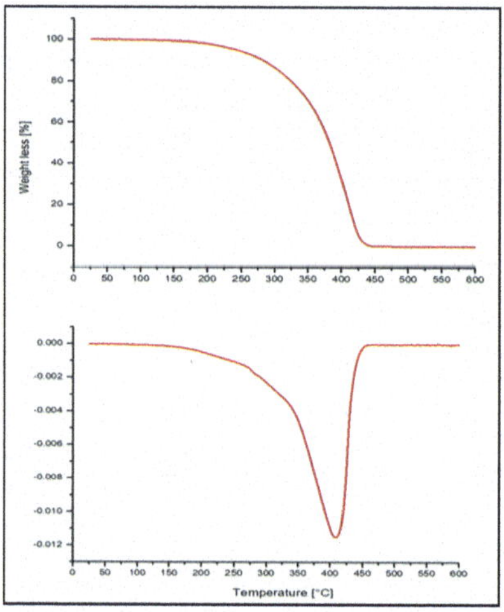

Fig. 1. TGA curve of Sample 1

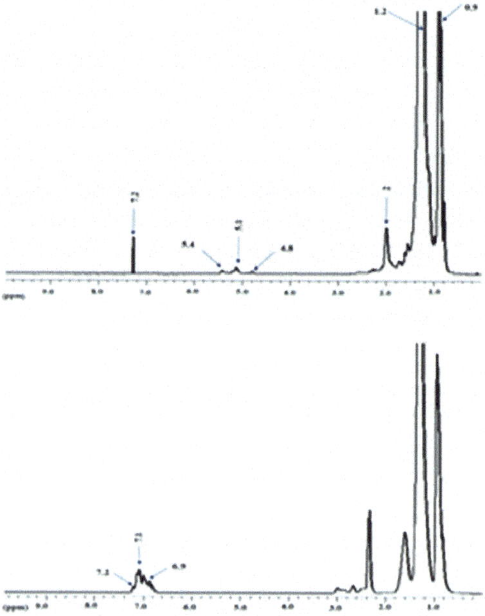

Fig. 2. ^{1}H NMR spectra for: (a) oligomer in the absence of toluene (sample 1) and (b) oligomer in the presence of toluene (sample 2)

Table 2. TGA results for sample 1

Thermal degradation temperature (°C)	Sample 1 (oligomer in the absence of toluene)
IDT[a]	230
$T^{b}_{10\,\%}$	290
T^{c}_{max}	410

[a]Initial degradation temperature, [b]temperature at which 10% weight loss occurs, [c]maximum degradation temperature

maximum degradation temperature (T_{max}) were observed at 270, 320, and 410 °C, respectively. Further-more, residual weight (%) was only 2% at 600 °C, due to the hydrocarbon nature of the synthesized PAO.

4 Conclusion

GPC results demonstrated the reduction of molecular weight of the samples by increasing the solvent amount. TGA analysis proved the oligomer obtained in the absence of the toluene solvent has high thermal stability. ^{1}H NMR analysis affirmed the presence of aromatic ring in the oligomer chains synthesized in the presence of toluene.

References

1. Rudnick, L.R., Shubkin, R.L.: Synthetic Lubricants and High-Performance Functional Fluids, Revised and Expanded, CRC, Boca Raton (1999)
2. Wight, A., Davis, M.: Chem. Rev. **102**(10), 3589–3614 (2002)
3. http://www.chemguide.co.uk/mechanisms/elsub/indalkyl.html#top

Predicting Droplet Size by Rheometry Analysis for PBAT/EVOH Blend

Mehdi Elhamnia and Ghodratollah Hashemi Motlagh(✉)

Advanced Polymer Materials and Processing Laboratory,
School of Chemical Engineering, College of Engineering, University of Tehran,
P.O. Box 11155-4563, Tehran, Iran
ghmotlagh@ut.ac.ir

Abstract. The aim of the present article is to predict the morphology and droplet size of the PBAT/EVOH blend. PBAT is a biodegradable polymer which is usually used in packaging, however to improve its barrier properties, the blending with EVOH was proposed. Thus, predicting the final morphology can achieve a good look. By combination of the theories and rheological data, the volume average radius of droplets was calculated and compared with those extracted from the FE-SEM images. The models of Palierne, Gramespacher and Meissner, and Bousmina fitted the rheological data and the volume average radius calculated by these models were 0.29, 1.01, and 0.40 μm, respectively. On the other hand, the FE-SEM images showed a volume average radius of about 1.1 ± 0.3 μm. Thus, the good compatibility between theory and experimental results was observed which could be useful to adjust the final morphology.

Keywords: Biodegradability · Surface tension · Solubility parameter · Morphology

1 Introduction

Poly(butylene adipate-*co*-terephthalate) (PBAT) is an aliphatic-aromatic co-polyester which due to its excellent flexibility and biodegradability, can be a proper alternative to conventional polymers in the packaging and agricultural application. However, PBAT has high permeability to gases, thus improving the permeability resistance of the polymer causes widespread applications. Poly(ethylene-co-vinyl alcohol) (EVOH) as a polymer which has good barrier properties is a suitable candidate for blending with PBAT [1] and can form good interfacial interaction due to enormous hydroxyl groups on its chain [2]. The morphology obtained from the blending of these two polymers controls the final properties of the alloy. Therefore, studying the morphology of the PBAT/EVOH alloy is of great significance. In this study, the morphology and droplet size of EVOH within PBAT as the matrix is predicted by theory and then compared with SEM images of alloy. By doing this, an important step is taken to control the morphology of this alloy.

H. Mirzadeh and A. A. Katbab (Eds.): ISPST 2018, *Eco-friendly and Smart Polymer Systems*, pp. 694–697, 2020.
https://doi.org/10.1007/978-3-030-45085-4_167

2 Experimental

2.1 Materials and Methods

PBAT (Ecoflex C1200) and EVOH (F101A) were provided by BASF Co., (Germany), and EVAL, Kuraray Co., (Belgium), respectively. The alloy consists of 20 wt% of EVOH was prepared at 200 °C for 8 min by a melt internal mixer. The rheometry analysis was done at 190 °C at Amirkabir University of Technology in Iran. The FE-SEM images were taken at University of Tehran in Iran.

3 Results and Discussion

The main step for predicting the morphology and droplet size of dispersed phase is determination of polymers interaction. Helfand and Tagami [3] proposed a self-consistent field model that was then modified by Helfand and Sapse [4] in which, the surface tension between two incompressible-structurally symmetric polymers has been presented in terms of the solubility parameter (Eq. (1)) [5].

$$\gamma = kT\alpha^{\frac{1}{2}}\left(\frac{\beta_A+\beta_B}{2}+\frac{1}{6}\times\frac{(\beta_A-\beta_B)^2}{\beta_A+\beta_B}\right) \quad (1)$$

Where, $\beta_k^2 = 1/6 \times \rho_{0k} b_k^2$, ρ_{0k} is the density of component k, b_k is Kuhn statistical segment length and $\alpha = 1/kT \times (\delta_A - \delta_B)^2$. The required parameters have been listed in Table 1. The approximate surface tension can be obtained by calculating the solubility parameters of the PBAT and EVOH co-polymers.

To obtain the solubility parameters of each part of each co-polymer, group contribution method according to Hoftyzer & Van Krevelen method [6] was employed. Thereupon, using the relationship proposed by Schneier (Eq. (2)) [7], the solubility parameter of the random co-polymer was calculated.

$$\delta_x = \frac{\left(\frac{n_1M_1^2}{\rho_1}\right)\delta_1 + \left(\frac{n_2M_2^2}{\rho_2}\right)\delta_2}{(n_1M_1+n_2M_2)\left\{\left[\left(\frac{n_1M_1^2}{\rho_1}\right)+\left(\frac{n_2M_2^2}{\rho_2}\right)\right]\left[\frac{1}{\rho_x}\right]\right\}^{\frac{1}{2}}} \quad (2)$$

Where, n is molar ratio, M and ρ are molecular weight of monomer and density of polymer, respectively and ρ_x is density of co-polymer that have been listed in Table 2.

Table 1. Parameters of Eq. (1)

	Density (g/cm^3)*	b_k (Å)**	$\beta_k \times 10^{-4}$ (m$^{-1/2}$)
PBAT	1.29	7	1.74
EVOH	1.19	7	3.88

* producer data
** is about 7 for many random coil polymers [12]

Table 2. Solubility parameter of discriminating polymers calculated by Hoftyzer & Van Krevelen method

	n^* (mol%)	M_0 (g/mol)	ρ (g/cm^3)	δ (MPa$^{1/2}$)
PE	32	28	0.94	18.12
PVA	68	44	1.19	25.77
PBT	44.4	220	1.31	21.57
PBA	55.6	200	1.02	17.92

* reported by Witt et al. [13] and Kuraray company

According to Table 2 and Eq. (2) the solubility parameter of PBAT and EVOH are calculated 20.47 and 25.25 MPa$^{1/2}$, respectively. So, by substitution of these value in Eq. (1), Υ = 14.5 mN/m was achieved.

Now, by fitting the rheometric data to the Palierne [8, 9], Gramespacher and Meissner [9, 10], and Bousmina [9, 11] models and placing the obtained surface tension in each model, the droplet radius of dispersed phase within the PBAT matrix can be achieved. The values of $\frac{\gamma}{R_v}$ and fitting curves have been showed in Fig. 1. Then, the volume average radius of dispersed phase was calculated and listed in Table 3.

The SEM images of the PBAT/EVOH 80/20 alloy are shown in Fig. 2. As seen in Fig. 2, the volume average radius of droplets is 1.1 ± 0.3 μm.

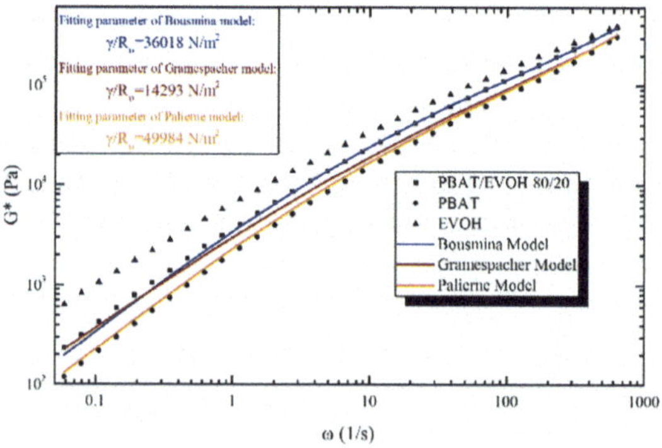

Fig. 1. Complex modulus of PBAT, EVOH and their blend with 20% EVOH. Also, three models was fitted to the complex modulus of the blend.

Table 3. Calculated volume average radius of dispersed phase

	Palierne	Gramespacher	Bousmina
R_v (μm)	0.29	1.01	0.40

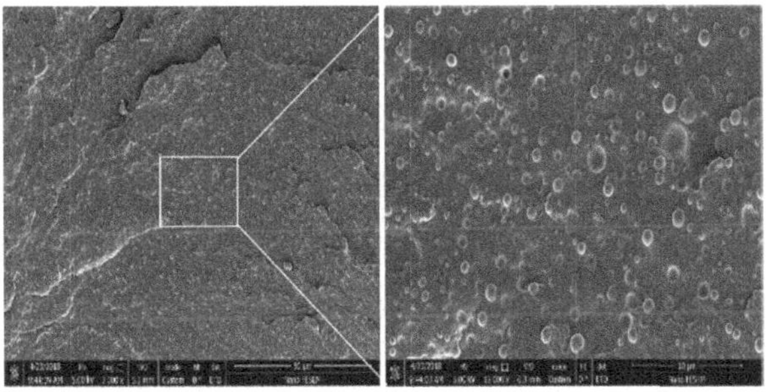

Fig. 2. SEM images of PBAT/EVOH 80/20 blend

4 Conclusion

The volume average radius obtained from the models is comparable with that obtained from the SEM images and shows that they have good compatibility. The difference between the models and the experiment can be contributed to the effect of temperature and the composition on the solubility parameter and considering the approximate value of the Kuhn length. Also, the results showed that Bousmina model is more appropriate for the present blend.

References

1. Lagaron, J.M., Powell, A.K., Bonner, G.: Polym. Test **20**(5), 569–577 (2001)
2. Gui, Z., Zhang, W., Lu, C., Cheng, S.: J. Macromol. Sci. Part B **52**(5), 685–700 (2013)
3. Helfand, E., Tagami, Y.: J. Polym. Sci. Part B: Polym. Lett. **9**(10), 741–746 (1971)
4. Helfand, E., Sapse, A.M.: J. Chem. Phys. **62**(4), 1327–1331 (1975)
5. Miquelard-Garnier, G., Roland, S.: Eur. Polym. J. **84**, 111–124 (2016)
6. Van Krevelen, D.W., Te Nijenhuis, K.: Properties of Polymers: Their Correlation with Chemical Structure; Their Numerical Estimation and Prediction from Additive Group Contributions, 4th edn., ch. 7, pp. 189–221. Elsevier (2009)
7. Schneier, B.: J. Polym. Sci. Part B: Polym. Lett. **10**(4), 245–251 (1972)
8. Pallerne, J.F.: Rheol. Acta **29**(3), 204–214 (1990)
9. Hassanpour Asl, F., et al.: J. Appl. Polym. Sci. **135**(16), 1–10 (2018)
10. Gramespacher, H., Meissner, J.: J. Rheol. (N. Y. N. Y) **36**(6), 1127–1141 (1992)
11. Xing, P., Bousmina, M., Rodrigue, D., Kamal, M.R.: Macromolecules **33**(21), 8020–8034 (2000)
12. Sperling, L.H.: Introduction to Physical Polymer Science, 4th edn., ch. 5, pp. 212–213. Wiley (2005)
13. Witt, U., Einig, T., Yamamoto, M., Kleeberg, I., Deckwer, W.D., Muller, R.J.: Chemosphere **44**(2), 289–299 (2001)

Appendix I. Plenary Lecture

The main message of ISPST 2018

Hamid Mirzadeh
Seminar Chairman
mirzadeh@aut.ac.ir

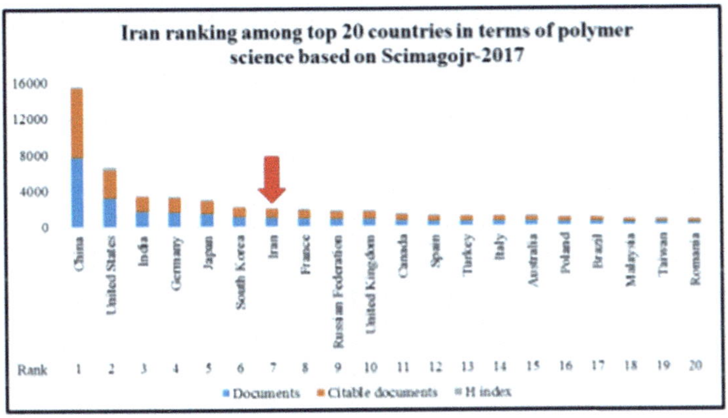

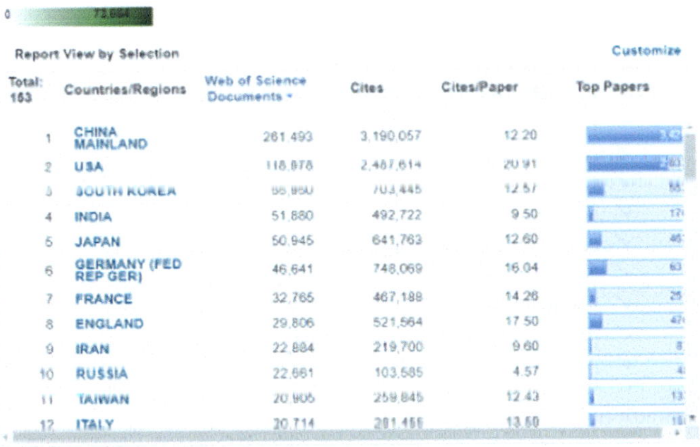

Essential Science Indicators (ESI)

Report View by Selection — Customize

Total: 153	Countries/Regions	Web of Science Documents	Cites	Cites/Paper	Top Papers
1	CHINA MAINLAND	261,493	3,190,057	12.20	[illegible]
2	USA	118,878	2,487,614	20.91	[illegible]
3	SOUTH KOREA	55,950	703,445	12.57	[illegible]
4	INDIA	51,880	492,722	9.50	[illegible]
5	JAPAN	50,945	641,763	12.60	[illegible]
6	GERMANY (FED REP GER)	46,641	748,069	16.04	63
7	FRANCE	32,765	467,188	14.26	25
8	ENGLAND	29,806	521,564	17.50	[illegible]
9	IRAN	22,884	219,700	9.60	8
10	RUSSIA	22,661	103,585	4.57	[illegible]
11	TAIWAN	20,905	259,845	12.43	[illegible]
12	ITALY	20,714	281,455	13.60	[illegible]

https://www.scimagojr.com/countryrank.php?cztegory=2507&year=2017

H. Mirzadeh and A. A. Katbab (Eds.): ISPST 2018, *Eco-friendly and Smart Polymer Systems*, pp. 701–709, 2020.
https://doi.org/10.1007/978-3-030-45085-4

111 years have passed since the inception of the production of synthetic polymers.

1907 Bakelite: Phenol-formaldehyde resin.
1924 Poly(vinyl alcohol)
1927 Poly(vinyl chloride)
1929 Urea-formaldehyde Resin
1933 Poly(vinylidene chloride)
1931 Poly(methyl methacrylate) (PMMA)
1938 Poly(tetrafluoro ethylene)
1939 Nylon
1941 Poly(ethylene Terephthalate) (PET)
1942 Polyethylene
1943 Poly(dimethylsiloxane) (PDMS)
1943 Polyurethane
1956 Poly(ethylene-co-vinyl acetate), PEVA
1957 Polypropylene
1957 Polycarbonate
1958 Polyacrylamide
1958 Carbon fiber
1960 Poly(hydroxyethyl Methacrylate) (PHEMA)
1964 Poly(p-phenylene terephthalamide) (Kevlar)
1964 Polyimide
1966 Poly(acrylic acid)
1970 thermopolymers polyester
1978 LLDPE
1985 Liquid crystal polymers
1986 Nanocomposites and advanced polymers

The growth of polymers production since 1950s has drastically developed fast than any other manufactured material.

https://www.bytheoceanweunite.org/plastic-facts/

Now, polymers have become an essential part in our life. Especially over the la 60 years polymers raised the standard of living and life without polymers exhausting.

You see polymers everywhere

- ✓ computer,
- ✓ cell phones,
- ✓ modern medicine and health,
- ✓ insulation,
- ✓ transportation,
- ✓ conduction,
- ✓ energy, heat and light,
- ✓ composites,
- ✓ membranes,
- ✓ packaging and storage, and
- ✓ housing

All are made from polymers

"I think that the development of polymerization is perhaps the biggest thing th chemistry has done, where it has had the biggest effect on everyday life" —Lor Todd, a Nobel prize laureate 1980.

https://www.sciencehistory.org/the-history-and-future-of-polymers

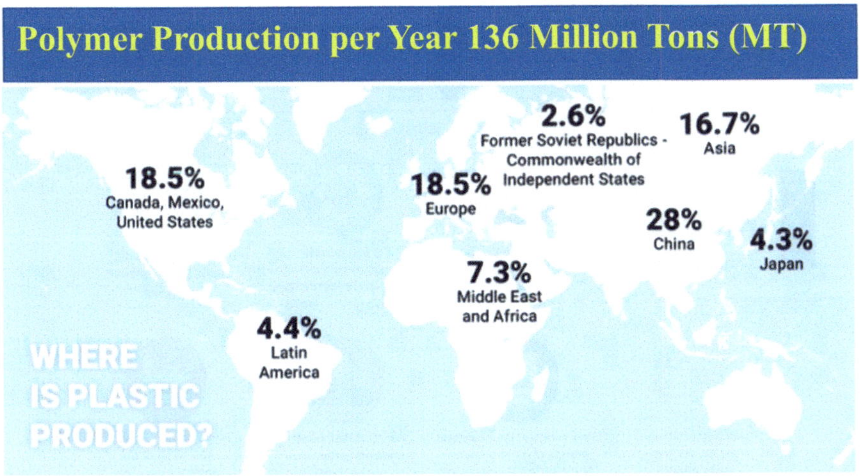

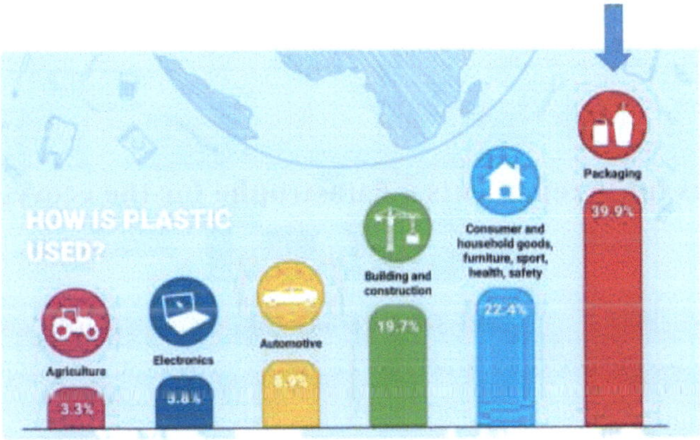

http://worldenvironmentday.global/en/news-category/beat-polymers-pollution

http://worldenvironmentday.global/en/news-category/beat-polymers-pollution

This table represents a catastrophe for the ecosystem

It is estimated:

Total polymer produced till (2017). (Mt)	Polymers waste Since 2015 (Mt)	Recycled % (2015).	Incinerated % (2015)	Accumulated in ecosystems (2015) %	Accumulated in ecosystems by 2050 (Mt)
8300	6300	9	12	79	12000

http://advances.sciencemag.org/content/3/7/e1700782.full

http://worldenvironmentday.global/en/news-category/beat-polymers-pollution

It is estimated that <u>over 80% of polymers in the oceans come from land based-sources</u>

https://www.bytheoceanweunite.org/polymers-facts/
<u>https://helpsavenature.com/polymers-pollution-in-ocean</u>

It was discovered that in the middle of the Pacific Ocean a circular pattern of water currents has accumulated enormous amounts of floating garbage. This huge amount is often referred as the "Eastern Pacific Garbage Patch" !

A British diver in this video shows what's happening on the beaches of the Balinese island of Indonesia. A plastic soup kitchen!

As polymers flow around the oceans, they are mistaken for food items by many marine animals.

Various polymers pieces floating in the water are responsible for the death of millions of fish and birds as they mistake these small pieces of polymers for food.

https://www.bytheoceanweunite.org/polymers-facts/

Plastics floating in marine environment are now commonly called microplastics which cause a lot of trouble for fish, bird and animal life.
The consumption of polymers can lead to a blockage of the digestive system leading to starvation as has been found in many seabirds, but it may also lead to the leaching of toxic chemicals from the polymers into the animals body.
As the polymers particles are eaten by marine life, these toxins enter the ocean's food webs including the fish we like to eat. These chemicals may well have detrimental effects on basic biological processes regulated through genes and hormones.

https://www.sciencehistory.org/the-history-and-future-of-polymers
https://www.sciencehistory.org/the-history-and-future-of-polymers
https://www.bytheoceanweunite.org/polymers-facts/

The problem of polymers pollution in the land and ocean is very serious and requires attention of all the countries in the world.

The ugly view that can be seen everywhere in the world.

Worldwide movements for making polymers safer.

More than 26 million tons (MT) of polymers waste materials are produced annually in the European countries.
Since October 2018, the European Union has decided to ban the use of four types of polymers to reduce polymer waste in EU member states.

France committed to use only recycled polymers in packaging by the end of 2025.

https://www.google.com/search?q=%D8%AF%D9%88%DB%8C%DA%86%D9%87+%D9%88%D9%84%D9%87+%D9%81%D8%A7%D8%B1%D8%B3%DB%8C&oq=%D8%AF%D9%88%DB%8C%DA%86%D9%87+%D9%88%D9%84%D9%87+%D9%81%D8%A7%D8%B1%D8%B3%DB%8C&aqs=chrome..69i57j0l5.29951j1j8&sourceid=chrome&ie=UTF-8

Polymers News Europes

8 African countries have seriously passed strict laws and regulations on the prohibition of disposable polymer materials.

There is a conflict between the two important issues:

- ✓ **Firstly: polymers are an important and necessary part of our current and future life**
- ✓ **Secondly: polymers pollution in the ecosystems is very serious also for our life.**

What shall we do?

Please:

- **Make biodegradable polymers**
- **Develop natural based polymers,**
- **Create polymers that are more ecosystems friendly**
- **Work to make polymers that are truly biodegradable.**

3 R should be considered as a global culture

Reduce: less packaging
Reuse: use reusable products
Recycle: items that can be recycled

We should consider that:

Any action on the use of polymers and the restriction of their use should be made according to the environmental and economic realities of different regions of the world.

Appendix II. Keynote Lectures

Novel Structural Changes During Temperature-Induced Self-Assembling and Gelation of PCLA-PEG-PCLA Triblock Copolymer in Aqueous Media

Josefine Eilsø Nielsen[1], Kaizheng Zhu[1], Sverre Arne Sande[2], Kenneth D. Knudsen[3], Lubomír Kováčik[4], Dušan Cmarko[4], and Bo Nyström[1(✉)]

[1] Department of Chemistry, University of Oslo, P.O. Box 1033, Blindern, 0315, Oslo, Norway
bo.nystrom@kjemi.uio.no
[2] School of Pharmacy, Department of Pharmaceutics, University of Oslo, P.O. Box 1068, Blindern, 0316, Oslo, Norway
[3] Department of Physics, Institute for Energy Technology, P.O. Box 40, 2027, Kjeller, Norway
[4] Institute of Biology and Medical Genetics, First Faculty of Medicine, Charles University and General University Hospital in Prague, Albertov 4, 128 01 Prague, Czech Republic

Abstract. Thermoresponsive amphiphilic biodegradable block copolymers of the type poly(ε-caprolactone-*co*-lactide)-poly(ethylene glycol)-poly(ε-caprolactone-*co*-lactide) (PCLA-PEG$_m$-PCLA) have great potential for various biomedical applications. In the present work, influences of PEG-spacer length (m = 1000 and 1500), temperature, and polymer concentration on the self-assembling process to form supramolecular structures in aqueous solutions of the PCLA-PEG$_m$-PCLA copolymer have been scrutinized. The rheological properties of both dilute and semidilute samples were characterized; especially the sol-to-gel transition was examined. Small-angle neutron scattering (SANS) experiments reveal fundamental structural differences between the two copolymers for both dilute and semidilute samples. The scattered intensity profiles from SANS for the copolymer with the long PEG spacer could be described by a spherical core-shell model over a broad temperature domain, whereas the copolymer with the short hydrophilic spacer forms rod-like species over an extended temperature range. This finding is supported by cryo-TEM images.

Introduction

Biodegradable and thermogelling polymer systems that can undergo sol-to-gel transition upon a temperature raise have attracted a great deal of attention in recent years for their potential in biomedical applications, 1–3 including drug delivery, cell therapy, and tissue engineering. In this study, triblock copolymers of the type PCLA-PEGm-PCLA (m = 1000 and m = 1500 are number average molecular weights) were

H. Mirzadeh and A. A. Katbab (Eds.): ISPST 2018, *Eco-friendly and Smart Polymer Systems*, pp. 713–715, 2020.
https://doi.org/10.1007/978-3-030-45085-4

synthesized by ring opening polymerization. Aqueous solutions of these copolymers were prepared, and the polymers were characterized at different temperatures in both dilute and semidilute aqueous solutions with the aid of various experimental techniques, such as turbidimetry, oscillatory shear, SANS, and cryo-TEM. The aim of this study is to acquire knowledge about how these features are influenced by type of hydrophobic block, temperature, and polymer concentration. This understanding is necessary to be able to tailor-make systems for specific biomedical applications.

Experimental

PCLA-PEGn-PCLA triblock copolymers were prepared by ring-opening polymerization (ROP) of ε-caprolactone and D,L-lactide with PEG as the initiator and stannous octoate as the catalyst. Feed ratios of the PEG/CL/LA were used to adjust the composition and molecular weight. The behavior of these samples was studied with different experimental methods such as turbidity, rheology, small-angle neutron scattering (SANS).4

Results and Discussion

All the hydrogels presented three basic physical states: solution, gel, and precipitate. We further found that the gel window was composed of two regions, transparent gels and opaque gels. The phase diagram reveals that the gelation region depends on temperature, polymer concentration and length of the PEG length.

Figure 1 shows a schematic illustration of the structures, based on the SANS results, for the two amphiphilic copolymers in dilute and semidilute solution. The copolymer with a long PEG-spacer form flower-like core-shell micelles in dilute solution in the temperature range 10–35 °C, but at 40 °C cylindrical structures are formed. In the semidilute concentration regime, the spherical micelles are closed-packed and the appearances of correlation peaks in the scattering profile suggest strong interactions between the micelles. In the case of the copolymer with the short PEG-spacer, extended ellipsoidal structures are formed over the considered temperature range. At 20 wt%, we have a close-packing of cylindrical structures and strings of micelles are probably developed

Conclusions

The SANS results disclosed some very interesting features of the systems at both dilute and semidilute conditions. In dilute solutions of the copolymer with a long PEG-spacer, the scattered intensity profiles could be fitted very well with a spherical core-shell model at temperatures up to 35 °C; the overall size (ca. 16 nm) of the moieties is virtually unaffected by temperature. The micelles are stable up to this temperature, and no hysteresis effects are detected upon cooling the samples. At 40 °C, a structural transition to cylinder-like species is observed and probably because of the high sticking probability, the entities tend to self-assemble in the form of strings. These structures are not stable and they exhibit hysteresis effects.

In the case of the copolymer with the short PEG-spacer, the scattered intensity profile can be described by an ellipsoidal model over the whole temperature region (10–35 °C), and these entities are not stable and marked hysteresis effects are observed upon cooling. The conjecture is that because of the short hydrophilic block it cannot loop back to the same micelle core, but the PCLA block will protrude out into the solution and at higher temperatures the enhanced stickiness will facilitate for the PCLA block to be accommodated in the core of a neighboring micelle. In this way a string of micelles will be formed and this self-assembling will lead to extended structures.

The type of copolymer system investigated here has great potential as drug carrier in drug delivery applications. For its proper design, it is important to understand how spacer length, composition, polymer concentration, and temperature will affect the properties of the system.

Acknowledgements

We acknowledge financial support from Norwegian Financial Mechanism 2009-2014 under Project Contract MSMT-28477/2014 (project 7F14009). DC thanks the Grant Agency of Czech Republic (P302/12/G157), and Charles University (Progress Q28 and UNCE 204022) for support.

References

1. Huang, K., Lee, B.P., Ingram, D.R., Messersmith, P.B.: Biomacromolecule **3**, 397–406 (2002)
2. Suh, J.M., Bae, S.J., Jeong, B.: Adv. Mater. **17**, 118–120 (2005)
3. Khademhosseini, A., Peppas, N.P.: Advanced Materials **21**, 3307–3329 (2009)
4. Nielsen, J.F., Zhu, K., Sande, S.A., Kováčik, L., Cmarko, D., Knudsen, K.D., Nyström, B.: J. Phys. Chem. B **121**, 4885–4899 (2017)

Simulation of Heat-Sealing Process in Multilayer Films

Zahra Kanani Aghkand, Abdellah Ajj, and Charles Dubois(✉)

3SPack, CREPEC, Department of Chemical Engineering, École Polytechnique de Montréal, Montréal, Canada
charles.dubois@polymtl.ca

Abstract. Over the last few years, flexible packaging industry has been in a continuous evolution to address novel applications and end-users requirements. Multilayer structures are often seen as the only way to bring in all the properties required in a single package. In every package, sealing is one of the most delicate operation, since failure of sealing also means failure of the whole package and product loss. The sealing operation becomes particularly challenging for multilayer film as the exhibit a wider range on physical properties such as melting point, glass transition temperature, etc. In this work, COMSOL Multiphysics software was used for simulating heat transfer in a heat-sealing process of multilayer polymeric films. To have a more accurate description of the process, the temperature dependence of material properties such as thermal conductivity, specific heat and density were measured over a wide temperature range and were used in COMSOL as interpolation functions. The temperature at the interface of two sealing films was measured by a fast-response thermocouple linked to a data acquisition system. The simulation results show a very good agreement with the experimental measurements and revealed the importance of considering thermal contact resistance (TCR) as the boundary condition between the jaw and the abuse layer. These results show that the simulation, for the first time, can exactly predict temperature profile at each point of the multilayer film in a heat sealing process without any fitting parameter borrowed from experimental measurements. Moreover, the obtained results showed that the simulation can successfully predict the effect of jaw temperature; dwell time, pressure, film thickness and different sealant materials in temperatures ranging from below to much over than the melting point of the sealant. The resulting COMSOL model proves to be a time and energy saving tool for industrial usage in selecting appropriate sealant materials and designing optimum package structures.

H. Mirzadeh and A. A. Katbab (Eds.): ISPST 2018, *Eco-friendly and Smart Polymer Systems*, p. 716, 2020.
https://doi.org/10.1007/978-3-030-45085-4

Semi-crystalline Polymer Foams with Enhanced Morphology and Mechanical Properties Nucleated by Trisamide Based Supramolecular Additives

Volker Altstädt(✉)

Polymer Engineering, University Bayreuth, 95447 Bayreuth, Germany
altstaedt@uni-bayreuth.de

Abstract. Semi-crystalline polymer foams are quite attractive for numerous applications such as acoustic and thermal insulators, packaging, automotive components due to their outstanding functional properties [1]. Nevertheless, obtaining foams with a fine cell structure is restricted due to the weak melt strength and melt elasticity of polymers, such as isotactic PP and PBT [2, 3]. To address this, new generation 1,3,5-benzene-trisamide based nucleating agents (BTAs), which can be dissolved in polymer melt at the processing temperature, are used. The dissolved NAs can crystallize upon cooling in finely dispersed, one-dimensional, highly ordered nano-fibrillar objects. Self-assembly of these nano-fibrils into three-dimensional objects, which act as nucleating sites for the foam cells, takes place [4]. In this study, we investigated the influence of BTAs on the nucleation behavior and foamability of extruded iPP and PBT. In the presence BTAs, iPP exhibited excellent foaming performance, yielding smaller mean cell sizes (around 30) and narrower cell size distributions than those of neat iPP. Only 0.02 wt% of 1,3,5-benzenetrisamide NA led to improve compression modulus (100%) than those of neat iPP foam. Initial results from PBT foams showed that the strategy to improve foam morphology of iPP with BTAs is also successfully applicable for PBT foams to achieve enhanced foam morphology and lower foam densities.

H. Mirzadeh and A. A. Katbab (Eds.): ISPST 2018, *Eco-friendly and Smart Polymer Systems*, p. 717, 2020.
https://doi.org/10.1007/978-3-030-45085-4

Novel Naturally-Derived Polymers for Medical Applications

Esmaiel Jabbari(✉)

Department of Chemical Engineering, University of South Carolina, Swearingen, Room 2C11, 301 Main Street, Columbia, SC 29208, USA
jabbari@engr.sc.edu

Abstract. There is a need to develop novel polymers for medical applications. The rapidly growing poultry industry worldwide produces more than 5 million tons of feather and other cornified skin products every year. The feather waste from poultry industry is in general processed into low-value-added products like fertilizer, animal food, and packaging and insulation materials. A major component of poultry feather is keratin, a relatively low-molecular matrix protein with high cysteine content. The disulfide crosslinks in combination with other structural features like crystallinity and physical interaction between keratin β-sheets impart high strength to feather. Partial alkylation of sulfhydryl groups, freeze drying, and chemical crosslinking have been used to produce membranes, sponges, and fiber meshes from feather keratin. There is a crucial need for novel injectable, cell-compatible polymers for soft tissue regeneration with tunable degradation that match the rate of tissue formation. In the first part, I will present the development and production of an injectable and photo-crosslinkable hydrogel from feather keratin which can be used as a matrix for cell encapsulation and delivery in soft tissue regeneration. The degradation of the feather keratin-based hydrogel can be tuned enzymatically. The cattle industry worldwide produces more than 62 million tons of beef each year. The meat processing plants also produce an enormous amount of bone remains which is generally disposed in landfills. The cartilage tissue attached to the ends of long bones is organized into layers which leads to the formation of gradients to minimize friction during articulation and transfer load to the underlying bone tissue. The layered structure of the articular cartilage is highly chondrogenic to mesenchymal stem cells. There is a critical need for the development of novel materials for stem cell delivery in cartilage regeneration for patients who suffer from post-traumatic osteoarthritis. In the second part, I will present the development and production of novel high-value biopolymers based on bovine articular cartilage tissue which can be used as a matrix for the delivery of mesenchymal stem cells in articular cartilage regeneration.

H. Mirzadeh and A. A. Katbab (Eds.): ISPST 2018, *Eco-friendly and Smart Polymer Systems*, p. 718, 2020.
https://doi.org/10.1007/978-3-030-45085-4

From Biomass to the Bio-Based Monomers and Polymers

M. J. Zohuriaan-Mehr(✉)

Biobased Monomers and Polymers (BIOBASED) Division & Adhesive and Resin Department, Iran Polymer and Petrochemical Institute (IPPI), P.O. Box 14965-115, Tehran, Iran
m.zohuriaan@ippi.ac.ir, bcst.ippi@gmail.com

Abstract. Nowadays, growing attention is being paid to the preparation and application of bio-based (bio-derived) polymers because of both environmental concerns and unsustainability of the fossil resources. The demand for crude oil grows increasingly, while its source is going to be finished up to the end of century. At the heart of this sustainable approach to new polymers is the synthesis and use of biomass originated building blocks of chemicals, so called platform molecules, e.g., glycerol and itaconic acid. Biomass, annually produced as high as 400,000 million tons, are materials of biological origin excluding the fossil sources (i.e., coal, oil and gas, which are not renewable). Since the bioresources are the unique alternative to the fossil resources (both of them give us either "energy" or "material"), the petro-refineries are being replaced by bio-refineries progressively. In direction of the aforementioned context, this presentation will highlight some particular bio-resourced systems; as representative instances of the recent research in the BIOBASED Division, IPPI. Furan-based self-healing networks, furan-based epoxy toughener, glycerol/lactic acid/itaconic acid- and tannic acid-based thermosetting resins, lignin-based reactive reinforcing agent for epoxy biocomposites, furan- and plant oil-based reactive diluents for epoxy and vinyl ester alloys, and high-performance bio-based crosslinkers for making improved superabsorbent materials are some of the typical materials exemplified as the materials developed in our research teams. Improved catalysts to convert furfural to furfuryl alcohol (a bio-monomer) is another promising case on the way toward scale-up to a pilot plant.

H. Mirzadeh and A. A. Katbab (Eds.): ISPST 2018, *Eco-friendly and Smart Polymer Systems*, p. 719, 2020.
https://doi.org/10.1007/978-3-030-45085-4

Upcycling of Technical Polymers from Post-Consumer Sources

Clemens Holzer[1(✉)], Matthias Katschnig[1], Uta Jenull-Halver[1], and Günther Höggerl[2]

[1] Chair of Polymer Processing, Department of Polymer Engineering and Science, Montan Universitaet Leoben, Leoben, Austria
kv@unileoben.ac.at
[2] Müller-Guttenbrunn Group, Müller-Guttenbrunn GmbH, Amstetten, Austria

Abstract. Moving towards sustainability is a social challenge that obliges society and politics the same way. Resource efficiency is thereby an important approach. Regarding plastics, a material flow from cradle to cradle (C2C) would be a desirable solution consecutively. C2C already exists, but only in producing simple products out of down-cycled plastic waste. Reasons for this are for instance rigorous material specifications, demanding RoHS (Restriction of Hazardous Substances) compliance and the need for reproducible supply. Hence, high-quality products made of engineering plastics are still far away from implementing recycled polymers on a larger scale and we mostly incinerate engineering plastic waste in Europe, which is neither resource-efficient nor saving CO_2. To establish C2C also in the demanding technical field we started the project Rec2TecPart (Recyclates to Technical Parts) based on a network of small and medium-sized companies in Austria. We accomplished a process design that upgrades engineering plastic waste (mostly POM from post-industrial sources) by compounding to fabricate tailor-made materials ("RecHQ") for technical components using the so-called Rec2TecPart process. This process is separated in modules (supply logistics, input management, formulation, producing and component) which simplifies the implementation afterwards. Moreover, we conducted a life cycle assessment of implemented RecHQ. The results confirmed the new approach, POM RecHQ save at least 25% CO_2 eq. and 30% costs providing technical equal material properties compared to virgin plastics and the successful roll-out of products made of 100% POM RecHQ has been made in 2017. In the next step we wanted to expand the Rec2TecPart concept to post-consumer-sources, which is more challenging due to the heterogenous and undefined waste input. As case study a PC/ABS system was chosen. Waste streams were derived completely from WEEE (Waste Electrical and Electronic Equipment), sorting/separation and material testing was conducted by a local company and the upcycling process and was done in Leoben. Sorting and separation included mechanical, electrical and optical technologies. A compounding step including twin-screw extrusion employed the upcycling. Design of Experiments (DoE) on screening level and thereafter as optimization modelling established a systematical analysis of

H. Mirzadeh and A. A. Katbab (Eds.): ISPST 2018, *Eco-friendly and Smart Polymer Systems*, pp. 720–721, 2020.
https://doi.org/10.1007/978-3-030-45085-4

compound recipes including the factors melt temperature, screw number of revolutions, separation quality and impact modifier. Target response comprised mechanical performance, which was evaluated by standard tensile and impact testing. Virgin material dealt as reference. Results displayed a potential cross impact of former impact modifier contamination with newly added impact modifier and a significant impact of separation selectivity on mechanical performance.

Study, Design, and Development of Printed Organic and Hybrid Optoelectronic Devices for Energy Application

Tayebeh Ameri(✉)

Chair of Functional Nanosystems, Research area of Physical Chemistry, Department of Chemistry, University of Munich (LMU), Butenandtstr. 11 (Haus E), 81377 Munich, Germany
tayebeh.ameri@lmu.de

Abstract. All the attractive and beneficial features of the organic materials such as their high absorption coefficient, the ability to tailor the bandgap, printability potential, and unique designs have brought them in the center of attention for the photovoltaic application. An important milestone towards OPV commercialization has been surpassed by reaching a power conversion efficiency (PCE) of over 13%. However, despite the significant improvements achieved, the PCE of OPV has to be increased to compete with commercial inorganic solar cells. The main limitation is due to the intrinsic narrow absorption window ($\sim$100–200 nm) of polymers compared to inorganic semiconductors such as Si, which makes it challenging to fully cover the solar spectrum with a single junction device. Furthermore, due to the low mobility of the photoactive materials, organic solar cells present a limited thickness, typically around 100 nm that dramatically limits light harvesting. To overcome the absorption limitation, ternary blend organic solar cells represent one of the dominant strategies that has been explored in the last decade. The outstanding advantage of ternary blends consists of maintaining the simplicity of the processing conditions used for single active layer devices. Moreover, all the optimization strategies developed for binary cells can be also effectively applied for ternary solar cells. In this elegant configuration all three photovoltaic parameters (VOC, JSC, FF) can be tackled simultaneously or individually by optimizing the ratio between the three materials used in the photoactive layer. In this presentation, we will discuss the achieved results in details for various prototype organic ternary systems with a central focus on the fundamental complexity of microstructure and charge transport mechanisms. Importantly, we will explain the key issues that result in modified recombination mechanisms and consequently improved FF and VOC in high efficiency ternary solar cells.

H. Mirzadeh and A. A. Katbab (Eds.): ISPST 2018, *Eco-friendly and Smart Polymer Systems*, p. 722, 2020.
https://doi.org/10.1007/978-3-030-45085-4

The Membrane Science Universality and the Rationalizing of Apparent Stagnation Regarding the Novel Polymer's Application Despite a Flood of Successful Reports

E. Ostadi[1,2], N. Mohammadi[1,2(✉)], and N. Yousefimehr[1,2]

[1] Polymer Engineering and Color Technology Department, Amirkabir University of Technology, P.O. Box 15475-4413, Tehran, Iran
mohamadi@aut.ac.ir
[2] Loghman Polymer Fanavari Gostar, 1599687143 Tehran, Iran

Abstract. By the late 1970s, reverse osmosis and the related membrane water treatment process of ultrafiltration were off the ground, commercialized and ready to be applied to gas separation. On the other hand, the bulk of today's market is supplied by essentially the same handful of polymers used for these applications in 1990s. There are several reasons for this lack of success or apparent stagnation. One reason is that the big and relatively easy applications are mostly solved problems. Another reason is that developers of new membrane materials have not fully appreciated that when they measure the permeability of these new candidates as a 100 μm thick films with pure gases and get encouraging results they are at the beginning of the development program, not at the end. In other words, all of the trade-off plots described in the literature are for pure gases. Few industrial companies are ready to invest the millions of dollars required to commercialize a new membrane application based on a simple laboratory film test result. Real world phenomena like plasticization and physical aging can change the relative positions of different membrane materials when tested with gas mixtures under application conditions. The fundamental and universal membrane science is going to be discernible and complemented with the practical routes toward university- industry cooperation for facilitated end use exploitation.

H. Mirzadeh and A. A. Katbab (Eds.): ISPST 2018, *Eco-friendly and Smart Polymer Systems*, p. 723, 2020.
https://doi.org/10.1007/978-3-030-45085-4

Halloysite Nanotube as Promising Carrier for the Controlled Release of Drugs

V. Haddadi-Asl(✉)

Department of Polymer Engineering and Color Technology, Amirkabir University of Technology, Tehran, Iran
haddadi@aut.ac.ir

Abstract. Halloysite (HNT) is a natural nanoscale clay mineral ($Al_2S_5O_5(OH)_4$. nH_2O) with tubular morphology containing a Al-OH sheet forming the inside and Si-O sheet the outside that belongs to the kaolin group formed by rolling kaolinite layers in tubes which is driven by a mismatch in the periodicity between the oxygen sharing tetrahedral SiO_4 sheets and adjacent octahedral AlO_6 sheets in the 1:1 layer and consisting a slender hallow structure which is between 300–1500 nm in length, 40–120 nm in outer and 15–100 nm in inner diameter depending on its deposit source. These morphology and chemical structure resulted in a negatively charged outer surface and positively charged inner surface that would be the main difference between the layered silicate and halloysite. In addition, they possess good biocompatibility, low cost, and environmental friendliness. Therefore, HNTs represent a promising candidate for drug delivery. Halloysite nanotubes were successfully employed for several hydrophobic drugs encapsulation. Two methods of drug formulation for halloysite based nanocontainer, inner lumen-loaded drug molecules and surface-absorbed, which can be combined for allowing initial drug burst release from the outer surface followed by slower discharge of inner trapped drug molecules distinguishes halloysite amongst other nanoparticles as a controllable sustained drug release carrier. Considering these properties, halloysite with good biocompatibility and low cytotoxicity can be used as an entrapment system for storage and controlled release of drugs. Generally, HNTs are modified by different polymers for sustained drug release. Therefore, the polymer modified-HNTs nanocomposites have shown many potential applications in drug and DNA delivery. Surface modification with polymer chains is commonly carried out by physical or chemical methods. In the physical methods, adsorption of polymer chains on the surface is implemented by physical bonds. However, in chemical methods, polymer chains are covalently grafted to the surface via three common ways of "grafting from", "grafting through", and "grafting to". The "grafting from" method is based on the modification of a surface by an initiator and subsequent propagation of polymer chains from the surface. To control the grafting density and molecular weight of the surface grafted polymer chains in grafting from reactions, one can use controlled radical polymerization (CRP) techniques. Control radical polymerizations (such as atom transfer radical polymerization (ATRP)) have been frequently used in combination with the grafting from technique to attach various polymer chains on the surface of HNT. Stimuli responsive polymers are macromolecules which their conformation varies upon a change in light, temperature, pH, and etc. Covering special substrates with these types of polymers in the form of brushes results in specific

H. Mirzadeh and A. A. Katbab (Eds.): ISPST 2018, *Eco-friendly and Smart Polymer Systems*, pp. 724–726, 2020.
https://doi.org/10.1007/978-3-030-45085-4

materials with lots of applications in the fields of biosensors, scaffolds for tissue engineering, drug delivery systems, and etc. The pH sensitive polymers respond to the pH variation and have been considered largely because of their application in medicine. Variation of pH in different body organs and also in certain cancers, inflamed, or wound tissues compared with the healthy cells makes these materials so applicable in the targeted drug release systems. The pH responsive polymers contain ionizable groups with a pKa value between 3 and 10.

References

1. Jafarzadeh, Sh., Haddadi-Asl, V., Roghani-Mamaqani, H.: Nanofibers of poly (hydroxyethyl methacrylate)-grafted halloysite nantoubes and polycaprolactone by combination of RAFT polymerization and electrospinning. J. Polym. Res. **22**, 123 (2015)
2. Zargarian, S.S., Haddadi-Asl, V., Hematpour, H.: Carboxylic acid functionalization of halloysite nanotubes for sustained release of diphenhydramine hydrochloride. J. Nanopart. Res. **17**, 218 (2015)
3. Hematpour, H., Haddadi-Asl, V., Roghani-Mamaqani, H.: Synthesis of pH-sensitive poly (N,N-dimethylaminoethyl methacrylate)-grafted halloysite nanotubes for adsorption and controlled release of DPH and DS drugs. Polymer **65**, 143–153 (2015)
4. Mirhosseini, M. M., Haddadi-Asl, V., Zargarian, S.S.: Fabrication and characterization of hydrophilic poly (e-caprolactone)/Pluronic P123 electrospun fibers. SymposiumJ. Appl. Polym. Sci. **133**(17) (2016)
5. Mirhosseini, M.M., Haddadi-Asl, V., Zargarian, S.S.: Fabrication and Characterization of Polymer–Ceramic Nanocomposites Containing Pluronic F127 Immobilized on Hydroxyapatite Nanoparticles. RSC Adv. **6**(84), 80564–80575 (2016)
6. Zargarian, S.S., Haddadi-Asl, V.: Surfactant-assisted water exposed electrospinning of novel super hydrophilic polycaprolactone based fibers. Artif. Cells Nanomed. Biotechnol. Int. J. **45**(5) (2017)
7. Zargarian, S.S., Haddadi-Asl, V.: Facile fabrication of novel polycaprolactone based electrospun fibers using in-process water exposure. Int. J. Polym. Anal. Charact. **21**(7) (2016)
8. Noein, L., Haddadi-Asl, V., Salami-Kalajahi, M.: Grafting of pH-sensitive Poly (N, N-dimethylaminoethyl methacrylate-co-2-hydroxyethyl methacrylate) onto halloysite nanotubes via surface-initiated atom transfer radical polymerization for controllable drug release. Int. J. Polym. Mater. Polymeric Biomater. **66**(3) (2017). http://www.tandfonline.com/toc/gpom20/current
9. Esmaili, Z., Bayrami, S., Dorkoosh, F., Akbari Javar, H., Seyedjafari, E., Zargarian, S.Sh., Haddadi-Asl, V.: Development and characterization of electrosprayed nanoparticles for encapsulation of curcumin. J. Biomed. Mater. Res. Part A **106A**(1), 285–292 (2018)
10. Ghaderi, M., Haddadi-Asl, V., Zargarian, S.Sh.: Fabrication and characterization of polymer ceramic nanocomposites containing drug loaded modified halloysite nanotubes. J. Biomed. Mater. Res. Part A (2018)

11. Sahebi Jouibari, I., Haddadi-Asl, V., Mirhossaini, M.M.: Effect of Nanofiller Content and Confined Crystalization on the Microphase Sepration Kinetics of Polyurethane Nano Composites. Polym. Compos. (2018)
12. Zeinali, E., Haddadi-Asl, V., Roghani-Mamaqani, H.: Synthesis of dubl thermo- and pH-sensitive poly(N-isopropylacrylalamide-*co*-acrylic acid)- grafted cellulose nanocrystals by reversible addition-fragmentation chain transfer polymerization. J. Biomed. Mater. Res. Part A **106A**(1) (2017)
13. Modarresi-Saryazdi, S.M., Haddadi-Asl, V., Salami-Kalajahi, M.: N, N-methylenebis(acrylamide)-crosslinked poly(acrylic acid) particles as doxorubicin carriers: a comparison between release behavior of physically loaded drug and conjugated drug via acid-labile hydrazone linkage. J. Biomed. Mater. Res. Part A **106**(2) (2017)

Green Tire Rubber Compounds: From Vulcanization to Viscoelastic Energy Dissipation

Mehdi Razzaghi-Kashani(✉)

Polymer Engineering Department, Faculty of Chemical Engineering, Tarbiat Modares University, P.O. Box 14115-114, Tehran, Iran
mehdi.razzaghi@modares.ac.ir

Abstract. Green tire technology has a strategic role in the modern automotive industry to reduce the environmental issues through reducing fuel consumption and CO2 emission as well as replacing materials with hazardous side-effects such as fillers, aromatic oils, and other chemical ingredients. This technology owes its success to the application of modified nano-silica in the rubber compounds. Reduction in rolling resistance of green tires occurs due to the controlled non-linear viscoelastic dissipation resulting from modifications in the interfacial effects in such highly-filled rubber vulcanizates. These effects not only impact the dynamic-mechanical dissipative behavior of these vulcanizates, but also influence the vulcanization of rubber compounds to a great extent. We have shown for the first time that there is a correlation between the dynamic behavior of rubber chains resulting from formation of filler networks and their tendency to crosslink during the vulcanization process. Fuel consumption and tire rolling resistance can result from both dynamic and vulcanization heterogeneities in tire compounds. Although this correlation is more pronounced in nano fillers such as nano silica, it is a universal behavior originating from the physical interactions in all active fillers rather than their chemical nature.

H. Mirzadeh and A. A. Katbab (Eds.): ISPST 2018, *Eco-friendly and Smart Polymer Systems*, p. 727, 2020.
https://doi.org/10.1007/978-3-030-45085-4

Author Index

H. Mirzadeh and A. A. Katbab (Eds.): ISPST 2018, *Eco-friendly and Smart Polymer Systems*, pp. 729–733, 2020.
https://doi.org/10.1007/978-3-030-45085-4

The manufacturer's authorised representative in the EU is Springer Nature Customer Service Centre GmbH, Europaplatz 3, 69115 Heidelberg, Germany. If you have any concerns regarding our products, please contact ProductSafety@springernature.com

Printed and bound by CPI Group (UK) Ltd, Croydon, CR0 4YY
30/09/2026
02225648-0020